TELLURITE GLASSES

HANDBOOK

Physical Properties and Data

Second Edition

TELLURITE GLASSES

GLASSES

HANDBOOK

Physical Properties and Data

Second Edition

Raouf A. H. El-Mallawany

Professor of Solid State Physics,
Science College, Northern Borders University,
Kingdom of Saudi Arabia

CRC Press
Taylor & Francis Group
Boca Raton London New York

CRC Press is an imprint of the
Taylor & Francis Group, an **informa** business

CRC Press
Taylor & Francis Group
6000 Broken Sound Parkway NW, Suite 300
Boca Raton, FL 33487-2742

First issued in paperback 2017

© 2012 by Taylor & Francis Group, LLC
CRC Press is an imprint of Taylor & Francis Group, an Informa business

No claim to original U.S. Government works

ISBN-13: 978-1-4398-4983-5 (hbk)
ISBN-13: 978-1-138-07576-4 (pbk)

Library of Congress Cataloging-in-Publication Data

El-Mallawany, Raouf A. H.
   Tellurite glasses handbook : physical properties and data / Raouf A.H. El-Mallawany. -- 2nd ed.
     p. cm.
   Includes bibliographical references and index.
   ISBN 978-1-4398-4983-5 (hardback)
   1. Fiber optics--Materials--Handbooks, manuals, etc. 2. Tellurites--Handbooks, manuals, etc. 3. Metallic glasses--Handbooks, manuals, etc. I. Title.

TA1800.E23 2011
621.36'92--dc23   2011039858

Visit the Taylor & Francis Web site at
http://www.taylorandfrancis.com

and the CRC Press Web site at
http://www.crcpress.com

To my family

Contents

Preface xiii
Acknowledgment xvii
About the Author xix

Chapter 1   Introduction to and Formation of Tellurite Glasses 1

    1.1   Crystal Structure and General Properties of Tellurium Oxide (TeO_2).............1
        1.1.1   Early Observations 1
        1.1.2   Analyses of Three-Dimensional Structure...............................1
        1.1.3   Analyses of Properties 2
    1.2   Definition of Tellurite Glasses and Review of Early Research 2
    1.3   Preparation of Pure Tellurium Oxide Glass 7
    1.4   Glass-Forming Ranges of Binary Tellurium Oxide Glasses...............8
        1.4.1   Phase Diagram and Immiscibility...11
        1.4.2   Structure Models 13
    1.5   Glass-Forming Ranges of Multicomponent Tellurium Oxide Glasses...........17
        1.5.1   Tellurite Glasses Prepared by Conventional Methods 17
        1.5.2   Tellurite Glasses Prepared by the Sol-Gel Technique...........22
    1.6   Nonoxide-Tellurite Glasses...23
        1.6.1   Chalcogenide Tellurite Glasses 23
        1.6.2   Fiber Preparation..28
        1.6.3   Halide-Tellurite Glasses (Chalcohalide Glasses)..................30
    1.7   Mixed Oxyhalide and Oxysulfate Tellurite Glasses.........................32
    1.8   General Physical Characteristics and Structure of Tellurite Glasses 37
    1.9   Structure and Bonding Nature of Tellurite Glasses..........................42
        1.9.1   Assignment of the Shoulder in the Oxygen-First Peak 45
        1.9.2   Profile of Oxygen-First Photoelectron Spectra 45
        1.9.3   Chemical Shifts of the Core Electron-Binding Energies 47
        1.9.4   Valence Band Spectra..47
        1.9.5   Intermediate-Range Order (as Determined by NMR, Neutron,
            XRD, X-Ray Absorption Fine Structure, Mossbauer Spectra,
            XPS, and X-Ray Absorption Near-Edge Structure Analyses) in
            Tellurite Glasses 49
    1.10   New Tellurite Glasses 53
    1.11   Applications of Tellurite Glass, Tellurite Glass Ceramics and Tellurite
        Glass Containing Nanoparticles...66
    References 72

PART I   Elastic and Acoustic Relaxation Properties

Chapter 2   Elastic Moduli of Tellurite Glasses 81

    2.1   Elastic Properties of Glass..81
        2.1.1   Terminology of Elasticity..82

		2.1.2	Semiempirical Formulae for Calculating Constants of Elasticity	85
			2.1.2.1 Mackashima–Mackenzie Model	86
			2.1.2.2 Bulk Compression Model	87
			2.1.2.3 Ring Deformation Model	90
			2.1.2.4 Central Force Model	91
	2.2	Experimental Techniques		92
		2.2.1	Pulse-Echo Technique	92
		2.2.2	Couplings	93
		2.2.3	Transducers	93
			2.2.3.1 Piezoelectric Transducers	93
			2.2.3.2 Characteristics of a Transducer	94
		2.2.4	Sample Holders	94
		2.2.5	Measurements of Elastic Moduli under Uniaxial and Hydrostatic Pressure	95
		2.2.6	Hardness Measurements	95
	2.3	Elasticity Modulus Data of TeO_2 Crystal		97
	2.4	Elastic Modulus Data of Pure TeO_2 Glass		97
		2.4.1	SOEC of Pure TeO_2 Glass	97
		2.4.2	TOEC and Vibrational Anharmonicity of Pure TeO_2 Glasses	97
		2.4.3	Physical Significance of SOEC	108
	2.5	Constants of Elasticity of Binary and Ternary Transition Metal Tellurite Glasses		109
		2.5.1	TeO_2–WO_3 and TeO_2–$ZnCl_2$ Glasses	109
		2.5.2	TeO_2–MoO_3 Glasses	111
		2.5.3	TeO_2–ZnO Glasses	112
		2.5.4	TeO_2–V_2O_2 Glasses	115
		2.5.5	TeO_2–V_2O_5–Ag_2O, TeO_2–V_2O_5–CeO_2, and TeO_2–V_2O_5–ZnO Glasses	117
		2.5.6	Effect of Gamma-Radiation on the Elasticity Moduli of Tricomponent Tellurite Glass System TeO_2–V_2O_5–Ag_2O	120
	2.6	Comparison between MoO_3 and V_2O_5 in Tellurite and Phosphate Glasses and K–V Relations		121
	2.7	Application of Makishima–Mackenzie Model to Pure TeO_2, $TeO2$–$V2O5$, and TeO_2–MoO_3 Glasses		130
	2.8	Elastic Moduli and Vickers Hardness of Binary, Ternary, and Quaternary Rare-Earth Tellurite Glasses and Glass-Ceramics		135
	2.9	Quantitative Analysis of the Elasticity Moduli of Rare-Earth Tellurite Glasses		139
	2.10	Elastic Properties of Te Glasses		143
	2.11	Elastic Properties of New Tellurite Glasses		147

Chapter 3	Acoustic Relaxation Properties of Tellurite Glasses		153
	3.1	Introduction		153
	3.2	Ultrasonic Attenuation of Oxide–Tellurite Glasses at Low Temperature		155
	3.3	Properties of Ultrasonic Attenuation in Nonoxide–Tellurite Glasses		162
	3.4	Radiation Effect on Ultrasonic Attenuation Coefficient and Internal Friction of Tellurite Glasses		162
	3.5	Structural Analysis of Ultrasonic Attenuation and Relaxation Phenomena		165
		3.5.1	Thermal Diffusion (Thermoelastic Relaxation)	165
		3.5.2	Direct Interaction of Acoustic Phonons with Thermal Phonons	166

3.6   Correlations between Low-Temperature Ultrasonic Attenuation and Room-Temperature Elastic Moduli 168
3.7   Acousto-Optical Properties of Tellurite Glasses 171
     3.7.1   Frequency Dependence of Ultrasonic Attenuation 172
     3.7.2   Temperature Dependence of Ultrasonic Attenuation 173
     3.7.3   Dependence of Ultrasonic Attenuation on Glass Composition 173
     3.7.4   Dependence of Specific Heat Capacity on Glass Composition 173
     3.7.5   Mechanism of Ultrasonic Attenuation at Room Temperature 173
     3.7.6   Figure of Merit of Tellurite Glass 174

Chapter 4   Applications of Ultrasonics on Tellurite Glasses 177
4.1   Introduction 177
4.2   Ultrasonic Detection of Microphase Separation in Tellurite Glasses 179
     4.2.1   Theoretical Considerations 179
     4.2.2   Application to Binary Tellurite Glasses 180
4.3   Debye Temperature of Oxide and Nonoxide–Tellurite Glasses 181
     4.3.1   Experimental Acoustic Debye Temperature 181
     4.3.2   Correlations between Experimental Acoustic and Calculated Optical Debye Temperatures 190
     4.3.3   Radiation Effect on Debye Temperatures 193
References 194

PART II   *Thermal Properties*

Chapter 5   Thermal Properties of Tellurite Glasses 201
5.1   Introduction 201
5.2   Experimental Techniques of Measuring Thermal Properties of Glass 204
5.3   Data of the Thermal Properties of Tellurite Glasses 208
     5.3.1   Glass Transformation, Crystallization, Melting Temperatures, and Thermal Expansion Coefficients 208
     5.3.2   Glass Stability against Crystallization and Glass-Forming Factor (Tendency) 214
     5.3.3   Viscosity and Fragility 220
     5.3.4   Specific Heat Capacity 222
5.4   Glass Transformation and Crystallization Activation Energies 228
     5.4.1   Glass Transformation Activation Energies 228
         5.4.1.1   Lasocka Formula 229
         5.4.1.2   Kissinger Formula 229
         5.4.1.3   Moynihan et al. Formula 230
     5.4.2   Crystallization Activation Energy 231
5.5   Correlations between Glass Transformation Temperature and Structure Parameters 234
5.6   Correlations between Thermal Expansion Coefficient and Vibrational Properties 238
5.7   Tellurite Glass-Ceramics 239
5.8   Nonoxide-Tellurite Glasses 245
5.9   Thermal Properties of New Types of Tellurite Glass 248
References 262

PART III   Electrical Properties

Chapter 6   Electrical Conductivity of Tellurite Glasses 269

6.1   Introduction to Current-Voltage Drop and Semiconducting
        Characteristics of Tellurite Glasses 269
6.2   Experimental Procedure to Measure Electrical Conductivity 271
        6.2.1   Preparation of the Sample 272
        6.2.2   DC Electrical Conductivity Measurements at Different
                   Temperatures 272
        6.2.3   AC Electrical Conductivity Measurements at Different
                   Temperatures 274
        6.2.4   Electrical Conductivity Measurements at Different Pressures 276
        6.2.5   Thermoelectric Power 276
6.3   Theoretical Considerations in the Electrical Properties of Glasses 277
        6.3.1   DC Conductivity 277
                   6.3.1.1   DC Conductivity of Oxide Glasses at High, Room,
                                 and Low Temperatures 277
                   6.3.1.2   DC Electrical Conductivity in Chalcogenide
                                 Glasses and Switching-Phenomenon Mechanisms 280
                   6.3.1.3   DC Electrical Conductivity in Glassy Electrolytes 282
                   6.3.1.4   Thermoelectric Power at High to Low Temperatures 287
        6.3.2   AC Electrical Conductivity in Semiconducting and Electrolyte
                   Glasses 287
6.4   DC Electrical-Conductivity Data of Tellurite Glasses at Different
        Temperatures 291
        6.4.1   DC Electrical Conductivity Data of Oxide–Tellurite Glasses
                   Containing Transition Metal Ions or Rare-Earth Oxides 292
                   6.4.1.1   DC Electrical Conductivity Data of Oxide–Tellurite
                                 Glasses at High Temperatures 292
                   6.4.1.2   DC Electrical Conductivity Data of Oxide–Tellurite
                                 Glasses at Room Temperature 296
                   6.4.1.3   DC Electrical Conductivity Data of Oxide–Tellurite
                                 Glasses at Low Temperatures 297
        6.4.2   DC Electrical Conductivity Data of Oxide–Tellurite Glasses
                   Containing Alkalis 306
        6.4.3   DC of Nonoxide–Tellurite Glasses 308
6.5   AC Electrical Conductivity Data of Tellurite Glasses 310
6.6   Electrical Conductivity Data of Tellurite Glass-Ceramics 315
6.7   Electrical Conductivity Data of New Tellurite Glasses 318

Chapter 7   Dielectric Properties of Tellurite Glasses 327

7.1   Introduction 327
7.2   Experimental Measurement of Dielectric Constants 328
        7.2.1   Capacitance Bridge Methods 328
        7.2.2   Establishing the Equivalent Circuit 329
        7.2.3   Low-Frequency Dielectric Constants 330
        7.2.4   Measurement of Dielectric Constants under Hydrostatic
                   Pressure and Different Temperatures 330

7.3   Dielectric Constant Models 331
       7.3.1   Dielectric Losses in Glass 332
               7.3.1.1   Conduction Losses 332
               7.3.1.2   Dipole Relaxation Losses 332
               7.3.1.3   Deformation and Vibrational Losses 332
       7.3.2   Dielectric Relaxation Phenomena 333
       7.3.3   Theory of Polarization and Relaxation Process 335
       7.3.4   Dielectric Dependence on Temperature and Composition 336
       7.3.5   Dielectric Constant Dependence on Pressure Models 336
7.4   Dielectric Constant Data of Oxide–Tellurite Glasses 337
       7.4.1   Dependence of Dielectric Constant on Frequency,
               Temperature, and Composition 337
               7.4.1.1   Establishing the Equivalent Circuit 339
               7.4.1.2   Low-Frequency Dielectric Constants 340
       7.4.2   Dielectric Constant Data under Hydrostatic Pressure and
               Different Temperatures 346
               7.4.2.1   Combined Effects of Pressure and Temperature on the
                         Dielectric Constant 348
       7.4.3   Dielectric Constant and Loss Data in Tellurite Glasses 351
7.5   Dielectric Constant Data of Nonoxide-Tellurite Glasses 354
7.6   Dielectric and Magnetic Data of New Tellurite Glasses and Ceramics 354
References 360

PART IV   Optical Properties

Chapter 8   Linear and Nonlinear Optical Properties of Tellurite Glasses in the Visible
             Region 367

       8.1   Introduction to Optical Constants 367
              8.1.1   What Is Optical Nonlinearity? 370
              8.1.2   Acousto-Optical Materials 371
       8.2   Experimental Measurements of Optical Constants 372
              8.2.1   Experimental Measurements of Linear Refractive Index and
                      Dispersion of Glass 372
              8.2.2   Experimental Measurements of Nonlinear Refractive Index of
                      Bulk Glass 373
              8.2.3   Experimental Measurements of Fluorescence and Thermal
                      Luminescence 375
       8.3   Theoretical Analysis of Optical Constants 377
              8.3.1   Quantitative Analysis of the Linear Refractive Index 377
                      8.3.1.1   Refractive Index and Polarization 377
       8.4   Linear Refractive Index Data of Tellurite Glasses 383
              8.4.1   Tellurium Oxide Bulk Glasses and Glass-Ceramics 383
              8.4.2   Linear Refractive Index Data of Tellurium Oxide Thin-Film Glasses ... 392
              8.4.3   Linear Refractive Index Data of Tellurium Nonoxide Bulk and
                      Thin-Film Glasses 393
       8.5   Nonlinear Refractive Index Data of Tellurite Glasses 394
              8.5.1   Nonlinear Refractive Indices of Tellurium Oxide Bulk, Thin-
                      Film Glasses, and Glass-Ceramics 394
              8.5.2   Nonlinear Refraction Index of Tellurium Nonoxide Glasses 403

8.6   Optical Applications of Oxide-Tellurite Glass and Glass-Ceramics
       (Thermal Luminescence Fluorescence Spectra) 404
       8.6.1   Mechanism 1 408
       8.6.2   Mechanism 2 409
8.7   Optical Properties of New Tellurium Glass 413

Chapter 9   Optical Properties of Tellurite Glasses in the Ultraviolet Region 423
9.1   Introduction—Absorption, Transmission, and Reflectance 423
9.2   Experimental Procedure to Measure UV Absorption and Transmission
       Spectra 424
       9.2.1   Spectrophotometers 424
9.3   Theoretical Absorption Spectra, Optical Energy Gap, and Tail Width 427
9.4   UV Properties of Tellurite Glasses (Absorption, Transmission, and
       Spectra) 430
       9.4.1   UV–Properties of Oxide–Tellurite Glasses (Bulk and Thin Film)......430
       9.4.2   Data of the UV Properties of Oxide–Tellurite Glass Ceramics 444
       9.4.3   UV Properties of Halide–Tellurite Glasses 450
       9.4.4   UV Properties of Nonoxide–Tellurite Glasses 451
9.5   UV Properties of New Tellurite Glasses 454

Chapter 10 Infrared and Raman Spectra of Tellurite Glasses 459
10.1   Introduction 459
10.2   Experimental Procedure to Identify Infrared and Raman Spectra of
       Tellurite Glasses 459
10.3   Theoretical Considerations for Infrared and Raman Spectra of Glasses 460
10.4   Infrared Spectra of Tellurite Glasses 466
       10.4.1   Infrared Transmission Spectra of Tellurite Glasses and Glass-
                Ceramics 466
       10.4.2   IR Spectral Data of Oxyhalide–Tellurite Glasses 477
       10.4.3   IR Spectra of Halide–Telluride Glasses 480
       10.4.4   IR Spectra of Chalcogenide Glasses 482
10.5   Raman Spectra of Tellurite Glasses 482
       10.5.1   Raman Spectra of Oxide–Tellurite Glasses and Glass-Ceramics 482
10.6   Infrared and Raman Spectra of New Tellurite Glasses 489
References 494

Index 501

Preface

The first edition of *Tellurite Glasses Handbook* presented a textbook on materials science for those at junior- or senior-level in an engineering curriculum and practical fields. The book gave an understanding of the physical properties of these new materials, which are academically known as the "physics of noncrystalline solids." Also, it covered the topic starting with the published references in the 1950s up to year 2000. The present edition will cover the dominant physical properties of this prototype glass system up to 2011. It will combine the scientific data and the updated practical applications of the strategic solid material, which is tellurite glass, including nanocomposites, and presents the three-dimensional structural models. The second edition of the *Tellurite Glasses Handbook* includes the new glass-forming systems, thermal, elastic, anelastic, electrical, and optical properties, together with the potential applications. In addition, the book contains the correlations and the new ways to combine the whole physical properties simultaneously to open new points for research. The present edition will gather the data in the main four major physical diretions: thermal, mechanical, electrical, and optical. In fact, the emphasis is on understanding and predicting the physics and technology of twenty-first century processing, fabrication, behavior, and the properties of tellurite glass and glass-ceramic materials. Some of the correlations and constants are described for the first time in book form, and data are often combined in novel ways to suggest new research directions.

This textbook, geared for junior- and senior-level materials science courses within engineering and other appropriate departments, is produced in part to address the perceived gap among specialists, such as physicists, chemists, and material scientists, in their understanding of the properties of tellurite glasses. Four groups of these properties—elastic and anelastic, thermal, electrical, and optical—have received the most attention. After an introductory chapter, the bulk of the book is organized into four parts based on the above property groups. Tellurite glasses themselves are differentiated by their compositions; for example, pure tellurite glass and binary-transitional, rare-earth metal oxide, and multicomponent tellurite glasses, including halides and oxyhalides. Each of the remaining nine chapters covers basic theories regarding a particular physical property, related experimental techniques, and representative data. The coverage of tellurite glasses in this book is unique in providing both a compilation of scientific data and views on practical and strategic applications based on the properties of these glasses.

Chapter 1 introduces the crystal structure of tellurium oxide (TeO_2), the precursor of tellurite glasses. It defines tellurite glasses and describes their general physical characteristics, touching on color, density, molar volume, and short- and intermediate-range structural properties, including bonding. The chapter summarizes the main systems for producing and analyzing pure and binary glasses, as well as many key concepts developed later in the book: phase diagramming and immiscibility, structure models, fiber preparation, and glass-forming ranges for multicomponent tellurite glasses, nonoxide-tellurite glasses (chalcogenide glasses), halide-tellurite glasses (chalcohalide glasses), mixed oxyhalide-tellurite glasses, and recent tellurite glass systems and their applications.

Part I: Chapter 2 covers the terminology of elasticity, semi-empirical formulae for calculating constants of elasticity, and experimental techniques for measuring factors that affect elasticity. Glasses have only two independent constants of elasticity. The rest, including both second- and third-order constants, must be deduced by various means. The elasticity moduli of TeO_2 crystal and of pure, binary transitional, rare-earth, and multicomponent tellurite glasses are summarized, as well as the hydrostatic and uniaxial pressure dependencies of ultrasonic waves in these glasses. The elastic properties of the nonoxide Te glasses are also examined. The elastic properties of new tellurite glass systems are included. Chapter 3 examines longitudinal ultrasonic attenuation at various frequencies and temperatures in both oxide- and nonoxide-tellurite glasses containing different modifiers in their binary and ternary forms. Experimental ultrasonic attenuation and acoustic activation energies of the oxide forms of these glasses and correlations among acoustic activation energy, temperature, bulk moduli, and mean cation-anion stretching force constants are discussed, as well as the effects of radiation and nonbridging oxygen atoms on ultrasonic attenuation and internal friction in tellurite glasses. Some tellurite glasses with useful acousto-optical properties for modulators and deflectors are highlighted. Phase separation, acoustic and optical Debye temperatures, and γ-radiation, and tests on nondestructive elastic moduli are summarized in Chapter 4 for analysis and manipulation of both oxide- and nonoxide-tellurite glasses in pure and multicomponent forms. The optical Debye temperatures have been calculated from the infrared spectra of tellurite glasses modified with rare-earth oxides. The effect of γ-radiation on acoustic Debye temperatures is examined.

Part II: In Chapter 5, experimental techniques are explained for measuring selected thermal properties of glass; for example, transformation temperature (T_g), crystallization temperature, melting temperature, and thermal expansion coefficient. Experimentally derived data for oxide-, nonoxide-, glass-, and ceramic-tellurite forms are compared with data calculated with different models. Correlations between thermal properties and average values for cross-link density and stretching force constants are summarized. Viscosity, fragility, and specific-heat capacities are compared between tellurite glasses and super-cooled liquids at T_g. Processing, properties, and structures of tellurite glass-ceramic composites are also discussed.

Part III: Chapter 6 examines the conduction mechanisms of pure tellurite glass; tellurite glasses containing transition metal, rare-earth, or alkaline components—oxide and nonoxide forms; and tellurite glass-ceramics. The effects of different temperature ranges, pressures, energy frequencies, and modifiers on AC and DC electrical conductivity in tellurite-based materials are summarized. Theoretical considerations and analyses of the electrical properties and conductivity of tellurite glasses and glass-ceramics based on their "hopping" mechanism are compared in high-, room-, and low-temperature conditions. This chapter explores the dependence of the semiconducting behavior of tellurite glasses on the ratio of low- to high-valence states in their modifiers, on activation energy, and on electron-phonon coupling. Ionic properties of tellurite glasses are also summarized. Chapter 6 clearly shows that electrical conduction parameters in tellurite-based materials are directly affected by temperature, frequency, and pressure, as well as the kinds and percentages of modifiers, together with electrical conductivity data of new tellurite glasses. The electric properties of tellurite glasses are explained in Chapter 7. Dielectric constant (ε) and loss factor data are summarized for both oxide- and nonoxide-tellurite glasses. These values vary inversely with frequency (f) and directly with temperature (T). The rates of change of ε with f and T, complex dielectric constants, and polarizability depend on the types and percentages of modifiers present in tellurite glasses. Data on the electric modulus and relaxation behavior of tellurite glasses are reviewed according to their stretching exponents. The pressure dependence of the ε is also examined. Quantitative analysis of the ε is discussed in terms of the number of polarizable atoms per unit volume, and data on the polarization of these atoms are related to the electrical properties of tellurite glasses and both dielectric and magnetic data of recent types of tellurite glasses and ceramics.

Part IV: Up-to-date linear and nonlinear refractive indices for oxide- and nonoxide-tellurite glasses and glass-ceramics (bulk and thin film) are discussed in Chapter 8, along with experimental measurements of the optical constants in these glasses. The refractive indices and dispersion values at different frequencies and temperatures are summarized for most tellurite glasses. Reflection, absorption, and scattering are related to dielectric theory. The relationship between refractive indices and numbers of ions/unit volume (N/V) along with the values of polarizability are explored. The reduced N/V of polarizable ions is primarily responsible for reductions in both the dielectric constant and refractive index, although reductions in electronic polarization also affect optical properties. Data on reductions in refractive indices, densities, and dielectric constants that occur with halogen substitution are also discussed, along with thermal luminescence, fluorescence, phonon sideband spectra, and the optical applications of tellurite glasses. Experimental procedures to measure ultraviolet (UV) absorption and transmission spectra in bulk and thin-film forms of tellurite glasses are summarized in Chapter 9, along with theoretical concepts related to absorption spectra, optical energy gaps, and energy band tail width. The UV spectrum data discussed here have been collected in the wavelength range from 200 to 600 nm at room temperature. Data are provided for the UV properties of oxide-tellurite glasses (bulk and thin film), oxide-tellurite glass-ceramics (bulk and thin film), halide-tellurite glasses, and nonoxide-tellurite glasses. From these experimental absorption spectra, the energy gap and band tail data are also summarized for these glasses. Analysis of these optical parameters is based on the Urbach rule, which is also explained. Chapter 10 describes infrared (IR) and Raman spectroscopy—two complementary, nondestructive characterization techniques, both of which provide extensive information about the structure and vibrational properties of tellurite glasses. The description begins with some brief background information on, and experimental procedures for, both methods. Collection of these data for tellurite glasses in their pure, binary, and ternary forms is nearly complete. IR spectral data for oxyhalide, chalcogenide, and chalcogenide halide glasses are now available. The basis for quantitative interpretation of absorption bands in the IR spectra is provided, using values of the stretching force constants and the reduced mass of vibrating cations-anions. Such interpretation shows that coordination numbers determine the primary forms of these spectra. Raman spectral data of tellurite glasses and glass-ceramics are also collected and summarized, and suggestions for physical correlations are made.

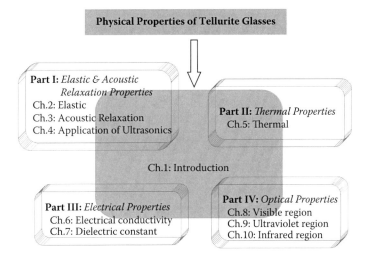

Acknowledgment

I would like to thank my colleagues in the international glass science community who asked and encouraged me to write this book, and for granting me permission to reproduce their data. I would also like to thank the publishers of the scientific journals, the officers of the companies providing specialized information, and the authors of the articles cited by references herein.

Raouf El-Mallawany
Northern Borders University
April, 2011

About the Author

Raouf H. El-Mallawany, PhD, professor of solid state physics in the Physics Department, Science College, Northern Borders University, Kingdom of Saudi Arabia. He has been a professor of solid state physics in the Physics Department, Faculty of Science, Menofia University, Egypt since 1994. He began his career in solid state physics in 1973, and earned his PhD in the discipline in 1986. He has received the following fellowships: the British Council Research Grant (1988), visiting scholarship at the University of California, USA (1990), and the ICTP Solid State Workshop, Italy (1995). He was head of the physics department at Qatar University, Qatar, and is also the former head of the physics department at Northern Borders University, Kingdom of Saudi Arabia.

He has taught physics to under- and post-graduate students for nearly four decades and has served as advisor and examiner for 25 PhD and MSc theses at Egyptian and Arab universities. Among the European and American universities he has visited are: Brunel University, UK; Bath University, UK; University of California, USA; International Center for Theoretical Physics, Italy; Renn University, France; Otto Schott Institute, Germany; Lehigh University, USA; Institute for Laser, Buffalo University, USA; the Materials Research Institute; and Penn State University, USA.

Professor El-Mallawany has presented papers and served as session chair, member of the International Advisory Committee, Egyptian delegate, and invited speaker for many annual international conferences and congresses on noncrystalline solids and glass. He has been invited to give several seminars in his area of expertise by the European Office of Aerospace Research and Development (EOARD); US Air Force Facilities (1993 and 1998); Rome Lab at Massachusetts Institute; the Center for Glass Research, Alfred University, New York; and the Materials Science and Engineering Department at the University of Florida, USA. He has also presented several talks at the National Science Foundation NSF-USA, and has organized and co-organized international workshops and conferences on materials science, such as the "International Conference on Materials Research and Education: Future Trends and Opportunities," State of Qatar, 2005, by National Science Foundation (NSF), Materials Research Institute at Northwestern University, USA, International Union of Materials Research Societies, University of Qatar, Qatar Foundation: Science and Technology Park and Texas A&M University at Qatar. He is a member of the Materials Research Society, USA, and the Arab Materials Science Society.

He is the author of several books, including *Physics of Solid Material* (1996, Arabic); *Tellurite Glasses Handbook* (1st Ed., 2002); he is the co-author of *Physics and Chemistry of Rarth-Earth Ions Doped Glasses*, (TTP publishers, Switzerland, 2008) and co-translated the Arabic copy of *The World Book Encyclopedia*, as well as the Arabic copy of *Scientific American*. Invitation by the International Materials Institute "IMI-USA" for New Functionality in Glass, 2005 to present five video seminars for international educational purposes entitled; "An Introduction to Tellurite Glasses," www.lehigh.edu/imi/resources.htm. He is a reviewer and contributing editor for several journals, and evaluates for Egyptian and Arab universities for the scientific work leading to professorship, assistant professorship in physics, and materials science. He is a member of the universities strategic planes and awarded: Egyptian Science Academy Award "Glass Science," 1990, Egyptian Encouragement State Award "Physics," 1994, Menofia University Merit certificate, 1995, 2006, and Egyptian Excellent Award "Advanced Technological Sciences," 2006.

1 Introduction to and Formation of Tellurite Glasses

This chapter introduces the crystal structure of tellurium dioxide (TeO_2). It defines the formation ranges for tellurite glasses and describes their general physical characteristics, touching on color, density, molar volume, and short- and intermediate-range structural properties, including bonding. The chapter summarizes the main systems for producing and analyzing pure and binary glasses, as well as many key concepts developed later in the book: phase diagramming and immiscibility, structure models, fiber preparation, and glass-forming ranges for multicomponent tellurite glasses, nonoxide-tellurite glasses (chalcogenide glasses), halide tellurite glasses (chalcohalide glasses), mixed oxyhalide tellurite glasses, new tellurite glass systems, density and molar volume, and applications of tellurite glasses.

1.1 CRYSTAL STRUCTURE AND GENERAL PROPERTIES OF TELLURIUM OXIDE (TeO_2)

1.1.1 EARLY OBSERVATIONS

Tellurium dioxide (TeO_2) is the most stable oxide of tellurium (Te), with a melting point of 733°C (Dutton and Cooper 1966). From the viewpoint of fundamental chemistry, the transitional position of Te between metals and nonmetals has long held special significance. The stability of tellurium oxides is one of the properties that originally attracted researchers, first to the crystalline solids and then to tellurite glasses. Arlt and Schweppe (1968) and Uchida and Ohmachi (1969), while analyzing the piezoelectric and photoelastic properties of paratellurite, a colorless tetragonal form of TeO_2, suggested the potential usefulness of these compounds in ultrasonic-light deflectors. Podmaniczky (1976) and Warner et al. (1972) noted the extremely slow-shear wave propagation velocity of these crystals along the (Lin et al. 2009) direction, their low acoustic losses, and their high refractive index (*n*). These researchers suggested that these properties could be put to good use in laser light modulators.

1.1.2 ANALYSES OF THREE-DIMENSIONAL STRUCTURE

As early as 1946, Stehlik and Balak, using qualitatively estimated x-ray intensities, reported on the crystal structure of the tetragonal tellurium dioxide α-TeO_2 (paratellurite). By 1961, Leciejewicz had undertaken analyses based on 13 reflections observed by neutron diffraction analysis. A detailed comparison of the structures of α-TeO_2 and orthorhombic β-TeO_2 (tellurite) was also provided by Beyer (1967). Lanqvist (1968) further refined data on the structure of α-TeO_2 with more extensive x-ray analyses. Crystals of α-TeO_2 were prepared by dissolving metallic tellurium in concentrated nitric acid. From this mixture, α-TeO_2 crystallized as colorless tetragonal bipyramids, the basal planes of which had the crystallographic *a*- and *b*-axes as edges. The cell dimensions of this α-TeO_2 were determined using the Guinier powder method (CuKα₁ λ radiation = 1.54050 Å), with KCl as the standard. Indexing of the cell constants was performed, with the following results: (1) $a - 4.8122 \pm 0.0006$ Å, (2) $c - 7.6157 \pm 0.0006$ Å, and (3) $V - 176.6$ Å³, where *c* is the speed of light and *V* is the volume of the unit cell. The structure of α-TeO_2 was defined in terms of a three-dimensional (3D) network built up from TeO_4 subunits, with each oxygen atom shared by two units, bonded in the equatorial position to

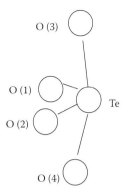

FIGURE 1.1 Schematic picture of the TeO$_2$ unit in the structure of α-TeO$_2$. (From Lanqvist, O., *Acta chem. Scand.*, 22, 977, 1968. With permission.)

TABLE 1.1

Distances between Components in Structure of α-TeO$_2$

Components	Distance (Å)
TeO$_1$ ↔ TeO$_2$	1.903
TeO$_3$ ↔ TeO$_4$	2.082
Te ↔ Te	3.740, 3.827, 4.070
O$_1$ ↔ O$_2$	2.959
O$_1$-O$_3$ ↔ O$_2$-O$_4$	2.686
O$_1$-O$_4$ ↔ O$_2$-O$_3$	2.775
O$_3$ ↔ O$_4$	4.144

Source: From O. Lanqvist, *Acta Chem. Scand.*, 22, 977, 1968.

one tellurium atom and in the axial position to another, as illustrated in Figure 1.1, and with distances between the atoms as summarized in Table 1.1 (Lanqvist 1968).

1.1.3 ANALYSES OF PROPERTIES

Arlt and Schweppe (1968) measured the dielectric constant of the tellurium dioxide crystal, and Samsonov (1973) estimated its band gap (~3.0 eV). Electrical conduction measurements on paratellurite TeO$_2$ single crystals were performed by Jain and Nowick (1981) for both parallel and normal orientations, with activation energy $E_{\parallel} = 0.54$ and $E_{\perp} = 0.42$ eV, respectively. Jain and Nowick (1981) concluded that the conductivity of TeO$_2$ crystal at temperatures below 400°C is ionic, falling into the extrinsic-dissociation range of behavior, possibly owing to oxygen ion vacancies. At high temperatures, the conductivity of TeO$_2$ increases sharply due to reduction of the sample and onset of electronic carriers. In that case, the TeO$_2$ crystal becomes slightly reduced with the corresponding introduction of N-type electronic conductors.

1.2 DEFINITION OF TELLURITE GLASSES AND REVIEW OF EARLY RESEARCH

The properties of tellurium oxides that give them their stability proved to be transferrable to their glass derivatives, thereby allowing experimentation with a wider selection of elements in the composition of tellurite glasses, and thus affording greater control over variations in performance characteristics. The first reports on tellurite glass were by Barady (1956, 1957), who showed that

TeO$_2$ forms a glass when fused with a small amount of Li$_2$O. In his later report, Barady had used x-ray analysis to investigate the structure and radial distribution of electrons within this glass; for Barady the most interesting feature of the distribution function was two well-resolved peaks—one at about 1.95 Å and the other at about 2.55 Å. The area of the larger peak in electrons was 2220, and the second peak had an area of 1340 electrons. The interatomic distances and areas of the peaks were consistent only with Te–O nearest-neighbor coordinations. In the 1956 report, Barady had assumed 5.6 and 57.0 as the effective numbers of O and Te electrons, respectively, in tellurite glass molecules; in his 1957 report, Barady had measured the areas of the two peaks at values of 3.8 and 2.3, respectively, which, within experimental error, correspond to Te–O nearest-neighbor areas of 4 and 2. Thus, the basic coordination scheme of the crystal was closely reproduced in the glass, although resolution of the peaks was not adequate to assign individual values for each Te–O interatomic distance. The distribution function showed a large increase in electron density (ρ) at radial distances beyond about 3.5 Å, which results from the Te–Te interatomic distribution. The high ρ of Te and the closeness of the peaks caused them to overlap, but there were indications of maxima at about 3.8 Å and 4.6 Å. In his seminal 1956 report, Barady had concluded that the calculated shapes of the inner part of the distribution function are not caused by errors in intensity measurements at large angles; Barady continued the x-ray study of tellurite glass to obtain better separation of the peaks of the distribution function at larger radial distances, the results of which were reported in the second article (1957). He concluded that tellurite glass is unusual because it is octahedral, with faces that exhibit a close-order configuration such that four oxygen atoms are found uniformly at a distance of 1.95 Å and two others at 2.75 Å. This configuration is similar to that found in the crystal. Barady's discussion of an octahedral link and other observations on the "edge-opening" process suggested the possible presence of crystallites in the glass itself.

To arrive at the later measurements, Barady used an x-ray diffractometer equipped with a scintillation counter, amplifying equipment, and a pulse-height analyzer. Each reading took at least 150 s, and the maximum probable error was 1%. The sample used for the 1957 report was prepared by fusing TeO$_2$ with Li$_2$CO$_3$ as flux. Microchemical analysis showed that the final composition was 98.15% TeO$_2$ and 1.84% Li$_2$O by weight. The molten glass was poured onto a flat steel surface and cooled rapidly. The diffraction pattern of Barady's TeO$_2$ glass is shown in Figures 1.2 and 1.3 (from Barady 1957). The radial distribution function (RDF) (Figure 1.3) shows two well-defined primary peaks, one at 1.95 Å and the other at 2.75 Å, as indicated above. Two additional well-resolved peaks are observed at 3.63 Å and 4.38 Å. Beyond these points, the distribution function becomes indefinite, and no additional features of any significance can be distinguished. The first two peaks have areas that are consistent only with Te–O distances. Barady proceeded directly from these

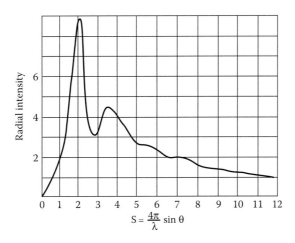

FIGURE 1.2 Diffraction pattern of TeO$_2$ glass. (From G. Barady, *J. Chem. Phys.*, 27, 300, 1957. With permission.)

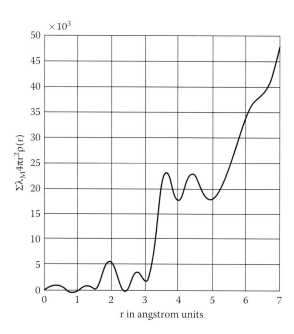

FIGURE 1.3 RDF in TeO_2 glass. (From G. Barady, *J. Chem. Phys.*, 27, 300, 1957. With permission.)

observations to an evaluation of the peak areas, which returned the number of atoms at selected distances from an atom placed centrally (radius $[r] = 0$). From these results, Barady concluded that, in tellurite glass with proportions of 98.15% TeO_2 and 1.84% Li_2O by weight, the basic coordination scheme of the parent crystal is maintained; that is, there are four nearest-neighbor O atoms surrounding a central Te atom at an average distance of 1.95 Å and two other O atoms situated at an average distance of 2.75 Å. Barady also noticed a rapid increase in ρ, which resulted from Te–Te interatomic distribution. The two peaks at 3.63 Å and 4.38 Å overlapped, and Barady could not be as definite about their maxima as about those of the inner, well-resolved peaks. Because various interatomic distances occur in this range in a crystalline system, the accuracy of attempts to resolve them as discrete peaks and compute their areas is not assured. However, it can be said that, as in the crystalline system, there are an average of two preferred sets of Te–Te distances.

The presence of the two peaks at equivalent distances and with the same nearest-neighbor numbers as those in the crystal strongly suggests that the same octahedral coordination is present in the glass structure derived from the crystalline material. Barady (1957) also visualized other close-order configurations that would exhibit two nearest-neighbor peaks, but it would have been fortuitous indeed if these had corresponded as closely to the octahedral coordination scheme of the crystal as do our findings. It seems preferable to conclude that the octahedral structure is preserved relatively unchanged when the crystalline material is transformed into glass. This conclusion agrees with earlier experimental findings by Warren (1942). Working with the SiO_2 glass system, Warren found that tetrahedral SiO_4 groups are present in all the known crystalline forms, and the same tetrahedral structure is found in the glassy state. These tetrahedral building blocks are assembled in different ways to accomplish various modifications, and it is their external arrangements that distinguish one form from another.

In 1957, Winter related glass formation—that is, the ability to form bonds leading to a vitreous network—to the periodic table of the elements. Thus, only four elements of group VIa of the periodic table (O, S, Se, and Te) are known to form monatomic (primary) glasses (i.e., simple glasses containing one kind of atom). These elements all retain the ability to form a vitreous network when mixed or chemically bound to one another, and they can also form binary glasses (i.e., glasses containing two kinds of atoms), by combining with elements from group III, IV, or

TABLE 1.2

Melting Point (T_m) and Forming Temperature (T_g) of Glass

Element	T_m (°C)	T_g (°C)
Oxygen	55	37
Sulfur	393	262
Selenium	493	328
Tellurium	725	484

Source: From A. Winter, *J. Am. Ceram. Soc.,* 40, 54, 1957. With permission.

V of the periodic table. Glasses that include group VIa elements are known to be very viscous in the liquid state and undercool easily; the temperature at which glass becomes a solid (T_g) and the corresponding melting points were given by Winter (1957) (Table 1.2), who prepared about 20 new glasses. Other binary glasses are known to be composed of an element from group VIIa (F, Cl, I, or Br) and either an element from group II, III, or IV or a transition element. Winter called these "self-vitrifying elements."

The existence of TeO_2 glass seems to conflict with one of the geometrical conditions for glass formation postulated by Zachariasen (1932), who stated that only compounds with oxygen triangles (tetrahedrals) can form a glass with energy comparable to that of the crystalline form. This rule was formulated based on empirical evidence. Goldschmidt (1926) computed that the radius ratio of cations to oxygen atoms lies between 0.2 and 0.4 for all glass-forming oxides, which corresponds in general to a tetrahedral arrangement of oxygen atoms. Furthermore, all attempts to make glasses from TiO_2 or Al_2O_3—both of which are octahedral in the crystalline state—were unsuccessful. It was concluded, therefore, that if the coordination complex of a compound comprises greater than four atoms, the resultant edge or face sharing of the polyhedral will fix the symmetry in too many directions in space. A shared face introduces high rigidity to a lattice; a shared edge has only the angle at the edge undetermined; and a shared corner allows the angle at the common corner of the polyhedron to be varied in all directions. Any transformation to a glass can be accomplished quite easily in the latter case, since any amount of disorder can be introduced into the network by simply varying this angle; in the other two cases, however, a large distortion of the octahedral structure is required in order to disrupt the symmetry sufficiently to produce the random structure that is characteristic of glass.

TiO_2 and Al_2O_3 are ionic compounds, and their octahedral structures are quite regular in the crystalline state. TeO_2 is covalent and has a highly deformed octahedron in its structure, because the valence characteristic of Te results in two sets of Te–O distances. Since each oxygen atom must be shared with three tellurium atoms, symmetry requirements force a distortion of the octahedrals to accommodate them into a regularly repeating lattice; thus, there are 6 different Te–O distances, 4 Te–Te distances, and 12 O–O distances. It is possible that this distortion produces a structure that is energetically similar to that of the vitreous state—in which there is only short-range order (SRO)—and, furthermore, that because Li^+ ions are added, some O atoms are not joined with three Te atoms, but instead form ionic bonds with Li^+. We have already seen from Barady's early analyses that the first peak in the RDF is at 1.95 Å, which is a decrease from that of the corresponding group of Te–O distance measurements in the crystal, and the Te–Te distance at 4.38 Å is somewhat greater than the equivalent spacing of 4.18 Å along the *b*-axis of the crystal. This indicates qualitatively that in the glass there is a less rigid arrangement of the octahedral than there is in the crystal, which would be expected if the octahedral assumes a partly ionic character to balance the charge of the Li^+ ions present in the network. The process of transformation to glass would be in some way related to breakdown of edge sharing, because it is precisely this property that introduces rigidity into the lattice.

Barady (1957) noticed that it is necessary to add a modifier (M) to TeO_2—such as Na_2O or Li_2O—before the material solidifies into a glass, and that attempts to melt the pure material and quickly quench it to a glass were unsuccessful because the melted material quickly recrystallized. Barady also noted that a concentration of about 10 mol% of the M is necessary before there is any tendency toward glass formation. It is simple to prepare SiO_2 glass from the pure material because the mechanism of formation is simply one of distortion of the corner tetrahedral Si–O–Si angle. To break one edge in TeO_2, however, an oxygen atom must be supplied in order to complete the octets of the two cations involved. The addition of Li_2O supplies ions for this process. When Li^+ ions are introduced, the disrupted-edge (O) ions may coordinate around them in much the same way as in the silicate system, in which Li^+ or Na^+ ions find positions in the holes in the network. Also, Li^+ ions are surrounded by the O ions, which are bonded only to one Si. The only difference for TeO_2 is that the O atom from the added oxide plays a more essential role—that of breaking an edge.

In crystalline TeO_2, four octahedrals share three edges. If the glass consists of a network of octahedrals in which all the shared edges are broken, the minimum molecular-mass ratio (mole ratio) of Li_2O to TeO_2 would be 3:4. Barady (1957) observed that the minimum mole ratio actually required is approximately 1:10. He concluded that only a fraction of the edges are opened—therefore, the glass cannot be made up completely of octahedrals linked together only by their corners. At first glance, this result might appear to be in conflict with Zachariasen's picture of the atomic arrangement in glass. In TeO_2 crystal, each unit cell has eight molecules. Zachariasen assumed that, for a crystalline size of this order to occur in glass, the effective mole ratio of Li_2O to TeO_2 should be divided by 8; that is, $3:4 \times 1:8 = 1:10$. That calculation was shown to produce the approximate minimum mole ratio of compounds required to form the glass. Barady (1957) concluded that the SRO structure in tellurite glass differs little from that of the crystal, and that this close order does not have to be tetrahedral. Stanworth (1952) has added excellent glass formers to the Pauling scale with electronegativites of 1.7 to 2.1. According to Pauling, these are values in the same range as arsenic, antimony (which form, for example, As_2O_3 and Sb_2O_3 glasses), and tellurium, which have an electronegativity value of 2.1; that is, the same as for phosphorus. These facts lead one to consider whether tellurium oxides or the tellurites or tellurates form glasses.

According to the data presented by Wells (1975), there are two crystalline forms of TeO_2, including a yellow orthorhombic form (the mineral tellurite) and a colorless tetragonal form (paratellurite). There is "four coordination" of Te in both forms, the nearest neighbors being arranged at four of the vertices of a trigonal bipyramid, which suggests a considerable covalent character to the Te–O bonds. Tellurite has a layered structure in which TeO_4 groups form edge-sharing pairs, which then form a layer by sharing their remaining vertices. The short Te–Te distance, 3.17 Å (compared with the shortest distance in paratellurite, 3.47 Å), may account for its color. In paratellurite, very similar TeO_4 groups share all vertices to form a 3D structure with 4:2 coordination in which the O-bond angle is 140°, distances of the two axial bonds are 2.08 Å, and distances of the two equatorial bonds are 1.9 Å.

Tellurite glasses exhibit a range of unique properties that give them potential applications in pressure sensors or as new laser hosts. These glasses are now under consideration in many other applications. Although the physical properties and structure of crystalline solids are now understood in essence, this is not the case for amorphous materials, including glass. The considerable theoretical difficulties in understanding the properties and structures of amorphous solids are amplified by the lack of precise experimental information. Research should be accelerated to fill this gap. The benefits will include providing the fundamental bases of new optical glasses with many new applications, especially tellurite glass-based optical fibers. Greater research attention must be given to new materials for optical switches, second harmonic generation (SHG), third-order nonlinear optical materials, unconventional glasses, and optical amplifiers. There are, however, many cases not discussed here in which researchers have studied and attempted to identify other uses of new glasses.

Great expectations have been placed on the development of new glasses as indispensable materials in developing the vital industries of the near future. This can clearly be seen in such fields as optoelectronics, multimedia, and energy development. "New glasses" are those that have novel functions and properties—such as a higher light regulation, extraordinary strength, or excellent heat, and chemical durability—which are far beyond the characteristics of conventional glasses. These functions and properties are realized through high technologies, such as super-high purification and ultra-precise processing, controlled production processes, utilization of new materials as composites in glass, and full exploitation of various specific characteristics of conventional glasses, including the following:

1. Optical homogeneousness and transparency to light
2. Excellent solubility, which enables almost all elements to mix with the original crystalline material, resulting in a wide range of composites with diverse functions and properties
3. Excellent hardness and chemical durability, and relatively high strength
4. Malleability, which allows formation into various shapes (i.e., the ability to form various shapes easily)
5. Adaptability to perform more specific functions as additional properties are provided to the glass by various surface treatments

The development of new glasses will add new functions and properties to the list above, often in response to requests from industries like electronics and optoelectronics. Various types of new glasses will continue to be needed in major industries of the twenty-first century. The relationships between new glasses and these industries can be configured into groups by the basic types of functions of glass: optical, thermal, electronic, mechanical, magnetic, and chemical.

1.3 PREPARATION OF PURE TELLURIUM OXIDE GLASS

In 1962, Cheremisinov and Zalomanov succeeded in melting white crystalline TeO_2 powder by chemical synthesis. They ascertained by x-ray analysis that this melt had a tetragonal lattice. The powder was nearly pure, containing 99.0% to 99.5% tellurium dioxide. An aluminum crucible containing the powdered tellurium dioxide was placed in a furnace, where it was heated to between 800°C and 850°C and kept at this temperature for 30 to 40 min until complete fusion was assured. The melt was then cooled to 400°C at a rate of 100°C/h and left at this temperature for about 1 h. Then, with the furnace turned off and closed, the glass was allowed to cool naturally. A transparent specimen was prepared in the form of a cylinder 10 mm in diameter by 10 mm in length. The resulting glass had a greenish hue and contained up to 6% Al_2O_3, which came from the walls of the crucible during the fusion process. This attempt fell short of producing pure tellurite glass.

Lambson et al. (1984) used white crystalline tetragonal TeO_2 powder (British Drug House grade 99%) to prepare their pure tellurium oxide glass. The tellurium dioxide was placed in an electric furnace preheated to 800°C and was kept at this temperature for 30 min, by which time complete fusion had occurred. The melt was then cooled at a rate of 20°C/min to a temperature of 700°C, at which it remained highly viscous. This melt was cast in a cube-shaped split mold of milled steel, which had been preheated to 400°C, and the volume of the molded sample was 1.5 cm³; this product was annealed in a second furnace at 300°C for 1 h, after which the furnace was switched off and the glass was allowed to cool *in situ* for 24 h. A large number of alternative thermal cycles were also tried, but they failed to produce glass. For example, when the TeO_2 melt was cast at 800°C directly in a mold that had been preheated to 400°C, a white, polycrystalline porcelain-like substance was obtained. The same result occurred when melts were cast at various temperatures between 800°C and the melting point of tetragonal TeO_2. The interim stage of cooling from 800°C to about 700°C is evidently crucial to glass formation. Lambson et al. (1984) conceptualized this result qualitatively into two principles:

- To form a glass rather than a polycrystal, a melt must be very viscous, which requires casting (i.e., supercooling) the melt at the lowest temperature possible, which permits the necessary flow.
- However, the starting material must initially be heated to well above its melting point so that any residual crystallinity in the melt is removed.

The glass produced by Lambson et al. had a greenish hue and contained up to 1.6% Al_2O_3, which was proportionately less than that occurring with the previous technique.

1.4 GLASS-FORMING RANGES OF BINARY TELLURIUM OXIDE GLASSES

Properties and structure of binary tellurite glasses containing monovalent and divalent cations were reported by Mochida et al. (1978). In the same year, Kozhokarov et al. examined the formation of binary tellurite glasses containing transition metal oxides (TMOs). Glass formation ranges were measured and investigated in the following binary tellurite systems (where M is the modifier): $MO_{1/2}$-TeO_2 (where M is Li, Na, K, Rb, Cs, Ag, or Ti) and MO–TeO_2 (where M is Be, Mg, Ca, Sr, Ba, Zn, Cd, or Pb). The properties of these glasses, such as density ρ, refractive index n, thermal expansion, and infrared (IR) spectra, were recorded, and glass formation range data were collected (Table 1.3). The possibility of additional SiO_2 impurities from the quartz crucible itself was not considered in these data. Moreover, the data represent the starting component rather than analyzed values, and in view of the relatively low boiling point of TeO_2 compared with those of other constituents, the proportions of M in these glasses probably were somewhat higher than the quoted values. Variation in the glass formation phase depends on both the nature of the M itself and the type of corresponding phase diagram, but it mainly depends on the position of the first eutectic point, as well as the formation of binary compounds in the tellurium dioxide-rich area of the system. For example, in the MnO–TeO_2 system, compounds formed in ratios of 6:1, 2:1, 3:2, 1:1, 6:5, and 4:3 cause a narrowing of the glass formation range compared with the range in the MnO_2–TeO_2 system. Analogous relations are observed in the TeO_2–CoO system, in which compounds are formed in ratios of 6:1, 2:3, 6:5, and 4:3. The upper limit of glass formation in the TeO_2–ZnO system is correlated with the $Zn_2Te_3O_8$ compound from the corresponding phase diagram. Melts above the geometric point of the latter compound do not cool as glass.

Marinov et al. (1988) prepared amorphous films based on tellurium dioxide and rare-earth metal oxides and measured the changes in optical absorption. These authors showed that heating alters the absorption coefficient (α) and the n of these films. They reported that suboxide thin films are influenced by laser beams at λ = 830 nm and are fit for use as media for optical information recording. In the work of Marinov et al. (1988) on thin-film synthesis, some binary powder mixtures of TeO_2 and rare-earth metal oxides were used, including 85 mol% TeO_2–15 mol% R_nO_m (where R is a rare-earth metal) and 95 mol% TeO_2–5 mol% R_nO_m, as discussed in Table 1.3. The R metal oxides introduced in the batch were La_2O_3, CeO_2, Pr_6O_{11}, Nd_2O_3, Sm_2O_3, Eu_2O_3, Gd_2O_3, Tb_4O_7, Dy_2O_3, Ho_2O_3, Er_2O_7, Tm_2O_3, Yb_2O_3, Lu_2O_3, and Sc_2O_3. Powder mixtures were homogenized in an agate mortar and then thermally treated in air at 600°C for 48 h. The mixtures were heated in special quartz crucibles with a tungsten heater. The amorphous thin-film samples (≅100 nm) were prepared by electrical evaporation in a standard vacuum installation at 2×10^{-5} torr. The substrate was a sheet of ultrasonically purified glass with dimensions of 25 mm by 20 mm by 1.5 mm. The films were centrifugally deposited at a speed of 100 revolutions/min, and the following list summarizes the visual observations of the colors of the prepared films after an additional thermal treatment in air at 250°C for 5 min (at which point the colored samples darken): TeO_2–La_2O_3, transparent; TeO_2–CeO_2, transparent; TeO_2–Pr_6O_{11}, dark

TABLE 1.3

Glass-Forming Ranges of Binary Oxide-Tellurite Glasses

Second Component (Reference)	Lower Limit mol%	Upper Limit mol%	ρ (g/cm³)	Color
$LiO_{1/2}$ (Mochida et al. 1978)	20.0	46.3	5.307–4.645	Pale yellow
$NaO_{1/2}$	10.0	46.3	5.406–4.450	Pale yellow
$KO_{1/2}$	2.5	34.6	5.520–4.516	Pale yellow
$AgO_{1/2}$	10.0	20.0	5.723–5.896	Pale yellow
$Tl_{1/2}O$	5.0	59.6	5.700–7.410	Yellow
BeO	10.0	20.0	5.445–5.217	Pale yellow
MgO	10.1	40.4	5.482–4.765	Pale yellow to dark yellow to amber
SrO	11.0	4.8	5.57–5.4487	Pale yellow
BaO	2.5	35.8	5.596–5.382	Pale yellow
ZnO	45.0	2.5	5.602–5.408	Pale yellow
CdO	5.0	10.0	5.637–5.681	Pale yellow
PbO	20.0	5.0	5.731–6.196	Pale yellow
Sc_2O_3 (Kozhokarov et al. 1978)	8.0	20.0		
TiO_2	7.0	18.5		
V_2O_5	7.5	58.0		
Cr_2O_3				
CrO_3	4.0	7.5		
MnO	15.0	27.5		
MnO_2	15.0	33.5		
Fe_2O_3	2.5	20.0		
Fe_3O_4				
CoO	6.0	14.3		
Co_3O_4	4.2	12.5		
NiO, Ni_2O_3				
CuO	26.2	50.0		
Cu_2O	17.0	22.5		
	30.0	37.5		
ZnO	9.2	40.0		
MoO_3	12.5	58.5		
WO_3	11.0	33.3		
B_2O_3 (Burger et al. 1984)	19.7	24.9	4.937–4.689	Transparent
P_2O_5 (Neov et al. 1984)	8.0	26.0	5.428–3.987	Transparent
GeO_2 (Ahmed et al. 1984)	90.0	70.0	3.83–4.11	
TeO_5–V_2O_5–MoO_5 (Kozhokarov et al. 1981)[a]				
La_2O_3, CeO_2, Pr_6O_{11}, Nd_2O, Sm_2O_3, Eu_2O_3, Gd_2O_3, Tb_4O_7, Dy_2O_3, Ho_2O_3, Er_2O_3, Tm_2O_3, Yb_2O_3, Lu_2O_3, Sc_2O_3 (Marinov et al. 1988)	5.0	15.0		

[a] See Figure 1.12.

brown; TeO_2–Nd_2O, black; TeO_2–Sm_2O_3, transparent; TeO_2–Eu_2O_3, dark brown; TeO_2–Gd_2O_3, transparent; TeO_2–Tb_4O_7, dark brown; TeO_2–Dy_2O_3, transparent; TeO_2–Ho_2O_3, dark brown; TeO_2–Er_2O_3, transparent; TeO_2–Tm_2O_3, light brown; TeO_2–Yb_2O_3, brown; TeO_2–Lu_2O_3, brown; and TeO_2–Sc_2O_3, brown.

After the film deposition, an amorphism test was performed by x-ray diffraction (XRD). The films were amorphous in this analysis, but after thermal treatment some low-intensity diffraction peaks were observed. Burger et al. (1984) studied the phase equilibrium, glass-forming properties, and structures of the TeO$_2$–B$_2$O$_3$ system (Table 1.3). Glass produced in this system was transparent, and the authors investigated 30 closely spaced geometric points in this glass, depending on the composition and distribution in the G or S regions (Figure 1.4). The most probable disposition of the fields of preliminary crystallization is presented in Figure 1.4. The monotectic temperature was determined by DAT to be 934 K. An invariant point corresponding to a composition of 73.6 + 0.5% TeO$_2$ was found; from this location, the system could be treated as a composite of two quite different regions (marked G and S in Figure 1.4):

1. In the TeO$_2$-rich part of the system, an exact determination of the glass formation range was made. In the α subregion, at a cooling rate of about 10 K s^{-1}, glasses (100 g batches) were completely free of crystals, whereas crystals occurred in the β subregion at a cooling rate of about 1 K s^{-1}. At above an 80% concentration of TeO$_2$, partial crystallization of the melts occurred, producing α-TeO$_2$.

2. In the B$_2$O$_3$-rich part of the system (Figure 1.4, Region 5), there was a stable miscibility (MG) gap. The glass melts, which were prepared by different methods and frozen at different temperatures, showed a distinct separation into two vitreous phases—namely, a transparent-glass phase and an opaque-glass phase with low ρ. As the temperature was increased, the dense phase became steadily less stable.

Burger et al. (1984) also prepared a melt with a composition of 60 mol% TeO$_2$ and 40 mol% B$_2$O$_3$ by cooling with rotating copper rollers at a rate of >10^3 K s^{-1}. The distribution of elements in this

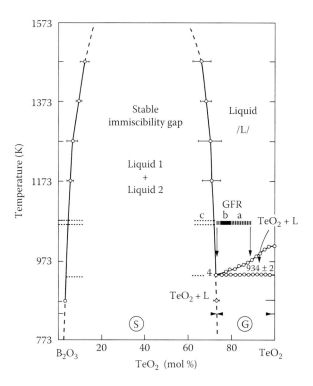

FIGURE 1.4 TeO$_2$–B$_2$O$_3$ system (see Table 1.3). The glass produced was transparent, and 30 geometric points were investigated in the system at small intervals, depending on the composition and its distribution in the G or S regions. GFR is the glass forming region. (From H. Burger, W. Vogel, V. Koshukharov, and M. Marinov, *J. Mater. Sci.*, 19, 403, 1984.)

sample, measured by x-ray microanalysis, showed an accumulation of 80% tellurium in the middle part, diminishing to about 4% at the edges.

1.4.1 Phase Diagram and Immiscibility

In relation to the glass former, it is known that the monotectic temperature (i.e., the isotemperature line at which phase equilibrium occurs between two liquids or a solid state occurs in a binary system) can be either above or below the melting point of the glass former. When it is above the melting point, an ideal two-component glass-forming system is present, in which a stable immiscibility gap occurs. In this case, the monotectic temperature lies between the melting point of the glass former and that of a second glass former beyond the immiscibility gap. Many borate and silicate systems with immiscibility gaps behave in this way. Figure 1.5 is a phase diagram for a TeO_2–B_2O_3 system, republished from Burger et al. (1984) and using their distributions applied to glass-forming systems below the monotectic temperature. The chemical incompatibility between the two mutual glass formers in this system favors bonding at points that simultaneously preserve the structure of the original networks; that is, a space-structural differentiation of the polyhedral occurs.

This co-effect of the two glass formers begins at a critical point, which for tellurite systems is when the concentration of the second glass former reaches $26 \pm 5\%$, as mentioned by Neov et al. (1980), who used neutron diffraction data to measure and describe the co-effects of two glass formers in the TeO_2–P_2O_5 system. The curves for the RDFs that they obtained show considerable

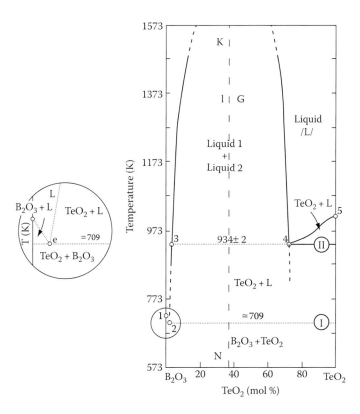

FIGURE 1.5 Phase distribution in the TeO_2–B_2O_3 system at between 573 K and 1573 K temperatures. Point 1, melting point of B_2O_3; point 2, eutecticum at <2 mol% TeO_2–98 mol% B_2O_3; points 3 and 4, nonvariant points; point 5, melting point of TeO_2 (1,006 K); line I, isotherm at ≈709 K; line II, monotectic temperature; K–N, isopleth at 38 mol% TeO_2–62 mol% B_2O_3. (From H. Burger, W. Vogel, V. Koshukharov, and M. Marinov, *J. Mater. Sci.*, 19, 403, 1984. With permission.)

destruction of the SRO in the tellurite matrix, whereas the basic coordination compound PO_4 poly-hedron remains unchanged. Their structural interpretation of the immiscibility is based on certain important factors:

- An essential influence on immiscibility in glasses is exerted by the structural interactions among its components.
- An important factor for developing a liquification process in glasses is the electrostatic bond strength, defined as Z/CN, where Z is the charge of the cation and CN is the coordination number. It is known that phase separation is typical for bond strengths of 3:8 and 1:2, which shows that a modifier must have a variable bond strength with oxygen to cause immiscibility. In liquification systems with an *M*, the stronger the *M*-oxygen bond, the stronger is the tendency toward phase separation. A necessary condition is that the electrostatic bond strength should be less than 1:2 and greater than 5:6. Irrespective of these criteria, however, the oxygen coordination polyhedron of the cation must be incompatible with the cation of the network former.
- Both cations in the studied system tend to keep their coordination polyhedra, chains, and layers stable. There can be no liquification if the cation-oxygen field strength is too great or, on the other hand, is negligible. It is therefore important to have, as a measure of the liquification process, a quantitative evaluation of the bonding of Te and P ions with oxygen. Such a quantitative measure is the difference in the ionic field strengths. The ionic field strength for this glass is 0.071 (i.e., the difference between the field strengths of Te–O and P–O), which satisfies the condition for liquification to proceed.

The experimental and theoretical interatomic distances for binary tellurite-phosphate glasses as measured and calculated by Neov et al. (1980) are given in Table 1.4. The structural characteristics of the two polyhedra are discussed above. Apart from the high electrostatic charge of the P^{5+} ions, the double P=O bond favors the closest packing of the polyhedron and, consequently, sharply increases the necessary minimum energy necessary to break the bonds—one reason that liquification in a binary phosphorous system is absent or extremely rare. Therefore, it is logical to seek the causes of liquification in the structural features of the TeO_4 polyhedra and their interactions with the

TABLE 1.4

Interatomic Distances in Binary Tellurite-Phosphate Glasses

Glass Components (mol%)[a]	Interatomic Distance (Å)			
	r (Te–O)	r (O–O)	r (Te–Te), (Te–O)	Coordination No.
TeO_2 (theor.)	2.00	2.85	3.85	4
$92\%TeO_2$–$8\%P_2O_5$ (exp.)	1.95	2.85	3.90	4
$92\%TeO_2$–$8\%P_2O_5$ (theor.)	2.00	2.90	3.90	
$84\%TeO_2$–$16\%P_2O_5$ (exp.)	1.90	2.70	3.70	
$84\%TeO_2$–$16\%P_2O_5$ (theor.)	2.00	2.90	3.90	
$80\%TeO_2$–$20\%P_2O_5$ (exp.)	1.80	2.75	3.90	
$80\%TeO_2$–$20\%P_2O_5$ (theor.)	2.00	2.90	3.85	
$74\%TeO_2$–$26\%P_2O_5$ (exp.)	1.57	2.60	3.70	
$74\%TeO_2$–$26\%P_2O_5$ (theor.)	2.00	2.85	3.85	

Source: S. Neov, I. Gerasimova, V. Kozhukarov, and M. Marinov, *J. Mater. Sci.*, 15, 1153, 1980.

[a] exp., experimental basis; theor., theoretical basis.

second glass former. The main structural changes that TeO_4 polyhedra undergo in the initial stages of liquification are summarized as follows:

1. The TeO_4 unit is a one-sided, coordinated polyhedron, with a free electron pair and highly mobile axial bonds.
2. Such a coordination pseudosphere of the Te atom favors secondary bonding; that is, an easy variability of the bond strength with respect to oxygen exists. The latter point follows from the high deformation of the Te–Te, Te–second-O, and O–second-O distances.
3. The active interaction of TeO_4 polyhedra on the side of their free electron pairs with the unsaturated oxygen ions cannot lead to the formation of strong Te–O–P ion covalent bond, as in the presence of an M.
4. With two glass formers present simultaneously and, because of the antagonistic tendencies for preservation of their own networks, a space-structural differentiation of the polyhedra takes place in the initial stages of phase separation.
5. The weaker acidity of TeO_2 (anhydride of the weak H_2TeO_3 acid) compared with that of P_2O_5 also favors easier attacking of the chains in α-TeO_2. In this aspect, TeO_2 probably plays the role of an M; that is, it appears more "alkaline" than the second glass former.

According to the above factors, the different glass-forming polyhedra have no inclination to associate among themselves, which provides grounds to consider that a space-structural differentiation of the polyhedra, groups, chains, and layers could take place; that is, a liquification process might occur. It should be noted, however, that the formation of low-polymerization phosphate groups at low P_2O_5 concentrations is possible. Their formation is influenced by two additional factors: first, a decrease in the extent of polymerization of phosphate structural patterns, because of the essential incorporation of the second component in the tellurite network; and second, the apparent incompatibility of the structural units of the two glass formers. At higher P_2O_5 concentrations, the polymerizing phosphate easily forms groups and chains from microscopic drop-like aggregates that are locked together by the defective tellurite matrix, which has considerably higher stereochemical instability.

1.4.2 Structure Models

Neov et al. (1980) discussed changes in the P_2O_5 amount and the behavior of the glasses in terms of two structural models based on RDF data they obtained. The first of these models presupposes microhomogeneity and is valid for glasses in the TeO_2-rich region. The second model describes the mechanism of the initial stage of microheterogeneity in glasses of the TeO_2–G_nO_m systems (in which G_nO_m is a glass former) and is valid in the G_nO_m-richer regions. For microhomogeneous glasses, the structural model illustrated in Figure 1.6 is considered to be reliable. The model can be represented by symbols (see Figure 1.7), but this means of representation does not permit the display of structural angular deformation or the stereo-ordering in tellurite glasses. The basis for this model is as follows: glass formation in tellurite systems takes place when the structure is highly defective—a fact that has been realized mainly through deformation of the bonds and angles and, in places, breaking of the Te–O–Te–O–Te chains and corresponding layers. Apart from this structural-rupture model, bonding rupture also occurs when PO_4 polyhedra are incorporated into the tellurite chains, according to the mechanisms illustrated in Figures 1.6 and 1.7. Independent of this, most of the PO_4 groups are implanted predominantly in the active tetrahedral holes between TeO_4 chains. This method of bonding increases the covalence of the wave-like chains and layers. The layers interact weakly among themselves through PO–Te bridging oxygen (BO) bonds. The PO_4 tetrahedrals positioned between the layers do not contribute to the strengthening and formation of stable ion-covalent structures because they tend toward coordinate saturation. In addition to connections of the type –Te–O–P–O–Te– (Figures 1.6 and 1.7), a certain number of PO_4 tetrahedrals have been

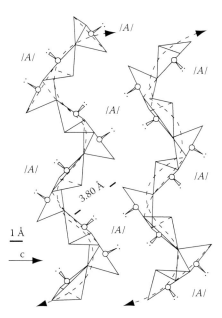

FIGURE 1.6 Model illustrating the structure of the atomic arrangement of the TeO$_2$-rich glass region. (From S. Neov, I. Gerasimova, V. Kozhukarov, and M. Marinov, *J. Mater. Sci.*, 15, 1153, 1980.)

shown to occupy sites by incorporation into the tellurite chains and layers. In this way the mobility of the polyhedra and chains sharply increases. With an increase in concentration of the second glass former, an increasing portion of the PO$_4$ units becomes distributed between the wave-like layers.

Thus, two types of bonding of the phosphorus ions are realized: that of P atoms in tellurite chains, and that of P atoms incorporated between chains and layers. The first type contributes to the breaking of the chains and high deformation of the structure—both of which take place as

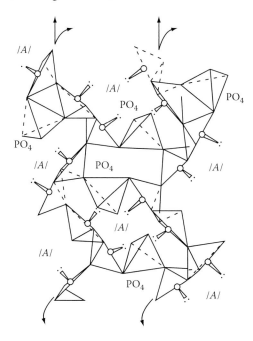

FIGURE 1.7 Representation of the model illustrated in Figure 1.6 by symbols. (From S. Neov, I. Gerasimova, V. Kozhukarov, and M. Marinov, *J. Mater. Sci.*, 15, 1153, 1980.)

early as the introduction of 2% P_2O_5, at which time the tellurite melt is easily overcooled. This result shows that, at low concentrations, the PO_4 polyhedra essentially contribute to the coordination asymmetry of the tellurite structural patterns. The second type of phosphorus atoms (denoted as P^x) are distributed between tellurite chains and layers and have a marked tendency to preserve their own coordination bonding. Both methods of bonding probably stimulate the creation of nonbridging oxygen (NBO) ions of –Te–O, whereas the NBO ions to which the PO_4 group contributes are predominantly of PO_4. On the basis of detailed neutron diffraction studies carried out in the P_2O_5-rich region, an SI model has been constructed (Figure 1.8). This model gives a satisfactory description of the mechanism of initial microheterogeneity in tellurite glasses, which accounts for the following main structural factors that stimulate stable liquification in the initial phases on the SRO level:

- With the increase in P_2O_5 concentration, more PO_4 polyhedra attack the tellurite chains and layers; notwithstanding the large deformation in structure, there is a tendency to preserve the original structural character and high covalence. Rupture of the tellurite in the phosphate chains and layers occurs at points where, because of their high stability, the PO_4 units have exerted their influence long before the addition of an equivalent amount of the second glass former. Regarding certain regularity, in tellurite glasses with a second glass former, there should be stable liquification starting with a critical composition within the limits $26 \pm 5\%$ G_nO_m. Variations in these limits depend on the conditions of glass formation (e.g., kinetic factors of the vitreous transition), the type of phase diagram, and the electrostatic bond strength, as well as the structural specificity, acidity, conservation, and other characteristics of the polyhedra.
- The SI model is a highly covalent, deformed structure between heavy layers that have very dynamic Te–P, Te–Te, Te–second-O, and O–second-O distances and BO bonds between them. By themselves, these qualities favor the weak interaction between groups and atoms belonging to different chains and layers. Parallel and perpendicular to the broken tellurite chains, fragments of chains of the second glass former are incorporated in addition to the polyhedra. This second glass former causes strong structural disorder because of its tendency to form its own network. Modifications to the functions of the two glass formers depend on their concentrations. The PO_4 units structurally differentiate even before the quimolecular composition is reached because of their stronger acidity and stability. Owing to their structural characteristics, the TeO_4 polyhedra cannot significantly resist the influence of the second glass former. Thus, each component forms its own microstructure. The strong antagonism in binding leads to special structural differentiation (Figure 1.8).

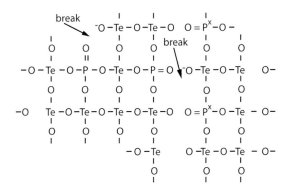

FIGURE 1.8 Structural incompatibility model illustrating the nature of the initial steps for immiscibility in the P_2O_5-rich glass region. (From S. Neov, I. Gerasimova, V. Kozhukarov, and M. Marinov, *J. Mater. Sci.*, 15, 1153, 1980.)

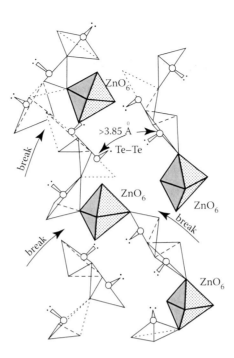

FIGURE 1.9   Model illustrating the nature of the atomic arrangement in 80 mol% TeO$_2$–20 mol% ZnO glass. (From V. Kozhukraov, H. Burger, S. Neov, and B. Sizhimov, *Polyhedron*, 5, 771, 1986).

The SI model describes the events in the initial stage of the structural regrouping of several polyhedra or chain fragments, which is the main prerequisite for the development of microheterogeneity.

The phenomenon of regrouping has not been observed by electron microscopy at this stage in structural differentiation; but in 1986, Kozhukharov et al. established, from the atomic arrangement in zinc-tellurite glass (80 mol% TeO$_2$–20 mol% ZnO), that atomic distances are most sensitive to transition to the glassy state at this stage, which may be explained by the fact that the shortest Te–Te distance in α-TeO$_2$ is 3.76 Å, as shown in Figure 1.9. Such a relatively long distance between the two heavy metal cations suggests that some weak metal bonds are present. Kozhukharov et al. (1986) compared the ion radius (r_k) and bond energy (D_{r-r}) in some common glass formers and found that they vary in the order shown in Table 1.5.

TABLE 1.5
Comparison of Ion Radius and Bond Energy of Selected Glass Formers

Parameter[a]	Characteristics of Selected Glass Formers[b]					
	B$_2$O$_3$	P$_2$O$_5$	SiO$_2$	GeO$_2$	As$_2$O$_3$	TeO$_2$
r_k (Å)	0.20	0.35	0.41	0.53	0.43	0.56
r_{k-O} (Å)	1.26	1.52	1.55	1.63	1.80	1.99
D_{R-R} (kcal/mol)		51.30	42.20	37.60	32.10	33.00

[a] r_k, ion radius; r_{k-O}, cation oxygen bond length; D_{R-R}, bond energy.
[b] Data adapted from V. Kozhukharov, H. Burger, S. Neov, and B. Sidzhimov, *J. Non-Cryst. Solids*, 5, 771, 1986.

It was established that Te atoms have the greatest r and mass, the highest value of the average Te–O distance, and the lowest Te–Te bond energy—nearly the same as those characteristics of As–As in As_2O_3. Kozjukharov et al. (1986) also noted that during glass transition, the bond r of Te and O (r_{Te-O}) of 3.8 Å between two tellurite-oxygen chains is not preserved because, along the c-axis in paratellurite, there are channels that can incorporate even the second-component polyhedra.

The TeO_2–GeO_2 binary glass system was prepared by Ahmed et al. (1984), who also reported the direct-current (DC) electrical conductivity, optical absorption edge, and IR optical absorption of these glasses. They found that conduction in TeO_2–GeO_2 glasses is electronic and the hopping of polarons seems to be the dominant process in the transport mechanism. The electrical activation energy E decreases with increases in tellurium content, and this decrease corresponds to a decrease in the optical energy gap. The optical gap is on the order of 2.74 eV—somewhat lower than that for many other oxide-based glasses. Most of the sharp absorption bands characteristic of the basic compounds GeO_2 and TeO_2 are modified with the formation of broad, strong absorption bands in the transition from a crystalline to amorphous state. Density ρ measurements show that the glasses have a compact structure.

The above text does not cover all binary tellurite glass systems; many of the more recently developed systems, most with very important uses, are discussed in later chapters.

1.5 GLASS-FORMING RANGES OF MULTICOMPONENT TELLURIUM OXIDE GLASSES

1.5.1 TELLURITE GLASSES PREPARED BY CONVENTIONAL METHODS

The glass formation ranges of 22 ternary tellurite systems were reported by Imoka and Yamazaki in 1968. Their experiments were similar to those of previous reports on the borate, silicate, and germinate systems. Their crucibles were made of an Au alloy containing 15% Pd. Except for TeO_2, the oxides were of 16 "a-group" elements—namely K, Na, Li, Ba, Sr, Ca, Mg, Be, La, Al, Tb, Zr, Ti, Ta, Nb, and W; and 5 "b-group" elements—namely Ti, Cd, Zn, Pb, and Bi. The work of Imoka and Yamazaki with ternary systems included all the combinations of these oxides (1968), except for systems with narrow glass formation ranges or with none at all. In the following list, optional third components that lead to such narrow glass formation ranges are listed parenthetically after selected second components of ternary tellurite glasses: CaO (La_2O_3, ThO_2, Ta_2O_5, CdO, PbO, or Bi_2O_3); Ta_2O_5 (Mg_2O, BeO, La_2O_3, Al_2O_3, ThO_2, TiO_2, Nb_2O_5, Tl_2O, CdO, ZnO, PbO, or Bi_2O_3); CdO (La_2O_3, Al_2O_3) Bi_2O_3 and (La_2O_3, Al_2O_3, ThO_2, or TiO_2).

According to Barady's (1957) data from x-ray analyses of TeO_2 glass, four O atoms are ranged around Te at a distance of 1.95 Å, and two others range at a distance of 2.75 Å. Therefore, the Te ion lies in an intermediate state between four- and six-atom coordination. TeO_2 itself is not vitrified. In the crystal state of TeO_2, the coordination number of Te^{4+} is six. If a small amount of an M ion is introduced, however, TeO_2 can be vitrified. Imoka and Yamazaki (1968) speculated that, in tellurite glasses, the Te–O distance shrinks somewhat; therefore, the four-atom coordination of Te^{4+} becomes more stable than the six-atom coordination. On the other hand, a series of ions that have no vitrifying range in any binary system with TeO_2 are produced. It is noteworthy that their r values are within a narrow range; as the valence of ions increases, their r range shifts somewhat to the shorter side. There are several possible explanations: if M has the structure of six-atom coordination, and if the size of MO_6 is nearly the same as that of TeO_6, then the six-atom coordination state of Te^{4+} might be stable.

Two remarkable features of the glass formation range of the tellurite system are that its range of potential M ions is wider than in the borate or the silicate system, and that it has no immiscible range in the tellurite system. These two properties very much resemble those of P_2O_5 systems. The first one can be explained by electronegativity (P = 2.1, Te = 2.1, B = 2.0, As = 2.0,

Si = 1.8, and Ge = 1.8). The electronegativity of Te is the same as that of P; therefore, the O–M bond may be ionic except for small, high-valence ions. The second property is probably a problem arising from the polymerization power of glass-forming oxides. Because many oxides are classified as Ms, ternary tellurite systems are classified largely as "A-type" (consisting of one glass former and two modifier components). Tungsten oxide (WO$_3$) cannot be considered an M component. In the tellurite systems generally, the glass formation ranges of WO$_3$-containing compounds are wide, very unlike those of borate, silicate, and so on. Imoka and Yamazaki (1968) reported that the tungsten coordination number is six in tellurite systems—the same as in borate, silicate, and other systems. Most of the actual glass formation range of WO$_3$ is above the A-type WO line. In this region the network structure contains WO$_3$ without M ions; instead, the regions within the A-to-D line contain a network of WO$_3$ with M ions. It is difficult to realize that Nb^{5+} is a network former with six-atom coordination in the four-atom-coordination network. The glass formation range of this A-type group resembles that of the B$_2$O$_3$–MgO–K$_2$O system. The glass formation ranges containing oxides of b-group elements (see above) are generally narrow. The few exceptions include the TeO$_2$–ZnO–La$_2$O$_3$ system. The narrow glass formation ranges of b-group elements containing oxides may result from the unstable self-network formation of b-group ions in tellurite systems.

Some properties of glasses obtained in the TeO$_2$–MoO$_3$–V$_2$O$_5$ system were studied by Kozhokarov et al. (1981). A good correlation between these properties and the phase diagram of the TeO$_2$–MoO$_3$ system was established. The glass resistance composition function varies between 6.85×10^9 Ω cm and 2.93×10^{10} Ωcm. The isolines of ρ properties for the glasses obtained from the TeO$_2$–MoO$_3$–V$_2$O$_5$ system are plotted as shown in Figure 1.10; the softening temperature is discussed in Chapter 5. Thermal properties and electrical resistance at and above room temperature along with the values of E are examined in Chapter 6. Electrical resistance is influenced by the concentrations of V$_2$O$_5$ and MoO$_3$, as well as by temperature. The glass absorption characteristics of thin layers have been determined in the visible range.

A topic that is of great interest to glass researchers and workers in the optics industry, as well as to those who observe trends in industrial development, is the preparation of glasses that have a high refractive index. In fact, such glasses possess optical characteristics, depending on the position, beyond the central region of the Abbe diagram. Glass forming in ternary tellurite systems was performed and studied by Marinov et al. (1983) and Kozhokarov et al. (1983), as shown in Table 1.6 and Figure 1.11. Tungsten oxide-tellurite glasses that show an extremely refractive n, low crystallization

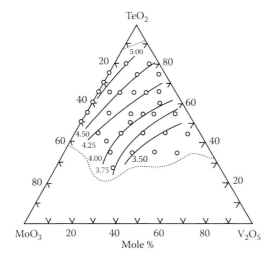

FIGURE 1.10 Density and glass formation of the ternary TeO$_2$–MoO$_3$–V$_2$O$_5$ glasses. (From V. Kozhukraov, S. Neov, I. Gerasimova, and P. Mikula, *J. Mater. Sci.*, 21, 1707, 1986.)

TABLE 1.6
Glass Formation Regions in Ternary Tellurite Glasses

Glass	Surface (%)	Immiscibility Gap (%)
$TeO_2-P_2O_5-BaO$	42.8	14.8
$TeO_2-P_2O_5-Tl_2O$	52.0	23.0
$TeO_2-P_2O_5-BaCl_2$	50.9	13.7
$TeO_2-P_2O_5-PbCl_2$	74.0	13.0
$TeO_2-B_2O_3-ZnO$	31.5	42.7
$TeO_2-B_2O_3-BaO$	27.0	36.0
$TeO_2-B_2O_3-NaF$	63.0	11.5
$TeO_2-B_2O_3-PbF_2$	50.6	38.8
$TeO_2-Cd(PO_3)-ZnO$	38.0	
$TeO_2.NaPO_3-BaCl_2$	53.0	
$TeO_2-Zn(P_2O_3)_2-Ba(PO_3)_2$	99.7	
$TeO_2-TlPO_3-Zn(PO_3)_2$	99.8	
$TeO_2-TlP_2O_3-Ba(PO_3)_2$	99.8	
$TeO_2-Ba(PO_3)_2-BaCl_2PO_5$	99.8	
$TeO_2-Cd(BO_2)_2-ZnO$	27.6	
$TeO_2-Pb(BO_2)_2-BiCl_2$	50.5	
$TeO_2-Zn(PO_3)_2-BiBO_3$	47.2	
$TeO_2-Ba(PO_3)_2-Zn(BO_2)_2$	75.0	
$TeO_2-BiBO_3-Pb(BO_2)_2$	98.2	
$TeO_2-NaFB_2O_3-PbF_2-B_2O_3$	44.0	

Source: V. Kozhokarov, M. Marinov, I. Gugov, H. Burger, W. Vogal, *J. Mater. Sci.*, 18, 1557, 1983.

ability, and good semiconducting and chemical resistance, as discussed in Chapters 6 and 8, have been formed. Glasses of this system possess a good biological effect against x-rays as well as γ-rays, as proved by Kozhokarov et al. (1977).

Glass formation in the vitreous ternary $TeO_2-MoO_3-CeO_2$ system was investigated by Dimitriev et al. (1988), who synthesized low-melting-point, stable glasses with up to 39 mol% CeO_2 (Figure 1.12). These researchers used IR-spectral investigations to develop their structural models for this system. CeO_2 mainly acts as an M without affecting appreciable changes to the glass network and coordination of the glass formers. Glasses in the molybdenum-rich compositional range are mainly composed of MoO_6 and TeO_3 polyhedra, whereas low-MoO_3-containing glasses consist of TeO_4 groups and isolated MoO_4 units. The basic structural polyhedra participating in the formation of the 3D glass-forming network are therefore TeO_4, TeO_3, MoO_6, MoO_4, and Mo_2O_8 (or MoO_5) units. The structural affinity of some ternary glasses for crystalline $Ce_4Mo_{11}Te_{10}O_{59}$ was pointed out by these researchers. The high electrical conductivity of the ternary glasses is interpreted as electron hopping between transition ions in different valence states with contributions from the Te (IV) network.

Glass formation in the quaternary $TeO_2-B_2O_3-MnO-Fe_2O_3$ system and in its ternary systems was investigated by Dimitriev et al. (1986a), as shown in Figure 1.13. A range of liquid immiscible phases located near the binary $TeO_2-B_2O_3$ and B_2O_3-MnO systems were determined. Using transmission electron microscopy, this group observed a trend toward metastable liquid-phase separation in the single-phase glasses near the boundary of immiscibility. An increase in the Fe_2O_3 and MnO contents during cooling of the melts enables a fine glassy crystalline structure to be formed. Dimitriev et al. (1986a) showed that by changing the upper limit of the melting temperature and adjusting the cooling rate, the glassy crystalline structure and the Fe_3O_4 content can be modified. The same group (Dimitriev et al. 1986b) also prepared the

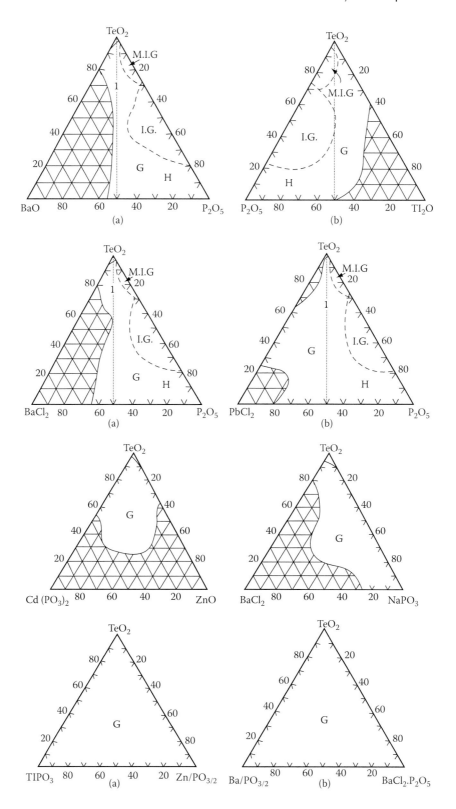

FIGURE 1.11 New family of tellurite glasses. (From V. Kozhokharov, M. Marinov, I. Gugov, H. Burger, and W. Vogal, *J. Mater. Sci.*, 18, 1557, 1983.)

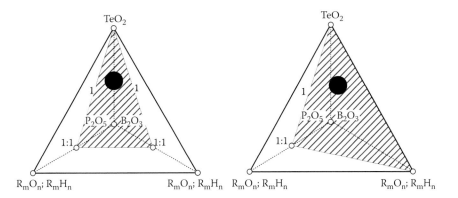

FIGURE 1.11 (Continued)

glass-forming region in the SeO_2–TeO_2–V_2O_5–MoO_3 quaternary system under increased oxygen pressure and at a slow melt cooling rate ($2°C$–$2.5°C$ min^{-1}), as shown in Figure 1.14. The stable glasses are located in the central part of the system, but nearer to the SeO_2–TeO_2 side. The structural units of these two glass formers are of decisive importance in building the glass lattice. Infrared spectra of selected compositions from the glass-forming region were taken. From the data obtained for the binary glasses in the TeO_2–V_2O_5, TeO_2–SeO_2, TeO_2–MoO_3, and V_2O_5–MoO_3 systems, and the spectra of these four component compositions, it was shown that the basic structural units participating in the glass lattice formation were the SeO_3, V_2O_5, TeO_4, and TeO_3 groups. A proposed structural model would show that glasses in the SeO_2 direction possess linear and chain structure, and that with an increase in TeO_2 concentration, a 3D structure is built. In 1986, Ivanova (unpublished data) performed phase image analysis of the ternary

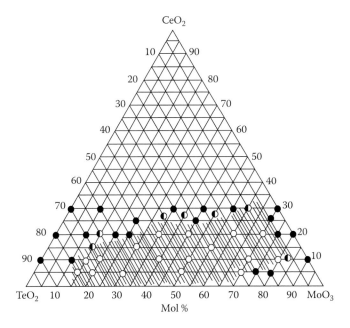

FIGURE 1.12 Glass formation range of TeO_2–MoO_3–CeO_2 glass. (From Y. Dimitriev, J. Ibart, I. Ivanov, and V. Idimtrov, *Z. Anorg. Allg. Chem.*, 562, 175, 1988. With permission.)

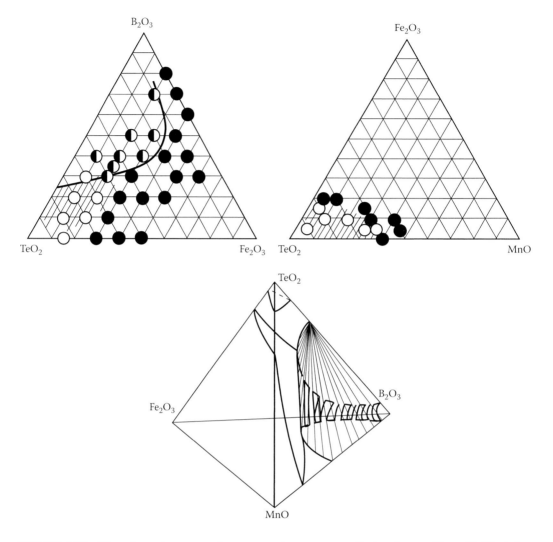

FIGURE 1.13 Glass formation range of ternary and quaternary tellurite glasses. (From Y. Dimitriev, I. Ivanova, V. Dimitrov, and L. Lackov, *J. Mater. Sci.*, 21, 142, 1986a.)

TeO_2–GeO_2–V_2O_5 system, as shown in Figure 1.15. Three binary diagrams forming the phase are described (Figure 1.15).

1.5.2 TELLURITE GLASSES PREPARED BY THE SOL-GEL TECHNIQUE

In 1999, Weng et al. prepared the first binary 90 mol% TeO_2–10 mol% TiO_2 thin films by a more highly homogeneous, pure, and controlled process at lower temperature—it is called the "sol-gel" process. Their work focused on developing this process for TeO_2–TiO_2 thin films, including stabilization of tellurium ethoxide ($Te[OEt]_4$), preparation of thin films of 90 mol% TeO_2–10 mol% TiO_2 by dip coating, and estimation of the values of n in these films (as shown in Chapter 8). Figure 1.16 represents the proposed precursor to hydrolysis of the derivative of tellurium 2-methyl 2,4-pentanediol. Although bulky ligands can prevent H_2O molecules from attacking the tellurium atom from one side, H_2O can always approach tellurium from the area around the lone pair of electrons, as shown in Figure 1.16. Weng et al. (1999) prepared the 90 mol% TeO_2–10 mol% TiO_2 thin films as follows:

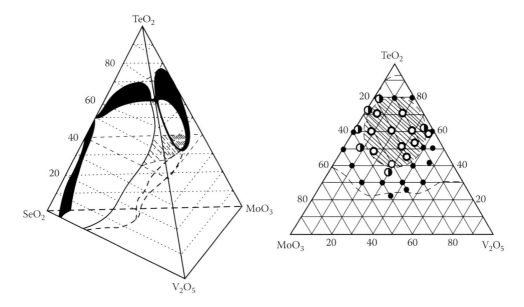

FIGURE 1.14 Glass formation range of ternary and quaternary tellurite glasses. (From Y. Dimitriev, E. Kashchiev, Y. Ivanova, and S. Jambazov, *J. Mater. Sci.*, 21, 3033, 1986b.)

1. Ethylene glycol was added to 0.1 M Te(OEt)$_4$ ethanol solution at a volume ratio of 0.02:1. The solution was refluxed for 2 h at 80°C under Ar protection.
2. Ti(OPr^i)$_4$ stabilized by acetylacetone with a molar ratio of 1:1 was added to this solution with a nominal composition of 90 mol% TeO$_2$–10 mol% TiO$_2$. The mixture was further refluxed at 80°C for 2 h.
3. Addition of 1,3-propanediol solution at a ratio of 0.1:1 (vol/vol) followed.
4. The thin film was made by dipping a glass substrate into the prepared solution and then withdrawing it at a speed of 0.6 mm/s.
5. The coated product was dried at 120°C in an oven and heat treated at 450°C for 10 min in static air.

1.6 NONOXIDE-TELLURITE GLASSES

1.6.1 CHALCOGENIDE TELLURITE GLASSES

Chalcogenide and halide glasses have received a great deal of interest as potential candidates for transmissions in the mid-IR region, as mentioned by Baldwin et al. (1981) and by Poulain (1983). The relatively poor chemical durability of halide glasses, together with their low T_g (especially for nonfluoride halide glasses), poses some serious problems in the development of practical applications for these glasses. On the other hand, chalcogenide glasses are well known for their high chemical durability and IR transmittance. However, their relatively high n values give rise to large intrinsic losses in the mid-IR range. It might be possible to improve the chemical durability of halide glasses by their incorporation into chalcogenide glass. It has been shown that the n of chalcogenide glass is decreased by combining this glass with members of the halogen group, which leads inevitably to reduced intrinsic scattering losses in the mid-IR region, as shown by Saghera et al. (1988). A new type of glasses prepared from mixtures of chalcogenide and halides are called "chalcohalides." Chalcohalides are potential candidates as materials for low-loss optical fibers that operate near the IR wavelength (NIR) and continue to wavelengths as long as 18–20 µm. Therefore, these glasses are also candidates to transport the CO$_2$ laser wavelength (10.6 µm) in

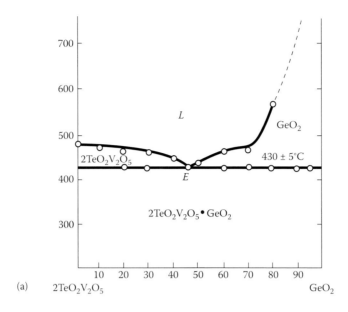

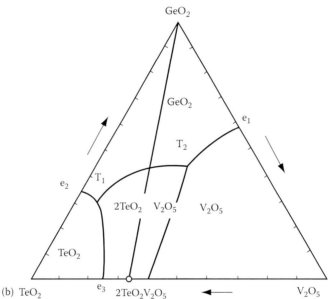

FIGURE 1.15 Phase diagram of $2TeO_2V_2O_5$–GeO_2 (a) and TeO_2–V_2O_5–GeO_2 (b). (From Y. Ivanova, *J. Mater. Sci. Lett.*, 5, 623, 1986.)

such applications as laser power delivery, remote spectroscopy, thermal imaging, and laser-assisted microsurgery, as indicated by Vogel (1985) and by Blanchetier et al. (1995).

Amorphous solids are most easily classified by the type of chemical bonding that is primarily responsible for their cohesive energy, as mentioned by Adler (1971). Van der Waals and hydrogen-bonded solids generally have low cohesive energies, and thus low melting temperatures; consequently, corresponding amorphous solids have not been studied to any great extent. Metals do not fall within the scope of a discussion of amorphous semiconductors.

The field of amorphous semiconductors can be broken into ionic and covalent materials. Ionic materials have been most studied for use in this field. These materials include the halide and oxide

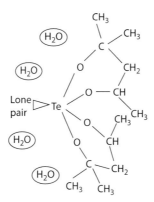

FIGURE 1.16 Proposed precursor to the hydrolysis of the derivative of tellurium 2-methyl-2,4-pentanediol. (From L. Weng, S. Hodgson, and J. Ma, *J. Mater. Sci. Lett.*, 18, 2037, 1999.)

glasses, particularly the TMO glasses. The compositions of these materials cannot be made to vary over a wide range, and the pure materials exhibit only positional disorder. On the other hand, the presence of impurities in the TMOs usually produces transition metal ions of two different valence states. For example, the introduction of P_2O_5 into V_2O_5 produces V^{4+} ions as well as V^{5+} ions. Thus, these materials are said to possess some degree of electronic disorder. Furthermore, the unsaturated transition metal ions generally contribute some spin disorder.

Covalent amorphous semiconductors can be divided into two classes: pure elemental material, which is perfectly covalently bonded and includes Si, Ge, S, Te, and Se among others (because all atoms are necessarily the same, these materials possess only positional disorder); and the binary materials As_2Se_3 and GeTe, as well as the multicomponent bride, arsenide, and chalcogenide glasses.

As_2Se_3 nominally contains 40 mol% As and 60 mol% Se, but there is no reason that an alloy of, for instance, 45 mol% As and 55 mol% Se (or these elements in any other proportion) cannot be made. In a similar manner, a combination of As, Se, Ge, and Te produces, for example, 31 mol% As, 21 mol% Se, 30 mol% Ge and −18 mol% Te is an example of a chalcogenide glass. This glass is significant because the covalent amorphous semiconductors always possess compositional and positional disorder. This joint disorder profoundly affects the electronic band structure and is responsible for the distinctive properties of these glasses.

Another classification scheme for amorphous semiconductors is to divide them into groups with the same short-range structural coordination. A general rule, stressed by Ioffe and Regel (1960), is that there must be preservation of the first coordination bond number of the corresponding crystal. Of course, for compositions that have no crystalline phase, this rule is meaningless. The classification scheme based on coordination bond number and, in general, normal oxide-based optical glasses does not apply to bonds beyond 3–5 μm in length, because of the strong absorption of chemical bonds in chalcogenide glasses. On the other hand, chalcogenide glasses are preferred for mid- and far-IR transmissions.

Hilton (1970) listed the qualitative results obtained for glasses made from systems based on the elements that are listed in the first column of Table 1.7. These glasses were evaluated for their softening points, *n* values, and suitability for application in the wavelength region of the two atmospheric windows. Hilton (1970) also compiled a list of properties measured for specific glass compositions (Table 1.8). These glasses represent the best compositions extant for their particular systems up to the time Hilton's report was published. Si and Ge for crystalline semiconductors, NaCl for alkali halides, and oxide-based optical glasses and other optical materials are included in Table 1.8 for comparison. Affifi (1991) used differential scanning calorimetry (DSC) data at different heating rates on Se–Te chalcogenide glass in order to find the glass transition temperature, crystallization temperature, and both the glass transition and crystallization *E*s. Moynihan et al. (1975)

TABLE 1.7

Qualitative Evaluation of Glasses from Nonoxide (Chalcogenide) Systems

System	Softening Point (°C)	n at ~5 μm	Absorption (dB/m)[a] 3–5	8–14
Si–P–Te	180	3.4	—	M
Si–As–Te	475	2.9–3.1	—	M
Ge–As–Te	270	3.5	—	—
Ge–P–Te	380	3.5	—	—
As–Se–Te	200	2.6–3.1	—	M
As–S–Se–Te	195	2.1–2.9	—	M
Si–Ge–As–Te	325	3.1	—	M

Source: A. Hilton, *J. Non-Cryst. Solids*, 2, 28, 1970.

[a] —, No appreciable absorption; M, medium absorption.

investigated IR absorption in the region of 250 to 4000 cm^{-1} in As_2Se_3 glasses doped with small amounts of As_2O_3 or purified by various procedures, with particular attention given to absorption in the wavelength regions of CO_2 and CO lasers. These authors listed the following steps to eliminate surface oxides:

1. Bake the melt tubes overnight under vacuum at 850°C–900°C to remove adsorbed water before loading the glass components.
2. Bake the components overnight at 100°C–120°C in the melt tubes under vacuum before sealing to remove surface moisture from the components.
3. Add to the melt components small amounts of metallic Al or Zr to act as oxide getters.

TABLE 1.8

Quantitative Evaluation of Glasses from Nonoxide Chalcogenide Systems

Chemical Composition[a]	Transmission Range (μm)	Softening Point (°C)
$Si_{25}As_{25}Te_{50}$	2.0–9.0	317
$Ge_{10}As_{20}Te_{70}$	2.0–20.0	178
$Si_{15}Ge_{10}As_{25}Te_{50}$	2.0–12.5	320
$As_{50}S_{20}Se_{20}Te_{10}$	1.0–13.0	195
$As_{35}S_{10}Se_{35}Te_{20}$	1.0–12.0	176
Si	1.2–15.0	1420[b]
Ge	2.0–23.0	942[b]
NaCl[c]	0.2–26.0	803
Optical glasses	0.2–3.0	700

[a] Subscript numbers are molecular percentages.

[b] Useful range much less than melting point.

[c] Very soluble in water.

[d] Rupture modulus can be increased by a factor of 3 by tempering.

4. Distill the glass as follows: after melting, seal the glass into one side of a quartz tube divided into sections by a coarsely porous quartz-fritted disk. With the distillation tube in a horizontal position, heat the glass to a suitable temperature so that it is distilled through the fritted disk and into the second side of the quartz tube, which is held at a lower temperature. Ordinarily the distillation is carried out with the distillation tube sealed under vacuum, but distillations can also be carried out with the tube sealed at room temperature under a one-third atmosphere of N_2, H_2, or 5% H_2–95% N_2. After distillation, the glass is remelted briefly in a quartz tube sealed under vacuum and is then annealed.

Katsuyama and Matsumura (1993) investigated the optical properties of Ge–Te and Ge–Se–Te chalcogenide glasses for CO_2 laser light transmission, obtaining the following results on bulk glasses:

- The absorption edge due to lattice vibration occurring below 700 cm^{-1} shifts to the shorter-wave number side as the Se content decreases and the Te content increases. In particular, this shift is prominent when the Te content exceeds 60 mol%.
- There is a small peak between wave frequencies of 760 and 775 cm^{-1} in the absorption spectrum, which affects transmission loss at 10.6 μm. The loss from this peak is proportional to the Se content; therefore, this peak originates from the Se bond. If the Se content is maintained below 20 mol%, the effect of this small peak on loss at 10.6 μm can be reduced.
- Scattering loss increases significantly as the Se content decreases below 5 mol%. Therefore, the appropriate Se content for obtaining low loss is 5 to 20 mol%.

In addition to this, Katsuyama and Matsumura (1993) also drew a 22 mol% Ge–20 mol% Se–58 mol% Te ternary glass block into fiber with the following results:

- Eliminating the small bumps existing on the surface of the fiber reduced wavelength-independent scattering loss.
- The absorption due to lattice vibration at 10.6 μm was about 0.4 dB/m. Increasing the Te content reduced the loss at 10.6 μm, but this required much more precise control of the drawing temperature.
- Transmission loss appearing at wave numbers above 1200 cm^{-1} produced about 1.0 dB/m loss at 10.6 μm. Suppressing the creation of the Te crystallites during the drawing process might reduce this loss.
- A small loss peak existed at 1100 cm^{-1} and was caused by Si impurities. This loss produced a 0.1-dB/m loss at 10.6 μm. Eliminating Si impurities might reduce this small loss peak.

Another investigation was done by Katsuyama and Matsumura (1994) on the transmission loss characteristics of Ge–Se–Te ternary chalcogenide glass optical fibers. They obtained the following results:

- A broadband transmission loss occurred at wavelengths between 2 and 10 μm, which greatly affected the loss at 10.6 μm.
- The broadband loss increased exponentially as the Te content increased. The loss α (in decibels per meter) at 2.5 μm is given by Te content x (mol%) as $\alpha = 6.80 \exp(0.0445x)$.
- The broadband loss was inversely proportional to the third power of the wavelength.
- Reflective high-energy electron diffraction measurement showed that the glass contains hexagonal Te microcrystals.

These results show that the broadband loss at wavelengths between 2 and 10 μm originates from Mie scattering due to hexagonal Te microcrystals. Katsuyama and Matsumura (1994) concluded, based on Mie-scattering theory, that the diameter of the hexagonal Te microcrystals is 0.2 μm. Therefore, they concluded that elimination of the hexagonal Te microcrystals is essential to obtaining low-loss chalcogenide optical fibers. Although the elements used for optical materials are ultrapure, further purification processes are required in order to eliminate surface oxidation. Hilton et al. (1975) were the first to use a distillation process to further purify raw materials, which were generally 6 N pure (not including surface oxides). Their purification procedure involved the use of a fused-quartz vessel with three interconnecting chambers. Before introduction of the raw materials, the chambers were etched, dried, and heated or flushed with "oxygen-free gas" to drive out moisture. Germanium was placed in the central chamber. One of the other chambers contained selenium, and the third contained arsenic (or antimony). The surface oxide removal of raw materials was carried out in temperatures at which the vapor pressure of the oxide exceeds that of the element in the presence of an inert gas flowing through the chambers (or under vacuum). After this step, the chambers were cooled using inert flowing gas (nitrogen or helium). The materials present in the two chambers on either side of the central chamber that is, selenium and arsenic (or antimony) were distilled or sublimated into the central chamber through porous quartz ferrites. The central chamber was kept cool to enhance the distillation process. After distillation, the two chambers were sealed, and the central chamber with the purified batch was placed in a rocking furnace for melting. This method has proved effective in reducing the oxide impurities from chalcogenide glasses, as shown by a greatly reduced α.

Nishii et al. (1987) prepared glasses in Ge–Se–Te and Ge–Se–Te–Tl systems. Raw materials were purified by H_2 reduction for Ge and Tl, and by distillation for Se and Te. The minimum loss of the Ge–Se–Te fibers was 0.6 dB/m at 8.2 μm (1.5 dB at 10.6 μm), at which the glass composition was 27 mol% Ge–18 mol% Se–55 mol% Te. The same authors found that, although the introduction of Tl above 4 mol% increases the intrinsic absorption loss, the best results—that is, 1.0 dB/m at 9.0 μm and 1.5 dB/m at 10.6 μm—are achieved for 22 mol% Ge–13 mol% Se–60 mol% Te–2 mol% Tl_2 glass fiber. Savage (1982) reported that the distillation process is very effective in germanium-arsenic sulfide, selenide, and telluride glasses. Inagawa et al. (1987) attributed the absence of impurity absorption in their Ge–As–Se–Te glasses to the distillation procedure. Reitter et al. (1992) introduced a modified technique for purification. Their results indicated that the oxide absorption bands in Ge–Sb–As–Se–Te glasses can be greatly reduced by a modified distillation procedure. El-Fouly et al. (1990) measured the temperature-induced transformations that are interesting characteristics of amorphous materials, including the x mol% Si–$(60 - x)$ mol% Te–30 mol% As–10 mol% Ge system, with $x = 5$, 10, 12, and 20 DTA, which was used to characterize the compositions. DTA traces of each glass composition at different heating rates from 5°C min^{-1} to 30°C min^{-1} were obtained and interpreted. Quick and slow cooling cycles were used to determine the rates of structure formation. Cycling studies of materials showed no memory effect, only ovonic-switching action. The compositional dependence of the crystallization E and the coefficient of the glass-forming tendency have been calculated. The transition temperature and associated changes in specific heat have been examined as functions of the Te/Si ratio by DSC. El-Fouly et al. (1990) found that both ρ (density) and E (activation energy) increase linearly with increasing tellurium content, whereas the heat capacity and glass factor tendency decrease with increasing tellurium content. The ρ of the system was changed from 4.9 to 5.79 g/cm by a decrease in Si from 20 to 5 mol% and a corresponding increase in Te from 40 to 55 mol%, with the rest of the components remaining the same. The relevant data for this system (x mol% Si–[60 – x] mol% Te–30 mol% As–10 mol% Ge) are shown in Table 1.9.

1.6.2 FIBER PREPARATION

In 1992, Nishii et al. made a significant advance in the fabrication of chalcogenide glass fiber for IR optical applications based on sulfide, selenide, and tellurite systems. They developed a new crucible drawing method for drawing fibers with glass cladding. The extrinsic losses caused by some oxide

TABLE 1.9

Effects of Changes in Te and Si Content in the x mol% Si–(60 – x) mol% Te–30 mol% As–10 mol% Ge System[a]

Si Content (x mol%)	Density (g/cm³)	T_g (°C)	T_{c1} (°C)[b]	T_{c2} (°C)[b]	T_m (°C)	K_{gl}^c
20	4.90	185	295		408	0.97
12	5.41	178	290	365		
10	5.47	172	285	388	405	0.94
5	5.79	165	202		360	0.23

[a] Data adapted from M. El-Fouly, A. Maged, H. Amer, and M. Morsy, *J. Mater. Sci.*, 25, 2264, 1990.

[b] T_{c1} and T_{c2}, temperatures at first and second crystal peaks, respectively.

[c] K_{gl}, glass-forming factor.

impurities were suppressed by purification of raw elements. The transition loss and mechanical (i.e., bending and tensile) strength of each fiber were investigated before and after the heat treatment under humid conditions. The fibers obtained were used for power delivery in a CO_2 laser (10.6 µm wavelength) and a Co laser (5.4 µm wavelength). Antireflection coating on fiber ends and cooling of fiber with gas or water were tested for improvements in power transmission efficiency. Nishii et al. (1992) described their method of fiber fabrication in detail. In their method, the purified elements are weighted in a glove box filled with Ar gas, and they are sealed in a dehydrated-silica ampoule under vacuum conditions at 1.3×10^{-5} Pa. The ampoule is then heated in a furnace at 850°C for 24 h.

The casting of chalcogenide glass melt is difficult because of its high vapor pressure. The process is summarized in the 1st Ed. The core rod is formed by rotating the ampoule vertically while cooling. The ampoule for the cladding tube is rotated horizontally. The temperature around the ampoule is carefully controlled to inhibit thermal-stress-induced cracking of the ampoule and chalcogenide glass. The internal diameter of the cladding tube is adjusted by varying the amount of glass melt in the ampoule. An aluminum abrasive is used to polish the core rod and the outside of the cladding tube. The space between the core and the cladding is narrowed to >3% of the diameter of the core rod. At the fiber-drawing temperature, chalcogenide glass is easily oxidized and crystallized by the residual oxygen and moisture surrounding the glass. Vaporization of the glass components also occurs, especially for glass with a high T_g. If a specified numerical aperture is not required, a Teflon cladding is useful. The fiber can be obtained by the conventional drawing method (i.e., a glass rod with a tightly contracted Teflon jacket is zonally heated and drawn into fiber). The transmission range of the fiber prepared by this method, however, is restricted to wavelengths shorter than 7.5 µm because strong absorption by Teflon appears between 7.5 and 11 µm.

A schematic diagram of this method of Nishii et al. (1992) is shown in the 1st Ed. In the first step, the core rod and cladding tube are dried at 130°C for 2 h under evacuation and placed in a silica crucible filled with an inert gas. The surrounding atmosphere of the nozzle is also replaced by an inert gas. Contact of the perform with air is absolutely avoided in order to prevent surface oxidation or hydration reactions between the chalcogenide glass and air. Second, the crucible is heated to the softening temperature of the glass only in the vicinity of the nozzle. After the cladding tube adheres uniformly around the inner surface of the crucible, the inside of the crucible is pressurized to 2×10^5 Pa with Ar gas, and gas in the space between the core rod and the cladding tube is evacuated down to 1.3 Pa. The fiber drawn from the nozzle is coated with an ultraviolet (UV)-curable acrylate polymer below the diameter monitor. The cladding diameter of the fiber prepared by this method is between 250 and 1000 µm. Nishii et al. (1992) also investigated the properties of Te–Ge–Se glass fiber, specifically for CO_2 laser power transmission. The glass-forming region of Te–Ge–Se

TABLE 1.10
Physical Properties of Chalcogenide (Te) Glass Fiber

Property	Glass System (Core [GeSeTe]/Clad [GeAsSeTe]) for Percentage of Te in Fiber	
	45%	**51.5%**
T_g (°C)	253/200	216/179
n (10.6 μm)	2.90/2.89	2.97/2.90
T Dependence of n (°C^{-1})	8×10^{-3}	14×10^{-3}
Numerical aperture	0.22	0.64
E (kpsi)	2970/—	2840/—
Vickers hardness (kpsi)	—	215/—

Source: J. Nishii, S. Morimoto, I. Inagawa, R. Hizuka, T. Yamashita, and T. Yamaguchi, *J. Non-Cryst. Solids*, 140, 199, 1992.

is narrow; therefore, the Te–Ge–Se–As glass system was chosen for its prominent stability against crystallization. The physical properties of the system were evaluated as shown in Table 1.10.

1.6.3 HALIDE-TELLURITE GLASSES (CHALCOHALIDE GLASSES)

Lacaus et al. (1986, 1987) and Zhang et al. (1988) reported the glass formation ability of new binary compositions based on tellurium halide (TeX) in the following systems: Te–Cl, Te–Br, Te–Cl–S, Te–Br–S, Te–Cl–Se, Te–Br–Se, Te–I–S, and Te–I–Se. The main characteristics of these glasses are the following:

- The ternary composition is extremely stable against devitrification, and most of these systems have no crystallization peak in the DSC analysis.
- Except for those systems very rich in halogen, these glasses show good resistance to corrosion by water and moisture, and some are totally inert in aqueous solution.
- The T_gs are in the range of 60°C–84°C, and the viscosity temperature dependence is such that fibering is easier than with fluoride glass.
- Analysis of the optical transmission domain indicates that all these vitreous materials can be classified into two large families: "light" TeX glasses contain a light element such as Cl or S, and "heavy" TeX glasses are based only on heavy elements such as I, Br, and Se.

Another new class of glasses in the binary system Te–Br was introduced in 1988 by Zhang et al.; the limits of this vitreous domain are formed by Te_2Br and TeBr. When calculated quantities of Te and Br_2 are heated to about 300°C in annealed glass tubes, the formation of a viscous melt occurs. When the tubes are cooled in air, the melt solidifies as a complete vitreous material with a Te/Br ratio between 1 and 2; under conditions of moderate quenching, Te_2Br and Te_3Br_2 form the limits of the vitreous area. The Te–Br glass is located in the middle of the diagram shown in the 1st Ed. Zhang et al. (1988) prepared a very pure glass, and the apparatus used for the preparation of the vitreous Te–Br–S and Te–Br–Se glasses and glass domain is shown in detail in the 1st Ed. The following experimental procedure was used by Zhang et al.:

1. The chips are first treated with HBr- and Br_2-containing solution to clean them of oxygen surface corrosion.
2. The Br_2 solution, always contaminated by water, is treated with P_2O_5 in the first container and transferred by condensation into a graduated tube where a calculated amount is stored.

3. This Br_2 is transferred again to a reaction tube containing tellurium; after sealing under vacuum, the tube is heated for 2 h in a rocking furnace at 300°C and is then cooled in air.

The black pieces of glass that result are not hygroscopic, and only those containing a large amount of Br, such as $TeBr_2$, show surface corrosion in normal atmosphere. The composition Te_3Br_2 is the most stable with regard to crystallization, and the addition of S or Se strongly decreases the devitrification rate. The T_gs are in the range 70°C–80°C and, for Se-doped glasses, there is no crystallization peak. Also in 1988, Lacaus et al. prepared the TeX system Te–Br–Se glasses and measured the optical-gap window corresponding to the band gap absorption mechanisms for different glasses of this system. These authors found that the band gap is shifted toward the shorter wavelength when (1) the content of the most electronegative element, Br, increases; and (2) the proportion of Se increases in a given family. They summarized their findings as follows:

1. Te 6 $(1 - x)$-Se6x Br4 glasses, $x \sim 0.15, 0.65$; $\rho = 4.95$–4.45 g/cm³
2. Te 7 $(1 - x)$-Se7x Br3 glasses, $x \sim 0.15, 0.85$; $\rho = 5.2$–4.30 g/cm³
3. Te 8 $(1 - x)$-Se8x Br2 glasses, $x \sim 0.38, 0.87$; $\rho = 5.2$–4.35 g/cm³
4. Te 9 $(1 - x)$-Se9x Br1 glasses, $x \sim 0.55, 0.88$; $\rho = 5.05$–4.42 g/cm³

In 1988, Saghera et al. reviewed the relatively uncommon (compared with chalcogenide and halide glasses), inorganic chalcohalide glasses. They collected data on the chemical, physical, and optical properties of known chalcohalide glass-forming systems, which possess interesting electrical and optical properties making them candidates in various applications; for example, electrical switching, memory functions, and transmission in the IR spectrum. Structural models were presented based on various spectroscopic studies, using techniques such as IR, Raman, and XRD analyses. The tellurite-chalcohalide glass systems examined were the following: As–Te–Br, As–Te–I, Ge–Te–I, Te–S–Cl, Te–S–Br, and Te–S–I.

Structural relaxation during such T_g annealing was reported by Ma et al. (1992), who recorded their observations while performing DSC monitoring of three chalcohalide glasses with low T_gs, including Te_3–I_3–Se_4 ($T_g = 49$°C), Te_3–Se_5–Br_2 ($T_g = 71$°C), and As_4–Se_3–Te_2–I ($T_g = 118$°C). On annealing at room temperature, the glass with the lowest T_g relaxed to equilibrium within a few days; the glass with an intermediate T_g showed substantial relaxation but not fully to equilibrium over a period of months; and the glass with the highest T_g showed no relaxation during a two-month period.

Optical properties of Ge–Te–I chalcogenide glasses for CO_2 laser light transmission were investigated by Katsuyama and Matsumura (1993). They found that Te content of more than about 60 mol% is essential to reduce the fundamental absorption caused by lattice vibration at 10.6 μm (the wavelength of CO_2 laser light). They also found that Se content above 5 mol% is required to eliminate wavelength-independent scattering loss due to glass imperfections. As a result, the transmission loss of optical fiber drawn from the 22 mol% Ge–20 mol% Te–58 mol% I glass block is reduced to 1.5 dB/m at 10.6 μm. Further loss reduction is accomplished by suppressing the creation of Te crystallites during the drawing process.

TeX glasses have been prepared in the form $Te_2Se_4As_3I$ and studied for their potential applications in such areas as thermal imaging, laser power delivery, and remote spectroscopy. These glasses are particularly interesting for their wide optical window, which ranges from 1 to 20 μm. Remote spectroscopy using IR optical fibers is being intensively studied because of the possibility of performing *in situ*, real-time, nonlinear analysis. TeX glass fiber offers a wide IR transmission region of 3.13 μm, at which many chemical species have their characteristic absorption. In 1995, Blanchetier et al. reported their preparation of TeX glass. They produced a glass rod with a core-cladding structure. Fiber attenuation was measured by using a Fourier transform IR spectrometer. The fiber losses, regularly measured, were <1 dB/m in the 5–9 μm range for mono-index fibers and about 1 dB/m for core–clad fibers.

One of the most important applications of TeX glass fiber is in CO_2 laser power delivery. More than 2.6 W has been generated through a 1 m long TeX fiber. These fibers have a wide optical window and have been used in remote spectroscopy to analyze chemical species that absorb at wavelengths between 3 and 13 µm. Other potential applications were discussed by Blanchetier et al. (1995). Laucas and Zhang (1990) demonstrated a new family of halide glasses based on tellurium, selenium, and the halogens Cl, Br, and I. They are called "TeX" glasses. IR optical fibers have been prepared from these glasses. The position and dependence of the multiphonon edge, as well as the band gap, were investigated by measuring the attenuation spectra, which showed that these glasses exhibit low optical losses in the 8–12 µm atmospheric window. Thin layers of TeX glasses are deposited by sputtering on silica or fluoride glass substrates, and antireflective coatings are deposited on bulk TeX glasses to reduce optical losses caused by reflection.

Chalcohalide glasses for IR transmission were investigated in the Ge–Te–As–I and Ge–Te–As–Se–I systems by Klaska et al. (1993). The glass-forming regions, thermal properties, and IR transmission properties were studied. Stable glasses containing up to 15% iodine (by weight) were shown to have a T_g ranging from 180°C to 250°C and to exhibit no peak of crystallization when heated at a rate of 10°C/min. Various purification techniques have been used to remove oxygen impurities. A series of new glasses in the As–Ge–Ag–Se–Te–I system were studied by Cheng et al. (1995). The glass-forming region of the ternary system involving As_2Se_3, GeTe, and AgI was obtained. The chalcohalide glasses studied have a wide range of IR transmission wavelengths (2–18 µm), relatively high transition temperatures, and low crystallization tendencies. Their excellent chemical durability and good glass-forming ability suggest that chalcohalide glasses are good candidates for optical fiber production. Jian et al. (1995) selected the Ge–As–Se system as the basic glass system for studying the effects of the addition of Te and I—separately and together—on the thermal, ρ, and optical properties of this system. The IR transmittance of Ge–As–Se–Te–I system glasses increases as iodine content increases—which, it is believed, is caused by increased homogeneity. The IR transmission edge decreases with the addition of I instead of Te, but shows little variation with the addition of I instead of Ge–As–Se–Te as a whole. A series of properties of the novel chalcohalide glasses in the As_2Se_3–GeTe–AgI system were studied by Chen et al. (1997). They showed that AgI acts as a network terminator in the chalcohalide glasses, making a more open network structure by producing a coordinated nonbridging iodine atom as well as some nonbridging Se and/or Te atoms. The AgI doping caused a more open network structure, which appears in a slight shift of the IR cutoff edge toward a longer wavelength.

1.7 MIXED OXYHALIDE AND OXYSULFATE TELLURITE GLASSES

In 1974, Vogel et al. prepared binary oxide-halide tellurite glasses and binary sulfate-oxide-tellurite glasses. They measured ρ, n, and the Abbe number (V_D) of numerous binary sulfate and TeX glasses, including all of the following: TeO_2, LiF, LiCl, LiBr, NaF, NaCl, NaBr, KF, KCl, KBr, RbF, RbCl, RbBr, BeF_2, $BeCl_2$, $BeBr_2$, MgF_2, $MgCl_2$, $MgBr_2$, CaF_2, $CaCl_2$, $CaBr_2$, SrF_2, $SrCl_2$, $SrBr_2$, BaF_2, $BaCl_2$, $BaBr_2$, LaF_2, $LaCl_2$, $LaBr_2$, $CdCl_2$, $CdBr_2$, PbF_2, $PbCl_2$, $PbBr_2$, ZnF_2, $ZnCl_2$, $ZnBr_2$, Li_2SO_4, Na_2SO_4, K_2SO_4, Rb_2SO_4, $BeSO_4$, $MgSO_4$, $CaSO_4$, $SrSO_4$, $BaSO_4$, $La_2(SO_4)$, and $PbSO_4$. Although the accommodation of fluorine is not unusual, that of significant amounts of halides and sulfates is remarkable. Property data like ρ, n, and V_D appear in Chapter 8. The data in Tables 1.11 and 1.12 relate to the lower and upper limits of glass formation in each system. These limits differ considerably. If halides or sulfates are melted with TeO_2, evaporation is more or less inevitable, so that ternaries of the volatile component result, such as $MeCl_2$–Me_2O–TeO_2. The compositions listed in the tables represent analytical values. The optical data show that a new valuable family of optical glasses have been discovered.

The results of a study of the glass-forming regions in 15 ternary TeX systems and information on their crystallizing tendencies, ρ values, and thermal-expansion coefficients were published by Yakhkind and Chebotarev (1980a, 1980b). The results of their chemical analysis show that there is

TABLE 1.11

Halide–Tellurite Glass Systems, Areas of Glass Formation, and ρ

Source and Glass System	Halide (mol%)	Metal Oxide (mol%)	ρ (g/cm^3)	Molar V (cm^3)
Vogel et al. 1974				
TeO_2–LiF				
TeO_2–LiCl	11.9–33.2	4.3–11.6	5.197–4.318	
TeO_2–LiBr	5.0–6.3	6.5–7.9	5.363–5.294	
TeO_2–NaF	19.0–32		5.207–4.901	
TeO_2–NaCl	16.6–22.0	0.8–1.1	5.001–4.802	
TeO_2–NaBr	8.6–26.4	4.6–2.9	5.196–4.724	
TeO_2–KF	16.1–42.0		5.103–4.241	
TeO_2–KCl	15.2–20.0	0.1–0.1	4.956–4.731	
TeO_2–KBr	10.8–23.8	2.7–2.5	5.056–4.594	
TeO_2–RbF	12.2–23.9	2.5–1.4	5.199–4.952	
TeO_2–RbCl	21.9–27.4	0.1–0.3	4.777–4.589	
TeO_2–RbBr	9.8–18.0	2.1–0.5	5.156–5.000	
TeO_2–BeF$_2$				
TeO_2–BeCl$_2$				
TeO_2–BeBr$_2$				
TeO_2–MgF$_2$				
TeO_2–MgCl$_2$	5.0–15.7	1.2–9.0	5.255–4.543	
TeO_2–MgBr$_2$	3.0–8.4	3.4–1.7	5.473–5.209	
TeO_2–CaF$_2$				
TeO_2–CaCl$_2$	6.6–15.5	0.8–4.5	5.221–4.801	
	3.2–4.8	2.9–5.2	5.383–5.323	
TeO_2–SrF$_2$				
TeO_2–SrCl$_2$	7.2–46.7	1.4–9.1	5.271–3.966	
TeO_2–SrBr$_2$	4.5–11.1	2.7–10.1	5.420–5.155	
TeO_2–BaF$_2$	6.2–15.4		5.573–5.494	
TeO_2–BaCl$_2$	7.2–19.7		5.334–4.988	
TeO_2–BaBr$_2$	6.3–37.3		5.463–4.973	
TeO_2–LaF$_2$				
TeO_2–LaCl$_2$	3.8–7.8	3.6–10.1	5.382–5.280	
TeO_2–LaBr$_2$	2.0–2.0	3.8 –14.0	5.537–5.632	
TeO_2–LaF$_2$	13.7–26.0		5.993–6.295	
TeO_2–LaCl$_2$	14.8–49.1	1.3–5.4	5.600–5.625	
TeO_2–LaBr$_2$	5.6–39.4	9.6–12.7	5.925–6.207	
TeO_2–CdCl$_2$	6.2–4.8	10.8–7.1	5.751–5.498	
TeO_2–CdBr$_2$	5.1–3.2	18.1–14.3	5.813–5.619	
TeO_2–PbF$_2$	13.7–26.0		5.993–6.295	
	14.8–49.1	1.3–5.4	5.600–5.625	
	5.6–39.4	9.6–12.7	5.925–6.207	
TeO_2–ZnF$_2$	4.3–36.0	16.2–28.8	5.538–5.308	
	2.5–33.6	9.3–14.2	5.202–4.316	
	0.3–29.2	9.9–39.9	5.343–4.694	
Yakhkind and Chebotarev 1980a				
TeO_2–WO$_3$–ZnCl–ZnO	52.3–6.3–39.3–2.1		5.180	29.54
	66.4–17.1–16.0–0.5		5.597	29.98

(continued)

TABLE 1.11 (Continued)
Halide–Tellurite Glass Systems, Areas of Glass Formation, and ρ

Source and Glass System	Halide (mol%)	Metal Oxide (mol%)	ρ (g/cm³)	Molar V (cm³)
TeO_2–BaO–ZnCl–ZnO	80.9–13.6–4.7–0.8		5.513	29.02
	65.0–10.9–22.6–1.5		5.140	29.62
TeO_2–Na_2O–ZnCl–ZnO	79.6–16.0–1.5–2.9		4.022	29.74
	75.1–15.4–6.9–2.6		4.856	28.89
Yakhkind and Chebotarev 1980b				
TeO_2–WO_3–$ZnCl_2$	$(80 - 20)$-x	$x = 0$		
		$x = 42.86$		
TeO_2–BaO–$ZnCl_2$	$(85.7 - 14.3)$-x	$x = 0$		
		$x = 36$		
TeO_2–Na_2O–$ZnCl_2$	$(83.3 - 16.7)$-x	$x = 0$		
		$x = 12$		
Sahar and Noordin 1995				
TeO_2–ZnO–$ZnCl_2$	60-10-30		5.18	
	50-30-20		5.11	
El-Mallawany et al. 1994				
$(1 - x)$ mol% TeO_2–x mol% MoO_3	$x = 20$		5.01	31.23
	$x = 50$		4.60	32.90
Sidky, M., El-Mallawany, R. et al. 1997				
$(1 - x)$ mol% TeO_2–x mol% V_2O_5	$x = 20$		4.9	33.48 cm³
	$x = 50$		3.99	42.73 cm³

a fairly close correspondence with the synthetic and analytical compositions stated by Vogel et al. (1974). The methods used for synthesis and chemical analysis of these glasses were described by Yakhkind and Chebotarev (1980a). The optical constants of TeO_2–WO_3–$ZnCl_2$, TeO_2–BaO–$ZnCl_2$, and TeO_2–Na_2O–$ZnCl_2$ glasses were measured in the visible region of the spectrum and the IR transmission spectrum. ZnCl reduced n and the average dispersion of tungsten–tellurite glasses less significantly than did ZnO, but did so more significantly than in barium–tellurite glass. The average refraction of the glass framework in all the systems did not depend on the composition, which is in good agreement with the observed constancy of electron polarizability among the average structural groups composing the framework of oxide–tellurite glasses. As the Br in the glass is replaced by Cl and F, the maximum IR absorption of the OH groups is shifted toward lower frequencies, which is in good agreement with the formation process of hydrogen bonds between the hydroxyl-1 groups in the glass structure and the halide atom.

In 1983, Kozhokarov et al. investigated glass-forming tendencies in two main groups of tellurite glasses, as shown in Figure 1.11 and Table 1.6:

1. Semiconducting tellurite glasses: TeO_2–R_mO_n–T_mX_n and TeO_2–R_mX_n–T_mX_n
2. Optical tellurite glasses: TeO_2–P_2O_5–R_nO_m, TeO_2–P_2O_5–R_nX_m, TeO_2–B_2O_3–R_nO_m, and TeO_2–P_2O_5–R_nX_m

where R_mO_n is any non-TMO and R_mX_n is any halide (fluorine, chlorine, or bromine) of nontransition metal. Figure 1.11 illustrates the glass formation tendencies when transition metal halides replace the respective metal oxides. Figure 1.11 also illustrates the glass formation of the ternary

TABLE 1.12

Tellurite–Sulfate Glasses, Halide–Tellurite Glass Systems, Area of Glass Formation, and ρ

System	Halide (mol%)	Metal Oxide (mol%)	ρ (g/cm³)
$TeO_2Li_2SO_4$			
$TeO_2Na_2SO_4$			
$TeO_2K_2SO_4$	3.4–6.8	0.6–1.1	5.283–4.931
$TeO_2Rb_2SO_4$	3.3–4.9	0.9–1.3	5.292–5.118
TeO_2BeSO_4	3.8–17.2	3.7–6.8	5.303–4.487
TeO_2MgSO_4	6.7–43.5	4.1–17.2	5.051–3.432
TeO_2CaSO_4			
TeO_2SrSO_4			
TeO_2BaSO_4	2.6–8.0	0.3–4.3	5.437–5.194
$TeO_2La_2(SO_4)_3$	4.0–0.7	7.4–2.1	5.476–5.591
TeO_2PbSO_4	5.0–12.4	13.6–23.0	6.020–6.271

Source: W. Vogel, H. Burger, G. Zerge, B. Muller, K. Forkel, G. Winterstein, A. Boxberger, and H. Bomhild, *Silikattechnelk* 25, 207, 1974.

tellurite–phosphate oxides and tellurite–phosphate halides, together with the glass formation of the ternary tellurite–borate oxides and tellurite–borate halides, and of the multitellurite–metaphosphate (metaborate)–metaphosphate (metaborate) section (K) from a four-component tellurite–phosphate or tellurite–borate system and metaphosphate (metaborate)–R_mO_n section (L) from a four-component tellurite–phosphate or tellurite–borate system. The results obtained provide a basis for determining the glass-forming tendencies in a number of tellurite–metaphosphate and tellurite–metaborate triple combinations. Hence, a new family of tellurite glasses have been produced.

It has been established that glass formation is mostly observed in alkaline, earth-alkaline-oxide, and halide-metaphosphate systems. The three-component tellurite-metaphosphate and tellurite-metaborate systems investigated have been treated as partial three-component sections of the main four-component phosphate and tellurite-borate systems. New families of optical tellurite glasses were obtained in 1985 by Burger et al., by melting TeO_2 with R_nO_m, R_nX_m, $R_n(SO_4)_m$, $R_n(PO_3)_m$, and B_2O_3. These new families exhibit a high transmission, with 86% in the visible, the NIR, and the mid-IR (MIR) regions. Glasses of the system TeO_2–R_nX_m (X = F, Cl, or Br) possess an entire transmission from 0.4 to 7 μm and do not have OH vibration absorption bands at 3.2 and 4.4 μm, respectively. It has been proved that heavy ions influence the absorption ability of glasses and shift the IR cutoff toward longer wavelengths. The same authors also investigated the n, dispersion, thermal-expansion coefficient, and ρ of the glasses. Discussion was based on the influence of the atomic mass, interatomic distances, and the cation stereochemical surroundings of the main building units in the vitreous matrix. It has been proved that Te–O stretching vibrations have a strong influence on multiphonon absorption.

Comparing the glass-forming regions of oxide–tellurite–phosphate and tellurite–borate three-component systems, it is clear that halides increase glass-forming ability. Introducing them into the glass, however, does not inhibit stable phase separation, which is characteristic of binary TeO_2–B_2O_3 and TeO_2–P_2O_5 systems as stated by Burger et al. (1984) and Kozhokarov et al. (1978). In analogy with tellurite–phosphate systems, tellurite–borate triple combinations along TeO_2–metaborate sections can easily give rise to the synthesis of phase-separation-free glasses. Kozhokarov et al. (1983) also prepared glasses of a "second generation," whose compositions are included in systems of the following types: TeO_2–metaphosphate–metal oxide, TeO_2–metaphosphate–metal halide, TeO$_2$–metaborate–metal oxide, TeO$_2$–metaborate–metal halide, TeO$_2$–metaphosphate–metaphosphate, TeO$_2$–metaborate–metaborate, and TeO$_2$–metaphosphate–metaborate.

In 1987, El-Mallawany and Saunders reported that they had prepared and measured both elastic moduli and compressibility of a series of binary TeO$_2$–ZnCl$_2$ (10, 20, and 33 mol%) glasses at room temperature for the first time. In 1988, Tanaka et al. prepared and investigated both structure and ionic conductivity for a series of TeO$_2$–Li$_2$O–LiCl glasses, as shown in the 1st Ed. It was observed that the Te–Oax bond, where ax is the axial position of the TeO$_4$ trigonal bipyramid, became weaker with the LiCl content in binary LiCl–TeO$_2$. In 1990, Ivanova studied glass formation in halide systems, including PbCl$_2$, NaCl, KCl, and BaCl$_2$; and in oxide-halide systems, including TeO$_2$. The ranges of glass formation in the systems TeO$_2$–PbCl$_2$–NaCl–KCl and TeO$_2$–BaCl$_2$–CdCl$_2$–NaCl were determined at different cooling rates as shown in Figure 1.17. Stable glasses possessing low melting temperatures and good chemical resistance have been achieved. Before Ivanova (1990), two problems of glass formation—chemical resistance and thermal stability—remained unsolved in relation to halide and oxyhalide glasses. El-Mallawany (1991, 1993) also studied the effect of chlorine on optical and elastic properties of tellurite glasses. Sahar and Noordin (1995) studied both thermal and optical properties of the oxyhalide glasses based on the TeO$_2$–ZnO–ZnCl$_2$ system. Zahar and Zahar (1995) measured the heat capacity changes that occur during the glass transition of TeO$_2$–TlO$_{0.5}$ glasses containing AgI or Ag$_{0.75}$–(Tl-I)$_{0.25}$. There was no interaction between the iodide in crystalline form and the host glass network.

Because tellurite glasses represent a compromise between the desire for a low-phonon-energy host and the need for mechanical strength and low processing temperatures, Sidebottom et al. (1997) studied the structural and optical properties of rare-earth metal oxide-doped zinc oxyhalide-tellurite glasses. The optical characterizations included optical absorption, Raman scattering,

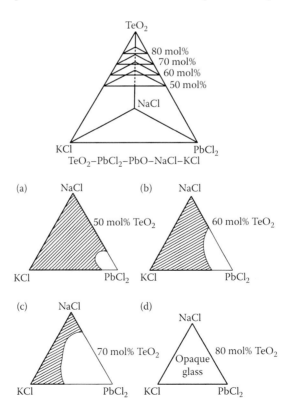

FIGURE 1.17 Investigated sections in the TeO$_2$–PbCl$_2$–KCl–NaCl systems. (From I. Ivanova, *J. Mater. Sci.*, 25, 2087, 1990.)

spectral analysis, temporal photoluminescence, and excitation (photon sideband) measurements (PSB). The goal of Sidebottom et al. (1997) was to develop insight into the manipulation of the local structural environment and its influence on both radiative and nonradiative excitation of the rare-earth metal oxide-doping population as discussed in Chapter 8. Also, these researchers studied the effects of structural depolymerization through ZnO addition and halide substitution for ZnF_2. Rare-earth ions possess optical transitions within the 4f level structure that are critical for the development of optical amplifiers and phosphors, as mentioned by France (1991). Reisfeld and Jorhensen (1977) measured both relative and nonrelative processes that mediate the excited-state decay within the 4f level and found that these processes are sensitive to the host material. A significant contribution to the measured emission lifetime—a key parameter of device performance—results from multiphonon nonradiative relaxation. Raman measurements provide an average picture of the vibrational ρ of the states in the glass. Depending on the local bonding environment of the rare-earth ion within the matrix, some of these vibrational modes may or may not be coupled to electronic excited states. Information about the local vibrational density of states (DOS) is conveyed by the phonon sideband (PSB) results. Sidebottom et al. (1997) reported that zinc-tellurite glasses appear to be excellent candidates for hosting rare-earth ions because they provide a low-phonon-energy environment to minimize nonradiative losses as well as possessing good chemical durability and optical properties. The emission lifetime, optical absorption, and vibrational ρ of states were examined in the glass system $xZnO–yZnF_2–([1-x-y]TeO_2)$ doped with 0.1 mol% Nd, Ho, Tm, or Er with a series of rare-earth metal oxides. Phonon sideband spectroscopy has been used to probe the vibrational structure in the intermediate vicinity of rare-earth ions. In particular, the lowest-energy PSB has about 40% less energy compared with the deformation mode seen in the glass. Also, the local vibrational modes obtained by PSB spectroscopy confirm the vibrational analogs seen in crystalline phases of zinc–tellurite. Fluorine substitution has been shown to affect the local crystal field symmetry of rare-earth ions and to result in substantial increases in the radiative lifetime.

1.8   GENERAL PHYSICAL CHARACTERISTICS AND STRUCTURE OF TELLURITE GLASSES

Stanworth (1952) studied the chemical durability of tellurite glasses and concluded that they have reasonable durability against water, acids, and alkaline. Glasses with compositions of 18 mol% PbO–82 mol% TeO_2 and 22 mol% PbO–78 mol% TeO_2 are very durable and show weight losses of only 5×10^{-7} and 20×10^{-7} g/cm² day⁻¹, respectively. The introduction of Li_2O and Na_2O reduces the water durability of lead–tellurite glasses, and there is a very severe increase in attack for tellurite glass containing B_2O_3. The water durability of 72.1 mol% TeO_2–27.9 mol% B_2O_3 is reflected in a loss of 228.000×10^{-7} g/cm² day⁻¹. Tellurite glasses containing oxides of phosphorus, tungsten, molybdenum, or zinc have good durability, indicated by weight losses of 20, 10, 40, or 0×10^{-7} g/cm² day⁻¹, respectively. Barium–tellurite glasses (e.g., 16 mol% BaO–84 mol% TeO_2), show only slight attack by citric acid, and only very slight attack by this same corrosive is shown in lead–tellurite glass, as well as tellurite glasses containing BaO, Li_2O, Na_2O, Cb_2O_5, P_2O_5, As_2O_3, MoO_3, and WO_3. The third test for chemical durability is resistance to attack by alkalis. Stanworth (1952) demonstrated only slight attack by alkalis against tellurite glasses containing 56.6 mol% TeO_2–19.8 mol% PbO–23.6% Cb_2O_5 and against tellurite glasses containing MoO_3 and WO_3. Heavy attack was shown in the lead–tellurite glasses, P_2O_5 and ZnF_2. The most heavily attacked were glasses containing BaO, Li_2O, Na_2O, and As_2O_5. Chen et al. (1997) studied the chemical durability of Te–Ge–Se–Se–I glasses. By comparing the behavior of glasses in different solutions, the authors concluded that the durability of glasses in acid solution is less than in water, while the lowest durability appears when glass samples are placed in a basic solution. The weight losses of glasses in HCl solution are proportional to, or the square root of, their immersion times.

Lambson et al. (1984) measured the physical characteristics of high-purity TeO$_2$ glass and found that it is semitransparent with a very pale lime green color. The binary tellurite glasses have different colors, ranging from light and dark brown to black or transparent, depending on the kind and percentage of *M*, as discussed in Section 1.4. For the pure oxide–tellurite glass, Lambson et al.'s x-ray exposure of the glasses using a Debye–Scherrer powder camera yielded diffuse halos typical of the vitreous state. Electron microscopy using the plastic replica technique presented no evidence of phase separation. Thermal analysis gave a transformation temperature of 320°C and a linear expansion coefficient over the temperature range 0°C–320°C of (15.5×10^{-6})°C^{-1}. Ultrasonic wave propagation is a method of establishing the vitreous nature of cast inorganic-oxide melts that is little known but is possibly the most rapid. Megacycle frequency ultrasonic echoes cannot be obtained from polycrystalline castings, whereas echoes are readily obtained from glasses, and this result can be obtained immediately after grinding two roughly parallel faces, whereas the x-ray method requires lengthy exposure times. Therefore, successful ultrasonic wave propagation gives the first indication that a high-purity TeO$_2$ glass has been prepared. Furthermore, the degree of homogeneity of glass samples can readily be estimated ultrasonically—if a velocity gradient occurs in the wave propagation direction, the interference maximum and minimum appear in the exponential-decay envelope of successive pulse echoes, even when all other causes of such interference patterns have been removed. Table 1.13 shows the collected ρ and calculated molar-volume data for TeO$_2$ crystal, TeO$_2$ glass, and binary and ternary tellurite glasses. The ρ of these glasses is measured at room

TABLE 1.13
Density and Molar Volume of Pure, Binary, and Ternary Tellurite Glasses

Glass System	Second Component (mol%)	ρ (g/cm³)	Molar V (cm³)	Reference
TeO$_2$ Crystal		5.70	26.60	Havinga 1961
TeO$_2$ Glass		5.11	31.29	Lambson et al. 1984
TeO$_2$–Li$_2$O	12.2–34.9	5.27–4.511		Vogel et al. 1974
TeO$_2$–Na$_2$O	5.5–37.8	5.43–4.143		
TeO$_2$–K$_2$O	6.5–19.5	5.27–4.607		
TeO$_2$–BeO	16.5–22.7	5.357–4.725		
TeO$_2$–MgO	13.5–23.1	5.347–5.182		
TeO$_2$–SrO	9.2–13.1	5.564–5.514		
TeO$_2$–BaO	8.0–23.1	5.626–5.533		
TeO$_2$–ZnO	17.4–37.2	5.544–5.66		
TeO$_2$–Tl$_2$O	13.0–38.4	6.262–7.203		
TeO$_2$–PbO	13.6–21.8	6.054–6.303		
TeO$_2$La$_2$O$_3$	4.0–9.9	5.662–5.707		
TeO$_2$Al$_2$O$_3$	7.8–16.8	5.287–4.850		
TeO$_2$TiO$_2$	7.8–18.9	5.580–5.242		
TeO$_2$WO$_3$	11.5–33.8	5.782–5.953		
TeO$_2$–Nb$_2$O$_5$	3.6–34.4	5.558–5.159		
TeO$_2$–Ta$_2$O$_5$	3.8–33.4	5.659–6.048		
TeO$_2$–ThO$_2$	8.4–17.1	5.762–5.981		
TeO$_2$–GeO$_2$	10.2–30.2	5.437–5.057		
TeO$_2$–P$_2$O$_5$	2.2–15.6	5.433–4.785		
TeO$_2$–B$_2$O$_3$	11.8–26.4	5.205–4.648		
(TeO$_2$–Al$_2$O$_3$) SiO$_2$ (5.2:1)	1.2–17.9	4.854–4.484		

temperature by the standard displacement method, using toluene as immersion liquid and following the formula:

$$\rho = \rho_0 \frac{\left(W - W_t\right)}{\left(W - W_t\right) - \left(W_1 - W_{1t}\right)} \tag{1.1}$$

where ρ is the density of toluene (0.805 g cm^{-3} at 398 K), W and W_1 are the weights of the glass samples in air and toluene, respectively, and Wt and W_1t are the weights of the suspended thread (0.01 mm in diameter) in air and toluene, respectively. Sensitive balance was used in these measurements. The sensitivity was 0.1 mg, and the maximum capacity was 160 g. The accuracy of the measurements was less than 0.001%.

The molar volume is calculated according to the following relation:

$$V = \frac{Mw_{\text{glass}}}{\rho_{\text{glass}}} \tag{1.2}$$

where ρ_{glass} is the density of the glass sample and Mw_{glass} is the molecular weight calculated from the relation:

$$Mw_{\text{glass}} = \left(1 - x - y\right) Mw_{\text{TeO}_2} + Mw_{\text{1st}} + y Mw_{\text{2nd}} \tag{1.3}$$

where, for the expression $(1 - x - y)$, x and y are the mole fractions of the first and second constituent oxides of TeO$_2$, respectively, and Mw_{TeO_2}, Mw_{1st}, and Mw_{2nd} are the molecular weights of modifying oxides.

The mutual benefits of a proposed cooperative research effort would include providing the fundamental bases for new optical glasses with new applications, especially tellurite-based glass optical fiber, which could benefit countries all over the world. Considerable international technological and fundamental interest in these materials is reflected in recent publications, including a 1992 article by El-Mallawany and several articles on the optical properties of tellurite glasses by Kim et al. (1993a, 1993b), Takabe et al. (1994), and Kim and Yako (1995). These articles report measurements of the nonlinear optical properties of pure tellurite glasses and glasses doped with La$_2$O$_3$. Among these findings, tellurite glasses have χ^3 (third-order, nonlinear optical-susceptibility) values that are approximately 50-fold greater that those of pure fused silica glass.

El-Mallawany (2000) has reported and interpreted the structures of pure tellurite glass and two different series of binary tellurite glasses, based on the ρ data for these glasses. El-Mallawany et al. (1994) and Sidky et al. (1997) calculated and analyzed the molar and mean atomic volumes for binary tellurite glasses doped with transition metals like MoO$_3$ and V$_2$O$_5$, respectively, as shown in Table 1.14. The molar volumes (V) of binary tellurite glasses of the form x mol% TeO$_2$–(100 – x) mol% M_nO$_m$ were calculated using the relation in Equation 1.2. M_nO$_m$ is the TMO MoO$_3$. V$_2$O$_5$ with x mol% Mw_{glass} is the glass molecular weight according to the mol% of every TMO present in the tellurite host glass, and ρ_g is the density of the glass. The calculated molar volumes of pure TeO$_2$ crystal and glass are 26.6 and 31.29 cm^3, respectively (ρ_{crystal} = 5.7 g/cm^3 [Havinga 1961]). This means that the ratio of $V_{\text{crystal}}/V_{\text{glass}}$ is 1.18; that is, the change is only 18% from crystalline solid to noncrystalline solid. Hence, V_{glass} is greater than V_{crystal}, which correlates extremely well with the longer number of TeO$_2$ units that can be accommodated in the more open structure of the vitreous state. For the first series, the binary TeO$_2$–MoO$_3$, the V increases from 31.23 to 32.9 cm^3. For the second series, TeO$_2$–V$_2$O$_5$ glass, V increases from 31.04 to 42.7 cm^3, as shown in Figure 1.18a. The change in relative concentrations within such structural groups with increasing TMO content should affect the physical properties of the binary TeO$_2$–TMO glasses.

TABLE 1.14

Molar V and Mean V_a of Binary TeO_2–V_2O_5 and TeO_2–MoO_3 Glasses for Different Percentages of the Modifier

Glass (Source) (mol%–mol%)	ρ (gm/cm³)	V (cm³)	V_a (cm³)
TeO_2 (Lambson et al. 1984; El-Mallawany 2000)	5.11	31.26	10.41
TeO_2–MoO_3 (El-Mallawany et al. 1994)			
80–20	5.01	31.23	9.76
70–30	4.90	31.61	9.58
55–45	4.75	32.20	9.33
50–50	4.60	32.90	9.40
TeO_2–V_2O_5 (Sidky et al. 1997)			
80–20	4.90	33.48	9.30
75–25	4.60	35.70	9.92
70–30	4.50	36.40	8.66
65–35	4.30	38.66	8.79
60–40	4.20	39.9	8.67
50–50	3.99	42.7	8.55

The mean atomic volume (V_a) equals $Mw_{glass}/q\rho_g$, where q is the number of atoms per formula unit. In the binary tellurite glasses just discussed, $q = (x)3 + (1 - x)$, where x is the number of atoms of the TMO (e.g., four for MoO_3, seven for V_2O_5, and three for TeO_2). From the simple model of compressibility by Mukherjee et al. (1992) of binary glass Ax–$B(1 - x)$ containing n_A formula units of type A, and n_B formula units of type B with the percentage $x = n_A/(n_A + n_B)$, the volume of the binary glass containing $n_A \times$ Avogadro's number formula units of type A and n_B formula units of type B can be easily determined from the ρ measurements using Equation 1.4:

$$V = \left(\frac{1}{\rho}\right)\left[Mw_A + \left(\frac{n_B}{n_A}\right)Mw_B\right] \qquad (1.4)$$

where ρ, Mw_A, and Mw_B are as previously defined, and Ax–$B(1 - x)$ and x are percentages also previously defined. The model supposes that the composition of binary glass AB changes from n_A and n_B formula units of type A and B, respectively, to $(n_A + 1)$ and $(n_B + 1)$ corresponding formula units. Although the total number of formula units of A and B taken together remains unchanged, the volume of the vitreous system changes by an amount called "the difference volume" (V_d) due to the exchange of one formula unit between A and B in the binary glass system. The compressibility model assumes that V_d and the mean volume (V_a) per formula unit of A in the binary glass Ax–$B(1 - x)$ are independent of the percentage of the M for a glass series and differ from one series to another. The model is also based on the assumption that the binary glass series has the same structure and no phase changes, which implies that the V of the binary system Ax–$B(1 - x)$ containing ($n_A \times$ Avogadro's number) of formula units of A and n_B formula units of B can be written as:

$$V = n_A V_A + n_B\left(V_d + V_A\right)$$

$$= V_o + \left(\frac{n_B}{n_A}\right)\left(n_A V_d + V_o\right) \qquad (1.5)$$

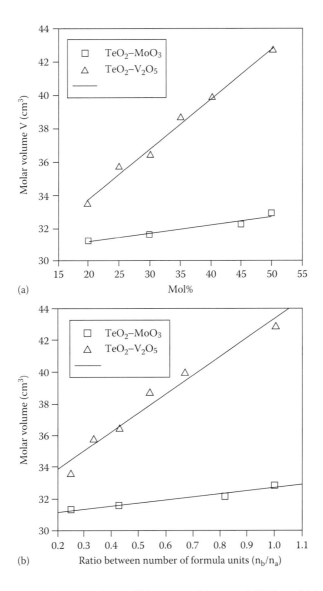

FIGURE 1.18 (a) Variation of molar volume of binary transition metal (V_2O_5 and MoO_3) tellurite glasses. (b) Relationship between molar volume and ratio of formula units (n_b/n_a), as calculated by El-Mallawany (2000).

where $V_o = n_A V_A$ represents the V of a vitreous system consisting of n_A formula units of type A only, with V_A the average volume per formula unit, and n_B/n_A is the composition ratio. When we envisage the molecular units in the glass network so that A stands for the "glass former TeO_2", and B stands for any M of V_2O_5 or MoO_3, Equation 1.5 clearly indicates that the plot of V against the composition ratio n_B/n_A follows a straight line from which the intersect with the y-axis gives V_o, and the slope gives $n_A V_d$. Figure 1.18b shows that for the two binary tellurite glasses studied in the present investigation, the value of V_o is 30.744 cm^3 for the line of TeO_2–MoO_3 and 31.419 cm^3 for the line of TeO_2–V_2O_5, respectively. The calculated values of the volume obtained from the simple model agree with the experimental values of the pure TeO_2 reported by Lambson et al. (1984). From the slope of both lines, the values of $n_A V_d$ and the V_d—due to the exchange of formula units between A and B times n_A—are −28.732 cm^3 and −19.494 cm^3 for the two binary glass series, respectively. These values are negative, whereas binary MoO_3–P_2O_5 and V_2O_5–P_2O_5 glasses have the respective values of 21.94 cm^3 and 9.538 cm^3, as stated by Mukherjee et al. (1992).

Understanding the properties of glasses sufficiently to offer predictions requires adequate knowledge of their atomic structure, which is a common requirement among those researching in this field. It seems clear from the coverage above that tellurite glasses are strategically important solid materials. The physics and chemistry of operating solids can sometimes stretch the limits of available materials. Glass, being thermodynamically metastable, can alter its structure and properties under various operating conditions. For example, long-term reactions with the fill materials in some high-intensity-discharge (HID) lamps can result in devitrification of the fused quartz arc tube. Under certain severe operating conditions, a poorly designed lamp can develop a thermally compacted glass structure resulting in high stresses, which are sufficient to literally rip apart an arc tube surface. High thermal loading of some developmental halogen lamps has caused glass "compaction" around the glass–metal seal, resulting in very high stresses, and ultimate seal failure. Careful analysis of glass–metal thermal-contraction stresses and understanding of the thermal limits of materials can help control thermal compaction stress and strain.

For every major type of glass, answers to questions on structure and properties are still needed. Much can be done, as has already been demonstrated, to understand those properties that depend on the local or long-range structure of glass; for example, optical properties and ionic diffusion in mesoscopic structures. Investigations of medium-range structure are more difficult. The difficulty of obtaining good structural data is matched by the limited usefulness of those data in predicting and controlling properties. When questions are properly posed, results can give valuable insights. Moreover, solutions to this part of the structural puzzle could lead to spectacular changes in fundamental concepts of the nature of glasses, other amorphous solids, and even liquids. An adequate understanding of the structures of amorphous solids such as tellurite glasses is a basic requirement for progress in affected fields in the next century.

However, the central questions remain the ones put forward by Gaskell (1997). What constitutes an adequate understanding of structure in relation to properties? It is not structural knowledge for its own sake, although few would argue that the structural questions posed by amorphous solids and liquids have immense intrinsic appeal for condensed-matter science. Gaskell (1997) posed two related questions whose answers could affect our understanding of structure–property relationships, along with our ability to better predict properties: just how far do we need to go? And how much structure knowledge do we need? Gaskell also provided an elaborate schematic representation of the properties of glasses with respect to their structural characteristics on various length scales, as shown in the 1st Ed. Experimental techniques for structure analysis include x-ray photoelectron spectroscopy (XPS), electron diffraction, electron probe microanalysis, Auger, electron paramagnetic resonance, and nuclear magnetic resonance (NMR), which is an effective technique for the analysis of the electronic structure of atoms. These leading-edge techniques have been explained in detail by Simmons and El-Bayoumi (1993).

1.9 STRUCTURE AND BONDING NATURE OF TELLURITE GLASSES

Considerable structural research has been done on tellurite glasses since 1982; for examples, see Kozhukharov et al. (1986), Burger et al. (1985), Kozhokharov et al. (1986), Neov et al. (1979, 1988, 1995a, 1995b), Bursukova et al. (1995), Zwanziger et al. (1995), Lefterova et al. (1996), Hiniei et al. (1997), Suehara et al. (1994, 1995), and Sekiya et al. (1992). The structure of α-TeO_2 tellurite glass and the atomic SRO of binary TeO_2–ZnO, as shown in Figure 1.9 by Kozhokharov et al. (1986), as well as the properties of TeO_2–WO_3 as shown in Figure 1.19 and described by Kozhokharov et al. (1986) and TeO_2–MoO_3, described by Neov et al. (1988), have all been described based on neutron diffraction. The main advantages of electron diffraction are the possibilities of determining the SRO in samples with small dimensions, such as microparticles and nanoparticles and thin films, and of identifying both light and heavy elements in the structures of these samples.

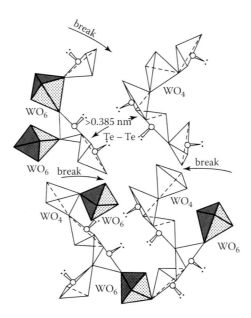

FIGURE 1.19 Model illustrating the manner of bonding of the nearest coordination polyhedra in binary TeO_2–WO_3 glasses. (From V. Kozhokharov, H. Burger, S. Neov, and B. Sidzhimov, *Polyhedron*, 5, 771, 1986).

Electron diffraction is suitable for investigating the SRO in model binary TeO_2–B_2O_3 glasses prepared as powders or thin films, as mentioned by Bursukova et al. (1995). These authors used electron diffraction to study the SRO in glasses of the TeO_2, B_2O_3 and TeO_2–B_2O_3 systems. The amorphous samples investigated were prepared using two methods. Powdered samples were produced either by rapid quenching using the roller technique (TeO_2 glass) or by slow cooling (B_2O_3 and TeO_2–B_2O_3 glasses), and amorphous TeO_2 and TeO_2–B_2O_3 thin films were deposited by vacuum evaporation with resistive heating. The 500 Å thick films were deposited onto NaCl pellet. Electron diffraction data indicate that the TeO_4 polyhedron is the main structural unit in TeO_2 glass. The boron atoms in the B_2O_3 glass are threefold coordinated with respect to oxygen, and the presence of B_3O_6 groups is indicated. In the binary glasses, mainly Te–O or B–O, distances are resolved, depending on their composition. The results for the powdered and thin-film samples have been compared, and a change in the Te–O distances for the films has been established. The interatomic distances for the binary tellurite–borate glasses are listed in Table 1.15.

Crystalline TeO_2 has two polymorphic forms: tetragonal α-TeO_2 (paratellurite), described by Leciejewicz (1961), and orthorhombic β-TeO_2 (tellurite), described by Beyer (1967), with fourfold and 4 + 2 coordination of the Te atom, respectively. On the basis of XRD data, Barady (1957) suggested a structural scheme for a TeO_2–Li_2O glass similar to that in β-TeO_2. Neutron diffraction studies of tellurite glasses by Neov et al. (1980) showed an atomic arrangement characteristic of the α-TeO_2 polyhedral structure, with a well-expressed tendency toward a transition in the Te coordination state from 4 to (3 + 1) to 3 with increasing M content. This transition is strongly dependent on the chemical nature and crystalline structure of the M.

Also in 1995, Neov et al. (1995a and 1995b) investigated binary tellurite–phosphate glasses and multicomponent halide–tellurite glasses by high-resolution SRO analysis. The aim of the first named study was to investigate the glass formation mechanism and SRO of tellurite glasses with compositions close to that of pure TeO_2. A small amount of P_2O_5 was used to prepare samples in the vitreous state. The changes in the basic unit of the glass network, the TeO_4 polyhedron, were studied by means of two independent neutron-scattering experiments—conventional and high-real-space-resolution time-of-flight (TOF) neutron diffraction. The atomic SROs in tellurite glasses containing 2 and 4 mol% P_2O_5 as a co-glass former have been investigated by two neutron diffraction

TABLE 1.15

Interatomic Distances and Coordination Numbers (CN) of Tellurite Glasses

Glass[a]	Reference	Distance I	Distance II	Distance II	CN
98 mol% TeO_2–2 mol% P_2O_5	Bursukova et al. 1995	0.19	2.08		
		CN = 1.994	CN = 2.30		
			4		
96 mol% TeO_2–4 mol% P_2O_5		0.19	0.21		
		CN = 1.92	CN = 2.11		
B_2O_3 (P)	Neov et al. 1995	1.40	2.35	3.6	BO = 3.3
4 mol% TeO_2–96 mol% B_2O_3 (P)		1.35	2.20	3.6	BO = 3.2
TeO_2 (P)		1.9	2.95	3.6	TeO = 3.6
TeO_2 (F)		2.05	2.80	3.75	TeO = 3.7
80 mol% TeO_2–20 mol% B_2O_3 (P)		1.90	2.55	3.60	TeO = 3.5
80 mol% TeO_2–20 mol% B_2O_3 (F)		2.00	2.75	3.80	TeO = 3.7

[a] P, powder; F, film.

techniques. The structure factor S_Q was measured up to 359 nm[1] by the TOF method, using a pulsed-neutron source. Conventional neutron diffraction measurements were carried out on a two-axis diffractometer. An asymmetric maximum located at 0.195 nm has been observed in the RDF of tellurite glass with 2 mol% P_2O_5. This maximum involves the Te–O pair distribution, and the calculated coordination bond number of the Te atoms is 3.8. Increasing the P_2O_5 content up to 4 mol% leads to a splitting of the first coordination maximum, with the two subpeaks at 0.199 and 0.214 nm, respectively. These experimental results are tabulated in Table 1.15 through a correlation with crystal-like model RDFs.

Neov et al. (1995b) studied the structures of complex tellurite glasses. The atomic SROs in multicomponent tellurite classes containing 80 mol% TeO_2 and 20 mol% CsCl, $MnCl_2$, $FeCl_2$, or $FeCl_3$ have been studied by the neutron-scattering method. Despite the easy deformation of the basic building unit of these glasses (i.e., the TeO_4 polyhedral), the first coordination maximum of the RDFs remains well separated. The coordination number of Te is four for all of the compositions studied. For the glasses with either of the 3d-metal chlorides as an M, the O–O distribution undergoes considerable change during transition to the vitreous state. The experimental RDF has been interpreted by comparison with model distribution functions composed of the experimental RDFs for pure TeO_2 glass and crystal-like RDFs for the modifier. Results for S_Q, where Q is the magnitude of the scattering vector, were obtained using a two-axis diffractometer. Analysis of the RDFs obtained by neutron-scattering experiments led to the following conclusions concerning the SROs in these glasses:

- The fourfold coordination of Te atoms is conserved for all the compositions studied.
- In TeO_2–$MnCl_2$, TeO_2–$FeCl_2$, and TeO_2–$FeCl_3$ glasses, the oxygen network is strongly influenced by chlorine-containing Ms.
- A strictly additive model of M incorporation into the TeO_2 matrix is applicable only for the SRO of TeO_2–$CsCl_3$ glass.
- For the isostructural Ms $MnCl_2$ and $FeCl_2$, the SROs in the corresponding TeO_2 glasses are similar.

Also in 1995, spectroscopic analyses (secondary ionization mass spectrometry, XPS, Mossbauer, IR, Auger, ERR, and NMR) of microstructures and modifications in borate and tellurite glasses were reported by Zwanziger et al. Through two-dimensional NMR experiments, a detailed picture of these chemical species and their local environments in glasses was obtained. The structural changes in borate glass on modification with rubidium oxide were monitored, including the relative

TABLE 1.16
Binding Energies of Various Core Level Electrons of
$(1 - x)(2TeO_2-V_2O_5)$-x mol% Ag_2O Glasses

x mol% Ag_2O	O1s	Ag3d$_{5/2}$	Te3d$_{5/2}$	V2p
0	531.1		576.9	517.7
33	530.1	367.7	575.8	516.4
50	531.1	368.7	575.8	516.8
67	530.3	368.4	575.8	516.6

Source: E. Lefterova, V. Krastev, P. Angelov, and Y. Demetriev, Ed., *12th Conference on Glass & Ceramics, 1996*, 150, 1997.

reactivity of different boron sites. Results were also presented on the composition dependence of the sodium environment in Na_2O–TeO_2 glasses. A qualitative change was observed in the average sodium coordination environment at a sodium oxide composition of 15%, which was interpreted in terms of the appearance of NBO ions, as shown in detail in the 1st Ed.

But in 1996, Lefterova et al. studied the XPS of Ag_2O–TeO_2–V_2O_5 glass as tabulated in Table 1.16. The XPS analysis proved that the addition of Ag_2O leads to the transformation of TeO_3 into TeO_4 groups and of VO_5 into VO_4 groups. At Ag_2O content above 25 mol%, NBO atoms have been observed, as shown in Figure 1.20.

Hiniei et al. (1997) measured the XPS of alkali–tellurite glasses of the forms R_2O–TeO_2 (where R is Li, Na, K, Rb, or Cs), using a fresh surface fractured in an ultrahigh vacuum (7×10^{-8} Pa) and irradiated with a monochromatic Al–Kα x-ray (hv = 1486.6 λ). In their report, the oxygen-first photoelectron spectra show only a single Gaussian–Lorentzian peak, which shifts toward smaller binding energies with increases in the Lewis basicity of oxide ions in the glasses. Two peaks attributed to BO (bridging oxygen) and NBO (nonbridgng oxygen) atoms are not observed. In the near-valence band spectra for the lithium–tellurite glasses, the spectral profile gradually becomes similar to that of Li_2–TeO_3 crystal as Li_2O content increases to 36 mol%. This variation of the profile is correlated with the change in the coordination structure of the tellurium atoms (TeO_4 trigonal bipyramids $\rightarrow$ TeO_3 trigonal pyramids) with the addition of the alkali oxides.

1.9.1 Assignment of the Shoulder in the Oxygen-First Peak

Hiniei et al. (1997) observed a shoulder in the oxygen-first (O-first) spectra for the glass surfaces exposed in air, which was not detected for the surfaces fractured in a vacuum. They assumed that this shoulder was due to some contamination on the glass surface. The amplitude of the O-first peak, around 284.6 eV, increased gradually with the passage of measuring time; this peak is attributed to a hydrocarbon accumulated on the glass surface in a vacuum. On the other hand, the chlorine O-first peak observed at about 289 eV for the surface exposed in air did not increase during measurement of the fresh surface.

1.9.2 Profile of Oxygen-First Photoelectron Spectra

In general, the O-first binding energy for alkali–tellurite glasses is small and close to that of the NBO for alkali–silicate glasses, which implies that the electronic ρ of the valence shell on the oxide ions in tellurite glasses is greater. Hiniei et al. (1997) assumed that some of the electrons of the lone pair on the Te atom in the TeO_2 polyhedron might be donated to the legend oxide ions through the Te–O σ-bonds, and thus that the electronic ρ of the valence shell on the oxide ions in the alkali–tellurite glass becomes greater. These authors suggested that the electronic ρ of the valence shell

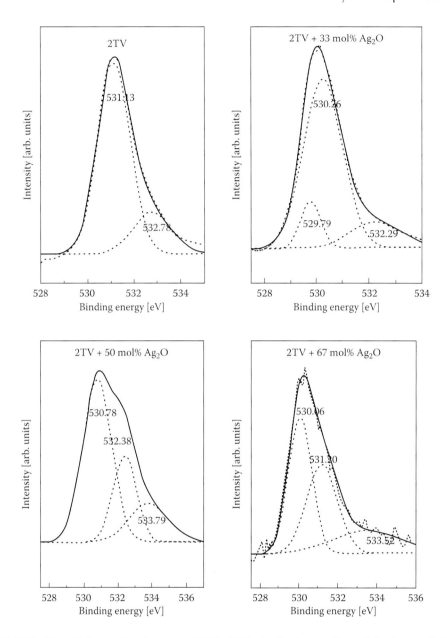

FIGURE 1.20 First-O deconvoluted spectra of TeO_2–V_2O_2–Ag_2O glasses. (From E. Lefterova, V. Krastev, P. Angelov, and Y. Dimitiev, eds., *12th Conference on Glass & Ceramics, 1996*, 150, 1997.)

on the NBO atom is almost equal to that on the BO atom; they theorized that equalization of electronic ρ occurs in the valence shell between BO and NBO atoms, because $p\pi$–$d\pi$ bonds are formed between O-2p and empty Te-5d orbitals. In contrast, two components attributed to BO and NBO atoms were observed in first-O photoelectron spectra for Na_2O–SiO_3 glasses, because the Si–O bonds in silicate glasses have less π-bonding character than the Te–O bonds in tellurite glasses, and lone-pair electrons are not donated to Si atoms but are localized on NBO atoms. The authors showed that Te–$3d_{5/2}$-binding energy decreases as well as that of the first O with increasing alkali oxide content. The shift in the Te $3d_{5/2}$-binding energy can be explained by an increase in the extent of the $p\pi$–$d\pi$ back donation to reduce the charge separation between tellurium and oxygen atoms in Te–O bonds as shown in the 1st Ed.

1.9.3　Chemical Shifts of the Core Electron-Binding Energies

Hiniei et al. (1997) showed that the first-O and the Te $3d_{5/2}$ peaks shift toward smaller binding energy as the ionic r of the alkali ions (Li $\rightarrow$ Na $\rightarrow$ K $\rightarrow$ Rh $\rightarrow$ Cs), as well as the alkali oxide content, increases. These chemical shifts are explained by a change in Lewis basicity of oxide ions in the glasses. The basicity of oxide ions is assumed to indicate the effective electronic density of their valence shells, which interact with cations. As mentioned before, the binding energy of the first-O atom is influenced by the electronic density of the valence shell on oxide ions, so some correlation between the binding energy of first Os and the basicity of oxide ions is expected. Hiniei et al. (1997) showed the binding energy of the first Os and Te $3d_{5/2}$ as a function of the optical basicity (Λcal), which represents the average Lewis basicity of oxide ions in a matrix. Hiniei et al. (1997) showed that a relatively linear correlation between the binding energy and the Λcal is found for the alkali–tellurite glasses. The increase in Λcal is assumed to indicate an increase in the effective electronic ρ of the valence shell on oxide ions, which can interact with a cation. Therefore, the increase in the covalency of Te–O bonds due to an increase in the Lewis basicity of oxide ions—that is, the extent of electron donation from oxygen to tellurium atoms—is reflected in the structural change from TeO_4 thp to more covalent TeO_3 tp with the addition of alkali oxides in the glasses as shown in the 1st Ed.

Hiniei et al. (1997) concluded that the calculated Λcal must be defined based on the assumption that all oxide ions have an identical chemical bonding state in the matrices and are independent of the local structure around the tellurium atoms. The experimental Λcal was obtained from frequency shifts of $^1S_0 \rightarrow {}^3P_1$ transition in the UV spectra of the probe ions Tl^+, Pb^{2+}, and Bi^{3+}, and this value reflects the basicity of individual oxide ions in the matrices.

This basicity, however, cannot be applied to the tellurite glasses because of their opaqueness to UV light. On the other hand, the binding energy of the first-O, which is determined by the electronic density of the valence shell on oxide ions, seems to be a better and more universal index of basicity. In Chapter 8, a detailed discussion of the n of tellurite glasses based on the change in Λcal of different oxides is given.

1.9.4　Valence Band Spectra

Hiniei et al. (1997) also measured the XPS spectra near the valence band for glasses and crystals in the TeO_2–Li_2O system. They concluded that variation in the valence band spectra may be due to hybridization in the bonding orbitals—that is, a change in coordination of tellurium atoms (TeO_4-tbp $\rightarrow$ TeO_3-tp) with the addition of the alkali oxide. However, Suehara et al. (1994) took no account of the Te-5d orbital. Thus, a more detailed understanding requires the molecular-orbital calculation involving Te-5d orbitals, which is being investigated.

In 1995, Suehara et al. reported on the electronic structure and nature of the chemical bond in tellurite glasses because, in a normal glass-forming system, M atoms are usually added to enhance glass formation, often as a result of "network breaking" (breaking the chains of structural units) and incremental entropy (decreasing liquidus temperature). In tellurite glasses, however, the M atoms play one more important role: variation of the structural unit itself, in contrast to the structural unit of silicate glasses (an SiO_2 tetrahedron), which is not affected by M atoms. Suehara et al. (1995) used this method of calculation based on the random-network model; glass has a structural unit similar to the SRO in its analogous crystalline compound. In general, a bond in a compound is partly covalent and partly ionic. In α-TeO_2, the charge states of a Te atom and an O atom are $Te^{4\delta-}$ and $O^{2\delta}$, where the parameter δ ($0 \leq \delta \leq 1$) is the fractional ionic character in a Te–O bond. Thus, the initial charge state for the TeO_6 octahedron should be $TeO_6^{8\delta-}$, but δ cannot be easily determined.

The ionicity of α-TeO_2 is estimated as follows: the net charge (n) for a $(TeO_6)^{n-}$ cluster can be defined as $8\delta + n_{CT}$, where n_{CT} is the amount of the charge transferred from M to TeO_6. The

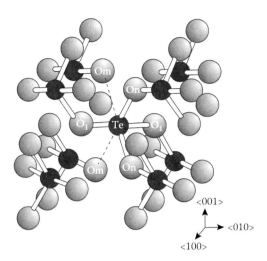

FIGURE 1.21   Schematic illustration of a TeO_6 cluster in paratellurite. (From S. Suehara, K. Yamamoto, S. Hishta, T. Aizawa, S. Inoue, and A. Nukui. *Phys. Rev.*, 51, 14919, 1995.) The TeO_6 cluster is made of one central Te atom and six octahedral O atoms, which can be divided into the following three types: axial type (O_I), found at a distance of 3.84 atomic units (a.u.) from the central Te atom; equatorial type (O_{II}), at a distance of 3.84 a.u.; and another equatorial type (O_{III}), found at a distance of 5.05 a.u.

cluster is made of a central Te atom and three kinds of octahedral O atoms: axial-type O_I, equatorial-type O_{II}, and O_{III}, as shown in Figure 1.21. Because no bonding is expected in the regions of the negative-overlap population, the bonds $Te–O_{III}$ and $Te–O_I$ should break, and consequently the coordination number of the Te atom should change from 6 via 4 to 2 as n increases from 6 to 8. Actual tellurite glasses are made not of one structural unit, but of a mixture of TeO_4, TeO_{3+1}, and/or TeO_3, as stated by Sekiya et al. (1992) and Neov et al. (1979). This variation is due to local inhomogenity; the electrons do not transfer from the *M* atoms to all the structure units; consequently, two or more states of the structural unit exist in the actual tellurite glasses. To understand the bond breaking, orbital-overlap populations for each Te–O bond in the neutral $(TeO_6)^0$ cluster must be examined. Also, energy originates from the O-2p states. In the rigid-band scheme, electrons occupy the levels marked 1, 2, 3, and 4 in order as *n* increases. Each level can contain two electrons. The occupation of levels 1, 2, and 3 hardly contributes to Te–O bonding, judging from the small orbital-overlap populations. These levels are of O-2p character and merely make lone pairs. Therefore, the bonding nature does not change up to *n* = 6. When *n* exceeds 7, electrons occupy marked level 4, and then the overlap populations of each bond decrease because of the large antibonding character at this level. In this case, the $Te–O_I$ bond is the most weakened. However, the $Te–O_{III}$ bond is the first broken, because this bond has a smaller overlap population than $Te–O_I$ from the beginning. The orbital-overlap population at marked level 4 for the $Te–O_{II}$ bond is the smallest antibonding characteristic, and therefore this bond remains in a bonding state. Thus, variation of the structural unit in tellurite glasses is most likely caused by the electrons in marked level 4.

   Suehara et al. (1995) also estimated the initial charge state and bond ionicity of α-TeO_2. As discussed above, the variation in structural units is caused by two or less electrons in marked level 4. This is consistent with the crystalline compounds; even assuming that all *M* atoms are completely ionized, the transfer charge (n_{CT} = 0) per cluster is 2 or less; for example, n_{CT} = 0 for α-TeO_2; 2 for Li_2TeO_3, $BaTeO_3$, and $CuTeO_3$; and 4/3 for $Zn_2Te_3O_8$. It follows from this estimation that the initial net charge (8δ) must be ~6 (i.e., $\delta \approx 0.75$) for a structural unit changing by n_{CT}. In α-TeO_2, therefore, the net charges of an α-Te atom and an O atom should be ~3 and ~1.5, respectively.

1.9.5 INTERMEDIATE-RANGE ORDER (AS DETERMINED BY NMR, NEUTRON, XRD, X-RAY ABSORPTION FINE STRUCTURE, MOSSBAUER SPECTRA, XPS, AND X-RAY ABSORPTION NEAR-EDGE STRUCTURE ANALYSES) IN TELLURITE GLASSES

Tagg et al. (1995) probed the sodium sites in tellurite glasses using a dynamic-angle-spinning NMR experiment in $(1 - x)$ mol% TeO_2–x mol% Na_2O glasses. They performed their experiments at two different magnetic field strengths in order to extract the chemical-shift and quadrupole-coupling parameters. The results suggested that sodium coordination changes from about six at low M concentrations to about five at high concentrations. This decrease is not monotonic. Berthereau et al. (1996) found the origin of the high nonlinear optical response in tellurite glasses of the forms TeO_2–Al_2O_3 and TeO_2–Nb_2O_5. The detailed structural investigations of these glasses were joined to a study of their electronic properties through *ab initio* calculations. Bulk glasses were prepared by the melt-quenching technique. Samples for x-ray absorption fine structure (XAFS) measurements were prepared by grinding and sieving the glasses to obtain fine powders of homogeneous granulosimetry (20 μm for the Te–L_{III} edge and 50 μm for the Te–K edge). The powdered glasses were mixed with a convenient ratio of boron nitride powder to fill a 1 mm thick and 1 cm² surface copper cell sandwiched between two x-ray-transparent Kapton adhesive tapes. Figure 1.22 represents the x-ray absorption near-edge structure (XANES) Te–L_{III} edge spectra of the x mol% TeO_2–$(1 - x)$ mol% Al_2O_3, where $x = 0$, 80, 85, 90, and 95, and also with reference crystals. Calculations of molecular orbitals have been performed in the *ab initio* restricted Hartree–Fock scheme (RHF). The core orbitals of tellurium atoms were reduced to a frozen orbital-effective core potential, and the clusters TeO_4^{4+} and TeO_3^{2-} were investigated. The molecular orbitals obtained in restricted Hartree–Fock calculations were subsequently used in a configuration interaction expansion, including single and/or double excitations from the Hartree–Fock reference. Sabadel et al. (1999), using x-ray absorption spectroscopy at the Te–L_{III} edge and a ¹²⁵Te Mossbauer analysis of TeO_2–BaO–TiO_2 glasses, also confirmed and completed the first structural results for the XANES study. The ¹²⁵Te Mossbauer spectra are characterized by isomer shift (δ) and quadrupole splitting (Δ). Elidrissi-Moubtaim et al. (1995) put the expression of the Mossbauer isomer shift as follows:

$$\delta = \left(\frac{Ze^2cR^2}{5\varepsilon_0 E_o} \right) \left(\frac{\Delta R}{R} \right) \left[\left| \Psi_a^0 \right|^2 - \left| \Psi_s^0 \right|^2 \right]$$

$$\delta = \alpha \Delta \rho (0) \tag{1.6}$$

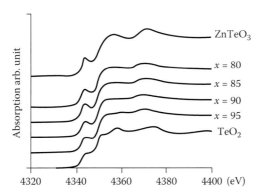

FIGURE 1.22   Te-L_{III} edge XANES spectra of x mol% TeO_2–$(1 - x)$ mol% Al_2O_3 glasses and reference crystals. (From A. Berthereau, E. Fargin, A. Villezusanne, R. Olazcuaga, G. Flem, and L. Ducasse, *J. Solid State Chem.*, 126, 143, 1996.)

where e is the electronic charge, c is the velocity of light, ε_o is the permittivity of the vacuum, Z is the electron number, E_o is the energy of the nuclear transition, and $\Delta R/R$ is the variation of the nuclear radius. The $|\Psi^0|^2$ is the relative ρ at the nucleus (a and s are absorber and source, respectively). According to the previous Equ, the δ varies linearly with $\rho\,(0)$ at the Mossbauer nucleus, which is mainly dependent on the s-type electron. The Δ is given by:

$$\Delta = \left(\frac{1}{2}\right)\left[eQV_{zz}\left(1+\frac{\eta^2}{3}\right)^{1/2}\right] \qquad (1.7)$$

where Q is the electric quadrupole moment, Vzz is the principal component of the diagonalized tensor of the electric field gradient, and $\acute{\eta}$ is the asymmetry parameter. Sabadel et al. (1999) found that for ^{125}Te, the excited nuclear level (3/2) splits into two sublevels ($\pm1/2$ and $\pm3/2$), and the existence of an electric field leads to a doublet structure for the absorption ($\pm1/2\ \pm1/2$ and $\pm1/2\ \pm3/2$). Equation 1.7 shows that Δ gives information on the charge distribution around the Te atoms, which can be related to Te coordination and chemical bonding.

Sincair et al. (1998) performed a high-resolution neutron diffraction study of binary tellurite–vanadate glasses of the form 95 mol% TeO_2–5 mol% V_2O_5. Reciprocal space data were obtained for high-scattering vectors, Q, and have been Fourier transformed to yield the real-space correlation function $T(r)$. The first-neighbor Te–O and O–O peaks in $T(r)$ were found, which suggests the presence of both TeO_3 and TeO_4 units. The shortest Te–O bond length is 1.91 Å, with two further contributions at 2.1 and 2.17 Å, whereas the average O–O distance within the TeO_n structural units is 2.76 Å, as shown in the 1st Ed. A composition with the crystalline polymorphs of TeO_2 indicates that the structure of the glass is nearer to α-TeO_2 than to β-TeO_2.

Ten years before Johnson et al. (1986) investigated TeO_2–V_2O_5 glasses by neutron diffraction, they found that, unlike the vanadate glasses, the structure of vitreous $Te_{1.298}$–$V_{0.295}$–$O_{3.407}$ (89.8 mol% TeO_2) is dominated by the tellurite component. The high resolution of the TOF diffraction data suggests that the tellurium atoms are predominantly fourfold coordinated. Two more Te–O bonds were inferred at a distance of ~2.85 Å, completing the octahedron, although the basic coordination scheme should be considered a distorted trigonal bipyramidal. A quasicrystalline model of α-TeO_2 bears no resemblance to the experimental correlation function. β-TeO_2 gives a better fit, but these results are inconclusive.

Rojo et al. (1990), using second-moment NMR signals, proposed two different distributions of lithium: one in which Li^+ ions are dispersed in TeO_2–Li_2O glasses, and another in which Li^+ and F ions are associated in the network of TeO_2–LiF glasses as shown in the 1st Ed. Rojo et al. (1992) also reported the substitution of fluorine for oxygen in lithium–tellurite glasses produced from LiF and α-TeO_2 nanocrystallites inside an amorphous matrix, as determined by NMR analysis as shown in the 1st Ed. The ternary TeO_2–Li_2O–LiF samples are actually combined glass-crystal materials, although their appearance and XRD patterns are typical of glass. In samples with higher fluorine contents (F/Te $\geq$ 0.5), the formation of the crystalline domains progresses, and a decrease in ionic conductivity is observed. The association of Li^+ and F ions induces the formation of α-TeO_2 domains, which promotes phase separation and devitrification. This result is in agreement with the difficulty encountered in preparation of amorphous LiF-rich samples.

In two very recent articles on tellurite crystals and TeO_2–M_2O (M: Li, Na, K, Rb, or Cs) glasses, Sakida et al. (1999a and 1999b) reported shortening the relaxation time of the Te nuclei during the preparation of 21 crystals by doping the crystals with small amounts of Fe_2O_3. Reagent-grade β-TeO_2, Cs_2CO_3, Nb_2O_5, Fe_2O_3, Li_2CO_3, $AgNO_3$, PbO, MgO, V_2O_5, ZrO_2, K_2CO_3, HfO_2, SnO_2, BaO_3, ZnO, and TiO_2 were used as starting materials. Crystals of Li_2TeO_3, Na_2TeO_3, $PbTeO_3$, $BaTeO_3$, $ZnTeO_3$, $MgTe_2O_5$, $Cs_2Te_2O_5$, $Te_2V_2O_5$, α-$Li_2Te_2O_5$, α-TeO_2, $Zn_2Te_3O_8$, $NaVTeO_5$, and $KVTeO_5$ were synthesized by crystallization from the melt and crystals of $TiTe_3O_8$, $ZrTe_3O_8$,

$HfTe_3O_8$, $SnTe_3O_8$, and $Te_3Nb_2O_{11}$, which were synthesized by solid-state reaction of the starting powder mixture of TeO_2 containing a trace of Fe_2O_3 and corresponding reagent chemicals. The Ag_2TeO_3 crystal was prepared by drying the precipitates obtained by adding a saturated solution of $AgNO_3$ to a saturated solution of $TeNa_2O_3$, containing a trace of Fe_2O_3, whereas the β-$Li_2Te_2O_5$ crystal was prepared by a phase transition of α-TeO_2 containing a trace of Fe_2O_3 on heating. The Li_2TeO_3 and α-TeO_2 crystals without Fe_2O_3 were synthesized in the same manner to examine the effect on ^{125}Te isotropic chemical shift by adding small amounts of Fe_2O_3 to the TeO_2-related compounds. The isotropic chemical shift derived from ^{125}Te-static NMR spectra [$\delta_{iso(static)}$], the chemical-shift anisotropy ($|\Delta\delta|$), and the asymmetry parameter ($\acute{\eta}$) were calculated. $\acute{\eta}$ is the measure of the deviation of the chemical-shift tensors from axial symmetry; $\acute{\eta} = 0$ for an axially symmetric electronic distribution around a tellurium atom, and $\acute{\eta} = 1$ for an axially asymmetric distribution. Sakida et al. (1999a) concluded that the Te atoms in tellurite crystals have three or four oxygen neighbors in the range 0.18–0.22 nm and one or two oxygen neighbors in the range 0.22–0.31 nm. Therefore, when a Te atom has n oxygen neighbors in the range >0.22 and m oxygen neighbors in the range 0.22–0.25 nm, the coordination bond number of the Te atom is calculated to be $N = (n + M)$, ignoring the weak bonds longer than 0.25 nm. Sakida et al. (1999b), as shown in the 1st Ed, classified tellurite crystals as follows:

1. TeO_3 types:
   a. Isolated TeO_3 types: Li_2TeO_3, Na_2TeO_3, Ag_2TeO_3, $PbTeO_3$, and $ZnTeO_3$
   b. Terminal TeO_3 types: $Cs_2Te_2O_5$, $V_2Te_2O_9$, and $Nb_2Te_3O_{11}$
2. TeO_{3+1} types: $MgTe_2O_5$, α-$Li_2Te_2O_5$, β-$Li_2Te_2O_5$, $Zn_2Te_3O_8$, and $Mg_2Te_3O_8$
3. TeO_4 types:
   a. α-TeO_2 types: α-$TiTe_3O_8$, -$ZrTe_3O_8$, -$HfTe_3O_8$, -$SnTe_3O_8$, -$Zn_2Te_3O_8$, -$Mg_2Te_3O_8$, and -$Nb_2Te_3O_{11}$
   b. β-TeO_2 types: β-TeO_2, -$NaVTeO_5$, and -$KVTeO_5$

Sakida et al. (1999b) concluded that without affecting the δ_{iso}, the addition of 0.3 mol% Fe_2O_3 to various tellurite crystals makes it possible to measure ^{125}Te NMR spectra at a pulse delay of 2.5 s, shortened from 20 s without Fe_2O_3. The $\acute{\eta}$-$|\Delta\delta|$ diagam is useful for examining the structure of tellurite crystals and glasses. The magnetic-angle-spinning (MAS) NMR results for $\delta_{iso(MAS)}$, $|\Delta\delta|$, and $\acute{\eta}$ of various tellurite crystals and glasses can be described by the following relations:

- $\delta_{iso(MAS)}$: $TeO_3 \geq TeO_{3+1} \approx \beta$-$TeO_2 > \alpha$-$TeO_2$
- $|\Delta\delta|$: β-$TeO_2 \geq \alpha$-$TeO_2 > TeO_{3+1} >$ terminal $TeO_3 \geq$ isolated TeO_3
- $\acute{\eta}$: $TeO_4 \geq TeO_{3+1} \geq TeO_4$

Sakida et al. (1999a) used the ^{125}Te-static NMR to study binary TeO_2–M_2O (M: Li, Na, K, Rb, or Cs) glasses and suggested a new model of vitrification and structural change for TeO_2–M_2O glasses. The MAS spectra of TeO_2–M_2O glasses were broad to obtain an δ_{iso}, based on the NMR data and the classification of structural units of TeO_3 type, TeO_{3+1} type, and TeO_4 type, as in the 1st Ed by Sakida et al. (1999b).

Sakida et al. (1999a) represented the relation between the $|\Delta\delta|$ and $\acute{\eta}$ as shown in the 1st Ed. Profile 1 is the peak for TeO_2 glasses, and profile 2 is determined from TeO_2–M_2O (M: Li, Na, K, Rb, or Cs) glasses. These authors also plotted the fraction of TeO_4-tbp (N_4) and TeO_3-tp (N_3) as a function of M_2O–TeO_2 glasses, as represented in the 1st Ed. They described the vitrification reaction as:

$$\frac{[TeO_3]}{[TeO_4]} = \frac{(2b+c)y}{1-(2b+c)y} \tag{1.8}$$

where a, b, and c are the fractions ($1 + b = 1$) in Equation 1.9 that hold only when $0 \leq y \leq 0.5$,

$$yO^{2-} + TeO_{4/2} \Rightarrow 2y \left\{ a \left[O_{3/2}Te - O^- \right] + b \left[O_{1/2}Te(=O) - O^- \right] \right\}$$
$$+ cy \left[O_{3/2}Te = O \right] + (1 - 2y - cy) TeO_{4/3}$$

$$(1.9)$$

Variations of N_4^0, N_4^-, and N_3^- with the Li_2O in the Li_2O–TeO_2 glass are shown in the 1st Ed. It is clear that N_4^0 decreases with increasing Li_2O content. It is interesting to note that the addition of M_2O to TeO_2 glass results in the formation not of N_4^- but of N_3^-, and the N_3^- increases rapidly above 20 mol% M_2O. The $O_{3/2^-}$–Te–O unit has two variations with NBO, at an axial or an equatorial position, because these variations are formed with equal probability on addition of M_2O to TeO_2, as shown in the 1st Ed. Uchino and Yanko (1996) reported that the trio of three-centered orbitals (bonding, nonbonding, and antibonding orbitals), of which the electronic configurations are as shown in the 1st Ed., are formed in the $O-_{ax.}Te-_{ax.}O$ bond of the TeO_4-tbp unit.

Sakida et al. (1999a) illustrated the way bonding of $2M_2O$ to TeO_2 glass creates deformed spirals by sharing the corners of TeO_4-tbps, as shown for α-TeO_2 in the 1st Ed. These M oxides break the $O-_{eq}Te_{ax}$–O linkage to form two $O_{3/2^-}$–Te–O units having $Te-_{eq}O-$ and $Te-_{ax}O-$ bonds. In 1999, Blanchandin et al. carried out a new investigation within the TeO_2–WO_3 glass system, by XRD and DSC. The investigated samples were prepared by the air quenching of totally or partially melted mixes of TeO_2 and WO_3. This investigation identified two new metastable compounds, which appeared during glass crystallization. In addition, use of a phase equilibrium diagram indicated that this system is a true binary eutectic one. Tellurite glass–ceramics are discussed in detail in Chapter 5.

In 2000, Mirgorodsky et al. compared lattice dynamic-model studies of the vibrational and elastic properties of both paratellurite (α-TeO_2) and tellurite (β-TeO_2). Emphasis was on the crystal chemistry aspects of the Raman spectra for these lattices. Results were used to interpret the Raman spectra of two new polymorphs, γ and δ, of tellurium dioxide and to clarify their relationships with the spectrum of pure TeO_2 glass. Lattice projections on the x,y planes of α-TeO_2 and β-TeO_2 were carried out as shown in Figure 1.23 (arrows indicate the positions of lone electron pairs of Te). The γ phase represents a new structural type, which is different from the α and β types, and the δ phase seems to exist as a superposition of domains of the α, β, and γ phases. Mirgorodsky et al. (2000) have used the potential function of two-body diagonal-force constants K_{Te-O} and K_{O-O}; the three-body O–Te–O and Te–O–Te diagonal-bending constants K_A and K_D, respectively; and the Te–O–Te–O stretching–stretching nondiagonal-force constants H (Te–O_{eq}–Te–O_{eq} via Te) and h (Te–O_{eq}–Te–O_{eq} via O and Te). The dependence of the force constant K_{Te-O} on interatomic l_{Te-O} was found in view of the smooth curve as represented in Figure 1.24a. The relevant curve K_{O-O} (l_{O-O}) is shown in Figure 1.24b.

Blanchandin et al. (1999) investigated the well-crystallized γ-TeO_2 obtained by slowly heating pure TeO_2 glass to 390°C and then annealing it for 24 h at this temperature, as in Figure 1.25. Its XRD pattern could be indexed in an orthorhombic cell with the following parameters: $a = 8.45$ Å; $b = 4.99$ Å; $c = 4.30$ Å; and $Z = 4$, whereas the δ-TeO_2 was detected in samples containing a small amount of WO_3. It was prepared as the unique crystallized phase, mixed with a small quantity of glass, by annealing it for 24 h at 350°C in a glassy sample containing 5–10 mol% of WO_3. The compound has a fluorite-related structure. Its XRD pattern can be unambiguously indexed in the $Fm3 - m(O_h^5)$ cubic space group, where $a = 5.69$ Å and $Z = 4$.

Bahgat et al. (1987) used the Mossbauer analysis to detect the structure of $[1 - (2x + 0.05)]$ mol% TeO_2–x mol% Fe_2O_3–$(x + 0.05)$ mol% Ln_2O_3, where $x = 0.0$ and 0.05, and $Ln =$ lanthanum, neodymium, samarium, europium, or gadolinium. These glasses were prepared by fusing a mixture of their respective reagent-grade oxides in a platinum crucible at 800°C for 1 h. Mossbauer parameters such as δ, Δ, and line width were found to be functions of the polarizing power (charge/radius) of the rare-earth cations. The Mossbauer parameters were not affected by heat treatment of the glass

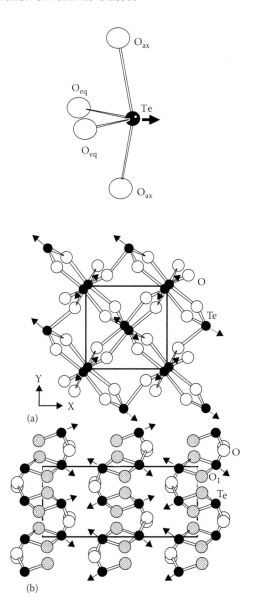

FIGURE 1.23 Structural model of TeO$_4$ unit present in α-TeO$_2$, β-TeO$_2$, and lattice projections on the x,y plane of (a) α-TeO$_2$, and (b) β-TeO$_2$. (From A. Mirgorodsky, T. Merle-Mejean, and B. Frit, *J. Phys. Chem. Solids*, 61, 501, 2000.)

samples, as shown in Figure 1.26. Both the Te–O–*Ln* and Te–O–Fe stretching vibrations are discussed in Chapter 10.

1.10 NEW TELLURITE GLASSES

Quaternary tellurite glass systems $(70 - x)$TeO$_2$–20WO$_3$–10Li$_2$O–xLn$_2$O$_3$, where $x = 0$, 1, 3, and 5 mol% have been prepared by the melt-quenching technique, and *Ln* are La, Pr, Nd, Sm, Er and Yb respectively. Densities of the obtained glasses and the molar volume have been measured and calculated by Hager and El-Mallawany (2010). The densities were: 5.78–5.840, 5.823–5.944, 5.788–5.853, 5.792–5.832, 5.792–5.888, 5.816–5.968, and 5.834–5.944 g/cm^3 for La, Pr, Nd, Sm, Er and Yb,

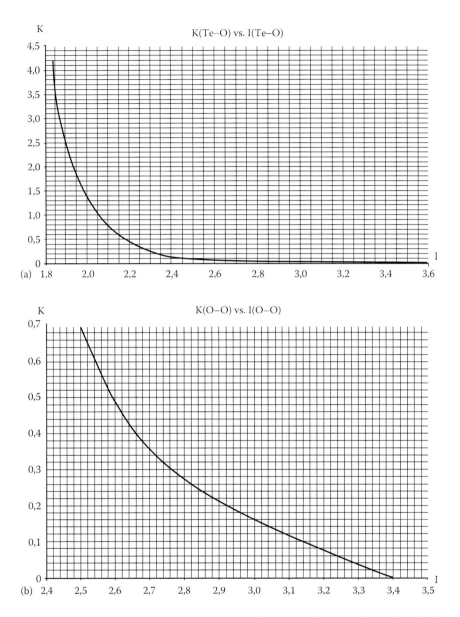

FIGURE 1.24 Dependence of the force constant K_{Te-O} on interatomic l_{Te-O} was determined by the smooth curve represented in panel a. The relevant curve K_{O-O} (l_{O-O}) (panel b) is from A. Mirgorodsky, T. Merle-Mejean, and B. Frit, *J. Phys. Chem. Solids*, 61, 501, 2000.

respectively. The values of molar volume were: 28.96–29.00, 29.14–34.35, 28.14–29.03, 28.1–28.96, 28.08–28.86, and 28.01–29.07 cm³/mol, respectively. Also, the quantitative interpretations were based on the concentration of ions per unit volume of Te, Ln, and O; and on the short distance in nanometer between ions for (Te–O) of TeO_4 and TeO_3 groups, (W–O) of WO_4, WO_6 groups, and calculated wave number, ν, for TeO_4 and TeO_3, respectively. The average stretching force constant that was present in these quaternary glasses has been calculated in order to interpret the data obtained. New ternary tellurite glasses and crystalline phases in the Bi_2O_3–CaO–TeO_2 systems have been synthesized and characterized by Chagraoui et al. (2010). The investigation in this system using XRD revealed new phases of γTeO_2. The glass composition was of the form TeO_2–Bi_2O_3–CaO, with the percentages of: 95–0–5, 90–0–10, and 85–0–15 mol%. The density was 5.44, 5.32, and 5.29 g/cm³;

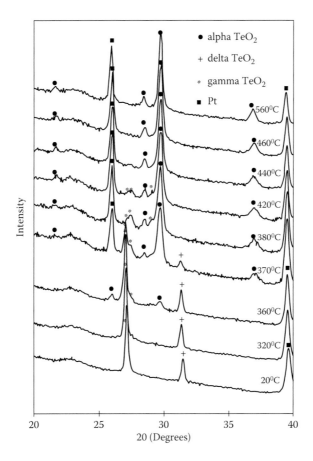

FIGURE 1.25 XRD patterns of the well-crystallized γ-TeO₂ and pure TeO₂ glass. (From Blanchandin, S., Marchet, P., Thomas, P., Champarnaud, J., Frit, B., and Chagraoui, A., *J. Mater. Sci.,* 34, 1, 1999.)

and the molar volume was 28.39, 28.054, and 27.235 cm³/mol respectively. The crystalline phases of glasses in TeO₂–CaO revealed γTeO₂ phase, which transforms into the stable αTeO₂ phase up to 500°C. Also, from the IR and Raman studies, it has been shown that transition of TeO₄, TeO₃₊₁, and TeO₃ units with increasing CaO content. Phases after crystallization in 5CaO–95TeO₂ were TeO₂(α), TeO₂(γ), αCaTe₂O₅, 10CaO–90TeO₂ with TeO₂(α), TeO2(γ), αCaTe₂O₅, and 15CaO–85 TeO₂ with TeO₂(α), TeO₂ (γ), and αCaTe₂O₅. Also, El-Mallawany et al. (2010) have prepared binary tellurite glass systems of the forms TeO₂(100 – x)–x AₙOₘ, where AₙOₘ = La₂O₃ or V₂O₅ and x = 5, 7.5, 10, 12.5, 15, 17.5, and 20 mol% for La₂O₃; and 10, 20, 25, 30, 35, 40, 45, and 50 mol% for V₂O₅. The densities and molar volumes of each glass were measured and calculated:

(100 – x)TeO₂–xLa₂O₃, (x = 5–0 mol%): 5.180–5.640 (g/cm³) and 32.42–34.21 (cm³/mol),
(100 – x)TeO₂–xV₂O₅, (x = 10–50 mol%): 5.040–4.010 (g/cm³) and 32.11–42.58 (cm³/mol)

The compressibility model has been used to find the difference volume V_d (as explained in section 1.8) due to the exchange of one formula unit between Te and both of La and V in the binary glass system and the mean volume V_A per formula unit in the present binary glass to check whether or not it is independent of the percentage of the modifier for a glass series and also different from series to another.

Inabaw et al. (2010) have discussed the density of silicate, borate, phosphate, tellurite, and germinate glasses by using an empirical equation for calculating the density of oxide glasses. The authors found that the density varies considerably, depending upon chemical composition and increases approximately in the order of tellurite > germanate > phosphate > silicate > borate glass system.

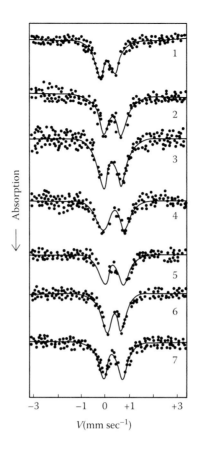

FIGURE 1.26 The Mossbauer effect spectra for tellurite glasses doped with rare-earth oxides. (From A. Bahgat, E. Shaisha, and A. Sabry, *J. Mater. Sci.*, 22, 1323, 1987.)

Reasonable linearity between the measured density and calculated value was obtained for all the oxide glasses investigated in this study. They have classified the oxides as network former oxides (NWF), intermediate oxides (IMO), and network modifier oxides (NWM), and have ordered the molar weight of NWM: $SiO_2 = 60.08$, $B_2O_3 = 69.64$, $GeO_2 = 104.59$, $P_2O_5 = 141.95$, and $TeO_2 = 159.60$ (10^{-3} kg/mol). The packing density parameter of SiO_2 (tetrahedral, hexavelent) = 13.9, 14.0; B_2O_3 (trivalent, tetrahedral) = 15.2, 20.8; GeO_2 (tetrahedral, hexahedral) = 14.0, 14.2; P_2O_5 (trivalent, tetrahedral) = 34.6, 34.8; TeO_2 (hexahedral) = 14.7 (10^{-6} m/mol).

Kaur et al. (2010) have prepared glasses of two systems: $xPbO – (100 – x)$ TeO_2 ($x = 13$, 15, 17, 19, and 21 mol%) and $yZnO – (100 – y)$ TeO_2, $y = 18$, 20, 22, 25, 30, 33, and 35 mol% at two melt-cooling rates (normal and splat-quenched samples, respectively) and characterized by density, UV-visible, Raman spectroscopy, DSC, and XRD measurements. Lead tellurites prepared at higher cooling rates were mostly amorphous, while samples containing 19 and 21 mol% of PbO, prepared at slower cooling rates, were translucent glass ceramics containing crystals of $PbTeO_3$, $Pb_2Te_3O_8$, and TeO_2. Also, Ozdanova et al. (2010) have prepared $Nb_2O_5–TeO_2$, $PbO–Nb_2O_5–$ TeO_2, and $PbO–TeO_2$ glasses from very pure oxides. Raman scattering measurements indicated that TeO4 trigonal bipyramids, TeO_{3+1}, TeO_3 trigonal unit, and NbO_6 octahedra, were the basic structural units of $Nb_2O_5–TeO_2$ and $PbO–Nb_2O_5–TeO_2$ glasses. The density values of $0.1Nb_2O_5–$ $0.9TeO_2$, $0.2Nb_2O_5–0.8TeO_2$ glasses were 5.41–5.24 g/cm³, while values of molar volume were 31.46–34.51 cm³/mol. The molar volumes for the $PbO–Nb_2O_5–TeO_2$ glasses series were 5.72, 5.53, 6.02 g/cm³; and 30.87, 33.85, and 30.39 cm³/mol for the compositions 0.1–0.1–0.8 mol%, 0.1–0.2–0.7 mol% and 0.2–0.1–0.7 mol%, respectively. While for binary $PbO–TeO_2$ glasses, the density was 6.27 g/cm^3 and the molar volume was 27.48 cm^3/mol. Upendera et al. (2010) have studied $(90 - x)$TeO$_2$–10GeO$_2$–xWO$_3$ glasses with $7.5 \leq x \geq 30$ mol% and showed that the glass system contains (TeO$_4$), (TeO$_{3+1}$)/(TeO$_3$), (GeO$_6$), (WO$_4$), and (WO$_6$) groups as basic structural units. The densities and molar volumes of the glasses were 5.680, 5.785, 5.909, and 6.011 g/cm^3; and 28.084, 28.510, 28.829, and 29.241(cm^3/mol) for $x = 7.5$, 15, 22.5, and 30 mol%, respectively.

Experimental results by Desirena et al. (2009) have indicateed that alkali metals with small ionic radii improve chemical durability (DR) and thermal properties like thermal expansion coefficient and glass transformation temperature (α, Tg) in alkali metal oxides R$_2$O (R = Li, Na, K, Rb and Cs), and network intermediate MO (M = Zn, Mg, Ba and Pb) in tellurite glasses of the form R$_2$O–MO–TeO$_2$. The density of the glasses 70Te–20Pb–10Cs, 10Rb, 10K, 10Na, and 10Li were 5.554, 5.554, 5.354, 5.454, and 5.574 g/cm^3, respectively; for 70Te–20Zn–10Cs 10Rb, 10K, 10Na, 10Li, they were 4.984, 4.874, 4.736, 4.806, and 5.079 g/cm^3, respectively; for 70Te–20Ba–10Cs, 10Rb, 10K, 10Na, 10Li, they were 4.733, 4.625, 4.479, 4.727, and 4.794 g/cm^3, respectively; and for 70Te–20Mg–10Cs, 10Rb, 10K, 10Na, 10Li, they were 5.035, 4.928, 4.784, 4.990, and 5.097 (g/cm^3), respectively. The researchers have also indicated that increased refractive index, third-order nonlinear susceptibility, and the lower thermal expansion coefficient were obtained for Zn and Mg while better chemical durability corresponds to Pb and Zn. Desirena et al. have stated that "it is important to know the chemical elements that should be added into the host in order to design glasses with desirable optical properties and high performance. The right selection of glass composition depends on the application and often is a compromise among many factors." From the conduction point of view, on adding BaO to telluirte glasses it creates nonbridging oxygen (NBO), which in turn affects the electronic paths and gets progressively blocked, thereby causing a decrease in electronic conductivity, as mentioned by Szu and Chang (2005). The Ba ion acts as a glass modifier, creates NBO, and breaks the glass network. Among the various types of oxide glasses, the tellurite-based glasses exhibit relatively high dielectric constant and electrical conductivity, which, it has been argued, is due to the unshared pair of electrons of the TeO$_4$ group that take part in bonding as by Chakraborty et al. (1997). Moawad et al. (2009) have measured the DC conductivity of silver vanadium tellurite glasses of the form $0.5[x$Ag$_2$O $- (1 - x)$ V$_2$O$_5]$–0.5TeO$_2$ glasses with $x = 0.1$–0.8 mol% over the wide range of temperatures 70–425 K. The densities of the glasses were: 4.132, 4.423, 4.650, 4.920, and 5.251 g/cm^3 for the percentages of V$_2$O$_2$ = 45, 40, 35, 30, and 25 mol%, respectively.

Kabalci et al. (2006) have prepared $(1 - x)$ TeO$_2$–xPbF$_2$ ($x = 0.10$, 0.15, and 0.25 mol) by quenching the melt. Thermal properties of $(1 - x)$ TeO$_2$–xPbF$_2$ binary glasses were studied in order to examine the effect of PbF$_2$ content. The optical band gaps of TeO$_2$–PbF$_2$ glasses decreased from 2.02 eV to 1.90 eV, and Urbach energy values increased from 0.34 eV to 0.36 eV when the PbF$_2$ content was increased from 0.10 mol% to 0.25 mol%, as we will see in Chapter 9. Also, Ghosh et al. (2009) have showed that the luminescence properties of a series of Er^{3+} ions-doped lead-free fluoro-tellurite glasses, up to a maximum of 1 mol% of Er$_2$O$_3$, can be incorporated in the glass to increase the luminescence efficiency of the glass. Eyzaguirre et al. (2007) have shown that the potential applications of tellurite glasses for broadband optical amplifiers can be amplified by increasing their glass stability ranges to produce better optical fibers. The authors have also shown that the addition of CsCl to the original composition can enhance the glass stability range GSR. The authors produced single-mode and multimode optical fibers using highly homogeneous $(78 - x)$ TeO$_2$–15.5ZnO–5Li$_2$O–1.5Bi$_2$O$_3$–xCsCl mol% and $(7 - x)$TeO$_2$–15.5ZnO–5Li$_2$O–xBi$_2$O$_3$–9.5CsCl mol% glasses and investigated the role of the CsCl in their GSR. The authors observed that CsCl acts like a network modifier in glass systems, weakening the network by forming Te–Cl bonds. Finally, the authors have shown that the thermal expansion coefficient mismatch is in the right direction for optical fiber fabrication purposes; that is, the core coefficient is larger than the clad one. It has also been shown that the Bi$_2$O$_3$ content can be used to control the refractive index of clad and core glasses, and has produced both single-mode and multimode optical fibers with Er^{3+}-doped tellurite glasses. Also, Donnell et al. (2003) have prepared ternary fluorotellurite glasses with improved mid-infrared transmission of the form: system $(90 - x)$ TeO_2-$xZnF_2$-$10Na_2O$ ($x = 5, 10, 15, 20, 25,$ and 30 mol%). Volatilization during the glass melting of the series is shown to increase, as expected, with ZnF_2 addition and melting time, resulting in lower optical loss. A core/clad glass pair is also proposed from the series, giving a suitable numerical aperture for guiding light. Identification of OH bands and other mid-infrared bands were discussed. Nazabal et al. (2003) have prepared and studied the glass-forming capability, glass-structural organization, and optical properties have been examined in rare-earth-doped oxyfluoride zinc tellurite glass system of the forms: $75TeO_2$-$20.1ZnO$-4.9 Na_2O, $75TeO_2$-$25ZnO$, $75TeO_2$-$18ZnO$-$7ZnF_2$, $69TeO_2$-$9ZnO$-$22ZnF_2$, $61TeO_2$-$12ZnO$-$27ZnF_2$, and $47TeO_2$-$18ZnO$-$35ZnF_2$ mol%. As a function of composition, the differential thermal analysis, vibrational spectra, optical absorption, spontaneous emission, and lifetime measurements have been analyzed in terms of fluorine influence. The authors concluded that the addition of fluoride compound results in a reduction of T–O–Te linkages due to a gradual transformation of trigonal bipyramid TeO_4 through TeO_{3+1} to trigonal pyramid TeO_3. The former network connectivity decrease is strengthened by a possible high ratio $(ZnO-ZnF_2)/TeO_2$ compared with pure oxide–tellurite glasses.

Erbium-doped tellurite glass of the form MoO_3-Bi_3O_3-TeO_2 have been fabricated and characterized optically by Li and Man (2009) and was found to be suitable for broadband optical amplifier applications. The upconversion luminescence intensity of Er^{3+}-doped MoO_3-Bi_3O_3-TeO_2 glasses is much weaker than that of the Na_2O-ZnO-TeO_2 glasses, and the luminescence intensity at 550 nm in Er^{3+}-doped MoO_3-Bi_3O_3-TeO_2 glasses is about 30 times weaker than that in Er^{3+}-doped Na_2O-ZnO-TeO_2 glasses. Erbium-doped fiber amplifiers (EDFAs) have been extensively studied as key devices for wavelength-division-multiplexing (WDM) networks. The results show that Er^{3+}-doped MoO_3-Bi_3O_3-TeO_2 glasses are excellent materials for broadband optical amplifiers in the WDM systems. The effect of Nb_2O_5 and Bi_2O_3 on the refractive index of new tellurite glasses within the $(1 - x - y)$ TeO_2-xNb_2O_5-yBi_2O_3, where $x = 0.05, 0.10$ and $y = 0.05, 0.10,$ and 0.15 mol% system has been examined Wang et al. (2009). The densities and molar volumes were 5.324, 5.713, 6.143 for $x = 0.05$ mol%; 5.633, 6.033, and 6.218 for $x = 0.1$ g/cm^3; 33.852, 34.229, and 35.956 for $x = 0.05$ mol%; and 32.938, 33.294, and 34.767 for $x = 0.10$ cm^3/mol, respectively. Also, the authors have found that a maximum refractive index value of 2.1927 (at 632.8 nm) was obtained for $75TeO_2$-$10Nb_2O_5$-$15Bi_2O_3$ glass. These glasses have good thermal stability against crystallization for $Bi_2O_3 < 10$ mol%. Also, Tokuda et al. (2003) have obtained a temporal and permanent refractive index change in niobium tellurite glasses by using ps- and fs-pulses, respectively. The He–Ne laser beam was guided by a line waveguide structure drawn by the fs-pulses, indicating a positive refractive index change. It is suggested that the refractive index change by the ps-pulses is due to an increase in the absorption coefficient, while that caused by the fs-pulses results from the rearrangement of the glass structure. Chen et al. (2008) have studied the nonlinear optical properties in the prepared tellurite glasses of the form: $0.2Bi_2O_3$-xWO_3-$(1 - x)$ TeO_2 and $x = 0.2, 0.25,$ and 0.3 mol%. It is seen that the densities increase from 6.370, 6.425, and 6.510 g/cm^3 by increasing the WO_3 content to 0.20, 0.25, and 0.3 with fixed $0.2Bi_2O_3$, respectively.

Tellurite glasses doped with europium ions have been prepared by Dehelean et al. (2009) using the sol-gel method to measure the magnetic susceptibility. The obtained data for the xEu_2O_3 $(1 - x)$ TeO_2 (where $x = 0.07, 0.11, 0.15$) system show that paramagnetic Curie temperatures were 40, 160, and 165 K, respectively. The magnetic europium ions are present in the host glass matrix in both their 2+ and 3+ valence states. The magnetic interactions between the europium ions are antiferomagnetic. Qin et al. (2009) have prepared TeO_2 nanoparticles in an acid medium at room temperature. The TeO_2 nanoparticles prepared in gallic acid were the orthorhombic phase β-TeO_2 spheres in the range of 30–200 nm. The TeO_2 nanoparticles prepared in acetum were the tetragonal phase α-TeO_2 irregular flakes in the range of 40–400 nm. This preparation method was mild and green, which might be suitable for the convenient preparation and mass production of TeO_2 nanoparticles.

Lakshminarayana et al. (2009) have prepared novel tellurite glasses with Pr^{3+}, Nd^{3+}, and Ni^{2+} ions as dopants in TeO_2–ZnO–WO_3–TiO_2–Na_2O glasses and studied the photoluminescence. The density of the $70TeO_2$–$10ZnO$–$10WO_3$–$5TiO_2$–$5Na_2O$ glass was 5.397 g/cm^3. These glasses had better thermal stability and strong visible luminescence and should have potential applications in optoelectronic materials. Sidek et al. (2009) have synthesized and studied the optical properties of $(1 - x)$ TeO_2–$xZnO$, x = 0.1 to 0.4 with an interval of 0.05 mol% glass systems have been achieved by Sidek et al. (2009). The density and molar volume ranged from 4.806 to 5.283 g/cm^3, and from 33.21 to 24.29 cm^3/mol. The increases in density were probably because of the decreases in the average interatomic spacing. The refractive index of the TeO_2–ZnO glasses increases with the substitution of ZnO oxides into TeO_2, bridging Te–O–Te bonds are altered and nonbridging Te–O–Zn^{2+} bonds are formed. The nonbridging oxygen (NBO) bonds have a much greater ionic character and much lower bond energies. Chen et al. (2009) have measured the spectroscopic properties and energy transfer of Tm^{3+}/Ho^{3+}-co-doped TeO_2–WO_3–ZnO glasses for a 1.47 μm amplifier. Fluorescence spectra and the analysis of energy transfer indicate that Ho^{3+} is an excellent co-dopant for a 1.47 μm emission. Although the pump efficiency of tungsten tellurite amplifiers is ≈50% less than that of fluoride glass, the figure of merit for bandwidth is approximately three times larger in tungsten tellurite glass than it is in fluoride glass. These characteristics indicate that Tm^{3+}/Ho^{3+}-co-doped tungsten tellurite glass is attractive for use as a broadband amplifier. Eraiah (2006) has prepared tellurite glasses with the composition of (Sm_2O_3) $x(ZnO)(40 - x)$ $(TeO_2)(60)$, where x = 0.1–0.5 mol% by the conventional melt-quenching method. The density, molar volume, and optical energy band gap of these glasses have been measured. The refractive index and the nonlinear variation of the above optical parameters with respect to samarium dopant have been explained. The density values were 10.645, 5.846, 7.464, 11.373, 13.872, and 7.393 g/cm^3 for the Sm_2O_3, with the values of 0, 0.1, 0.2, 0.3, 0.4, and 0.5 mol%, respectively. Caricato et al. (2003) have studied the optical properties of erbium-doped zinc–tellurite $60TeO_2$–$20ZnO$–$20ZnCl_2$:$1ErCl_3$ oxyhalide glass waveguides, deposited by reactive pulsed laser deposition (RPLD) on silica substrates. Er^{3+}-doped zinc–tellurite glass (ZT) targets were ablated in oxygen-dynamical flow at two different pressure values, 5 and 10 Pa, by ArF excimer laser at the fluence of 3.7 J/cm^2. The waveguiding properties of the deposited films were investigated by the m-line technique. The TE_0 mode excitation was used for photoluminescence (PL) and Raman measurements, to study the Erbium ion $^4I_{13/2} \rightarrow {}^4I_{15/2}$ transition and structural properties of the deposited films, respectively. Optical band gap and wavelength dependence of the real and imaginary parts of the refractive index were estimated from transmission spectra. Also, Duclere et al. (2009) estimated the value of Kerr electro-optical sensitivity of several tellurite glasses. They have found that the highest value of Kerr coefficient B ≈190 × 10^{-16} mV^{-2} was registered for $0.6TeO_2$–$0.3TlO$–0.5–$0.1ZnO$ glass. This evidences the prospects of a thallium–tellurite glass system for electro-optical applications. A very important investigation of glass formation and color properties in the P_2O_5–TeO_2–ZnO system has been conducted by Konishi et al. (2003). The color of the samples varied from clear to reddish, depending on the composition, so that the color became deeper with increasing TeO_2 content. The coloring was considered to be due to colloidal coloration by precipitated metallic Te particles.

Munoz et al. (2009) have studied nonlinear materials in the $(1 - x-y)$ TeO_2–xWO_3–$yPbO$, with x = 0.0 to 40 and y = 0.0 to 30 mol%, with an interval of 0.005 mol% system as potential candidates for photonic devices and prepared by conventional melting at temperatures ranging between 710°C and 750°C. The density was 5.798 g/cm^3 for the composition 0.9–0.1–0.1 and 7.018 g/cm^3 for the composition 0.40–0.40–0.20 mol%, respectively. Some compositions were cystallized like: 0.9–0.0–0.1, 0.6–0.1–0.3, 0.5–0.4–0.1, and 0.4–0.3–0.3 mol%. The main glass former oxide is TeO_2, which arranges TeO_4 groups with tetrahedral coordination, while PbO plays as a glass modifier oxide. Tungsten oxide is incorporated as a network former, alternating with TeO_2 and forming mixed linkages Te–O–W and W–O–W. The WO_3 is the component that contributes the most to increasing the glass transition temperature and to decreasing both the oxygen molar volume and the thermal expansion coefficient. Lasbrugnas et al. (2009) have studied the SHG of thermally poled tungsten tellurite glass of the form $85TeO_2-15WO_3$ mol%. Two complementary hypotheses have been proposed to explain the origin of the second-order nonlinearity property of this tellurite glass:

   i. Reorientation of the TeO_4 glass structural entities under electric field
   ii. Formation of an anodic depletion region of sodium ions

El-Mallawany and Ahmed (2008) have prepared the quaternary tellurite glass systems of the form $80TeO_2-5TiO_2-(15 - x) WO_3-xA_nO_m$, where A_nO_m is Nb_2O_5, Nd_2O_2, and Er_2O_3, $x = 0.01$, 1, 3, and 5 mol% for Nb_2O_5, and $x = 0.01$, 0.1, 1, 3, 5, and 7 mol% for Nd_2O_3 and Er_2O_3 by the melt quenching and shown in Figure 1.27. The color of the glass changes from light yellow to reddish brown due to the increase of the neodymium concentration in the glass. Density values were: 5.455 and 5.673 g/cm^3 for 0.01 and 5.0 Nb_2O_5 mol%; 5.178, and 5.791 g/cm^3 for Nd_2O_3: 0.01 and 7.0 mol%; and 5.754 and 5.989 g/cm^3 for Nb_2O_5: 0.01 and 7.0 mol%. While molar volumes were 30.51–29.63, 32.19–29.99, and 28.82–29.54 cm^3/mol for the three tellurite glass series doped with Nb_2O_5, Nd_2O_3, and Er_2O_3, respectively. The decrease in molar volumes indicates that Nb_2O_5 and Nd_2O_3 modifiers have been accommodated in the glass structure and have created a more compact glass network. By contrast, the molar volume increases as the content of the modifier Er_2O_3 increases. Although Er_2O_3 has the highest density glass among the studied modifiers, it is accommodated in the more open structure of the glassy state. Glass densities were found to vary according to the type and content of the modifier, and they are higher than that of the silicate samples. The density increased with increases in the transition metal oxide modifier Nb_2O_5. The same behavior in density accompanied the addition of the rare-earth oxide modifiers Nd_2O_3 and Er_2O_3. They showed an increase in density as their contents increased. One of the clear factors in controlling the density of the three glass series is the direct substitution of WO_3 with the higher molecular mass modifiers Nb_2O_5, Nd_2O_3, and Er_2O_3, resulting in high glass density, and, consequently, partial control of the bulk properties of the glass such as in the linear refractive index and thermal expansion coefficient. The molar volume has been found to decrease as the content of the modifiers Nb_2O_5 and Nd_2O_3 increase. This decrease in molar volumes indicates that Nb_2O_5 and Nd_2O_3 modifiers have been accommodated in the glass structure and have created a more compact glass network. By contrast, the molar volume increases as the content of the modifier Er_2O_3 increases. Although Er_2O_3 has the highest density glass among the studied modifiers, it is accommodated in the more open structure of the glassy state.

Golis et al. (2008) have synthesized tellurite glasses from the $TeO_2-WO_3-PbO-La_2O_3$ system for optoelectronics devices. The glass compositions were 60–30–10–0, 59–30–10–1, 58–30–10–2 and 57–30–10–3. The effect of lanthanum oxide content on the tendency toward the crystallization of glassy matrix was investigated. The authors investigated tellurite glasses using the Faraday method, which is a rotation of polarization of linearly polarized light in an isotropic transparent material under the magnetic field. It has been stated that the addition of lanthanum to tellurite glass from the TeO_2-WO_3-PbO system hinders the crystallization process of glass, which is very important during fiber drawing. Tellurite glasses have physical and chemical properties that make this class of materials a potential candidate in different applications, such as electrochemical and optoelectronic devices. In the field of photonics, these glasses may be used as precursors for infrared fibers or windows due to the possibility of changing their phonon energies and therefore the domain of wavelength transparency as a function of compositions. Also, Xu et al. (2006) succeeded in preparing Er^{3+}/Yb^{3+}-co-doped tungsten–tellurite glasses of the forms: $(100 - x)TeO_2-xWO_3-10BaO$, $(100 - x)TeO_2-xWO_3-10La_2O_3$, $(100 - x)TeO_2-xWO_3-10Bi_2O_3$ and $65TeO_2-25WO_3-10PbO$, with $x = 15$, 20, 25, and 30 mol%. The density of the glasses was: 6.31, 6.20, and 6.17 g/cm^3 for the first series with $x = 30$, 25, and 20 mol%; 6.3, 6.42, and 6.28 g/cm^3 for the second series with $x = 25$, 20, and 15 mol%; 6.7, 6.66, and 6.75 g/cm^3 for the third series; and 6.42 g/cm^3 for the fourth glass,

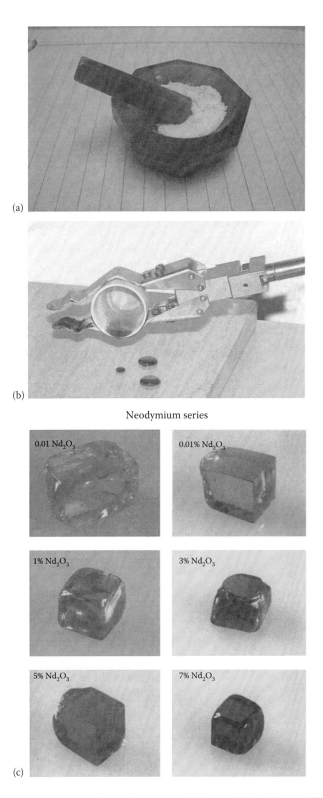

FIGURE 1.27 **(See color insert)** (a–c) Color of a series of $80TeO_2$–$5TiO_2$–$(15 - x)$ WO_3–xNd_2O_3 glass with $x = 0.01, 0.1, 1, 3, 5,$ and 7 mol% for Nd_2O_3 (From I. Abbas, Ph.D. thesis at Sudan University of Science and Technology, 2006).

with $x = 25$ mol%, respectively. The Er^{3+}/Yb^{3+}-co-doped tungsten–tellurite glasses exhibit good thermal stability, and most of these samples show no crystallization tendencies upon heating, which is very important when they would be drawn into optical fibers. The results show that the tungsten tellurite glasses would be the promising host materials for 1.55 μm. Also, Chillcce et al. (2006) have reported a method for producing, from the raw materials, high optical and geometrical quality glass tubes and photonic crystal fiber (PCF) preforms, without using extrusion or drilling at any stage. A thermal glass study has been carried out in order to choose the appropriate glass composition {$66TeO_2$–$18WO_3$–$7Na_2O$–$9Nb_2O_5$, $70TeO_2$–$19WO_3$–$7Na_2O$–$4Nb_2O_5$, $71TeO_2$–$22.5WO_3$–$5Na_2O$–$1.5Nb_2O_5$} to avoid crystallization problems during the tube, preform, and fiber fabrication. A two-period PCF was fabricated in addition to a co-doped erbium and thulium photonic crystal fiber. In the latter, a 187 nm wide amplified spontaneous emission (ASE) spectrum was obtained when pumping a 15 cm long fiber at a wavelength of 790 nm. Jose et al. (2006) have systematically added WO_3 (up to 10 mol%) and P_2O_5 (up to 16 mol%) to a TeO_2–BaO–SrO–Nb_2O_5 (TBSN)-based glass system and studied the thermal and optical properties of the resultant glasses. The primary series of glasses takes the general formula $(78 - x - y)$ TeO_2–$3.5BaO$–$10.5SrO$–$8Nb_2O_5$–xWO_3–yP_2O_5, ($x = 0, 2, 4, 6, 8, 10$ and $y = 0, 4, 8, 12, 16$). The first composition in the primary series was $78TeO_2$–$3.5BaO$–$10.5SrO$–$8Nb_2O_5$, and the second was $76TeO_2$–$3.5BaO$–$10.5SrO$–$8Nb_2O_5$–$2WO_3$. The seventh composition in this series was $74TeO_2$–$3.5BaO$–$10.5SrO$–$8Nb_2O_5$–$4P_2O_2$, the eighth was $72TeO_2$–$3.5BaO$–$10.5SrO$–$8Nb_2O_5$–$2WO_3$–$4P_2O_5$, and so on. The primary series consisted of 28 glasses. The secondary series of glasses takes the general formula $(74 - y)TeO_2$–$3.5BaO$–$10.5SrO$–mNb_2O_5–$(12 - m)$ WO_3–yP_2O_5 ($m = 2, 4, 6, 8$). These doped glasses are characterized by higher thermal stability and wider Raman spectra. These features make them promising new candidates for photonic device materials for optical systems as a new broadband Raman gain media.

Shivachev et al. (2009) prepared $90TeO_2$–$5Bi_2O_2$–$5GeO_2$, $80TeO_2$–$10Bi_2O_2$–$10GeO_2$, and $60TeO_2$–$20Bi_2O_2$–$20GeO_2$ glasses by the conventional melt-quenching technique, and the value of the refractive index of the glasses was approximately 2.13. Also, the values of the density were 5.84, 6.01, and 6.28 g/cm^3, respectively. El-Hagry et al. (2009) studied the effects of rare-earth oxides Sm_2O_3/Yb_2O_3 co-doping upon optical properties of $80TeO_2$–$10GeO_2$–$7K_2O$–$(3 - x)$ Sm_2O_3–xYb_2O_3 glasses and $x = 1$ and 2 mol% as a thin film. The absorption coefficient has been determined from the transparency and reflectivity spectrum in the strong absorption region. Hussain et al. (2008) have developed a spectral analysis of absorption and emission of Sm^{3+} and Dy^{3+} in lithium boro tellurite glasses in the form: $20TeO_2$–$74B_2O_2$–$5Li_2O$–$1RE_2O_3$, where RE = Sm^{3+} and Dy^{3+}. The densities were 2.388 and 2.383 g/cm^3 for glass doped with Sm^{3+} and Dy^{3+}, respectively. Feng et al (2008) successfully fabricated a very large mode area (3000 μm²) tellurite holey fiber for infrared applications and have shown effective single-mode guidance in it. The fiber has a propagation loss of 2.9 dB/m at 1.55 μm and a zero dispersion wavelength at 2.15 μm. Broadband SC spectra extending from 0.9 μm to beyond 2.5 μm were generated in a 9 cm long piece of the fiber, with a maximum output power of 6 mW. The resistance to the high peak pump powers required for generating single-mode broadband supercontinuum SC in this large mode area fiber suggests that the scaling to even higher average powers is likely to be viable. Villegas and Navarro (2007) have prepared twelve glasses in the $(100 - x - y)TeO_2$–TiO_2–Nb_2O_5, where $x = 0.0, 5, 10, 15$, and $y = 0.0, 5, 10, 15$ mol%, and determined the glass-forming region. The glass structure is mainly built by TeO_4 groups, while Nb^{5+} and Ti^{4+} ions play as network modifiers. As the Nb_2O_5 and TiO_2 concentration increases, TeO_4 groups progressively change to TeO_3 groups as a consequence of the network opening. The contribution of the three ions to the oxygen molar volume follows the order: Te4+ > Ti4+ > Nb5+. The TiO_2 incorporation and even increase in the Nb_2O_5 content improve the thermal stability of the glasses and the network reinforcement. The density values were: 5.518 g/cm^3 for 95–0–5, 95–5–0 (crystalline and density not determined), 5.413 g/cm^3 for 90–0.0–10, 5.429 g/cm^3 for 90–5–5, 5.455 g/cm^3 for (90–10–0), 5.327 g/cm^3 for 85–0–15, 5.334 g/cm^3 for 85–5–10, 5.345 g/cm^3 for 85–10–5, 5.361 g/cm^3 for 85–15–0 (crystalline), 5.241 g/cm^3 for 80–5–15, 5.255 g/cm^3 for 80–10–10, and 5.170 g/cm^3 for 75–10–15 (opal glass).

Tellurite glasses were generally applied in rare-earth optical materials due to their excellent physical and chemical properties. Jihong et al. (2007) have prepared tellurite glasses in the form of the following compositions $(93 - x)TeO_2–xTiO_2–5La_2O_3–2Er_2O_3$ $(3x = 0, 2, 5, 10)$ in mol%. In E^{3+}-doped $TeO_2–TiO_2–La_2O_3$ glasses, changes in Raman spectra were found that the intensity ratio of 672 to 760 cm^{-1} increased, which meant that trigonal bypiramidal TeO_4 groups transform to trigonal pyramid TeO_3/TeO_{3+1} units due to the addition of TiO_2. It has been concluded that the fourth glass could be applied as a potential host material for upconversion laser and broadband amplifier. Nandi et al. (2006) made and characterized a new, erbium-doped tellurite glass of the form $67TeO_2–30P_2O_5–1Al_2O_3–1.75La_2O_3–0.25Er_2O_3$ (all in mol%) that has a high glass transition temperature. The addition of phosphate is found to increase the phonon energy. Many channels have been written inside this glass with focused, 45 femtosecond laser pulses at 806 nm wavelength, using different energies and by varying substrate translation speeds.

Lin et al. (2006) fabricated Sm^{3+}-doped alkali–barium–bismuth–tellurite glass of the following molar host composition: 5% Li_2O, 5% K_2O, 5% BaO, 10% Bi_2O_3, and 75% TeO_2 and 2 wt% Sm_2O_3 was added to dope the glass and measure the high refractive index (which at sodium wavelength is 2.344) and its visible fluorescence spectra. The density of the glass samples was measured to be 5.88 g/cm^3 and the number density of Sm^{3+} ions in 2 wt% Sm_2O_3-doped glass was estimated as equal to 4.06×10^{20} cm^{-3}. Rivara et al. (2006) reported the preparation of planar waveguides by $Ag^+ \rightarrow Na^+$ ion exchange in Er^{3+}-doped tellurite glass with a composition of $75TeO_2–2GeO_2–10Na_2O–12ZnO–1Er_2O_3$ mol%. The thermal measurements have indicated that the glass has good thermal stability. Measurements of the refractive index, absorption spectrum, luminescence, and lifetime were made. The glass was chemically stable during the ion exchange process. Monomode and multimode planar waveguides in the tellurite glasses were prepared. The authors determined the depth of the guides, effective diffusion coefficient, and the activation energy. The depths of the waveguides could be controlled by varying the ion exchange temperatures and times (250°C–288°C and 3–12 h). Hart and Zwanziger (2005) interpreted the chemical shifts of tetravalent inorganic oxide network formers (SiO_2, GeO_2, and TeO_2) in terms of the local paramagnetic shielding effect. DiGiulio et al. (2005) studied the optical properties of Rf-sputtering growth of stoichiometric amorphous TeO_2 thin films ($TeO_2–GeO_2–K_2O$) co-doped with rare-earth oxides Sm_2O_3/Yb_2O_3.

Udovic et al. (2006) prepared the $Bi_2O_3–XO_2–TeO_2$ (X = Ti, Zr) systems. A large glass-forming domain was found for X = Ti, but no glass formation was evidenced for X = Zr. Densities, glass transition (T_g), crystallization temperatures (T_c), and Raman spectra of the relevant glasses were studied as functions of the composition. Densities of $Bi_2O_3–TO_2–TeO_2$ ranged from 5.33 to 6.44 g/cm^3 for the compositions of 0.0–18–82 and 25–10–65 mol%, respectively. The Raman spectra of the glasses were interpreted in terms of the structural transformations produced by the modifiers. It was established that the addition of Bi_2O_3 and TiO_2 content to TeO_2 glass influences the T_g temperature in a similar manner: this value progressively increases with the increase of the modifier concentration. However, the structural evolutions are different:

a. Addition of TiO_2 to TeO_2 glass keeps the polymerized framework structure in transforming a number of Te–O–Te bridges into the Te–O–Ti ones without producing any tellurite anions (i.e., the $[TeO_3^{-2}]$ groups)

b. On the contrary, the addition of Bi_2O_3 destroys the glass framework by giving rise to the island-type $[Te_nO_m]^{(m-2n)-2}$ complex tellurites anions, thus causing a depolymerization of the glass

Lima et al. (2006) studied the thermal lens, thermal relaxation calorimetry, and interferometric methods that are applied to investigate the thermo-optical properties of tellurite glasses (in mol%: $80TeO_2–20Li_2O$ [TeLi], $80TeO_2–15Li_2O–5TiO_2$ [TeLiTi-5], and $80TeO_2–10Li_2O–10TiO_2$ [TeLiTi-10]). While, the crystalline phase of $15(K_2O)–15(Nb_2O_5)–70(TeO_2)$ glass ceramic is a polymorph of $K_2Te_4O_9$ by Hart et al. (2004). The authors proposed a new model for the crystalline phase of 15(K_2O)–15(Nb_2O_5)–70 (TeO_2) glass ceramic. A polymorph of $K_2Te_4O_9$ in space group P21 = c with a tetragonal lattice (a = 7:821 Å, b = 16:590 Å) is suggested, which explains all the features of the observed diffraction patterns. This model of the ceramic phase has the cubic ordering of the cations, which is indicated by the x-ray diffraction patterns, but has reasonable bond lengths and angles for the TeO_x polyhedra and predicts the additional features that appear in the neutron diffraction data. The composition of the crystalline phase also explains the observed phase separation behavior. This new description of crystallization in the potassium–niobate–tellurite system has implications for understanding the nonlinear optical response of the glass ceramic.

A series of new glasses of $70TeO_2$–$(20 - x)$ ZnO–xWO_3–$5La_2O_3$–$2.5K_2O$–$2.5Na_2O$ mol%, doped with Yb^{3+}, has been prepared by Wang et al. (2004a) and the density values were: 5.387, 5.443, 5.516, 5.573, and 5.624 g/cm^3 for x = 0, 5, 10, 15, and 20 mol%, respectively. Thermal stability and spectroscopic properties of Yb^{3+} ions have also been measured. It was found that $70TeO_2$–$15WO_3$–$5ZnO$–$5La_2O_3$– $2.5K_2O$–$2.5Na_2O$ composition glass had better thermal stability T_x–T_g > 160°C than $75TeO_2$– $20ZnO$–$5Na_2O$ glass. Evaluated from the good potential laser parameters, this glass is promising for miniature solid fiber lasers or high peak power and high average power lasers, also for waveguides or tunable lasers. Glasses in the $70TeO_2$–$(20-x)$ ZnO– $xGeO_2$–$5La_2O_3$–$2:5Na_2O$–$2:5K_2O$ (x = 0, 5, 10, 15, and 20 mol%; Yb_2O_3 1.0 mol%) systems were prepared by the conventional method of power fusion and quenching have been studied by Wang et al. (2004b). The densities were: 5.387, 5.397, 5.405, 5.415, and 5.430 g/cm^3 for the samples with x = 0, 5, 10, 15, and 20 mol%, respectively. All of the glasses have better thermal stability than $75TeO_2$–$20ZnO$–$5Na_2O$ glass by Wang et al. (1993). Also, evaluated from the good potential laser parameters, this glass is a promising host of miniature solid lasers such as short-pulse generation in diode-pumped lasers, short-pulse generation tunable lasers, high peak power and high average power lasers, and so on.

Lim et al. (2004) determined the electronic structure of two alkali–tungsten–tellurite glass series: (a) $10Li_2O$–xWO_3–$(90 - x)$ TeO_2, and (b) $xK_2O(10 - x)$–Li_2O–$25WO_3$–$65TeO_2$ by x-ray photoelectron spectroscopy. In contrast to alkali–tellurite glasses, the core level spectra of the various elements appear unaffected when TeO_2 is replaced by WO_3, with the alkali–oxide concentration remaining constant. The O1s spectra do not indicate a clear separation of the bridging and nonbridging oxygen contributions. Thus, the authors have concluded that WO_3 behaves essentially as a network former in tellurite glasses. The authors also obtained valence band spectra of the glasses and observed major changes with the addition of WO_3. In particular, the intensity corresponding to the lowest energy part of the spectra, attributed to the O2p bonding orbital, decreases significantly with increasing WO_3, indicating the appearance of a new bonding configuration. Noguera et al. (2004) have studied the stable $[TeO_2]_n$ polymers (n < 7) in the form of chains, rings, and three-dimensional entities are studied through the *ab initio*. The density functional theory realized in the Beck's three-parameter hybrid method using the Lee-Yang-Parr correlation functional (B3LYP) calculations including their Raman spectra, polarizabilities, and hyperpolarizabilities. The 3D-polymers are found to be the most stable, and can be proposed as the model fragments of TeO_2 glass. According to the calculations, the huge hyperpolarizability of TeO_2-based compounds is likely to be related to the essentially delocalized electron states (inherent to polymerized systems) and not to the electron lone pairs of tellurium atoms.

Tellurite thin films have the potential to be applied in optical and electronic areas, but it is difficult to prepare them from tellurium(IV) alkoxides by sol-gel processing due to the instability of tellurium(IV) alkoxides and their derivatives. The precursor of tellurium(VI) complex was synthesized and applied to make tellurium oxide thin films without the addition of any modifiers in the study. Weng et al. (2004) have achieved controllable sol-gel processing of tellurite glasses through the use of Te(VI) precursors. The decomposition process of the complex was investigated, and the thin films derived from this precursor were characterized structurally, morphologically, and optically.

New opportunities for photonic components resulted from the discovery of the SHG in thermally poled glasses in the early 1990s. This has attracted a lot of interest because it offers new developments in the optical glassy material research. Ferreira at al. (2003) have prepared a new tellurite glass with $70\%TeO_2$–25% $Pb(PO_3)_2$–$5\%Sb_2O_3$ with a density of 5.28 g/cm^3. The structure of the unpoled glass has been studied by IR, Raman, and XAFS spectroscopy. Optical properties of transmission, such as the linear refractive index and third-order nonlinear optical susceptibility χ^3, as well as dielectric susceptibilities, have also been investigated. A second harmonic signal was observed for the poled glass, which is an order of magnitude better than the silica glass efficiency.

Sundaram et al. (2003) used femtosecond laser pulses to produce localized damage in the bulk and near the surface of the baseline, Al_2O_3-doped and La_2O_3-doped sodium tellurite glasses. Single or multiple laser pulses were nonlinearly absorbed in the focal volume by the glass, leading to permanent changes in the material in the focal volume. These changes were caused by an explosive expansion of the ionized material in the focal volume into the surrounding material—that is, a microexplosion. The writing of simple structures (periodic array of voxels, as well as lines) was demonstrated. The regions of microexplosion and writing were subsequently characterized using scanning electron microscopy (SEM), energy-dispersive spectrometry (EDS), and atomic force microscopy (AFM). Fingerprints of the microexplosions (concentric lines within the region and a concentric ring outside the region), which were caused by the shockwave generated during the microexplosions, were evident. In the case of the baseline glass, no chemistry change was observed within the region of the microexplosion. However, Al_2O_3-doped and La_2O_3-doped glasses showed depletion of the dopant from the edge to the center of the region of the microexplosions, thereby indicating a chemistry gradient within the regions. Interrogation of the bulk- and laser-treated regions using micro-Raman spectroscopy revealed no structural change due to the microexplosions and writing within these glasses.

The correlation between the specific heat and the refractive index change formed by laser spot heating on tellurite glass surfaces has been investigated by Inoue et al. (2003). The ternary tellurite glasses of TeO_2–Na_2O–Al_2O_3, TeO_2–Na_2O–GeO_2, and TeO_2–Na_2O–TiO_2 doped with 2 mol% of CoO were irradiated by a green lightbeam spot (532 nm) from a second harmonic generator of a Q switch pulse YAG laser. The density of the glasses were 5.025, 4.854, and 4.694 g/cm^3 for $85TeO_2$–$15Na_2O$, $80.0TeO_2$–$14.2Na_2O$–$5Al_2O_3$, and $76.5TeO_2$–$13.5Na_2O$–$10Al_2O_3$, with 2 CoO mol%; 4.948, 4.983, and 4.923 g/cm^3 for the systems 80.8 TeO_2–$14.2Na_2O$–$5GeO_2$, 76.5 TeO_2–$13.5Na_2O$–$10GeO_2$, and $72.3TeO_2$–$12.7Na_2O$–$15GeO_2$ with 2 mol% CoO; and 4.89 and 4.847 g/cm^3 for the systems $80.8TeO_2$–$14.2Na_2O$–$5TiO_2$ and $76.5TeO_2$–$13.5Na_2O$–$10TiO_2$ doped with 2 mol% of CoO, respectively. Rolli et al. (2003) have investigated the erbium-doped tellurite glasses with high quantum efficiency and a broadband-stimulated emission cross section at 1.5 µm in two tellurite glasses of molar composition $75TeO_2$:$12ZnO$:$10Na_2O$:$2PbO$:$1Er_2O_3$ and $75TeO_2$:$12ZnO$:$10Na_2O$:$2GeO_2$:$1Er_2O_3$. The measured absorption and emission spectra were analyzed by Judd–Ofelt and McCumber theories to obtain the radiative transition rates and stimulated emission cross sections. It was found that these glasses have high and broadband absorption and stimulated emission cross sections at 1.5 µm. For the metastable $^4I_{13/2}$ level, a quantum efficiency of above 80% was found by comparing the measured lifetime with the calculated radiative decay time. Because tellurite glasses were promising candidates for optical fiber laser and amplifier applications due to their excellent optical and chemical properties, Marjanovic et al. (2003) studied the emission spectrum from erbium in tellurite glasses and fibers and found it was was almost twice as broad as the corresponding spectrum in silica. The nine glass compositions were of the form: $(100 - x - y)$ TeO_2–x ZnO–yNa_2O and $x = 20, 27.5, 35, 20, 27.5, 35, 15, 22.5,$ and 30; and $y = 0.0, 0.0, 0.0, 5, 5, 5, 10, 10,$ and 10 mol%, respectively. The authors concluded that broad erbium emission spectra have been observed and a small signal net gain of ≈30 dB was demonstrated in 1 m long fiber.

Nayaka et al. (2003) studied the optical waveguiding characteristics of amorphous TeO_{2-x} (20:80, 35:65, 50:50, 70:30, and 100:0) films deposited by reactive sputtering under different O_2: Ar gas mixtures are investigated on fused quartz and Corning glass substrates. Infrared absorption band in the range $641-658 cm^{-1}$ confirmed the formation of a Te–O bond, and a $20O_2$:80Ar gas mixture ratio was found to be optimum for achieving highly uniform and transparent films at a high deposition rate. The grown amorphous films exhibited a large band gap (3.76 eV) and a high refractive index value (2.042–2.052) with low dispersion over a wide wavelength range of 500–2000 nm. Optical waveguiding with a low propagation loss of 0.26 dB/cm at 633 nm was observed on films subjected to a postdeposition annealing treatment at 200°C. Packing density and etch rates were determined and correlated with the lowering of optical propagation loss in the annealed films. The density values of the thin films were 5.58, 5.6, and 5.55 g/cm^3 for the TeO_2 (20:80) at annealing temperatures of 100°C, 200°C, and 300°C with thicknesses of 1.32, 1.29, and 1.15 μm, respectively. Jhaa et al. (2001) have studied the high-refractive index chalcogenide, and heavy metal oxide glasses have been developed recently for their application as all-optical switching devices in telecommunication networks. Depending upon the nature of electromagnetic phenomena, the switching speed may vary between milliseconds in a rare-earth-doped glass and femtoseconds for surface plasmon relaxation in a nanoscale dispersion of metal particles in glasses.

Prasad et al. (2003) prepared a highly transparent and yellow-colored Ho^{3+}-doped TeO_2–B_2O_3–Li_2O glass {density = 2.315 g/cm^3} to study its optical properties from the measured UV–VIS–NIR absorption, emission, and upconversion emission spectra. The authors observed a prominent bright red color when this optical glass was brought under a UV lamp. The recorded upconversion emissions revealed a strong green color along with blue and violet colors upon excitation with $\lambda_{exc} = 643$ nm. The mechanism was explained by an energy level diagram as a three-photon absorption process. Roychoudhury et al. (2002) studied the ultrasonic properties of five different samples of the $(Li_2O)0.2x$–$(Na_2O)x$–$(TeO_2)0.8$ glass system within the temperature range 140–420 K for longitudinal polarization and 140–300 K for transverse polarization. The densities were: 5.2407, 4.9759, 4.9685, 5.0413, and 5.0977 g/cm^3 for $x = 0.0, 0.05, 0.10, 0.15,$ and 0.20, respectively. The physical quantities such as ultrasonic velocity, optical energy gap, refractive index, and optical dielectric constant were evaluated from the spectra. The theoretical fitting of the optical absorption indicates that the present glass system behaves as an indirect gap semiconductor. Also, Prakash et al. (2001) prepared $10Na_2O$–xNb_2O_5–$(90 - x)TeO_2$ and $x = 0.0, 5, 10,$ and 20 mol% glasses, and investigated some of their physical and optical properties. The densities of the glasses were 4.91, 5.41, 5.17, and 5.11 g/cm^3, respectively. The refractive index variation with wavelength is measured using a novel technique—white light interferometry. Optical band gaps and Urbach energies are estimated from the optical absorption spectra. Optical parameters—namely, dispersion energy, average oscillator energy, Abbe's number, and third-order nonlinear susceptibility values—were estimated from the dispersion of the refractive index. Respective refractive index and estimated third-order nonlinear optical susceptibility values are larger than for the tellurite glasses reported earlier. Relatively large refractive index values obtained for the present glasses are attributed to the hyperpolarizability of the Nb–O bands. Also, study of the optical band gap and the dispersion energies of these glasses suggest the decrease in covalency of these glasses with the increase in niobium (Nb) content.

1.11 APPLICATIONS OF TELLURITE GLASS, TELLURITE GLASS CERAMICS AND TELLURITE GLASS CONTAINING NANOPARTICLES

At the Seventh International Conference on Amorphous and Liquid Semiconductors in Edinburgh, UK, Flynn et al. (1977) announced that binary TeO_2–V_2O_5 glass is a semiconductor. Above 200 K, the DC conductivity has a constant activation energy E of about 0.25–0.34 eV, depending on composition. In comparison with the phosphate glasses, the conductivity of tellurites is 2.5 to 3 orders of magnitude greater for similar vanadium concentrations, and, perhaps more importantly, the variation in conductivity with composition in tellurites is due mainly to variations in the pre-exponential constant, rather than the E, as in the phosphate glasses. In the 1980s, Matsushita Electric Industrial Co. announced what is believed to be the first optical disk system on which stored information can be erased and rewritten, as first described by Garner (1983). Erasability is achieved by the addition of several substances including germanium and indium—to the tellurium suboxide layer, which forms the basic recording surface of conventional "record-and-playback" optical disk systems. By passing a laser beam recording "shots" on the surface, it is possible to change the state of surface layers from a crystalline to an amorphous (noncrystalline) phase, and vice versa, as shown in the 1st Ed of this book. These phases offer high and low light reflectivity, respectively; information can be "read" by a laser beam of a certain power and wavelength as a disk revolves at a speed of 1800 rpm. The successful development of an "erasable" system is likely to enhance the image of the optical disk as an alternative medium to magnetic devices such as the floppy disk for mass information storage, most notably in office filing-system applications. It could also have important long-term implications for the future prospects of optical disk-based consumer products, which currently include the laser videodisk and the digital audio disk. A 12 cm diameter optical disk can store up to 15,000 color pictures, or 10,000 A4 size documents, which is thousands of times greater capacity than that of an 8-in-1 MB capacity floppy disk. Information on the optical disk can be erased and rewritten up to a million times. Also in the 1980s, tellurite glasses were shown by Burger et al. (1985) to be good transmitters in the visible spectral region and also as windows for IR transmission in the MIR region. The optical glasses in the systems comprising TeO_2 and R_nO_m, R_nX_m, $R_n(SO_4)_m$, $R_n(PO_3)_m$, or B_2O_3 transmit in the NIR and MIR regions at wavelengths up to 7 μm. It has been established that absorption bands in the MIR region result from R–O influence on the second glass former. Tellurite glasses have high n (>1.80), low dispersion coefficients (v < 30), and high ρ (> 4.5 g/cm³). The T_g varies from 220°C to 450°C, with coefficients of thermal expansion from 120 to 220×10^{-7} K^{-1}. Burger (1985) proved that n decreases strongly in the order RBr, R'Br$_2$>RCl, R'Cl$_2$>RF, R'F$_2$. As for their optical characteristics, tellurite glasses are in the superheavy optical flint class.

In 1992, Mizumo et al. reported magnetic recording devices that have become smaller in size but boast greater memory. The metal in-gap head is now sufficiently developed to keep up with the high-coercive-force medium and to enable high-density recording. The composition of this tellurite glass system is 85 mol% [TeO_2–xPbO–y(B_2O_3)]–5 mol% ZnO–10 mol% CdO. This glass system has been used as a bonding material in magnetic heads because it offers a low thermal expansion coefficient, good water resistance, and low interaction with the amorphous alloy, as shown in the 1st Ed of the book.

To develop and use TeO_2-based glasses, an understanding of their thermal stability is necessary. Studies of the thermal stability and crystallization behaviors in TeO_2-based glass are limited and very recent. For example, in 1995, Shioya et al. studied the optical properties of TeO_2–Nb_2O_5–K_2O glass and its ceramic forms and found that this system exhibits good optical transparency at the wavelength of visible light. Such transparent glass–ceramic material is considered a new type of nonlinear optical ceramic materials made by the controlled crystallization of glasses. The ceramic materials thus produced have outstanding mechanical, thermal, and electrical properties. The most notable characteristic of glass–ceramic materials is the extremely fine grain size, and it is likely that this feature is responsible in large measure for the valuable properties of these materials. It is to be expected that a glass–ceramic material would have an almost ideal polycrystalline structure since, in addition to its fine texture, the crystals are fairly uniform in size and are randomly oriented.

The electronics company TNN announced in year 2000 the use of tellurite single-mode fiber. The potential for this new fiber is creating much interest in the fiber optics industry as shown in the 1st Ed of the book. It was expected that different combinations of the glass elements would result in different performance characteristics. Tellurite's resistance to moisture provides better reliability than fluoride glass in telecommunication applications. Another advantage of tellurite fiber is its ability to be pumped at 980 nm as well as 1480 nm. The Japanese manufacturer KDD expects a broader photoluminescence spectra and new wavelengths that are not available from silica and fluoride fiber lasers. The bend strength has been measured for this fiber. The mean value of the breaking stress is 2.1%, which corresponds to a breaking-bend radius of 3.7 mm. To examine tellurite's resistance to moisture, this fiber has been maintained at 80°C and 80% humidity for 400 h. There is no evidence of degradation of the glass surface. The gain bandwidth of the Er-doped tellurite is about 80 nm, whereas the conventional Er-doped silica fiber amplifier is between 35 and 40 nm. One of the most promising applications of tellurite glass fiber is in Er-doped tellurite fiber amplifiers for wavelength division-multiplexing (WDM) optical systems, being developed by NTT (Japan) as shown in the 1st Ed of the book. The advantage of the tellurite amplifier is its high gain and large gain bandwidth. Tellurite glass is generally more stable than fluoride glass, which allows for a wider selection of elements in the composition of the preform.

In fiber optic communications, WDM is a technology that multiplexes multiple optical carrier signals on a single optical fiber by using different wavelengths (colors) of laser light to carry different signals. Chryssou (2001) has established WDM systems in the C-Band using Er^{3+}-doped tellurite optical waveguide amplifiers. It was 16-channel and 2.5 Gb/s. The WDM system is analyzed with its channels allocated in the 1.52–1.56 μm wavelength region in order to increase the usable amplifier bandwidth to ≈45 nm. To avoid ASE noise and the nonuniform signal gain in the wavelength region, an amplifier module consisting of an Er^{3+}-doped tellurite waveguide amplifier, an ASE filter, and two concatenated long-period grating filters are proposed. A tellurite-based amplifier was chosen as the amplifying element because of its broad emission bandwidth (~80 nm), its high emission cross section (6.44×10^{-25} m²), and its high rare-earth ion solubility. The amplifier model is based on propagation and population rate equations and includes both uniform and pair-induced upconversion mechanisms. It is solved numerically by combining finite elements and a Runge–Kutta algorithm. The analysis predicts that by using the proposed amplifier module, the channels may be transmitted to a maximum distance of 1800 km, thereby finding applications in large optical networks where either many wavelengths are required or channel spacing must be large. Mori et al. (2003) have also established the first ultra-wideband tellurite-based fiber Raman amplifier (T-FRA) for application to seamless ultra-large capacity dense WDM systems and confirmed that the Raman scattering characteristics of the tellurite-based fiber has so large a gain coefficient and Stokes shift to achieve a wideband T-FRA with a shorter fiber length than when using silica-based fiber. In 2004, Mori and Masuda started their article with the question, "what is a fiber Raman amplifier FRA?" and answered it as follows: the FRA differs in principle from rare-earth-doped fiber amplifiers in that it uses molecular vibration and an optical-scattering phenomenon in the fiber material that acts as an amplification medium. Mori and Masuda (2004) then mentioned the advantages of T-FRAs and successfully extended the amplification bandwidth to 170 nm by using tellurite fiber (with tellurium dioxide as its main component) as the amplification medium while the bandwidth is of 100 nm in the 1500 nm band for silica fiber. This is possible because the vibration frequency Ω, caused by the thermal vibration of a Te–O molecule, is larger than the thermal vibration Ω of a Si–O (or Ge–O) molecule. Also, Mori and Masuda (2004) compared this spectrum with that of Raman scattering in a dispersion compensation fiber (DCF). Silica fiber, which has a Stokes shift of 100 nm in the 1500 nm band, has a spectrum with a unimodal shape. By contrast, the tellurite fiber has a Stokes shift 1.7 times larger (=170 nm) and a bimodal spectrum. Moreover, tellurite glass has a nonlinear susceptibility about one order of magnitude greater than that of silica glass, thus its Raman gain coefficient (scattering intensity) is about 16 times larger. These features show that tellurite fiber is more promising than silica fiber for making an ultra-wideband FRA with a shorter fiber length and fewer pump wavelengths. Mori and Masuda (2004) show the configuration of a tellurite FRA using the multiwavelength pumping technique. This amplifier pumps a 250 m tellurite fiber module using laser diodes (LDs) operating at four different wavelengths. The fiber length here is about one-tenth that of commonly used silica DCF (about 3 km).

Murugan et al. (2005) have prepared tellurite glasses that are optimized for higher Raman gain and broad bandwidth. These glasses were found to have improved thermal stabilities, which make them suitable for fiber device applications. While maintaining the Raman gains at higher values, the Raman bandwidths could be broadened by the proper addition of alkaline earth oxides and heavy metal oxides to the tellurite glasses. The relative Raman gain and Raman cross sections of the present glasses are better than that of the tellurite-based glasses reported earlier. Thus, higher Raman gain and broader Raman amplifications could be possible by using this tellurite glass system as a gain medium compared with the conventional T-FRAs. Stegeman et al. (2005) have fabricated and tested several different compositions of tellurium–thallium oxide glasses ($TeO_2–TlO_{0.5}$ and $TeO_2–TlO_{0.5}–PbO$) for their Raman gain performance. The addition of PbO to the glass matrix increased the surface optical damage threshold by 60%–230%. The maximum material Raman gain coefficient that was experimentally obtained was (58 ± 3) times higher than the peak Raman gain of a 3.18 mm thick Corning 7980-2F fused silica sample ($\Delta v = 13.2$ THz). The highest peak in the Raman gain spectrum of the tellurium–thallium glass is attributed to the presence of TeO_3 and TeO_{3+1} structural units with thallium ions in the vicinity at a frequency shift near 21.3 THz.

Wang et al. (2006) have studied the source of optical loss in tellurite glass fibers. Nonsilica-based optical fibers generally suffer from high optical loss and low strength, whereas they have other optical properties superior to that of silica for certain device applications. Tellurite glasses offer one of the best compromises among optical, mechanical, and processing properties. Wang et al (2006) have achieved low-loss tellurite fibers for active and nonlinear applications. The authors also explained that single-mode tellurite fibers doped with $KNbO_3$ were made with losses varying from 1.3 to 6 dB/m. The sources of loss were striation, dust particles, and bubbles. The decrease of striation was observed by employing a lower pouring temperature, and the increase of fiber strength was achieved by the hydrochloric acid etching of the preform.

Jose at al. (2006) have studied new tellurite-based glasses and broadband Raman gain media to realize high performance FRAs. These performances were achieved by WO_3 and P_2O_5 doping in $TeO_2–BaO–SrO–Nb_2O_5$ (TBSN) glass. The maximum gain coefficient obtained in the present study was ≈ 50 times that of silica glass for 532 nm excitation. The bandwidth (i.e., full width at half the maximum gain) of these doped glasses was more than twice that of a conventional tellurite-based glass, and was 70% larger than that of silica glass. Tellurite and fluorotellurite glasses for fiber optic Raman amplifiers have been investigated by Donnell et al. (2006). In their study the authors reported the glass characterization, optical properties, Raman gain, preliminary fiberization, fiber characterization, and spontaneous Raman scattering spectra of nine oxide–tellurite and fluorotellurite glasses from three glass systems: sodium–zinc–tellurite (TZN), tungsten–tellurite, and fluorotellurite. A Raman gain and surface damage threshold of 1064 nm were also shown for a selection of these glasses, all of which exhibited high gain and damage resistance. Raman gain spectra were directly measured and accurately calculated for selected TZN and fluorotellurite glasses after Fresnel, internal solid angle, and Bose–Einstein corrections. The calculated gain showed good fits to the Raman gain measurements made using a calibrated nonlinear optics apparatus. Infrared and UV-Vis absorption spectra, characteristic temperatures obtained by differential thermal analysis, densities acquired by the Archimedes principle, and refractive indices measured by spectroscopic ellipsometry were also given. The ternary systems $TeO_2–WO_3–Bi_2O_3$ and $TeO_2–Na_2O–ZnF_2$ and the quaternary system $TeO_2–Na_2O–ZnO–PbO$ show promise as Raman amplifiers, as they are relatively easy to draw into optical fibers and, to the authors' collective knowledge, this is the first time that Raman gain has been presented on halide-containing tellurite glasses. The oxyfluoride system studied here, $TeO–Na_2O–ZnF_2$, exhibited a dependence on the peak Raman intensity with ZnF_2 addition. The calculations of preform geometry for monomode and multimode guidance and stresses in similar and dissimilar (core suction) core–clad pairs were shown. Dispersion in the mid-infrared and initial fiber drawing studies was also reported, with fibers showing reasonable unclad losses.

Masuda et al (2006) have studied the design and the spectral gain and noise figure (NF) characteristics of FRAs, which are T-FRAs or hybrid tellurite/silica FRAs (hybrid FRAs). The propagation equations for a multiwavelength (multi-λ)-pumped T-FRA that included the pump interaction terms are presented, with which the gain and NF of the T-FRA could be calculated. Tellurite fiber (TF) length dependences of gain and the NF of a T-FRA are clarified experimentally and theoretically. Numerical calculations on the gain and NF spectra of the tellurite-based FRAs show that a T-FRA with a two-stage configuration and a hybrid FRA with a three-stage configuration can provide seamless gain bands with widths of more than 130 nm over the S-, C-, and L-bands. The two-stage T-FRA has a couple of two-λ-pumped T-FRA stages, a gain equalizer (GEQ), and a DCF between the T-FRA stages; while the three-stage hybrid FRA has a couple of two-λ-pumped T-FRA stages, a GEQ, and a two-λ-pumped DCF Raman amplifier (DCF-RA) stage between the T-FRA stages. The numerical calculations also showed that the two-stage T-FRA and the three-stage hybrid FRA achieved top gains with regard to their flattened gain spectra of 9.7 and 24 dB, and maximum NFs in their gain bands of 11.8 and 8.2 dB, respectively. The measured gain and NF spectra of the tellurite-based FRAs coincide well with the corresponding calculated spectra.

Qin et al. (2007-A) have demonstrated stimulated Brillouin amplification in a tellurite fiber as a potential system for slow light generation. A Brillouin gain of 29 dB is achieved in a 100 m tellurite fiber with a pump power of 10 mW at 1550 nm. Stimulated Brillouin scattering (SBS)-induced time delay per unit power and per unit length is also calculated using the measured data of Brillouin gain coefficients. A peak value of 0.09246 ns $mW^{-1}m^{-1}$ and a time delay slope efficiency of 1.75 ns/dB were obtained for this tellurite fiber. The potential performance of a tellurite fiber for slow light generation was clarified on the base of the Brillouin gain characteristic. The results of Qin et al. (2007b) showed that tellurite fiber is a potential candidate for Brillouin fiber amplifiers and slow light generation. Jose et al (2007) have studied the nonlinear susceptibility in TeO_2–BaO–SrO–Nb_2O_5 tellurite glasses by systematically adding WO_3 and P_2O_5 in a TeO_2–BaO–SrO–Nb_2O_5 TBSN glass system to study it for ultra-broadband Raman amplifiers. The response in nonlinear indices to this addition was studied and reported herewith. The third-order optical susceptibility (χ^3) measured using the Maker-fringe analysis increased with an increase in WO_3 content and decreased with an increase in P_2O_5 content. When these components were added simultaneously, the χ^3 reached a value similar to that of pure TeO_2 glass. In view of their higher Raman gain coefficient and amplification bandwidth, and with the present result of higher nonlinear indices, these glasses are likely to be suitable for photonics applications.

Ohishi et al. (2007) have studied novel photonics materials for broadband lightwave processing and glass-based photonics materials research for practical functional devices in future optical networks. As new fiber Raman gain media, the TBSN glass system containing WO_3 and P_2O_5 was systematically studied. The TBSN glass doped with WO_3 and P_2O_5 showed high stability against crystallization. New Raman bands due to WO_4 and PO_4 tetrahedra occurred and broadened the Raman spectrum of the glass system. The Raman gain coefficient and bandwidth of the TBSN tellurite glass have been tailored by systematically adding WO_3 and P_2O_5. The glass system showed the broadest gain bandwidth so far achieved in tellurite glasses while maintaining higher gain coefficients. The gain bandwidths of these glasses were more than twice that of a conventional tellurite-based glass and were 70% larger than that of the silica glass. These glasses are promising candidates for photonics devices in future photonic systems. In the same article, the authors also announced the design of ultimate gain-flattened O, E, and S+ C+ L ultra-broadband fiber amplifiers using a new fiber Raman gain medium. By solving the inverse amplifier design problem, gain-flattened O (~17.5 THz), E (~15.1 THz), and S+ C+ L (~20.9 THz) ultra-broadband FRAs are designed using a new TeO_2–BaO–SrO–Nb_2O_5–P_2O_5–WO_3 (TBSNWP) tellurite fiber. When increasing the numbers of pump wavelengths from two to eight, the gain profiles become flatter and the effective bandwidth becomes larger. The relative gain flatness of ~1% could be achieved over bandwidths of up to 15.1 THz (which corresponds to E-band) without any gain equalization devices. When narrowing the gain bandwidth from the S+ C+ L-band (20.9 THz) to the E-band (15.1 THz), the relative gain flatness is reduced from 4.51 to 4.12%. The effects of the shape of the Raman gain spectra on the relative gain flatness along with the effective bandwidth are also investigated using the TBSNWP glass with one broad Raman shift peak (full width at half maximum ~11 THz) and TeO_2–Bi_2O_2–ZnO–Na_2O (TBZN) glass with twin peaks. The simulation results show that the relative gain flatness and the effective bandwidth of TBSNWP FRA are better and larger than those of TBZN FRA, respectively. Our results suggest that the TBSNWP glasses are promising candidates for broadband FRA in photonic systems.

Ma et al. (2008) have designed and presented a complete analysis on improving the gain flatness of ultra-broadband T-FRAs by optimizing the parameters of multiwavelength pumps. Jose et al. (2008) have studied the issue of widening the Raman spectral bandwidth of tellurite glasses, while maintaining higher scattering intensity is also addressed. Raman spectral bandwidths of tellurite glasses were widened by using a single Raman active component of suitable concentration in appropriate base glasses. It was observed that the MoO_2 octahedra, due to their high octahedral distortion, have high Raman polarizability compared with WO_6, NbO_6, and TaO_6 octahedra, and PO_4 tetrahedra. This high Raman polarizability enabled the broadening of the spectral width up to 350 cm^{-1} while maintaining high Raman scattering intensities. Although similar bandwidths could be achieved using the combined generation of WO_6 octahedra and PO_4 tetrahedra, the resultant Raman scattering intensity of such glasses is only half of what would be achievable using MoO_2. It is evident that the simplest tellurite glass showing wide spectral broadening is a quaternary system comprising a network modifier (BaO or Bi_2O_3) and two Raman oscillators (NbO_6 and MoO_6 octahedra).

In 2009, Liao et al. prepared new fluorotellurite glasses for photonics applications for fabricating mid-infrared optical fiber lasers. The reduced absorption loss would allow them to be good candidate materials for this application. The prepared glasses were in the form $(85 - x)$ TeO_2–$xZnF_2$–$12PbO$–$3Nb_2O_5$, $x = 0, 10, 20, 30$, and 40 mol%, and the densities = 5.9, 5.81, 5.76, 5.71, and 5.65 g/cm^3. The increase of the ZnF_2/TeO_2 ratio resulted in a significant change of the Raman spectra of the fluorotellurite glass. Also, it has been concluded by Gandhi et al. (2009) that there is an increase in the degree of disorder of the octahedral in the glass network of the form $20ZnF_2$–$30As_2O_3$–$(50 - x)$ TeO_2 with an increase in the concentration of x = 0.0, 0.1, 0.2, 0.3, 0.4, 0.5, and 0.6 mol% V_2O_5 oxide. The density values of these glasses were 5.304, 5.299, 5.295, 5.288, 5.283, 5.277, and 5.269 g/cm^3, respectively. Also, Carlie et al. (2009) studied tellurite-based glasses for high Raman gain amplification. A series of lead phosphotellurite glasses: namely, $48TeO_2$–$17PbO$–$17P_2O_5$–$18Sb_2O_3$, $56TeO_2$–$20PbO$–$20P_2O_5$–$4Sb_2O_3$, and $76.5TeO_2$–$9PbO$–$9P_2O_5$–$5.5Sb_2O_3$, $85TeO_2$–$10Nb_2O_5$–$5MgO$ and $85TeO_2$–$15WO_3$ were prepared to evaluate the impact of network former type TeO_2/P_2O_5 ratio and the influence of other heavy metal oxide additives, such as PbO and Sb_2O_3. The authors concluded that:

1. Tellurite and pyrophosphate glass network former-based compositions appear to be attractive candidates for Raman gain application purposes. The tellurite glasses exhibit an absolute Raman gain coefficient of up to 30 times higher than silica, and pyrophosphate glasses with d0 ions exhibit Raman vibration up to 1300 cm^{-1}. The intermediate and modifier glass constituents influence the glass network and the overall Raman gain performance of the glass. Borophosphate glass networks also appear to be a good host for rare-earth ions and can be drawn into fiber with a similar structure and luminescence properties as the preform.

2. Chalcogenide-based materials possess high nonlinear refractive indices, which increase with an increase of Se content up to 400 times the n_2 of fused silica for $As_{24}S_{38}Se_{38}$ glass. These glasses can be successfully deposited into films using the thermal evaporation technique. Waveguides can be written in the surface using IR fs-laser irradiation, thereby inducing a change in the linear refractive index. This has been attributed to different

mechanisms, such as the creation of As-As and S-S bonds in As-based glasses, and to changes in interconnections between GeS_4 units in Ge-based glasses.

3. Oxysulfide films, deposited from a sulfinated target using the RF sputtering technique, are also sensitive to laser exposure with a change of the refractive index and a red shift of the absorption band gap.

Such modification opens the pathway toward the laser writing of photonic devices in the surface of the chalcogenide and oxysulfide materials. In 2009, Lin et al. studied the optimizing glass composition and, using a multistage dehydration process, a ternary $80TeO_2–10ZnO–10Na_2O$ glass was obtained, which showed excellent transparency in the wavelength range from 0.38 up to 6.10 µm. Based on this optimized composition, the authors reported on the fabrication of a single-mode solid-core tellurite glass fiber with large mode area (effective area = A_{eff}) of 103 µm^2 and low loss of 0.24 ~ 0.7 dB/m at 1550 nm. By using the continuous-wave self-phase modulation method, the nonresonant nonlinear refractive index n_2 and the effective nonlinear parameter $\gamma = 2\pi\, n_2/\lambda A_{eff}$ of this tellurite glass fiber were estimated to be 3.8×10^{-19} m^2/W and 10.6 W^{-1}km^{-1} at 1550 nm, respectively.

Massera et al. (2010) reported a method to produce new core–clad tellurite-based preforms with the compositions $70TeO_2–10Bi_2O_3–20ZnO$ and $72.5TeO_2–10Bi_2O_3–17.5ZnO$ for the clad and core compositions using a rotational caster. The authors developed preform-preparation and fiber-processing conditions that allowed the authors to successfully draw multimode fiber with a core radius of 57 ± 1 µm. No variation in the composition, as quantified by energy dispersive spectroscopy (EDS), has been induced by the drawing process. Micro-Raman spectroscopy has confirmed the presence of an increased number of TeO_4 units and Te–O–Te bridging units in the resulting fiber as compared with the structure of the preform, which has been attributed to subtle molecular unit reorientation during the new thermal history of the drawing process. The extent of this reorganization is more pronounced in the fiber cladding, which sees a more rapid cooling rate during the fiber drawing. The propagation losses in the fiber have been measured at (3.2 ± 0.1) dB/m at 632 nm and 2.1 ± 0.1 dB/m at 1.5 µm, and are believed to be dominated by residual impurities or moisture within the bulk glass. The losses at 1.5 µm have have been related to OH groups in the glass estimated, from the absorption spectrum in the IR region, to be 5.10×10^{19} ions/cm^3. In 2010, Mizuno et al. studied tellurite glass fiber with a high Brillouin gain employed for the distributed strain measurement with Brillouin optical correlation-domain reflectometry (BOCDR), as shown in Chapter 2. Recently, Rivera et al. (2011) studied the growth of silver nanoparticles embedded in tellurite glass in the form $75TeO_2–2GeO_2–15Na_2O–7ZnO–1Er_2O_3–xAgCl$ (mol%). It has been shown that the nanoparticles carried out an energy transfer process between the Ag nanoparticles and the Er^{3+} ions in the tellurite glasses. Also, Kassab et al. (2011) studied the effects of gold nanoparticles in the green and red emissions of $TeO_2–PbO–GeO_2$ glasses doped with Er^{3+}–Yb^{3+}. It has been shown that the combined effects of gold nanoparticles and the efficient Yb^{3+} to Er^{3+} energy transfer mechanism change the upconversion visible spectrum.

REFERENCES

Abbas, I., Ph.D. thesis at Sudan University of Science and Technology, 2006.

Adler, D., *Amorphous Semiconductors*, CRC Press: Boca Raton, FL, 5, 1971.

Affifi, N., *J. Non-Cryst. Solids*, 136, 67, 1991.

Ahmed, A., Hogarth, C., and Khan, M., *J. Mater. Sci.*, 19, 4040, 1984.

Arlt, G., and Schweppe, H., *Solid State Commun.*, 6, 78, 1968.

Bahgat, A., Shaisha, E., and Sabry, A., *J. Mater. Sci.*, 22, 1323, 1987.

Baldwin, C. M., Almeida, R. M., and Mackenzie, J. D., *J. Non-Cryst. Solids,* 43, 309, 1981.

Barady, G., *J. Chem. Phys.*, 24, 477, 1956.

Barady, G., *J. Chem. Phys.*, 27, 300, 1957.

Berthereau, A., Fargin, E., Villezusanne, A., Olazcuaga, R., Flem, G., and Ducasse, L., *J. Solid State Chem.*, 126, 143, 1996.

Beyer, H., *Z. Krist.*, 124, 228, 1967.

Blanchandin, S., Marchet, P., Thomas, P., Champarnaud, J., Frit, B., and Chagraoui, A., *J. Mater. Sci.*, 34, 1, 1999.

Blanchetier, C., Foulgoc, K., Ma, H., Zhang, X., and Lucas, J., *J. Non-Cryst. Solids*, 184, 200, 1995.

Burger, H., Vogel W., Koshukharov, V., and Marinov, M., *J. Mater. Sci.*, 19, 403, 1984.

Burger, H., Vogel, W., and Kozhokarov, V., *Infrared Phys.*, 25, 395, 1985.

Bursukova, M., Kashchieva, E., and Dimitriev, Y., *J. Non-Cryst. Solids*, 192/193, 40, 1995.

Caricato, A., Fernandez, M., Ferrari, M., Leggieri, G., Martino, M., Mattarelli, M., Montagna, M., Resta, V., Zampedri, L., Almeida, R., Conçalves, M., Fortes, L., and Santos, L., *Mater. Sci. Eng. B*, 105, 65, 2003.

Carlie, N., Petit, L., and Richardson, K., *J. Eng. Fiber Fabr.*, 4, 21, 2009.

Chagraoui, A., Tairi, A., Ajebli, K., Bensaid, H., and Moussaoui, A., *J. Alloy Compd.*, 495, 67, 2010.

Chakraborty, S., Satou, H., and Sakata, H., *J. Appl. Phys.*, 82, 5520, 1997.

Chen, G., Zhang, Q., Cheng, Y., Zhao, C., Qian, Q., Yang, Z., and Jiang, Z., *Spectrochim. Acta Part A*, 72, 734, 2009.

Chen, W., Chen, G., and Cheng, J., *Phys. Chem. Glasses*, 38, 156, 1997.

Chen, Y., Nie, Q., Xu, T., Dai, S., Wang, X., and Shen, X., *J. Non-Cryst. Solids*, 354, 3468, 2008.

Cheng, J., Chen, W., and Ye, D., *J. Non-Cryst. Solids*, 184, 124, 1995.

Chereminsinov, V., and Zalomanov, V., *Opt. Spectrosc.*, 12, 110, 1962.

Chillcce, E., Cordeiro, C., Barbosa, L., and Cruz, C., *J. Non-Cryst. Solids,* 352, 3423, 2006.

Chryssou, C., *Fiber Integrated Opt.*, 20, 581, 2001.

Dehelean, A., and Culea, E., *J. Phys. Conf. Ser.*, 182, 012064, 2009.

Desirena, H., Schulzgen, A., Sabet, S., Ramos-Ortiz G., Rosa E., and Peyghambarian N., *Opt. Mater.,* 31, 787, 2009.

DiGiulio, M., Zappettini, A., Nasi, L., and Pietralung, S., *Cryst. Res. Technol.*, 40, 1023, 2005.

Dimitriev, Y., Ibart, J., Ivanova, I., and Dimitrov, V., *Z. Anorg. Allg. Chem.*, 562, 175, 1988.

Dimitriev, Y., Ivanova, I., Dimitrov, V., and Lackov, L., *J. Mater. Sci.*, 21, 142, 1986a.

Dimitriev, Y., Kashchiev, E., Ivanova, Y., and Jambazov, S., *J. Mater. Sci.*, 21, 3033, 1986b.

Donnell, M., Miller, C., Furniss, D., Tikhomirov, V., and Seddon, A., *J. Non-Cryst. Solids,* 331, 48, 2003.

Donnell, M., Richardson, K., Stolen, R., Seddon, A., Furniss, D., Tikhomirov, V., Rivero, C., Ramme, M., Stegeman, R., Stegeman, G., Couzi, M., and Cardinal, T., *Presented at the American Ceramic Societies Glass and Optical Materials Division (GOMD) Meeting,* Hyatt Regency: Greenville, SC, May 16–19, 2006.

Duclere, J., Lipovskii, A., Mirgorodsky, A., Thomas, P., Tagantsev, D., and Zhurikhina, V., *J. Non-Cryst. Solids,* 355, 2195, 2009.

Dutton, W., and Cooper, W., *Chem. Rev.*, 66, 657, 1966.

El-Fouly, M., Maged, A., Amer, H., and Morsy, M., *J. Mater. Sci.,* 25, 2264, 1990.

El-Hagary, M., Emam-Ismaila, M., Shaabana, E., and Shaltout, I., *J. Alloy Compd.*, 485, 519, 2009.

Elidrissi-Moubtaim, M., Aldon, L., Lippens, P., Oliver-Fourcade, J., Juma, J., Zegbe, G., and Langouche, G., *J. Alloy Compd.*, 228, 137, 1995.

El-Mallawany, R., Abdel-Kader, A., El-Hawary, M., and El-Khoshkhany, N., *J. Mater. Sci.*, 45, 871, 2010.

El-Mallawany, R., and Ahmed, I., *J. Mater. Sci.*, 43, 5131, 2008.

El-Mallawany, R., and Saunders, G., *J. Mater. Sci. Lett.*, 6, 443, 1987.

El-Mallawany, R., *J. Appl. Phys.*, 72, 1774, 1992.

El-Mallawany, R., *J. Appl. Phys.*, 73, 4878, 1993.

El-Mallawany, R., *Mater. Chem. Phys.*, 63, 109, 2000.

El-Mallawany, R., *Mater. Sci. Forum*, 67/68, 149, 1991.

El-Mallawany, R., Sidky, M., Kafagy, A., and Affif, H., *Mater. Chem. Phys.*, 3, 295, 1994.

Eraiah, B., *Bull. Mater. Sci.*, 29, 375, 2006.

Eyzaguirre, C., Eugenio Rodriguez, E., Chillcce, E., Osorio, S., Cesar, C., Barbosaw, L., Mazali, I., and Alves, O., *J. Am. Ceram. Soc.*, 90, 1822, 2007.

Feng, X., Loh, W., Flanagan, J., Camerlingo, A., Dasgupta, S., Petropoulos, P., Horak, P., Frampton, K., White, N., Price, J., Rutt, H., and Richardson, D., *Opt. Express*, 16, 13651, 2008.

Ferreira, B., Fargin, E., Guillaume, B., Le Flem, G., Rodriguez, V., Couzi, M., Buffeteau, T., Canioni, L., Sarger, L., Martinelli, G., Quiquempois, Y., Zeghlache, H., and Carpentier, L., *J. Non-Cryst. Solids*, 332, 207, 2003.

Flynn, B., Owen, A., and Robertson, J., *7th Conference on Amorphous and Liquid Semiconductors*, ed. Spear, W. E., University of Edinburgh, Cent. 89, 69664 1978, 469664 CAPLUS.

France, P. W., ed., *Optical Fiber Lasers and Amplifiers*, CRC Press: Boca Raton, FL, 1991.

Gandhi, Y., Venkatramaiah, N., RaviKumar, V., and Veeraiah, N., *Physica B*, 404, 1450, 2009.

Garner, R., *Financial Times*, p. 12, April 19, 1983.

Gaskell, P., *J. Non-Cryst. Solids*, 222, 1, 1997.

Ghosh, A., and Debnath, R., *Opt. Mater.*, 31, 604, 2009.

Goldschmidt, V., *Skrifter Norske Videnskaps-Akad, Oslo*, 8, 137, 1926.

Golis, E., Reben, M., Wasylak, J., and Filipecki, J., *Opt. Appl.*, vol. XXXVIII, no. 1, 2008.

Hager, I. Z., and El-Mallawany, R., *J. Mater. Sci.*, 45, 897, 2010.

Hart, R., Jr. and Zwanziger, J., *J. Am. Ceram. Soc.*, 88, 2325, 2005.

Hart, R., Zwanziger, J., and Lee, P., *J. Non-Cryst. Solids*, 334, 48, 2004.

Havinga, E., *J. Phys. Chem. Solids*, 18/23, 253, 1961.

Hilton, A. R., Hayes, D. J., and Rechtin, M. D., *J. Non-Cryst. Solids*, 17, 319, 1975.

Hilton, A. R., *J. Non-Cryst. Solids*, 2, 28, 1970.

Hiniei, Y., Miura, Y., Nanba, T., and Osaka, A., *J. Non-Cryst. Solids*, 211, 64, 1997.

Hussain, S., Hungerford, G., El-Mallawany, R., Gomes, M., Lopes, M., Ali, N., Santos, J., and Buddhudu, S., *J. of Nanosci. and Nanotechnol.*, 8, 1, 2008.

Imoka, M., and Yamazaki, T., *J. Ceram. Assoc. Japan*, 76, 5, 1968.

Inabaw, S., and Fujino, S., *J. Am. Ceram. Soc.*, 93, 217, 2010.

Inagawa, I., Iizuka, R., Yamagishi, T., and Yokota, R., *J. Non-Cryst. Solids*, 95/96, 801, 1987.

Inoue, S., Nukui, A., Yamamoto, K., Yano, T., Shibata, S., and Yamane, M., *J. Non-Cryst. Solids*, 324, 133, 2003.

Ioffe, A. F., and Regel, A. R., *Prog. Semicond.*, 4, 237, 1960.

Ivanova, I. *J. Mater. Sci.*, 25, 2087, 1990.

Ivanova, Y., *J. Mater. Sci. Lett.*, 5, 623, 1986.

Jain, H., and Nowick, A., *Phys. Stat. Solids*, 67, 701, 1981.

Jhaa, A., Liua, X., Kar, A., and Bookey, H., *Curr. Opin. Solid State and Mater. Sci.*, 5, 475, 2001.

Jihong, Z., Haizheng, T., Yu, C., and Xiujian, Z., *J. Rare Earths*, 25, 108, 2007.

Johnson, P., Wright, A., Yarker, C., and Sinclair, R., *J. Non-Cryst. Solids*, 81, 163, 1986.

Jose, R., Qin, G., Arai, Y., and Ohishi, Y., *J. Appl. Phys.*, 46, L651, 2007.

Jose, R., Qin, G., Arai, Y., and Ohishi, Y., *J. Opt. Soc. Am. B: Opt. Phys.*, 25, 373, 2008.

Jose, R., Suzuki, T., and Ohishi, Y., *J. Non-Cryst. Solids*, 352, 5564, 2006.

Kabalci, I., Ozen, G., Oglu, M., and Sennaroglu, A., *J. Alloys Compd.*, 419, 294, 2006.

Kassab, L., Camilo, M., Amancio, C., Da Silva D., and Martinelli, J., *Opt. Mater.*, 33, 1948, 2011.

Katsuyama, T., and Matsumura, H., *J. Appl. Phys.*, 75, 2743, 1993.

Katsuyama, T., and Matsumura, H., *J. Appl. Phys.*, 76, 2036, 1994.

Kaur, A., Pesquera, A., Gonzalez, F., and Sathe, V., *J. Non-Cryst. Solids*, 356, 864, 2010.

Kim, S., and Yako, T., *J. Am. Ceram. Soc.*, 78, 1061, 1995.

Kim, S., Yako, T., and Saka, S., *J. Am. Ceram. Soc.*, 76, 2486, 1993.

Kim, S., Yako, T., and Saka, S., *J. Appl. Phys.*, 76, 865, 1993.

Kingry, W., Bowen, H., and Uhlman, D., *Introduction to Ceramics*, 2nd ed., John Wiley & Sons, 1960.

Klaska, P., Zhang, X., and Lucas, J., *J. Non-Cryst. Solids*, 161, 297, 1993.

Konishi, T., Hondo, T., Araki, T., Nishio, K., Tsuchiya, T., Matsumoto, T., Suehara, S., Todoroki, S., and Inoue, S., *J. Non-Cryst. Solids*, 324, 58, 2003.

Kozhokarov, V., Kerezhov, K., Marinov, M., Todorov, V., Neov, S., Sidzhimov, B., and Gerasimova, I., Bulgarian patent 24719, C 03CS/24, 1977.

Kozhokarov, V., Marinov, M., and Pavlova, J., *J. Mater. Sci.*, 13, 977, 1978.

Kozhokarov, V., Marinov, M., Gugov, I., Burger, H., and Vogal, W., *J. Mater. Sci.*, 18, 1557, 1983.

Kozhokarov, V., Marinov, M., Nikolov, S., Bliznakov, G., and Klissurski, D., *Z. Anorg. Allg. Chem.*, 476, 179, 1981.

Kozhokarov, V., Neov, S., Gerasimova, I., and Mikula, P., *J. Mater. Sci.*, 21, 1707, 1986.

Kozhukarov, K., Marinov, M., and Grigorova, G., *J. Non-Cryst. Solids*, 28, 429, 1978.

Kozhukharov, V., Burger, H., Neov, S., and Sidzhimov, B., *Polyhedron*, 5, 771, 1986.

Lacaus, J., Chiaruttini, I., Zhang, X., and Fonteneau, G., *Mater. Sci. Forum*, 32/33, 437, 1988.

Lakshminarayana, G., Yang, H., and Qiu, J., *J. Alloys Compd.*, 475, 569, 2009.

Lambson, E., Saunders, S., Bridge, B., and El-Mallawany, R., *J. Non-Cryst. Solids*, 69, 117, 1984.

Lanqvist, O., *Acta Chem. Scand.*, 22, 977, 1968.

Lasbrugnas, C., Thomas, P., Masson, O., Champarnaud-Mesjard, J., Fargin, E., Rodriguez, V., and Lahaye, M., *Opt. Mater.*, 31, 775, 2009.

Leciejewicz, J., *Z. Krist.*, 116, 345, 1961.

Lefterova, E., Krastev, V., Angelov, P., and Dimitiev, Y., Ed., *12th Conference on Glass & Ceramics, 1996,* Science Invest.: Sofia, 1997, 150.

Li, H., and Man, S., *Opt. Commun.*, 282, 1579, 2009.

Liao, G., Chen, Q., Xing, J., Gebavi, H., Milanese, D., Fokine, M., and Ferraris, M., *J. Non-Cryst. Solids,* 355, 447, 2009.

Lim, J., Jain, H., Toulouse, J., Marjanovic, S., Sanghera, J., Miklos, R., and Aggarwa, I., *J. Non-Cryst. Solids,* 349, 60, 2004.

Lima, S., Falco, W., Bannwart, E., Andrade, L., Oliveira, R., Moraes, J., Yukimitu, K., Araujo, E., Falcao, E., Steimacher, A., Astrath, N., Bento, A., Medina, A., and Baesso, M., *J. Non-Cryst. Solids,* 352, 3603, 2006.

Lin, A., Elizabeth, A., Bushong, J., and Toulouse, J., *Opt. Express*, 17, 16716, 2009.

Lin, H., Wang, X., Lin, L., Yang, D., Xu, T., Yu, J., and Pun, E., *J. Lumin.*, 116, 139, 2006.

Lucas, J., and Zhang, X., *Mater. Res. Bull.*, 21, 871, 1986.

Lucas, J., and Zhang, X., *J. Non-Cryst. Solids*, 125, 1, 1990.

Ma, H., Zhang, X., Lucas, J., and Moynihan, C., *J. Non-Cryst. Solids*, 140, 209, 1992.

Ma, J., Jiang, C., and Wu, J., *Opt. Eng.,* 47, 045007, 2008.

Marinov, M., Kozhokarov, V., and Dimitriev, V., *J. Mater. Sci. Lett.*, 7, 91, 1988.

Marinov, M., Kozhukarov, V., and Burger, H., *Report on the 2nd Otto-Schott Colloqium*, 17, 253, 1983.

Marjanovic, S., Toulouse, J., Jain, H., Sandmann, C., Dierolf, V., Kortan, A., Kopylov, N., and Ahrens, R., *J. Non-Cryst. Solids,* 322, 311, 2003.

Massera, J., Haldeman, A., Milanese, D., Gebavi, H., Ferraris, M., Foy, P., Hawkins, W., Ballato, J., Stolen, R., Petit, L., and Richardson, K., *Opt. Mater.*, 32, 582, 2010.

Masuda, H., Mori, A., Shikano, K., and Shimizu, M. *J. Lightwave Technol.*, 24, 504 2006.

Mirgorodsky, A., Merle-Mejean, T., and Frit, B., *J. Phys. Chem. Solids*, 61, 501, 2000.

Mizumo, Y., Ikeda, M., and Youshida, A., *J. Mater. Sci. Lett.*, 11, 1653, 1992.

Mizuno, Y., He, Z., and Hotate, K., *Opt. Commun.*, 283, 2438, 2010.

Moawad, H., Jain, H., and El-Mallawany, R., *J. Phys. Chem. Solids*, 70, 224, 2009.

Mochida, M., Takashi, K., Nakata, K., and Shibusawa, S., *J. Ceramic Assoc. Japan*, 86, 317, 1978.

Mori, A., and Masuda, H., *NTT Technical Review*, 2, 51, 2004.

Mori, A., Masuda, H., Shikano, K., and Shimizu, M., *J. Lightwave Technol.*, 2, 1300, 2003.

Moynihan, C. T., Macedo, P. B., Maklad, M. S., Mohr. R. K., and Howard, R. E., *J. Non-Cryst. Solids*, 17, 369, 1975.

Mukherjee, S., Ghosh, U., and Basu, C., *J. Mater. Sci. Lett.*, 11, 985, 1992.

Munoz-Martin, D., Villegas, M., Gonzalo, J., and Fernandez-Navarro, J., *J. Euro. Ceram. Soc.*, 29, 2903, 2009.

Murugan, G., Suzuki, T., and Ohishi, Y., *Appl. Phys. Lett.,* 86, 161109, 2005.

Nandi, P., Jose, G., Jayakrishnan, C., Debbarma, S., Chalapathi, K., Alti, K., Dharmadhikari, A., Dharmadhikari, J., and Mathur, D., *Opt. Express,* 14, 12145, 2006.

Nayaka, R., Guptaa, V., Dawarb, A., and Sreenivasa, K., *Thin Solid Films*, 445, 118, 2003.

Nazabal, V., Todoroki, S., Inoue, S., Matsumoto, T., Suehara, S., Hondo, T., Araki, T., and Cardinal, T., *J. Non-Cryst. Solids,* 326/327, 359, 2003.

Neov, S., Gerasimova, I., Kozhukarov, V., and Marinov, M., *J. Mater. Sci.*, 15, 1153, 1980.

Neov, S., Gerasimova, I., Kozhukharov, V., Mikula, H., and Lukas, C., *J. Non-Cryst. Solids,* 192/193, 53, 1995.

Neov, S., Gerasimova, I., Sidzhimov, B., Kozhukarov, V., and Mikula, P., *J. Mater. Sci.,* 23, 347, 1988.

Neov, S., Ishmaev, S., and Kozhukharov, V., *J. Non-Cryst. Solids*, 192/193, 61, 1995.

Neov, S., Kozhukharov, V., Gerasimova, I., Krezhov, K., and Sidzhimov, B., *J. Phys. Chem.* 12, 2475, 1979.

Nishii, J., Morimoto, S., Inagawa, I., Hizuka, R., Yamashita, T., and Yamagishi, T., *J. Non-Cryst. Solids*, 140, 199, 1992.

Nishii, J., Morimoto, S., Yokota, R., and Yamagishi, T., *J. Non-Cryst. Solids*, 95/96, 641, 1987.

Noguera, O., Smirnov, M., Mirgorodsky, A., Merle-Mejean, T., Thomas, P., and Champarnaud-Mesjard, J., *J. Non-Cryst. Solids,* 345/346, 734, 2004.

NTT Electronics, "Optical fiber modules," http://www.nel.co.jp/photo/fa/tfm.html, 2000.

Ohishi, Y., *Proc. SPIE,* 6469, 646908, 2007.

Ozdanova, J., Ticha, H., and Tichy, L., *Opt. Mater.*, 32, 950, 2010.

Podmaniczky, A., *Opt. Commun.*, 16, 161, 1976.

Poulain, M., *J. Non-Cryst. Solids,* 54, 1, 1983.

Prakash, G., Narayana Rao, D., and Bhatnagar, A., *Solid State Commun.*, 119, 39, 2001.

Prasad, N., Annapurna, K., Hussain, N., and Buddhudu, S., *Mater. Lett.,* 57, 2071, 2003.

Qin, B., Bai, Y., Zhou, Y., Liu, J., Xie, X., and Zheng, W., *Mater. Lett.*, 63, 1949, 2009.

Qin, G., Jose, R., and Ohishi Y., *J. Lightwave Technol.*, 25, 2727, 2007b.

Qin, G., Sotobayashi, H., Tsuchiya, M., Mori, A., and Ohishi, Y., *J. Appl. Phys.* 46, L810, 2007a.

Reisfeld, R., and Jorhensen, C., *Lasers & Excited States of Rare Earth,* Springer, Berlin, 1977.

Reitter, A. M., Sreeram, A. N., Varshneya, A. K., and Swiler, D. R., *J. Non-Cryst. Solids*, 139, 121, 1992.

Rivera, V., Chillcce, E., Rodriguez, E., Cesar, C., and Barbosa, L., *J. Non-Cryst. Solids,* 352, 363, 2006.

Rivera, V., Osorio, S., Manzani, D., Messaddeq, Y., Nunes, L., and Marega Jr., E., *Opt. Mater.,* 33, 888, 2011.

Rojo, J., Herrero, P., Sanz, J., Tanguy, B., Portier, J., and Reau, J., *J. Non-Cryst. Solids*, 146, 50, 1992.

Rojo, J., Sanz, J., Reau, J., and Tanguy, B., *J. Non-Cryst. Solids*, 116, 167, 1990.

Rolli, R., Montagna, M., Chaussedent, S., Monteil, A., Tikhomirov, V., and Ferrari, M., *Opt. Mater.*, 21, 743, 2003.

Roychoudhury, R., Batabyal, S., Paul, A., Basu, C., Mukherjee, S., and Goswami, K., *J. Appl. Phys.*, 92, 3530, 2002.

Sabadel, J., Arman, P., Lippens, P., Herreillat, C., and Philippot, E., *J. Non-Cryst. Solids*, 244, 143, 1999.

Saghera, J. S., Heo, J., and Mackenzie, J. D., *J. Non-Cryst. Solids*, 103, 155, 1988.

Sahar, M., and Noordin, N., *J. Non-Cryst. Solids*, 184, 137, 1995.

Sakida, S., Hayakawa, S., and Yoko, T., *J. Non-Cryst. Solids*, 243, 13, 1999a.

Sakida, S., Hayakawa, S., and Yoko, T., *J. Non-Cryst. Solids*, 243, 1, 1999b.

Samsonov, G., ed., *The Oxide Handbook*, IFJ/Plenum Press: New York, 1973.

Sanghera, J., Heo, J., and Mackenzi, J., *J. Non-Cryst. Solids*, 103, 155, 1988.

Savage, J. A., *J. Non-Cryst. Solids*, 47, 101, 1982.

Sekiya, T., Mochida, N., Ohtsuka, A., and Tonokawa, M., *J. Non-Cryst. Solids*, 144, 128, 1992.

Shioya, K., Komatsu, T., Kim, H., Sato, R., and Matusita, K., *J. Non-Cryst. Solids*, 189, 16, 1995.

Shivachev, B., Petrov, T., Yoneda, H., Titorenkova, R., and Mihailova, B., *Scripta Mater.*, 61, 493, 2009.

Sidebottom, D., Hruschka, M., Potter, B., and Brow, R., *J. Non-Cryst. Solids*, 222, 282, 1997.

Sidek, H., Rosmawati, S., Talib, Z., Halimah, M., and Daud, W., *Am. J. Appl. Sci.*, 6, 1489, 2009.

Sidky, M., El-Mallawany, R., Nakhala, R., and Moneim, A., *J. Non-Cryst. Solids*, 215, 75, 1997.

Simmons, C., and El-Bayoumi, O., ed., *Experimental Techniques of Glass Science*, The American Ceramic Society: Westerville, OH, 1993, 129.

Sincair, R., Wright, A., Bachra, B., Dimitriev, Y., Dimitrov, V., and Arnaudov, M., *J. Non-Cryst. Solids*, 232, 234, 38, 1998.

Stanworth, J., *Nature*, 169, 581, 1952.

Stegeman, R., Rivero, C., Richardson, K., Stegeman, G., Delfyett Jr., P., Guo, Y., Pope, A., Schulte, A., Cardinal, T., Thomas, P., and Champarnaud-Mesjard, J., *Opt. Express*, 13, 1144, 2005.

Stehlik, B., and Balak, L., *Collect. Czach. Chem. Commun.*, 14, 595, 1946.

Suehara, S., Yamamoto, K., Hishta, S., Aizawa, T., Inoue, S., and Nukui, A., *Phys. Rev.*, 51, 14919, 1995.

Suehara, S., Yamamoto, K., Hishta, S., and Nukui, A., *Phys. Rev.*, 50, 7981, 1994.

Sundaram, S., Schaffer, C., and Mazur, E., *Appl. Phys. A*, 76, 379, 2003.

Szu, S., and Chang, F., *Solid State Ionics*, 176, 2695, 2005.

Tagg, S., Youngman, R., and Zwanziger, J., *J. Phys. Chem.*, 99, 5111, 1995.

Takabe, H., Fujino, S., and Morinaga, K., *J. Am. Ceram. Soc.*, 2455, 1994.

Tanaka, K., Yoko, T., Yamada, H., and Kamiya, K., *J. Non-Cryst. Solids*, 103, 250, 1988.

Tokuda, Y., Saito, M., Takahashi, M., Yamada, K., Watanabe, W., Itoh, K., and Yoko, T., *J. Non-Cryst. Solids,* 326/327, 472, 2003.

Uchida, N., and Ohmachi, Y., *J. Appl. Phys.*, 40, 4692, 1969.

Uchino, T., and Yanko, T., *J. Non-Cryst. Solids*, 204, 243, 1996.

Udovic, M., Thomas, P., Mirgorodsky, A., Durand, O., Soulis, M., Masson, O., Merle-Mejean, T., and Champarnaud-Mesjard, J., *J. Solid State Chem.*, 179, 3252, 2006.

Upendera, G., Vardhania, C., Suresha, S., Awasthib, A., and Chandra, V., *Mater. Chem. Phys.,* 121, 335, 2010.

Villegas, M., and Navarro, J., *J. Euro. Ceram. Soc.*, 27, 2715, 2007.

Vogel, W., Burger, H., Zerge, G., Muller, B., Forkel, K., Winterstein, G., Boxberger, A., and Bomhild, H., *Silikattechnelk*, 25, 207, 1974.

Vogel, W., *Chemistry of Glass,* American Ceramics Society: Westerville, OH, 1985.

Wang, G., Zhang, J., Xu, S., Dai, S., Hu, L., and Jiang, Z., *J. Lumin.*, 109, 1, 2004b.

Wang, G., Xu, S., Dai, S., Zhang, J., and Jiang, Z., *J. Alloys Compd.*, 373, 246, 2004a.

Wang, J., Prasad, S., Kiang, K., Pattnaik, R., Toulouse, J., and Jain, H., *J. Non-Cryst. Solids,* 352, 510, 2006.

Wang, J., Vogel, E., and Snitzer, E., *Opt. Mater.*, 3, 187, 1993.

Wang, Y., Dai, S., Chen, F., Xu, T., and Nie, Q., *Mater. Chem. Phys.*, 113, 407, 2009.

Warner, A., White, D., and Bonther, V., *J. Appl. Phys.*, 43, 4489, 1972.

Warren, B., *J. Appl. Phys.*, 13, 602, 1942.

Wells, A. F., *Structure of Inorganic Chemistry*, 4th ed., Oxford University Press: Oxford, U.K., 581, 1975.

Weng, L., Hodgson, S., and Ma, J., *J. Mater. Sci. Lett.*, 18, 2037, 1999.

Weng, L., Hodgson, S., Baob, X., and Sagoe-Crentsil, K., *Mater. Sci. Eng. B*, 107, 89, 2004.

Winter, A., *J. Am. Ceram. Soc.*, 40, 54, 1957.

Xu, J., Yang, R., Chen, Q., Jiang, W., and Ye, H., *J. Non-Cryst. Solids*, 184, 302, 1995.

Xu, T., Shen, X., Nie, Q., and Gao, Y., *Optical Materials*, 28, 241, 2006.

Yakhkind, A., and Chebotarev, S., *Fiz. Khim. Stekla*, 6, 164, 1980.

Yakhkind, A., and Chebotarev, S., *Fiz. Khim. Stekla*, 6, 485, 1980.

Yasui, I., and Utsuno, F., *Amer. Ceramic Soc. Bull.*, 72, 65, 1993.

Zachariasen, W., *J. Am. Chem. Soc.*, 54, 541, 1932.

Zahar, C., and Zahar, A., *J. Non-Cryst. Solids*, 190, 251, 1995.

Zhang, X., Fonteneau, G., and Laucas, J., *J. Non-Cryst. Solids*, 104, 38, 1988.

Zhang, X., Fonteneau, G., and Laucas, J., *J. Mater. Sci. Forum*, 19/20, 67, 1987.

Zwanziger, J., Youngman, Y., and Tagg, S., *J. Non-Cryst. Solids*, 192/193, 157, 1995.

Part I

Elastic and Acoustic Relaxation Properties

Chapter 2: Elastic Moduli of Tellurite Glasses

Chapter 3: Acoustic Relaxation Properties
 of Tellurite Glasses

Chapter 4: Applications of Ultrasonics on
 Tellurite Glasses

2 Elastic Moduli of Tellurite Glasses

Elastic properties provide considerable information about the structures and interatomic potentials of solids. Chapter 2 covers the terminology of elasticity, semiempirical formulae for calculating constants of elasticity, and experimental techniques for measuring factors that affect elasticity. Glasses have only two independent constants of elasticity. The rest, including both second- and third-order constants, must be deduced by various means. The elasticity moduli of TeO_2 crystal and of pure, binary transitional, rare-earth, and multicomponent tellurite glasses are summarized, as well as the hydrostatic and uniaxial pressure dependencies of ultrasonic waves in these glasses. The elastic properties of the nonoxide Te glasses are also examined. Data on new tellurite glasses have been included.

2.1 ELASTIC PROPERTIES OF GLASS

Elastic properties differentiate solids from liquids. The application of a shearing force to a solid is met with considerable resistance. The magnitude of the resulting deformation of the solid is proportional to the amount of force applied and remains constant with t if the force is constant. The deformation instantaneously recovers upon removal of the force, as long as the magnitude of the applied force is below the fracture strength of the solid. Liquids, on the other hand, flow or deform at a constant rate over t if a constant shear force is applied, and the shear force relaxes over t if a constant strain is applied by the liquid. Thus, t must be kept in proper perspective; for example, solids recover "instantaneously," whereas liquids "deform over time."

The elastic properties of glass are important because the uses of most glass products critically depend on the solid-like behavior of the glass. At temperatures well below the glass transition range, glass can reasonably be considered a linear elastic solid, obeying Hooke's law when the applied stress magnitudes are low relative to the fracture strength. This means that, upon application of a stress (force per unit area), glass undergoes instantaneous deformation such that the ratio of the stress to the resulting strain (change in length per unit length) is a constant, which is called the "modulus of elasticity" (measured in pascals in the international system [1 Pa = 1 N/m^2], in dynes per square centimeter in the centimeter-gram-second system, and in pounds per inch in the English system), which is independent of the magnitude of the strain. An example of non-Hookean elastic behavior that is commonly observed is the stretching of rubber bands, in which the modulus of elasticity varies significantly with deformation. Non-Hookean behavior, such as nonconstancy of elasticity moduli in glass, is often observed when the applied stresses are very high; that is, perhaps two-thirds of the fracture strength. This situation occurs, for instance, when glass is loaded by an indenter for a microhardness test to measure its abrasion resistance. Because of the small contact area in such a test, the applied load translates to a very large local stress that causes yielding and plastic deformation of the glass network. Other examples of plastic behavior (such as irreversible compaction under hydrostatic compression) and viscoelastic behavior (such as delayed elasticity) also occur in glass. The magnitude of such nonlinear or nonelastic behaviors generally increases with the magnitude of the applied stresses and as temperatures approach the glass range. The second-order constants of elasticity (SOEC) are of central importance to any study of the vibrational properties of a solid because they determine the slope of the dispersion curves at the long-wavelength limit; their pressure dependencies provide information on the shift of these vibrational energies with compression. Although the vibrational spectra of crystalline solids are now understood in essence, this is not the case for amorphous materials. To fill this gap—at least in our understanding of the long-wavelength limit of glass—will require measurements of the pressure and temperature dependencies of ultrasonic wave velocities (v), which determine the SOEC and third-order constants of elasticity (TOEC) of semiconducting tellurite glasses. The TOEC are of interest because they characterize anharmonic properties; that is, the nonlinearity of the atomic displacements. Brassington et al. (1980, 1981) found that the higher-order elasticity of glasses falls into two quite distinct regimes. On one hand, the nonlinear elastic properties of silica-based glasses, as described by Bogardus (1965), Kurkjian et al. (1972), and Maynell et al. (1973); and those of BeF_2 glass, as described by Kurkjian et al. (1972), are quite different from those of most materials. In particular, the pressure derivatives of the bulk and shear moduli (K and S, respectively) are negative, whereas their temperature derivatives are positive and their TOEC are anomalously positive. On the other hand, the elastic behaviors of many glasses, including amorphous arsenic and As_2O_3 glasses (Brassington et al. 1980 and 1981, respectively), chalcogenide glasses (Thompson and Bailey 1978), and a fluorozirconate glass (Brassington et al. 1981), resemble that of crystalline solids (i.e., those that do not exhibit some form of lattice instability such as acoustic-mode softening).

In 1984, Lambson et al. measured the elastic properties of tellurite glass in its pure form. At about the same time, Lambson et al. and Hart also measured the SOEC and TOEC of tellurite glasses of the following forms: 50 mol% TeO_2–20 mol% PbO–30 mol% WO_3 and 75 mol% TeO_2–21 mol% PbO–4 mol% WO_3 (Hart 1983); 66.7 mol% TeO_2–20.7 mol% PbO–12.4 mol% Nb_2O_5 and 76.82 mol% TeO_2–13.95 mol% WO_3–9.2 mol% BaO (Lambson et al. 1985). In addition, Bridge (1987) reported the elastic properties of the TeO_2–$ZnCl_2$ system; El-Mallawany and Saunders (1987, 1988) measured the elasticity moduli of binary TeO_2–WO_3 glasses and of the binary rare-earth tellurite glasses, respectively. In 1990, El-Mallawany reported the quantitative analysis of elasticity moduli in tellurite glasses and later used these data to find the Debye temperature (θ_D, discussed below) and the phase separation in these glasses (El-Mallawany 1992a, 1992b). In 1993, El-Mallawany reported a comparison of the roles of ZnO and $ZnCl_2$ in tellurite glasses. With coworkers, El-Mallawany later reported on the elasticity moduli of binary tellurite glass of the forms TeO_2–MoO_3 and TeO_2–V_2O_5 (El-Mallawany et al. 1994a, 1994b; Sidky et al. 1997a, 1997b).

In the late 1990s, El-Mallawany and El-Moneima (1998) compared the elasticity moduli of tellurite and phosphate glasses, and El-Mallawany (1998) reviewed the research to date on the elasticity moduli of tellurite glasses. In 1999, the application of ultrasonic wave velocity to tellurite glasses was reviewed based on previous data such as the θ_D and phase separation on tellurite glasses (e.g., El-Mallawany 1999). More recently, El-Mallawany (2000a) published reports on the elasticity moduli of tricomponent TeO_2–V_2O_5–Ag_2O glasses and the radiation effects on the same glasses (El-Mallawany 2000b), as well as the elasticity moduli of TeO_2–V_2O_5–CeO_2 and TeO_2–V_2O_5–ZnO (El-Mallawany et al. 2000c). El-Mallawany (2000a) also reported a structural analysis of the elastic moduli of tellurite glasses and the relation of these moduli to compressibility.

2.1.1 TERMINOLOGY OF ELASTICITY

In discussing the stress–strain relationships in crystals, it is convenient to first consider the forces acting on a small cube (e.g., with dimensions dx by dy by dz) that forms part of a crystal under stress, as presented by Dekker (1981). The force exerted on the cube by the surrounding material can be represented by three components on each of the six faces of the cube. However, when the cube is in equilibrium, the forces on opposite faces must be equal in magnitude and must be of opposite signs. Thus, the stress condition of the cube can be described by nine "couples." Three such couples have been indicated in Figure 2.1—for example, those for which the forces are parallel to the x-axis. One of these corresponds to a compression or tensile stress σ_{xx} (force per square centimeter); the other two are the shearing stresses (τ_{xz} and τ_{xy}) which, respectively, tend to rotate the cube in the y and

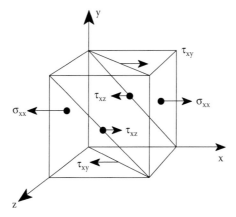

FIGURE 2.1 Illustration of the three couples of forces acting along the x direction; σ_{xx} is a tensile stress; τ_{xy} and τ_{xz} are shear stresses; τ_{xy} represents a force acting along the x-axis in a plane perpendicular to the y-axis, and so on. Similar forces act along the y- and z-axes.

z directions. Extending this reasoning to the forces parallel to the y- and z-axes, one thus finds the following stress tensor configuration:

$$
\begin{array}{ccc}
\sigma_{xx} & \tau_{xy} & \tau_{xz} \\
\tau_{yx} & \sigma_{yy} & \tau_{yx} \\
\tau_{zx} & \tau_{zy} & \sigma_{zz}
\end{array}
$$

However, the reader will be readily convinced that if rotation is absent, the tensor must be symmetrical (i.e., τ_{yx} must equal τ_{xy}). The stress condition may thus be specified by six independent stresses: σ_{xx}, σ_{yy}, σ_{zz}, τ_{xy}, τ_{xx}, and τ_{yx}.

As a result of these stresses, the crystal is strained; that is, an atom that in the unrestrained crystal occupies the position x,y,z will occupy the position x',y',z' in the strained crystal. When the distortion is homogeneous, the displacements are proportional to x,y,z, and we have, in analogy with the definition of the strain, the more general expressions:

$$
\begin{aligned}
x' - x &= \varepsilon_{xx} x + \gamma_{xy} y + \gamma_{xz} z \\
y' - y &= \gamma_{yx} x + \varepsilon_{yy} y + \gamma_{yz} z \\
z' - z &= \gamma_{zx} x + \gamma_{zy} y + \varepsilon_{zz} z
\end{aligned}
\tag{2.1}
$$

where ε and γ refer to normal and shearing strains, respectively. The strain tensor is again symmetrical if rotation is absent, and the strain condition of the cube may be specified by the six strain components ε_{xx}, ε_{yy}, ε_{zz}, γ_{yz}, γ_{zx}, and γ_{xy}. When Hooke's law is satisfied, the strain and stress components are linearly related as follows:

$$
\varepsilon = \sigma_x / E
\tag{2.2}
$$

where E is Young's modulus. Thus, in analogy with Equation 2.2, we have, for example,

$$
\sigma_{xx} = C_{11}\varepsilon_{xx} + C_{12}\varepsilon_{yy} + C_{13}\varepsilon_{zx} + C_{14}\varepsilon_{yz} + C_{15}\varepsilon_{zx} + C_{16}\varepsilon_{xy}
\tag{2.3}
$$

There are six such equations and hence 36 moduli of elasticity or elastic stiffness constants (C_{ij}). These relationships, which are the inverse of Equation 2.3, express strains in terms of stresses; for example,

$$\varepsilon_{xx} = S_{11}\sigma_{xx} + S_{12}\sigma_{yy} + S_{13}\sigma_{zx} + S_{14}\tau_{yz} + S_{15}\tau_{zx} + S_{16}\tau_{xy} \tag{2.4}$$

The six equations of this type define 36 constants (S_{ij}), which are called the "constants of elasticity." It can be shown that the matrices of C_{ij} and S_{ij} are symmetrical; hence, a material without symmetrical elements has 21 independent S_{ij} or moduli. Due to the symmetry of crystals, several of these constants may vanish. In cubic crystals, three independent elasticity moduli are usually chosen (C_{11}, C_{12}, and C_{44}). The semiempirical approach to determining the elasticity of solids is based primarily on conclusions from various experiments that have been conducted on most engineering materials. Solids are usually subjected to three types of stressing conditions in these experiments: uniaxial stress, triaxial stress, and pure shear (Figure 2.2). These tests assume that a normal stress does not produce shear strain in all directions and that a shear stress produces only one shear strain in its direction.

Up to the proportional limit, stress is directly proportional to strain (Hooke's law):

$$\sigma = E\varepsilon \tag{2.5}$$

where σ is the normal (tensile) stress and E and ε are as previously defined. Similarly, the shear stress τ is directly proportional to γ, as follows:

$$\tau = G\gamma \tag{2.6}$$

where G is the modulus of rigidity or the modulus of elasticity in shear. When a sample is extended in tension, there is an accompanying decrease in thickness; the ratio of the thickness decrease ($\nabla d/d$) to the length increase ($\nabla l/l$) is Poisson's ratio (σ):

$$\sigma = \left[\frac{(\nabla d/d)}{(\nabla l/l)} \right] \tag{2.7}$$

For plastic flow, viscous flow, and creep, the volume remains constant so that $\sigma = 0.5$. For elastic deformation, σ is found to vary between 0.2 and 0.3, with most materials having a value of

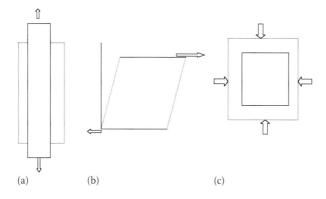

(a)   (b)   (c)

FIGURE 2.2   Illustrated definitions of constants of elasticity. (a) Young's modulus; (b) shear modulus; and (c) bulk modulus.

approximately 0.2 to 0.25. σ relates the modulus of elasticity and modulus of rigidity by the following equation:

$$\sigma = \left[\left(\frac{E}{2G} \right) - 1 \right]$$

(2.8)

This relationship is applicable only to an isotropic body in which there is one value for the Sij and that value is independent of direction. Generally, this is not the case for single crystals, but the relationship represented by these equations is a good approximation for glasses and for most polycrystalline–ceramic materials.

Under conditions of isotropic pressure, the applied pressure (P) is equivalent to a stress of $-P$ in each of the principal directions. In each principal direction, we have a relative strain

$$\varepsilon = \left(\frac{-P}{E} \right) + \sigma \left(\frac{P}{E} \right) + \sigma$$

$$\varepsilon = \left(\frac{P}{E} \right) \left(2\sigma - 1 \right)$$

(2.9)

The relative volume change is given by

$$\left(\frac{\Delta V}{V} \right) = 3\varepsilon$$

$$= \left(\frac{3P}{E} \right) \left(2\sigma - 1 \right)$$

(2.10)

K, defined as the isotropic pressure divided by the relative volume change, is given by

$$K = -\left(\frac{PV}{\Delta V} \right)$$

$$= \frac{E}{3\left(1 - 2\sigma \right)}$$

(2.11)

The stresses and strains corresponding to these relationships are illustrated in Figure 2.2.

2.1.2 SEMIEMPIRICAL FORMULAE FOR CALCULATING CONSTANTS OF ELASTICITY

Elastic properties are very informative about the structures of solids, and they are directly related to the interatomic potentials. According to Gilman (1961, 1963, and 1969), the E of ionic crystals can be approximately derived as follows: for a pair of ions of opposite sign with the spacing r_o, the electrostatic energy of attraction (U) is equal to

$$U = \left(\frac{-e^2}{r_o} \right)$$

(2.12)

To account for the many interactions between ions within a crystal, this value is multiplied by the Madelung constant (α), giving the Madelung energy:

$$U_m = \alpha U$$

(2.13)

The force between ions is $(\partial U_m / \partial r)$, and so the stress ($\sigma$) is

$$\sigma \approx \left(\frac{1}{r^2}\right)\left(\frac{\partial U_m}{\partial r}\right) \tag{2.14}$$

Then the change in σ for a change in r is $d\sigma/dr$, and therefore

$$d\sigma = \left(\frac{dr}{r^2}\right)\left(\frac{\partial^2 U_m}{\partial r^2}\right) \tag{2.15}$$

but this is just $E\,d\varepsilon$ where the strain $d\varepsilon = dr/r_o$. Thus,

$$\begin{aligned} E &= \left(\frac{d\sigma}{d\varepsilon}\right) \\ &\approx \left(\frac{1}{r_o}\right)\left(\frac{\partial U_m}{\partial r^2}\right) \\ &\approx \left(\frac{2\alpha e^2}{r^4}\right) \end{aligned} \tag{2.16}$$

Equation 2.16 shows that E of ionic crystals is inversely proportional to the fourth power of r_o, and this relationship has been confirmed for many ionic and covalent crystals, although it has not been evaluated for glasses. Equation 2.16 can now be rewritten as follows:

$$\begin{aligned} E &= \left(\frac{2\alpha}{r^3}\right)\left(\frac{e^2}{r_o}\right) \\ &= \left(\frac{2\alpha U_m}{r^3}\right) \end{aligned} \tag{2.17}$$

From Equations 2.17 and 2.13, E is twice the U_m per cubic volume r^3. The single-bond strength of oxides has been determined by Sun (1947) from the ratio of the dissociation energy, and the coordination bond number (n) and has been measured by such methods as x-ray or neutron diffraction or x-ray photoelectron spectroscopy, as explained in Chapter 1.

2.1.2.1 Mackashima–Mackenzie Model

This section summarizes the available models that are used to understand the structure of glasses. The E of glasses has been empirically studied by many authors, including, in the 1970s, Makishima and Mackenzie (1973, 1975), and Soga et al. (1976); in the 1980s, Higazy and Bridge (1985), Bridge et al. (1983), Bridge and Higazy (1986), and Bridge (1989a, 1989b); and from 1990 through 2000 (either to derive the relationship between chemical composition and E or to obtain high-E glass compositions for the manufacture of strong glass fiber), El-Mallawany (1990, 1998, 1999, 2000a), and El-Mallawany et al. (2000a, 2000b, and 2000c).

Makishima and Mackenzie (1973, 1975) derived a semiempirical formula for the theoretical calculation of E, including that of tellurite glass, based on chemical compositions of the glass, the packing density of atoms, and the bond energy (U_o) per unit volume. These authors assumed that, for a pair of ions of opposite sign with spacing r_o, $U = -e^2/r_o$, and for many interactions between ions within a crystal, this energy is multiplied by the α, thus giving the Madelung energy constant.

If the A–O U_o in one molecule of oxide (A_xO_y) is similar for the crystal and the glass forms, then—provided that the n values are the same—it is reasonable to apply the above treatment to oxide glasses. However, because of the disordered structure of glass, it is difficult to adopt a meaningful α as we can for crystalline oxides. In place of U_m per cubic volume (r^3), Makishima and Mackenzie (1973) considered that a more appropriate U_m of glass (U_m') is given by the product of the dissociation energy per unit volume (G) and the packing density of ions (V_t). For example, in a single-component glass such as silica,

$$E = 2V_tG \tag{2.18}$$

For polycomponent glasses,

$$E = 2V_t \sum_i G_i X_i \tag{2.19}$$

The V_t is defined by

$$V_t = \left(\frac{\rho}{Mw} \right) \sum_i V_i X_i \tag{2.20}$$

where Mw is the effective molecular weight, ρ is the density, X_i is the mole fraction of component I, and V_i is a packing factor obtained, for example, from the following equation for oxide A_xO_y:

$$V_i = 6.023 \times 10^{23} \left[\left(\frac{4\pi}{3} \right) \left(xR_A^3 + yR_O^3 \right) \right] \tag{2.21}$$

R_A and R_O are the respective ionic radius of metal and oxygen (Pauling's ionic radii [Pauling, 1940]). Thus, the E of glass is theoretically given by

$$E = 83.6V_t \sum_i G_i X_i \tag{2.22}$$

This expression gives E in kilobars if units of G_i are in kilocalories per cubic centimeter.

The simple division of pure vitreous oxides into two distinct categories according to their Poisson's ratios—$\sigma(P_2O_5, B_2O_3,$ or $As_2O_3) \approx 0.3$ and $\sigma(GeO_2$ or $SiO_2) \approx 0.15$—suggests the following straightforward theoretical interpretation as presented by Bridge and Higazy (1986): the three hypothetical chain networks of Figure 2.3 are identical except for their cross-link densities (defined as the number of bridging bonds per cation less two $[n_c]$) of 0, 1, and 2, respectively. Cross-link density also seems to be key to understanding the radically different melting points and melt viscosities of vitreous oxides (whose bond strengths are relatively constant).

2.1.2.2 Bulk Compression Model

The bulk compression model was first proposed by Bridge and Higazy (1986), who computed a theoretical K (K_{bc}) for a glass by using available network bond-stretching force constants (f) on the assumption that an isotropic deformation merely changes the network bond lengths (l) without changing the bond angles. From this definition of K calculated by Equation 2.11, where P is a uniform applied pressure and $\Delta V/V$ is the fractional volume change, the elastic strain per unit volume for an elastically compressed block is given by

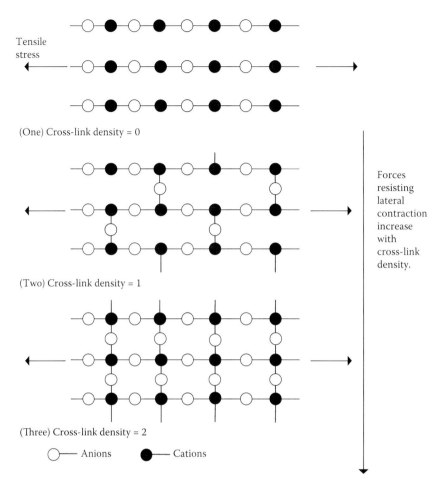

FIGURE 2.3 Variation of Poisson's ratio (lateral strain/longitudinal strain) with cross-link density for tensile stresses applied in parallel to oriented chains. (From B. Bridge, N. Patel, and D. Waters, *Phys. Stat. Solids*, 77, 655, 1983. With permission.)

$$U = \frac{1}{2}\left(\frac{P^2}{K}\right)$$

(2.23)

For any glass, uniform compression results in a shortening of every network bond by a fractional amount, given as

$$\left(\frac{\Delta r}{r}\right) = \left(\frac{\Delta V}{V}\right)$$

(2.24)

without any concomitant change in bond angles. The stored energy per bond is then simply $(1/2)$ $f(\Delta r)^2$, and, neglecting the effect of the bond–bond interaction force constant and hence according to Balta and Balta (1976),

$$\left(\frac{1}{2}\right)f(\Delta r)^2 n_b = \left(\frac{1}{2}\right)\left(\frac{P^2}{K_{bc}}\right)$$

(2.25)

where n_b is the number of network bonds per unit volume. Bridge and Higazy (1986) modeled the value of K_{bc} and n_b as

$$K_{bc} = \frac{n_b r^2 f}{9}$$ (2.26)

and

$$n_b = n n_f = \left(\frac{N_f N_A \rho}{M} \right)$$ (2.27)

where r is the bond length and f (newtons/meter) is the first-order f. Finally, n is the coordination bond number, n_f is the number of network bonds per unit formula, N_A is Avogadro's number, and ρ and M are as previously defined. The K for a polycomponent oxide glass based on this bond compression model is given by the equation

$$K_{bc} = \sum_i \left(X_i n_i l_i^2 f_i \right) \frac{\rho N_A}{9M}$$ (2.28)

where X_i is the mole fraction of the ith oxide and $f = 1.7/r^3$, as mentioned by Bridge and Higazy (1986). The bond compression model gives the calculated σ by an equation containing the average cross-link density per cation in the glass (n_c) in the form

$$\sigma = 0.28 \left(n_c' \right)^{-1/4}$$ (2.29)

where

$$n_c' = \left(\frac{1}{\eta} \right) \sum_i \left(n_c \right)_i \left(N_c \right)_i$$ (2.30)

and

$$\eta = \sum_i \left(N_c \right)_i$$

and where n'_c is the average cross-link density per unit formula, n_c equals the number of bonds less 2, N_c is the number of cations per glass formula unit, and η is the total number of cations per glass formula unit. The other calculated M (E_{th}) follows readily by combining the K_{bc} and σ for each glass system as follows:

$$G_{cal} = \left(\frac{3}{2} \right) K_{bc} \left(\frac{1 - 2\sigma_{cal}}{1 + \sigma_{cal}} \right)$$

$$L_{cal} = K_{bc} + \left(\frac{4}{3} \right) G_{cal}$$ (2.31)

$$E_{cal} = 2G \left(1 + \sigma_{cal} \right)$$

The ratio K_{bc}/K_e, where K_e is an experimental value, is a measure of the extent to which bond bending is governed by the configuration of the network bonds; that is, K_{bc}/K_e is assumed to increase with ring size (l). The bond compression model also assumes a value of K_{bc}/K_e such that 1 indicates a relatively open (i.e., large-ringed) three-dimensional network with l tending to increase with K_{bc}/K_e. Alternatively, a high value of K_{bc}/K_e could imply the existence of a layer or chain network, thus indicating that a network bond-bending process predominates when these materials are subjected to bulk compression.

2.1.2.3 Ring Deformation Model

In 1983, Bridge et al. observed that vitreous ls increases systematically with the ratio K_{bc}/K_e. The authors proposed, therefore, a role for K_{bc}/K_e as an indicator of l. An alternative approach to the question of whether the range of ls for oxide glasses can explain the orders of magnitude of observed moduli examines a material's macroscopic elastic behavior; the deformation of a loaded-beam assembly is strongly dependent on beam lengths. For example, the central depression of a uniformly loaded beam clamped at both ends and of length l is proportional to l^{-4}. Taking a ring of atoms subjected to uniformly applied pressure, Bridge et al. (1983) proposed that K will show a high-power dependence on ring diameter. Since the atomic rings are puckered (i.e., nonplanar), it is convenient to define the atomic ring size in terms of an external diameter (l), defined as the ring perimeter (i.e., n times the bond r) divided by π, as shown in Figure 2.4. Assuming that K is roughly proportional to $f_b l^{-4}$, where f_b is a bond-bending force constant, which—according to the initial postulate—is proportional to f. Bridge et al. (1983) performed a least-squares linear regression on the quantities $\ln(K_{bc}/f)$ and l, using the l values for many glasses (i.e., B_2O_3, As_2O_3, SiO_2, GeO_2, and P_2O_5), which yielded the equation (with a correlation coefficient of 93%):

$$K_{rd} = 0.0128\ f\left(l^{-3.31}\right)(\text{GPa})$$

(2.32)

where l is in nanometers, f is in newtons per square meter, and K_{rd} is K according to the ring deformation model. l is recomputed from Equation 2.32 by substituting K_e for K_{rd} to agree to be within an average of 5% with the values determined by the bulk compression model. However, the correlation coefficient improves to 99% with the equation:

$$K_e = 0.0106\ f\left(l^{-3.84}\right)\text{GPa}$$

and

$$f' = \frac{\sum_i f_i' n_i \left(N_c\right)_i}{\sum_i n_i \left(N_c\right)_i}\left(\text{N/m}^2\right)$$

(2.33)

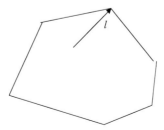

FIGURE 2.4 Ring deformation model l = (number of bonds × bond length)/2π. (From B. Bridge, N. Patel, and D. Waters, *Phys. Stat. Solids*, 77, 655, 1983. With permission).

where f' is in newtons per square meter. In conclusion, Bridge et al. (1983) established that the elastic behavior of inorganic oxide glasses can be understood in terms of their three-dimensional ring structure.

2.1.2.4 Central Force Model

Bridge and Patel (1986b) presented a model of the magnitude of the two-well barrier heights and deformation potentials that would occur based on the phenomenological theory illustrated in Figure 2.5a. To a first-order approximation, the mutual potential energy of two atoms in a diatomic molecule, for longitudinal vibrations, takes the form

$$U = \left(\frac{-a}{r}\right) + \left(\frac{b}{r^m}\right) \tag{2.34}$$

where $6 < m < 12$ and a and b are constants for a given molecular type, which can be obtained from the relationship

$$U_o = \left(\frac{a}{r_o}\right)\left[\left(1 - \frac{1}{m}\right)\right] \tag{2.35}$$

where

$$\left(dU/dr\right)_{r=r_o} = 0$$

and where $b = [(ar_o^{m-1})/m]$ U_o is the bond energy, and r_o is the equilibrium interatomic separation. Considering a linear arrangement of three atoms consisting of an anion in the middle of two cations (or vice versa) separated by a distance R, and assuming that the potential energy of the system is given by a superposition of two potentials of the form in Equation 2.34; that is,

$$U = \left(\frac{-a}{r}\right) + \left(\frac{b}{r^m}\right) + \left[\left(\frac{-a}{(R-r)}\right) + \left(\frac{b}{(R-r)^m}\right)\right] \tag{2.36}$$

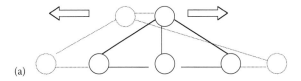

(a)

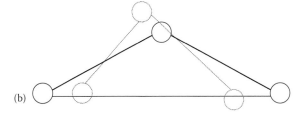

(b)

FIGURE 2.5 Anion vibration: (a) Longitudinal vibration; (b) transverse vibration. (From B. Bridge and N. Patel, *J. Mater. Sci.*, 21, 3783, 1986. With permission.)

then

$$U = \left(-a\right)\left[\left(\frac{1}{r}\right)+\left(\frac{1}{2er_o - r}\right)\right] + b\left[\frac{1}{r^m} + \left(\frac{1}{\left(2er_o - r\right)^m}\right)\right]$$ (2.37)

where $e = R/2r_o$ is the elongation factor—that is, the A–A separation divided by the equilibrium separation ($2r_o$). The quantity $U/2$ can be regarded as the mutual potential energy of half the oxygen atom plus that of one of the A– atoms. And is taken as the potential with which the O– atom moves each A– atom being considered infinitely heavy. The variable that accounts for the direct interaction between A atoms has been ignored in this relationship because it is a function of R only, and it does not affect the degree of variation of U with r. For the system P–O–P, the authors calculated the values $U_o = 6.18$ eV and $r_o = 0.156$ nm, so that with $m = 9$, they obtained values of 1.085 eV/nm for the constant a and 4.229×10^{-8} eV/nm^9 for the constant b.

For the transverse vibrations, the potential energy of a linear arrangement of three atoms A–O–A when the O– atom is transversely displaced by an amount d, as shown in Figure 2.5b, is, using previously described notation,

$$U = \left(-2a\right)\left[\frac{1}{\left(e^2 r_o^2 + d^2\right)^{1/2}}\right] + \left(2b\right)\left[\left(e^2 r_o^2 + d^2\right)^{m/2}\right]$$ (2.38)

Taking the same values of a, b, and m as in the calculation of longitudinal vibration, Bridge and Patel (1986b) showed that for $e > 1$, only a single minimum in the potential for the vibrating oxygen occurs. However, for $e < 1$, potential wells of the order of magnitude required to explain low-temperature acoustic loss occur, as is further discussed in Chapter 3. The model predicts a two-well system and associated low-temperature acoustic-loss peaks in all amorphous materials. The central force model shows that for any single two-well system A–O–A, further O– atoms to the left and right of the A– atoms will generally be situated at slightly different distances from A–. In 1989, Bridge (1989a) estimated the K of polycomponent inorganic oxide glasses and improved the model for calculating the K by using an arithmetic mean n between the cation and anion (Bridge, 1989b).

2.2 EXPERIMENTAL TECHNIQUES

2.2.1 Pulse-Echo Technique

Ultrasonic techniques have been widely used for a number of different types of investigations, as stated by Truell et al. (1969). Recently, the most commonly used method has been the pulse-echo technique, in which a short sinusoidal electrical wave activates an ultrasonic transducer cemented to the sample that is being analyzed. The transducer then introduces a soundwave train into the sample. An advantage of this method is that the sound v can be measured at the same time as the attenuation. Polarized shear waves as well as longitudinal waves can be used, and a wide range of sound frequencies can be used.

Errors in the pulse-echo method may result in the appearance of a nonexponential peak decrement and of spurious peaks on the oscilloscope screen. These errors could occur as a result of nonparallelism of the two surfaces of the sample, or due to diffraction of the soundwaves inside the specimen. Further inaccuracies may be introduced by a phase change on reflection at the transducer–specimen interface. These errors, however, depend only on wavelength and dimensions of the specimen and transducer; the results remain constant for such determinations as variations of absorption with temperature, provided that the wavelength changes are negligible. The block diagram of the pulse-echo system has been shown in the 1st Ed. The principal purpose of this equipment is to produce high frequencies, in the megahertz range. The transducers responsible for converting electric signals to ultrasonic vibrations are used simultaneously as both transmitters and receivers. Due to the great difference between voltage on the initial pulse (9 V) and that of the reflected echoes (measured in millivolts), a protection bridge is introduced to attenuate the initial pulse so that it does not load the amplifier. Therefore, the reflected echo can be measured with high accuracy (± 0.003) on the cathode ray tube.

2.2.2 COUPLINGS

One of the practical problems in ultrasonic testing is the transmission of ultrasonic energy from the source into the test material. If a transducer is placed in contact with the surface of a dry part, very little energy is transmitted through the interface into the material because of the presence of an air layer between the transducer and the sample. The air causes a great difference in acoustic impedance (impedance mismatch) at the interface.

A coupling is used between the transducer face and the test surface to ensure efficient sound transmission from the transducer to the test surface. A coupling, as the name implies, couples the transducer ultrasonically to the surface of the test specimen by smoothing out the irregularities of the test surface and by excluding all air that otherwise might be present between the transducer and the test surface. The type of coupling can vary from a large variety of liquids to semiliquids, and even to some solids that satisfy the following requirements:

1. The coupling must wet (fully contact) both the surface of the test specimen and the face of the transducer, and must exclude all air between them.
2. The coupling must be easy to apply and have a tendency to stay on the test surface but also be easy to remove.
3. The coupling must be homogeneous and free of air bubbles or solid particles for a nonsolid.
4. The coupling must be harmless to the test specimen and transducer.
5. The coupling must have an acoustic impedance value between the impedance value of the transducer and that of the test specimen, preferably approaching that of the test surface.

2.2.3 TRANSDUCERS

The key role of transducers in ultrasonic systems is well known. They are the elements by which ultrasonic waves and pulses are transmitted and detected in materials. Their principle of operation may be used on mechanical, electromagnetic, or thermal phenomenon, or on combinations of these.

2.2.3.1 Piezoelectric Transducers

The most common way of generating ultrasonic waves in crystals is to make use of their piezoelectric properties. Piezoelectricity, regarded as pressure electricity, is produced as positive and negative charges in certain crystals that are subjected to pressure forces along certain axes. Among all the piezoelectric materials, quartz has been most extensively applied. It is strong, resistive to chemical attack, and impervious to moisture.

From larger crystals, the piezoelectric crystal can be cut in various orientations to generate the types of waves that are required. The type of cut most frequently used in ultrasonic applications is the so-called "X-cut" plate, which is cut out perpendicularly with respect to the electrical axis (x-axis) of the crystal. In addition to the X-cut, ultrasonic applications make frequent use of the "Y-cut." A Y-cut plate is made so that its lateral faces are perpendicular to the mechanical axis (y-axis) of the quartz crystal, and the electrodes of such a plate are placed over these faces.

2.2.3.2   Characteristics of a Transducer

Whereas the exact relation between the input and output is known absolutely for an "ideal" transducer, this is not the case for a "good" transducer. Rather, such a device has response characteristics that are sufficient to meet the requirements of any material-testing situation. Although the basic requirements of an ultrasonic transducer are good sensitivity and resolution and a controlled radiation pattern, a more complete list by which a good transducer may be defined would include:

1. Controlled frequency response
2. High power as surface
3. High sensitivity as receiver
4. Wide dynamic range
5. Linear electromechanical acoustic response
6. Controlled geometric radiation field effects

The ultrasonic flaw detector (catalog number USM 2; Krautkramer, Germany) is the main instrument used in these measurements. This apparatus is capable of producing high-frequency pulses in the specimen, in the range of 0.5–10 MHz, and it usually operates simultaneously with the same transducer as both its transmitter and receiver as shown in the 1st Ed. Applying the ultrasonic pulses to a sample under investigation results in a train of echoes that are visualized on a cathode ray tube fitted to the unit. The received echoes are transmitted to an oscilloscope (PM 3055; Philips), which is a dual-channel, dual-delay sweep with a built-in 100 MHz crystal-controlled counter. The oscilloscope is capable of measuring t intervals with an accuracy of 0.002% of the reading.

The time interval t, which is the time elapsed between the first and second selected echoes, can be read directly on the display counter in units of microseconds. The ultrasonic wave velocity v is therefore obtained from the relation

$$v = \frac{2X}{\Delta t} \tag{2.39}$$

where X is the thickness of the glass sample and Δt is the time interval. All measurements of v are carried out at a 6 MHz frequency at room temperature (300 K). The measurements are repeated several times to check the reproducibility of the data. With an accuracy of 0.003% for Δt measurements, we find that the percentage error for measurements of v in each glass series is also 0.003%.

2.2.4   Sample Holders

A special sample holder is shown in (as shown in the 1st Ed.) for mounting the transducer. It consists of four brass rods, which provide a rigid mount for the holder body. Three copper disks are introduced through the rods to hold the transducer–sample combination together. A spring-loaded rod connected to a small plate acts as housing for the transducer and maintains the connection to high-frequency voltage. The v values of longitudinal and shear (v_l and v_s, respectively) ultrasonic waves propagated through the glasses are determined at room temperature by the pulse-echo overlap technique. Ultrasonic pulses at a frequency of 10 MHz are generated and received by X- and Y-cut quartz transducers. The following equations measure the velocities and the indicated technical elasticity moduli:

$$\text{Longitudinal modulus L: } C_{11}^s = \rho v_l^2 \tag{2.40a}$$

$$\text{S: } C_{44}^s = \rho v_s^2 \tag{2.40b}$$

$$K : K^s = \rho\left(3v_l^2 - 4v_s^2\right)\big/3 \qquad (2.40c)$$

$$E : E^s = \rho v^2 \left(3v_l^2 - 4v_s^2\right)\big/\left(v_l^2 - v_s^2\right) \qquad (2.40d)$$

$$\sigma : \mu_s = \left(v_l^2 - 2v_s^2\right)\big/2\left(v_l^2 - v_s^2\right) \qquad (2.40e)$$

In addition, $\theta_D = h/k[(3N/4\pi)^{1/3}\, v_m]$; $v_m = [(1/3)[(1/v_l^3) + (2/v_s^3)]]^{-1/3}$; and v_m is the mean v that can be calculated from the longitudinal and shear v values. The v_m for tellurite glasses are summarized in Chapter 4.

2.2.5 Measurements of Elastic Moduli under Uniaxial and Hydrostatic Pressure

The hydrostatic pressure dependencies of the ultrasonic wave transit times (t_p) have been measured in a piston-and-cylinder apparatus that uses castor oil as the liquid pressure medium. The hydrostatic pressure is found by measuring the change in electrical resistivity of a managing wire coil within the pressure chamber. To account for pressure-induced changes in crystal dimensions and ρ, the "natural velocity" $W(l_oT_p)$, where l_o is the path length at atmospheric pressure, is computed. The natural-velocity technique is used as presented by Yogurtcu et al. (1980). Experimental data for the change in T_p with pressure are compared with the change in natural velocity [$(W/W_o) - 1$]. The pressure dependencies of the relative changes in natural velocity of the longitudinal and shear waves have been found to be linear up to the maximum pressure applied (1.4 k bar). The values of the present derivatives $[d(\rho W)/dP]_{p=0}$ obtained from these data are 8.03 (± 1.5)% and 1.46 (± 1.5)% for L and S, respectively. To obtain sufficient information to determine all the TOEC, the change in one of the v values with uniaxial pressure must be measured in addition to the hydrostatic pressure data. Thus, the sample is loaded under uniaxial composition in a screw press, and the uniaxial stress is measured by a calibrated proving ring. Changes in the ultrasonic longitudinal wave v are measured using automatic frequency-controlled, gated-carrier pulse superposition apparatus as described by Thurston and Brugger (1964) and Thompson and Bailey (1978), as shown in the 1st Ed. The uniaxial pressure dependence is linear up to the maximum pressure applied (50 bar), the pressure derivative $[d(\rho W)/dP]_{p=0}$ for the longitudinal mode being $-1.05 \pm 2\%$.

2.2.6 Hardness Measurements

The temperature dependence of the Vickers hardness (H) measurement of glasses can be examined by using high-temperature-type H equipment, such as Nikon QM-2, in temperatures ranging from room temperature to around the glass transition temperature in a vacuum. The applied load is about 490 mN, and the loading time is about 15 s. The temperature of the diamond pyramid indenter should be kept at that of the sample ($\pm 1°C$). Vickers H at room temperature can be measured by using AkashiMVK-100 in air. The applied load can be 245, 490, or 980 mN, and the time required for loading is about 15 s. As shown schematically in Figure 2.6, in the Vickers indenter test, under an applied load "P," a deformation/fracture pattern is observed in which "a" and "C" (see Figure 2.6) are the characteristic indentation diagonal and crack lengths, respectively. Equation 2.41.1 evaluates Vickers H according to the relation:

$$H = \left[\frac{P}{\alpha_o a^2}\right] \qquad (2.41.1)$$

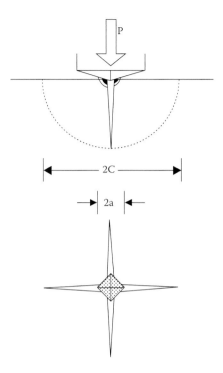

FIGURE 2.6 Deformation/fracture pattern showing medium crack in Vickers indentation test as demonstrated by Watanabe et al. (From P. Watanabe, Y. Benino, K. Ishizaki, and T. Komatsu, *J. Ceramic Soc. Japan*, 107, 1140, 1999. With permission.)

where α_o is an indenter constant (calculated as 2.15 in the experiment by Watanabe et al. [1999], when used in a diamond pyramid indenter). The fracture toughness K, which is the measure of the resistance to fracture, can be calculated from the relation:

$$K = \left[\frac{P}{\beta C^{3/2}} \right] \tag{2.41.2}$$

where β is the function of E and H. Anstis et al. (1981) proposed a relation of the form:

$$K = 0.016 \left[\frac{P}{C^{3/2}} \right] \sqrt{\frac{E}{H}} \tag{2.42.1}$$

The ratio C/a is an important parameter for measuring the brittleness of a material. The relation between C/a and H/K is as follows:

$$\frac{C}{a} = A \left[\frac{E^{1/3}}{H^{1/2}} \right] \left(\frac{H}{K} \right)^{3/2} \left(P^{1/6} \right) \tag{2.42.2}$$

The H can be calculated from the elasticity moduli and v by using the relation:

$$H = \frac{(1-2\sigma)E}{6(1+\sigma)} \tag{2.42.3}$$

where E and σ are as previously defined.

2.3 ELASTICITY MODULUS DATA OF TeO$_2$ CRYSTAL

In 1968, Arlt and Schweppe measured the elastic and piezoelectric properties of paratellurite. A large single crystal of paratellurite was grown by the Czochrlski method, with a maximum exterior dimension of 60 mm and a maximum diameter of 25 mm. The crystal belonged to point group D$_4$ and was characterized by a piezoelectric tensor consisting of only two nonvanishing components. Torsional-resonance modes, which were almost free of spurious responses, could be excited. The Sij of TeO$_2$ crystal determined by Arlt and Schweppe (1968) were:

- $C_{11} = 5.6 \pm 3 \times 10^{11}$ dyn/cm^2
- $C_{12} = 5.16 \pm 3 \times 10^{11}$ dyn/cm^2
- $C_{13} = 2.72 \pm 6 \times 10^{11}$ dyn/cm^2
- $C_{44} = 2.7 \pm 3 \times 10^{11}$ dyn/cm^2
- $C_{33} = 10.51 \pm 2 \times 10^{11}$ dyn/cm^2
- $C_{66} = 6.68 \pm 2 \times 10^{11}$ dyn/cm^2

The elastic tensor equals that of polarized ferroelectric ceramics except the constant C_{66}.

Schweppe (1970) measured the elastic and piezoelectric properties of a single crystal of paratellurite TeO$_2$. In the $\langle 110 \rangle$ direction, shear waves may propagate with the extremely low-phase velocity of 0.6×10^3 m/s. The coupling coefficient for shear waves in the $\langle 110 \rangle$ direction is on the order of 10%. Direct excitation of torsional vibrations is possible by using a very simple electrode configuration. Also, there are certain directions in which the phase velocity of shear waves is independent of temperature.

2.4 ELASTIC MODULUS DATA OF PURE TeO$_2$ GLASS

Lambson et al. (1984) measured the elasticity moduli (SOEC and TOEC) of pure tellurite glass at room temperature; these data are listed in Table 2.1 and compared with measurements for other pure oxide glasses, such as SiO$_2$, GeO$_2$, P$_2$O$_5$, As$_2$O$_3$, and B$_2$O$_3$, in Table 2.2A, B, and C. El-Mallawany and Saunders (1987) measured the compressibility of pure and binary tellurite glasses (Table 2.3). The quantitative analytical data for these tellurite glasses are presented in Tables 2.4, 2.6, and 2.7.

2.4.1 SOEC of Pure TeO$_2$ Glass

No attempt was made to assemble the SOEC data for pure TeO$_2$ glass until the comprehensive effort by Lambson et al. (1984). About the same time, Hart (1983) and Lambson et al. (1985) included much systematic property data as well. The measured v values of longitudinal and shear ultrasonic waves propagated through the TeO$_2$ glass are summarized in Table 2.1. From the ρ data, together with the velocity data, it is easy to calculate the values of the technical elastic moduli; L, S, K, and E moduli are shown in Table 2.1, as well as the SOEC and σ of TeO$_2$ glass and other main glass formers, including SiO$_2$, P$_2$O$_5$, GeO$_2$, As$_2$O$_3$, and B$_2$O$_3$. The physical significance of the SOEC of tellurite glasses is discussed in detail in Section 2.4.3, whereas that of the compressibility data is covered in Section 2.5.

2.4.2 TOEC and Vibrational Anharmonicity of Pure TeO$_2$ Glasses

To obtain sufficient information to determine the TOEC of pure tellurite glasses, the effect of a change in uniaxial pressure on one of the v values must be measured in addition to the hydrostatic

TABLE 2.1

Second Order Elastic Constants of Transition Metal Tellurite Glasses

Glass (Reference)	ρ(g/cm³)	v_l(m/s)	v_s(m/s)	E_s(GPa)	K_s(GPa)	σ^s	C_{11}(GPa)	C_{12}(GPa)	C_{44}(GPa)
TeO₂ (Lambson, Saunders, Bridge, and El-Mallawany 1984)	5.101	3,403	2,007	50.7	31.7	0.233	59.1	18.00	20.6
85 mol% TeO₂–15 mol% WO₃ (El-Mallawany and Saunders 1987)	5.250	3,532	2,031	54.3	36.6	0.253	65.5	22.18	21.7
79 mol% TeO₂–21 mol% WO₃	5.390	3,561	2,080	57.9	37.0	0.241	68.35	21.71	23.3
67 mol% TeO₂–33 mol% WO₃	5.700	3,555	2,098	61.9	38.6	0.230	72.04	25.09	25.1
90 mol% TeO₂–10 mol% ZnCl₂	5.000	3,362	1,879	44.9	32.9	0.273	56.50	21.20	17.7
80 mol% TeO₂–20 mol% ZnCl₂	4.870	3,324	1,812	41.2	32.5	0.289	53.8	21.8	15.9
67 mol% TeO₂–33 mol% ZnCl₂	4.63	3,312	1,807	44.0	30.6	0.289	50.8	20.6	15.1
80 mol% TeO₂–20 mol% MoO₃ (El-Mallawany et al. 1994)	5.01	3,272	1,870	44.0	30.28	0.257	53.6		17.5
70 mol% TeO₂–30 mol% MoO₃	4.90	3,190	1,823	41.0	28.15	0.258	49.9		16.3
55 mol% TeO₂–45 mol% MoO₃	4.75	3,147	1,798	38.7	26.56	0.257	47.0		15.4
50 mol% TeO₂–50 mol% MoO₃	4.60	3,137	1,793	37.2	25.55	0.257	45.3		14.8
90 mol% TeO₂–10 mol% ZnO (El-Mallawany 1993)	5.46	3,468					65.7		
75 mol% TeO₂–25 ZnO	5.47	3,775					77.9		
60 mol% TeO₂–40 mol% ZnO	5.50	3,819					80.2		
90 mol% TeO₂–10 mol% V₂O₅ (Sidky et al. 1997)	5.312	3,210	1,650	37.4	34.8	0.320	53.7		14.2
80 mol% TeO₂–20 mol% V₂O₅	4.900	2,810	1,330	23.6	27.2	0.360	38.7		8.7
75 mol% TeO₂–25 mol% V₂O₅	4.620	3,080	1,470	27.0	30.5	0.350	43.8		10.0
70 mol% TeO₂–30 mol% V₂O₅	4.564	3,110	1,590	30.5	28.8	0.323	44.1		11.5
65 mol% TeO₂–35 mol% V₂O₅	4.330	3,410	1,690	33.1	33.9	0.337	50.4		12.4
60 mol% TeO₂–40 mol% V₂O₅	4.225	3,620	1,790	36.2	37.3	0.338	55.4		13.5
55 mol% TeO₂–45 mol% V₂O₅	4.100	3,560	1,630	29.8	37.5	0.367	52.0		10.9
50 mol% TeO₂–50 mol% V₂O₅	3.996	3,790	1,840	36.4	39.4	0.346	57.4		13.5
50 mol% TeO₂–45 mol% V₂O₅–5 mol% Ag₂O (El-Mallawany et al. 2000a)	4.475	4,100	1,850	44.0	53.1	0.375	76		16.0
50 mol% TeO₂–40 mol% V₂O₅–10 mol% Ag₂O	4.775	3,850	1,750	39.3	52.1	0.374	71.2		14.3
50 mol% TeO₂–35 mol% V₂O₅–15 mol% Ag₂O	5.075	3,600	1,475	32.3	51.0	0.394	66.1		11.6
50 mol% TeO₂–30 mol% V₂O₅–20 mol% Ag₂O	5.375	3,350	1,375	31.3	47.0	0.388	62.1		11.3

Composition								
50 mol% TeO$_2$–27.5 mol% V$_2$O$_5$–22.5 mol% Ag$_2$O	5.450	3,300	1,350	30.1	46.3	0.392	60.1	10.8
50 mol% TeO$_2$–25 mol% V$_2$O$_5$–25 mol% Ag$_2$O	5.700	3,150	1,225	29.4	44.3	0.388	58.4	10.6
50 mol% TeO$_2$–22.5 mol% V$_2$O$_5$–27.5 mol% Ag$_2$O	5.800	3,100	1,150	25.8	38.3	0.402	56.7	9.2
50 mol% TeO$_2$–20 mol% PbO–30 mol% WO$_3$ (Hart 1983)	6.680	3,169	1,786	54.0	38.7	0.267	67.1	21.3
75 mol% TeO$_2$–21 mol% PbO–4 mol% La$_2$O$_3$	6.145	3,038	1,711	45.6	32.7	0.267	56.7	18.0
66.9 mol% TeO$_2$–12.4 mol% Nb$_2$O$_5$–20.7 mol% PbO (Lambson et al. 1985)	5.888	3,295	1,908	53.5	35.3	0.248	63.9	21.4
76.82 mol% TeO$_2$–13.95 mol% WO$_3$–9.23 mol% BaO	5.669	3,378	1,951	53.9	35.9	0.267	64.7	21.6
99 mol% TeO$_2$–1 mol% Al$_2$O$_3$ (Bridge 1987)	5.365	3,365	1,970	50.4	32.2	0.239	59.3	20.3
80.7 mol% TeO$_2$–15.8 mol% ZnCl$_2$–3.5 mol% Al$_2$O$_3$	4.870	3,093	1,757	37.9	26.5	0.262	46.6	15.0
64.7 mol% TeO$_2$–36.6 mol% ZnCl$_2$	4.370	2,872	1,569	27.7	21.7	0.287	36.1	10.8
70 mol% TeO$_2$–(30 − x) mol% V$_2$O$_5$–x mol% CeO$_2$ (El-Mallawany et al. 2000c)								
x = 3	6.090	3,460	1,429	35.7	56.0	0.395	72.9	12.7
x = 10,	6.760	3,760	1,567	46.3	73.5	0.391	95.6	16.6
70 mol% TeO$_2$–(30 − x) mol% V$_2$O$_5$–x mol% ZnO								
x = 3	5.070	4,129	1,721	42.0	66.5	0.390	86.6	15.04
x = 10,	5.260	2,641	1,100	17.7	28.2	0.395	36.7	6.36
SiO$_2$ (Bridge et al. 1983)	2.200			73.0	36.1	0.162	78.0	31.4
SiO$_2$ (cross-link = 2)								
GeO$_2$ (cross-link = 2)	3.629			43.3	23.9	0.192	48.0	18.1
P$_2$O$_5$ (cross-link = 1)	2.520			31.1	25.3	0.290	41.4	12.1
B$_2$O$_3$ (cross-link = 1)	1.834			17.4	12.1	0.260	21.3	6.9
As$_2$O$_3$ (cross-link = 1)	3.704			11.9	10.9	0.320	16.9	4.5
70 mol% TeO$_2$–20 mol% V$_2$O$_5$–10 mol% CeO$_2$ (El-Mallawany et al. 2000c)	6.760	3,760	1,567	46.3	73.5	0.391	95.6	16.6
70 mol% TeO$_2$–20 mol% V$_2$O$_5$–10 mol% ZnO (El-Mallawany et al. 2000c)	5.258	2,641	1,100	17.7	28.2	0.395	36.7	6.4

TABLE 2.2A

Third-Order Constants of Elasticity of Tellurite Glasses: Hydrostatic Pressure Derivatives $[d(\rho W_{1,s}^2)/dP]_{p=0}$; $[d(\rho W_1^2)/dP]^{uniaxial}_{p=0}$

Glass (Reference)	TOEC for Mode				
	1	2	3	4	5
TeO$_2$ (Lambson, Saunders, Bridge, and El-Mallawany 1984)	8.03	1.46	−1.05		
85 mol% TeO$_2$–15 mol% WO$_3$ (El-Mallawany and Saunders 1987)	7.07	1.48		0.42	1.20
79 mol% TeO$_2$–21 mol% WO$_3$	8.16	1.34			
67 mol% TeO$_2$–33 mol% WO$_3$	8.26	1.39	0.75	1.14	
90 mol% TeO$_2$–10 mol% ZnCl$_2$	7.85	1.41		0.48	1.23
80 mol% TeO$_2$–20 mol% ZnCl$_2$	8.27	1.50		0.29	1.29
67 mol% TeO$_2$–33 mol% ZnCl$_2$	7.44	1.38	−1.45		

pressure data. Thus, the sample is loaded under uniaxial compression in a screw press, the uniaxial stress being measured by a calibrated proving ring. Changes in v are then measured. The uniaxial pressure dependence is linear up to the maximum pressure applied (50 bar) (Lambson et al. 1984). The pressure derivative $\{d[\rho W^2(1~\&~s)]/dP\}_{p=0}$ and the derivative $\{d[\rho W^2(l)/dP]\}^{uniaxial,~P=0}$ collected for TeO$_2$ glass are summarized in Table 2.2A.

The TOEC of an isotropic material are $C_{111}=C_{222}=C_{333}$, C_{123}, $C_{144}=C_{255}=C_{366}$, C_{456}, $C_{112}=C_{223}=C_{133}=C_{122}=C_{233}$, and $C_{155}=C_{244}=C_{344}=C_{166}=C_{266}=C_{355}$. However, only three of these are independent. If these are taken as $C_{123}=u_1$, $C_{144}=u_2$, and $C_{456}=u_3$, then the others are given by the linear combinations:

$$C_{113} = u_1 + 2u_2, C_{155} = u_2 + 2u_3, \text{ and } C_{111} = u_1 + 6u_2 + 8u_3 \tag{2.43}$$

These three independent S_{ij} have been obtained from the measurements of the hydrostatic pressure derivatives $[d(\rho W_1^2)/dP]_{p=0}$ according to the method of Thurston and Brugger (1964):

$$\left[\frac{d(\rho W_1^2)}{dP}\right]_{P=0} = -1 - \left(\frac{1}{3B^T}\right)(2C_{11} + 3u_1 + 10u_2 + 8u_3) \tag{2.44}$$

$$= 8.03 \pm 1.5\%$$

$$\left[\frac{d(\rho W_s^2)}{dP}\right]_{P=0} = -1 - \left(\frac{1}{3B^T}\right)[2C_{44} + 3u_2 + 4u_2] \tag{2.45}$$

$$= 1.46 \pm 1.5\%$$

$$\left[\frac{d(\rho W_1^2)}{dP}\right]^{Uniaxial}_{P=0} = \left(\frac{1}{E^T}\right)[\sigma^T(2C_{11} + 8u_3) + u_1(2\sigma^T - 2)] \tag{2.46}$$

$$= 1.05 \pm 2\%$$

The values obtained for u_1, u_2, and u_3—and hence of C_{ijk}—are given in comparison with those of other glasses in Table 2.2B. The hydrostatic pressure derivatives of K

$$\left(\frac{\partial K}{\partial P}\right)_{P=0} = -\left(\frac{1}{9B}\right)\left[C_{111}+6C_{121}+2C_{123}\right] \qquad (2.47)$$

and S are also given in Table 2.2B, together with the Gruneisen parameters (γ_s, γ_l, and γ_{el}). Glasses fall into two quite different categories as far as the temperature and pressure dependence of their elastic properties is concerned. For TeO_2 glass, the pressure derivatives of the bulk $(\partial K/\partial P)_{p=0}$ and shear $(\partial \mu/\partial P)_{p=0}$ are positive, and the TOEC are negative. The physical principles underlying the TOEC of a material can be understood by considering the force acting between pairs of atoms vibrating longitudinally in a linear chain; that is,

$$F = -(aX)+bX^2+cX^3 \qquad (2.48)$$

where a, b, and c are all positive and X is the displacement of the interatomic separation about the equilibrium value. The first term is the harmonic one in its energy $F = -\partial U/\partial x$ and is the Hooke's law approximation. The second term is asymmetric in its potential energy; its effect is to increase the force less rapidly (than is expected on the basis of Hooke's law), as the displacement x is increased in the positive direction, but to increase the force less rapidly as x is made more negative, thus reflecting the fact that interatomic repulsive forces have a shorter range than interatomic attractive forces. The third term is symmetric with respect to x as far as the potential is concerned and causes F to increase less rapidly with x at large vibrational amplitudes, which has a pronounced influence during phonon mode softening in materials that show incipient acoustic-mode instabilities. For a steady uniaxial pressure applied to the chain causing the mean value of x to become X (the latter being negative), the pressure can be written as:

$$P = a|X|+b|X|^2 - c|X|^3 \qquad (2.49)$$

Hence, the effective elasticity modulus for a wave motion along the chain of amplitude ($\Delta x \ll X$) is

$$dM = \frac{dP}{d|X|}$$

$$= a + 2b|X|-6c|X|^2$$

and

$$\frac{dM}{dP} = \left(\frac{dM}{d|X|}\right)\left(\frac{d|X|}{dP}\right)$$

so

$$= \left(\frac{dM}{d|X|}\right)\left(\frac{1}{M}\right)$$

$$= \frac{\left(2b-6c|X|\right)}{M}$$

TABLE 2.2B
Third-Order Constants of Elasticity of Tellurite and Other Glasses

Reference	C_{111} (GPa)	C_{112} (GPa)	C_{123} (GPa)	C_{144} (GPa)	C_{155} (GPa)	C_{456} (GPa)	$(\partial K/\partial P)_{p=o}$	$(\partial\mu/\partial P)_{p=o}$	γl	γs	γel
TeO₂ (Lambson et al. 1984)	−732	−120	−186	−33	−153	−94	6.40	1.70	2.14	1.11	1.45
85 mol% TeO₂–15 mol% WO₃ (El-Mallawany and Saunders 1987)	−685	−166	−39	−63	−130	−33	5.42	1.68	1.98	1.30	1.52
79 mol% TeO₂–21 mol% WO₃	−770	−224	−130	−47	−136	−45	6.90	1.60	2.28	1.09	1.49
67 mol% TeO₂–33 mol% WO₃	−655	−166	−90	−31	−122	−46	6.74	1.60	2.22	1.02	1.42
90 mol% TeO₂–10 mol% ZnCl₂	−644	−180	−92	−46	−115	−35	6.28	1.59	2.29	1.33	1.85
80 mol% TeO₂–20 mol% ZnCl₂	−616	−129	−121	−38	−122	−59	6.52	1.67	2.50	1.50	1.80
67 mol% TeO₂–33 mol% ZnCl₂							5.94	1.54	2.24	1.38	1.67
GeO₂*							−6.30	−4.10	−2.80	−2.36	−0.88
B₂O₃*							−4.72	−2.39	−1.74	−1.50	+0.28
Fused Silica*	530	240	50	90	70	−10					−2.50
Pyrex*	400	30	260	−120	90	105					−1.58
62 mol% P₂O₅–38 mol% Fe₂O₃*	−450	−200	−160	−18	−62	−22	+4.73	−0.16	+1.10	−0.30	+0.70
BeF₂*											−1.4
As₂S₃*	−267	−78	−26	−26	−47	−11	+6.52	+1.87	+2.61	+2.49	+2.53

* Data from E. Lambson, G. Saunders, B. Bridge, and R. El-Mallawany, *J. Non-Cryst. Solids*, 69, 117, 1984.

For $X \to 0$, $|X| \to (P/a)$; hence

$$\left(\frac{dM}{dP}\right) = \left(\frac{1}{M}\right)\left[2b - \left(\frac{6cP}{a}\right)\right] \tag{2.50}$$

A positive pressure gradient for the elasticity modulus M usually arises from the third-order constant b, although at a sufficiently high pressure (P), a negative gradient can arise through the fourth-order constant (c). However, this simple approach predicts that $[(dM/dP)P \to 0]$ is always positive, whereas for some materials—in particular vitreous silica (Table 2.2C), the pressure derivatives of the elasticity moduli for $P \to 0$ are negative. For real materials, the physical interpretation of TOEC is of course far more complex because the interatomic forces are functions of angle as well as of atomic separation, and, in general, the application of uniaxial pressure involves changes of both bond angles and bond l values. In summary, when a material is subjected to pressure, collapse is resisted by the interatomic repulsive forces, which have a shorter range than the attractive forces and tend to dominate the TOEC. Thus, the TOEC are negative, and the hydrostatic pressure derivatives of the S_{ij} are positive in a material that shows the normal effect of stiffening under the influence of stress—as TeO_2 glass does. There have been reports by Brassington et al. (1981) of a good correlation between values of $[(1/\mu)(\partial\mu/\partial P)_{p=0}]$ and σ for a substantial number of glass–TeO_2–glass combinations fit into this correlation, which, for the glasses in Table 2.2C, yields a correlation coefficient of 86%. A linear regression performed on $[(1/K)(\partial K/\partial P)_{p=0}]$ against σ gives an even better correlation coefficient of 96%.

The effect of hydrostatic pressure on mode vibrational frequencies can be quantified by consideration of the Gruneisen mode γs, which expresses the volume (or strain) dependence of the normal mode frequency (ω_i)

$$\gamma_i = -\left(\frac{d \ln \omega}{d \ln V}\right) \tag{2.51}$$

For an isotropic solid there are only two components of the Gruneisen parameter for acoustic modes at the long-wavelength limit; namely, γ_l and γ_s, which refer to longitudinal and shear elastic waves, respectively. These two parameters can be obtained from the SOEC and TOEC by using the relations provided by Brugger and Fritz (1967):

$$\gamma_{L,S} = -\left(\frac{1}{6\omega_{L,S}}\right)\left[3K^T + 2\omega_{L,S} + K_{L,S}\right] \tag{2.52}$$

where $\omega_l = C_{11}$, $\omega_s = C_{44}$, and k is a well-defined wave vector $[k_L = C_{111} + 2C_{112}; k_S = 0.5(C_{111} - C_{123})]$.

TABLE 2.2C
Third-Order Constants of Elasticity and α Values of Tellurite and Other Glasses

Glass	$K^{-1}(\partial K/\partial P)_{p=o}$ $(10^{-10}$ Pa$^{-1})$	$\mu^{-1}(\partial\mu/\partial P)_{p=o}$ $(10^{-10}$ Pa$^{-1})$	α $(10^{-6}\,°C^{-1})$
TeO_2	+2.00	+0.83	+15.500
As_2S_3	+5.11	+2.92	22.400
62 mol% P_2O_5–38 mol% Fe_2O_3	+1.03	−0.06	+7.514
Pyrex	−1.23	−0.87	+3.200
Fused Silica	−1.69	−1.04	+0.450
B_2O_3	+3.88	+0.52	+13.515
GeO_2	−0.03	−0.61	+7.500
BeF_2	−1.67	−1.52	+7.500

Source: E. Lambson, G. Saunders, B. Bridge, and R. El-Mallawany, *J. Non-Cryst. Solids*, 69, 117, 1984.

It can be shown that for an isotropic solid

$$\gamma_i = -\left(\frac{K}{C_{11}}\right)\left\{3 - \left(2\frac{C_{12}}{K}\right)\left[3 - \left(\frac{dK}{dP}\right) - \left(\frac{4d\mu}{dP}\right)\right]\right\} \tag{2.53}$$

$$\gamma_s = -\left(\frac{1}{6\mu}\right)\left[2\mu - \left(2\frac{Kd\mu}{dP}\right) - \left(\frac{3}{2}\right)(K) + \left(\frac{3}{2}\right)C_{12}\right] \tag{2.54}$$

where the Lame constant (μ) is C_{44} and

$$\left(\frac{d\mu}{dP}\right) = \frac{\left(3C_{11} + 3C_{12} + C_{111} - C_{123}\right)}{6K} \tag{2.55}$$

The mean long-wavelength acoustic-mode Gruneisen parameter (γ_{el}) is given by

$$\gamma_{el} = \left(\frac{\gamma_L + 2\gamma_s}{3}\right) \tag{2.56}$$

The L and S Gruneisen parameters, which measure the shift of the long-wavelength mode frequencies with volume change, and γ_{el} of TeO_2 glass, are compared with those of other glasses in Table 2.2B. The positive signs of γ_l and γ_s found for TeO_2 glass show that the application of hydrostatic pressure causes the long-wavelength acoustic-mode frequencies to increase. This is a normal behavior in that the energies associated with those modes are raised if the glass is subjected to volumetric strain; that is, the acoustic modes stiffen. By contrast, in silica and Pyrex, both the acoustic-L and -S Gruneisen parameters are negative; the acoustic modes soften under pressure as their frequencies and energies decrease. The linear thermal expansion coefficient (α) measured at room temperature for TeO_2 is large compared with those temperatures obtained in vitreous material, which exhibit anomalous negative pressure derivatives of the S_{ij} and positive TOEC. In fact, for TeO_2 glass the α is somewhat larger than that of B_2O_3, an oxide glass known to show normal elastic behavior under pressure (Table 2.2C) and temperature. In addition to determining the elastic behavior of a material under pressure, vibrational anharmonicity is responsible for thermal expansion, although that of a glass has not been described quantitatively at a microscopic level, in the absence of sufficiently detailed knowledge of the vibrational density of states. Experimentally, an average thermal Gruneisen parameter (γ^{th}) can be evaluated from the expressions:

$$\gamma^{th} = \frac{3\alpha V K_S}{C_p}$$

$$= \left(\frac{3\alpha V K_T}{C_v}\right)$$

$$= \left(\frac{\sum_i C_i\gamma_i}{\sum_i C_i}\right) \tag{2.57}$$

where C_i is the mode contribution to specific heat ($C_v = \Sigma C_i$), and the mode gamma (γ_l) is given by Equation 2.57. Since long-wavelength acoustic modes must contribute substantially to the collective summations (Σ) over other modes, there should be a strong correlation between γ^{th} and γ_{el}, which, in principle, would enable an assessment of the relative contributions to γ^{th} of the long-wavelength acoustic phonons and the shorter-wavelength excitations, which do not have a well-defined k. However, thermodynamic data are sparse on glasses whose elastic behavior under pressure is known, so a comparison has been made between the room temperature α and γ_{el}, as shown in Figure 2.7a for the glasses listed in Table 2.2C. A systematic trend can be seen between these two properties—which are determined by vibrational anharmonicity—for this wide variety of glassy materials. At one end of the scale, vitreous silica has anomalously negative elasticity-constant pressure derivatives and hence Equations 2.51 through 2.57 are negative (γ_{el}); α is small because of the existence of an individual mode (γ_1) with both negative and positive signs, depending on the nature of the atomic displacements associated with the mode (Phillips 1981). TeO_2 glass is positioned in the range of glasses showing the more usual positive value of γ_{el} coupled with a moderate thermal expansion as shown in Figure 2.7a for the glasses listed in Tables 2.2B and 2.2C.

Equations 2.51 through 2.57 show that negative values of the pressure derivatives of the elastic modulus ($P \rightarrow 0$) for a wave motion along a linear chain and negative values of the corresponding Gruneisen mode γ cannot arise from longitudinal atomic vibrations alone. It is therefore natural to consider whether transverse components of atomic motions are responsible for TOEC anomalies. It is intuitively easy to understand that pure transverse vibrations can lead to small or even negative α and therefore, according to Equation 2.57 (which is derived solely from thermodynamical arguments), to negative-mode γ. However, as to the mechanisms by which transverse vibrational components give rise to these effects, several qualitative models have been given in the literature; for example, by Phillips (1981). The following discussion differs from these accounts in that it predicts negative-mode gammas from a consideration of pure transverse vibrations alone on central force theory, rather than from an analysis of the simultaneous effects of transverse and compressional vibrations of directional bonds. Consider a chain of atoms with equilibrium separation r_o reducing to r on application of uniaxial pressure along the chain. If an atom vibrates transversely with instantaneous displacement d from the axis, then assuming central forces the atom will move in a double-well potential having minima at positions defined by the expression $d^2 + r^2 = r_o$ and a central maximum on the chain axis as mentioned by Bridge and Patel (1986). For a sufficiently small pressure, the total energy of the atom (E) (even in the ground state) will exceed the barrier height V (eV) so that the atom may be considered (classically) to vibrate across the barrier with kinetic energy ($E - V$). As P increases from zero, V increases, and so, for a given E, the kinetic energy and the vibrational frequency (v) both decrease. Thus, dP/dv is negative, making it plausible that negative Gruneisen mode γ values can be associated with transverse components of atomic motions. The arguments above predict TOEC anomalies in crystals. In glasses the same mechanism takes effect, but it is complicated by the fact that applied pressure now perturbs the two-well systems that are already present because of the spread of r_o, which occurs in the amorphous state. It is therefore not surprising that a given material, available as both a crystal and a glass, exhibits different TOEC behaviors in each of the two forms.

In general, the vibrational anharmonicity of the acoustic modes of TeO_2 glass does not exhibit the unusual features of silica-based glasses, which can be ascribed to their low coordination and relative ease of bending vibrations as stated by Hart (1983). The anomalous decreases of K and S induced by hydrostatic pressure in silica and Pyrex are not a necessary consequence of the glassy state, but are part of systematic behavior, which is controlled by the nature of the anion, the n, and the structure. Silica-type glasses comprise fourfold coordinated cations. The negative pressure dependencies and corresponding positive temperature dependencies identified by Sato and Anderson (1980) for the elasticity moduli in these glasses stem from this open structure, with the low n based on SiO_4 tetrahedra, which allows bending vibrations of the bridging oxygen (BO) ions—corresponding to transverse motion against small-force constants. Although the n of the tellurium atoms in TeO_2 is also small (~4, as shown in Chapter 1), their elastic behavior under pressure differs entirely from that of silica-based glasses, resembling rather the chalcogenide

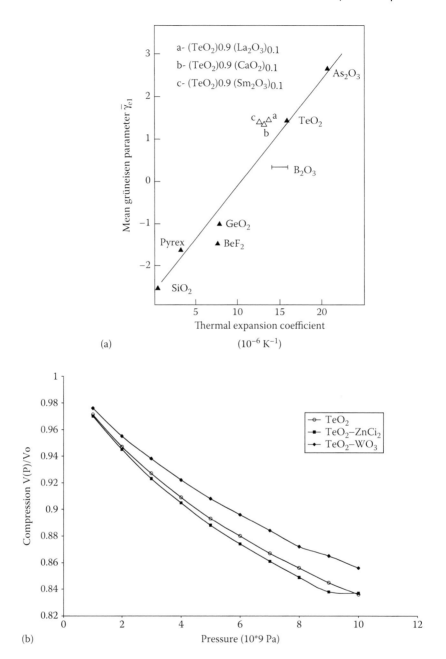

FIGURE 2.7 (a) Correlation between the mean long-wavelength, acoustic-mode Gruneisen parameter γ_{el} and the linear coefficient of thermal expansion for a number of glasses. (From El-Mallawany, *J. Mater. Res.*, 5, 2218, 1990.) (b) Compressibility of pure TeO_2 glass, binary 67 mol% TeO_2–33 mol% WO_3 glass, and 67 mol% TeO_2–33 mol% $ZnCl_2$ glass. (Data from El-Mallawany and Saunders 1987, and representation above by the author).

glasses and amorphous arsenic and arsenic trisulfide glasses (Table 2.2B). Clearly, low coordination alone is not sufficient to produce the anomalous elastic behavior. However, there are structural differences between TeO_2 and SiO_2 glasses, which account plausibly for their different elastic responses to applied pressures. The neutron diffraction studies of TeO_2 glasses by Warren (1942) and Zacharaisen (1932) showed that there are well-defined peaks at 2 Å in the radial distribution function. These closely correspond to Te–O and O–O distances in crystalline paratellurite and tellurite, and both are characterized by a fourfold coordination of the tellurium atom. The good correlation between the Te–O and O–O distances within the two tetragonal crystalline forms indicates that the basic coordination polyhedra are trigonal bipyramids (Figure 1.1), which are connected vertex-to-vertex so that the equatorial and axial bonding of oxygen is preserved in the glass. A certain freedom of rotation of these TeO_4 units is permitted in the glass. However, the anomalous elastic behavior of silica glasses under pressure arises from the motion of the BO atoms, which are free to move individually. Inspection of the structure of the TeO_2 glass shows that a similar motion of individual oxygen atoms is not permitted because they are bound rigidly in the TeO_4 trigonal bipyramid. More explicitly, this rigidity stems from the fact that each oxygen atom is bound in an axial position to one Te atom and in an equatorial position to another. Thus a transverse motion of a Te–O_{ax} bond involves a fairly direct compression of a Te–O_{eq} and vice versa; a pure transverse vibrational mode is therefore possible only if the entire structural grouping is broken up, which involves a large amount of energy compared with that available in elastic waves. It is plausible that the improbability of pure transverse vibrations could be linked with the low value of K_{bc}/K_e, which has been attributed to the relatively small amount of bond bending taking place under isotropic compression.

Hart (1983) measured the elasticity moduli of tellurite joined with other compounds in two forms: 20 wt% PbO–30 wt% WO_3 and 21 wt% PbO–4 wt% La_2O_3. He found that K and S decrease with temperature and increase with pressure in the normal way. This normal elastic behavior of these heavily modified tellurite glasses might well be construed as arising from the inhibition of bending vibrations by the modifying ions, as happens in silica-based glasses. Lambson et al. (1985) also measured the elasticity moduli of tellurite glasses of the forms 66.9 mol% TeO_2–12.4 mol% Nb_2O_5–20.7 mol% PbO and 76.82 mol% TeO_2–13.95 mol% WO_3–9.23 mol% BaO. Sato and Anderson (1980) found for a soda–lime silica glass (70.5 wt% SiO_2–11.6 wt% CaO–8.7 wt% Na_2O%–77 wt% K_2O) and for a lead–silica glass (46.0 wt% SiO_2–45.32 wt% PbO–5.62 wt% K_2O) that, while the longitudinal v increases under pressure, that of the shear wave decreases. The pressure derivatives of the K values for these two glasses are +2.6 and +2.80, respectively, and those of S are −0.235 and +0.0057, respectively. The elastic behaviors of these two silica-based glasses represent a middle position between that of silica itself and those of the glasses, such as arsenic and arsenic trisulfide, which are normal. In the silica-based glasses, the pressure of alkali metal ions in the silica interstices inhibits the transverse vibrations responsible for the anomalous elastic behavior. A similar situation occurs for the iron phosphate glass with composition 38 mol% Fe_2O_3–62 mol% P_2O_5 for which $(\partial Ks/\partial P)_{p=0}$ is strongly positive but $(\partial \mu/\partial P)_{p=0}$ is slightly negative, as stated by Murnaghan (1944). Nevertheless, the results obtained on the parent glass allow a defined statement to be made concerning TeO–PbO and TeO_2–La_2O_3; unlike silica glasses, there is no need to postulate that the modifier inhibits bending vibrations and produces normal elastic behavior, because pure TeO_2 glass itself does not show the anomalous effects found in vitreous silica.

Inaba et al. (1999) identified the E and compositional parameters of oxide glasses for silicate, borate, phosphate, and tellurite. E was measured using an ultrasonic method after its values were predicted using the empirical compositional parameters G_i and V_i, based on the Makishima–Mackenzie theory, where G_i is the dissociation energy and V_i is the packing density parameter of a single-component oxide. The relationship between the calculated E from the compositional parameters and the measured E was investigated. Experimental results indicated that the Es of phosphate and tellurite glasses could not be predicted using the compositional parameters. Thus, it was necessary to modify G_i by considering P_2O_5 and TeO_2 as glass network formers. As for the phosphate glass, it exhibited a layered structure that consisted of a P=O double bond and three chains of a P–O bond. In their report, Inaba et al. (1999) calculated the modified G_i of P_2O_5 from the assumption that the P=O double bond is a nonbridging bond and does not contribute to E. For tellurite glass, the glass structure is mainly composed of TeO_4 trigonal pyramids, and the addition of other oxides results in structural changes to the TeO_3 trigonal pyramid. However, the mechanisms of such structural changes have not yet been clarified. Therefore, the modified G_i of TeO$_2$ is calculated from the measured value using the ρ and E of pure TeO$_2$ glass. The results reveal that the calculated values from our proposed parameter are in agreement with the measured values all through the oxide glasses.

2.4.3 PHYSICAL SIGNIFICANCE OF SOEC

Bridge et al. (1983) pointed out that the pure vitreous inorganic oxides SiO$_2$, GeO$_2$, P$_2$O$_5$, As$_2$O$_3$, and B$_2$O$_3$ can be divided into two groups according to the magnitude of their cross-link density and Poisson's ratio. Thus, SiO$_2$ and GeO$_2$, with two cross-links per cation, have σ of 0.15 and 0.19, respectively; whereas for P$_2$O$_5$, As$_2$O$_3$, and B$_2$O$_3$, each having one cross-link per cation, the ratios are 0.29, 0.30, and 0.32, respectively. Lambson et al. (1984) observed that vitreous TeO$_2$ continues this pattern; with its cross-link density of 2 per cation and σ of 0.233, it belongs to the first group of oxides. It seems that bond-bending distortions do not play a significant role in the elastic behavior of this material. A useful guide to the structure of simple covalently bonded glasses and compounds containing only one type of bond can be obtained by calculating K on the assumption that isotropic compression of the structure results in uniform reductions in l without any changes in bond angles, as stated by Bridge et al. (1983) and by Bridge and Higazy (1986). The K obtained on this bond compression model is given by Equations 2.27 and 2.28. For structures that closely approximate microscopic isotropy (diamond, for example), Equations 2.27 and 2.28 give a result of the right order—typically within about 30% of the experimental value K_e. Usually, however, K_{bc} is found to be greater than K_e (typically by a factor of 3–10) and the ratio K_{bc}/K_e forms a rough measure of the degree to which bond-bending processes are involved in isotropic elastic deformation of the structure. It has been argued further that, at least for the pure inorganic oxides, K_{bc}/K_e increases systematically with atomic l (i.e., the shortest closed circuits of network bonds), provided the ratios of f to f_b for the different glasses are the same. Lambson et al. (1984) used an l value of 2.08 Å for Te–O$_{ax}$, as indicated by Wells (1975). The value of $f = 2.1 \times 10^2$ N/m was obtained by Bridge et al. (1983) and by Bridge and Higazy (1986), by extrapolating a curve of known force constant versus l for a number of simple oxides and assuming fourfold coordination for tellurium, $n_b = 7.74 \times 10^{28}$ m^{-3}. Substitution of these data into Equation 2.27 gives $K_{bc} = 7.15 \times 10^{10}$ N/m^2. Comparison of this result with $K_e = 3.17 \times 10^{10}$ N/m^2 shows that the ratio K_{bc}/K_e is about 2.3, which is considerably smaller than that obtained by Bridge et al. (1983) for all of the other pure inorganic oxide glass formers. This suggests that isotropic deformation results in much less bond-bending in the TeO$_2$ network compared with, for SiO$_2$ glass $K_{bc}/K_e = 3$ and for GeO$_2$ glass $K_{bc}/K_e = 4.4$.

It is therefore plausible that TeO$_2$ can be considered a three-dimensional network composed of (TeO)$_n$ rings, in which the average number of $2n$ cations and anions in a ring is somewhat smaller than that in silica glass, in which n averages 6. Alternatively, the result could indicate that the ratio of first-order f_b and f for the Te–O bond is substantially larger than the ratio of the Si–O and Ge–O bonds.

Lambson et al. (1984) drew the similar conclusion that the average ls in simple oxide glasses can be related to K by an empirical equation (Equation 2.33), as stated by Bridge et al. (1983) and by Bridge and Higazy (1986). Assuming f_b to be proportional to f, Equation 2.33 can be refitted to the relation stated by Bridge et al. (1983) to assume values of K, f, and l for the pure inorganic oxide glasses (SiO$_2$, GeO$_2$, As$_2$O$_3$, B$_2$O$_3$, and P$_2$O$_5$) and for diamond, by means of linear regression. Lambson et al. (1984) applied this result to TeO$_2$ and found a ring diameter of $l = 0.5$ nm, which (rounded to the nearest whole number) corresponds to the Te$_4$O$_4$ (8-atom) rings. *This result suggests that TeO$_2$ glass is a disordered version of paratellurite, which is indeed a three-dimensional network of Te$_4$O$_4$ rings.*

On the other hand, if glass is to be envisaged as a disordered, three-dimensional version of tellurite with (on average) Te$_6$O$_6$ rings, then the Te–O f values must be unusually strong (i.e., compared with these constants in other vitreous oxides), which would make the constant that was used in Equation 2.33 to obtain our *l* estimate erroneous. This conclusion is certainly consistent with the evident difficulties of producing the vitreous TeO_2 network by deformation of Te–O–Te angles.

2.5 CONSTANTS OF ELASTICITY OF BINARY AND TERNARY TRANSITION METAL TELLURITE GLASSES

Many transition metal oxides (TMO) form glasses when melted with TeO_2, as shown in Chapter 1. It is well known to every scientist in the field that tellurite glasses containing relatively large concentrations of TMO are electronic semiconductors, as shown in Chapter 6. The rigidity, and hence the response of this glass as a new noncrystalline solid, is determined by the sizes of its rings, which in turn are influenced by the chemical nature of the modifying oxide.

From 1987 to 2000, El-Mallawany (1993, 1998, 1999, 2000), and El-Mallawany et al. (1988, 1994, 2000 a,b,c), published many articles on the elastic properties of tellurite glasses. Therefore, the purpose of this section is to collect the previous experimentally measured elastic moduli with the calculated elastic moduli SOEC and TOEC (as listed in Table 2.1) for binary tellurite glass systems in the forms: TeO_2–WO_3, TeO_2–$ZnCl_2$, TeO_2–MoO_3, TeO_2–ZnO, TeO_2–V_2O_2, TeO_2–V_2O_5–Ag_2O, and the effect of Gamma radiation on the elastic moduli of the tricomponent tellurite glasses TeO_2–V_2O_5–Ag_2O glasses through the hardness values. The compressibility of tellurite glasses will also be reviewed.

2.5.1 TeO_2–WO_3 AND TeO_2–$ZnCl_2$ GLASSES

The objective of El-Mallawany and Saunders (1987) was to study the compositional dependence of the elastic properties of two selected tellurite glass systems, each with modifiers that were expected to produce quite different changes in the physical properties. One modifier that was examined, the TMO WO_3, is of interest based on measurements of its capacitance, dielectric loss, and AC conductivity, and for the effects of pressure on its dielectric molar volume (*V*), temperature, thermal expansion, and infrared spectra, as discussed in Chapters 6, 9, and 10. The effects of $ZnCl_2$ on the elastic properties of these selected systems were also examined.

For the SOEC of the binary TeO_2–WO_3 system, the ρ of the glass has been observed to increase with increasing WO_3, as shown in Table 2.2A; whereas $ZnCl_2$ has had the opposite influence in the binary TeO_2–$ZnCl_2$ system (Table 2.2A). It was anticipated that changes in the elastic properties of glass that were induced by these modifiers correlate with their different effects on ρ. Increased $ZnCl_2$ content decreases *v* through glass, and elastic stiffness is also reduced. By contrast, the addition of WO_3 stiffens glass elastically; hence, both sets of glasses exhibit the usual relationships between ρ and elastic stiffness that occur across glass systems. Measurements of *v* in these two glass systems, as reported in the 1987 work by El-Mallawany and Saunders, exactly repeat and are indistinguishable from the results with the same TeO_2 specimen that were reported earlier by Lambson et al. (1984), despite three additional years during which the specimen was exposed to the atmosphere. The agreement of these results despite the elapsed time between them is significant. The unchanged elastic properties and ρ establish that vitreous TeO_2—the very existence of which has been doubted—is stable at room temperature, and also that it could not have absorbed any significant amount of water vapor (further explained in Chapter 10).

The comparative changes in acoustic-mode anharmonicity in these two glass systems were ascertained by determining the hydrostatic and uniaxial stress dependencies of their *v* values. The hydrostatic pressure dependencies of their elasticity moduli and the TOEC of these glasses are summarized in Table 2.2C; the latter can be represented as $(\partial K/\partial P)_{p=0}$ and $(\partial \mu/\partial P)_{p=0}$ of *K* and *S*, respectively. The negative signs of the TOEC and the positive hydrostatic pressure derivatives of the SOEC show that these binary tellurite glasses, like vitreous TeO_2, behave normally when they

TABLE 2.3
Compressibility of Tellurite Glasses

Pressure ($\times 10^9$ Pa)	Compression $[V(P)/V_o]$ in Glass		
	TeO$_2$	67 mol% TeO$_2$–33 mol% ZnCl$_2$	67 mol% TeO$_2$–33 mol% WO$_3$
1	0.971	0.970	0.976
2	0.947	0.945	0.955
3	0.927	0.923	0.938
4	0.909	0.905	0.922
5	0.893	0.888	0.908
6	0.880	0.874	0.896
7	0.867	0.861	0.884
8	0.856	0.849	0.872
9	0.845	0.838	0.865
10	0.836	0.837	0.856

Source: R. El-Mallawany and G. Saunders, *Mater Sci Lett*, 6, 443, 1987.

stiffen under the influence of external stress. The increase of ZnCl$_2$ reduces the glass ρ, and stiffness extends to the pressure derivative of bulk modulus $(\partial K/\partial P)_{p=0}$. The addition of either modifier has little influence on the pressure derivative $(\partial \mu/\partial P)_{p=0}$.

Knowledge of compression $[V(P)/V_o]$ (the ratio of the volume V at a pressure P to volume at atmospheric pressure $[V_o]$) is useful in both experimental and theoretical studies now underway on the behavior of these materials when under pressure. These data are collected in Table 2.3. Compression of solids has been calculated by the Murnaghan "equation of state" (Murnaghan 1944), which is expressed in logarithmic form as:

$$\ln\left(\frac{V_o}{V(P)}\right) = \left(\frac{1}{K_o^T}\right)\ln\left(K'^T_o\left(\frac{P}{K_o^T}+1\right)\right) \quad (2.58)$$

and describes well the compression of many solids. Ultrasonic measurements give adiabatic (S) moduli, therefore it is necessary to transform the data to isothermal (T) quantities using the transformation described by Overton (1962):

$$K'^S_o = \frac{K'^S_o}{(1+\alpha\gamma T)}$$

$$K'^T_o = \left(\frac{\partial K^T}{\partial P}\right)_{P=0} \quad (2.59)$$

$$= K'^S_o + \alpha\gamma T\left(\frac{K_o^T}{K_o^S}\right)\left[1 - \frac{2}{\alpha K_o^T}\left(\frac{K_o^T}{\partial T}\right)_P - 2K'^S_o\right]$$

The volume α of TeO$_2$ is 46.5 × ∂10^{-6} °C^{-1}, with α-TeO$_2$–33 mol% WO$_3$ being about 37 × (10^{-6}/°C), as calculated by Lambson et al. (1984). The temperature dependencies of the K values for $(\partial K/\partial T)_{p=0}$

have been obtained and are equal to -7×10^7 Pa K^{-1}. By replacing the Gruneisen parameter γ with γ_{el}, El-Mallawany and Saunders (1987) found the following:

- For pure TeO_2 glass, $K_o^T = 31.1$ GPa and $K'^T_o = 6.0$
- For TeO_2–33 mol% WO_3 glass, $K_o^T = 37.8$ and $K'^T_o = 6.3$
- For TeO_2–33 mol% $ZnCl_2$ glass, $K_o^T = 3.0$ Gpa and $K'^T_o = 5.5$

These isothermal quantities have been inserted into Murnaghan's equation of state (1944) (Equation 2.59) to obtain compressions for these glasses as shown in Table 2.3 and Figure 2.7b. The reduction of stiffness achieved by modifying glass with $ZnCl_2$ has the effect of reducing $V(P)/V_o$ from that in TeO_2 itself, while WO_3 does the opposite. Insight into the anharmonicity of the long-wavelength acoustic modes can be gained by considering the Gruneisen γ values, which quantify the volume dependence $(\partial ln\omega_i/\partial lnV)$ of the normal mode frequencies (ω_i). The γ_l and γ_s Gruneisen parameters for these glasses are summarized in Table 2.2B and are given, respectively, by:

$$\gamma_L = \left(\frac{1}{6C_{11}}\right)\left(3K^T + 2C_{11} + C_{111} + 2C_{112}\right)$$

and

$$\gamma_S = \left(\frac{-1}{6C_{44}}\right)\left[3K^T + 2C_{44} + \frac{1}{2}\left(C_{111} + 2C_{123}\right)\right] \qquad (2.60)$$

The positive signs show that the application of hydrostatic pressure to tellurite glasses leads to an increase in the frequencies of the long-wavelength acoustic modes. This is normal behavior, corresponding to an increase in vibrational energy when the glasses are subjected to volumetric strain.

The elastic behavior of tellurite glasses is in accord with their structure. The basic coordination polyhedra in TeO_2 glasses are composed of trigonal bipyramids, which are joined vertex-to-vertex, as described in Chapter 1. In silica-based glasses, the n is reduced to 4. The elastic behavior of tellurite glasses under pressure indicates that this small coordination number does not, in itself, lead to anomalous behavior. In silica glasses, the anomalous elastic and thermal properties are associated with the motion of BO atoms. However, a similar bending motion is not possible in the tellurite glasses because the oxygen atoms are rigidly bound in the TeO_4 trigonal bipyramids. The binary TeO_2–WO_3 and TeO_2–$ZnCl_2$ glasses, like those of the parent material TeO_2, follow normal trends in their elastic behavior under pressure.

2.5.2 TeO_2–MoO_3 GLASSES

Since 1990, binary molybdenum–tellurite glasses of the form MoO_3–TeO_2 and of various compositions have been reported to function as semiconductors over the temperature range 100–500 K (as discussed further in Chapter 6), and results have been published by Neov et al. (1988) about tendencies toward glass formation and properties of these glasses (discussed in Chapter 1). Previously, El-Mallawany et al. (1994) prepared binary molybdenum–tellurite glasses of various compositions. The ρ's of the glasses were shown to decrease with the substitution of more MoO_3 for TeO_2, whereas the V's of these glasses had the opposite behavior, as shown in Table 2.1A. These observed variations in the molybdenum-substituted glasses are consistent with differences in the M, the molecular weight of the TeO_2 and MoO_3 oxides in the glass. The values of v (both longitudinal and shear) decrease with increasing MoO_3 concentration in these glasses. The measured elastic properties of different molybdenum–tellurite glasses are listed in Table 2.1A. Among the various elasticity

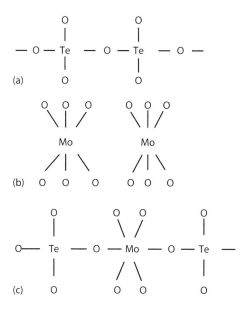

FIGURE 2.8 Schematic two-dimensional representation of (a) crystalline TeO$_2$, (b) crystalline MoO$_3$, and (c) binary TeO$_2$–MoO$_3$ glass. (Reprinted from *Mater. Chem. Phys.* 39, R. El-Mallawany, 161 [1994], with permission from Elsevier Science.)

moduli, El-Mallawany et al. (1994) first considered the composition dependence of K. With increasing MoO$_3$ concentration in the glass, K decreases from 31.7 to 25.6 GPa. The quantitative interpretation of these experimental results is that the larger molar volume of the glass unit cell results in a decrease in the average number of bonds per unit volume. When the modifying oxide has six-bond coordination for MoO$_3$ in a medium with four-bond coordination for TeO$_2$, the values of f are nearly equal as calculated using the crystal structure shown in Figure 2.8; that is, equal connectivity in binary MoO$_3$–TeO$_2$ glass has less effect on the level of K.

The previous quantitative interpretation of the experimental elastic behavior was based on the K_{rd} by Bridge et al. (1983) as given in Equation 2.32 ($K = 0.0106\, f_b/l^{3.84}$). Using this formula, El-Mallawany et al. (1994) estimated the ls for a number of modified tellurite glasses. The value of l is 5.3 Å for pure TeO$_2$ glass and increases to 5.7 Å for 50 mol% TeO$_2$–50 mol% MoO$_3$ glass. As shown in Chapter 1, the neutron diffraction results described by Neov et al. (1988) led to the conclusion that the cause of increased l is some edge sharing of MoO$_6$. Interpretation of other behavior of the other elasticity moduli led to the same conclusions. The composition dependence of σ is clear in Table 2.1: it increases from 0.233 for pure TeO$_2$ glass to 0.257 for binary glass with 50% MoO$_3$. This slight increase in σ could be attributed to the breaking of network linkages. Figure 2.9 is a schematic two-dimensional representation of binary TeO$_2$–MoO$_2$ glasses proposed by El-Mallawany et al. (1994).

2.5.3   TeO$_2$–ZnO Glasses

In this section, the longitudinal elastic moduli of binary zinc oxide–tellurite glasses are discussed, along with the effect of halogen substitution in a tellurite glass network. Zinc–tellurium glasses are very stable and can serve as superheavy optical flint glasses, as discussed in Chapter 8. Earlier analysis of elasticity moduli by El-Mallawany (1993) depended on the most crucial quantities for the moduli of the very important TeO$_2$–ZnO and TeO$_2$–ZnCl$_2$ systems. Use of both glass systems has been reported as a basis for multicomponent optical glass synthesis, and as useful media for ultra-low-loss (1 dB/1000 m) optical fibers transmitting wavelengths in the 3.5–4 µm

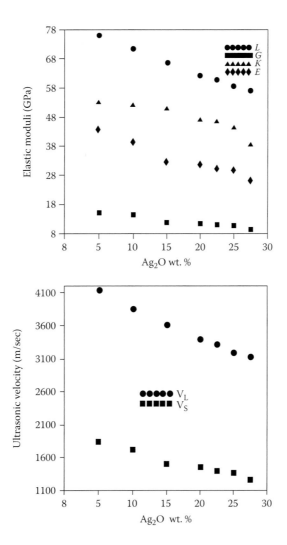

FIGURE 2.9 The measured longitudinal and shear ultrasonic wave velocities v_l and v_s and the longitudinal (L), shear (G), bulk (K), and Young's (E) elasticity moduli for 50 mol% TeO_2–(50 − x) mol% V_2O_5–x mol% Ag_2O glasses. (From R. El-Mallawany, A. Aboushely, and E. Yousef, *J. Mater. Sci. Lett.*, 19, 409, 2000. With permission.)

region, as discussed in Chapter 1. Variations of ρ and V with ZnO concentration are shown in Table 2.1. The ρ of all glasses increases with the substitution of more ZnO for TeO_2, while V decreases. From Table 2.1, it is clear that the behavior of ZnO is opposite to that of $ZnCl_2$ in tellurite glasses. Variations in longitudinal sound velocity and Sij, with variations in the composition of these prepared binary tellurite glasses, also are shown in Table 2.1. Among the structural variations that occur in these binary systems, we next consider the composition dependence of the longitudinal modulus.

First, the addition of ZnO (n = 6) results in higher network rigidity, which in turn results in an increase of the longitudinal modulus. Second, the strength of the modifying bonds and their disposition are considered. "Disposition" here means the type of ring structure formed by the bonds. Thus, bond strength and greater connectivity give rise to higher strength in the modifying bonds of the moduli. The values of *f* are *f*(Zn-O) = 219 N/m and *f*(Zn-Cl) = 185 N/m. This behavior is opposite to that of binary zinc chloride–tellurite glasses, as reported by El-Mallawany and Saunders (1987).

TABLE 2.4

n_b **and Average Cross-Link Density of Binary Zinc Oxide and Zinc Chloride Glasses, and Crystal Structure of Zinc Oxide and Zinc Chloride**

Sample Type	Reference	r (nm)	F (N/m)	n_f	Modifier (mol%)	n_b (10^{22} cm^{-3})	n_c
			Oxide				
TeO$_2$	Walls 1975	0.1990	216	4			
ZnO		9.1988	219	6			
ZnCl$_2$		0.2320	185	4			
			Glass				
TeO$_2$	Lambson et al. 1984					7.74	2.0
TeO$_2$–ZnO	El-Mallawany 1993				10.0	9.11	2.2
					25.0	10.06	2.5
					40.0	12.28	2.8
TeO$_2$–ZnCl$_2$	El-Mallawany 1993				10.0	7.65	2.0
					20.0	7.51	2.0
					33.0	7.34	2.0

These two opposite behaviors are interesting and are discussed quantitatively according to these structural changes from which the role of the halogen could be determined.

A quantitative analysis of the above experimental results was based on the bond compression model (Equations 2.27 and 2.28) proposed by Bridge et al. (1983) for the multicomponent glass of the formula $xA_{n2}O_{m1}(1 - x)\, G_{n2}O_{m2}$, which depends on the crystal structure of each oxide in the glass, as shown in Table 2.4. The parameters are the bond length l (in nanometers), the n_f, and the f, calculated as $f = 17/r^3$, as stated by Bridge and Higazy (1986). The structures of these two tellurite glass systems have been modeled based on the calculations of the number of bonds per unit volume (n_b) and the average number of n_c in each glass (Equation 2.30). The values of both quantities for each glass sample are summarized in Table 2.4.

The calculated n_b and n_c (Table 2.4) are the two main effective factors necessary to describe the modulus. TeO$_2$ glass is considered a disordered paratellurite with a mean ring diameter of eight atoms (TeO$_4$ rings). Possible structures for pure ZnCl$_2$ and TeO$_2$–ZnCl$_2$ glasses based on elasticity data have been discussed. Since our experimental data suggest the complete absence of phase separation (El-Mallawany 1992b), it is very important to consider how TeO$_2$ and ZnO can combine into a single homogeneous structure and how chlorine and oxygen affect the network. The atomic arrangement of zinc–tellurite glasses has been examined and described, both experimentally and theoretically, by using neutron diffraction patterns (see Figure 1.11) as measured by Kozhukarov et al. (1986). Studies of elasticity moduli by El-Mallawany and Saunders (1987) in monocomponent and polycomponent three-dimensional vitreous tellurite networks determined that: (1) elastic moduli increase linearly with the force constants for a constant mean ring size, and (2) that elastic moduli increase linearly with the value of cross-links for f; that is, they are inversely proportional to the mean ring diameter, which equals the mean number of atoms in a ring times the mean bond l divided by π.

In this light, there are no coordination changes to substitute Cl atoms for O atoms in TeO$_2$–ZnCl$_2$ glasses. The structural parameters in Table 2.4 confirm this, because the average n'_c is 2.0 and the n_b changes only from 7.74×10^{22} cm^{-3} for pure TeO$_2$ glass to 7.34×10^{22} cm^{-3} for TeO$_2$–ZnO glasses.

The number of bonds changes to 12.28×10^{22} cm^{-3} for 40 mol% ZnO in glass, because the average n for ZnO is 6, which in turn increases the n_c' from 2.0 for TeO glass to 2.8 for 40 mol% ZnO. The change in the force constant obtained by changing one Te–O bond into a Te–Cl bond, and one Zn–Cl bond into a Zn–O bond, is 10%. The most likely cause of the higher modulus is therefore a reduction in the size of rings. Since we have proposed rings of eight atoms for the basic TeO$_2$ glass, this implies, at the most, the introduction of rings with six atoms.

2.5.4 TeO$_2$–V$_2$O$_2$ GLASSES

The first switching phenomena in tellurite–vanadate glasses were measured by Flynn et al. (1979). The bipolar threshold and memory switching of these glasses have been measured and are summarized in Chapter 6. Tellurite–vanadate glasses have been known as semiconductors for a long time, but in recent years fast ion-conducting glasses containing silver and lithium have been the most widely studied because of their applications in solid batteries, which has enabled progress at an international level in our understanding and use of these glasses. In 1997, Sidky et al. measured the variations in v (shear and longitudinal) and density with different percentages of vanadium oxide in tellurite glasses. The change in ρ (Table 2.1) accompanying the addition of V$_2$O$_5$ was related to the changes in the atomic mass and atomic volume of the constituent elements. The atomic masses of Te and V are 127.6 and 50.94, respectively, and their respective atomic radii are 1.6 and 1.34 Å.

This explains the observed linear decrease in ρ with increasing V$_2$O$_5$. Table 2.1A presents the elasticity moduli of V$_2$O$_5$. It is clear that the oxygen ion V of tellurite–vanadate glasses decreases rapidly in the range from 0 to 20 mol% V$_2$O$_5$. Above 20 mol% V$_2$O$_5$, the oxygen ion V decreases gradually up to 50 mol% V$_2$O$_5$. On the contrary, the variation in V with increasing V$_2$O$_5$ concentration is not sensitive up to 10 mol%, but it increases linearly and rapidly with increasing V$_2$O$_5$ in the range from 10 to 50 mol%.

The variation in longitudinal and shear v values with changes in V$_2$O$_5$ concentrations is given in Table 2.1. The addition of up to ~20 mol% V$_2$O$_5$ to vitreous TeO$_2$ decreases both longitudinal and shear v values, whereas further addition of the V$_2$O$_5$ modifier above 20 mol% to 50 mol% causes a noticeable increase in v values (longitudinal and shear). Based on this compositional dependence of v, the studied glass system can be divided into two composition regions: one in which $0 < $ V$_2$O$_5 < 20$ mol% and another in which 20 mol% $< $ V$_2$O$_5 < 50$ mol%. It is worth mentioning that all values of elasticity moduli are smaller than those corresponding to the vitreous TeO$_2$. All the composition-related variations of elasticity moduli are similar to those of v values. The addition of $<$20 mol% V$_2$O$_5$ to pure TeO$_2$ decreases both v and ρ values, and this in turn decreases elasticity moduli. When V$_2$O$_5$ is increased above 20 mol%, ρ decreases linearly while v increases rapidly; therefore, all elasticity moduli values increase gradually. The variation of σ with composition for the investigated tellurite–vanadate glasses is nearly opposite to that of the elasticity moduli. σ increases in the first composition region and then decreases variably in the second composition region. σ is defined as the ratio between lateral and longitudinal strains produced when σ force is applied. For σ stresses applied parallel to the chains, the longitudinal strain produced will be the same and is unaffected by cross-link density, whereas lateral strain is greatly decreased with cross-link density (n_c). These cross-links generate a strong covalent force resisting lateral contraction, as shown previously. The values of mean n_c in pure TeO$_2$ and 50 mol% TeO$_2$–50 mol% V$_2$O$_5$ are 2.0 and 2.67, respectively, as reported by Wells (1975). Moreover, σ increases with increasing atomic l, and in the first composition region, the elasticity moduli and σ are more affected by increases in the atomic l. In the second composition region, increasing V$_2$O$_5$ content increases n_c', and the average atomic l slightly decreases with a subsequent decrease in σ.

In pure TeO$_2$ glasses, the basic coordination polyhedra are trigonal bipyramids, as mentioned before, and are joined vertex-to-vertex. The replacement of TeO$_2$ by V$_2$O$_5$ causes a change in the atomic l of the network. The average atomic l is 0.535 nm for 20 mol% V$_2$O$_5$ in the glass, whereas it is slightly less for all other mol% V$_2$O$_5$. For TeO$_2$ glasses, the average atomic l equals 0.500 nm.

In the first composition region (20 mol% V_2O_5), the three-dimensional tellurite network is partly broken by the formation of TeO_3 trigonal pyramids. This in turn leads to reduced glass rigidity, as evidenced by an increase in l and a decrease in elasticity moduli.

Moreover, for each glass, Sidky et al. (1997) computed the K_{bc} according to the bond compression model of Bridge and Higazy (1986). According to them, as the content of V_2O_5 is increased, the glass M increases and ρ decreases; that is, a value variation represented as ρ/M_g. Therefore, the n_b increases from 7.74×10^{28} m^{-3} for pure TeO_2 to 7.96×10^{28} m^{-3} for glasses containing 10 mol% V_2O_5. Above 10 mol% V_2O_5, n_b decreases steadily up to a 50 mol% V_2O_5 concentration, as shown in Table 2.5. The dependence of the calculated and experimental K on V_2O_5 content in binary V_2O_5–TeO_2 glass has been shown in the 1st Ed. It is clear from the figure that K_{bc} and K_e have the same trend up to 20 mol% V_2O_5 and, above this concentration, K_{bc} continues decreasing while K_e increases gradually. The important parameter in K_e calculation is the measured as v and ρ, whereas for theoretically calculated K, the parameters are ρ and M. The observed agreement between the behaviors of K_{bc} and K_e through the first composition region can be explained by the fact that values of both ρ and v decrease with an increase in V_2O_5 content up to 20 mol%, and that above 20 mol% V_2O_5, v increases rapidly, which causes an increase in K_e. At the same time, the continuous decrease in ρ influences calculations of K_{bc}, and its value consequently decreases. It is clear from Table 2.5 that the ratio of calculated K to experimental K_e is 2.3 for pure TeO_2 glass, and ranges between 2.7 and 1.6 for the other tellurite glasses. These values are considerably smaller than those for other pure inorganic oxide glass formers ($K_{bc}/K_e = 3.08$ for P_2O_5, 4.39 for GeO_2, and 10.1 for B_2O_3). The relation between K_{bc}/K_e and the calculated atomic l is illustrated in the 1st Ed. A value of $K_{bc}/K_e \gg 1$

TABLE 2.5

Calculated K and σ of Binary $(100 - x)$ mol% TeO_2–x mol% V_2O_5 Glass and Ternary 70 mol% TeO_2–$(30 - x)$ mol% V_2O_5–x mol% CeO_2 and 70 mol% TeO_2–$(30 - x)$ mol% V_2O_5–x mol% ZnO Glasses

Glass (Reference)	ρ/M (10^6 m³)	n_b (10^{28} m³)	K_{bc} (GPa)	K_{bc}/K_e
$(100 - x)$ mol% TeO_2–x mol% V_2O_5 (Sidky et al. 1997) for the following values of x				
00.0	0.031	7.74	71.5	2.3
10.0	0.032	7.96	76.4	2.2
20.0	0.030	7.56	73.3	2.7
25.0	0.028	7.16	69.7	2.3
30.0	0.027	7.11	69.6	2.4
35.0	0.026	6.78	66.6	2.0
40.0	0.025	6.65	65.6	1.8
45.0	0.024	6.48	64.2	1.7
50.0	0.023	6.34	63.1	1.6
70 mol% TeO_2–$(30 - x)$ mol% V_2O_5–x mol% CeO_2 (El-Mallawany et al. 2000c) for the following values of x				
03.0	0.273	10.0	98.4	1.76
10.0	0.245	18.4	104.2	1.42
70 mol% TeO_2–$(30 - x)$ mol% V_2O_5–x mol% ZnO (El-Mallawany et al. 2000c) for the following values of x				
03.0	0.322	8.06	82.5	1.24
10.0	0.297	8.8	83.2	2.95

indicates a relatively open three-dimensional network with l's tending to increase with K_{bc}/K_e. The relatively high value of $K_{bc}/K_e = 2.7$ for tellurite glasses containing 20 mol% V_2O_5 is attributed to the very open three-dimensional structure (0.535 nm) at that composition. σ can be calculated with the knowledge of n_c', according to the formula in Equation 2.30, where n_c is the number of cross-links per cation. The average cross-link ρ and the calculated σ are given in Table 2.5. It is clear that the calculated ρ decreases steadily with the increase in V_2O_5 mol% concentration, and this steady decrease is mainly caused by the mean cross-link ρ in these glasses, which increases from 2.0 for pure TeO_2 glass to 2.67 for 50% TeO_2 glass.

A variation in the microhardness of tellurite glasses of the form TeO_2–V_2O_5 has been shown in the 1st Ed. The microhardness value of pure TeO_2 glass is 3.67 GPa and decreases to 1.39 with the introduction of 50 mol% V_2O_5, as measured by Sidky et al. (1997). This change clearly shows a minimum at about 20 mol% V_2O_5 as obtained for the elasticity moduli. In the first region, the reduction of glass rigidity is evidenced by a decrease in microhardness values. In the second composition region, the glass rigidity increases because microhardness increases.

2.5.5 TeO_2–V_2O_5–Ag_2O, TeO_2–V_2O_5–CeO_2, and TeO_2–V_2O_5–ZnO Glasses

El-Mallawany et al. (2000a) prepared ternary tellurite glasses of the form 55 wt% TeO_2–[(50 – x) wt% V_2O_5]–xAg_2O, where $x = 5$, 10, 15, 20, 22.5, 25, and 27.5 wt%. Also, El-Mallawany et al. (2000c) measured the elasticity moduli of 70 wt% TeO_2–[(30–x) wt% V_2O_5]–$xCeO_2$ or $xZnO$ glasses at room temperature. Longitudinal and shear v values have been measured at room temperature. These values, plus E and both θ_D and σ, have been calculated. θ_D is discussed in Chapter 4, according to the number of vibrating atoms per unit volume in the present glass.

Measured and calculated physical properties have been found to be sensitive to the Ag_2O wt% in glass. Determination of absolute v by the pulse-echo technique is done by the same method as for determination of the v. The technique involves a very precise measurement of the time intervals (t) between two echoes and also an accurate measurement of sample thickness. The accuracy of t measurements (Δt) is within $\pm 0.002\%$ of the reading. The measured longitudinal and shear ultasonic waves v_l and v_s for the present glasses are represented in Figure 2.9. Variations in the velocities of both waves occur as follows: v_l and v_s decrease with increasing Ag_2O content; for example, v_l decreases from 64,135 to 3130 m/s, and v_s decreases from 1846 to 1261 m/s. In Table 2.1, ρ data have been used to obtain values of the longitudinal (γ_1), shear (γ_2), K, and E elasticity moduli, which are also represented in Figure 2.9. The figure shows that the ρ values of all glasses examined increased with the gradual substitution of Ag_2O for V_2O_5. The ρ of the glass sample modified with 5 wt% Ag_2O has higher value than that for 50 mol% TeO_2–(50 – x) mol% V_2O_5, as measured by Sidky et al. (1997), which was 4.0 g/cm³. The decrease in ρ caused by the addition of Ag_2O is related to the change in atomic mass and atomic volume of constituent elements. The atomic masses of Te, V, and Ag atoms are 127.6, 50.942, and 107.87, respectively, and their atomic radii are 1.6, 1.34, and 1.44 Å, respectively. Variations of the L, G, K, and E for these glasses are shown in Figure 2.9. All the composition-related variations in elasticity moduli are similar to the composition-related variation in v. An inspection of the L, G, and E elasticity moduli show that the addition of Ag_2O decreases elastic stiffness. The increase of Ag_2O in binary tellurite–vanadate glass up to 27.5% causes a decrease in the K of these glasses, as found by Sidky et al. (1997). The effect of adding Ag_2O from $x = 5$ to $x = 27.5$ wt% is explained as follows. First, the addition of x% Ag_2O such that $5 < x < 27.5$ wt% in TeO_2–V_2O_5 glass causes splitting of the glassy network, composed of O–Te–O groups and formatted nonbridging oxygen (NBO) adjacent to Ag_2O. Second, more ions begin to open up the interstices of cages within the network—thus, weakening of the glass structure or a reduction in the rigidity of the glasses for $5 < x < 27.5$ wt% results in a decrease in velocity.

A quantitative interpretation of experimental elasticity moduli is based on the n_b calculated from Equation 2.28, which is 4 for TeO_2 and 5 for V_2O_5. n_b decreases from 6.6×10^{28} m⁻³ to

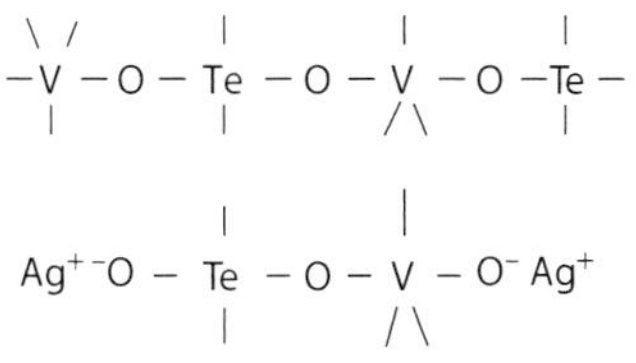

FIGURE 2.10 Two-dimensional representation of 50 mol% TeO_2–$(50 - x)$ mol% V_2O_5–x mol% Ag_2O glasses. (From R. El-Mallawany, A. Aboushely, and E. Yousef, *J. Mater. Sci. Lett.*, 19, 409, 2000. With permission.)

6.11×10^{28} m^{-3} for tellurite-vanadate glasses containing from 5 to 27.5 wt% Ag_2O. These are represented in Table 2.5. Figure 2.10 is a two-dimensional representation of 55 mol% TeO_2–$(50 - x)$ mol% V_2O_5-x mol% Ag_2O. The role of Ag_2O oxide is to break some part of the above covalent bonds (BO) related to both TeO_2 and V_2O_5. The decrease in numbers of BO is due to the creation of NBO and also explains a decrease in the thermal properties of this glass, as considered in Chapter 5. One conclusion drawn from the results obtained over the whole glass series is that, by increasing the percentage of the ionic modifier Ag_2O instead of V_2O_5 in these glasses, the values of v and the elasticity moduli are decreased due to the transformation of some BO to NBO.

The values of longitudinal and shear v, and of L and S moduli of the ternary tellurite glasses 70 mol% TeO_2–$(30 - x)$ mol% V_2O_5–x mol% A_nO_m, where A_nO_m is CeO_2 or ZnO and $x = 3, 5, 7$, or 10 mol%, have been measured by El-Mallawany et al. (2000c), as shown in Table 2.1. The L, G, K, and E elastic properties are calculated together with σ and θ_D. A qualitative analysis has been carried out after calculating the ring diameter of the glass network. Quantitatively, the estimated K and σ of the tricomponent transition metal rare-earth oxide–tellurite glasses have been calculated using the bond compression model, and according to the cation–anion bond of each oxide present in the glass, to analyze room temperature elasticity moduli data. Information about the structure of the glass can be deduced after calculating the n_b, the average f, the average l, and the n_c'.

Table 2.1 shows variation in both ρ and V for all glasses as a function of their modifiers. The calculation of V of all glasses has been carried out in agreement with the following definition: $V = M_g/\rho_g$, where M_g and ρ_g are the glass M_w and ρ, respectively. The results show that ρ_g increased from 6.09 to 6.76 g/cm^3 with increasing CeO_2 content (3–10 mol%), and that it also increased with an increase in ZnO (3–10 mol%) from 5.08 to 5.26 g/cm^3. The observed increase in ρ_g with increasing CeO_2 and ZnO contents can be attributed to the substitution of Ce and Zn atoms—which have atomic weights of 140 and 65.4, respectively—for vanadium atoms, which have an atomic weight of 50.94. ρ of the tricomponent tellurite glasses is higher than that of the binary 70 mol% TeO_2–30 mol% V_2O_5 glasses ($\rho = 4.6$ g/cm^3) according to Sidky et al. (1997). The change in V of these ternary glasses results from a change in their structure caused by the decrease of r_o inside the glass network, which results in a more compact and dense glass. The V values of these systems decrease from 27.3 to 24.3 cm^3, and from 32.2 to 29.7 cm^3, when modified with CeO_2 and ZnO, respectively, as shown in Table 2.1. From this, one can conclude that the V of the corresponding structure with its surrounding space decreases with the introduction of transition metal and rare-earth oxides.

The longitudinal and shear v values in four cerium and four zinc tellurovanadate glasses with compositions of 3, 5, 7, and 10 mol% for each modifier are given in Table 2.1. It can be seen that in the ternary tellurite–vanadate–cerium system, both longitudinal and shear v values increase from 3460 to 3760 m/s, and from 1442 to 1567 m/s, respectively, as the cerium oxide content increases and both velocities follow the same trend. The continuous increase in v with the increase in the molar percentage of CeO_2 is the result of increasing the rigidity of the glass system. Consequently, cerium atoms enter the glass network with n values of 8, which provide an octahedral form. Octahedral groups are more rigid compared with tetrahedral groups of vanadium. The increase in v with the increased molar percentage of CeO_2, in light of the above discussion, suggests that the fraction consisting of octahedral groups increases at the expense of the tetrahedral groups. On the other hand, both longitudinal and shear v decrease from 4129 to 2641 m/s, and from 1721 to 1100 m/s, respectively, as the zinc oxide content increases.

Table 2.1 summarizes the elastic properties of the different cerium or zinc tellurovanadate glasses, including values for L, G, K, E, and σ. Elasticity moduli increased with higher percentages of CeO_2 in the 70 mol% TeO_2–$(30 - x)$ mol% V_2O_5–x mol% CeO_2 glass system, while they decreased for the second tricomponent tellurite vanadate glass system in which ZnO replaced CeO_2. θ_D is explained in detail in Chapter 4.

Once K data are qualitatively interpreted, based on the relationship of K to the other moduli, G, L, and E follow automatically, according to the bond compression models by Bridge et al. (1983) and by Bridge and Higazy (1986). Table 2.5 provides the calculated values of the ring diameters for both glass series. In the first case (3–10 mol% CeO_2), K increases from 56 to 7305 GPa as the V=O bonds are replaced by V–O–Ce. This increase in K is caused by a gradual increase in the cross-linking density of the glass structure, which is attributed to all the added cerium needed to form octahedral CeO_8 groups. This increase tends to cause the network ring size to decrease from 0.444 to 0.405 nm with increasing cerium oxide content; consequently, K tends to increase. On the other hand, as ZnO enters the glass network as a third component, K decreases from 66.5 to 28.2 GPa. El-Mallawany et al. (2000c) believed that Zn^{2+} ions in this composition reside in the glass interstitial layer as network modifiers, which break the network bonds and replace them with ionic bonds while interstitially modifying Zn^{2+} ions and the oxygen atoms of the network. This alteration in structure is accompanied by a change in the l's. It seems that glasses with large atomic l's, from 0.427 to 0.530 nm, cause a decrease in K.

In ternary glasses in which both the degree of cross-linking and the relative proportions of different types of bonds can be changed with composition, it is clear that both types of mechanisms are present because (1) σ decreases with increasing cross-link density (at a constant ratio of f_b/f), and (2) σ decreases with an increasing ratio of f_b/f (at constant cross-link density).

The first mechanism was found to be most favorable for results of elasticity moduli of the tricomponent tellurite-vanadate-cerium or -zinc glasses. El-Mallawany et al. (2000c) found that the variation of σ with composition ought to be exactly the reverse of the K variations described earlier; that is, σ should decrease as the ratio of cross-link density with CeO_2 content increases. In the second group, σ must increase with decreasing cross-link density as ZnO increases (Table 2.5). It can be seen that when the CeO_2 concentration increases in the glass network, θ_D increases from 218 to 240 K, which implies an increase in the rigidity of the glass network that continues as long as V decreases and the mean v increases from 1131 to 1229 m/s. On the other hand, as ZnO increases, θ_D decreases from 246 to 156 K, thus implying a decrease in the rigidity of the glass network, accompanied by a corresponding fall in all other parameters. These values are close to the previous values for $(100 - x)$ mol% TeO_2–x mol% V_2O_5 as determined by Sidky et al. (1997).

To understand and analyze the experimental values of the elasticity moduli obtained in the tricomponent tellurite–vanadate–cerium/zinc glasses, El-Mallawany et al. (2000c) used the bond compression model devised by Bridge and Higazy (1986) to calculate the K and K_{bc} of polycomponent oxide glasses, as calculated in Equations 2.28 through 2.31. The coordination numbers, cation–anion bond l values, and average stretching force constants (f') for the TeO_2, V_2O_5, CeO_2, and ZnO oxides are summarized in Table 2.6.

Table 2.5 provides the calculated K_{bc}, n_b, and f', together with the measured K_e, of the tricomponent tellurite glasses discussed in this section, doped with CeO_2 and ZnO. The total calculated K values are higher than those of binary vanadium–tellurite glasses as determined by Sidky et al. (1997). The difference with those previous results can be explained by the increase in n_b from 1.0×10^{28} m^{-3} to 1.8×10^{28} m^{-3}, and from 8.06×10^{28} m^{-3} to 8.80×10^{28} m^{-3} when the glasses are modified with CeO_2 and ZnO, respectively, although the f' is reduced from 234 to 216 N/m, and from 239 to 233 N/m for each series. It was concluded by El-Mallawany et al. (2000c) that if

TABLE 2.6
Property Data for the TeO_2, V_2O_5, CeO_2, and ZnO Oxides

Property	TeO_2	V_2O_5	CeO_2	ZnO
n coordination number	4	5	8	6
Cation–anion l (nm)	0.199	0.180	0.241	0.215
Reference	Wells 1975a	16b	16c	16d
f' (N/m)	216	291	121	171

K_e is less than K_{bc}, this implies that, for these materials, compression proceeds via a mechanism requiring much less energy than is required for the pure compression of network bonds. When the ratio of K_{bc}/K_e is close to 1.0, it means that an interaction between neighboring bonds is neglected in the model, as in the case of CeO_2 contents. On the other hand, when the ratio of K_{bc}/K_e is >1.0, this could indicate a relatively open three-dimensional network, with l tending to increase with increasing K_{bc}/K_e. Alternatively, a high value of K_{bc}/K_e could imply the existence of a layer or chain of networks, as occurs in the ZnO series. The σ of the tricomponent tellurite-vanadate-cerium/zinc glasses has been discussed according to the cross-link density present in these multicomponent glasses (Equation 2.30). Table 2.5 also provides the values of the calculated σ for both glass series, which decreases from 0.222 to 0.219 with the increase of CeO_2 concentration in the CeO_2–V_2O_5–TeO_2 glasses. This decrease is due to an increase in the mean cross-link density from 2.53 to 2.66 with increases in CeO_2 content. On the other hand, σ increases from 0.223 to 0.227, implying decreases of the cross-link density from 2.48 to 2.32 as the ZnO concentration increases.

After calculating both K_{bc} and σ_{th}, it is easy to calculate the theoretical values for the rest of the elasticity moduli from Equation 2.31, including G_{th}, L_{th}, and E_{th}. Table 2.5 collects the calculated values of G, L, and E for the present tricomponent tellurite glasses. It is clear that the behavior of all the calculated elasticity moduli (G_{th}, L_{th}, and E_{th}) has the same trend as that of K_{bc}. El-Mallawany et al. (2000c) concluded that their v measurements of the tricomponent tellurite glass system revealed the following facts:

1. The variation in v (both longitudinal and transverse) with composition indicates an increasing or decreasing linear variation with increases in the third component, either CeO_2 or ZnO, respectively.
2. The results for the elasticity moduli L, G, K, E, θ_D, and σ indicate an opposite behavior for both series. The trend in the values of the elasticity moduli and both θ_D and σ indicate a tightening in the binding of the network with increases in either third component, where $A_nO_m = CeO_2$ or ZnO.
3. Information about the structure of these noncrystalline solids has been deduced after calculating the n_b, the values of f', the average l, and the mean cross-link density.

2.5.6 EFFECT OF GAMMA-RADIATION ON THE ELASTICITY MODULI OF TRICOMPONENT TELLURITE GLASS SYSTEM TeO_2–V_2O_5–AG_2O

El-Mallawany et al. (2000b) were the first to measure the effects of radiation on v (longitudinal and shear) in tellurite glasses. They carried out their measurements in the semiconducting ternary tellurite glasses of the form 50 wt% TeO_2–(50 − x) wt% V_2O_5–x wt% Ag_2O, where x = 5, 10, 15, 20, 25, and 27.5 wt%. This glass was irradiated with γ-rays in doses of 0.2, 0.5, 1.0, 2.0, 3.0, 4.0, and 5.0 Gy. The elasticity moduli of the glasses discussed in this section have been found to be sensitive to these doses. The interaction of energetic radiation with matter is a complex phenomenon; as γ-rays penetrate a substance, they interact with its atoms. There are two principle interactions of γ-rays with glasses: the ionization of electrons, and the direct displacement of atoms by elastic scattering. Irradiating these glasses with radiation such as γ-rays causes changes in the physical properties of the glass.

Glass specimens, prepared as described by El-Mallawany et al. (2000b), were exposed to different γ-ray dosages from a ^{60}CO γ cell as a source of gamma radiation in air. The exposure rate of 2×10^2 rad/h was applied at room temperature. Different doses were achieved by exposing the sample to the source for different periods of time. Variations in longitudinal and shear v and elasticity moduli (L, G, and E) for the present ternary tellurite glasses as a function of each dose of radiation are represented in Figure 2.11 (a through f). Each figure represents a variation of the above physical properties with the Ag_2O wt% for each (γ-dose, 0.2, 0.5, 1.0, 2.0, 3.0, 4.0, and 5.0 Gy). For all v values, it is clear that both longitudinal and shear v values are decreased by increasing the dose; that is, due to the exposure of samples to γ-rays, the number of broken bonds increases and the elasticity moduli decrease.

The early part of the work reported by El-Mallawany et al. (2000b) is a quantitative analysis of the elasticity moduli based on n_b, as determined by Equation 2.28. The value of n_b decreases from 6.6×10^{28} m^{-3} to 6.11×10^{28} m^{-3} for the present glasses containing 5–25 wt% Ag_2O. Variations in v and the elasticity moduli for all prepared glasses that are subjected to increasing γ-ray doses (e.g., from 0.2 to 5.0 Gy) can be summarized as follows:

1. When glass is subjected to ionizing radiation, electrons ionized from the valence band move throughout the glass matrix and are either trapped by preexisting flaws to form defect centers in the glass structure or they recombine with positively charged holes.
2. If NBO atoms in the glass lose electrons, the interstitial cations change their positions in the matrix, and, consequently, the NBOs trap a hole, giving rise to color centers.

The effects produced in glass by irradiation can be represented by the general expression: defect in glass + h v → positive hole + e^-.

When silver tellurite glasses are irradiated by γ-rays, the silver ion Ag^+ acts as a trap for the electrons and positive holes, leading to the formation of Ag^o (4 d^{10} 5 s^1) and Ag^{2+} (4 d^9), respectively; that is, $Ag^+ + e^- \rightarrow Ag^o$ and $Ag^+ + e^+ \rightarrow Ag^{2+}$ γ-rays change the structure of these glasses and produce free silver atoms or form charged double-silver ions, and the solid medium becomes less elastic. Evidence of decreased v and room temperature elasticity moduli due to irradiation by γ-rays of the tricomponent silver–vanadate–tellurite glasses reported in this volume shows the following:

1. Elastic moduli decrease with increasing γ-ray dosage.
2. γ-ray irradiation of silver ion exchange glasses accelerates decreases in elastic moduli.

2.6 COMPARISON BETWEEN MOO$_3$ AND V$_2$O$_5$ IN TELLURITE AND PHOSPHATE GLASSES AND K–V RELATIONS

In 1998, El-Mallawany and El-Moneim reported the results of their comparison of the elasticity moduli of binary tellurite and binary phosphate glasses containing MoO_3 and V_2O_5, which they analyzed according to the bond compression model of Bridge et al. (1983). Table 2.7 gives the experimental values of the elasticity moduli of both glasses, together with the complete set of parameters adopted from the crystal structure described by Wells (1975), which is needed to calculate both K (K_{bc}) and $σ_{bc}$ by the bond compression model. Parameters are the bond l values between cations and anions, the value of the f of every type of bond present (calculated according to the empirical relation $f = 17/r^3$ [Bridge et al. 1983]), and the n for every cation. The K_e values for pure TeO_2 and P_2O_5 glasses were 31.7 GPa, as given by Lambson et al. (1984), and 25.3 GPa, from Bridge et al. (1983),

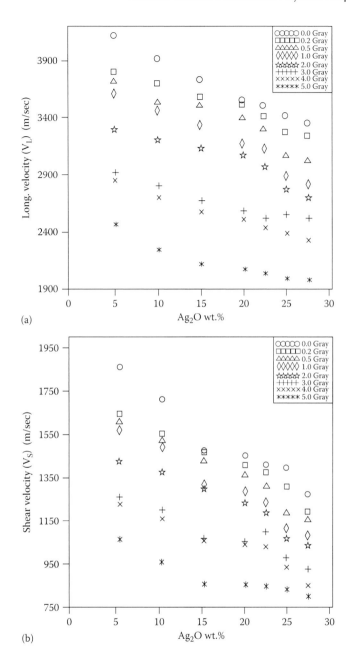

FIGURE 2.11 Variations of (a) longitudinal and (b) shear v values and elasticity moduli ([c] longitudinal, [d] shear, [e] Young's, and [f] bulk) for the ternary tellurite glasses 50 mol% TeO$_2$–(50 − x) mol% V$_2$O$_5$–x mol% Ag$_2$O as a function of γ-ray dose. (From R. El-Mallawany, A. Aboushely, A. Rahamani, and E. Yousef, *J. Mater. Sci. Lett.*, 19, 413, 2000. With permission.)

respectively; when modified with MoO$_3$ or V$_2$O$_5$, these K values increase or decrease, depending on the modifier. Based on the parameters for calculating K included in Table 2.7, it seems clear that the TeO$_2$ and P$_2$O$_5$ n values are both 4, but also that P$_2$O$_5$ has one major difference—namely, that one bond in phosphate glass is a double bond. Vitreous P$_2$O$_5$ is built up of an infinite three-dimensional network of PO$_4$ tetrahedrals, with each tetrahedron joined at three corners to other tetrahedrals and with the remaining oxygen atom linked to the phosphorus atom by a double bond with a cross-link,

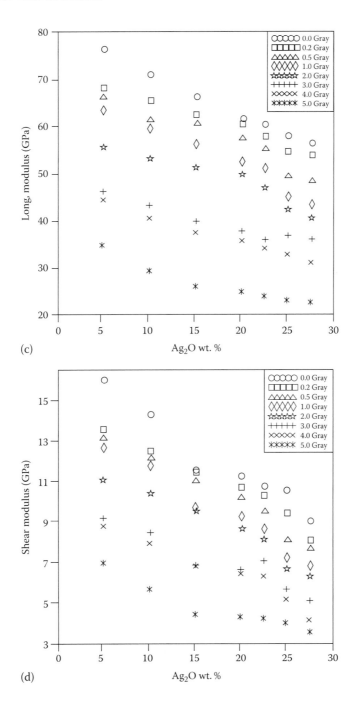

FIGURE 2.11   (Continued)

as mentioned by Bridge et al. (1983). In pure TeO_2, the network has an n of 4 and a cross-link number of 2, as mentioned by Lambson et al. (1984). The value of n_b in pure TeO_2 is 7.74×10^{28} m^{-3}, whereas it is 6.42×10^{28} m^{-3} in pure P_2O_5 glass. The f' of the bonds Te–O and P–O are 216 and 450 N/m, respectively.

From Table 2.7, the calculated values of K for both end-member glasses are as follows: K_{bc} of $TeO_2 = 73.25$ GPa, and K_{bc} of $P_2O_5 = 78$ GPa. From these values, the ratio K_{bc}/K_e can be calculated

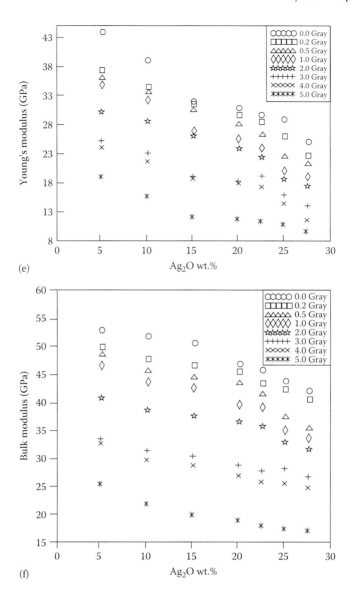

FIGURE 2.11 (Continued)

for each glass (2.3 for TeO_2 glass and 3.08 for P_2O_5 glass). Lambson et al. (1984) argued that higher values of the ratio K_{bc}/K_e (e.g., from 3 to 10, indicate that network bond bending or nonnetwork bond compression processes predominate when these materials are subjected to bulk compression. Figure 2.15a represents changes in the ratio K_{bc}/K_e of pure TeO_2 and P_2O_5 that occur when both tellurite and phosphate glasses are separately doped with either MoO_3 or V_2O_5. Both binary tellurite and phosphate glasses become more closely linked as a result of modifications with MoO_3 ($n = 6$) or V_2O_5 ($n = 5$), as indicated by Wells (1975). This in turn increases both the n_b and the average n_c (n_c') of the glass. Consequently, the calculated K_{bc} of the present tellurite and phosphate glasses increases. The experimental elasticity moduli of binary TeO_2–MoO_3 glasses decrease for higher MoO_3, whereas they increase for higher V_2O_5. In binary phosphate glasses, with higher MoO_3, K increases, whereas it decreases with higher V_2O_5. From Table 2.7, the K_{bc} for every glass series is directly proportional to n_b and the value of f', except for TeO_2–V_2O_2 and P_2O_5–MoO_3 glasses. In

TABLE 2.7

Effects of Various Modifier Percentages on Calculated and Experimental Ks of TeO$_2$– V$_2$O$_5$ and TeO$_2$–MoO$_3$ Transition Metal Tellurite Glasses

Glass (mol% TeO$_2$–mol% Modifier)	K_e (GPa)	n_b (10^{28} m^{-3})	K_{bc}	K_{bc}/K_e	f (N/m)	l (nm)
TeO$_2$	31.70	7.74	73.27	2.30	216	0.500
TeO$_2$–MoO$_3$[a]						
80–20	30.28	8.6	82.35	2.72	219	0.512
70–30	28.15	8.8	83.55	2.97	220	0.522
55–45	26.56	9.12	87.74	3.30	221	0.532
50–50	25.55	9.12	87.27	3.40	221	0.537
TeO$_2$–V$_2$O$_5$						
80–20	27.20	7.56	73.3	2.7	230.5	0.535
75–25	30.50	7.16	69.7	2.3	233.9	0.521
70–30	28.80	7.11	69.6	2.4	237.3	0.531
65–35	33.90	6.78	66.6	2.0	240.5	0.510
60–40	37.30	6.65	65.6	1.8	243.7	0.499
50–50	39.40	6.34	63.1	1.6	249.9	0.486

Source: R. El-Mallawany and A. El-Moneim, *Phys. Stat. Solids,* 166, 829, 1998.

[a] Properties of MoO$_3$ oxide: r, 1.96 Å; f, 225 N/m; n_b, 6 (Wells, A. F., *Structure of Inorganic Chemistry*, 4th ed., Oxford University Press: Oxford, 1975, p. 581.)

binary TeO$_2$–MoO$_3$ glasses, the increase in K_{bc} is attributed to a gradual increase in both n_b and f'; but in TeO$_2$–V$_2$O$_5$ glasses, f' increases with increasing V$_2$O$_5$, and the number of bonds decreases, which consequently decreases the calculated K. In phosphate glass, when modified with MoO$_3$ or V$_2$O$_5$, the force constant decreases but n_b increases more rapidly, so that the calculated values of K gradually increase.

Figure 2.12a shows the composition dependence of the ratio K_{bc}/K_e for both tellurite and phosphate families. It is clear that the ratio has higher values for doping with V$_2$O$_5$ oxide than doping the same glass with MoO$_3$. The ratio also increases for the glass series TeO$_2$–MoO$_3$ and P$_2$O$_5$–V$_2$O$_5$, and decreases for TeO$_2$–V$_2$O$_5$ and P$_2$O$_5$–MoO$_3$. This means that the two TMOs, V$_2$O$_5$ and MoO$_3$, have opposite influences on the elastic properties of both glasses. If K_{bc}/K_e is >1, a relatively open three-dimensional network structure with l directly proportional to K_{bc}/K_e is indicated. Figure 2.12b illustrates the variation of the estimated atomic l with the modifier concentration for the same studied glass systems, by using the K_{rd} of Bridge et al. (1983). Figure 2.15b shows that the average atomic l behaves similarly to the ratio K_{bc}/K_e for both glass families. The relatively high values for the K_{bc}/K_e of phosphate glasses are attributed to the very open three-dimensional structure of the glasses at the studied composition range. To correlate this behavior to other oxide glasses, it is very important to study the relation between the ratio K_{bc}/K_e and the calculated l of the other pure oxide glasses as shown in Figure 2.12c. From this figure, it is clear that the behavior of both tellurite and phosphate glasses is in the range of the systematic behavior. The σ of both tellurite and phosphate glasses is calculated according to Equation 2.31. The bond compression model by Bridge (1986) states that σ is inversely proportional to n'_c, which generates a strong covalent force to resist lateral contractions. The n'_c of pure tellurite glasses is 2, as mentioned by Lambson et al. (1984), and that of pure phosphate glasses is 1, according to Bridge et al. (1983), and increases for both binary transition metal tellurite and phosphate glasses, as shown in Table 2.7. The calculated σ decreases steadily for these

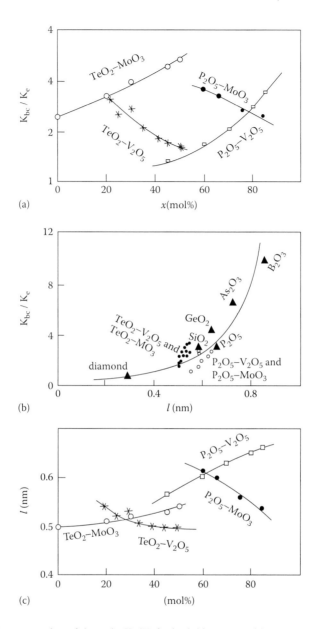

FIGURE 2.12 (a) Representation of the ratio K_{bc}/K_e for both binary transition metal tellurite and phosphate glasses. (b) Variation of the calculated ring diameter with composition for binary tellurite and phosphate glasses doped with TMOs. (c) Representation of both binary transition metal tellurite and phosphate glasses in comparison with the pure oxide glasses. (From R. El-Mallawany and A. El-Moneim, *Phys. Stat. Solid.*, 166, 829, 1998. With permission.)

families of tellurite and phosphate glasses, as concluded by El-Mallawany and El-Moneim (1998) and summarized as follows:

1. The theoretical K (K_{bc}) of both tellurite and phosphate glasses is governed by the forces interlocking the network and is increased or decreased by placing transition metal ions of high or low field strengths in the interstices of the network.
2. The ratio K_{bc}/K_e has been calculated for both glass series when doped with V_2O_5 or MoO_3. The behavior of binary glass is dependent on the kind of both glass former and glass modifier.

3. The calculated σ indicates that binary molybdenum–tellurite glass is more rigid than vanadium–tellurite glass, but the opposite is true in phosphate glasses; that is, binary molybdenum–phosphate glass is less rigid than binary vanadium–phosphate glass.
4. An estimation of the average l of tellurite or phosphate has been calculated and found to depend on the type of modifier.

El-Mallawany (2000a) interpreted the structures of tellurite glasses from the point of view of their elasticity. This analysis of the compositional dependence of elastic properties of the binary tellurite glasses, including K–V and bond compression model relationships, provided a basis to discuss the interatomic bonding of network modifiers and network formers and the correlation of these bonding properties with the elastic properties of noncrystalline solids. In that study, elastic-property data were collected and analyzed based on volume changes in each kind of tellurite glass, because these elastic properties depend on bonding forces between atoms in solids and because different kinds of tellurite glass differ in their types of bonding, in both network-former and -modifier molecules.

Of the elasticity moduli, K is a thermodynamic property defined by Schreiber et al. (1973) as the second derivative of the internal energy, as computed by Equation 2.61. By assuming the Born potential of ionic solids to calculate the internal energy, a simple relationship between K and V can be given for ionic solids of similar structure, based on the calculation of Soga (1961) and Brown (1967), as in Equation 2.61:

$$K = V \left(\frac{\partial^2 U}{\partial V^2} \right)$$

$$K = \left(\frac{r^2}{9V_{\text{equilib}}} \right) \left(\frac{\partial^2 U}{\partial r^2} \right)$$

$$K = \left(\frac{U}{9V_{\text{equilib}}} \right) (n-1)$$

$$K = \left[\frac{A \left(Z_1 Z_2 e^2 \right)(n-1)c}{9V^{4/3}} \right]$$

(2.61)

where U denotes internal energy, V is the volume, V_{equilib} is the volume at equilibrium, r is the effective atomic radius at equilibrium, and n is the Born exponent as described by Brown (1967). α is the Madelung constant; $Z_1 e$ and $Z_2 e$ are the charges of the cation and anion, respectively; n is the power of the repulsive term in the Born potential; and c is a constant related to the packing condition of ions in solids. The K values were measured previously from the relation $K = L - (4/3)$ G, where $L = \rho v_l^2$, $G = \rho v_s^2$, and v_l and v_s are the longitudinal and shear v values, as measured by El-Mallawany et al. (1994) and Sidky et al. (1997). Figure 2.13a represents the relations between K and V for the binary tellurite glasses, which is discussed in this section. From Figure 2.13a, it is clear that, in the V_2O_5–TeO_2 glasses, K increases with an increase in V, whereas for the binary MoO_3–TeO_2 glasses, K is inversely proportional to V.

Because glass is elastically isotropic and usually has a lower ρ than its crystalline state, it is considered a mutual solution to the random distribution of its constituent components, and any excess volume existing in the glassy state shows a very large change in K, as stated by Soga (1961). However, K decreases (i.e., is negatively correlated) with increasing mean atomic volume in many crystalline and glassy material, (such as silicate glasses), giving the materials the relation:

$$K = V_a^{-m}$$

(2.62)

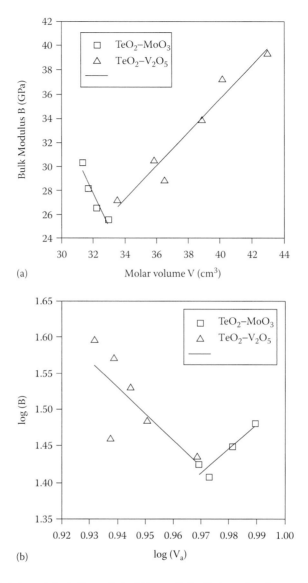

(a)

(b)

FIGURE 2.13 (a) Variation of the bulk moduli of binary TeO_2–V_2O_5 and TeO_2–MoO_3 glasses with volumes (V) of the modifiers (mol%). (b) Variation of the bulk moduli of binary TeO_2–V_2O_5 and TeO_2–MoO_3 glasses with mean atomic volumes (V_a) of the glass in log-log representation. (c) Variation of the ratio K_{bc}/K_e of binary TeO_2–V_2O_5 and TeO_2–MoO_3 glasses with different percentages of the modifier (mol%). (d) Ring diameter (l) of binary TeO_2–V_2O_5 and TeO_2–MoO_3 glasses with different percentages of the modifier of V_2O_5 and MoO_3 (mol%).

where $V_a = M_g/q\rho_g$ is the mean atomic volume and q is the number of atoms per formula unit. The above relation holds well for $m = 4/3$, as determined by Soga (1961). Other values than 4/3 have been reported by Ota et al. (1978), and m has been found to be negative by Mahadeven et al. (1983). The variation of m in K–V relationships is determined by the nature of the bonding on one hand, and by the nature of the coordination polyhedral on the other. When a volume change occurs without a change in the nature of the bonding or in the coordination polyhedral, log K–log V_a plots are generally linear and possess a slope of −4/3, as reported by Damodaran and Rao (1989). In the binary tellurite glasses covered here, the number of atoms per formula unit $q = (x)3 + [(1 − x)(\text{no. of atoms}$ of the TMO)]. The q values are 4 for MoO_3, 7 for V_2O_5, and 3 for TeO_2. The volume dependencies of K for the present binary transition metal tellurite glasses V_2O_5–TeO_2 and MoO_3–TeO_2 are shown

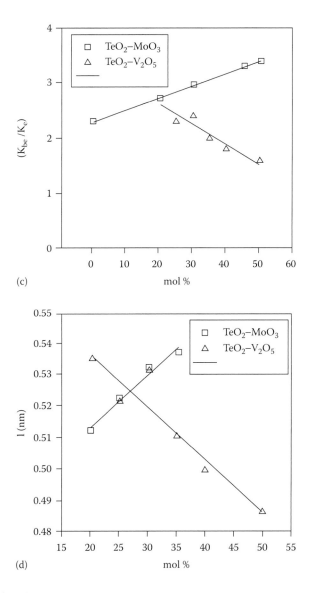

(c)

(d)

FIGURE 2.13 (Continued)

in Figure 2.13b as log–log plots. The log K–log V_a relations are linear with positive $m = +3.594$ for the glass series MoO_3–TeO_2 and negative $m = -3.335$ for the second series V_2O_5–TeO_2. In the transition metal-substituted tellurite glasses V_2O_5–TeO_2 and MoO_3–TeO_2, the K–V and K–V_a are opposite (it is clear from Figure 2.13a and Figure 2.13b that their behaviors and positions are opposite). Considering that highly modified glasses are governed by Coulombic interactions, the tendency of m to decrease toward 4/3 appears to be related to the validity of Equation 2.56 for more ionic materials, and its derivation is based on the Born–Lande type of potential. The composition dependence of K can be discussed in terms of glass structure as follows:

1. The addition of TMO increases connections and network rigidity (K).
2. K of a covalent network is determined by n_b.
3. The cationic field strength of the modifier plays a role in determining f' (high field strength cations polarize their environment strongly and enhance ion-dipole interactions).

4. The cationic field strength increases V_t because of a local contraction of the network around such cations, together with the effects of increasing the K.
5. Difference in volume (V_d) due to the exchange of one formula unit of the tetrahedral TeO_2 and the modifiers MoO_3 (n, 6) and V_2O_5 (n, 5) has been calculated for the modification.
6. For binary transition metal tellurite glasses, the relation between K_es and V_a satisfied the relation $K = V-m$, where the power $m = +3.335$ for the TeO_2–MoO_3 glass series and $m = -3.594$ for TeO_2–V_2O_5 glasses.
7. The K_{bc} of tellurite glasses is governed by the forces interlocking the network. The ratio K_{bc}/K_e has been calculated for both glass series when doped with V_2O_5 or MoO_3. The behavior of binary glass is dependent on both the kind of glass former and the kind of glass modifier. Estimation of the average l of tellurite is calculated as shown in Figure 2.13c and Figure 2.13d. (Note the K_{bc}/K_e values of the following modifiers: $P_2O_5 = 3.08$, $GeO_2 = 4.39$, $SiO_2 = 3.05$, and $B_2O_3 = 10.1$.)

Values of $K_{bc}/K_e \gg 1$ indicate a relatively open three-dimensional network with l tending to increase with K_{bc}/K_e. The relatively high value of K_{bc}/K_e for 80 TeO_2–20 V_2O_5, 2.7, is attributed to a very open three-dimensional structure ($l = 5.35$ nm).

2.7 APPLICATION OF MAKISHIMA–MACKENZIE MODEL TO PURE TeO$_2$, TeO$_2$–V$_2$O$_5$, AND TeO$_2$–MoO$_3$ GLASSES

It has long been known that certain glass properties, including E, are related in an approximately additive manner to composition. By consideration of a large amount of data on composition and property, it is possible to derive a best-fit set of coefficients that can be interpolated or extrapolated to compositions on which no data exist, by applying the previous Equations 2.18 through 2.22 in Section 2.1 from Makishima and Mackenzie (1973 and 1975) to binary tellurite glasses (e.g., TeO_2–V_2O_5 and TeO_2–MoO_3). The different factors necessary to calculate E are listed in Table 2.8, which also contains the calculated values according to the Makishima and Mackenzie (1973 and 1975) model for dissociation energy (G_t), V_t, E, K, S, and σ.

From these important factors, consider first the occupied volume of glass basic unit volume (V_1), which is 31×10^{-3} m^3 for TeO_2. By introducing modifiers with smaller volumes than TeO_2, the occupied volume of glass becomes lower in the binary 50 mol% TeO_2–50 mol% V_2O_5 system to reach the value of 23×10^{-3} m^3, whereas in the binary TeO_2–MoO_3 tellurite glass system, the V_1 increases to 33×10^{-3} m^3 for the composition 50 mol% TeO_2–50 mol% MoO_3. The packing factor V_i (m^3) of ions in the present tellurite glass has an opposite behavior due to the introduction of V_2O_5 oxide or MoO_3 oxide, whereas the V_t of binary transition metal tellurite glasses behaves like the packing factor. The dissociation energy U in kilocalories per cubic centimeter in the present binary tellurite glasses increases from 0.225 to 0.262 kcal/cm^3 for 50 mol% TeO_2–50 mol% V_2O_5, and decreases from 0.225 to reach 0.22397 kcal/cm^3 for the binary 50 mol% TeO_2–50 mol% MoO_3 glass.

Calculation of the elastic moduli by this model gives us the values of E, K, S, and σ. As shown in Table 2.8 and represented in Figure 2.14 (a, b, c, and d) from El-Mallawany (2001), it is clear that the behavior of tellurite glass is opposite in these different series. In the binary TeO_2–V_2O_5 glasses, by increasing the modifier V_2O_5, E increases from 24.88 to 36.57 GPa, K increases from 39.43 to 72.96 GPa, and S increases from 9.45 to 13.67 GPa. σ increases from 0.395 to 0.417, although for the binary 50 mol% TeO_2–50 mol% MoO_3 glass, the behavior is opposite.

Applying the Makishima–Mackenzie models (1973 and 1975) for these tellurite glass systems, the values of the elastic moduli are lower than those measured experimentally, perhaps resulting from an decrease/increase in packing density, which mainly affects the elasticity moduli in this model. On the other hand, the reciprocal of the V reduces the G, which also affects the elasticity moduli by affecting the packing density in this model. This leads to the conclusion that the effects of V on the elasticity moduli through the packing density and on G are pronounced and clear.

TABLE 2.8

Calculated Elasticity Values of the Binary Glass Systems $(100 - x)$ mol% TeO_2–x mol% V_2O_5 and $(100 - x)$ mol% TeO_2–x mol% MoO_3

Glass Composition (mol% TeO_2–mol% Modifier)	S (GPa)	K (GPa)	E (GPa)	$G_t \times 10^9$ (kcal/cm^3)	σ	V_t	V (m^3)	Molar Volume (m^3)	Molecular Weight (g)
TeO_2–V_2O_5									
100–0	9.45	39.43	24.88	0.225	0.395	1.325	0.0411	0.031	159.6
90–10	9.37	37.24	24.58	0.232	0.390	1.267	0.0405	0.032	161.8
80–20	10.14	42.59	26.71	0.240	0.396	1.333	0.0400	0.030	164.1
75–25	10.91	48.98	28.87	0.243	0.402	1.419	0.0397	0.028	165.2
70–30	11.39	52.77	30.19	0.247	0.405	1.461	0.0394	0.027	166.3
65–35	11.90	56.98	31.61	0.251	0.408	1.507	0.0392	0.026	167.4
60–40	12.44	61.69	33.14	0.255	0.411	1.556	0.0389	0.025	168.5
55–45	13.03	66.98	34.79	0.258	0.414	1.610	0.0386	0.024	169.6
50–50	13.67	72.96	36.57	0.262	0.417	1.668	0.0384	0.023	170.7
TeO_2–MoO_3									
80–20	8.56	31.78	22.32	0.224	0.380	1.190	0.037	0.031	156.5
70–30	8.05	27.77	20.86	0.224	0.375	1.110	0.035	0.032	154.9
55–45	7.29	22.28	18.68	0.224	0.361	0.997	0.032	0.032	152.6
50–50	6.95	20.05	17.72	0.224	0.353	0.095	0.031	0.033	151.8

Values were calculated according to the Makishima–Mackenzie model (El-Mallawany, in preparation). Pauling ionic radii (nm): Te, 0.221; O, 0.14; V, 0.059; Mo, 0.062. G_t per n (kcal/mol): TeO_2, 22.54 × 10^7; V_2O_5, 29.99 × 10^7; MoO_3, 22.34 × 10^7.

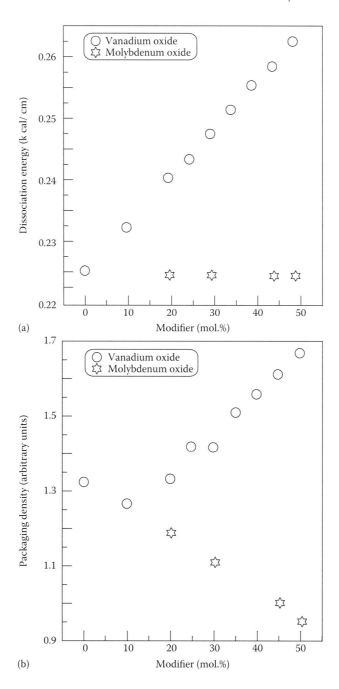

(a)

(b)

FIGURE 2.14 Dependence of the calculated dissociation energy, (a) packing density, (b) bulk and Young's moduli, (c and d) Poisson's ratio, (e) and shear modulus (f) of binary $(100 - x)$ mol% TeO_2–x mol% V_2O_5 and $(100 - x)$ mol% TeO_2–x mol% MoO_3, on the modifier percentage. (From El-Mallawany, R., Aboushely, A., Rahamani, A., and Yousef, E., *J. Mater. Sci. Lett.*, 19, 413, 2000a. With permission.)

The value of σ in this model lies near the experimental one. This means that the structure has some ionic features rather than covalent features, and that this model can be applied successfully for the ternary rather than the binary system.

For a perfect crystalline arrangement, all A–A and A–O separations (bond l values) are the same. In this study, the tellurite glass is regarded as a three-dimensional network of A–O–A bonds

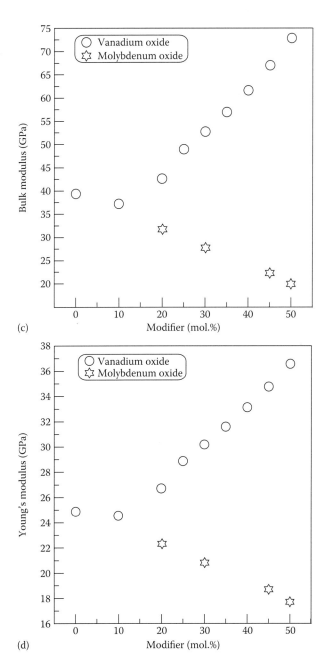

FIGURE 2.14 (Continued)

(A = cation; O = anion and oxygen atom), and A–O–A angles have a spread of values around the fixed values in the corresponding crystalline tellurite, as given by Bridge and Patel (1986). The spread of cation–cation spacing is smaller than the equilibrium (crystalline) values for bond angles that are more acute than normal, and larger for bond angles that are more obtuse than normal. The longitudinal and transverse double-well potentials are associated with a spread of bond l values and of cation–cation spacing, respectively. In this case, central force theory predicts that all anions will move in identical, symmetric interatomic wells. The wells have a single central minimum corresponding to the equilibrium positions of the anions, and they are harmonic for sufficiently small

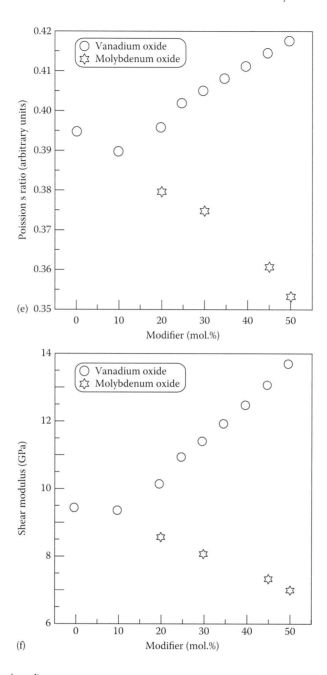

FIGURE 2.14 (Continued)

oxygen vibrations, although at larger amplitudes anharmonic effects appear as the wells become flat-bottomed.

The values of the oxygen density [O], the r_o, and the corresponding mutual potential energy of a studied glass system are found by using Equations 2.34 through 2.38. Also, the [O] that is present in the glass is calculated according to Equation 2.63.

$$[O] = \left(\frac{C}{D} \right) \left(\frac{N_A}{16} \right) \tag{2.63}$$

TABLE 2.9
Oxygen Density of Pure and Binary Tellurite Glasses

Glass and Composition (mol% TeO$_2$–mol% Modifier)	[O] × 10^{28} m^{-3}
TeO$_2$–V$_2$O$_5$	
100–0	3.89
90–10	4.33
80–20	5.22
75–25	5.92
70–30	6.47
65–35	7.07
60–40	7.71
50–50	9.17
TeO$_2$–MoO$_3$	
80–20	4.25
70–30	4.38
55–45	4.57
50–50	4.56

The bond r of TeO$_2$ is 0.199 nm, and that of V$_2$O$_5$ is 0.180 nm.
(R. El-Mallawany, in preparation.)

---

where C is the total amount of oxygen in 100 g of the glass, and D is the volume of 100 g of the glass.

For the pure tellurite glass, it is clear from Table 2.9 that the [O] is 3.89×10^{28} m^{-3}. For the binary tellurite glass system $(100 - x)$ mol% TeO$_2$–x mol% MoO$_3$, the [O] increases from 4.25×10^{28} m^{-3} to 4.56×10^{28} m^{-3}. In terms of oxygen atoms vibrating between static heavier atoms, the total number of acoustically active two-well systems is proportional to [O]. Table 2.9 also summarizes the behavior of $(100 - x)$ mol% TeO$_2$–x mol% V$_2$O$_5$ glass. The values of the barrier height are clarified in Chapter 3.

For the tellurite glass in question, the variability of the Te–O–Te, Mo–O–Mo, and Te–O–Mo bond angles means that there is a spread of atomic l values in these glasses. Of course, the average l in a given tellurite glass is larger than the l value occurring in the nearest equivalent crystalline tellurite, although there are some rings smaller than the crystalline one. In the last rings, the bonds have few 180° angles. Thus, one can expect the amount of distorted cation–cation spacing and the average degree of elongation to increase with average l. For a given l, variations in the bond strength do not affect the total amount of distorted cation–cation spacing or the degree of distortion.

2.8 ELASTIC MODULI AND VICKERS HARDNESS OF BINARY, TERNARY, AND QUATERNARY RARE-EARTH TELLURITE GLASSES AND GLASS-CERAMICS

Glasses containing rare-earth ions in high concentrations are potentially useful for optical data transmission and in laser systems. Earlier studies of the effect of incorporating rare-earth cations, such as lanthanum, neodymium, samarium, europium, or cerium, on the structure and physical properties of tellurite glasses have been reported by El-Mallawany and Saunders (1988). Interest in erbium-containing tellurite glasses developed because the photochromic properties of erbium-doped tellurite glasses are substantially more pronounced than those of silicate glasses containing

TABLE 2.10

Composition, ρ, and Color of Binary, Ternary, and Quaternary Rare-Earth Tellurite Glasses

Glass Formula (mol%)	Notation	ρ (g/cm^3)	Color[a]
90 TeO$_2$–10 La$_2$O$_3$	A	5.685	Pale Lime (T)
90 TeO$_2$–10 CeO$_2$	B	5.706	Dark Reddish–Brown (O)
90 TeO$_2$–10 Sm$_2$O$_3$	C	5.782	Yellow (O)
60 TeO$_2$–30 WO$_3$–10 Er$_2$O$_3$	D	6.713	Pink (T)
77 TeO$_2$–20 WO$_3$–3 Y$_2$O$_3$	E	6.018	Lime Yellow (T)
77 TeO$_2$–20 WO$_3$–3 La$_2$O$_3$	F	6.027	Lime Yellow (T)
77 TeO$_2$–20 WO$_3$–3 Sm$_2$O$_3$	G	6.110	Yellow (T)
74 TeO$_2$–21 WO$_3$–5 CeO$_2$	H	5.781	Very Dark Reddish–Brown
49 TeO$_2$–29 WO$_3$–2 Er$_2$O$_3$–20 PbO	I	6.813	Pink (T)

Source: R. El-Mallawany and G. Saunders, *J. Mater. Sci. Lett.*, 6, 443, 1988.

[a] T, transparent; O, opaque.

an equivalent erbium ion concentration, as reported by Hockroodt and Res (1975). Table 2.10 provides the densities and colors of rare-earth oxide–tellurite glasses in the binary, ternary, and quaternary forms from El-Mallawany and Saunders (1988).

Data on room temperature v values, the SOEC, the adiabatic K and E, and σ for rare-earth oxide–tellurite glasses are given in Table 2.11, according to El-Mallawany and Saunders (1988). El-Mallawany (1990) analyzed the elasticity modulus data of rare-earth oxide–tellurite glasses. El-Mallawany and Saunders (1988) inspected the constants of elasticity for the binary glasses and showed that the inclusion of 10 mol% La$_2$O$_3$, CeO$_2$, or Sm$_2$O$_3$ in tellurite glasses increases the

TABLE 2.11

Experimental Second-Order Constants of Elasticity of Binary, Ternary, and Quaternary Rare-Earth Oxide–Tellurite Glasses

Elastic Property	Values of Properties for Glass[a]										
	A	B	C	X	Y	D	E	F	G	H	I
v_l (m/s)	3,415	3,429	3,446	4,642	4,439	3,548	3,471	3,481	3,515	3,408	3,566
v_s (m/s)	2,093	2,102	2,148	2,740	2,493	2,139	2,031	2,035	2,068	2,011	2,137
C^s_{11} (GPa)	66.3	67.1	68.7	70.3	66.9	84.6	72.5	73.0	75.5	67.2	86.7
C^s_{12} (GPa)	16.5	16.7	15.3	21.2	24.8	23.1	22.8	23.1	23.3	20.4	24.4
C^s_{44} (GPa)	24.9	25.2	26.7	24.5	21.1	30.7	24.8	25.0	26.1	23.4	31.1
K^s (GPa)	33.1	33.5	33.1	38.5	38.8	43.6	39.4	39.7	40.7	36.0	45.1
E^s (GPa)	59.8	60.5	63.1	60.4	53.6	74.7	61.6	58.8	64.5	72.6	75.9
σ^s	0.199	0.199	0.182	0.230	0.270	0.214	0.239	0.240	0.235	0.233	0.220
N'_c	2.65	2.68	2.71			3.09	2.64	2.54	2.67	2.16	3.3

Source: R. El-Mallawany and G. Saunders, *J. Mater. Sci. Lett.*, 6, 443, 1988.

[a] One-letter designations for glass types correspond to Columns 1 and 2 of Table 2.10.

elastic stiffness: both S and K are larger in rare-earth oxide–tellurite composites than in vitreous TeO_2 itself. The introduction of an ionic binding component due to the presence of the rare-earth ion stiffens the structure. It is interesting to note that, although the rare-earth phosphate glasses are much less dense, they have much the same elastic stiffness as the tellurite glasses. It is plausible that this arises from substantially greater ionic contributions to the binding in the phosphates than in the tellurites. The elastic properties of the vitreous samarium–phosphates in relation to structure and binding have been discussed elsewhere by Sidek et al. (1988) and by Mierzejewski et al. (1988). The v values of both longitudinal and shear ultrasonic waves in the binary, ternary, and quaternary tellurite glasses are closely similar to those of vitreous TeO_2 itself, as measured by Lambson et al. (1984). This is an unusual feature; v values in other glass systems, such as silicates, borates, or phosphates, are usually quite sensitive to the inclusion of other oxides whether as formers or modifiers. It appears that the tellurite matrix determines the long-wavelength acoustic-mode velocities (v_l or v_s). Because the elasticity moduli are defined by the appropriate value ρv^2, they become larger as the ρ of the oxide that is added to the tellurite former is increased (Table 2.11).

For samarium–phosphate glasses, it has been found that both K and S decrease with application of hydrostatic pressure; these materials show the extraordinary property of becoming easier to compress as the pressure on them is increased (Sidek et al. 1988). The similarity of the Raman spectra of these samarium glasses to those of other phosphate glasses, including lanthanum, suggests that the samarium glasses have similar structural features to those of other phosphate glasses as measured by Mierzejewski et al. (1988). The unusual elastic behavior under pressure could therefore be caused by the variable valence of the samarium ion. To test this, it is useful to measure the elastic properties of samarium ions in a glass under pressure, based on a different glass former; tellurium dioxide is well studied for this purpose. A lanthanum–tellurite glass has also been studied: in contrast to samarium, whose ions can be 2+ or 3+, lanthanum ions can only be 3+. To seek other possible valence effects, cerium, which can have 3+ or 4+ ions, has also been included in the study.

The pressure dependencies of the relative changes in natural velocity of the longitudinal and shear v_l and v_s values for the rare-earth tellurite glasses are found to increase linearly up to the maximum pressure (3 kbar). This is a good indication that the glasses are homogeneous on a microscopic scale. The hydrostatic pressure derivatives $(\partial C_{11}/\partial P)_{p=0}$, $(\partial K/\partial P)_{p=0}$, and $(\partial C44/\partial P)_{p=0}$ of the moduli obtained from the experimental measurements for rare-earth oxide–tellurite glasses by El-Mallawany and Saunders (1988) are given for each glass in Table 2.12. The positive values obtained for the hydrostatic pressure derivatives of the SOEC show that these rare-earth oxide–tellurite glasses, like vitreous TeO_2 itself, behave normally in that they stiffen under the influence of external stress. The hydrostatic pressure derivatives of the SOEC are combinations of the TOEC and hence correspond to cubic terms in the Hamiltonian with respect to strain. They measure the anharmonicity of the long-wavelength vibrational modes and thus relate to the nonlinearity of the atomic forces with respect to atomic displacements. Insight into the mode anharmonicity can be gained by considering the acoustic-mode Gruneisen γ values, which represent the quasiharmonic approximation of the volume V-dependent $(-\partial \ln \omega_i / \partial \ln V)$ of the normal mode frequency ω_i. The longitudinal (γ_l) and transverse (γ_s) acoustic-mode Gruneisen parameters in the long-wavelength limit are given in Table 2.12 from El-Mallawany and Saunders (1988). Their positive signs show that the application of hydrostatic pressure to rare-earth oxide-tellurite glasses leads to an increase in the frequencies of long-wavelength acoustic modes, which is normal behavior, corresponding to an increase in the vibrational energy of the acoustic modes when the glass is subjected to volumetric strain. By contrast, when a hydrostatic pressure is applied to samarium–phosphate glasses, their moduli have the anomalous property of decreasing, as measured by Sidek et al. (1988). Although uncommon, such acoustic-mode softening behavior is also known in glass that is based on silica, as stated by Hughes and Killy (1953), for which it

TABLE 2.12

Experimental Third-Order Constants of Elasticity of Binary, Ternary, and Quaternary Rare-Earth Tellurite Glasses

Glass Composition (mol%)[a]	$(\partial C_{11}/\partial P)_{p=0}$	$(\partial C_{44}/\partial P)_{p=0}$	$(\partial K/\partial P)_{p=0}$	γ_l	γ_s	γ_{el}
90 TeO$_2$–10 La$_2$O$_3$	+9.29	+1.79	+6.80	+2.15	+1.03	+1.40
90 TeO$_2$–10 CeO$_2$	+8.19	+1.87	+5.70	+1.88	+1.07	+1.34
90 TeO$_2$–10 Sm$_2$O$_3$	+8.52	+1.71	+6.24	+1.89	+0.89	+1.23
X = 90 P$_2$O$_5$–10 Sm$_2$O$_3$	−0.88	−0.69	+0.05	−0.40	−0.70	−0.60
Y = 90 P$_2$O$_5$–10 La$_2$O$_3$	+0.22	+0.11	+2.07	+0.48	-0.06	+0.12
60 TeO$_2$–30 WO$_3$–10 Er$_2$O$_3$	+11.64	+1.77	+9.28	+2.83	+1.09	+1.67
77 TeO$_2$–20 WO$_3$–3 Y$_2$O$_3$	+10.19	+1.06	+8.78	+2.60	+0.67	+1.31
77 TeO$_2$–20 WO$_3$–3 La$_2$O$_3$	+10.10	+2.46	+6.83	+2.58	+1.79	+2.05
77 TeO$_2$–20 WO$_3$–3 Sm$_2$O$_3$	+7.41	+1.83	+4.97	+1.83	+1.26	+1.45
74 TeO$_2$–21 WO$_3$–5 CeO$_2$	+6.67	+1.95	+4.07	+1.62	+1.33	+1.43
49 TeO$_2$–29 WO$_3$–2 Er$_2$O$_3$–20 PbO	+9.99	+3.28	+5.62	+2.44	+2.21	+2.29

[a] Data for all glasses from El-Mallawany, R., and Saunders, G., *J. Mater. Sci. Lett.*, 7, 870, 1988; data for glass X are from Sidek, H., Saunders, G., Hampton, R., Draper, R., and Bridge, B., *Phil. Mag. Lett.*, 57, 49, 1988, and for glass Y are from Mierzejewski, A., Saunders, G., Sidek, H., and Bridge, B., *J. Non-Cryst. Solids*, 104, 323, 1988.

has been attributed to the open fourfold coordination structure, which enables bending vibrations of the BO ions, corresponding to transverse motion against a small force constant as described by Brassington et al. (1981). If P–O bridging and P bending were to cause the anomalous elastic behavior of samarium–phosphate glasses under pressure, then other phosphate glasses would be expected to have negative values of *dB/dP* and *dμ/dP*.

The valence transition from 2+ to 3+ involves size collapse of the samarium ion Sm^{3+}(4f^5), which is about 20% smaller than Sm^{2+}(4f^6), and could lead to an observed reduction under pressure of *K* and *S*. One objective of the study by El-Mallawany and Saunders (1988) was to find out whether samarium in a glassy tellurite matrix glass also shows this extraordinary behavior. If it does not, then either the pressure effects on the valence state or the way in which the samarium ion is bound would seem to be different in the tellurite than in the phosphate glasses. These opening studies of such a fundamental property as elastic stiffness and its behavior under pressure show that there is much to learn about how rare-earth ions in high concentrations are bound in a glassy matrix and how they behave under an applied stress.

In 1999, Sidky et al. measured the elastic moduli and microhardness of tricomponent tellurite glass of the form TeO$_2$–V$_2$O$_5$–Sm$_2$O$_3$. The effect of adding Sm$_2$O$_3$ on the expanse of V$_2$O$_5$ was investigated in terms of the n_b of the glass. The results obtained by Sidky et al. (1999) showed that these glasses become stable and compact when modified with the rare-earth oxide Sm$_2$O$_3$. This, in turn, increases the n_b and decreases the average *l* of the network, consequently increasing the elastic properties of these glasses as shown in Figure 2.15 and Table 2.13. Watanabe et al. (1999) measured the temperature dependence of Vickers *H* for TeO$_2$-based glasses and glass-ceramics as shown in Figure 2.16. The temperature dependence of the *H* of the K–Nb–Te glass and glass-ceramics is indicated in Figure 2.16. The glass-ceramic has a *H* of 4.8 GPa at room temperature, which is much larger than that of precursor glasses (*H* = 3.3 GPa), and it has a value of *dH/dT* = −6.3 × 10^{-3} GPa/K, which is very close to that of soda–lime silicate glass. The temperature dependence of relative change in *H* for the transport of glass–ceramic material is shown in Table 2.13. Watanabe et al. (1999) concluded that the *H* of TeO$_2$-based glasses is largely improved by crystallization.

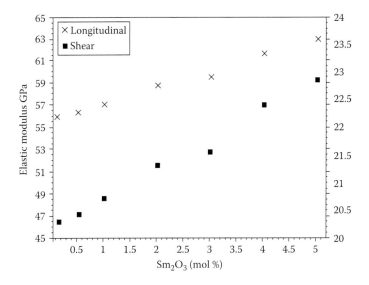

FIGURE 2.15 Elastic moduli and microhardness of ternary tellurite glasses of the form Te–V–Sm. (Reprinted from *Mat. Chem. Phys.*, 61, M. Sidky, A. El-Moneim, and L. El-Latif, 103, 1999. With permission from Elsevier Science.)

2.9 QUANTITATIVE ANALYSIS OF THE ELASTICITY MODULI OF RARE-EARTH TELLURITE GLASSES

El-Mallawany (1990) analyzed the elasticity modulus data for rare-earth oxide–tellurite glasses. The estimated K and σ values of binary, ternary, and quaternary rare-earth oxide–tellurite glasses have been calculated using the bond compression model of Bridge and Higazy (1986), according to the cation–anion bond of each oxide present in the glass as stated in Equations 2.27 through 2.31. Information about the structure of the glass can be deduced after calculating the n_b, the value of the f', the average l, the structure sensitivity factor, and the mean cross-link density. Comparisons between the calculated and experimental elasticity moduli and σ have been carried out. The longitudinal and shear elastic-stiffness values of binary rare-earth oxide–tellurite glasses have been compared and analyzed with those of other binary rare-earth glasses.

Table 2.14 gives the complete set of variables needed to calculate K and σ using Equations 2.27 through 2.31. To calculate K with ratio K_{bc}/K_e, where K_e is the experimental K, the important variable for the K calculation is the number of n_bs per formula unit (n_f), which equals 4 for TeO_2. By introducing a modifier with a higher value of n_f, the structure becomes more linked. For example, n_f equals 6 for WO_3; 7 for La_2O_3, Sm_2O_3, Y_2O_3, and Er_2O_3; and 8 for CeO_2. After introducing PbO, which has an n_f of 4, no significant change in the structure occurs. Thus, a more linked structure occurs by introducing a modifier with a higher n_f. The value of average cross-links per cation in the pure TeO_2 glass is 2.0, which changes to 2.55, 2.4, and 2.54 for different binary systems, and reaches 3.09 and 3.3 for the ternary and quaternary systems, respectively. Consequently, after increasing the number of cross-links per cation of the glass, the n_b increases. After calculating these two main parameters, it can be seen that the n_b changes from 7.74×10^{28} m^{-3} for pure TeO_2 to 8.25×10^{28} m^{-1} for binary glass A; 9.1×10^{28} m^{-3} for ternary system F; and 9.64×10^{28} m^{-3} for quaternary system I. An increase in n_b increases K to 76.9, 87.1, and 91.1 GPa for glasses A, F, and I, respectively. The high K_{bc} values (GPa) are 25.3 (P_2O_5), 23.9 (GeO_2), 12.1 (B_2O_3), and 36.1 (SiO_2) for pure or multicomponent glasses, due to the high n_b in the glass-forming structures. Previously, Lambson et al. (1984) proved that the ratio of calculated and experimental K is 2.3. This value is considerably smaller than that for other pure inorganic oxide glass formers as reported by Bridge et al. (1983),

TABLE 2.13

Microhardness of Tellurite Glasses

Glass (Reference)	H (GPa)	dH/dT (GPa/K) (10^{-3})	Mean Atomic V $(10^{-6}$ m^3/g/Atom)
70 mol% TeO$_2$–20 mol% WO$_3$–10 mol% K$_2$O (Watanbe et al. 1999)	2.5	−4.6	9.25
70 mol% TeO$_2$–15 mol% Na$_3$O–15 mol% ZnO	2.6	−4.5	9.46
70 mol% TeO$_2$–15 mol% Nb$_2$O$_5$–15 mol% K$_2$O	3.3	−5.0	9.82
70 mol% TeO$_2$–15 mol% Nb$_2$O$_5$–15 mol% K$_2$O (Crystallized)	4.8	−6.3	9.19
73 mol% SiO$_2$–14 mol% Na$_2$O–13 mol% CaO	6.0	−6.6	8.27
65 mol% TeO$_2$–(35 − x) mol% V$_2$O$_5$–x mol% Sm$_2$O$_3$ (Sidky et al. 1999) for the following values of x			
0.1	3.9		
3.0	4.1		
5.0	4.3		
TeO$_2$	3.5		
100 mol% TeO$_2$–(Sidky et al. 1997)	3.7		
90 mol% TeO$_2$–10 mol% V$_2$O$_5$	1.7		
80 mol% TeO$_2$–20 mol% V$_2$O$_5$	0.9		
75 mol% TeO$_2$–25 mol% V$_2$O$_5$	1.0		
70 mol% TeO$_2$–30 mol% V$_2$O$_5$	1.4		
65 mol% TeO$_2$–35 mol% V$_2$O$_5$	1.3		
60 mol% TeO$_2$–40 mol% V$_2$O$_5$	1.5		
55 mol% TeO$_2$–45 mol% V$_2$O$_5$	1.0		
50 mol% TeO$_2$–50 mol% V$_2$O$_5$	1.4		
10 mol% Te–10 mol% Ge–(80 − x) mol% Se–x mol% Sb (El-Shafie 1997) for the following values of x			
3	1.2		
6	1.0		
12	0.8		

and by Bridge and Higazy (1986) (i.e., $K_{bc}/K_e = 3.08$ [P$_2$O$_5$], 4.39 [GeO$_2$], 10.1 [B$_2$O$_3$], and 3.05 [SiO$_2$]). From Table 2.15, the ratio of K_{bc}/K_e is in the range of 2.3 ± 6% for all systems, which suggests that the elastic properties of tellurite glass are mainly caused by the Te–O bond rather than by the modifier bond. The relation between K_{bc}/K_e and the calculated atomic l ($K_e = 0.0106\ F\ l^{-3.84}$), where l is in nanometers and F is in newtons per meter, is illustrated by the systematic relationship as shown in the 1st Ed.

The values of l are in the same range as for pure TeO$_2$ glass, which has an l of 0.5 nm, as calculated by Lambson et al. (1984). The structure-sensitive factor has been found to be 0.43 ± 10% for this family of tellurite glasses. Finally, the calculated σ decreases steadily for this family of tellurite glasses. The steady decrease is interesting because it is caused as the mean cross-link density per cation increases from 2.0 for pure TeO$_2$ to 3.3 for glass I.

El-Mallawany (1990) described the basis of the variation in estimated atomic l with the values K_{bc}/K_e, as shown in the 1st Ed. With the calculated values of K and σ, it is possible to do the following:

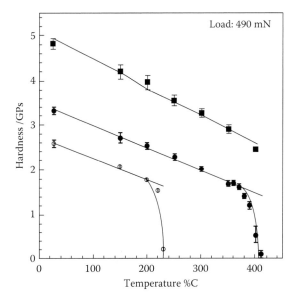

FIGURE 2.16 Temperature dependence of the hardness of K–Nb–Te glass and glass-ceramics. (From P. Watanabe, Y. Benino, K. Ishizaki, and T. Komatsu, *J. Ceramic Soc. Japan*, 107, 1140, 1999. With permission.)

1. Compute the rest of the elasticity moduli (γ_l, γ_s, and E) using Equation 2.31, as shown in Table 2.16 (the calculated moduli are in the range of the experimental values, which have been measured before by El-Mallawany and Saunders [1988]).

2. Compare tellurite and other glass formers; for example, binary tellurite and binary phosphate glasses modified with samarium and lanthanum rare-earth oxides (the following data for pure P_2O_5, 85 mol% P_2O_5–15 mol% Sm_2O_3, and 85 mol% P_2O_5–15 mol% La_2O_3 glasses are from Sidek et al. [1988]: $\rho = 2.52$, 3.28, and 3.413 g/cm³; longitudinal modulus = 41.1, 66.4, and 67.6 GPa; $S = 12.1$, 23.6, and 23.1 GPa; $K = 25.3$, 34.9, and 36.9 GPa; and $E = 31.4$, 57.8, and 57.2 GPa [all data listed in order with respect to the named glasses]).

TABLE 2.14

Parameters Adopted from Crystal Structure of Each Rare-Earth Oxide

Oxide	r (nm)	F (N/m)	n_f
WO_3	0.187	261	6
PbO	0.230	139	4
La_2O_3	0.253	105	7
CeO_2	0.248	112	8
Sm_2O_3	0.249	110	7
Y_2O_3	0.228	143	7
Er_2O_3	0.225	149	7

Source: A. F. Wells, *Structure of Inorganic Chemistry,* 4th ed., Oxford University Press, 1975.

TABLE 2.15

Calculated K and σ of Rare-Earth Oxide–Tellurite Glasses

Property (Unit)	Value of Indicated Property for Glass[a]								
	A	B	C	D	E	F	G	H	I
M_{glass}	176.23	160.86	176.53	203.58	176.02	179.04	176.3	173.99	197.77
η	1.10	1.00	1.10	1.10	1.03	1.03	1.05	1.00	1.38
n_c	2.55	2.40	2.54	3.09	2.64	2.64	2.48	2.17	3.30
n_b (10^{28} m^{-3})	8.25	9.39	8.38	9.73	9.24	9.10	9.08	8.92	9.64
K_{bc} (GPa)	76.9	86.2	77.1	93.1	88.8	87.1	85.6	80.6	91.1
K_{bc}/K_c	2.3	2.5	2.3	2.1	2.3	2.2	2.1	2.2	2.0
F (N/m)	194	189	193	264	221	217	203	182	316
l (nm)	0.515	0.51	0.514	0.519	0.509	0.506	0.495	0.496	0.538
S	0.43	0.40	0.43	0.48	0.43	0.45	0.48	0.45	0.50
$\sigma_{calculated}$	0.222	0.225	0.216	0.211	0.219	0.219	0.223	0.231	0.207

Source: R. El-Mallawany, *J. Mater. Sci.*, 5, 2218, 1990.

[a] One-letter designations for glass types correspond to Columns 1 and 2 of Table 2.10.

TABLE 2.16
Calculated Elasticity Moduli of Rare-Earth Oxide–Tellurite Glasses

Glass[a]	K_e (GPa)	G_e (GPa)	L_e (GPa)	E_e (GPa)
A	33.1	24.9	66.3	59.8
B	33.5	25.2	67.1	60.5
C	33.1	26.7	67.6	63.1
D	43.6	30.7	84.6	74.7
E	39.4	24.8	72.5	61.6
F	39.7	25.0	73.0	58.8
G	40.7	26.1	75.5	64.5
H	36.0	23.4	67.2	72.6
I	45.1	31.1	86.7	75.9

Source: R. El-Mallawany, *J. Mater. Sci.,* 5, 2218, 1990.

[a] One-letter designations for glass types correspond to Columns 1 and 2 of Table 2.10.

A comparison of these elastic properties reveals a rather surprising fact: tellurite glasses, when modified with different rare-earth oxides, undergo a small change in elasticity moduli compared with the phosphate glasses, which undergo a large change in elasticity moduli when modified with different rare-earth oxides. As an example, the measured K of tellurite glass containing 10 mol% Sm_2O_3 is 33.1 GPa, while that of pure TeO_2 glass is 31.7 GPa. In binary rare-earth phosphate glasses, the situation is the opposite; the addition of 15 mol% Sm_2O_3 causes a change in K from 25.3 to 34.9 GPa. In phosphate glasses, the difficulty in interpreting trends of elasticity moduli is that the substitution of another rare-earth atom for a phosphate atom produces simultaneous changes in n, force constant, number of P=O bonds replaced by bridging bonds, and proportions of pyrophosphate and metaphosphate structures. In tellurite glasses, the calculated variables affecting the moduli (i.e., n and average force constant) are nearly the same.

In 1996, El-Adawi and El-Mallawany discussed and analyzed the elasticity moduli of rare-earth oxide–tellurite glasses according to the Makishima–Mackenzie models (1973, 1975) by using Equations 2.18 through 2.22, as shown in Table 2.17. From Table 2.17, the value of the occupied volume of glass in pure TeO_2 is 41×10^{-3} m^3 and changes to 38.7×10^{-3} m^3 for binary cerium–tellurite glass. El-Adawi and El-Mallawany first found that the values of the occupied volume decrease for the tricomponent glass and also for the tetracomponent systems. Second, the dissociation energy of these glasses increases for multisystems compared with that of pure tellurite glass (Table 2.17), with the important variable affecting the dissociation energy being the n_f. Third, the packing density of the glass depends on the kind of modifier and its ionic radius. El-Adawi and El-Mallawany (1996) also compared the calculated elasticity moduli from the bond compression model and the Makisima–Mackenzie model, which showed good agreement (Figure 2.17).

2.10   ELASTIC PROPERTIES OF Te GLASSES

Farley and Saunders (1975) measured the v values (and ultrasonic attenuation as shown in Chapter 3) for several chalcogenide glasses, including 48 mol% Te–10 mol% Ge–12 mol% Si–30 mol% As, 48 mol% Te–20 mol% Si–32 mol% As, 49 mol% Te–10 mol% Ge–12 mol% Si–29 mol% As, and 49 mol% Te–12 mol% Ge–14 mol% S–35 mol% As. The glasses had low v values and low θ_D. The K, G, and E moduli and σ were insensitive to glass composition (Table 2.18). As shown in Table 2.18, Farley and Saunders noted that the interatomic binding forces are much weaker in the chalcogenide

TABLE 2.17
Calculated Elasticity Moduli of Rare-Earth Oxide–Tellurite Glasses According to Mackishima–Mackenzie Model

Glass[a]	M_g	P (g/cm³)	X_i	V_i (m³ × 10⁻³)	Uo (kJ/ mol × 10⁹)	G_i (10⁹)	V_t (10⁹)	G_t (10⁹)	C_t (10⁹)	E (GPa)
TeO₂	159.6	5.101	1.5961	41.08	146.85	4.69	0.2095	7.48	65.57	37.15
A	176.23	5.685	1.7623	39.80	142.817	4.61	0.2262	8.12	70.14	47.60
B	160.86	5.607	1.6086	38.70	144.584	5.13	0.1958	8.25	62.25	42.90
C	176.53	5.782	1.7653	39.60	143.972	4.72	0.2289	8.33	69.90	48.70
D	203.58	5.713	2.0358	33.60	158.983	4.46	0.1916	9.08	68.40	51.90
E	176.02	6.018	1.7602	36.70	156.132	5.34	0.2209	9.40	64.60	50.80
F	179.04	6.027	1.7904	36.80	155.137	5.22	0.2218	9.35	65.90	51.50
G	176.3	6.110	1.7630	36.40	154.890	5.37	0.2224	9.47	64.20	50.83
H	173.99	5.781	1.7399	35.80	155.628	5.17	0.2069	9.00	62.28	46.90
I	197.77	6.813	1.9777	29.10	152.282	5.25	0.1983	10.38	57.55	49.90

Source: A. El-Adawi and R. El-Mallawany, *J. Mat. Sci. Lett.*, 15, 2065, 1996.

[a] One-letter designations for glass types correspond to Columns 1 and 2 of Table 2.10.

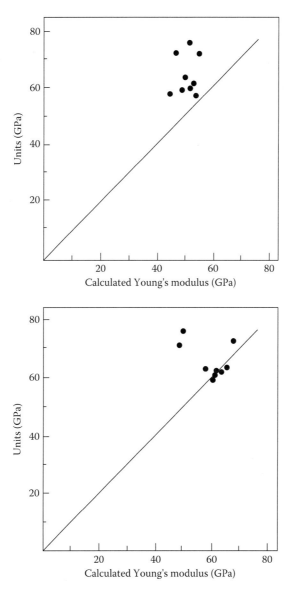

FIGURE 2.17 Experimental and calculated Young's modulus of rare-earth oxide–tellurite glasses according to the Makishima–Mackenzie model. (From A. El-Adawi and R. El-Mallawany, *J. Mater. Sci. Lett.*, 15, 2065, 1996.)

glasses than in covalently bound crystals. The S_{ij} of 10 mol% Ge–12 mol% Si–30 mol% As, 48 mol% Te–20 mol% Si–32 mol% As, 49 mol% Te–10 mol% Ge–12 mol% Si–29 mol% As, and 49 mol% Te–12 mol% Ge–14 mol% S–35 mol% As glasses are not consistent with covalent binding as the only binding force. Plausibly, the glasses comprise covalently bound, multiply branched chains and rings that are held together by much weaker binding forces. The temperature dependencies of the v values are also negative, as shown in the 1st Ed., and no evidence has been found for the existence of two-well systems in these chalcogenide glasses, as will be explained in Chapter 3. Thompson and Bailey (1978) measured both shear and longitudinal v values for 10 MHz in amorphous semi-conducting chalcogenide glasses of the form Te–Se–Ge_2–As_2, Te_{15}–Ge_3–As_2, and As–Ge–Te over the pressure range 0–8 kbar and the temperature range 25°C–100°C. Thompson and Bailey (1978)

TABLE 2.18
Elasticity Constants of Te Glasses

| Glass Composition (mol%) (Reference) | ρ (g/cm^3) | Elastic Modulus (10^9 N/m^2) | | | |
		K	S	E	σ
10 Ge–12 Si–30 As–48 Te (Farley and Saunders 1975)	4.94	15.2	8.35	21.8	0.27
20 Si–32 As–48 Te	4.90	14.1	8.15	20.5	0.26
10 Ge–12 Si–29 As–49 Te	5.15	14.8	9.66	23.8	0.23
12 Ge–14 Si–35 As–38 Te	4.92	15.7	9.78	24.3	0.24
x Te–(1 − x) Se (Carini et al. 1984) for the following values of x					
06.8	4.41	98.7 (10^9 d/cm^2)	38.90	102.9	
14.0	4.54	99.7	41.60	109.6	
21	4.68	105.7	43.80	115.4	
Te–Se–Ge$_2$–As$_2$ (Thompson and Bailey 1978)	5.08	2.021 (10^{11} d/cm^2)	11.98 (10^{10} d/cm^2)	3.001 (10^{11} d/cm^2)	0.25
Te$_{15}$–Ge$_3$–As$_2$	5.80	1.7	9.67	2.4	0.27
As–Ge–Te	4.90	1.3	7.63	1.9	0.26
10 Ge–77 Si–3 S–10 Te (Elshafie 1997)	3.81	10.5	6.73	16.5	0.23
10 Ge–74 Si–6 Sb–10 Te	4.03	14.0	5.64	14.0	0.24
10 Ge–68 Si–12 Sb–10 Te	4.28	8.4	4.63	11.7	0.26

concluded that those glasses with larger values of the connections were stiffer and showed less pressure effect, as shown in Table 2.18.

In 1984, Carini et al. measured the S_{ij} of x mol% Te–(1 − x) mol% Se, where x = 0, 6.8, 14, and 21 at 15 MHz, as in Table 2.18. Carini et al. (1984) used the results of optical spectroscopy and shear velocity tests to explain the role of tellurium in these alloys, and they found a double action played through the covalent bonds on the inside of the copolymer chains and between them by means of van der Waals forces. Sreeram et al. (1991) measured both the V and elastic properties of multicomponent chalcogenide glasses of the form Te–Ge–Sb–Se–As, as represented in the 1st Ed. These glasses were prepared by vacuum melting of previously distilled 5–6 N pure raw materials from which the surface oxide was also removed in some cases. The V and elastic data were interpreted by the average n and, near the tie-line, a chemically ordered covalent-network model for the atomic arrangement in these glasses was found to be preferable over the chance coordination that was predicted by the random covalent network model. Also in 1991, Domoryad used the effects of γ irradiation (^{60}Co) on the S, microhardness, and internal friction (see Chapter 3) of Te$_3$As$_2$–As$_2$Se$_3$ glass. The relative changes in S ($\Delta G/G$) and microhardness ($\Delta H/H$) increase linearly with the logarithm of the radiation dose up to saturation. γ radiation stimulates densification of these glasses.

A more significant consideration by Elshafie (1997) is also more peculiar to the glass composed of 10 mol% Te–(80 − x) mol% Se–x mol% As–10 mol% Ge and regards the v, ρ, elasticity modulus, θ_D, and σ values, together with the microhardness data, summarized in Tables 2.13 and 2.17. The change in the physical properties of this glass were attributed to a change of very low ordering character in the average force constant of the unit cell within the formation of (80 − x) mol% Se–x mol% As. Also, from the results concerning bulk samples, it was concluded that the decrease in physical properties is due to a decrease in the polymeric chains and a lower connectivity character with increasing Sb content.

2.11 ELASTIC PROPERTIES OF NEW TELLURITE GLASSES

It is important to review the progress that has been made in the research on the elastic moduli of the new tellurite glass which has been published in the past ten years. Acoustic and optical properties of xNa$_2$O–(0.2–x)Li$_2$O–0.8 TeO$_2$ glasses with x = 0.0, 0.05, 0.10, 0.15, and 0.20 mol.% have been measured by Roychoudhury et al. (2002a). The initial mixture of the starting materials was taken in the ratio $(0.22x)$ M$_1$:xM$_2$:0.8M$_3$ by weight and M$_1$, M$_2$, and M$_3$ are the molecular weights of Li$_2$CO$_3$, Na$_2$CO$_3$.H$_2$O, and TeO$_2$, respectively. The study of the ultrasonic properties of five different samples of Na$_2$O–Li$_2$O–TeO$_2$ glass systems within the temperature range 140–420 K for longitudinal polarization and 140–300 K for transverse polarization was carried out. The measurements of ultrasonic velocity and attenuation for both polarizations were performed at a 10 MHz frequency using a quartz transducer. The variation of the ultrasonic velocity of all the samples showed a negative gradient. In contrast to CuO–TeO$_2$ or PbO–TeO$_2$ glasses (Paul et al. [2000] and Roychoudhury et al. [2002b]) where the temperature variation of ultrasonic velocity shows an anomaly, the temperature variation of velocity for both transverse and longitudinal polarizations in the present ternary system shows the usual behavior as observed in many oxide glasses (Mukherjee et al. [1992], Mukherjee [1993], Paul et al. [1997, 1999]); that is, they decrease with an increase in temperature over the entire range and for all compositions. Roychoudhury et al. (2002a) have found, from the velocity and density data, that the values of longitudinal, L, shear, G, and bulk, B moduli and Poisson's ratio at different temperatures for the glass system (Li$_2$O)$_{0.2-x}$ (Na$_2$O)$_x$ (TeO$_2$)$_{0.8}$ were as follows:

For x = 0.0 mol%, (ρ = 5.2407 g/cm^3) at temperature 140 K, L = 1.968(10^{10} Nm^{-2}), G = 1.809(10^{10} Nm^{-2}), B = 3.174(10^{10} Nm^{-2}), and σ = 0.260; at temperature 300 K, L = 1.915(10^{10} Nm^{-2}), G = 1.700(10^{10} Nm^{-2}), B = 3.049(10^{10} Nm^{-2}), σ = 0.265.

For x = 20.0 mol% $\approx$ (ρ = 5.0977 g/cm^3) at temperature 140 K, L = 2.181(10^{10} Nm^{-2}), G = 1.8556(10^{10} Nm^{-2}), B = 3.4186(10^{10} Nm^{-2}), and σ = 0.270; at temperature 300 K, L = 2.125(10^{10} Nm^{-2}), G = 1.7636(10^{10} Nm^{-2}), B = 3.3016(10^{10} Nm^{-2}), and σ = 0.273.

The longitudinal ultrasonic velocity slightly decreases and then slightly increases with increasing Na$_2$O mol%, while the shear ultrasonic velocity slightly increases and then slightly decreases with increasing Na$_2$O mol%. The data of Debye temperature θ_D at different temperatures will be given in Chapter 4. It is interesting to note that the temperature of Na–Li–Te glass samples could not be lowered below 140 K as the samples developed cracks irrespective of composition when the ultrasonic waves passed through them. Unlike the Cu–tellurite or Pb–tellurite glass system, there is a monotonous decrease in the ultrasonic velocity with temperature for both longitudinal and transverse polarizations.

In 2003, valence force field (VFF) theory, elastic, acoustic and thermodynamic properties, and a potential function (PF) of TeO$_2$ glass were studied by Gopala Rao et al. The authors correlated the pressure derivatives to some properties of the substances. Thus, some equations have been derived to correlate with the Gruneisen parameter, which is evaluated from Schofield's equations and Bhatia–Singh's (BS) parameters. They have been used to compute the longitudinal γ^L and transverse γ^T Gruneisen constants. γ^L calculated by different methods agree well with experiment. γ^T obtained from BS parameters give rather higher values, while Schofield's equations give results that are in agreement with the experiment. The DeLaunay–Nath–Smith (DNS) equation has been used to derive a relation to compute γ^{el} (elastic). A method has been extended to calculate the third-order elastic constants (TOEC) and has been found to give excellent values of TOECs that are in agreement with experiment results. The absorption band position of TeO$_2$ has been predicted to occur at 276 cm^{-1}. The phonon dispersion curves have been calculated through BS equations for TeO$_2$. Several other properties of TeO$_2$ have been computed, such as the thermal Gruneisen parameter γ^{Th}, its pressure derivatives $(\gamma^{Th})' = (\gamma^{Th}/dP)$, the pressure variation of bulk modulus C$_1$=(dK$_T$/dP$_T$), and its pressure derivatives (dC$_1$/dP)T, which are in turn related to $(\gamma^{Th})'$, the heat capacity at constant volume C$_v$, and the second Gruneisen constant, Q. In some cases, the authors have calculated these quantities using different methods and the agreement between them is good. The authors have evaluated γ^{AG}_T, the Anderson–Gruneisen parameter. A very important aspect of this article was the formulation of the PF of TeO_2 from which the authors have calculated the SOEC. These were found to be in excellent agreement with the experiment. All other properties mentioned already have also been calculated through the use of the newly formulated PF, and the calculated values that were obtained through the various other equations are in good agreement with those obtained via PF. According to valence force field theory, all atomic forces can be resolved into bond-bending β and bond-stretching α forces. It is shown that TeO_2 does not satisfy Martin's unity rule (1970); hence, it is concluded that there is an effective dynamic charge on Te in TeO_2. Using the experimental elastic constants, the bond-bending force β and the bond-stretching force α, along with their pressure derivatives, have been evaluated. In addition, the reststrahlen optic frequency ω has also been calculated. A self-consistent check has been made by evaluating C_{44} through the calculated values of α and β. The authors have used Mie-type bireciprocal PF, as in the equation:

$$\phi(r) = \left(\frac{\varepsilon}{n-m}\right)\left(\frac{n^n}{m^m}\right)^{1/n-m}\left[\left(\frac{\sigma}{r}\right)^n - \left(\frac{\sigma}{r}\right)^m\right] \tag{2.64}$$

For any amorphous substance, $V = KR^3$, where V is the volume, R is the length and K is proportionality constant. The authors have taken $m = 6$ and have fit n and ε through the use of experimental values of K_T, the bulk modulus, and $C_1 = (dK_T/dP)$. Therefore, the SOEC have been calculated after dividing $\phi''(R)$ by 2, as it is the interaction between two particles. Also, the authors have used Martin, R. (1970) to calculate β:

$$\beta = \frac{L}{\sqrt{3}}(C_{11} - C_{12}) - 0.053SC_o \tag{2.65}$$

where, L = the bond length, C_{11} = the longitudinal modulus, C_{44} = the shear modulus, $C_{12} = C_{11} - 2C_{44}$, $S = (Z^*)^2/\varepsilon$ and $C_o = e^2/L^4$, and Z^* is the ration of the effective charge on the atom to the free electron charge, e, and α by Ziman (1977) to calculate $\alpha = -C_v/K_2$. Rao and Venkatesh (2003) used the previous experiment data of TeO_2 glass, such as: density = 5105 kgm^{-3}, molecular weight = 159.6 g/mol, V_L = 3403 ms^{-1}, V_T = 2007 ms^{-1}, C_{11} = 59.1 GPa, C_{44} = 20.6 GPa, C_{12} = 17.9 GPa, Y = 50.7 GPa, α_T = 15.5 × 10^{-6} deg^{-1}, nearest neighbor distance = 2.0A°, number of nearest neighbor (n_c) = 4.03, C_1 = 6.4, C_{11} = 8.03, C_{44} = 1.7, θ_D = 177.6 K, K_T = 31.7 GPa, V_m = 31.26 cm^3/mol. to prove that the potential function model is good. The force constants and their pressure derivatives were: bond stretching (α, $dynes/cm^2$) = 24 × 10^3, bond bending (β, $dynes/cm^2$) = 3.7 × 10^3, $d\alpha/dP$(cm) = 28.3 × 10^{-8}, $d\beta/dP$ = 393 × 10^{-8}, and (α/β) = 6.6, respectively. Also, Wang et al. (2003) have mentioned that all metallic glasses have positive pressure dependence of both the longitudinal and shear velocities, while for most nonmetallic glasses, both velocities decrease with increasing pressure. Thus, the Gruneisen parameters evaluated for the mean mode and the shear mode for metallic glasses and nonmetallic glasses have opposite signs. This indicates that the influence of short-range-order structure and bonding, according closely with the atomic configurations in the vitreous amorphous materials, plays an important role in the properties determined by vibrational anharmonicity. The negative and positive Gruneisen parameters exhibit mode softening and stiffness, respectively, under high pressure. Narita et al. (2003) investigated the Vickers nanoindentation measurements that were carried out for transparent nanocrystallized glasses of $10BaO–10Er_2O_3–80TeO_2$. The universal hardness H_u was found to be H_u = 4.0–9.4 GPa and increased with decreasing penetration depth. The elastic recoveries during unloading were 0.53–0.61. The elastic U_e and plastic U_p deformation energies for the penetration depth of 500 nm in the nanocrystallized glasses were estimated to be U_u = 106 and U_p = 91 kJ/mol, respectively. These values are much larger than those of the base glass (U_e = 83 and U_p = 57 kJ/mol), indicating that a transparent nanocrystallized glass has high resistance against deformation during

Vickers loading compared with the base glass. Quantitative estimations of deformation energies from load/unload displacement curves using a Vickers nanoindentation technique seem to be very informative in nanoscaled glass science and technology. On the interpretation and significance of the Gruneisen parameter in thermoelastic stress analysis, Dulieu–Smith and Stanley (1998) collected the Gruniesen parameter of glasses and ceramics according to Morrell (1985) as follows:

- γ pyrex = 0.06
- γ soda–lime glass = 0.14
- γ vitreous silica glass = 0.14

Tellurite glasses containing vanadium $(50 - x)V_2O_5–xBi_2O_3–50TeO_2$ with different bismuth $(x = 0, 5, 10, 15, 20,$ and 25 wt%) contents have been prepared by Rajendran et al. (2003) using the rapid quenching method. Ultrasonic velocities (both longitudinal and shear) and attenuation (for longitudinal waves only) measurements have been made using a transducer operated at the fundamental frequency of 5 MHz in the temperature range 150–480 K. A continuous decrease in velocities, elastic moduli, and an increase in attenuation and acoustic loss as a function of temperature in all the glass samples exhibit without showing any anomalies, unlike many other transition metal oxides glasses. The bulk and Young's moduli increased from $\approx$30 to 37 GPa. Also, Afifi and Marzuk (2003) measured the ultrasonic velocity of the studied glass, which revealed the fact that adding PbO content to the tungsten tellurite network causes an easy movement for the ultrasonic wave inside the network of the glass structure and, hence, the tendency of the ultrasonic velocity to increase as the mol% of PbO increases. The ternary tellurite glasses were of the form: $(80 - x)$ $TeO_2–xPbO–20WO_3$, $x = 10, 12.5, 15, 17.5,$ and 20 mol%. The results of the elastic moduli, along with those of the Poisson ratio, microhardness, softening temperature, and the Debye temperature indicate the presence of a tightening in the bonding of the glass structure while PbO content increases from 10 to 20 mol% in intervals of 2.5 mol%. Also, the authors did a comparison between the elastic moduli that were observed experimentally and that were calculated theoretically by the bond compression model and the Makishima–Mackenzie model. The data were tabulated and interpreted after calculating the average force constant, ring diameter, the mean cross-link density, and the packing density. The bulk and Young's moduli were in the ranges: 31.98–49.21 and 39.98–51.77 GPa, while the Poisson ratio was in the range 0.367–0.342.

Longitudinal and shear ultrasonic velocities were measured by Sidky and Gaffer (2004) in different compositions of the glass system $0.80 (TeO_2)–(20 - x)WO_3–x(K_2O)$, $x = 0.0, 0.5, 0.10, 0.15,$ and 0.20 mol% by using the pulse-echo technique. Measurements were carried out at 6 MHz frequency and at room temperature. Elastic moduli and some other physical parameters such as the glass transition temperature, Debye temperature, softening temperature, and fugacity were calculated. Results indicated that these parameters depend upon the alkali–oxide modifier (K_2O) content; that is, glass composition. All of the elastic moduli decreased as follows: Young's modulus (E) = 54.73–32.62 GPa, shear modulus S = 21.94–12.71 GPa, bulk modulus K = 36.09–25.12 GPa, microhardness H = 3.696–1.834 GPa, but the Poisson ration increased from 0.2472 to 0.2836. Quantitative analysis has been carried out in order to obtain more information about the structure of these glasses, based on the bond compression model and the ring deformation model (i.e., the cation–anion bond of each oxide). The authors concluded that:

1. The addition of K_2O at the expense of WO_3 in the glass system studied resulted in the 4-coordination polyhedra being transformed into 3-coordination polyhedra and increase in the ratio $N_3=N_4$.
2. The decrease in the elastic constants that were observed in this study, with the increase in K_2O mol% content is mainly due to the increasing number of cations per glass formula unit and the decrease in the average cross-link density. This indicates the breaking of network linkage and the weakness of the network structure.

In 2005, Saddeek has measured the velocity of longitudinal and transverse ultrasonic waves in different compositions of the glass system $0.65TeO_2$–$(0.35 - x)V_2O_5$–xGd_2O_3, $(0 \leq x \leq 0.1)$ and x in mol fraction by conventional quenching method. The authors have measured the ultrasonic velocities at room temperature using the ultrasonic pulse-echo technique at 4 MHz. The velocity data of the glass system have been used to find the elastic moduli, Poisson's ratio, and the Debye temperature. The Young's and bulk moduli were in the range: 41.9–45.8 GPa and 29.3–35.5 GPa, respectively. The compositional variation of the elastic moduli of the studied glass system, along with some ternary tellurovanadate glass systems, can be explained in terms of the values of the average cross-link density, the number of the network bonds per unit volume, and the average ring diameter.

El-Mallawany et al. (2006) measured both the longitudinal and shear ultrasonic velocities in different compositions of the glass system by using the pulse-echo method at 5 MHz frequency and at room temperature. The elastic properties of ternary tellurite glasses $0.50(TeO_2)$–$(50 - x)(V_2O_5)$–$x(TiO_2)$, $x = 0.02, 0.05, 0.10, 0.125$, and 0.15 mol% were measured as a function of composition. The ultrasonic velocity data, density, calculated elastic moduli, microhardness, softening temperature, and Debye temperature all depend on the glass composition.

Longitudinal, shear, Young's, and bulk moduli increased as follows: L = 56.53–64.46, G = 18.80–21.97, E = 47.03–54.55, and K = 31.47–35.17, and microhardness H = 3.121–3.786 GPa, respectively, while the Poisson ratio decreased from 2.502 to 2.483. By calculating the number of network bonds per unit volume, the average stretching force constant, and the average ring size, information about the structure of the glass can be deduced. The theoretical value of the bulk modulus $K_{bc} = 64.08$–74.23 GPa and the calculated number of bond per unit volume $n_b = 6.44$–$7.53 \times 10^{31} m^{-3}$ increased. The ultrasonic velocity measurements of the tricomponent tellurite glasses show a linear increase in the density, ultrasonic velocities elastic moduli, Debye temperature, and microhardness, with the addition of TiO_2 mol%, while the Poisson ratio and softening temperature decrease linearly with the addition of TiO_2. Information about the structure of the present glass has been deduced after calculating the number of network bonds per unit volume, the value of the average stretching force constant, the average ring diameter, and the mean cross-link density. The experimental nondestructive ultrasonic velocity measurement, calculated elastic moduli, and the theoretical elastic moduli give insights into the mechanistic role of titanium in addition to the present semiconducting glass (TeO_2)–(V_2O_5). The elastic moduli of the tellurite glass $0.80 (TeO_2)$–$(20 - x) WO_3$–$x(K_2O)$ by Sidky and Gaffer (2004) and the other tellurite glass series of the form $0.50(TeO_2)$–$(50 - x)(V_2O_5)$–$x(TiO_2)$ by Elmallawany et al. (2006) were opposite. The elastic moduli decreased by increasing $x(K_2O)$ in the first tellurite glass series, while it increased by increasing $x(TiO_2)$ in the second series. The structure and elastic properties of TeO_2–BaF_2 glasses have been studied by Begum and Rajendran (2006). Binary tellurite $(100 - x)TeO_2$–$xBaF_2$ glasses for different compositions of BaF_2 ($x = 8, 10, 12, 15, 18$, and 20 wt%) have been prepared using the rapid quenching method. The velocities and attenuation during the propagation of the ultrasonic waves in all glasses were measured using a transducer operated at a fundamental frequency of 5 MHz at room temperature. The ultrasonic velocity and attenuation have been measured in BFT glasses at room temperature. The decrease in velocity, density, and elastic moduli, and the increase in attenuation (as shown in Chapter 3) indicate the structural role of TeO_2, depending on BaF_2 content. The bulk and Young's moduli decreased from ≈ 34 to 32.7 and from 48.3 to 44.3 GPa, respectively. The incorporation of BaF_2 in the tellurite glasses results in the loose packing of the network, which induces a depolymerization of the TeO_2 (TeO_4–TeO_3) network with an increase in NBO.

In 2007, Saddeek et al. studied the mechanical properties and structure of TeO_2–Li_2O–B_2O_3. The glass compositions were of the form $Li_{0.6}$–$Te_x B_{1.4-2x}$–$O_{2.4-x}$ and $x = 0, 0.1, 0.2, 0.3$, and 0.35 mol.%, and were prepared by the conventional rapid quenching method over a wide range of compositions. Ultrasonic velocities (longitudinal and shear) were measured in these glasses at room temperature. The elastic moduli and the Debye temperature were calculated and discussed quantitatively in terms of the glass transition temperature, the cross-link density, and the packing density. The monotonic decrease in the velocities, the glass transition temperature, and the elastic moduli (bulk modulus = 54.62–40.69 and shear modulus = 26.45–24.79 GPa) as a function of TeO_2 modifier content reveals the loose packing structure, which is attributed to the increase in the molar volume and the reduction in the vibrations of the borate lattice. The compositional dependence of these parameters suggested that TeO_2 changes the rigid character of $Li_{0.6}B_{1.4}O_{2.4}$ to a matrix of ionic behavior bonds by breaking down the lithium borate structure. This was attributed to the creation of more discontinuities and defects in the glasses. Also, Gowda et al., in 2007, conducted a very important study on the elastic moduli of three kinds of glasses. The authors investigated the pseudobinary sodium borate glasses containing $(1 - y)Na_2B_4O_7 - yM_aO_b$ (where M_aO_b = PbO, Bi_2O_3, and TeO_2; and y = 0.25, 0.5, 0.67, and 0.79). Sound velocities (longitudinal and shear) have been measured at 10 MHz frequency using quartz transducers. Density increases with increases the percentage y, and the molar volume decreases. Sound velocities also decrease with increasing y until y ≈ 0.66, above which it increases slightly. Steeper decreases in velocity have been observed in TeO_2-containing glasses. Elastic moduli and Poisson's ratio have been calculated. The bulk modulus was in the ranges of 35.59–38.08, 41.81–43.59, 40.81–32.99 GPa for PbO, Bi_2O_3, and TeO_2 glass series, respectively; Young's modulus was in the ranges of 61.51–54.79, 62.92–51.61–62.92, 55.18–43.94 GPa for PbO, Bi_2O_3, and TeO_2 glass series, respectively; and the Poisson ratio was in the ranges of 0.21–0.26, 0.25–0.27, and 0.27–0.28 PbO, Bi_2O_3, and TeO_2 glass series, respectively. Gowda et al. (2007) also calculated the average ring size, which is also a determinant of the observed molar volume since larger ring sizes are expected to give rise to larger V_m. In other words, the number of bonds n may also be related to V_m. Thus, the authors have deduced that $K \approx V_m^{(S-t-1.3)}$ and found that for the three glass systems, PbO, Bi_2O_3, and TeO_2 glass series, $K \approx V_m^{(1.05-t)}$, $K \approx V_m^{(2.09-t)}$, and $K \approx V_m^{(-0.17-t)}$, and t = −0.61 for the tellurite glass series, and t is positive in both PbO and Bi_2O_3.

Studies of some of the mechanical and optical properties of $(70 - x)TeO_2 + 15B_2O_3 + 15P_2O_5 + xLi_2O$ glasses have been conducted by Khafagy et al. (2008). Specimens of the glassy system: $(70 - x)TeO_2 + 15B_2O_3 + 15P_2O_5 + xLi_2O$, where x = 5, 10, 15, 20, 25, and 30 mol% were prepared by the melt-quenching method. An ultrasonic pulse-echo technique was employed at 5 MHz for measuring: the ultrasonic attenuation, longitudinal and shear wave velocities, elastic moduli, Poisson ratio, Debye temperature, and the hardness of the present glasses. It was found that the gradual replacement of TeO_2 by Li_2O in the glass matrix up to 30 mol% leads to a decrease in the average cross-link density and the rigidity of prepared samples, which affects certain properties; that is the hardness, ultrasonic wave velocities, and elastic moduli are decreased, while the Poisson ratio and the ultrasonic attenuation are increased. Elastic moduli are all rapidly decreased (longitudinal modulus = 72.12–27.04 GPa, shear modulus = 24.63–9.03 GPa, bulk modulus = 39.29–15.00 GPa, and Young's modulus = 61.11–22.56 GPa) upon the replacement of tellurium oxide by lithium oxide in the glass matrix over all the investigated range of x from 5 to 30 mol%. The variation of the Poisson ratio, σ, has a gradual increase from 0.2470 to 0.2495 with the gradual replacement of TeO_2 oxide by Li_2O. Debye temperature decreased gradually from 430 to 265 K due to the substantial increase in Li_2O in the glass matrix up to 30 mol% and will be discussed in Chapter 4 in detail. Abdel Al et al. (2009) measured the elastic properties of vanadium tellurite glasses, $65TeO_2 - (35 - x) V_2O_5 - xCuO$, with different compositions of Copper and x = 0, 7.5, 10, 12.5, 15, and 17.5 mol% at room temperature (300 K). The bulk modulus decreased from 33.97 to 29.37 GPa, and the Young's modulus decreased from 54.86 to 42.87 GPa, while the Poisson ratio increased from 0.231 to 0.257 for the increase of CuO from 0 to 17.5 mol%. The Debye temperature decreased from 0.231 to 242 K for the same increase of CuO. The data of ultrasonic velocities, elastic moduli, the Poisson ratio, cross-link density, microhardnes, the Debye temperature, and the softening temperature of the glasses reveal a decrease in the rigidity of the glass network when the CuO concentration increases from 7.5 to 20 mol%.

The effects of the concurrent reduction of TeO_2 and the addition of ZnO on elastic properties of $(90 - x)TeO_2 - 10Nb_2O_5 - (x)ZnO$ for x = 0, 5, 10, and 15 mol% have been studied by Mohamed et al. (2010). The values of longitudinal modulus (C_L) were decreased from 75.25 to 70.62 GPa, the shear modulus (μ) increased from 24.58 to 25.13 GPa, the bulk modulus (K) decreased from

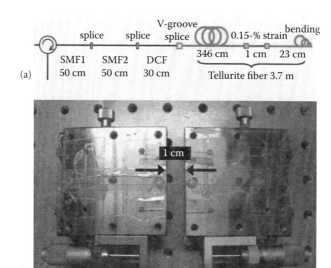

FIGURE 2.18 (a) Structure of the fiber under test for a distributed measurement, (b) photo for the strain-applied section. (by Mizuno, Y., He, Z., and Hotate, K., *Optics Commun.*, 283, 2438, 2010. With permission.)

42.5 to 37.1 GPa, Young's modulus (Y) decreased from 61.8 and 58.6 GPa, the Debye temperature (θ_D) decreased from 283 to 276 K, hardness H increased from 3.9 to 4.7 GPa, and the Poisson ratio decreased from 0.26 to 0.22. A reduction of TeO_2 caused detrimental effects on the elastic properties of the glass system while the addition of ZnO produced a partial recovery of the elastic properties. At a ZnO addition of $x = 5$ mol% and a TeO_2 reduction by 5 mol%, the longitudinal modulus (C_L), shear modulus (μ), bulk modulus (K), Young's modulus (Y), and Debye temperature (θ_D) were observed to drop. Further increases in ZnO caused the shear velocity (v_S), shear modulus (μ), and Debye temperature (θ_D) to recover steadily to higher values, with a full recovery achieved at $x = 15$ mol%. However, the longitudinal (C_L) and bulk (K) moduli did not recover significantly. Far transmission infrared (FTIR) showed the formation of both BO and NBO, but the former was found to be more dominant. It was suggested that recovery of the rigidity of the glass system was due to the formation of BO, which resulted from the addition of ZnO.

A tellurite glass fiber with a high Brillouin gain was employed for distributed strain measurement with Brillouin optical correlation-domain reflectometry (BOCDR) by Mizuno, et al, 2010. Figure 2.18a and b shows the structure of the fiber under test and the photo of the strain-applied section. The investigation on the dependence of the Brillouin frequency shift (BFS) on strain in the tellurite fiber shows that the BFS in this kind of fiber shifted toward lower frequency with increasing strain, and its coefficient was found to be –0.023 MHz/$\mu\varepsilon$. This negative dependence seems to originate from the negative dependence of the Young's modulus of the tellurite fiber on strain. First, the spatial resolution of BOCDR was evaluated using the tellurite fiber. With the high Brillouin gain of the fiber, it was clearly confirmed in the experiment that the spatial resolution is limited by the Rayleigh scattering-induced noise. Using this tellurite fiber, the distribution of the BFS around a 1 cm strain-applied section was successfully measured with BOCDR of a nominal spatial resolution of 6 mm.

3 Acoustic Relaxation Properties of Tellurite Glasses

This chapter examines longitudinal ultrasonic attenuation at various frequencies and temperatures in both oxide– and nonoxide–tellurite glasses containing different modifiers in their binary and ternary forms, together with new data. Experimental ultrasonic attenuation and acoustic activation energies of the oxide forms of these glasses, and correlations among acoustic activation energy, temperature, bulk moduli, and mean cation–anion stretching force constants, as well as the effects of radiation and nonbridging oxygen atoms on ultrasonic attenuation and internal friction in tellurite glasses, are discussed. Some tellurite glasses with useful acousto-optical properties for modulators and deflectors are also highlighted.

3.1 INTRODUCTION

Chapter 2 demonstrated that the elastic properties of covalent networks are very sensitive to average coordination numbers; that is, high-coordination-bond networks form relatively hard glasses, and their elastic constants are determined by covalent forces; whereas low-coordination-bond networks form relatively soft glasses, and their elastic constants are determined by longer-range forces. Chapter 2 also examined both second- and third-order elastic constants (SOEC and TOEC, respectively) of tellurite glasses and, in the process, shear- and longitudinal-acoustic-mode Gruneisen parameters. Estimated bulk moduli and Poisson's ratios are calculated using a bond compression model that focuses on the cation–anion bond of each oxide present in the glass. Information about the structure of the glass is deduced after calculating the number of network bonds per unit volume (n_b) and the values of the average stretching force constant (f'), average ring size, and mean cross-link density. The role of halogen inside the glass network was also discussed in Chapter 2.

To complement and capitalize on the information gained from ultrasonic (including both longitudinal- and shear-velocity) studies of tellurite glasses at room temperature, it is also very important to examine the data on ultrasonic attenuation at low temperatures and correlate these with previous data—that is, earlier room-temperature and low-temperature results. The low-temperature transport, thermodynamic, and acoustic properties of glasses are radically different from those of crystalline solids. Anderson and Bommel (1955) showed that characteristics of the loss peak in fused silica are consistent with a structural-relaxation mechanism with a range of activation energies. In the 1960s and early 1970s, Strakena and Savage (1964), Kurkjian and Krause (1966), and Krause (1971) studied ultrasonic spectra and the variables affecting acoustic loss in silicate, germinate, borate, and arsenic glasses. In 1973, Maynell et al. stated that, because the ultrasonic-loss characteristics in $Na_2O–B_2O_3–SiO_2$ glasses are not consistent with a thermal phonon-damping mechanism, the effects of glass composition and phase separation—including changes during heat treatment—on ultrasonic attenuation and velocity (v) must be interpreted on the basis of a structural-relaxation model. Jackle (1972) and Jackle et al. (1976) completed a theoretical description of the effect of relaxation of two-state structural defects on the elastic properties of glasses in temperatures ranging from 0.3 and 4 K to 100 K.

In this chapter, we describe the effects of room temperature and lower temperatures on the acoustic properties of tellurite glasses. Various models are examined in light of experimental evidence, and their benefits and limitations are identified. Ultrasonic attenuation characteristics are measured from propagation of ultrasonic waves, as described in Chapter 2; for the binary glass systems TeO_2–MoO_3 and TeO_2–V_2O_5, the methods of El-Mallawany et al. (1994a) and Sidky et al. (1997b) are examined; for the ternary glass system consisting of 70 mol% TeO_2–(30 − x) mol% V_2O_5–x mol% A_nO_m in the temperature range 100–300 K, where A_nO_m is CeO_2 or ZnO and $x = 0.03$–0.10 mol%, the method described by El-Mallawany et al. (2000d) is examined. Ultrasonic attenuation properties of the following nonoxide–tellurite glasses are also discussed: 48 mol% Te–10 mol% Ge–12 mol% Si–30 mol% As and 48 mol% Te–20 mol% Si–32 mol% As are examined using the method of Farley et al. (1975); 7 mol% Te–93 mol% Se and 21 mol% Te–79 mol% Se are examined using the method of Carini et al. (1984); and Te–Ge–Se–Sb systems are examined using the method of Elshafie (1997).

The importance of network modifiers has already been emphasized, but some emphasis in this chapter is directed toward the effects of these constituents on ultrasonic attenuation in tellurite glasses. Ultrasound absorption is governed below room temperature by a broad loss peak, which increases in amplitude and shifts to a higher temperature as the ultrasound frequency (f) increases. Section 3.2 provides an analysis of low-temperature acoustic attenuation ($\alpha_{acoustic}$) in tellurite glasses, in the framework of a two-well potential model. Section 3.3 correlates low-temperature ultrasonic attenuation data and room-temperature elastic properties data, drawing on a 1994 report by El-Mallawany. Section 3.4 deals with the effects of γ-radiation and the presence of nonbinding oxygen atoms (NBO) on internal friction (Q^{-1}) and the ultrasonic wave velocity (v) in silver–vanadate–tellurite glasses (El-Mallawany et al. 1998) because radiation (such as γ-rays) changes the physical and chemical properties of solid materials while passing through them.

Section 3.5 deals with relaxation phenomena in solids. I refer the reader to several excellent articles and books that have also reviewed the subject: Zener (1948), Mason (1965), Berry and Nowick (1966), and Nowick (1977). Since the early 1980s, there has been almost explosive activity in the area of ultrasonic attenuation in phosphate glasses, and many papers have been written on this subject, particularly by Bridge and Patel (e.g., 1986a, 1986b, and 1987). These are also reviewed. At the end of the chapter, new directions in the studies of ultrasonic attenuation are described, such as low-f Raman and Brillouin scattering, and stress relaxation.

Many solid materials have been studied for use in acousto-optical devices such as light modulators and deflectors. The following are the main criteria of acceptable materials for these applications:

1. High acousto-optical interaction efficiency
2. Low acoustic loss
3. Minimal temperature effect on optical and acoustical parameters
4. Availability in large quantities
5. Ease of fabrication

A glass can be used in acousto-optical devices only if it has a high acousto-optical "figure of merit" (Me) and low acoustic loss; examples of such glasses include fused quartz, lead glass, extra-dense flint (as shown in Chapter 8), and As_2O_3 glass. The Me equals $n^6P^2/\rho v^3$, which is measured in cubic seconds per gram for light-scattering efficiency. For example, when the four glasses named above are used as acousto-optical deflectors for longitudinal soundwaves, their Me values are equal to 1.51, 5.48, 19, and 433×10^{-18} s³/g, respectively. The Me values for fused quartz and lead glass are too low for practical acousto-optical applications. Although As_2O_3 glass exhibits a high Me value, it shows strong absorption of visible light. For this reason, extra-dense flint glass has been the best material for practical acousto-optical applications to date. Both Yano et al. (1974) and Izumitani and Masuda (1974) have measured the Me of tellurite glasses, as discussed in Section 3.7. Yano et al. (1974) measured Me values both for a tellurite glass known as TeFD5, which was developed by Hoya Glass Works, Ltd., Tokyo, Japan, and for TeO_2 crystal. Izumitani and Masuda (1974) measured Me values for the TeO_2–WO_3–Li_2O system, while varying the ratios of its components.

3.2 ULTRASONIC ATTENUATION OF OXIDE–TELLURITE GLASSES AT LOW TEMPERATURE

In 1977, Sakamura and Imoka reported their measurements for the Q^{-1} of network-forming oxide glasses. They used two binary network-forming oxide glasses with different coordination bond numbers, network-forming oxides, and modifier oxides with divalent cations. A broad peak was observed in every system except the binary tellurite system. Sakamura and Imoka concluded that the peak appears when strong and weak parts coexist in the network structure, and that increasing the weak parts raises the height of the background echo, as observed on the screen of a flaw detector, as for chalcogenide glasses. A peak was not observed in the system binary tellurite system, probably due to each aggregation of weak or strong parts in the network structure.

Tellurite glasses used to measure ultrasonic properties have been prepared from high-purity oxides, and their melting schedules (times and temperatures for melting, casting, and annealing) were described by El-Mallawany et al. (1994b), Sidky et al. (1997b), and El-Mallawany et al. (2000b). Longitudinal ultrasonic attenuation in these glasses has been measured at frequencies of 2, 4, 6, and 8 MHz, and in the temperature range 100–300 K. Figures 3.1a through 3.1d represent the variation in ultrasonic attenuation with different temperatures in tellurite glasses. These figures also show a very well-defined peak that shifts to a higher temperature as f increases, suggesting a kind of relaxation process. The acoustic activation energy (E), as well as the relaxation f_o, are calculated and correlated with the relaxation strength (A) for each glass composition, as shown in Section 3.3. Measurements of the change in ultrasonic attenuation with temperature can be made by measuring the change of height of a particular echo, as observed on the screen of a flaw detector (explained in Chapter 2). The general equation for obtaining the α_{acoustic} is given in the form:

$$\alpha_{\text{acoustic}} = \left(\frac{20}{2x}\right)\log\left[\frac{A_n}{A_{n+1}}\right] \tag{3.1}$$

where x is the thickness of the glass sample and A_n and A_{n+1} are the heights of two successive echoes. The Q^{-1} is calculated in Equation 3.2, in which $f = 4$ MHz and ω is the angular frequency ($\omega = 2\Pi f$) that is used for the generation and detection of longitudinal ultrasonic waves traveling with a velocity C:

$$Q^{-1} = \left(\frac{\alpha\lambda}{\pi}\right)$$

or

$$Q^{-1} = \left(\frac{2\alpha C}{\omega}\right) \tag{3.2}$$

From values of both the ultrasonic attenuation and ultrasonic wave velocity (C), the Q^{-1} is calculated according to Equation 3.2. The temperature dependence of the total attenuation coefficient (α) for longitudinal waves at four frequencies, 2, 4, 6, and 8 MHz in the binary forms TeO_2–MoO_3 and TeO_2–V_2O_5, and in the ternary forms TeO_2–V_2O_5–ZnO and TeO_2–V_2O_5–CeO_2, respectively, are shown in Figures 3.1a through 3.1d. The temperatures (T_p) at which α is maximally shifted toward higher temperature as f increases from 2 to 8 MHz for the investigated glasses are tabulated in Table 3.1. Furthermore, the height of the peak loss increases with increasing f from 2 to 8 MHz at constant composition. This behavior is similar to that observed earlier in other glasses by Bridge and Patel (1986a, 1986b). Plots of log f against the inverse peak temperature T_p^{-1} are shown in

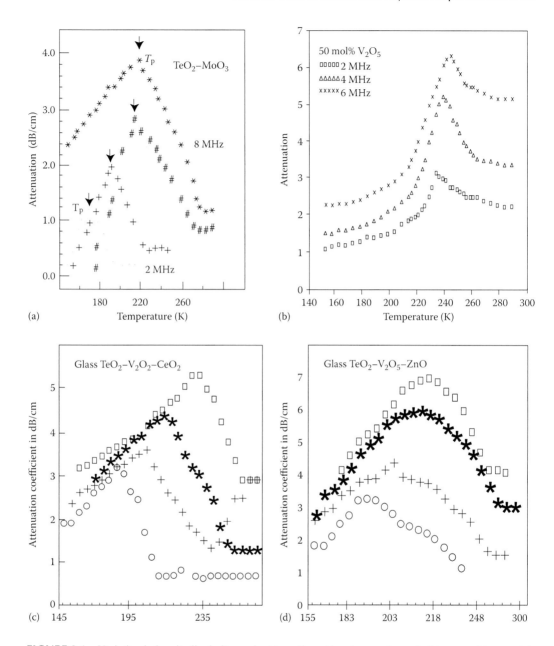

FIGURE 3.1 Variation in longitudinal ultrasonic attenuation at low temperatures for binary and ternary tellurite glasses. (a) TeO$_2$–MoO$_3$ (From El-Mallawany et al., *Mater. Chem. Phys.*, 37, 197, 1994b), (b) TeO$_2$–V$_2$O$_5$ (From Sidky et al., *Phys. Stat. Sol. A*, 159, 397, 1997b), (c) TeO$_2$–V$_2$O$_2$–CeO$_2$ (From El-Mallawany et al., *J. Appl. phys.*, 107, 053523, 2010), (d) TeO$_2$–V$_2$O$_5$–ZnO (From El-Mallawany et al., *J. Appl. phys.*, 107, 053523, 2010), (e) 90TeO$_2$–5Nb$_2$O$_5$–5Li$_2$O (From Sidky et al., *Mater. Chem. Phys.*, 74, 222, 2002).

Figures 3.2a through 3.2d; they yield a straight line for all glasses examined, and these data can be fitted to an equation of the form:

$$f = f_o \exp\left(\frac{-E}{kT}\right) \tag{3.3}$$

where f_o is the attempt frequency and determined from the intercept of the slope of the lines, E is the acoustic activation energy (in electron volts), k is the Boltzman constant, and T is the absolute

TABLE 3.1

Experimental Low-Temperature Properties of Binary and Ternary Tellurite Glasses

Glass and mol% Composition (Reference)	Frequency (MHz)[a]	α (dB/cm)	T_p(K)	f_o (s⁻¹)	E(eV)
TeO$_2$–MoO$_3$ (El-Mallawany et al. 1994b)					
80–20	2	1.22	168		
	4	2.00	190		
	6	2.84	214		
	10	3.90	219	5.04×10^{-11}	0.110
70–30	2	2.64	178		
	4	4.22	198		
	6	5.30	218		
	10	5.40	228	2.83×10^{-11}	0.124
50–50	2	3.96	198		
	4	4.38	218		
	6	5.40	232		
	10	5.62	250	1.59×10^{-11}	0.139
TeO$_2$–V$_2$O$_5$ (Sidky et al. 1997b)					
80–20	6	5.40	238	1.27×10^{-10}	0.068
75–25	6	6.75	255	1.9×10^{-14}	0.166
70–30	6	7.00	238	4.4×10^{-17}	0.223
TeO$_2$–V$_2$O$_5$–CeO$_2$ (El-Mallawany et al. 2010)					
70–27–3	2	2.06	180		
	4	2.22	188		
	6	3.20	193		
	8	4.40	200	2.66×10^{-11}	0.095
70–25–5	2	2.68	188		
	4	3.06	198		
	6	3.32	205		
	8	4.70	210	1.05×10^{-11}	0.092
70–20–10	2	3.00	203		
	4	4.26	218		
	6	5.13	228		
	8	6.24	238	0.2×10^{-11}	0.049
TeO$_2$–V$_2$O$_5$–ZnO (El-Mallawany et al. 2010)					
70–27–3	2	1.70	153		
	4	2.20	170		
	6	2.60	185		
	8	3.50	195	0.12×10^{-10}	0.036
70–25–5	2	2.74	163		
	4	3.08	180		
	6	4.10	187		
	8	4.30	198	0.62×10^{-10}	0.049
70–20–10	2	3.30	198		
	4	4.40	203		
	6	6.00	210		
	8	7.00	220	13.8×10^{-10}	0.098

(continued)

TABLE 3.1   (Continued)

Glass and mol% Composition (Reference)	Frequency (MHz)[a]	α (dB/cm)	T_p(K)	f_o (s⁻¹)	E(eV)
T–Ge–Se–Sb (Elshafie 1997)				Internal Friction[d]	
10–10–77–3	2[a]			3,989	
10–10–74–6	2[a]			1,649	
10–10–68–12	2[a]			872	
T–Ge–Si–As (Farely and Saunders 1975)					
48–10–12–30	20[b]	0.15–0.32[c]			
	60[b]	0.6–1.0[c]			
	100[b]	1.6–1.9[c]			

[a] Room-temperature frequency.
[b] Frequency at 200–300 K.
[c] α(dB/μs).
[d] Calculated from $Q^{-1} \times 10^{-6}$ (see text).

temperature. The values of f_o and E for every kind of tellurite glass are shown in Table 3.1. From Table 3.1, for example, the binary 80 mol% TeO_2–20 mol% MoO_3 glass shows a shift from $T_p \cong 168$ K to $T_p \cong 219$ K due to an increase in the ultrasonic f from 2 to 10 MHz, and the α at the T_p increases rapidly with f. The ternary tellurite glasses of the systems TeO_2–V_2O_5–ZnO and TeO_2–V_2O_5–CeO_2 show similar behavior. E increases from 0.11 to 0.139 eV for the binary TeO_2–MoO_3 and from 0.07 to 0.242 eV for binary vanadate glasses. However, the values of the activation energies of ternary tellurite glass systems increase from 0.036 to 0.098 eV with changes in the Te–V–ZnO contents The value of E for binary and ternary tellurite glasses is less than the activation energy of binary molybdenum–phosphate glasses, as reported by Bridge and Patel (1986a). Also, tellurite glasses have less acoustic attenuation and activation energy than vitreous silica glass [$E(SiO_2) \cong 0.05$ eV], as reported by Strakana and Savage (1964); but from Table 3.1, it is clear that the behavior of tellurite glasses is opposite to that of phosphate glasses in one respect, because in both binary molybdenum– and vanadium–tellurite glasses, the T_p shifts to lower values at higher frequencies (El-Mallawany et al. 1994b; Sidky et al. 1997b). For higher percentages of the transition metal modifier in binary tellurite glasses and in ternary transition metal rare-earth oxide–tellurite glasses, the T_p values become higher. Previously, Strakena and Savage (1964) found that the effect of temperature broadening on relaxation loss curves was evident in GeO_2, SiO_2, and B_2O_3 glasses. In a second observation in this analysis, these authors found that while the integrated A of SiO_2 is sixfold that of B_2O_3, their relaxation curves appear to be comparable.

Paul et al. (2000) measured v and attenuation in copper–tellurite glasses of different compositions in the temperature range 80–300 K. The longitudinal ultrasonic absorption in binary $(100 - x)TeO_2$–xNb_2O_5 and ternary $(100 - x)$ TeO_2–$0.5xNb_2O_5$–$0.5xLi_2O$ tellurite glass systems have been measured by Sidky et al. (2002) using the pulse-echo technique at ultrasonic frequencies of 2, 4, 6 and 8 MHz in the temperature range 200–280 K. The relaxation spectra or the shape of maximum peaks showed the presence of well-defined broad peaks at various temperatures depending upon the glass composition and operating frequency. The maximum peaks shift to higher temperatures with increasing frequency, suggesting some kind of relaxation process. This process has been interpreted as a thermally activated relaxation process, which arises when ultrasonic waves disturb the equilibrium of an atom moving in a double-well potential in the glass network. Results showed that the mean activation energy of the process is strongly dependent on the modifier content. The dependence of the maximum peaks on the composition was analyzed in terms of an assumed loss of standard linear solid type, with low dispersion and a broad distribution of Arrhenius-type relaxation with temperature-independent relaxation strength. The relaxation strength and deformation potential were determined experimentally and theoretically. The acoustic activation energy and deformation potential of the binary tellurite glasses $96TeO_2-4Nb_2O_5$, $90TeO_2-10Nb_2O_5$, $85TeO_2-15Nb_2O_5$ and $77.5TeO_2-22.5Nb_2O_5$ were: 0.196, 0.193, 0.184, and 0.175 eV; and for $90TeO_2-5Nb_2O_5-5Li_2O$, $80TeO_2-10Nb_2O_5-10Li_2O$, $75TeO_2-12.5Nb_2O_5-12.5Li_2O$, $70TeO_2-15Nb_2O_5-15Li_2O$, they were 0.127, 0.12, 0.118, and 0.096 eV, respectively. From the study above, longitudinal ultrasonic absorption at low temperatures showed the presence of well-defined peaks whose heights increase as the applied frequency increases. This was attributed to a

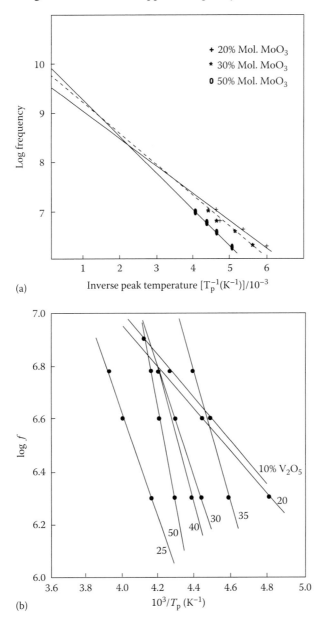

(a)

(b)

FIGURE 3.2 Variation of the logarithm of f versus inverse T_p (K^{-1}) for binary and ternary tellurite glasses. (a) TeO_2-MoO_3 (From El-Mallawany et al., *Mater. Chem. Phys.*, 37, 197, 1994b), (b) $TeO_2-V_2O_5$ (From Sidky et al., *Phys. Stat. Sol. A*, 159, 397, 1997b), (c) $TeO_2-V_2O_2-CeO_2$ (From El-Mallawany et al., *J. Appl. phys.*, 107, 053523, 2010), (d) $TeO_2-V_2O_5-ZnO$ (From El-Mallawany et al., *J. Appl. phys.*, 107, 053523, 2010).

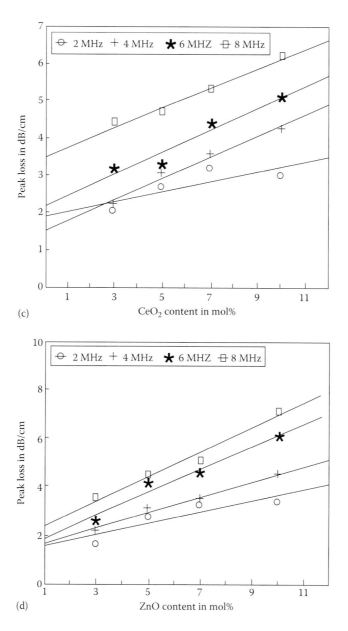

FIGURE 3.2 (Continued)

thermally activated relaxation governed by Arrhenius relationship. The absorption increases linearly with the applied frequency and decreases with the modifier. The peak temperature increases as the modifier content increases. The activation energy and the attempt frequency of this type of relaxation decrease with an increase in the modifier content in both systems. The theoretical treatment of the longitudinal vibrations of two-well systems revealed that the equilibrium interatomic separation decreases with an increase in the mutual potential energy and the oxygen density for the binary and ternary systems. The number of loss centers decreases with the activation energy. The area under the curve of the longitudinal ultrasonic absorption versus temperature decreases as the modifier increases. The number of loss centers is related to the compositional dependence of the elastic moduli and the oxygen density. The relaxation strength decreases with frequency for the same modifier content. Although the experimental deformation potential decreases with increasing modifier content in the two systems due to the decrease in experimental activation energy, the corresponding theoretical values did not show the same trend.

Acoustic and optical properties of x.Na$_2$O–0.2 Li$_2$O–0.8 TeO$_2$, x = 0.0, 0.05, 0.10, 0.15, and 0.20 glasses have been measured by Roychoudhury et al. (2002a). No well-defined broad peak has been observed in the temperature variation of attenuation, and a mixed alkali-type peak starts growing at temperature 350 K. No well-defined broad relaxation peak as observed in usual oxide glasses (by Mukherjee et al. 1992, 1993; Paul et al. 1997, 1998, 1999, and 2000; and Roychoudhury 2002) was obtained in the present mixed alkali–tellurite glass system. However, for the longitudinal attenuation, we observe a rise in the curves at temperature 350 K. From the very nature of the graphs, one may expect to get a new peak at a higher temperature. For the transverse waves, we did not have any data beyond 300 K. The expression for internal friction $Q^{-1} = (2\alpha C/\omega)$, as derived from attenuation considering thermal relaxation in an asymmetric double-well potential only, is given by Paul et al. (2000), Roychoudhury (2002), and Roychoudhury et al. (1994). Values of the longitudinal and transverse internal frictions of x.Na$_2$O–0.2 Li$_2$O–0.8 TeO$_2$, x = 0.0, 0.05, 0.10, 0.15, and 0.20 were in the ranges 0.991, 0.994, 0.991, 0.980, 0.989, 0.974, 0.928, 0.989, 0.987, and 0.976, respectively. Relaxation parameters were explained by Paul et al. (1998 and 1999) and Jain et al. (1983). Roychoudhury et al. (2002) have taken the relaxation parameter τ_o to be approximately 10^{-13} sec. Roychoudhury et al. (2002a) could not able to fit the attenuation curves due to the absence of any well-defined peak. Tellurite-containing vanadate $(50 - x)$V$_2$O$_5$–xBi$_2$O$_3$–50TeO$_2$ glasses with different bismuth (x = 0, 5, 10, 15, 20, and 25 wt%) contents have been prepared by Rajendran et al. (2003) using the rapid-quenching method. Ultrasonic velocities (both longitudinal and shear) and attenuation (for longitudinal waves only) measurements have been made using a transducer that was operated at the fundamental frequency of 5 MHz in the temperature range 150–480 K. The attenuation of ultrasound waves in binary vanadium–tellurite glass is slightly less than that of Bi$_2$O$_3$ containing vanadium–tellurite glass where Bi$_2$O$_3$ is added into the network as a network modifier. In Bi$_2$O$_3$-containing ternary vanadate glasses, a gradual increase in attenuation with the addition of increasing amounts of Bi$_2$O$_2$ content has been noticed. This may be due to the following facts: the increase in Bi$_2$O$_3$ content in TeO$_4$ network results due to two types of NBO (stated by Paul et al. 2000); that is, Te–O$_{eq}$ and Te–O$_{ax}$ (where O$_{eq}$ and O$_{ax}$ refer to the oxygen in an equatorial plane and the oxygen in an axial position, respectively) bonds as assumed in alkali–tellurite glasses as stated by Sekiya et al. (1992) and also due to the formation of point defects and dislocations. In binary copper oxide (CuO)x–(TeO$_2$)1 $- x$ glasses, a similar increase in attenuation with the addition of CuO due to the breaking of bonds was studied by Paul et al. (2000).

The ultrasonic attenuation at room temperature that propagated in the binary tellurite glasses of the form $(100 - x)$ TeO$_2$ – xBaF$_2$ for different compositions of BaF$_2$ (x = 8, 10, 12, 15, 18, and 20 wt%) have been studied by Begum and Rajendran (2006). The velocities and attenuation during the propagation of the ultrasonic waves in all glasses were measured using a transducer that was operated at a fundamental frequency of 5 MHz at room temperature. The ultrasonic longitudinal attenuation coefficient α_L was 0.57, 0.62, 0.65, 0.75, 0.79, and 0.81 (dB cm^{-1}), and the ultrasonic shear attenuation α_s was 0.36, 0.40, 0.41, 0.44, 0.49, and 0.50, respectively. Afifi et al. (2007) measured the longitudinal ultrasonic attenuation measurements using the pulse-echo method at fundamental frequencies of 2, 4, 6, and 8 MHz in 20WO$_3$–$(80 - x)$ TeO$_2$–xPbO ternary tellurite glasses (x = 10, 12.5, 15, 17.5, and 20 mol%) in the temperature range 160–280 K. The results showed the presence of a broad peak, which shifts to a higher temperature with increasing frequency. The ultrasonic attenuation peaks suggest that the experimental behavior is controlled by thermally activated structural relaxations. The internal friction, acoustic activation energy, deformation potential, relaxation strength, number of loss centers, and density of state have been calculated, as a function of both temperature and PbO content. The acoustic activation energy was found to decrease from 0.156 to 0.135 eV with the increase in PbO content. The results showed that both the number of loss centers and their activation energy decrease with the atomic ring size. An increase in the density of state is observed with the addition of PbO content at the same frequency in the whole range of temperatures, which is associated with structural units formed when PbO is added. Also, the ultrasonic attenuation of the tellurite glasses of the form: $(70 - x)TeO_2 + 15B_2O_3 + 15P_2O_5 + xLi_2O$, where $x = 5, 10, 15, 20, 25$, and 30 mol%, has been studied by Khafagy et al. (2008). The attenuation coefficient increased from ≈ 4.9 to ≈ 6.2 dB/cm for the present tellurite glasses with $5Li_2O$ to $30Li_2O$ mol%, respectively.

3.3 PROPERTIES OF ULTRASONIC ATTENUATION IN NONOXIDE–TELLURITE GLASSES

Farely and Saunders (1975) measured the α in IVb–Vb–VIb chalcogenide glasses in the temperature range 0–300 K. The chalcogenide glasses they used were 48 mol% Te–30 mol% As–10 mol% Ge–12 mol% Si; 48 mol% Te–32 mol% As–20 mol% Si; and 50 mol% Se–30 mol% As–20 mol% Ge. Farely and Saunders (1975) concluded that at lower temperatures in many oxide glasses, the v values exhibit minima that correspond to the relaxation of elastic moduli accompanying the broad loss peak in ultrasound attenuation. There is no such relaxation in the elastic moduli of chalcogenide glasses, and no temperature range in which the temperature coefficient of velocity is positive. Farely and Saunders (1975) added that the temperature dependence of ultrasonic attenuation in the chalcogenide glasses is different from that in the oxide glasses, and there is no attenuation peak in the range from 4.2 to 300 K for either longitudinal or transverse ultrasonic waves (10–100 MHz frequencies) in any of the chalcogenide glasses listed above.

Carini et al. (1984) used acoustic measurements to calculate the low activation energies in Te–Se glasses, including both 21 mol% Te–79 mol% Se and 7 mol% Te–93 mol% Se glasses. These authors took measurements in temperatures ranging from 4.2 to 300 K for ultrasonic wave frequencies ranging from 15 to 75 MHz. They found low-temperature peaks connected to thermally activated relaxation processes with low activation energies, as shown in the 1.st Ed. Carini et al. discussed these peaks and the overall acoustic behavior from the viewpoint of the polymeric structures of these glasses.

Elshafie (1997) measured the α values of nonoxide–tellurite glasses of the form 10 mol% Te–10 mol% Ge–$(80 - x)$ mol% Se–x mol% Sb at room temperature and at different frequencies. Elshafie also calculated the Q^{-1} for these glasses, with different values of x (i.e., 3, 6, and 12) as shown in Table 3.1. It was clear that α increases as f increases; whereas at the same frequency, the ultrasonic attenuation decreases due to the substitution of Sb for Se, with the proportions of Te and Ge remaining unchanged. The Q^{-1} of the 10 mol% Te–10 mol% Ge–$(80 - x)$ mol% Se–x mol% Sb glasses at room temperature and at an f of 2 MHz decreased when Sb was substituted for Se.

3.4 RADIATION EFFECT ON ULTRASONIC ATTENUATION COEFFICIENT AND INTERNAL FRICTION OF TELLURITE GLASSES

Glasses and their properties are subject to a variety of changes resulting from high-energy radiation. In general, these effects extend from the reduction of specific ions to metal to collapse of the entire network. Less energetic ultraviolet (UV) radiation can produce different effects. Some glasses, called photochromic glasses, are partially transformed to glass with crystalline regions by UV irradiation. Photochromic glasses are defined as glasses that are sensitive to changes of the glass system from state A to state B. The transmission percentages of some glasses also change to various degrees after irradiation by various ^{60}Co-γ radiation doses. Since the area of interaction of high energy and glass is extraordinarily large, dosimeter glasses must exhibit quite different sensitivities to radiation doses of various intensities.

El-Mallawany et al. (1998) measured the α of a glass with composition 50 mol% TeO_2–$(50 - x)$ mol% V_2O_5–x mol% Ag_2O at room temperature. With an increase in Ag_2O content from 5 to 25

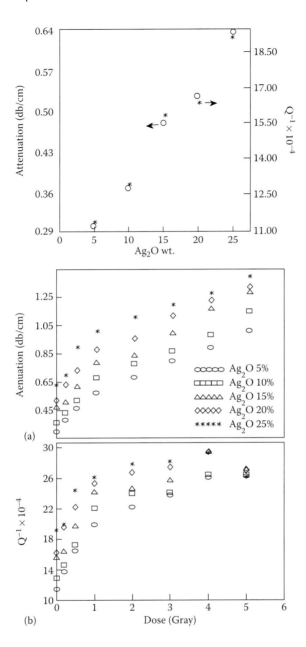

FIGURE 3.3 (a, b) Effect of γ-radiation on α and the Q^{-1} of 50 mol% TeO_2–(50 – x) mol% V_2O_5–x mol% Ag_2O. (From R. El-Mallawany et al., *Mater. Chem. Phys.*, 53, 93, 1998.)

wt%, α increases from 0.3 to 0.64 dB/cm (Figure 3.3a). The increase in attenuation when Ag_2O content is increased from 5 to 27.5 wt% is explained as follows: the addition of Ag_2O to TeO_2–V_2O_5 forms an NBO, which causes splitting of the glassy network. These NBO (oxygen atoms that are free on one side) can absorb more ultrasonic waves than bounded oxygen atoms (BO); that is, the glass with greater Ag_2O has an increased α. The variations in both Q^{-1} and attenuation with changes in Ag_2O content (on a percentage weight basis) are similar. The variation of Q^{-1} has been discussed in terms of its relation to variations in ultrasonic attenuation; v decreases rapidly with increases in Ag_2O content from 5 to 25 wt%, as shown in Figure 3.3b. Q^{-1} increases with increasing Ag_2O content.

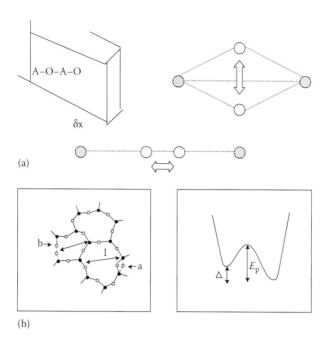

FIGURE 3.4 (a) Schematic two-dimensional representation of the noncrystalline solid material and its l in both the longitudinal and transverse modes, (b) the two-well system with barrier height (V) and asymmetry (Δ).

A quantitative interpretation of experimental Q^{-1} data is based on the number of covalent n_b [$n_b = \Sigma(n_f N_A \rho_i)/M_i$], as stated previously by Lambson et al. (1984) and cited in Chapter 2, where n_f is number of n_b (i.e., coordination bond number), which is four for TeO_2 and five for V_2O_5; N_A is Avogadro's number; ρ is the density; and M is the molecular weight of the glass. The number of n_b decreases from 6.6×10^{28} m^{-3} to 6.11×10^{28} m^{-3} as Ag_2O content increases from 5% to 25% for tricomponent tellurite glasses. The following is a two-dimensional representation of a binary vanadium–tellurite glass before the addition of Ag_2O oxide: $-O-Te-O-V-O-$. The following is a representation of the binary vanadium–tellurite glass after the addition of Ag_2O oxide; that is, such that Ag_2O breaks some part of the above covalent BO related to both TeO_2 and V_2O_5: $Ag^+-O-Te-O-V-O-Ag^+$. The decrease in the number of BOs is due to the creation of NBOs, which also explains changes in thermal properties as mentioned by El-Mallawany et al. (1997) and discussed in Chapter 5.

Variations of α and Q^{-1} for all $TeO_2-V_2O_5-Ag_2O$ glasses with increasing γ-ray doses from 0.2 to 5.0 Gy are shown in Figure 3.4. It is well known that when glass (a noncrystalline solid) is subjected to ionizing radiation, electrons are ionized from the valency band, move throughout the glass matrix, and either become trapped by preexisting flaws to form defect centers in the glass structure or recombine with the positively charged holes. It also has been suggested that if NBO in glasses lose electrons, the interstitial cations change their positions in the matrix; consequently, an NBO traps a hole, giving rise to color centers. The effects produced in glass by irradiation can be represented by the following general rule: (defect in glass) + $hv \rightarrow$ (positive hole) + e^-.

When silver tellurite glasses are irradiated by γ-rays, the silver ion Ag^+ acts as a trap for the electrons and positive holes leading to the formation of Ag^o ($4d^{10}\ 5\ s^1$) and $Ag^{2+}(4d^9)$, respectively; that is, $Ag^+ + e^- \rightarrow Ag^o$ and $Ag^+ + e^+ \rightarrow Ag^{2+}$.

So γ-rays change the structure of the glasses and produce free silver atoms or form charged double-silver ions, which absorb more ultrasonic waves. Based on the results obtained from the ultrasonic techniques in the 50 mol% $TeO_2-(50-x)$ mol% V_2O_5-x mol% Ag_2O glasses, the following conclusions are drawn:

1. Ternary glasses have higher α and Q^{-1} compared with binary glasses due to the modification with the ionic oxide Ag_2O, which causes a reduction in BO and an increase in NBO.
2. Irradiating either glass with γ-rays increases both α and Q^{-1}.

3.5 STRUCTURAL ANALYSIS OF ULTRASONIC ATTENUATION AND RELAXATION PHENOMENA

In general, relaxation is considered the gradual approach to equilibrium of a system in response to rapid changes in external constraints on the system. It is impossible here to review all the models that have been produced to study relaxation phenomena in solids. Some include the Maxwell (Mason 1965), Voigt (Berry and Nowick 1966), and Zener (Nowick 1977) models. The Zener model dates to 1948 and is also known as the standard linear-solid model. Indeed all known loss mechanisms of dynamic linear or nonlinear types might be satisfactorily approximated by these three models with the exception of phase variation effects and resonant absorption mechanisms.

Phenomenologically, a resonant acoustic-absorption model of a solid can be constructed by adding an internal term to the time-dependent stress–strain relations of the Maxwell, Voigt, or Zener models. In a material exhibiting ultrasonic absorption, Hooke's law cannot hold, and $\sigma = M*\varepsilon$, where $M*$ is the complex elastic modulus. The loss mechanisms likely to be present in glass are discussed next.

3.5.1 THERMAL DIFFUSION (THERMOELASTIC RELAXATION)

In a compression wave propagation, the alternating compressions and refractions generally take place too rapidly for isothermal conditions to apply, and in the extreme case ($\omega \Rightarrow \infty$), deformations take place adiabatically. From ordinary thermodynamic analysis, the formula for the quantity $\Delta M/M'$, where M' is the mean modulus [$M' = (\text{Im}.M \times \text{Re}.M*)^{1/2}$], ΔM is the difference between both parts of the modulus, and $\tan \infty = (\text{Im}.M*/\text{Re}.M)$. The loss factor or Q^{-1} is given by:

$$Q^{-1} = \left[\frac{\tan \delta}{\left(1 + \tan^2 \delta\right)^{1/2}} \right] \quad (3.4)$$

$$Q^{-1} = \left(\frac{\Delta M}{M} \right) \left(\frac{\omega^2 \tau}{\left(1 + \omega^2 \tau^2\right)} \right) \left[1 + \left(\frac{\Delta M}{M} \right)^2 \left(\frac{\omega^2 \tau^2}{\left(1 + \omega^2 \tau^2\right)^2} \right) \right]^{-1/2} \quad (3.5)$$

Because the loss per unit length is related to the loss factor by $\alpha = Q^{-1} \omega/2c$, where $c = \omega L$ and L is the mean length of the free path, or $c \cong (M/\rho)^{1/2}$, where ρ is the density of the solid material, the following holds:

$$\alpha = \left(\frac{1}{2c} \right) \left(\frac{\Delta M}{M} \right) \left(\frac{\omega^2 \tau}{\left(1 + \omega^2 \tau^2\right)} \right) \left\{ 1 + \left(\frac{\Delta M}{M} \right)^2 \left[\frac{\omega^2 \tau^2}{\left(1 + \omega^2 \tau^2\right)^2} \right] \right\}^{-1/2} \quad (3.6)$$

So, for $\Delta M/M' \ll 1$, $(\Delta M/M)^2$ is neglected, and the ultrasonic attenuation is of the form:

$$\alpha \approx \left(\frac{1}{2} \right) \left(\frac{\Delta M}{M} \right) \left(\frac{\rho}{M} \right)^{1/2} \left(\frac{\omega^2 \tau}{\left(1 + \omega^2 \tau^2\right)} \right) \quad (3.7)$$

3.5.2 Direct Interaction of Acoustic Phonons with Thermal Phonons

Several authors have used different quasiclassical approaches to estimate theoretical loss, but, in all cases, standard linear solid behavior is obtained, and loss takes the form described by Mason (1965):

$$\alpha = \left[\frac{C_V T \gamma^2 \omega^2 \tau_{th}}{2 \rho c^3 \left(1 + \omega^2 \tau_{th}^2\right)} \right] (np/cm) \qquad (3.8)$$

where results of the calculation are in nanophonons per centimeter and C_v is the heat capacity per unit volume, c is the velocity, and γ^2 is a structurally dependent arbitrary constant representing the nonlinearity of the elastic moduli; that is, the Gruneisen parameter for various phonon modes of the material (Gruneisen parameters determined experimentally from the TOEC are discussed in Chapter 2). But the observed behavior could indicate the presence of anelastic effects due to mobile particles subjected to a thermally activated relaxation process, as demonstrated by Berry (1966) and Nowick (1977). Correlations between low-temperature and room-temperature α that exist in a double-well system (Figure 3.4) with a distribution of barrier heights (E) for both longitudinal and transverse motions of the anion are described in Chapter 2 and have been reported by Bridge and Patel (1986a and 1986b), El-Mallawany et al. (1994 and 2000d), and El-Mallawany (1994). Both the number of two-well systems and the average E (in electron volts) decrease with decreasing mean atomic ring diameter (l). Bridge and Patel (1986a), using the central-force model, suggested that the possible origin of two-well systems is the mutual potential energy of two atoms (separated by a distance r) in a diatomic molecule (discussed in Chapter 2).

Bridge and Patel (1986a) also proved a theoretical expression for the deformation potential (D) according to the definition proposed by Jackle et al. (1976)—that the shift in the minima of the two wells due to acoustic interaction is $\delta U = D\varepsilon$. The number of cation–anion–cation (A–O–A) units per cubic meter of glass is shown by $n_b N_A / M$, where n_b is the number of A–O–A units per formula unit, N_A is Avogadro's number, and M is the mass of 1 kmol of glass, as in Figure 3.5; and the number of these units intersecting and the cross-section of the unit area in square meters $= n_b N_A \rho \delta \times M$, where δx is the A–A unit. D is given by the expression:

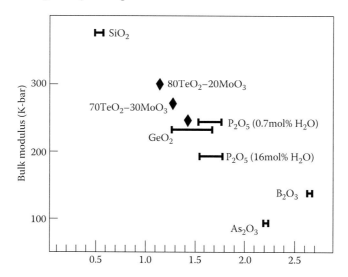

FIGURE 3.5 Relation between K (at room temperature) and E (at low temperature) of TeO$_2$–MoO$_3$ glass (From R. El-Mallawany, *Mater. Chem. Phys.*, 39, 161, 1994), together with values of the pure oxide glasses SiO$_2$, GeO$_2$, P$_2$O$_5$, and B$_2$O$_3$. (From B. Bridge and M. Patel, *J. Mater. Sci.*, 29, 3783, 1986a.)

$$D' = \left(\frac{q}{2}\right)\left[\frac{M}{n_b N_A \rho}\right]\left(\frac{\delta y}{\delta x}\right) \tag{3.9}$$

where q = longitudinal elastic modulus, δx = normal cation-cation separation, and δy is the separation of minima in the two-well potentials. The above model proposes that only the motion of the oxygen atoms is responsible for acoustic loss in glasses. The factors that affect the above model include the following:

1. The total number of acoustically active double-well systems is proportional to the oxygen density [O].
2. Both the number of double-well (acoustic) loss centers and their average activation energy increase with l.
3. The barrier height is directly proportional to the bond strength.

The basic representation of the α at any temperature T and $f(\omega)$ was given by Carini et al. (1982) and incorporates the model for $AgI–Ag_2O–B_2O_3$ glasses in Equation 3.9:

$$\alpha(T,\omega) = \left(\frac{n(A)}{2v}\right)\left[\frac{\omega^2 \tau_{th}}{(1+\omega^2 \tau_{th}^2)}\right] \tag{3.10}$$

where n is the number of loss centers (i.e., the number of oxygen atoms that absorb the acoustic waves), v and ω are the ultrasonic velocity and angular frequency, respectively, and A is the relaxation strength defined by the equation:

$$A = \left(\frac{2\alpha_{MRL}v}{2\pi f}\right) \tag{3.11}$$
$$= \left(\frac{D^2}{4\rho vkT}\right)$$

where α_{MRL} is the maximum relaxation loss at T_p, f is the frequency, ρ is the density of the material, and D is the deformation potential that expresses the energy shift ΔE of the relaxing states in a strain field of unit strength ε, which are correlated to the relation $\Delta E = D\varepsilon$. From Chapter 2, if the product of $\omega\tau$ equals 1, then the maximum acoustic absorption, obtained values of the relaxation strength A, and D for the tellurite glasses discussed here have been achieved. Table 3.2 provides data on ρ and v at room temperature for different tellurite glass series. El-Mallawany et al. (1994a and 2000d) and Sidky et al. (1997a). The deformation potential of the binary tellurite glasses $96TeO_2–4Nb_2O_5$, $90TeO_2–10Nb_2O_5$, $85TeO_2–5Nb_2O_5$, and $77.5TeO_2–22.5Nb_2O_5$ and $90TeO_2–5Nb_2O_5–5Li_2O$, $80TeO_2–10Nb_2O_5–10Li_2O$, $75TeO_2–12.5Nb_2O_5–12.5Li_2O$, $70TeO_2–15Nb_2O_5–15Li_2O$ glasses, were 0.507, 0.501, 0.486, 0.378, and 0.378, 0.364, 0.361, 0.315eV respectively by Sidky, et al. 2002.

On the basis of the calculated A and D values of tellurite glass in the binary forms $TeO_2–MoO_3$ and $TeO_2–V_2O_5$, values of both the A and D increase with increasing ultrasonic f and with increasing proportions of the modifiers MoO_3 and V_2O_5. Two significant conclusions can be drawn from the experimental measurements of ultrasonic attenuation with different temperatures and different frequencies: first, that the curves of losses against temperatures tend to be skewed to the right; second, that, in a set of loss curves versus temperatures whose tails overlap each other, the effect of increasing f is greater amounts of overlap due to thermal brooding.

TABLE 3.2

Theoretical Low-Temperature Acoustic Relaxation Phenomena for Binary Tellurite Glasses

Glass and mol% Composition	$\rho(g/cm^3)$	$V_1(m/s)$ (Source)	Relaxation Strength	Deformation Potential (eV)
TeO$_2$–MoO$_3$			P.W.	P.W.
80–20	5.010	3,272	0.379 (2 MHz)	0.66
		(El-Mallawany et al.	0.76 (4 MHz)	0.90
		1994a)	1.08 (6 MHz)	1.14
			1.48 (10 MHz)	1.35
70–30	4.900	3,190	1.49 (2 MHz)	1.19
			2.38 (4 MHz)	1.59
			2.99 (6 MHz)	1.87
			3.05 (10 MHz)	1.93
50–50	4.600	3,137	2.25 (2 MHz)	1.49
			2.49 (4 MHz)	1.64
			3.07 (6 MHz)	1.88
			3.20 (10 MHz)	1.99
TeO$_2$–V$_2$O$_5$				
80–20	4.900	2,810 (Sidky et al. 1997a)	5.38×10^{-8} (6 MHz)	0.25
70–30	4.564	3,110	1.8×10^{-5} (6 MHz)	2.87
60–40	4.225	3,620	1.79×10^{-4} (6 MHz)	14.97
50–50	3,996	3,790	3.83×10^{-4} (6 MHz)	22.02
				P.W.

Thus, the resulting ordinate due to superposition of loss curves becomes larger as the f increases. Quantitatively, E can be correlated to both the f' and the number of oxygen atoms (loss centers) present in these noncrystalline solids, as discussed in the next section.

3.6 CORRELATIONS BETWEEN LOW-TEMPERATURE ULTRASONIC ATTENUATION AND ROOM-TEMPERATURE ELASTIC MODULI

Quantitative analysis of the increase in absorption of ultrasonic waves with wave frequencies for all glasses tested indicates that this phenomenon is caused by the presence of interatomic two-well potential systems. The corresponding variations in bulk moduli (K) of the same binary or ternary tellurite glasses are correlated with changes in the values of the f' of all the different bonds and also with the values of l. Figure 3.5 represents the l of the glass network, which is defined as the number of bonds times bond length divided by π. The correlation between ultrasonic attenuation at low temperatures and the elastic modulus is based on the fact that in all noncrystalline solids (e.g., glass systems), there is a distribution of thermally average cation–anion–cation spacing about the equilibrium values, and a corresponding distribution of cation–anion–cation angles.

Previously, Bridge and Patel (1986b) evaluated the peak loss or relaxation spectra and K for the present oxide networks based on the fact that the total number of two-well systems per unit volume (N) is proportional to the product of [O], which is explained in Chapter 2, Equation 2.63, and l as follows:

TABLE 3.3
Room-Temperature K, f, l, [O], and ln[O], ln (E/K), $(f',K)^{0.576}$, $f'(f', K)^{0.076}$ of Molybdenum–Tellurite Glasses

	Results for Glass		
Property	20 mol% MoO$_3$–80 mol% TeO$_2$	30 mol% MoO$_3$–70 mol% TeO$_2$	50 mol% MoO$_3$–50 mol% TeO$_2$
K (kPa)	303	282	265
f' (N/m)	218	220	222
l (nm)	53	55	57
[O] 10^{28} (m^{-3})	4.24	4.38	4.563
ln [O]	65.916	65.949	65.99
ln (E/K)	−7.92	−7.73	−7.52
$(f'/K)^{0.576}$	2.68/10^{-5}	2.64/10^{-5}	2.518/10^{-5}
$f'(f'/K)^{0.076}$	5.84/10^{-3}	6.14/10^{-3}	6.44/10^{-3}

Source: Reprinted from *Mater Chem Phys.*, 39, El-Mallawany, R., 161, 1994, with permission from Elsevier.

$$N\alpha[\text{O}](l^m) \tag{3.12}$$

where

$$[\text{O}] = \left(\frac{C}{D}\right)\left(\frac{N}{16}\right) \tag{3.13}$$

and C is the total amount of oxygen in 100 g of glass, as explained in Chapter 2, D is the volume of 100 g of the glass, and N_A is Avogadro's number. In 1994, El-Mallawany computed the parameters for correlations between low-temperature ultrasonic attenuation and room-temperature elastic moduli of tellurite glasses. The values of the oxygen densities are given in Table 3.3. It is also worthwhile to understand that the average E (average barrier height between the two wells) is proportional to the product of the mean first-order f and l, as reported by Bridge and Patel (1986b); that is,

$$E \, \alpha \, f' \, l^m \tag{3.14}$$

where m is equal to 2.21, as stated by Lambson et al. (1984), and f' is the average stretching force constant (explained in detail in Chapter 5). Earlier in 1984, Lambson et al. argued that, for the K of oxide networks, $K = 0.106 \, f/l^n$, where l is the ring diameter and n is equal to 3.84. Quantitatively, Equations 3.11 and 3.12 are rewritten in the following form by eliminating l, using Equation 3.13:

$$N = (\text{constr.}1)\left\{\frac{F'}{K}\right\}^{m/n} \tag{3.15}$$

and

$$E = (\text{constr.}2)(F')\left\{\frac{F'}{K}\right\} \tag{3.16}$$

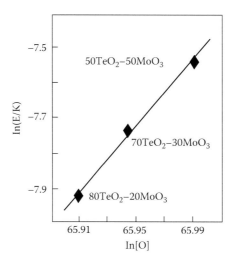

FIGURE 3.6 Relation between activation energy and *l*. (From R. El-Mallawany, *Mater. Chem. Phys.*, 39, 161, 1994.)

So Equations 3.14 and 3.15 take the forms:

$$N = (0.589)\left(\frac{F'}{K}\right)^{0.576}$$ (3.17)

and

$$E = (0.19)(F')\left(\frac{F'}{K}\right)^{0.576}$$ (3.18)

Eliminating *f'* from Equations 3.16 and 3.17,

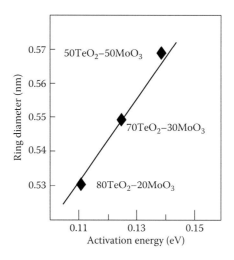

FIGURE 3.7 Relation between $f'(f'/K)^{0.5}$ and the activation energy of TeO_2–MoO_3 glass (regression 97%). (Reprinted from *Mater. Chem. Phys.*, 39, El-Mallawany, R., 161, 1994, with permission from Elsevier.)

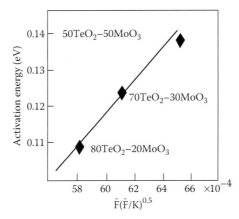

FIGURE 3.8 The variation between $\ln(E/K)$ and $\ln[\text{O}]$ of TeO_2–MoO_3 glass. Linear regression performed on $\ln(E/K)$ and $\ln[\text{O}]$ yielded a correlation of 100%. (Reprinted from *Mater. Chem. Phys.*, 39, El-Mallawany, R., 161, 1994, with permission from Elsevier.)

$$E = (\text{constr.3})\, KN^{\left(1 + \frac{m}{n}\right)} \tag{3.19}$$

The linear regression performed on $\ln(E/K)$ and $\ln(N)$ yields the result:

$$E = KN^{1.576} \tag{3.20}$$

In 1994, El-Mallawany calculated the structural parameters for binary transition metal tellurite glasses as presented in Table 3.3, which shows room-temperature ultrasonic properties of TeO_2–MoO_3 glasses, including K, l, f', [O], and number of loss centers and values of the above relations. In binary TeO_2–MoO_3 glass, both the calculated stretching-force constant and l increased with [O], while the experimental K decreased with higher percentages of the modifier MoO_3. The relationship between K (at room temperature) and E (at low temperature) is represented in Figure 3.5, together with the values of pure oxide glasses (SiO_2, GeO_2, P_2O_5, and B_2O_3). Tellurite glasses lie on the experimental hyperbolic curve and are considered normal glass.

Figure 3.6 represents the relationship between activation energy and l (calculated from the room-temperature elastic moduli), which is directly proportional to E. The level of [O] of the present glass increases from 4.24×10^{28} to 4.563×10^{28} m^{-3}, which means that the number of loss centers increases for higher percentages of MoO_3. The relationship between $f'(f'/K)^{0.5}$ and activation energy is represented in Figure 3.7, with a regression of 97%. Figure 3.8 represents the variation between $\ln(E/K)$ and $\ln[\text{O}]$. Linear regression performed on $\ln(E/K)$ and $\ln[\text{O}]$ yields a correlation of 100%.

3.7 ACOUSTO-OPTICAL PROPERTIES OF TELLURITE GLASSES

Both Yano et al. (1974) and Izumitani and Masuda (1974) measured the acousto-optical properties of tellurite glasses by using the Dixon–Cohen method. The Dixon–Cohen comparison includes the assumption that soundwave transmission through sample–referent bonding is reciprocal to a 6,328-Å laser light source (He–Ne). The reciprocity should be verified experimentally. The profiles of acoustic-wave flux intensity distribution are obtained by moving the optical beam parallel to the transducer plane. Profile P1 corresponds to the acoustic-wave flux through which the first-order diffracted light L1 is observed. Similarities between P1, P2, P3, and P4 indicate good sound transmission reciprocity through the bonding layer. A "Z-cut $LiNbO_3$" transducer is connected using an

TABLE 3.4
Me of Tellurite Glasses

Physical Property	TeO$_2$–WO$_3$–Li$_2$O (Izumitani and Masuda 1974)	Tellurite-Based Glass (Yano et al. 1974)	
		Long	Shear
Light transmission	0.43–2.7 μ		
Ultrasonic Velocity (10^5 cm/s)	3.33	3.4	1.96
Ultrasonic Attenuation (dB/cm at 100 MHz)	2.0	3.0	1.5
Me (10^{-18} s^3/g) at 6,328 Å	20.0	23.9	0.12
Refractive Index at 6,328 Å		2.089	
Density (g^3/cm)		5.87	
Optical Polarization		Parallel 0.257	
		Perpendicular 0.214	
		Parallel or perpendicular 0.0079	

α-cyanoacrilate bonding agent. Measurements are made in one of two ways: either the transducer is cemented to the sample, or the transducer is cemented to the referent.

The relationship between diffracted-light intensities and the square of the input voltage recorded for TeFD5-based tellurite glass, as examined by Yano et al. (1974), is linear. The Me and P that result from these values are listed in Table 3.4. Also, Yano et al. (1974) measured transmission as a function of wavelength, and acoustic absorption as a function of wave f. From Table 3.4, the value of the Me of TeFD5-based tellurite glasses measured by Yano et al. (1974) is 23.9 × 10^{-18} s^3/g, which is larger than that of extra-dense flint glass. The values of Me for both tellurite glass and crystal are close. The ratio of Me (P11) to Me (P12) equals 1.14. This polarization dependence of the Me for tellurite glass is less than the 1.35 ratio of Me (P33) and Me (P13) for TeO$_2$ crystal. This property is very useful for practical light modulators. The value of the acoustic loss for a longitudinal sound-wave in tellurite-based glass (α = 3.0 dB/cm at 100 MHz) is larger than that of TeO$_2$–001 crystal (α = 1.5 dB/cm at 100 MHz).

Izumitani and Masuda (1974) measured tellurite glasses with different compositions and classi-fied the reasons (see below) that these glasses all have high Me and low acoustic losses; that is, the mechanisms of ultrasonic attenuation at room temperature for ternary tellurite glasses of the form TeO$_2$–WO$_3$–LiO$_2$. The acoustic loss at 30 MHz, thermal conductivity, n, specific heat capacity (C_v), V, Me, and other values for two selected tellurite glasses—AOT-44 and AOT-5—are given in Table 3.4. Izumitani and Masuda (1974) discussed their data in terms of the following categories:

1. Frequency dependence of ultrasonic attenuation
2. Temperature dependence of ultrasonic attenuation
3. Dependence of ultrasonic attenuation on glass composition
4. Dependence of C_v capacity on glass composition
5. Discussion of the mechanism of ultrasonic attenuation at room temperature
6. Me of tellurite glass

3.7.1 Frequency Dependence of Ultrasonic Attenuation

Izumitani and Masuda (1974) measured the f dependence of α in the range 10–200 MHz at room temperature in glasses in the TeO$_2$–WO$_3$–Li$_2$O system. They found that

$$\alpha \propto f^x \qquad (3.21)$$

where α is the ultrasonic attenuation in decimals per centimeter. In the same report, these authors found that the value $x = 1.7$ is close to the f dependence of quartz. This value of x does not vary with changes of temperature in the range $-100°C$ to $+100°C$. α is independent of acoustic power in the range 0.1 to 1.3 W. The amplitude independence shows that changes in attenuation are not hysteresis-type acoustic losses.

3.7.2 TEMPERATURE DEPENDENCE OF ULTRASONIC ATTENUATION

The dependence of attenuation of longitudinal acoustic waves in fused quartz on temperature was measured at $3°C$ intervals starting from $-200°C$ by Izumitani and Masuda (1974) to check the experimental error caused in glass by the temperature dependence of bonded resin. As the temperature of liquid nitrogen surrounding tellurite glass in the TeO_2–WO_3–Li_2O system increases from $-200°C$ to $150°C$, α remains nearly unaffected; however, α increases gradually as the temperature increases from $150°C$ to the glass softening point. No attenuation peak due to deformation loss like those observed in silicate glasses is observed in tellurite glass.

3.7.3 DEPENDENCE OF ULTRASONIC ATTENUATION ON GLASS COMPOSITION

Izumitani and Masuda (1974) measured ultrasonic attenuation at 30 MHz f in seven glasses in the ternary TeO_2–WO_3–Li_2O system. They found that ultrasonic attenuation mainly depends on the TeO_2 content. TeO_2-rich and WO_3- or Li_2O-poor glasses have low values of α.

3.7.4 DEPENDENCE OF SPECIFIC HEAT CAPACITY ON GLASS COMPOSITION

From the compositional dependence of C_v, the relaxation time from thermal conductivity τ_{th} was calculated by Izumitani and Masuda (1974) using the equation:

$$\tau_{th} = \left(\frac{3K}{C_V V^2} \right) \qquad (3.22)$$

where K is the thermal conductivity, C_v is the specific heat capacity at constant volume, and V is the sound velocity.

3.7.5 MECHANISM OF ULTRASONIC ATTENUATION AT ROOM TEMPERATURE

From their experimental results on acoustic loss at room temperature in the range from 10–200 MHz frequencies, Izumitani and Masuda (1974) pointed out the following typical loss features of tellurite glasses:

1. Temperature independence of attenuation is observed in the range from $-200°C$ to $+150°C$
2. Frequency dependence is roughly subject to the quadratic law
3. Attenuation is proportional to τ_{th} derived from thermal conductivity

Ultrasonic attenuation in α-quartz in the f range of about 100 MHz at room temperature is attributed to the interaction between acoustic phonons and lattice phonons, as reported by Bommel and Dransfeld (1960). Such a process of ultrasonic attenuation is also expected in glass. In the phonon–phonon interaction attenuation for condition $\omega\tau \ll 1$—called the Akhieser mechanism

(Akhieser 1939)—ω is the angular f of acoustic waves, and τ_1 is the relaxation time of lattice phonons. In this case, the mean free path of lattice phonons is short enough compared with the wavelength of the acoustic wave. Lattice phonon distribution is modulated by acoustic waves. The deviation from equilibrium distribution is relaxed through the anharmonic interaction between lattice phonons. Akhieser acoustic loss (α in the following equation) is expressed by the Woodruff–Ehrenreich equation, as follows:

$$\alpha = \left[\left(\omega^2 \tau_1 \right) \left(\frac{C_V T \gamma^2}{3 \rho v^3} \right) \right] \tag{3.23}$$

where C_v is the specific heat, γ is an average Gruneisen constant, and τ_1 is relaxation time for lattice phonons. Equation 3.23 shows that acoustic loss varies with ω^2 and is independent of temperature ($\tau_1 = T^{-1}$).

Izumitani and Masuda (1974) found that in tellurite glasses, the f dependence of ultrasonic attenuation is given by $\omega^{1.7}$ and approximately agrees with the quadratic f dependence of the Woodruff–Ehrenreich equation. Temperature independence of attenuation in tellurite glass in the range from $-200°C$ to $-150°C$ supports the finding that the acoustic loss in tellurite glass is due to the Akhieser mechanism. The relaxation time τ_1 in the typical tellurite glass AOT-5 (TeO_2–WO_3–Li_2O) is calculated to be 5.68×10^{-11} s from Equation 3.23, using $\alpha = 3$ dB/cm at 100 MHz, $C_v = 0.340$ cal/cm^3/deg, $\rho = 5.87$ g/cm^3, $v = 3.48 \times 10^5$ cm/s, and $\gamma = 1.65$.

In a similar way, the calculated relaxation times of α-quartz and fused silica are as follows: τ_1 of quartz, 0.8×10^{-11} s; τ_1 of fused silica, 1.0×10^{-11} s. The calculated relaxation time of tellurite glass is roughly the same as the relaxation times of α-quartz and fused silica, whose acoustic losses are explained by phonon interaction. This suggests that the acoustic loss in tellurite glass is due to phonon–phonon interaction.

Attenuation is apparently proportional to the relaxation time of the thermal phonon τ_{th}, but the attenuation calculated from the Woodruff–Ehrenreich equation using τ_{th} is about 10^{-3}-fold smaller than the observed value. This means that τ_{th} of the thermal phonon applied to crystal can be applied to glass in Equation 3.23; although in glass material, τ_1 should be applied in place of τ_{th} in Equation 3.23. An approximately linear relationship is found between τ_1 and τ_{th}. Thus, the apparent relationship between attenuation and τ_{th} can be understood. From the above discussions, acoustic loss in the f range 10–200 MHz at room temperature is due to phonon–phonon interaction in the $\tau\omega << 1$ range.

The τ_1 (10^{-11} s) from ultrasonic attenuation is much longer than the τ_{th} (10^{-13} s) from thermal conductivity. Such a phenomenon is observed not only in tellurite glasses, but also in fused silica and chalcogenide glasses. The τ_{th} of amorphous materials is known to be determined by the irregularity of glass structures. The relationship between the τ_1 calculated from acoustic loss and glass structure is not yet clarified. Accordingly, Izumitani and Masuda (1974) could not explain the inconsistency between τ_1 and τ_{th} in glass; but the inconsistency between τ_1 and τ_{th} can be regarded as a characteristic of amorphous materials. The longer τ_1 in the vitreous state is the main reason that ultrasonic attenuation of amorphous material is generally larger than that of crystalline material.

3.7.6 Figure of Merit of Tellurite Glass

Izumitani and Masuda (1974) measured the composition dependence of the Me, V, n, ρ, and photo-elastic constant P in TeO_2–WO_3–Li_2O ternary glass systems according to the relation:

$$Me = \left(\frac{n^6 P^2}{\rho V^3} \right) \tag{3.24}$$

The main factor that determines Me in the TeO_2–WO_3–Li_2O ternary glass system is TeO_2 content; that is, Me increases with increasing TeO_2 content. The measured Me values range from 20 to 27×10^{-18} s^3/g. These values are very high in comparison with other high-n oxide glasses, whose Me values are, at most, 19×10^{-18} s^3/g. The n and V are also determined by TeO_2 content.

The relationship between Me and ρ, v, n, and P are summarized as follows:

1. n plays the most important role in determining Me values and also V, whereas ρ and P do not affect the Me.
2. The effect of acousto-optical properties of tellurite glasses on ultrasonic attenuation in these glasses is mainly determined by the relaxation time of the Woodruff–Ehrenreich equation and Me by n.

Izumitani and Masuda (1974) concluded that a high Me and low loss are not contradictory. The increase of n with increasing TeO_2 content is ascribed to the high polarizability of the Te^{4+} ion. Tellurite glass has a high Me and low acoustic loss because it contains a large amount of network-former Te ions, which have higher polarizability than the other network-forming cations, plus a small amount of modifier ions. High transmittance in the visible wavelength region of tellurite glass is also one of its useful properties for applications in acousto-optical material.

Still, we suggest that a low-f Raman and Brillouin scattering analysis be carried out on tellurite glasses over a wide range of temperatures—from that of liquid helium to the glass transition temperature—to give a deeper insight into the problem of low-temperature sound velocity and attenuation.

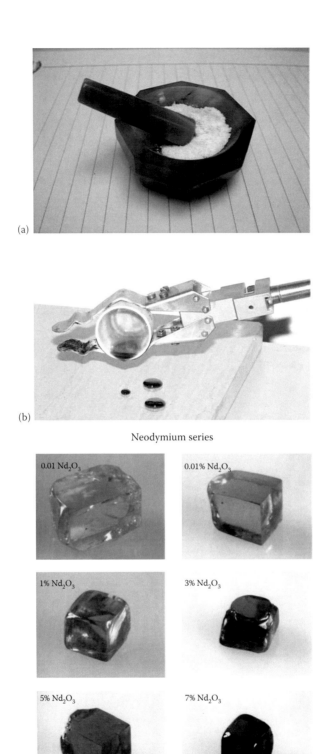

(a)

(b)

Neodymium series

0.01 Nd₂O₃   0.01% Nd₂O₃

1% Nd₂O₃   3% Nd₂O₃

5% Nd₂O₃   7% Nd₂O₃

(c)

FIGURE 1.27 (a–c) Color of a series of $80TeO_2$–$5TiO_2$–$(15 - x)$ WO_3– xNd_2O_3 glass with $x = 0.01$, 0.1, 1, 3, 5, and 7 mol% for Nd_2O_3 (From I. Abbas, Ph.D. thesis at Sudan University of Science and Technology, 2006).

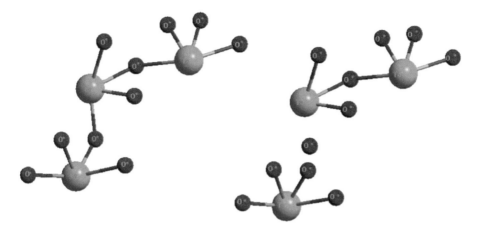

FIGURE 7.15 A mechanism for structural changes in TeO glass network from TeO_4 (tbp) to TeO_{3+1} polyhedra induced by modifiers. (Reprinted from Y. Gandhi, N. Venkatramaiah, V. Ravi Kumar, and N. Veeraiah, *Physica B: Condensed Matter*, 1450, 2009. With permission.)

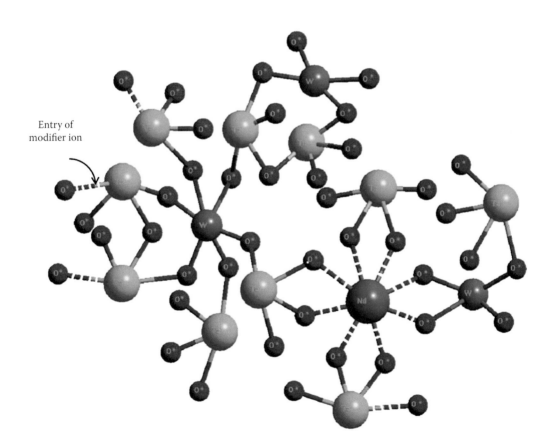

Entry of modifier ion

FIGURE 7.16 An illustration of a Nd^{3+} ion embedded in the TeO_2 glass network with tungsten ions in tetrahedral and octahedral positions. (Reprinted from Y. Gandhi, I.V. Kityk, M.G. Brik, P. Raghava Rao, and N. Veeraiah, *J. Alloys and Comp.*, 278, 2010. With permission.)

4 Applications of Ultrasonics on Tellurite Glasses

In this chapter, phase separation, acoustic and optical Debye temperatures, and γ-radiation tests on nondestructive elastic moduli are summarized for analysis and manipulation of both oxide– and nonoxide–tellurite glasses in pure, multicomponent forms and new tellurite glass systems. The optical Debye temperatures have been calculated from the infrared spectra of tellurite glasses modified with rare-earth oxides. The effect of γ-radiation on acoustic Debye temperatures is also examined. Other topics in passing include microhardness, calculated softening temperatures, calculated thermal expansion, and stress fatigue in tellurite glasses.

4.1 INTRODUCTION

Ultrasonic techniques continue to be favored to measure the elastic constants of glasses. These increasingly precise methods are evolving, no doubt, as a result of the higher complexity and lower regularity in new species of glasses and the great increase in sites performing such research and development. Some of these new tools have made it more profitable, for instance, to study pressure and temperature effects on elastic constants. For example, one such elegant new measurement technique, although not yet applied to tellurite glasses, is called Brillouin scattering, which is based on the fact that at any temperature above absolute zero, all elastic solids contain thermally activated density fluctuations that propagate within the solid in multiple directions at many velocities and frequencies. If a transparent elastic solid is probed with a laser and the laser satisfies Bragg's law, then Brillouin scattering, as well as Rayleigh scattering, occurs. In Brillouin scattering, the frequency and direction of wave train propagation produced by the laser provide data that, if accurately measured, enable precise calculations of both Young's modulus and the shear modulus—and from these, Poisson's ratio.

One new analytical method is based on the fact that not all structural elements of glass precursor crystals are destroyed immediately upon liquification. The influences of intermediate or transitional structures in a melt on the resulting glass structure are becoming increasingly evident. Proof for the existence of microinhomogeneities in glasses is inseparably associated with the development of electron microscopy as a structure research tool. Interest in these progressive changes in the structure in glasses, a process termed "phase separation," started in the 1950s but has become an essential topic of modern glass research. Several authors have published chapters on this phenomenon in glasses (e.g., Vogal 1985, Tomozawa 1979). Tomozawa (1979) addressed glass phase separation in terms of metastable immiscibility boundaries, kinetics of phase separation, and effects of phase separation on such properties as chemical durability, electrical properties, mechanical properties, thermal properties, density and refractive index, light scattering characteristics, and crystallization.

Levan (1970) presented a method for understanding the glass structure using the principles of crystal chemistry, in which cations are generally surrounded by several nonbridging oxygen (NBO) ions. Because the number of these ions is limited, it is necessary for them to bond to more than one cation. The requirement that two cations must bond to the same oxygen determines the "limiting composition." The limiting composition leading to immiscibility can be calculated by finding a cation concentration corresponding to the cation–oxygen–cation bond length. Thus, the immiscibility range of a glass is primarily controlled by the ionic field strength of its cations. As Tomozawa (1979) observed, the fact that the properties of glass change in general as phase separation progresses can be used to manipulate the properties of glass and to study the kinetics of its microstructure development. In addition, phase-separated glass offers a model system for a ceramic–ceramic composite material. Elastic constants appear insensitive to microstructural size; however, they are dependent on changes in the volume fractions as well as on the compositions of the phases.

Two-phase materials (or "composites") offer unique physical properties often unattainable from their individual constituents alone (Clarke 1992, Chawla 1987). Metal matrix composites and the electrical resistivity of composites have been reported by Everett (1991) and McLachian (1990). Oxide glasses can be regarded as solutions of oxides in which no definite combinations occur. Each oxide exerts its own influence so that any physical property of glass as a whole can reasonably be considered as the sum of the influences of its individual oxides. There have been a number of theoretical approaches to predict these properties as a function of volume fraction in second phase. In particular, predictions of elastic properties such as Young's modulus, shear modulus, and Poisson's ratio in two-phase materials have been made by Shaw (1971), Hashin (1962), Hashin and Strikman (1963), and Ross (1968). Expressions have been devised for upper and lower bounds on the elastic properties of two-phase systems, and these have been compared with experimental data. Of all the models proposed, the boundaries derived using principles of elasticity most accurately describe the variation of elastic properties with second-phase volume fractions, as mentioned by Hashin and Strikman (1963).

The Debye theory of specific heat has also proven very useful because it is a single-parameter theory, based on the Debye temperature (Θ_D), which predicts actual observations remarkably well. The Θ_D need not be determined by any heat capacity measurements but can be calculated from the elastic moduli. Once the Θ_D parameter has been determined from the elastic moduli, the Debye theory specifies the lattice contribution to specific heat with a standard error of only 10%–20% over most of the temperature range. The theoretical model assumes that the solid is an elastic continuum in which all the soundwaves travel at the same velocity, independently of their wavelength. Such a model is satisfactory within the limits of long wavelengths or low temperatures. Atoms in a solid are so tightly coupled that they most certainly oscillate together. To extend this fact into theory, Debye conceived of the solid as an elastic continuum and derived an expression for the frequencies that such a system could support.

In a well-known book, *Glass: Science and Technology*, Ernsberger (1980) discussed the hardness of glasses in detail, from definitions and experimental methods to the hardness values of common glasses. The familiar square-pyramid Vickers indentation test is the most widely used test of hardness. It was inevitable that someone would try this test on glass, although glass is a brittle material for which the concept of plastic deformation at first seems eminently unsuitable. An important problem with the hardness data for glass is the poor reproducibility of measurements. Ernsberger (1980) concluded that hardness testing of glass is not particularly fruitful. It is easy to produce data, but not so easy to establish what they mean; however, Ernsberger did reach two conclusions: hardness data have only a limited absolute significance in glass research and development, and the measurement itself is highly subjective. Ernsberger's first conclusion counsels caution in comparing data generated in different laboratories; the second suggests that even within a single laboratory, the data may unconsciously be manipulated by the experimenter unless double-blind procedures are followed.

In addition to failure caused by brittleness, another characteristic of the rupture of glass is that its measured strength depends on the length of time that a load is applied or on the loading rate. Termed "static fatigue," this characteristic has been associated with two phenomena: (1) a stress correction process, in which a sufficiently large stress enhances the rate of corrosion of the crack tip relative to that at the sides, leading to sharpening and deepening of the crack; and (2) eventual failure and lowering of the surface energy by the adsorption of active energy, which leads to a decrease in the fracture surface energy to perform fracture surface work.

Still another kind of fracture that should be mentioned is "creep failure," which is encountered with polycrystalline materials, under which conditions a significant part of plastic deformation results from grain/boundary sliding.

4.2 ULTRASONIC DETECTION OF MICROPHASE SEPARATION IN TELLURITE GLASSES

4.2.1 THEORETICAL CONSIDERATIONS

The ultrasonic detection method gives information on the interiors of bulk specimens, whereas an electron microscope probes only surface layers. It is possible to perform a test for the presence or absence of a two-phase system by using ultrasonically derived data and theoretical analysis of the composition dependence of elastic moduli compared with experimental values. In two-phase systems, molecules in general are intermediate between the high- and low-modulus components, a fact investigated by a number of researchers, including Shaw and Uhlamann (1971), Hashin (1962), Hashin and Strikman (1963), and Ross (1968). Generally, these researchers discussed the upper and lower bounds between which the various elastic properties are applicable. For a binary material consisting of a matrix of volume fractions (V_1 and V_2), the second phase (V_2) is given by:

$$V_2 = \left\{ \frac{\rho_1 (x - x_1)}{\left[(x - x_1)(\rho_1 - \rho_2) \right] + (x_2 - x_1)\rho_2} \right\} \tag{4.1}$$

where x is the weight fraction of the second phase ($x_1 < x < x_2$, and x_1 and x_2 are called the end-member components), and ρ_1 and ρ_2 are the densities of the end-member components. According to the first Voigt model by Shaw et al. (1971), which assumes uniform strain, the bulk modulus (K^*), the shear modulus (G^*), and Young's modulus (E^*) of the composite become:

$$K_H^* = (1 - V_2) K_1 + V_2 K_2$$

$$G_H^* = (1 - V_2) G_1 + V_2 G_2 \tag{4.2}$$

$$E_H^* = (1 - V_2) E_1 + V_2 E_2$$

where the subscripts 1 and 2 refer to the first and second end-member components.

The second model by Ross (1968) assumes uniform stress, and the moduli become:

$$\frac{1}{K_L^*} = \frac{(1 - V_2)}{K_1} + \frac{V_2}{K_2}$$

$$\frac{1}{G_L^*} = \frac{(1 - V_2)}{G_1} + \frac{V_2}{G_2} \tag{4.3}$$

$$\frac{1}{E_L^*} = \frac{(1 - V_2)}{E_1} + \frac{V_2}{E_2}$$

Equation 4.2 forms the upper limits and Equation 4.3 forms the lower limits of the various quantities. Hashin and Strikman's expression (1963) was for $K_2 > K_1$ and $G_1 > G_2$ as follows:

$$K_H^* = K_2 + \frac{(1-V_2)}{\left[\dfrac{1}{(K_2-K_1)}\right] + \left[\dfrac{3V_2}{(3K_2+4G_2)}\right]}$$

$$K_L^* = K_1 + \frac{(V_2)}{\left[\dfrac{1}{(K_2-K_1)}\right] + \left[\dfrac{3(1-V_2)}{(3K_2+4G_2)}\right]} \qquad (4.4)$$

$$G_H^* = G_2 + \frac{(1-V_2)}{\left[\dfrac{1}{(G_1-G_2)}\right] + \left[\dfrac{6(K_1+2G_1)V_2}{5G_2(3K_2+4G_2)}\right]}$$

where the subscripts H and L denote the upper and lower bounds, respectively; subscripts 1 and 2 denote the matrix and the particulate phase, respectively; and V is as previously defined. It should be noted that these bounds are not independent, since they were obtained by merely reversing the roles of the matrix before inclusion in the analysis. Equivalent upper and lower bounds for E^* are obtained simply by combining Equations 4.4 with the relation:

$$E^* = \frac{9K^*G^*}{3G^* + G^*} \qquad (4.5)$$

The slopes and curvatures of the relations described by Equation 4.4 show that the modulus-volume fraction curves depend only on the relative values of the end-member moduli; the curvatures of such plots are always positive (i.e., concave upward); and no maxima, minima, or points of inflection or discontinuities exist.

4.2.2 APPLICATION TO BINARY TELLURITE GLASSES

In 1992, El-Mallawany tested phase separation in binary zinc–tellurite glasses, as shown in Figure 4.1, using data collected by Kozhukarov et al. (1986) for the composition dependence of all the elastic moduli of the glass system TeO_2–ZnO through the entire vitreous range from 0 to 10% (weight basis [wt%]) ZnO. Examination of the figures shows no evidence for two-phase immiscibility gaps in the entire composition range of zinc–tellurite glasses, because all experimental values of bulk, shear, and Young's moduli lie well outside the upper and lower bounds of the calculated values of the K^*, G^*, and E^* moduli. There was reasonable agreement between the predictions of the above models and the electron-micrographic observations by Kozhukarov et al. (1986). The moduli in Figure 4.1 are plotted against wt% composition instead of volume fraction (mol%), and the curvatures can be positive, negative, or zero according to the criteria given in Table 4.1. In all cases, the slopes of the functions are controlled by the relative values of the end-member properties, and no maxima, minima, inflections, or discontinuities exist between the end members. The atomic arrangement in this kind of tellurite glass has been accurately described by Kozhukharov et al. (1986) by analysis of the theoretical and experimental radial-distribution functions for neutron diffraction. The atomic arrangement in the zinc–tellurite glass $(TeO_2)_{0.8}$–$(ZnO)_{0.2}$ shows that the main structural units in this glass are similar to those in paratellurite and zinc–tellurite, and they are predominantly those of the $Zn_2Te_3O_8$ crystal structure, as illustrated in Chapter 1.

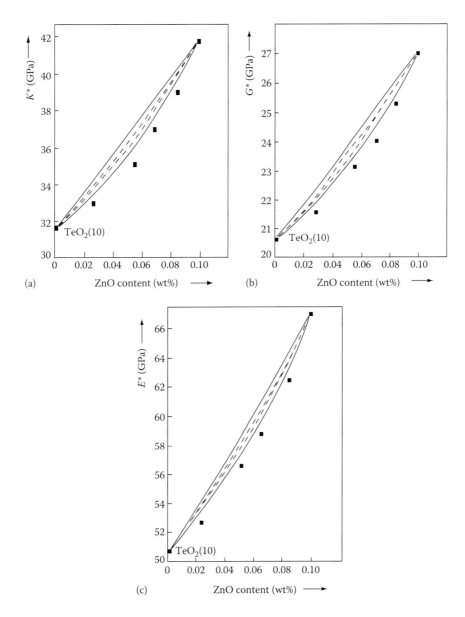

FIGURE 4.1   Variations (in gigapascals) of (a) bulk (K^*), (b) shear (G^*), and (c) Young's (E^*) moduli of TeO_2–ZnO glass with compositions (wt%). Filled squares represent experimental data, solid lines show Voigt and Reuss boundaries, and dashed lines represent Hashin and Strikman's upper and lower bounds, if the glasses are assumed to consist of separate TeO_2 and ZnO phases. (From R. El-Mallawany, *Phys. Stat. Sol.*, 133, 245, 1992. With permission.)

4.3   DEBYE TEMPERATURE OF OXIDE AND NONOXIDE–TELLURITE GLASSES

4.3.1   EXPERIMENTAL ACOUSTIC DEBYE TEMPERATURE

The Debye theory of specific heat, although based on a rather primitive model for the frequency spectrum of a solid, provides a remarkably useful device for comparing the thermal properties of solids in a fashion similar to the corresponding-states model applied to gases. According to the Debye model, one can relate a characteristic temperature (Θ_D) to the maximum frequency of vibration (ν_{max}) or to the velocity of sound in the solid by the following relations:

TABLE 4.1
Curvature of Hashin–Strikman/Kerner Elastic Modulia[a]

Effective Bulk Modulus (K^*)	Curvature	Effective Shear Modulus (G^*)
$K2 > K1$	Positive	$G2 > G1$
$\dfrac{\rho_1}{\rho_2} < \dfrac{3K_2 + 4G_1}{3K_1 + 4G_1}$		$\dfrac{\rho_1}{\rho_2} < \dfrac{G_1(7 - 5\sigma_1) + G_2(8 - 10\sigma_1)}{G_1(15 - 15\sigma_1)}$
$K2 < K1$	Positive	$G2 < G1$
$\dfrac{\rho_1}{\rho_2} < \dfrac{3K_2 + 4G_1}{3K_1 + 4G_1}$		$\dfrac{\rho_1}{\rho_2} < \dfrac{G_1(7 - 5\sigma_1) + G_2(8 - 10\sigma_1)}{G_1(15 - 15\sigma_1)}$
$K2 > K1$	Negative	$G2 > G1$
$\dfrac{\rho_1}{\rho_2} < \dfrac{3K_2 + 4G_1}{3K_1 + 4G_1}$		$\dfrac{\rho_1}{\rho_2} < \dfrac{G_1(7 - 5\sigma_1) + G_2(8 - 10\sigma_1)}{G_1(15 - 15\sigma_1)}$
$K2 < K1$	Negative	$G2 < G1$
$\dfrac{\rho_1}{\rho_2} < \dfrac{3K_2 + 4G_1}{3K_1 + 4G_1}$		$\dfrac{\rho_1}{\rho_2} < \dfrac{G_1(7 - 5\sigma_1) + G_2(8 - 10\sigma_1)}{G_1(15 - 15\sigma_1)}$
$K2 = K1$	Zero	$G2 = G1$
$\dfrac{\rho_1}{\rho_2} = \dfrac{3K_2 + 4G_1}{3K_1 + 4G_1}$		$\dfrac{\rho_1}{\rho_2} = \dfrac{G_1(7 - 5\sigma_1) + G_2(8 - 10\sigma_1)}{G_1(15 - 15\sigma_1)}$

Source: R. Shaw and D. Uhlmann, *J. Non-Cryst. Solids*, 5, 237, 1971.

[a] The effective Young's modulus (E^*) has the same curvature as the bulk and shear moduli over the same composition range. It should be noted that modulus relations are quite insensitive to variations in Poisson's ratio. ρ, density; σ, Poisson's ratio.

$$\Theta_D = \left(\frac{h}{K_B}\right) v_{max}$$

$$= \left(\frac{h}{K_B}\right)\left(\frac{9N}{4\pi V}\right)^{1/3}\left[\left(\frac{1}{v_L}\right) + \left(\frac{2}{v_s}\right)\right]^{-1/3} \tag{4.6}$$

where v_L and v_s are longitudinal and shear velocities of sound, h and K_B are the Planck and Boltzman constants, and N/V is the number of vibrating atoms per atomic volume (as explained in Chapter 1). The sound velocities were measured as explained in Chapter 2.

The measured values of Θ_D (acoustic Θ_D) for tellurite glasses (pure and modified) have been collected and summarized from the work of El-Mallawany and Saunders (1987, 1988), El-Mallawany et al. (1994a), Sidky et al. (1997a), El-Mallawany (1992b), El-Mallawany et al. (2000a,b,c), and Sidky et al. (1999). The acoustic Θ_D data were obtained from v (longitudinal and shear) measurements according to the relation in Equation 4.6. Optical Θ_Ds were calculated and correlated with the measured acoustic Θ_D for rare earth tellurite glass systems by El-Mallawany (1989). The optical Θ_Ds were calculated from infrared absorption spectra. Table 4.2 summarizes the experimental values of the acoustic Θ_D for binary and ternary tellurite glasses. The effect of radiation on the Θ_D has been discussed by El-Mallawany (2000b).

From Table 4.2, it is clear that acoustic Θ_D (elastic—hereinafter "Θ_D [el]") values of tellurite glasses measured from room temperature ρ and v values range from 250 to <300 K. The Θ_D of pure TeO$_2$ glass is 249 K, as given by El-Mallawany and Saunders (1988), which is lower than Θ_D (el) values for pure SiO$_2$ (495 K) and Θ_D (el) for B$_2$O$_3$ (276 K), as mentioned by Maynell and Saunders (1973),

TABLE 4.2

Experimental Values of the Acoustic Debye Temperatures of Oxide– and Nonoxide–Tellurite Glasses

Glass (Reference[s])	ρ (g/cm³)	v_m (m/s)	Θ_D (K)	L (nm)
Oxide Tellurite Glasses				
TeO_2 (El-Mallawany and Saunders 1987)	5.1050	2,065	249.0	0.530
85 mol% TeO_2–15 mol% WO_3 (El-Mallawany and Saunders 1987)	5.2500	2,094	253.0	
79 mol% TeO_2–21 mol% WO_3	5.3900	2,142	261.0	
67 mol% TeO_2–33 mol% WO_3	5.7000	2,158	266.0	
90 mol% TeO_2–10 mol% $ZnCl_2$	5.0000	1,944	233.0	
80 mol% TeO_2–20 mol% $ZnCl_2$	4.8700	1,879	225.0	
76 mol% TeO_2–33 mol% $ZnCl_2$	4.6300	1,873	292.0	
90 mol% TeO_2–10 mol% La_2O_3 (El-Mallawany and Saunders 1988)	5.6850	2,145	265.0	0.515
90 mol% TeO_2–10 mol% CeO_2	5.7060	2,154	268.0	0.510
90 mol% TeO_2–10 mol% Sm_2O_3	5.7820	2,197	271.0	0.514
60 mol% TeO_2–10 mol% WO_3–10 mol% Er_2O_3	6.7130	2,195	281.0	0.519
77 mol% TeO_2–20 mol% WO_3–3 mol% Y_2O_3	6.0180	2,091	259.0	0.509
49 mol% TeO_2–29 mol% WO_3–2 mol% La_2O_3	6.0270	2,095	264.0	0.506
75 mol% TeO_2–20 mol% WO_3–5 mol% Sm_2O_3	6.1100	2,128	268.0	0.495
74 mol% TeO_2–21 mol% WO_3–5 mol% CeO_2	5.7810	2,069	257.0	0.496
49 mol% TeO_2–29 mol% WO_3–20 mol% PbO–2 mol% Er_2O_3	6.8130	2,194	275.0	0.538
77 mol% TeO_2–20 mol% WO_3–3 mol% Y_2O_3	6.0180	2,091	259.0	0.509
49 mol% TeO_2–29 mol% WO_3–2 mol% La_2O_3	6.0270	2,095	264.0	0.506
75 mol% TeO_2–20 mol% WO_3–5 mol% Sm_2O_3	6.1100	2,128	268.0	0.495
74 mol% TeO_2–21 mol% WO_3–5 mol% CeO_2	5.7810	2,069	257.0	0.496
49 mol% TeO_2–29 mol% WO_3–20 mol% PbO–2 mol% Er_2O_3	6.8130	2,194	275.0	0.538
80 mol% TeO_2–20 mol% MoO_3 (El-Mallawany 1994)	5.0100	1,932	237.4	
70 mol% TeO_2–30 mol% MoO_3	4.9000	1,883	232.8	0.550
55 mol% TeO_2–45 mol% MoO_3	4.9000	1,858	231.5	
50 mol% TeO_2–50 mol% MoO_3	4.7500	1,853	230.2	0.570
65 mol% TeO_2–35 mol% V_2O_5 (Sidky et al. 1997)	4.3300	1,882	225.0	0.510
50 mol% TeO_2–50 mol% V_2O_5	3.9900	2,051	247.0	0.496
65 mol% TeO_2–(35 − x) mol% V_2O_5–x mol% Sm_2O_3 (Sidky et al. 1999)				
x = 0.1	4.3760		291.0	
x = 1.0	4.4360		293.0	
x = 3.0	4.5310		295.0	
x = 5.0	4.6890		299.0	
70 mol% TeO_2–27 mol% V_2O_5–3 mol% CeO_2 (El-Mallawany et al. 2000)	6.0880	1,131	218.0	0.444
70 mol% TeO_2–25 mol% V_2O_5–5 mol% CeO_2	6.1327	1,152	219.0	0.436
70 mol% TeO_2–27 mol% V_2O_5–7 mol% CeO_2	6.5380	1,182	228.0	0.421
70 mol% TeO_2–27 mol% V_2O_5–10 mol% CeO_2	6.7610	1,229	240.0	0.405
70 mol% TeO_2–27 mol% V_2O_5–3 mol% ZnO	5.0769	1,350	246.0	0.426
70 mol% TeO_2–27 mol% V_2O_5–5 mol% ZnO	5.1169	1,296	235.0	0.433
70 mol% TeO_2–27 mol% V_2O_5–7 mol% ZnO	5.1919	1,072	195.0	0.475
70 mol% TeO_2–27 mol% V_2O_5–10 mol% ZnO	5.2585	863	156.0	0.527
50 mol% TeO_2–45 mol% V_2O_5–5 mol% Ag_2O (El-Mallawany et al. 2000)	4.4680	1,935	288.5	

(continued)

TABLE 4.2 (Continued)
Experimental Values of the Acoustic Debye Temperatures of Oxide– and Nonoxide–Tellurite Glasses

Glass (Reference[s])	ρ (g/cm³)	v_m (m/s)	Θ_D (K)	L (nm)
50 mol% TeO₂–30 mol% V₂O₅–20 mol% Ag₂O	5.3676	1,527	285.8	
50 mol% TeO₂–22.5 mol% V₂O₅–27.5 mol% Ag₂O	5.7880	1,331	284.3	
Nonoxide–Tellurite Glasses				
48 mol% Te–10 mol% Ge–12 mol% Si–30 mol% As30 (Farley and Saunders 1975)	4.9400		135.0	
48 mol% Te–20 mol% Si–32 mol% As	4.9000		135.0	
49 mol% Te–10 mol% Ge–12 mol% Si–29 mol% As	5.1500		144.0	
39 mol% Te39–12 mol% Ge–14 mol% Si–35 mol% As	4.9200		150.0	
x mol% Te–(1 – x) mol% Se (Carini et al. 1984)				
$x = 0$	4.2600		101.2	
$x = 6.8$	4.4100		102.6	
$x = 14.0$	4.5400		103.8	
$x = 21.0$	4.6800		104.8	

and Θ_D (el) for pure P₂O₅ (307 K) from Bridge and Patel (1986). In binary transition metal (TM) or RE tellurite glasses, the Θ_D is increased by increasing the percentage of the TM oxide (e.g., WO₃, MoO₃, or V₂O₅), whereas for the ternary vanadium–tellurite glasses, the Θ_D decreases for higher percentages of V₂O₅. It is also clear from Table 4.2 that increasing the percentages of TM oxides (e.g., WO₃, V₂O₅, MoO₃ or ZnCl₂) either increases or decreases Θ_D. For example, by increasing the percentage of WO₃ from 15 to 33 mol%, the Θ_D is increased from 253 to 266 K. This increase in Θ_D is also seen in vanadium oxide– and zinc halide–tellurite glasses when oxides increase; but in binary tellurite–molybdenum glasses, the situation is opposite—increasing MoO₃ from 20 to 50 mol% in the glass decreases the value of Θ_D from 237.4 to 230.2 K, similar to the effect seen with binary TM phosphate glasses (e.g., vanadium– or molybdenum–phosphate glasses), as described by Farley and Saunders (1975). The Θ_D decreases for P₂O₅–V₂O₅ from 195.2 to 167.8 K as V₂O₅ increases from 47.7 to 82.3 mol%, as shown in Table 4.2. In the MoO₂–P₂O₅ system, the Θ_D increases from 307 K (pure P₂O₅ glass) to 358 K for 50 mol% P₂O₅ glass, and then decreases to 317 K at 62.2 mol% (Farley and Saunders 1975). The interpretation of this discontinuity was that a structural change occurs at 50 mol% MoO₃. For binary rare earth tellurite glasses, Θ_D values have been found to differ for the same percentages of RE oxide in the glass. The RE oxides in this case were La₂O₃, Sm₂O₃, and CeO₂. Table 4.2 also includes values of Θ_D for ternary and quaternary TM–RE–tellurite glasses of the forms TeO₂–WO₃–RE, TeO₂–V₂O₅–RE, TeO₂–V₂O₅–Ag₂O, and TeO₂–V₂O₅–Sm₂O₃.

From the above description of the experimental acoustic Θ_D, it could be concluded that the compositional dependence of Θ_D is determined by the mean value of v (longitudinal and shear) and the mean atomic volume. Therefore, if the modifier oxide is introduced into the TeO₂ network, then the mean atomic volume and the mean v both change, although they do so inversely. The creation of smaller rings increases the Debye frequency (the cage vibrational frequency) and also Θ_D, whereas more open rings in the glass network give rise to lower free energy; that is, lower Θ_D values. From Table 4.2, the values of the glass ring diameter (L) have been collected, as explained in Chapter 2, for every kind of tellurite glass, and it agrees with the above explanation. Using the simple lattice, the Einstein temperature ($\Theta_E = 0.75 \Theta_D$) could be calculated.

From the velocity data, Farley and Saunders (1975) measured the Θ_D of nonoxide–tellurite glasses in the following systems: 48 mol% Te–10 mol% Ge–12 mol% Si–30 mol% As, 48 mol% Te–20 mol% Si–32 mol% As, and 49 mol% Te–10 mol% Ge–12 mol% Si–29 mol% As. Their experimental error was <5 K. They concluded that the Θ_D of these chalcogenide glasses characterizes the total vibrational spectrum, including optical and acoustic spectra, as stated previously by Anderson (1959). Farley and Saunders also stated that these values were smaller than the Θ_D of other vitreous silica and many other oxide glasses. We can generalize from this fact by adding data for oxide–tellurite glass. Due to an increasing interest in the amorphous state, particularly in the chalcogenide amorphous semiconductors, attention has been devoted to Se–Te alloys by various authors during the past 25 years. For example, Carini et al. (1984) measured Θ_D for the $(1 - x)$ mol% Se–x mol% Te amorphous alloys by using v values. They also studied the role of tellurium in this amorphous solid. These values of the Θ_D are also in Table 4.2. It is clear that by increasing Te atoms from 0 to 21 mol%, the Θ_D is increased from 101.2 to 104.8 K.

Inaba et al. (2003) studied the heat capacity of oxide glasses at a high temperature region. Vitreous silica, some binary and ternary silicate, borate, and phosphate glasses were measured in the temperature range 300–840 K. One-dimensional Debye temperature dominated the compositional dependence of heat capacity. Thus, the authors derived the semiempirical equation for estimating the one-dimensional Debye temperature from glass composition. On that basis, the authors proposed the compositional parameters in predicting heat capacity, and suggested that taking the compositional parameters into account is one of the ways of designing a glass with a stable heat capacity for a specific purpose. The Debye temperature of xNa$_2$O–$(0.2 - x)$Li$_2$O–0.8 TeO$_2$ glasses with $x = 0.0, 0.05, 0.10, 0.15,$ and 0.20 mol% have been measured by Roychoudhury et al. (2002a) in the temperature range 140–300 K. Some of the data of the experimental values of the Debye temperature Θ_D are in Table 4.3. Due to the variation in ultrasonic velocity of all samples by a negative gradient with temperature, the Debye temperature (Θ_D) of the glass x.Na$_2$O–0.2 Li$_2$O–0.8 TeO$_2$ decreased from 250.6 to 243.1 K as the temperature increased from 140 to 300 K for the x.Na$_2$O–0.2 Li$_2$O–0.8 TeO$_2$ with $x = 0.0$ mol%. The rest of the values of the Debye temperature Θ_D of other percentages of Na$_2$O are in Table 4.3. Rajendran et al. (2003) measured the Debye temperature of tellurite glasses containing vanadium oxide in the form of $(50 - x)$ V$_2$O$_5$–xBi$_2$O$_3$–50TeO$_2$, with different bismuth ($x = 0, 5, 10, 15, 20,$ and 25 wt%) contents prepared using the rapid-quenching method and by measuring the ultrasonic velocities. Rajendran et al. (2003) measured the ultrasonic velocities (both longitudinal and shear) using a transducer operated at the fundamental frequency of 5 MHz in the temperature range 150–480 K. The temperature dependence of Θ_D shows a gradual decrease (negative gradient) in all Bi$_2$O$_3$ content without showing any anomalies over the entire range of temperatures from 150 to 480 K. The Debye temperature decreased from 190 to 173 K as the Bi$_2$O$_3$ content increased from 0.0 to 0.25 mol%, as in Table 4.3.

With the continuous addition of Bi$_2$O$_3$ content, the glass network becomes more discontinuous, as mentioned by Neov et al. (1979), and thus leads to an increase in glass density and lowers the Debye temperature. From these results, it can be inferred that the structure of these glasses is an independent combination of TeO$_4$ and Bi$_2$O$_3$ structures. Also, Affifi and Marzouk (2003) measured the Θ_D of heavy metal tellurite glasses in the form 20WO$_3$–$(80 - x)$TeO$_2$–xPbO, where $x = 10, 12.5, 15, 17.5,$ and 20 mol%. If the Debye model (Equation 4.6) represents all the vibration modes (acoustics and optics), then N must be taken as being equal to the number of atoms per unit volume. Table 4.3 indicates that as the mol% of PbO increases from 10 to 20 mol%, the Debye temperature shows a regular increase from 187.8 to 227.1 K. Since Θ_D represents the temperature at which nearly all modes of vibration are excited in the solid, its increase implies an increase in the rigidity of the glass system; that is, it indicates the presence of a tightening in the bonding of the glass structure while PbO content increases from 10 to 20 mol%, as stated by Farley (1975).

In 2004, Sidkey and Gaafer measured both the longitudinal and shear ultrasonic velocities in different compositions of the glass system (TeO$_2$)80–(WO$_3$)$(20 - x)$–(K$_2$O)x using pulse-echo technique and $x = 0, 5, 10, 15,$ and 20 mol%. Measurements were carried out at 6 MHz frequency and at room temperature, from which the Sidky and Gaafer 2004 have calculated the Debye temperature. The Θ_D decreased from 258 to 213 K for the change from 0.0 to 20.0 mol%, as in Table 4.3. Sidkey and Gaafer stated that Θ_D decreases gradually as the alkali modifier (potassium oxide) content increases, thus indicating the decrease in the glass rigidity of the glass. Saddeek and

TABLE 4.3

Debye Temperature of New Multicomponent Tellurite Glasses

Glass	Reference	Density (g/cm³)	Θ_D (K)	
(0.2-x) (Li₂O)–x (Na₂O)–0.8 (TeO₂)	Roychoudhury et al. 2002a			
X = 0.0 mol.%		5.2407	250.6	at 140 K
			243.1	at 300 K
X = 0.05		4.9759	248.5	at 140 K
			241.8	at 300 K
X = 0.10		4.968	251.2	at 140 K
			245.2	at 300 K
X = 0.15		5.0413	260.7	at 140 K
			253.1	at 300 K
X = 0.20		5.0977	259.3	at 140 K
			252.8	at 300 K
(50 – x)V₂O₅–xBi₂O₃–50TeO₂	Rajendran et al. 2003			
50–0–50		3.9956	190-	
35–15–50			180	
25–25–50		6.0314	173	
30WO₃– (70 – x)TeO₂–xPbO	Afifi and Marzouk 2003			
X = 10 mol.%		5.5875	187.8	$l = 0.477$ nm
X = 12.5		5.7179	198.9	0.463
X = 15		5.7379	208.9	0.455
X = 17.5		5.8851	217.3	0.449
X = 20		5.9012	227.1	0.442
(TeO₂)80–(WO₃)(20-x)– (K₂O)x	Sidkey and Gaafar 2004			
X = 0.0 mol.%		5.766	258	
X = 5		5.453	247	
X = 10		5.091	233	
X = 15		4.766	221	
X = 20		4.500	213	
Teₓ–Na₂₋₂ₓ–B₄.₄ₓ–O₇.₅ₓ	Saddeek and Abd El-Latif 2004			
X = 0.0		2.377	520.54	
X = 0.05		2.401	501.26	
X = 0.15		2.505	444.56	
X = 0.25		2.612	418.01	
X = 0.35		2.782	388.69	
0.65TeO₂– (0.35 − x)V₂O₅ –xGd₂O₃	Saddeek 2005			
X = 0.025 mol.%		4.167	264	
X = 0.050		4.311	274	
X = 0.075		4.471	280	
X = 0.100		4.572	260	
(TeO₂)50–(V₂O₅)50−x(TiO₂), X mol.%	El-Mallawany et al. 2006			
X = 2		3.988	295.9	0.5251 nm
X= 5		4.018	299.1	0.5211
X =10		4.062	303.0	0.5146
X= 12.5		4.131	309.7	0.5105

TABLE 4.3   (Continued)
Debye Temperature of New Multicomponent Tellurite Glasses

Glass	Reference	Density (g/cm³)	Θ_D (K)	
X =15		4.182	313.8	0.5059
$(100 - x)TeO_2 - xBaF_2$	Begum and Rajendran 2006			
X = 8		5.7437	168	
X= 10		5.6698	166	
X = 12		5.6436	164	
X = 15		5.6198	163	
X = 18		5.6028	161.5	
X = 20		5.6008	160	
$Na_2B_4O_7$–PbO	Gowda et al. 2007			
75 – 25		3.87	418	
21 – 79		5.35	271	
$Na_2B_2O_7$–Bi_2O_3				
75 – 25		3.50	399	
21 – 79		6.72	263	
$Na_2B_4O_7$–TeO_2				
75 – 25		2.71	428	
21 – 79		3.74	296	
$Li_{0.6}$–Te_x–$B_{1.4-2x}$–$O_{2.4-x}$	Saddeek et al. 2007			
X = 0		2.266	560	
X = 0.1		2.613	515	
X = 0.2		2.900	470	
X = 0.3		3.190	430	
X = 0.35		3.300	390	
$(70 - x)TeO_2 + 15B_2O_3 + 15P_2O_5 + xLi_2O$	Khafay et al. 2008			
X = 5 mol.%			430	
X = 10			390	
X = 15			350	
X = 20			325	
X = 25			275	
X = 30			260	
$(90 - x)TeO_2 - 10Nb_2O_5 - (x)ZnO$	Mohamed et al. 2010			
X = 0		5.513	283 ± 5	
X = 5		5.494	276 ± 5	
X = 10		5.431	279 ± 5	
X = 15		5.416	285 ± 5	

Abdel Latif (2004) measured the Debye temperature of sodium borate glass containing tellurite as Te_x–Na_{2-2x}–B_{4-4x}–O_{7-5x} with $x = 0$; 0.05, 0.15, 0.25, and 0.35 mol%. Ultrasonic velocity (both longitudinal and shear) measurements were made using a transducer operated at the fundamental frequency of 4 MHz at room temperature. The Debye temperature decreased from 520.54 to 388.69 K where $x = 0.0$ to 0.35 mol% of TeO_2, as in Table 4.3. The monotonic decrease in both the sound velocities, the elastic moduli, and the increase in the both ring diameter and the ratio (K_{bc}/K_c) as a function of TeO_2 modifier content reveal the loose packing structure. This was attributed to the increase in the molar volume and the reduction in the vibrations of the borate lattice. The observed results confirm that the addition of TeO_2 changes the rigid character of $Na_2B_4O_7$ to a matrix of ionic behavior bonds (NBOs). This is due to the creation of discontinuities and defects in the glasses, thus breaking down the borax structure.

In 2005, Saddeek measured the ultrasonic velocities of the Gd^{3+}-doped tellurovanadate glasses using the pulse-echo technique. The tellurovanadate glasses containing different gadolinium contents as $0.65TeO_2-(0.35-x)V_2O_5-xGd_2O_3$ (with $0 \leq x \leq 0.1$, x mol%) were prepared using the conventional quenching method. There was an observed increase in the ultrasonic velocity, the elastic moduli, and a decrease in the Debye temperature from 264 to 260 K, as in Table 4.3. The glass transition temperature showed an increase in Gd_2O_3 content up to $x = 0.075$, which was attributed to the decrease in the molar volume and the increase in the rigidity of the glasses. These parameters decreased when $x = 0.1$, which was attributed to the formation of NBO atoms in a direct result of the transformation of VO_5 to VO_4. There was agreement between the observed values of the Debye temperature of the tellurovanadate glass systems (El-Mallawany et al. 2000, Sidkey et al. 1999, Rajendran et al. 2003, and Rajendran 2002) and the values of the Debye temperature of the studied glass system. El-Mallawany et al. (2006) measured both the longitudinal and shear ultrasonic velocities of ternary tellurite glasses of the form: $0.50TeO_2 - (50-x) V_2O_5 - xTiO_2$, with $x = 0.02, 0.05, 0.10, 0.125, 0.15$ mol%. The elastic moduli, microhardness, softening temperature, and Debye temperature were calculated by using ultrasonic velocity and density. The Debye temperature increased from 259.9 to 313.8 K due to the increase of the TiO_2 mol% from 1 to 5 mol%, as in Table 4.3. The compositional dependence of Θ_D can be discussed on the basis of the variation in the number of atoms per volume and the mean ultrasonic velocity. The number of atoms per volume increased from 7.05×10^{28} m^{-3} for $50(TeO_2)-50(V_2O_5)$ to 7.53×10^{28} m^{-3} for $50 (TeO_2)-35(V_2O_5)-15(TiO_2)$, while the molar volume decreased from 42.3 to 37.17 cm^3; that is, the replacement of the V atom by the Ti atom explains the observed increase in density with increases in TiO_2. Also, the decrease in the ring diameter L from 0.5251 to 0.5059 nm increased the Debye temperature.

In 2006, Begum and Rajendran prepared binary tellurite $(100 - x)$ TeO_2-xBaF_2 glasses for different compositions of BaF_2 ($x = 8, 10, 12, 15, 18$, and 20 wt%). The Debye temperature (Table 4.3), which plays an important role in the determination of elastic moduli, decreased from 168 to 160 K as the percentage of BaF_2 increased from 8 to 20 wt%. The observed decrease in bulk modulus (as explained in Chapter 2) with the addition of BaF_2 content can be related to the changes in the coordination number of Te from 4 to 3, which are known to exist as a TeO_3 trigonal pyramid, and hence reduces the rigidity of the network. The continuous decrease in rigidity and elastic moduli with the addition of modifier has been observed in $TeO_2-BaO-TiO_2$ glasses (Sabadel et al. 1999, 2000). Furthermore, the monotonic decrease in rigidity, velocity and, hence, elastic modulus in $BaTiO_3$-doped lead bismuth glasses (Rahendran et al. 2001) with the addition of PbO also supports the above observation.

The Debye temperature of ternary tellurite glasses of the form: $TeO_2-Li_2O-B_2O_3$ has been studied by Saddeek et al. (2007). The glass compositions were of the form $Li_{0.6}-Te_x-B_{1.4-2x}-O_{2.4-x}$ where $x = 0, 0.1, 0.2$, and 0.3 mol%, and were prepared using the conventional rapid-quenching method over a wide range of compositions ($x = 0.0$ to 0.35 mol%). The Debye temperature decreased from 560 to 390 K as x increased from 0 to 0.35 mol%, as in Table 4.3. The observed continuous decrease in velocity with the addition of TeO_2 is ascribed to the change in the coordination number. When TeO_2 is added to the rigid framework glasses $Li_{0.6}-B_{1.4}-O_{2.4}$, distorted TeO_4 units, followed by the creation of regular TeO_3 sites, will be formed, besides the transformation of BO_4 into BO_3. This ionic character bond results in a monotonic decrease in velocity (as shown in Chapter 2). The observed decrease in the Debye temperature supports the claim that the addition of TeO_2 will lose the packed structure of the $Li_{0.6}-B_{1.4}-O_{2.4}$ glass, which reduces the vibrations of the lattice of the formed glasses, as observed earlier by Bridge and Higazy (1986).

Gowda et al. (2007) measured Θ_D of the pseudobinary sodium borate glasses containing $(1 - y)$ $Na_2B_4O_7-YM_aO_b$, where $M_aO_b = PbO$, Bi_2O_3, and TeO_2, and $y = 0.25, 0.5, 0.67$, and 0.79 mol%.

Sound velocities (longitudinal and shear) were measured at 10 MHz frequency using quartz transducers. The calculated Debye temperature Θ_D values as a function of y mol% were in Table 4.3. The Debye temperature represents the temperature at which all the low-frequency "lattice" vibrational modes are excited. The calculated Debye temperatures have been found to decrease in each of the series, as expected, because of the introduction of heavy elements. The emergence of low-frequency vibrational modes in the IR spectra was generally in evidence since there is a clear shift of absorption toward lower frequencies with increasing y. For glasses where PbO and Bi_2O_3 concentrations are 79 mol%, Θ_D values tend to increase marginally, thereby reflecting changes in structure. Also, Θ_D values are almost equal, which is consistent with the near-equal values of the reduced masses for Pb–O and Bi–O vibrations and similar force constant $[F = 17/(r_{M-O})^3]$ values. Furthermore, Θ_D $(33Na_2B_4O_7-67TeO_2)/\Theta_D$ $(33Na_2B_4O_7-67PbO) \approx 1.27$ and is close to the calculated ratio of Θ_D $(TeO_2)/\Theta_D$ $(PbO) \approx 1.36$. Khafagy et al. (2008) measured the Debye temperature of specimens of the glassy system: $(70 - x)$ $TeO_2-15B_2O_3-15P_2O_5-xLi_2O$, where $x = 5, 10, 15, 20, 25$, and 30 mol%, which were prepared using the melt-quenching method. An ultrasonic pulse-echo technique was employed at 5 MHz for measuring the longitudinal and shear ultrasonic velocities. The Debye temperature gradually decreased from 430 to 265 K, and there was a substantial increase in Li_2O in the glass matrix of up to 30 mol%. Equation 4.6 shows that Θ_D is proportional to the mean ultrasonic velocity, V_m, and the number of vibrating atoms per unit volume in the chemical formula unit of the glass. Therefore, the average calculated value of V_m decreased from 2606 to 1713 m/s, and that for the number of vibrating atoms per unit volume also decreased from 7.28×10^{28} to 6.17×10^{28} m^{-3} as the Li_2O content increased from 5 to 30 mol% in the glass matrix, which confirms the obtained decrease in the Θ_D of tested glasses and agrees with previous results (by El-Mallawany et al. 2006).

The composition dependence of Debye temperature of $(70 - x)TeO_2 + 15B_2O_3 + 15P_2O_5 + xLi_2O$ glasses was measured by Khafagy et al. (2008). The Debye temperature gradually increased from 430 to 265 K with the substantial increase in Li_2O in the glass matrix of up to 30 mol%. There was variation in Poisson's ratio, σ, which showed a gradual increase from 0.2470 to 0.2495 with the gradual replacement of TeO_2 oxide by Li_2O. Abd El-Aal et al. (2009) measured the elastic properties of vanadium tellurite glasses, $65TeO_2-(35 - x)$ V_2O_5-xCuO, with different compositions of copper, where $x = 0, 7.5, 10, 12.5, 15$, and 17.5 mol% at room temperature (300 K). The bulk modulus decreased from 33.97 to 29.37 GPa and the Young's modulus decreased from 54.86 to 42.87 GPa, while Poisson's ratio increased from 0.231 to 0.257 due to the increase of CuO from 0 to 17.5 mol%. The Debye temperature decreased from 0.231 to 242 K for the same increase of CuO. The data of ultrasonic velocities, elastic moduli, Poisson's ratio, cross-link density, microhardness, Debye temperature, and softening temperature of the glasses reveal a decrease in the rigidity of the glass network when CuO concentration increases from 7.5 to 20 mol%.

Mohamed et al. (2010) studied the niobium-containing tellurite glass with a starting composition of $(90 - x)TeO_2-10Nb_2O_5-(x)ZnO$, where $x = 0, 5, 10$, and 15 mol%, which was prepared using the melt-quenching method. The effect of the reduction of TeO_2 with a simultaneous increase in ZnO in elastic properties was also studied. The Debye temperature changed from 283 ($x = 0$ mol%) to 276 K ($x = 5$ mol%), due to the reduction of both shear and longitudinal moduli. The increase in Θ_D from 276 K ($x = 5$ mol%) to 285 K ($x = 15$ mol%) indicates an increase in the rigidity of the glass system. This is due to dominance of shear velocity recovery for $x > 5$ mol% compared with the longitudinal component. Poisson's ratio describes the expansion of a sample in a perpendicular direction to an applied stress, where a low Poisson's ratio of between 0.1 and 0.2 shows higher resistance toward lateral expansion when compressed compared with a Poisson's ratio of between 0.3 and 0.5. For glasses, the former has been associated with high cross-link density while the latter was associated with low cross-link density. In the present study, the drop in the Poisson's ratio from 0.260 to 0.220 with ZnO addition indicates an increase in the cross-link density with ZnO. This can be attributed to the positive role of ZnO as a network former in place of the reduced TeO_2.

4.3.2   CORRELATIONS BETWEEN EXPERIMENTAL ACOUSTIC AND CALCULATED OPTICAL DEBYE TEMPERATURES

Previously, El-Mallawany (1992b) reported that the lattice vibrational spectrum of a perfect crystal with P atoms per unit cell consists of (1) three acoustic phonon branches with acoustic phonon wave vectors spanning a quasi-continuous spectrum from zero up to a cutoff value v_{ac}^u (called the "reciprocal lattice vector"); (2) $3P - 3$ optical phonon branches with a quasi-continuous spectrum of phonon frequencies and an upper limiting frequency v_{op}^U (at K = 0) and a lower limiting frequency v_{op}^L (at K = π/2a); that is, being a reciprocal lattice vector.

These optical phonon branches give rise to an optical absorption band centered on the upper-limit frequencies, all of which lie essentially in the IR region. The frequencies v_{op}^U, v_{op}^L, and v_{ac}^U from Figure 4.2 are systematically related to one another since they are all functions of the cation–anion force constant and masses (the highest values of v_{op}^U could be identified with the upper limit of the experimental IR-absorption spectrum, while the lowest values of v_{op}^L would be associated with the lowest limit of the IR spectrum). For the diatomic chain, the theoretical values of the limiting phonon frequencies were given by Kittel (1966):

$$v_{op}^U = 2F\left(\frac{1}{M_1} + \frac{1}{M_2}\right)^{1/2}$$

$$v_{op}^L = \left(\frac{2F}{M_2}\right)^{1/2} \tag{4.7}$$

$$v_{ac} = \left(\frac{2F}{M_1}\right)^{1/2}$$

where M_1 and M_2 are the cation and anion masses ($M_1 > M_2$) and F is the anion–cation force constant. For a perfect crystal with P atoms per unit cell, there are ($3P - 3$) optical modes with different v_{op}^U and v_{op}^L values and three acoustic modes (v_{ac}^U and v_{ac}^L) with different end values. In Equation 4.6,

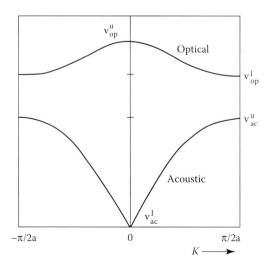

FIGURE 4.2 Schematic diagram of the dispersion relationships for optical and acoustic phonon modes in a vibrating lattice.

v_{ac} is the upper-limiting frequency for the acoustic phonon modes occurring in a real crystal lattice. In the polaron model of hopping, which was conducted in amorphous materials, another parameter was Θ_D (optical):

$$\Theta_D\left(optical\right) = \left(\frac{hv_{optical}}{K_B}\right) \tag{4.8}$$

where v_{op} is the average optical phonon frequency (i.e., the average over all v_{op}^U and v_{op}^L values). However, because $v_{op}^U - v_{op}^L$ is very small, then $\bar{v}_{op} \approx \bar{v}_{op}^{-U} \approx \bar{v}_{op}^{-L}$. Since $\bar{v}_{op}$, $\bar{v}_{op}^{-U}$, and $\bar{v}_{op}^{-L}$ values were systematically related, one might expect that Θ_D (ac) and Θ_D (op) exhibit a systematic relationship. From the Debye continuum of phonon states, the number of phonon states per unit volume in the frequency range v and $v + dv$ is given by $g(v)dv$, where:

$$g\left(v\right) = 4\pi\left[\left(\frac{1}{V_L^3}\right) + \left(\frac{2}{V_S^3}\right)\right] \tag{4.9}$$

where $0 < v < v_{ac}$; $g(v) = 0$; and $v > v_{ac}$.

However, the total number of phonon states was assumed to be $3N$ where N is the number of atoms per unit volume, so that from the relation $[g(v)dv = 3N]$, and Equations 4.6 and 4.9 would be of the form:

$$v_{ac}^3 = \left(\frac{9N}{4M}\right)\left[\left(\frac{1}{V_L^3}\right) + \left(\frac{2}{V_S^3}\right)\right] \tag{4.10}$$

and

$$\Theta_D = \left(\frac{h}{K_B}\right)\left(\frac{3\rho N_A P}{4\pi M}\right)\left[\left(\frac{1}{V_L^3}\right) + \left(\frac{2}{V_S^3}\right)\right]^{1/3} \tag{4.11}$$

where ρ is the density, N_A is Avogadro's number, P is the number of atoms in the chemical formula, Θ_D is determined by acoustic phonon velocities, and M is the molecular weight. Previously, El-Mallawany (1992b) reported a study of the Θ_D of tellurite glasses, the main purpose of which was to see whether there is any correlation between the Θ_D values for tellurite glasses determined by the ultrasonic methods of El-Mallawany and Saunders (1987) and the calculated optical Θ_D values from IR-characterized spectra (El-Mallawany 1989) (Figure 4.3). El-Mallawany (1989) computed the optical Θ_D (IR) by using the average frequency of the absorption band in the previous determination of IR spectra of pure and binary $TeO_2-A_nO_m$ ($A = La$, Ce, or Sm), as shown in Figure 4.3.

By adopting the values of the acoustic Θ_D from El-Mallawany and Saunders (1988) and comparing them with the present optical Θ_D, it is clear that Θ_D (IR) is related to Θ_D (el) for these glasses. In a step toward the possibility of correlating between Θ_{IR} and Θ_D (el), it is clear that the ratio of Θ_D (IR)/Θ_D (el) approximately equals 3.56 for all glasses. This ratio can be understood theoretically by considering the glasses as an assembly of linear diatomic chains of the types O–A–O and O–Te–O.

The composition dependence of Θ_D (el) was determined by variations in (1) the N/V and (2) the mean v (v_m), calculated as $v_m = [(1/v_L^3) + (2/v_T^3)]^{-1/3}$, where v_L is the longitudinal v and v_T is the transverse v.

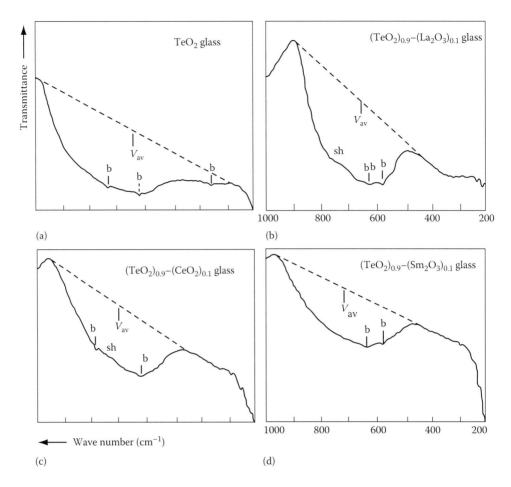

FIGURE 4.3 (a-d) Experimental IR spectra of pure and binary rare-earth tellurite glasses in the range 200–1000 waves cm^{-1}. b, band; bb, broadband; sh, shoulder; av, average wave number. (Reprinted from R. El-Mallawany, *Infrared Phys.*, *29*, 781, 1989. With permission from Elsevier.)

El-Mallawany (1992b) previously calculated the values of N/V, $[\Theta_D \ (\text{el}) = \Theta_D]$, $[\Theta_D \ (\text{IR})]$, and $[\Theta_D/\Theta_D \ (\text{el})]$, as shown in Table 4.3. If V changes without a change in the nature of bonding or coordination polyhedra, plots of log (M/P_ρ) against log $V(M/P_\rho)$, the mean atomic volume, where V is the molar volume, are generally linear and possess a negative slope. In the tellurite glasses discussed here, the V changes from 31.29 cm^3 for TeO$_2$ to 30.89, 28.19, and 30.87 cm^3 when modified with La$_2$O$_3$, CeO$_2$, and Sm$_2$O$_3$, respectively. The N/V is changed by each modifier—for example, 5.775 × 10^{28} for TeO$_2$ and 6.216, 6.409, and 6.25 × 10^{28} m^{-3} for La$_2$O$_3$-, CeO$_2$-, and Sm$_2$O$_3$-modified binary tellurite glasses, respectively, cause increases in Θ_D (el) from 249 K for TeO$_2$ to 265, 268, and 271 K for TeO$_2$ with the above glass modifiers, respectively. The replacement of a Te atom with La, Ce, or Sm (coordination numbers 7, 8, or 7, respectively) increases the cross-link density from 2.0 for pure TeO$_2$ glass to 2.55, 2.4, and 2.54 for the respective modified forms, which decreases the average ring sizes of the networks from 0.53 nm to 0.515, 0.51, and 0.514 nm, respectively. As mentioned in Chapter 2, the creation of smaller rings increases the average v (as seen in Tables 4.2, 4.3 and 4.4), which, in the same sequence, increases the Θ_D.

Calculations of the optical Θ_D and experimental values of the upper and average frequencies have been done for the tellurite glasses (El-Mallawany 1992). The optical Θ_D changes to higher values by modifying with RE oxides. Since the vibration frequency of the cations increases directly with the Θ_D, the Θ_D derived in this way characterizes the total vibrational spectrum in that optical and acoustic modes

TABLE 4.4
Calculated Properties and Optical Debye Temperatures of Tellurite Glasses

Glass	TeO_2	$(TeO_2)_{0.9}–(La_2O_3)_{0.1}$	$(TeO_2)_{0.9}–(Ce_2O_3)_{0.1}$	$(TeO_2)_{0.9}–(Sm_2O_3)_{0.1}$
Density(g/cm³)	5.101	5.685	5.706	5.782
Molar Volume	31.29	30.998	28.19	30.87
N/V (10^{28} m⁻³)	5.775	6.216	6.409	6.25
Average Cross-Link Density	2.0	2.65	2.68	2.71
Average Ring Size (nm)	0.53	0.515	0.51	0.514
Θ_D (el) (K)	249	265	268	271
Θ_D (IR) (K)	899	928	959	972
Θ_D (IR)/Θ_D (el)	3.61	3.50	3.58	3.58

Source: R. El-Mallawany, *Mater. Chem. Phys.*, 53, 93, 1998. With permission.

are separated. Because the thermal irregular network is governed by behavior of long-wavelength phonons, the irregular network of the amorphous material tellurite glass approximates well the elasticity continuum, and the thermal properties and propagation of soundwaves become closely interrelated.

4.3.3   RADIATION EFFECT ON DEBYE TEMPERATURES

El-Mallawany et al. (2000b) measured the effect of γ-rays on the acoustic Θ_D of the ternary silver–vanadium–tellurite glasses of the form TeO_2–V_2O_5–Ag_2O. The glass samples were exposed to different γ-ray doses using ^{60}Co γ-rays as a source of gamma radiation in the air. The exposure rate of 2×10^2 rad/h was applied at room temperature. The different doses—0.2, 0.5, 1.0, 2.0, 3.0, 4.0, and 5.0 Gy—were achieved by exposing the sample to the source for different periods of time.

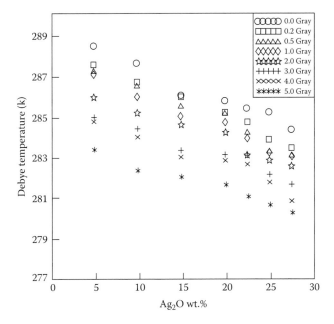

FIGURE 4.4   Effect of γ-radiation on Debye temperatures of tricomponent 50 mol% TeO2–(50 − x) mol% V_2O_5–x mol% Ag_2O glasses with various γ-ray doses, as indicated in the figure. (Data from R. El-Mallawany, A. Aboushely, and E. Yousef, *J. Mater. Sci. Lett.*, 19, 409, 2000. With permission.).

Variation of the Θ_D of the present tellurite glasses with γ-ray doses is shown in Figure 4.4. It is clear that Θ_D decreases with increasing γ-ray doses. The observed decrease in Θ_D is due to the decrease in v_m, which is caused by increasing the γ-ray doses. This analysis agrees with the results for ultrasonic wave attenuation and internal friction analyses of the same glasses, as previously published by El-Mallawany et al. (1998) and presented in Chapter 3 of this volume.

The interaction of energetic radiation with matter is a complex phenomenon because when γ-rays penetrate through a substance, they interact with its atoms. There are two principal interactions of γ-rays with glasses: ionization of electrons and direct displacement of atoms by elastic scattering.

Irradiating glass with γ-rays causes changes in its physical properties, based on the results obtained in ultrasonic attenuation analyses of TeO_2–V_2O_5–Ag_2O glasses, as explained in Chapter 3. For example, ternary silver–vanadium–tellurite glasses of the form TeO_2–V_2O_5–Ag_2O have higher absorption coefficients and internal frictions in comparison with binary tellurite–vanadate glasses. The higher ultrasonic absorption is attributed to the presence of the ionic oxide Ag_2O, which causes a reduction in the number of bridging oxygen atoms and creates more nonbridging atoms; radiating the glass with γ-rays increases the ultrasonic attenuation coefficient and the internal friction.

The above variation of Θ_D for the indicated tellurite glasses with γ-ray doses from 0.2 to 5.0 Gy is explained as follows:

- When tellurite glass is subjected to ionizing radiation, electrons are ionized from the valance band, move through the glass matrix, and are either trapped by preexisting flaws to form defect centers in the glass structure or recombine with positively charged holes.
- If NBO atoms in the glass lose electrons, the interstitial cations change their positions in the matrix and, consequently, the NBOs trap a hole, thus giving rise to color centers.

REFERENCES

Abd El-Aal, N., and Afifi, H., *Arc. Acoustics*, 34, 407, 2009.
Afifi, H., and Marzouk, S., *Mat. Chem. Phys.*, 80, 517, 2003.
Afifi, H., Marzouk, S., and Abd El-Aal, N., *Physica B*, 390, 65, 2007.
Akhieser, A., *J. Phys. (U.S.S.R.)*, 1, 277, 1939.
Anderson, O., and Bommel, H., *J. Am. Ceram. Soc.*, 38, 125, 1955.
Anderson, O., *J. Phys. Chem. Solids*, 12, 41, 1959.
Anstis, G., Chantikul, P., Lawn, B., and Marshall, D., *J. Am. Ceram. Soc.*, 64, 533, 1981.
Arlt, G., and Schweppe, H., *Solid State Commun.*, 6, 783, 1968.
Balta, P., and Balta, E., *Introduction to the Physical Chemistry of the Vitreous State*, Abacus Press: Kent, United Kingdom, 1976.
Begum, A. N., and Rajendran, V., *J. Phys. Chem. Solids*, 67, 11697, 2006.
Berry, B., and Nowick, A., In *Physical Acoustics*, Vol. IIIA, ed. by W. Mason, Academic Press: New York, 1966, p. 1.
Bogardus, E., *J. Appl. Phys.*, 36, 2504, 1965.
Bommel, H., and Dransfeld, K., *Phys. Rev.*, 117, 1245, 1960.
Brassington, M., Hailing, T., Miller, A., and Saunders, G., *Mat. Res. Bull.*, 16, 613, 1981.
Brassington, M., Lambson, W., Miller, A., Saunders, G., and Yogurtcu, Y., *Phil. Mag. B*, 42, 127, 1980.
Brassington, M., Miller, A., and Saunders, G., *Phil. Mag. B*, 43, 1049, 1981.
Brassington, M., Miller, A., Pelzl, J., and Saunders, G., *J. Non-Cryst. Solids*, 44, 157, 1981.
Bridge, B., and Higazy, A., *J. Mater. Sci.*, 20, 4484, 1985.
Bridge, B., and Higazy, A., *J. Mater. Sci.*, 21, 2385, 1986.
Bridge, B., and Higazy, A., *Phys. Chem. Glasses*, 27, 1, 1986.
Bridge, B., and Patel, N., *J. Mater. Sci.*, 21, 3783, 1986a.
Bridge, B., and Patel, N., *J. Mater. Sci. Lett.*, 5, 1255, 1986b.
Bridge, B., and Patel, N., *J. Mater. Sci.*, 21, 1187, 1986c.
Bridge, B., and Patel, N., *J. Mater. Sci.*, 22, 781, 1987.
Bridge, B., *J. Mater. Sci. Lett.*, 8, 1060, 1989a.
Bridge, B., *J. Mater. Sci.*, 24, 804, 1989b.

Bridge, B., Patel, N., and Waters, D., *Phys. Stat. Sol. A*, 77, 655, 1983.

Bridge, B., *Phys. Chem. Glasses*, 28, 70, 1987.

Brown, F., *The Physics of Solids*, Benjamin Cummings: New York, 1967, p. 105.

Brugger, K., and Fritz, T., *Phys. Rev.*, 157, 524, 1967.

Carini, G., Cutroni, M., Federico, M., and Galli, G., *J. Non-Cryst. Solids*, 64, 29, 1984.

Carini, G., Cutroni, M., Federico, M., and Galli, G., *Solid State Commun.*, 44, 1427, 1982.

Chawla, K., *Composite Materials, Science and Engineering*, Springer-Verlag: New York, 1987.

Clarke, D., *J. Am. Ceram. Soc.*, 75, 739, 1992.

Cutroni, M., and Pelous, P., *Solid State Ion.*, 28, 788, 1988.

Damodaran, K., and Rao, K., *J. Am. Ceram. Soc.*, 72, 533, 1989.

Dekker, A., *Solid State Physics*, Macmillan India Ltd.: Madras, 1995.

Dixon, R., and Cohen, M., *Appl. Phys. Lett.*, 8, 205, 1966.

Domoryad, I., *J. Non-Cryst. Solids*, 130, 243, 1991.

Dulieu-Smith, J., and Stanley, P., *J. Mater. Process. Technol.*, 78, 75, 1998.

El Mallawany, R., Abousehly, A., and Yousef, E., *J. Mater. Sci. Lett.,* 19, 409, 2000.

El Mallawany, R., Sidkey, M., and Afifi, H., *Glastech. Ber. Glass Sci., Technol.*, 73, 61, 2000.

El-Adawi, A., and El-Mallawany, R., *J. Mater. Sci. Lett.*, 15, 2065, 1996.

El-Mallawany, R., *J. Mater. Res.,* 5, 2218, 1990.

El-Mallawany, R., *Mater. Chem. Phys.*, 39, 161, 1994.

El-Mallawany, R., Aboushely, M., Rahamani, A., and Yousef, E., *Phys. Stat. Sol. A*, 163, 377, 1997.

El-Mallawany, R., Aboushely, A., Rahamani, A., and Yousef, E., *J. Mater. Sci. Lett.*, 19, 413, 2000a.

El-Mallawany, R., Aboushely, A., and Yousef, E., *J. Mater. Sci. Lett.*, 19, 409, 2000b.

El-Mallawany, R., Aboushely, M., Rahamani, A., and Yousef, E., *Mater. Chem. Phys.*, 52, 161, 1998.

El-Mallawany, R., and El-Moneim, A., *Phys. State Sol. A*, 166, 829, 1998.

El-Mallawany, R., and Saunders, G., *J. Mater. Sci. Lett.*, 6, 443, 1987.

El-Mallawany, R., and Saunders, G., *J. Mater. Sci. Lett.*, 7, 870, 1988.

El-Mallawany, R., El-Khoshkhany, N., and Afifi, H., *Mater. Chem. Phys.*, 95, 321, 2006.

El-Mallawany, R., *Infrared Phys.*, 29, 781, 1989.

El-Mallawany, R., *J. Appl. Phys.*, 73, 4878, 1993.

El-Mallawany, R., *Mater. Chem. Phys.*, 53, 93, 1998.

El-Mallawany, R., *Mater. Chem. Phys.*, 60, 103, 1999.

El-Mallawany, R., *Mater. Chem. Phys.*, 63, 109, 2000a.

El-Mallawany, R., *Phys. Stat. Sol. A*, 177, 439, 2000b.

El-Mallawany, R., *Phys. Stat. Sol. A*, 133, 245, 1992a.

El-Mallawany, R., *Phys. Stat. Sol. A*, 130, 103, 1992b.

El-Mallawany, R., Sidky, M. and Afifi, H., *J. Appl. phys.*, 107, 053523, 2010.

El-Mallawany, R., Sidky, M., and Affifi, H., *Glass Technol. Berl.*, 73, 61, 2000c.

El-Mallawany, R., Sidky, M., Kafagy, A., and Affifi, H., *Mater. Chem. Phys.,* 37, 295, 1994a.

El-Mallawany, R., Sidky, M., Kafagy, A., and Affifi, H., *Mater. Chem. Phys.*, 37, 197, 1994b.

El-Shafie, A., *Mater. Chem. Phys.*, 51, 182, 1997.

Ernsberger, F., *Glass: Science and Technology*, Vol. 5, no. 1, eds. by D. Uhlmann and N. Kreidl, Academic Press: New York, 1980, p. 1.

Everett, R., and Arsnault, R., *Metal Matrix Composites: Mechanisms and Properties,* Academic Press: New York, 1991.

Farley, J., and Saunders, G., *J. Non-Cryst. Solids*, 18, 417, 1975.

Farley, J., and Saunders, G., *Phys. Stat. Sol. A*, 28, 199, 1975.

Gilman, J., *Ceramic Sciences,* vol., eds. by I. J. Burke, Pergamon: Elmsford, New York, 1961.

Gilman, J., *Mechanical behaviour of crystalline solids, Proc. Am. Ceram. Soc., Symp.*, Monograph 59, National Bureau of Standards, Washington, DC, 1963, 79.

Gilman, J., *Micromechanics of Flow in Solids*, McGraw-Hill: New York, 1969, p. 29.

Gopala Rao, R., and Venkatesh, R., *J. Phys. Chem. Solids*, 6, 897, 2003.

Gowda, V., Reddy, C., Radha, K., Anavekar, R., Etourneau, J., and Rao, K., *J. Non-Cryst. Solids*, 353, 1150, 2007.

Hart, S., *J. Mater. Sci.*, 18, 1264, 1983.

Hashin, Z., and Strikman, S., *J. Mech. Phys. Solids*, 11, 127, 1963.

Hashin, Z., *J. Appl. Mech.*, 29, 143, 1962.

Higazy, A., and Bridge, B., *J. Non-Cryst. Solids*, 72, 81, 1985.

Hockroodt, R., and Res, M., *Phys. Chem. Glasses*, 17, 6, 1975.

Hughes, D., and Killy, J., *Phys. Rev.*, 92, 1145, 1953.

Inaba S., Oda, S., and Morinaga, K., *J. Non-Cryst. Solids*, 325, 258, 2003.

Inaba, S., Fujino, S., and Morinaga, K., *J. Am. Ceram. Soc.*, 82, 12, 1999.

Izumitani, T., and Masuda, I., Int. Congr. Glass [Pap]. 10th, 5-74-81, 1974.

Farley, J. M., and Saunders, G. A., *Phys. Stat. Sol. (a)*, 28, 199, 1975.

Jackle, J., Piche, L., Arnold, W., and Hunklinger, S., *J. Non-Cryst. Solids*, 20, 365, 1976.

Jackle, J., *Z. Physica*, 257, 212, 1972.

Jain, H., Peterson, N., and Downing, H., *J. Non-Cryst. Solids*, 55, 283, 1983.

Khafagy, A., El-Adawy, A., Higazy, A., El-Rabaie, S., and Eid, A., *J. Non-Cryst. Solids*, 354, 3152, 2008.

Kittle, C., *Introduction to Solid State Physics*, 4th Ed. by J. Welly & sons, Inc., N.Y., London, Sydney, Toronto, 1971.

Kozhukharov, V., Burger, H., Neov, S., and Sidzhinov, B., *Polyhedron*, 5, 771, 1986.

Krause, J., *J. Appl. Phys.*, 42, 3035, 1971.

Kurkjian, C., Krause, J., McSkimin, H., Andereatch, P., and Bateman, T., *Amorphous Materials*, ed. by R. Douglas and B. Ellis, Wiley: New York, 1972, p. 463.

Kurkjian, J., and Krause, J., *J. Am. Ceram. Soc.*, 49, 171, 1966.

Lambson, E., Saunders, G., and Hart, S., *J. Mater. Sci.*, 4, 669, 1985.

Lambson, E., Saunders, G., Bridge, B., and El-Mallawany, R., *J. Non-Cryst. Solids*, 69, 117, 1984.

Mahadevan, S., Giridhor, A., and Sing, A., *J. Non-Cryst. Solids*, 57, 423, 1983.

Makishima, A., and Mackenzie, J., *J. Non-Cryst. Solids*, 12, 35, 1973.

Makishima, A., and Mackenzie, J., *J. Non-Cryst. Solids*, 17, 147, 1975.

Martin, R., *Phys. Rev. B*, 1, 4005, 1970.

Mason, W., Ed., *Physical Acoustics*, Vol. IIIB, Academic Press: New York, 1965, p. 245.

Maynell, C., Saunders, G., and Scholes, S., *J. Non-Cryst. Solids*, 12, 271, 1973.

McLachian, D., Blaszkiewicz, M., and Newnham, R., *J. Am. Ceram. Soc.*, 73, 2187, 1990.

Mierzejewski, A., Saunders, G., Sidek, H., and Bridge, B., *J. Non-Cryst. Solids*, 104, 323, 1988.

Mizuno, Y., He, Z., and Hotate, K., *Optics Commun.*, 283, 2438, 2010.

Mochida, N., Takahshi, K., and Nakata, K., *Yogyo-Kyokai Shi*, 86, 317, 1978.

Mohamed, N., Yahya, A., Deni, M., Mohamed, S., Halimah, M., and Sidek, H., *J. Non-Cryst. Solids*, 356, 1626, 2010.

Morrell, R., *Handbook of Properties of Technical and Engineering Ceramics*, HMSO: London, 1985.

Mukherjee, S., Basu, C., and Ghosh, U., *J. Non-Cryst. Solids*, 144, 159, 1992.

Mukherjee, S., Maiti, A., Ghosh, U., and Basu, C., *Philos. Mag. B*, 67, 823,1993, Erratum, ibid., 68, 809, 1993.

Murnaghan, F., *Proc. Natl. Acad. Sci. USA*, 30, 244, 1944.

Muthupari, S., Raghavan, S., and Rao, K., *J. Mat. Res.*, 10, 2945, 1995.

Narita, K., Benino, Y., Fujiwara, T., and Komatsu, T., *J. Non-Cryst. Solids*, 316, 407, 2003.

Neov, S., Kozhukharov, V., Gerasimove, I., Krezhov, K., and Sidzhimov, B., *J. Phys(C): Solid State*, 12, 2475, 1979.

Neov, S., Gerasimova, I., Sidzhimov, B., Kozhukarov, V., and Mikula, P., *J. Mater. Sci.*, 23, 347, 1988.

Nowick, A., *Physical Acoustics*, Vol. XIII, eds. by W. Mason and R. Thurston, Academic Press: New York, 1977, p. 1.

Ota, R., Yamate, T., Soga, N., and Kunu, M., *J. Non-Cryst. Solids*, 29, 67, 1978.

Overton, W., *J. Chem. Phys.*, 37, 116, 1962.

Paul, A., Roychoudhury, P., Mukherjee, S., and Basu, C., *J. Non-Cryst. Solids*, 275, 83, 2000.

Paul, A., Maiti, A., and Basu, C., *J. Appl. Phys.*, 86, 3598, 1999.

Paul, A., Ghosh, U., and Basu, C., *J. Non-Cryst. Solids*, 221, 265, 1997.

Paul, A., Roychoudhury, P., Mukherjee, S., and Basu, C., *J. Non-Cryst. Solids*, 275, 83, 2000.

Paul, A., Roychoudhury, P., Mukherjee, S., and Basu, C., *J. Non-Cryst. Solids*, 275, 83, 2000.

Paul, A., Chattopadhyay, A., and Basu, C., *J. Appl. Phys.*, 84, 2513, 1998 and 85, 4265, 1999.

Paul, A., Ghosh, U., and Basu, C., *J. Non-Cryst. Solids*, 221, 265, 1997.

Paul, A., Roychoudhury, P., Mukherjee, S., and Basu, C., *J. Non-Cryst. Solids*, 275, 83, 2000.

Pauling, L., *Nature of Chemical Bond and Structure of Molecules and Crystals*, 2nd ed., Cornell University Press: Ithaca, NY, 1940.

Phyllips, W., *Amorphous Solids*, ed. by W. Phillips, Springer: Berlin, 1981, p. 53.

Rajendran, V., Palanivelu, N., Chaudhuri, B., and Goswami, K., *J. Mater. Sci. Lett.*, 21, 1691, 2002.

Rajendran, V., Palanivelu, N., Chaudhuri, B., and Goswami, K., *J. Non-Cryst. Solids*, 320, 195, 2003.

Rajendran, V., Palanivelu, N., Palanichamy, P., Jayakumar, T., Raj, B., and Chaudhuri, B., J. *Non-Cryst. Solids*, 296, 39, 2001.

Ross, R., *J. Am. Ceram. Soc.*, 51, 433, 1968.

Roychoudhury, P, Batabyal, S., Paul, A., Basu, C., Mukherjee, S., and Goswami, K., *J. Appl. Phys.,* 92, 3530, 2002a.

Roychoudhury, P., and Sunandana, C., *J. Non-Cryst. Solids*, 175, 51, 1994.

Roychoudhury, P., Paul, A., Basu, C., Mukherjee, S., and Goswami, K., *Phys. Chem. Glasses,* 43, 155, 2002.

Roychoudhury, P., Paul, A., Basu, C., Mukherjee, S., and Goswami, K., *Phys. Chem. Glasses,* 43, 155, 2002b.

Sabadel, J., Armand, P., E-Lippens, P., Cachau-Herreillat, D., and Philippot, E., *J. Non-Cryst. Solids,* 244, 143, 1999.

Sabadel, J., Armand, P., Terki, F., Pelous, J., Cachau-Herreillat, D., and Philippot, E., *J. Phys. Chem. Solids*, 61, 1745, 2000.

Saddeek, Y., Afifi, H., and Abd El-Aal, N., *Physica B*, 398, 1, 2007.

Saddeek, Y., and Abd El-Latif, L., *Physica B*, 348, 475, 2004.

Saddeek, Y., *Mater. Chem. Phys.*, 91, 146, 2005.

Sakamura, H., and Imoka, M., *J. Ceram. Soc. Japan*, 85, 121, 1977.

Sato, Y., and Anderson, O., *J. Phys. Chem. Solids*, 41, 401, 1980.

Schreiber, E., Anderson, O., and Soga, N., *Elastic Constants and Their Measurements*, McGraw-Hill: New York, 1973, p. 4.

Schweppe, H., *Ultrasonics*, 4, 84, 1970.

Sekiya, T., Mochida, N., Ohtsuka, A., and Tonokawa, M., *J. Non-Cryst. Solids*, 144, 128, 1992.

Shaw, R., and Uhlmann, D., *J. Non-Cryst. Solids*, 5, 237, 1971.

Sidek, H., Saunders, G., Hampton, R., Draper, R., and Bridge, B., *Phil. Mag. Lett.,* 57, 49, 1988.

Sidky, M., Abd El-Moneim, A., and Abd El-Latif, L., *Mater. Chem. Phys.*, 61, 103, 1999.

Sidky, M., and Gaafar, M., *Physica B*, 348, 46, 2004.

Sidky, M., El Mallawany, R., Abousehly, A., and Saddeek, Y., *Mater. Chem. Phys.*, 74, 222, 2002.

Sidky, M., El-Mallawany, R., Nakhala, R., and El-Moneim, A., *J. Non-Cryst. Solids*, 215, 75, 1997a.

Sidky, M., El-Mallawany, R., Nakhala, R., and El-Moneim, A., *Phys. Stat. Sol. A*, 159, 397, 1997b.

Sidky, M., El-Moneim, A., and El-Latif, L., *Mater. Chem. Phys.*, 61, 103, 1999.

Soga N., *Bull. Inst. Chem. Res.,* Kyoto University, 59, 1147, 1961.

Soga, N., Yamanaka, H., Hisamoto, C., and Kunugi, M., *J. Non-Cryst. Solids*, 22, 67, 1976.

Sreeram, A., Varshneya, A., and Swiler, D., *J. Non-Cryst. Solids*, 128, 294, 1991.

Strakana, R., and Savage, H., *J. Appl. Phys.*, 35, 1445, 1964.

Sun, K., *J. Am. Ceram. Soc.*, 30, 277, 1947.

Thompson, J., and Bailey, K., *J. Non-Cryst. Solids*, 27, 161, 1978.

Thurston, R., and Brugger, K., *Phys. Rev.*, 133, A1604, 1964.

Tomozawa, M., Phase separation in glass, *Treatises on Materials Science and Technology*, Vol. 12, 71, 1979.

Truell, R., Elbaum, C., and Chick, B., *Ultrasonic Methods in Solid State Physics*, Academic Press: New York, 1969.

Vogal, W., *Chemistry of Glass*, American Ceramic Society: Westerville, OH, 1985.

Wang, R., Wang, W., Wang, L., Zhang, Y., Wen, P., and Wang, J., *J. Phys. Condens. Mat.*, 15, 603, 2003.

Watanabe, T., Benino, Y., Ishizaki, K., and Komatsu, T., *J. Ceram. Soc. Japan*, 107, 1140, 1999.

Wells, A. F., *Structure of Inorganic Chemistry*, 4th ed., Oxford University Press: Oxford, 1975, p. 581.

Woodruff, T., and Ehrenrich, H., *Phys. Rev.*, 123, 1553, 1961.

Yano, T., Fukumoto, A., and Watanabe, A., *J. Appl. Phys.*, 42, 3674, 1974.

Yogurtcu, Y., Lambson, E., Miller, A., and Saunders, G., *Ultrasonics*, 18, 155, 1980.

Zener, C., ed., *Elasticity and Anelasticity of Metals*, University of Chicago Press: Chicago, IL, 1948.

Zeng, Z., *Kuei Suan Yen Hsueh Pao* (Chinese), 9, 228, 1981.

Ziman, J., *Principles of the Theory of Solids*, Vikas Publishing House: New Delhi, 1977, p. 67.

Part II

Thermal Properties

Chapter 5: Thermal Properties of Tellurite Glasses

5 Thermal Properties of Tellurite Glasses

In this chapter, the experimental techniques for measuring selected thermal properties of glass are explained; for example, glass transition temperature (T_g), crystallization temperature, melting temperature, and thermal expansion coefficient. Experimentally derived data for oxide, nonoxide, glass, and ceramic tellurite forms are compared with data calculated with different models. We examine the correlations between thermal properties and average values for cross-link density, and summarize stretching-force constants. Viscosity, fragility, and specific-heat capacities are compared between tellurite glasses and super-cooled liquids at T_g. Processing, properties, and structures of tellurite glass–ceramic composites are also discussed. Data of thermal properties for new and recent types of tellurite glass are also collected.

5.1 INTRODUCTION

The physical properties of tellurite glasses have attracted the attention of many researchers not only because of the numerous potential and realized technical applications for these properties, but because of a fundamental interest in their microscopic mechanisms. Although the thermal properties of crystalline solids are generally well understood, this is not the case for amorphous materials. As indicated by the schematic representation of volume (V)–temperature (T) relationships between a material's liquid, crystalline, and glass states (Figure 5.1), it is very important to characterize the temperatures at which these changes in structure occur. The melting T (T_m) is the T at which a material is melted. The crystallization T (T_c) is defined by the following process: if the rate of cooling of the liquid is sufficiently slow and the necessary nuclei are present in the melt, at the T_c a crystallization process begins and sudden change occurs in the V of this substance; if the liquid is cooled very rapidly, so that no crystallization at the same T can occur, the V of the new super-cooled liquid will continue to decrease without any discontinuity along a line BX, which represents the smooth continuation of the initial plot of change in V (AB). On further cooling, a region of T is reached in which a bend would appear in a plotted line of the data representing the V–T relationship. This region is termed the "glass transition region," and the T at which a change of shape occurs is called the "glass transition" T (T_g). In this region, the viscosity of a material increases to a sufficiently high value, typically about 10^{13} poise, so that its form changes, and at T values below the bend, the slope of the V–T curve becomes smaller than that of the liquid. If the T of the glass is held constant just below the T_g, and the cooling is stopped, then the glass will slowly contract further until its V returns to a point on the smooth continuation of the construction curve of the super-cooled liquid. This decrease in V is represented by a dotted line in Figure 5.1, and this process is known as "stabilization" (S). This characteristic is an important difference between glass and super-cooled liquid, which cannot achieve a stable state without crystallization. Stabilization is defined mathematically by $S = (T_c - T_g)$. Because of S, the properties of a glass also change with the length of time spent in the vicinity of the T_g; that is, the V of the glass depends on the cooling rate required to form the glass. Also, the glass-forming tendency K_g equals $(T_c - T_g)/(T_m - T_c)$, because $S = (T_c - T_g)$ is an indicator of stability against crystallization.

The physical properties of a glass can be characterized by its viscosity (represented by the formula P = Fd/vA, where F is the shearing force applied to a liquid of plane area A, d is the distance

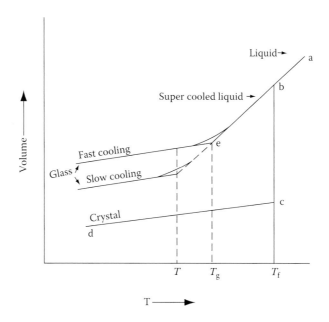

FIGURE 5.1   The volume–temperature relationship between the liquid and super-cooled liquid states.

apart, v is the resultant relative velocity of flow, and P is viscosity in poise (10^{-1} BS). Viscosity varies with T. The viscosity of a glass determines the melting conditions, the Ts of working and annealing, the upper T for use, and also the devitrification rate. Some glass Ts are defined in terms of relative viscosity; the working point, softening point, annealing point, and strain point of glass are defined as the Ts at which the viscosity of the glass is 10^4 Pa, $10^{7.6}$ Pa, $10^{13.4}$ Pa, and 10^4 Pa, respectively. The thermal expansion curve of annealed glass begins to deviate considerably from linearity at a viscosity of about 10^{14} or 10^{15} Pa. The physical properties of a glass can also be changed with variations in time at T_g. Variations of viscosity with time for quenched samples involve structural changes from states characteristic of higher Ts to those of lower Ts. If a glass exhibits a strong increase in viscosity with decreasing T near its liquidus point, then its K_g is high, whereas a weakly increased viscosity with decreasing T tends to lead to crystallization problems.

Heat capacity is a measure of the energy that is required to raise the T of a material or the increase in energy content per degree of T rise. It is normally measured at constant pressure (C_p [in calories per mole per degree Celsius]). The specific V of any given crystal increases with T, and the crystal tends to become more symmetrical. The general increase in V with T is mainly determined by the increased amplitude of atomic vibrations compared with mean amplitude. The repulsion between atoms changes more rapidly with atomic separation than does the attraction. Consequently, the minimum-energy trough is nonsymmetrical, as shown in the lattice energy as a function of atomic separation (Figure 5.2). As the lattice energy increases, the increased amplitude of vibration between equivalent energy positions leads to a higher value for the atomic separation, which corresponds to a lattice expansion.

Thermodynamically, structure energy increases as entropy decreases. The thermal expansion coefficient (α_{th}), which represents the relative change in length or V associated with T changes, may be linear (l)) ($\alpha_{th} = [dl/ldT]$) or volumetric ($\alpha_{th} = [dV/VdT]$) expansion. The energy that is required to raise the T of a material goes into its vibrational energy; rotational energy is required to raise the energy level of electrons and change their atomic positions. The T at which heat capacity becomes constant or varies only slightly with further changes in T depends on other properties of the material as well, including bond length, elastic constants, and T_m. The concept of fragility (m) in super-cooled liquids introduced in 1985 and 1991 by Angel gave new insights into glass transition,

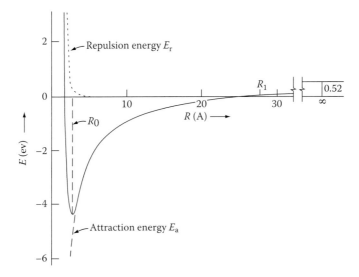

FIGURE 5.2   The minimum-energy trough is nonsymmetrical, as shown by the lattice energy as a function of atomic separation.

structural-relaxation phenomena, glass or super-cooled liquid structure, and other properties. As defined by Bohmer et al. (1993), the degree of m at T_g can be expressed as:

$$m = \left[\frac{d \log \langle \tau \rangle}{d(T_g - T)} \right]$$

where $\langle \tau \rangle$ is the average relaxation time. Furthermore, Komatsu (1995) and Komatsu et al. (1995) gave the expression of m at $T = T_g$ as:

$$m = E_\eta / 2.303 \, RT_g \tag{5.1}$$

where η is the shear viscosity, E_η is the activation energy for viscous flow at $\sim T_g$, and R is the gas constant. Therefore, the glasses with high E_η values at $\sim T_g$ or with low T_g values in a given glass system have a tendency to be more fragile than those with low E_η or high T_g values. Komatsu et al. (1995) stated that, as a general trend, heat capacity changes (ΔC_p) during glass transition in the so-called "fragile-glass-forming" liquids (those with large m values as estimated from the previous two equations) are much larger than those in "strong-glass-forming" liquids (those with small m values).

The conduction process for heat-energy transfer under the influence of a T gradient depends on the energy concentration present per unit V, its v of movement, and its rate of dissipation with the surroundings. The conduction of heat in dielectric solids can be considered either the propagation of anharmonic elastic waves through a continuum, or the interaction between quanta of thermal energy, called phonons. The frequency of these lattice waves covers a range of values, and catering mechanisms or wave interactions may depend on the frequency. The thermal conductivity (K) can be written in a general form:

$$K = \left(\frac{1}{3} \right) \int C(\omega) V l(\omega) \, d\omega \tag{5.2}$$

where the average rate at which molecules pass a unit area in the given direction is $Nv/3$, $C(\omega)$ is the contribution to specific heat per frequency interval for lattice waves of that frequency, v is the average velocity of N concentration of molecules past a unit area due to the T gradient, and $l(\omega)$ is the attenuation length for the lattice waves (ω).

The aims of this chapter are to describe and summarize the following:

1. Thermal-property data—that is, the temperature at which crystallization begins (T_x, T_g, T_c, T_m, α, S, and K_g) of tellurite glasses, as well as the energies of glass transformation and of crystallization with different compositions of elements and compounds in tellurite glasses according to different models, viscosity, and specific heat capacity data for oxide, nonoxide, glass, and ceramic forms of tellurite glasses.
2. Correlation data between thermal and structural properties; for example, between T_g and mean cross-link density (n_c') and the average force constant present in the glass network (these two parameters are used in Chapter 2 to interpret the elastic properties of glass.
3. C_p data of tellurite glasses.
4. Correlations between third-order elastic constants (TOEC) (the calculation of long-wavelength, acoustic-mode Gruneisen parameters [γ_{el}s]) and the thermal properties to discuss the atomic vibrations present in these noncrystalline solids.
5. Different methods of processing and different properties and structures of tellurite glass-ceramics.

5.2 EXPERIMENTAL TECHNIQUES OF MEASURING THERMAL PROPERTIES OF GLASS

The characterized glass T values (T_g, T_c, T_m, and C_p) have been determined by using instruments of differential thermal analysis (DTA) or differential scanning calorimetry (DSC). The values of S, K_g, and C_p can be calculated. The E to crystallize glass (E_c) and its viscosity and fragility can also be calculated. Both instruments are explained in this section. Figure 5.3 is a block diagram of the DTA analyzer. The basic features of the DTA analyzer consist of the following components:

- Sample holder assembly (incorporating sample and reference containers mounted in a suitable holder, thermocouples, etc.)
- Furnace or heating device (incorporating a T sensor)
- T programmer (with control system)
- Recording device (with amplifier)
- Atmosphere control
- Cooling control

Disk or plate thermocouples are used as platforms for the sample/reference container pans, which are flat. Exothermic or endothermic changes in a sample give rise to enthalpy changes—that is, changes in a sample's heat content. Under conditions of constant pressure that normally apply during thermal analysis, enthalpy changes can be assumed to correspond to a heat reaction and are usually written as ΔH. Changes in a thermodynamic property of a sample are expressed as the final value of that property minus its initial value. The general appearance of DTA curves is exothermic (plotted upward as a maximum) or endothermic (plotted downward as a minimum). The characterized glass T values (T_g, T_c, and T_m) can be determined at the point of change in the slope between endothermic and exothermic processes in the DTA chart, as shown next. The K_g and S against crystallization are calculated as $K_g = (T_c - T_g)/(T_m - T_c)$ and $S = (T_c - T_g)$, respectively.

Such curves are useful both qualitatively and quantitatively because the positions and profiles of the peaks are characteristic of the sample. The peak area (A) in DTA depends on the mass of the sample (M) used, the heat of the reaction or enthalpy change (ΔH) concerned, sample geometry, and K. The later two factors give rise to an empirical constant (K is previously used as a variable for "glass-forming tendency"). The similarity is potentially confusing.

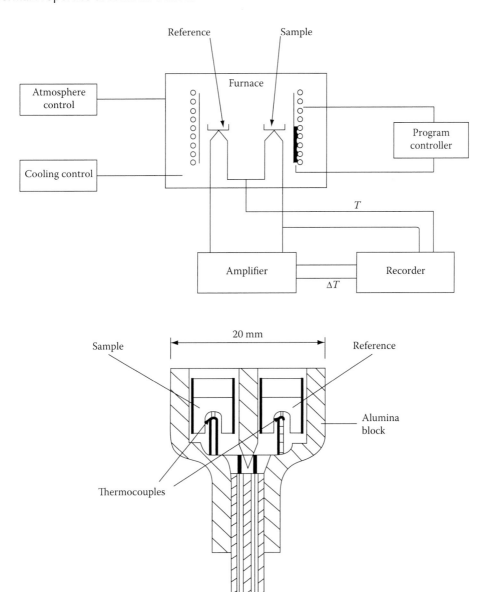

FIGURE 5.3 Block diagram of the differential thermal analysis (DTA) analyzer and details of head assembly.

$$(A) = \pm K\Delta\ HM \qquad (5.3)$$

The positive sign applies for endothermic reactions in which ΔH is >0, and the negative sign applies for exothermic reactions in which ΔH is <0. The qualitative practical application of this DTA instrument is in calculating the K factor as the calibration constant, which is T dependent. This means that peak areas cannot be converted into mass or energy units directly unless we know the value of K at the T concerned.

The DSC curves are used to record the differential heat input to a sample, expressed as the heating rate (dH/dT) in meters per joule per second, or meters per calorie per second on the ordinate against T or time (t) on the abscissa. The idealized representation of the three major processes observed by DSC is as follows:

1. Exothermal upward
2. Endothermic downward (as in DTA)
3. Displacement (d) for estimation of C_p

When the ΔH of a reaction is >0, the sample heater in the DSC instrument is energized and a corresponding signal is obtained. For an exothermic change ($\Delta H < 0$), the reference heater is energized to equilibrate the sample and reference Ts again and restore ΔT to zero, which gives a signal in the opposite direction. Because these energy inputs are proportional to the magnitude of the thermal energies involved in the transition, the records give calorimetric measurements directly. The peak areas in DSC are proportional to the thermal effects experienced by the sample as it is subjected to the T program. When the sample is subjected to a heating program in DSC (or in DTA), the rate of heat into the sample is proportional to its C_p. Virtually any chemical/physical process involves a change in the value of C_p for the sample. The technique of DSC is particularly sensitive to such change, which may be detected by the displacement of the base line from one nearly horizontal position to another. Such displacement occurs just to the right of the endothermic on the DSC record. The value of the C_p may be determined at a particular T by measuring this displacement with the equation:

$$C_p = \frac{change}{(heating\ rate)(M)}$$

$$= \frac{(dH/dt)}{M(dT/dt)} \tag{5.4}$$

$$= (mJ/tg)\left(1/\left(°C^{-1}\right)\right)$$

$$= mJg^{-1}°C^{-1}$$

In practice, we measure the baseline shift by reference to a baseline obtained for empty and reference pans. Furthermore, in order to minimize experimental error, we usually repeat the procedure with a standard sample of known C_p.

Of the DTA and DSC curves, the more quantitative interpretation (evaluation of specific heats and enthalpy change) is the evaluation of kinetic parameters from DTA curves. The theoretical treatments for the evaluation of the order of reaction, rate of reaction, and E from DTA curves invariably necessitate the use of simplifying assumptions that are hardly realizable in practice and restrict the application of the final equation to certain types of reactions. Applicable relationships have been reported, and the reader is referred to recent reviews for more detailed treatment. General lines of approach have been used for the comparison of DTA curves at different heating rates. Kissinger (1956) concluded that the peaks for high-order reactions are more symmetrical than those for lower-order reactions. He cited a relationship between the peak T (T_p) and the heating rate (Φ) for a series of DTA curves from which E may be derived using the equation:

$$\frac{d\left(\ln\Phi/T_p^2\right)}{d\left(1/T\right)} = -\left(E/R\right) \tag{5.5}$$

where R is the gas constant. This application of the DTA is known as the crystallization phenomenon. All of the present mechanisms and their applications to tellurite glasses are discussed in detail next. There are some physical and chemical phenomena that can be studied by DTA or DSC, such as melting and freezing transitions, purity determinations, and glass transition; therefore, the two techniques have similar applications. However, DSC instruments measure the E change in a sample directly, not as a T change, and they are consequently more suitable than DTA instruments for quantitative measurements of heat, including heats of reaction and transition, specific heats, and so forth.

To measure the linear α_{th} of a solid, it is important to measure the relative increase in length with respect to the increase in T. There are two different methods to measure α_{th}:

1. By using a thermal analyzer (TMA) to determine the increase in T to a value over the T_g, which gives another value of the quantity dl/l as shown in Figure 5.4 (another tool to find T_g)
2. By measuring the change in length of the solid at different Ts with a sensitive micrometer attached to the sample and a thermocouple (for measuring the T) to obtain the results as shown in Figure 5.5

Viscosity measurements in the range 10^9–10^6 Pa are performed using a computerized parallel plate viscometer. Senapati and Varshneya (1996) explained this method applied to the viscosity of

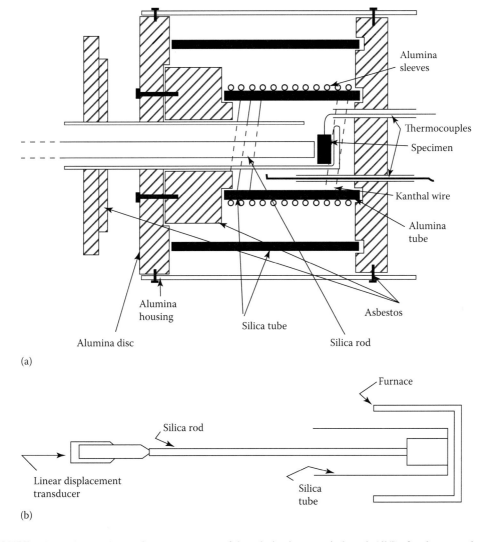

FIGURE 5.4 (a, b) Instrument for measurement of the relative increase in length (dl/l) of a glass sample with temperature up to a temperature lower than the glass transformation temperature. (From B. Bridge and N. Patel, *Phys. Chem. Glasses*, 27, 239, 1986. With permission.)

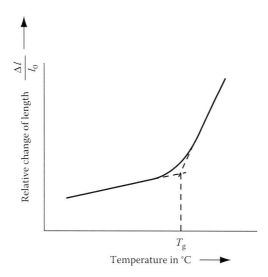

FIGURE 5.5 Change in length of a solid at different temperatures.

chalcogenide glass-forming liquids. The samples used for the viscosity measurements were typically around 6 mm in diameter and 5 mm in length, sliced from longer rods using a low-speed diamond saw. Most of these specimens had adequately parallel faces. Viscosity was calculated from the deformation rate (dh/dt) by using Gent's equation:

$$\eta = \frac{\left(2\pi Mgh^5\right)}{\left[3V\left(\dfrac{dh}{dt}\right)\left(2\pi h^3 + V\right)\right]} \tag{5.6}$$

where η is the viscosity, M is the mass placed on the load pan, g is the acceleration due to gravity, h is the sample thickness at time t, and V is the volume. The viscosity change in the glass transition region can also be measured using DTA. In 1995, Komatsu et al. measured the T dependence of viscosity at the glass transition point and m change in tellurite glasses. The T range for the viscosity measurements was about 50°C at near T_g. The viscosity range from 10^7 to $10^{10.5}$ Pa and the plot of log η against $1/T$ shows a straight line because of the narrow T range. The E_η for each glass was calculated by using the so-called Andrade equation from Gent (1960) ($\eta = A \exp[E_\eta/RT]$), where A is a constant. The thermal conductivity and specific C_p of oxide–tellurite glasses have been found by measuring the thermal diffusion coefficient (D) ($D = K/\rho C_v$). The thermal dissipation ratio $= \Delta(K \rho C_v)$, as described by Izumitani and Masuda (1974). These values were obtained by measuring the phase difference of periodical thermal current at the glass surface, or at a distance from the glass surface, using calibrated curves for the standard material.

5.3 DATA OF THE THERMAL PROPERTIES OF TELLURITE GLASSES

5.3.1 GLASS TRANSFORMATION, CRYSTALLIZATION, MELTING TEMPERATURES, AND THERMAL EXPANSION COEFFICIENTS

The tellurite glasses studied so far are remarkable for their refractive indices, which often lie in the range 2.0–2.3, as shown in Chapter 8. They usually have a marked yellow color although this may be considerably reduced by using very pure raw materials. These glasses are very easily melted, forming fluid melts at Ts below 1000°C, and they have low deformation Ts (250°C–350°C) and high α_{th}s (150–200 × 10^{-6} °C^{-1}) as shown in Table 5.1.

TABLE 5.1

Thermal Properties of Tellurite Glasses

Glass Composition (mol%) (Source)	Glass Property (Unit)		
	ρ (g/cm³)	T_g (°C)	α (10⁻⁶ °C⁻¹)
82 TeO$_2$–18 PbO (Stanworth1952)	6.05	295	18.5
78 TeO$_2$–22 PbO	6.15	280	17.7
80.4 TeO$_2$–13.5 PbO–6.1 BaO	5.93	305	17.5
88.2 TeO$_2$–9.8 PbO–2 Li$_2$O	5.56	270	19.7
86.4 TeO$_2$–9.6 PbO–4 Na$_2$O	5.46	275	21.5
42.5 TeO$_2$–29.6 PbO–27.9 B$_2$O$_3$	4.41	405	17.5
56.6 TeO$_2$–19.8 PbO–23.6 Cb$_2$O$_5$	5.78	410	11.5
44.1 TeO$_2$–22.5 PbO–13.4 P$_2$O$_5$	5.99	335	15.5
29.4 TeO$_2$–20.5 PbO–50.1 V$_2$O$_5$	4.31	255	14.0
31.4 TeO$_2$–15 BaO–53.6 V$_2$O$_5$	3.98	285	13.5
67.6 TeO$_2$–21.6 BaO–10.8 As$_2$O$_3$	5.19	415	14.5
63.5 TeO$_2$–22.2 PbO–14.3 MoO$_3$	5.93	315	18.0
41.1 TeO$_2$–19.1 PbO–39.8 WO$_3$	6.76	400	12.7
66.3 TeO$_2$–23.2 PbO–10.5 ZnF$_2$	6.08	280	17.5
85.9 TeO$_2$–12.0 PbO–2.1 MgO	5.77	315	17.0
73.25 TeO$_2$–22.7 PbO–4.05 TiO$_2$	6.03	320	16.5
72.3 TeO$_2$–22.4 PbO–5.3 GeO$_2$	6.07	315	17.0
74.7 TeO$_2$–21.4 PbO–3.9 La$_2$O$_3$	6.16	310	17.7
83.3 TeO$_2$–16.7 WO$_3$ (Stanworth 1954)	5.84	335	15.5
65.5 TeO$_2$–34.5 WO$_3$	6.07	380	13.9
82 TeO$_2$–18 PbO (Baynton et al. 1956)	6.05	295	18.5
20 TeO$_2$–80 PbO	7.24	260	17.1
83.3 TeO$_2$–16.7 WO$_3$	5.84	335	15.5
65.5 TeO$_2$–34.5 WO$_3$	6.07	380	13.8
69 TeO$_2$–31 MoO$_3$	5.03	315	16.5
78 TeO$_2$–22 Na$_2$O (Yakakind and Martyshenko 1970)			28.0
TeO$_2$–ZnCl$_2$–WO$_3$			13.0
or TeO$_2$–ZnCl$_2$–BaO			20.0
or TeO$_2$–ZnCl$_2$–Na$_2$O (Yakakind and Chebotarev 1980)			22.0
60 TeO$_2$–20 PbO–20 WO$_3$ (Heckroodt and Res 1976)	6.44	313	15.1
60 TeO$_2$–20 WO$_3$–20 Er$_2$O$_3$	6.45	432	11.2
TeO$_2$–MO$_{1/2}$ (M = Li, Na, K, Ag, and Tl) (Mochida et al. 1978)			28.2, 37.8, 43.8, 31.0, 40.6
TeO$_2$–MO (M = Be, Mg, Sr, Ba, Zn, Cd, and Pb) (Mochida et al. 1978)			11.0, 15.3, 19.0, 27.0, 15.3, 20.0, 22.0
TeO$_2$ (Lambson et al. 1984)	5.11	320	15.5
TeO$_2$, RnOm, RnXm–RnSO$_4$			
62.6 TeO$_2$–7.1 Al$_2$O$_3$–30.3 PbF$_2$	6.17	282	13.5

(continued)

TABLE 5.1   (Continued)
Thermal Properties of Tellurite Glasses

	Glass Property (Unit)		
Glass Composition (mol%) (Source)	**ρ (g/cm³)**	**T_g (°C)**	**α (10⁻⁶ °C⁻¹)**
92.4 TeO₂–7.6 Al₂O₃	5.29	343	16.3
50.5 TeO₂–49.5 P₂O₅ (Mochida et al. 1988)	4.14	399	12.3

Glass Composition (mol%) (Source)	**ρ (g/cm³)**	**T_g (°C)**	**T_c (°C)**
45 TeO₂–55 V₂O₅ (Hiroshima et al. 1986)		228	292, 328, 404
50 TeO₂–50 V₂O₅		232	299, 339
60 TeO₂–40 V₂O₅		242	389, 430

Glass Composition (mol%) (Source)	**ρ (g/cm³)**	**T_g (°K)**	**α (10⁻⁶ °C⁻¹)**
80 TeO₂–20 B₂O₃ (Burger et al. 1984)	4.94	605	15.6
77.5 TeO₂–22.5 B₂O₃	4.81	614	14.7
75 TeO₂–25 B₂O₃	4.69	619	14.2
80 TeO₂–20 B₂O₃ (Burger et al. 1985)	4.94	605	15.6
75 TeO₂–25 B₂O₃	4.69	610	14.2
TeO₂–rare-earth oxide (El-Mallawany 1992)			
90 TeO₂–10 La₂O₃	5.69	620	13.2
90 TeO₂–10 CeO₂	5.71	625	12.9
90 TeO₂–10 Sm₂O₃	5.78	635	12.5

Glass Composition (mol%) (Source)	**ρ (g/cm³)**	**T_g (°K)**	**T_g (°C)**	**T_m (°C)**	**K_g**
TeO₂ TMO (El-Mallawany 1995)					
80 TeO₂–20 MnO₂	4.80	620	740	1,000	0.45
70 TeO₂–30 MnO₂	4.60	640	740	1,000	0.40
95 TeO₂–5 Co₂O₄	4.75	625	725	1,025	0.33
92 TeO₂–8 Co₂O₄	4.73	635	730	1,035	0.31
80 TeO₂–20 MoO₃	4.90	620	740	1,015	0.46
70 TeO₂–30 MoO₃	4.75	625	745	1,035	0.41

Glass Composition (mol%) (Source)	**ρ (g/cm³)**	**T_g (°C)**
TeO₂–V₂O₅–Ag₂O (El-Mallawany et al. 1997)		
50–30–20	5.37	484
50–27.5–22.5	5.50	481
50–25–25	5.72	474
50–22.5–27.5	5.80	463

Glass Composition (mol%) (Source)	**ρ (g/cm³)**	**T_g (°C)**	**T_c (°C)**
TeO₂–ZnCl₂–ZnO (Sahar and Noordin 1995)			
90–10–0	5.30	314	380
40–60–0	4.91	298	465
60–30–10	5.18	290	495
50–20–30	5.11	315	450

TABLE 5.1 (Continued)
Thermal Properties of Tellurite Glasses

Glass Composition (mol%) (Source)	ρ (g/cm³)	T_g (°C)	T_C (°C)
TeO₂–TlO₀.₅ (Zahra and Zahra 1995)			
(TeO₂–0.66 TlO₀.₅)₀.₉₅ (AgI)₀.₀₅		195	
(TeO₂–0.86 TlO₀.₅)₀.₉ (AgI)₀.₁		163	
(TeO₂–0.86 TlO₀.₅)₀.₇₄ (AgI₀.₇₅TlI₀.₂₅)₀.₂₆		119	

TeO₂–KNbO₃ (Hu and Jain 1996)	ρ (g/cm³)	T_g (°C)	T_c	T_m	T_g/T_m	$T_c – T_g$
90 TeO₂–10 KNbO₃	4.50	395	446	667	0.71	51
82 TeO₂–9PbO–9TiO₂	5.20	365	532	672	0.67	167
83.7 TeO₂–4.7 PbO–9.3 TiO₂–2.3 La₂O₃	5.27	382	537	700	0.67	155
90 TeO₂–5 LiTaO₃₅–NbO₃	4.50	385	442	645	0.72	57
85 TeO₂–5LiTaO₃–10 NbO₃	4.75	396	467	641	0.73	71
TeO₂–WO₃–K₃O (Kosuge et al. 1998)						
80 TeO₂–10 WO₃–10 K₃O	5.31	308	456	619		
40 TeO₂–35 WO₃–25 K₃O	5.08	314	452	504		
20TeO₂–50WO₃–25K₃O	5.02	325	422	550		
(TeO₂–RTIO₀.₅).₍₁₋ₓ₎ (AgI)x (Rossignol et al. 1993)		105	148	244	43	
R = TeO₂/TlI₀.₅ = 1, x = 0.2		85	127	221	42	
R = 1.33, x = 0.25		135	202	260	64	
R = 0.86, x = 0.1						
TeO₂–PbO–CdO (Komatsu and Mohri 1999)						
100–0–0		336	400			
80–20–0		280	307			
0–20–20		3297	334			
0–20–30		293	340			
0–20–40		293	340			

Table 5.1 summarizes the T_g, T_c, T_ms, and α_{th}s of tellurite glasses, starting from the very old literature by Stanworth (1952 and 1954). In both references by Stanworth, the electronegativity values of elements led to the conclusion that lead oxide–tellurium oxide and barium oxide–tellurium oxide mixtures might well form glasses with low softening points and high refractive indices. Many three-component glasses have been reported containing TeO₂, PbO, or BaO combined with one of the following: Li₂O, Na₂O, Ba₂O₃, Cb₂O₅, P₂O₅, MoO₃, WO₃, ZnF₂, V₂O₅, MgO, CdO, TiO₂, GeO₂, ThO₂, Ta₂O₅, or La₂O₃. In 1956, Baynton et al. did research on tellurium, vanadium, molybdenum, and tungsten oxide-based tellurite glasses (Table 5.1). In the 1960s and 1970s, Yakakind et al. (1968, 1970, and 1980) studied the spectral-equilibrium diagrams and the glass formation, crystallization tendency, density, and thermal-expansion properties of tellurite glasses and tellurite–halide glasses (Table 5.1). Heckroodt and Res (1976) measured the T_g and α_{th} of 60 mol% TeO₂–20 mol% PbO–20 mol% WO₃ and 60 mol% TeO₂–20 mol% WO₃–20 mol% Er₂O₃ (Table 5.1). Mochida et al. (1978) studied the thermal properties of binary TeO₂–$MO_{1/2}$ glasses, where M = Li, Na, K, Rb, Cs, Ag, or Tl as in Table 5.1.

In 1984, Lambson et al. measured the T_g and α_{th} of pure tellurite glasses as 320°C and 15.5×10^{-6} °C^{-1}, respectively (Table 5.1). Also in 1984, Burger et al. (1985) measured the thermal properties of TeO_2–B_2O_3 glasses (Table 5.1), including both thermal and IR transmission in what were then new families of tellurite glasses of the forms TeO_2–$[R_nO_m, R_nX_m, R_n(SO_4)_m, R_n(PO_3)_m,$ or $B_2O_3]$, where X = F, Cl, or Br (Table 5.1). Hiroshima et al. (1986) measured the thermal and memory-switching properties of TeO_2–V_2O_5 glass (Table 5.1). Ten years after Mochida's first article in 1978, Mochida et al. (1988) measured the thermal expansion and transformation T of binary TeO_2–$PO_{5/2}$ glasses as shown in Table 5.1. In 1988, Zhang et al. used DSC to analyze the glasses Te_3–Br_2 and Te_3–Br_2–S, and they found no crystallization peak in the later tellurite glass, as shown in Figure 5.6. From the early 1990s, many researchers have been attracted to tellurite glasses for their semiconducting properties. Inoue and Nukui (1992) studied the phase transformation of binary alkali–tellurite glasses, as illustrated in Figure 5.7. Nishida et al. (1990) found the composition dependence of T_g for x mol% K_2O–$(95 - x)$ mol% TeO_2–5 mol% Fe_2O_3, x mol% MgO–$(95 - x)$ mol% TeO_2–5 mol% Fe_2O_3, and x mol% BaO–$(95 - x)$ mol% TeO_2–5 mol% Fe_2O_3 in comparison with x mol% Na_2O–$(95 - x)$ mol% TeO_2–5 mol% Fe_2O_3, as shown in Figure 5.8.

Also in 1992, El-Mallawany measured the thermal properties of rare-earth (RE) tellurite glasses and studied their structural and vibrational properties as summarized in Figure 5.9 and Table 5.1. Sekiya et al. (1992) measured both the T_g and α_{th} of TeO_2–$MO_{1/2}$ (M = Li, Na, K, Rb, Cs, or Tl) glasses, as shown in the 1st Ed. Rossignol et al. (1993) measured the T_g, T_c, T_m, and $T_c - T_g$ of the tellurite glass system $(1 - x)$ mol% $(TeO_2$–$RTlO_{0.5})$–x mol% AgI, where $R = (TeO_2/TlO_{0.5})$ and x is the percentage of AgI. Sekiya et al. (1995) measured both the T_g and α_{th} of TeO_2–WO_3 glasses and also studied the thermal properties of TeO_2–MoO_3, as shown in the 1st Ed. El-Mallawany (1995) studied both the devitrification and vitrification of binary transition metal (TM) tellurite glasses of the form $(1 - x)$ mol% TeO_2–x mol% A_nO_m, where A_nO_m = MnO_2, Co_3O_4, or MoO_3, as summarized in Table 5.1. Also in 1995, Elkholy studied the nonisothermal kinetics for binary TeO_2–P_2O_5, as discussed in Section 5.4.

The thermal properties of TeO_2–ZnO–$ZnCl_2$ tellurite glass systems, as described by Sahar and Noordin (1995), and of TeO_2–$TlO_{0.5}$ containing AgI or $(AgI)_{0.75}$–$(TlI)_{0.25}$ (Zahra and Zahra 1995) are tabulated in Table 5.1. Cuevas et al. (1995) measured both the T_g and α_{th} of TeO_2–Li_2O–TiO_2

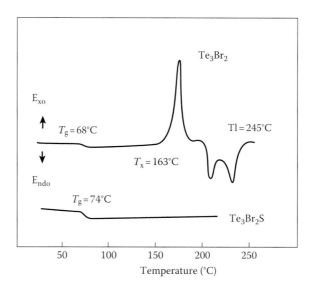

FIGURE 5.6 Results of differential scanning calorimetry analysis of Te_3–Br_2 and Te_3–Br_2–S glasses. (From X. Zhang, G. Fonteneau, and J. Lucas, *J. Mater. Res. Bull.*, 23, 59, 1988.)

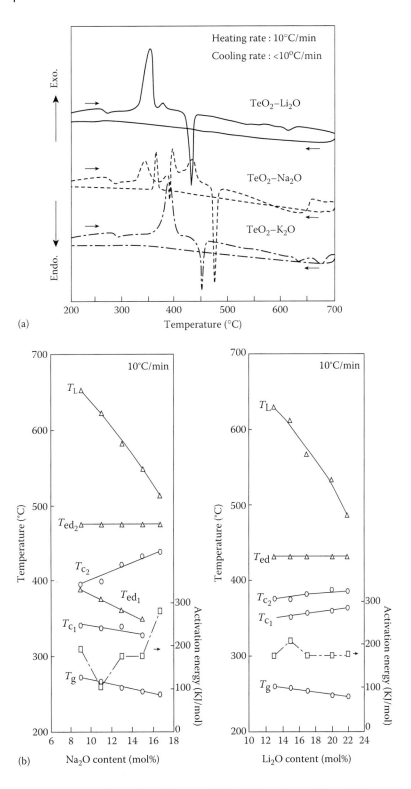

FIGURE 5.7 (a, b) Phase transformation of binary alkali–tellurite glasses. (From S. Inoue and A. Nukui, In: L. Pye, W. LaCourse, and H. Steven, eds., *Physics of Non-Crystalline Solids*, Taylor & Francis: London, 1992, p. 406. With permission.)

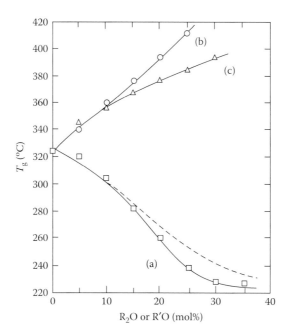

FIGURE 5.8 Glass transition of (a) x mol% K_2O–(95 − x) mol% TeO_2–5 mol% Fe_2O_3, (b) x mol% MgO–(95 − x) mol% TeO_2–5 mol%, (c) x mol% BaO–(95 − x) mol% TeO_2–5 mol% Fe_2O_3, in comparison with x mol% Na_2O–(95 − x) mol% TeO_2–5 mol% Fe_2O_3 (dashed line). (From T. Nishida, M. Yamada, H. Ide, and Y. Takashima. *J. Mater. Sci.*, 25, 3546–3550, 1990.)

glass systems, as shown in the 1st Ed., and the 80 mol% TeO_2–(20 − x) mol% Li_2O–x mol% Na_2O glasses were analyzed by Komatsu et al. (1995) as shown in Figure 5.10.

Hu and Jiang (1996) measured the thermoanalytical properties of TeO_2-based glasses containing ferroelectric components like $KNbO_3$, PbO–TiO_2, PbO–La_2O_3, $KNbO_3$–$LiTaO_3$, and $KNbO_3$–$LiTaO_3$, as listed in Table 5.1. The thermal properties of the glass systems Ag_2O–TeO_2–P_2O_5 described by Chowdari and Kumari (1996a), and 30 mol% Ag_2O–70 mol% [xB_2O_3–(1 − x) TeO_2] glasses by Chowdari and Kumari (1996b) were identified as shown in Figure 5.11.

In 1997, El-Mallawany et al. studied the tellurite glass system 50 mol% TeO_2–(50 − x) mol% V_2O_5–x mol% Ag_2O calorimetrically, as shown in Section 5.4. The thermal properties of the tellurite glass system TeO_2–V_2O_5–Na_2O were analyzed by Jayasinghe et al. (1997) as shown in Figure 5.12. The glass transformation temperature of the TeO_2–Fe_2O_3–MoO_3 glass system was measured by Qiu et al. (1997) as shown in Figure 5.13. Chowdari and Kumari (1997) identified the T_g of the glass TeO_2–Ag_2O–MoO_3 (Figure 5.14).

The thermal properties of TeO_2–Nb_2O_5–K_2O glasses were measured by Komatsu et al. (1997a) as shown in Figure 5.15. Muruganandam and Seshasayee (1997) measured the thermal properties of TeO_2–$LiPO_3$ glass (Table 5.1). In 1999, Komatsu and Mohri measured both the T_g and T_c of TeO_2–PbO–CaO (Table 5.1). In 2000, Damrawi measured the T_g and α_{th} of PbO–PbF_2–TeO_2 glasses (Table 5.1). The rest of the thermal properties of tellurite glasses—for example, glass stability against crystallization, glass-forming factors, viscosity, m, C_p, and K—are discussed under separate headings next.

5.3.2 Glass Stability against Crystallization and Glass-Forming Factor (Tendency)

El-Mallawany (1992, 1995) reported on thermal properties of tellurite glasses modified by RE oxides and TMOs, respectively, and El-Mallawany et al. (1997) reported on the thermal properties

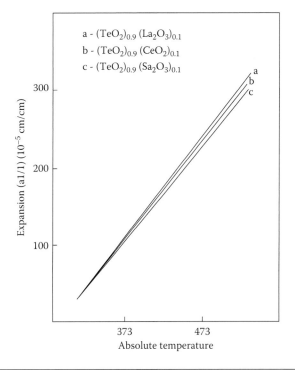

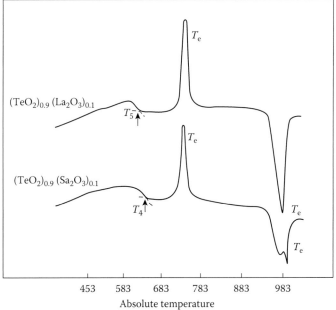

FIGURE 5.9 Thermal properties of rare-earth tellurite glasses. (From R. El-Mallawany, *J. Mater. Res.*, 7, 224–228, 1992. With permission.)

of ternary-form tellurite glasses containing either TMO or ionic oxide. These thermal properties included T_g, T_c, T_m, and α_{th} as shown in Figure 5.9 and listed in Table 5.1.

The α_{th} values were measured by using a dilatometer and the thermal analysis system described previously. The relative expansion ($\Delta L/L_o$) of the glass yielded the value of the thermal expansion between 300 and 500 K (below the T_g). The T_g was obtained by DTA. Figures 5.9a and b represent

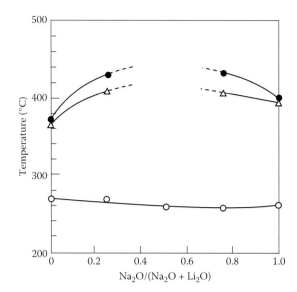

FIGURE 5.10 Glass transformation temperature of 80 mol% TeO_2–(20 − x)–mol% Li_2O–x–mol% Na_2O glasses. (From T. Komatsu, R. Ike, R. Sato, and K. Matusita, *Phys. Chem. Glasses*, 36, 216–221, 1995. With permission.)

the curves of both the TMA and DTA of some tellurite glasses (Table 5.1). From Figure 5.9b, it is clear that each curve has three peaks. The endothermic and exothermic peaks at high Ts represent the T_m and T_c values. The weak endothermic peak at low T corresponds to the T_g. The strong dependence of the T_g and α_{th} on the kind of modifier is evident in Table 5.1. In the range between room T and T_g, the α of the glass was often assumed to be independent of T and was defined as $\alpha = (\Delta L/L_0)/\Delta T$. In practice, the thermal coefficient depends on various parameters such as the thermal history of the glass and the T region. Thus, the glass must be well annealed; otherwise, relaxation occurs during heating as T_g is approached. The T_g increases from 593 K for pure TeO_2, to 635 K for samarium–tellurite glass. An understanding of how the α_{th} varies as a function of the glass composition is needed.

Table 5.1 also includes the ρ and molar V values of each glass. Changes in molar V are attributed to changes in structure caused by decreases in interatomic spacing, which can in turn be attributed to an increase in the stretching-force constant of the bonds inside the glassy network, resulting in a more compact, denser glass. For example, RE oxides increase their density from 5.105 g/cm³ for pure TeO_2, to 5.685, 5.706, and 5.782 g/cm³ for binary tellurite glasses when modified with La_2O_3, CeO_2, and Sm_2O_3, respectively, as measured by El-Mallawany (1992). When tellurite glasses are modified with a TMO such as MnO_2, Co_3O_4, or MoO_3, the density decreases as measured by El-Mallawany (1995). The molar V of the first series decreased from 31.26 cm³ for pure TeO_2, to 28.19 cm³ when modified with CeO_2. From these values of the molar V, one may understand V as including both the volume of the corresponding structural unit and its surrounding space, which is decreased by introducing the RE oxides. In the second series, the molar V (in cubic centimeters) is increased by modifying tellurite glasses with TMO, as seen in Table 5.1; that is, the basic structural units are linked more randomly. In the case of binary tellurite glass TeO_2–V_2O_5, the molar V increases from 31.29 to 42.73 cm³ for the 50%–50% glass reported by Sidky et al. (1997). Section 5.4 describes the work of El-Mallawany et al. (1997) with the thermal properties of the tricomponent silver–tellurite–vanadate glasses. The situation for these glasses is opposite that of the binary tellurite glasses cited above. An increasing percentage of Ag_2O added to 50–50 vanadate–tellurite glass caused the molar V to decrease from 42.73 to 32 cm³. This interpretation is based on the creation of a nonbridging oxygen (NBO) from Ag_2O in the glass, after which each unit V decreases as measured and discussed by El-Mallawany et al. (1997).

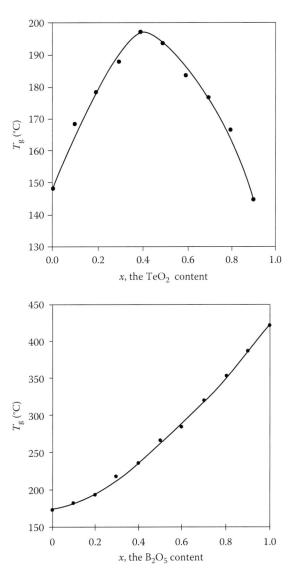

FIGURE 5.11 Glass transformation temperature of $Ag_2O–TeO_2–B_2O_5$. Materials consist of 30 mol% $Ag_2O–$ {70 mol%[x mol% B_2O_3–$(1 - x)$ mol% TeO_2]}. (From B. Chowdari and P. Kumari, *Solid State Ionics*, 86–88, 521–526, 1996b. With permission.)

Figure 5.9b (from El-Mallawany 1995) shows the values of T_g, T_x, T_c, and T_m for binary tellurite glasses containing TMO. These peaks were found in all curves of the DTA. It was apparent that the endothermic peak at higher T and the exothermic peak at intermediate T corresponded to melting and crystallization, respectively. All tellurite glasses showed the same trend, and no characteristic feature was observed. The dependence of T_g on the type of modifier is evident in Table 5.1. The glass transformation T_g changed from 598 K for pure TeO_2 glass, to 638, 633, and 623 K for binary transition metal tellurite glasses. The increase in the T_g that was induced by the addition of the modifier could be explained by the increased degree of polymerization. The T levels at which the crystallization processes started (T_x) for these glasses are also summarized in Table 5.1. The differences between both T_g and T_x were calculated to illustrate the size of the working range ($T_g - T_x$). For pure TeO_2 glass, the difference was 75 K, and changed to reach 90 K, 60 K, and 100 K for binary transition metal tellurite glasses. This difference might have arisen partly because the TMO

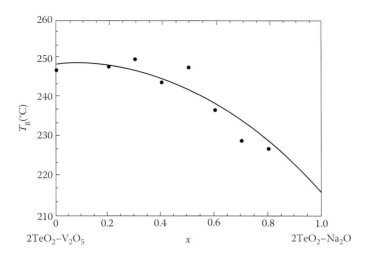

FIGURE 5.12 Thermal properties of tellurite glass systems of the form $TeO_2–V_2O_5–Na_2O$. (From G. Jayasinghe, M. Dissanayake, M. Careem, and J. Souquet, *Solid State Ionics,* 93, 291–295, 1997. With permission.)

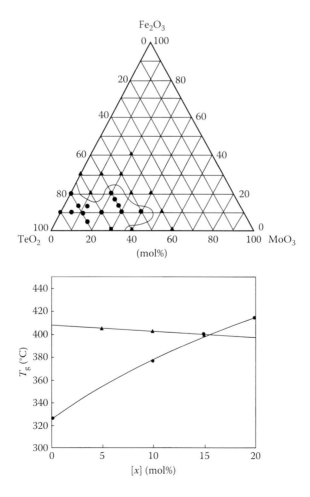

FIGURE 5.13 Glass transformation temperature of $TeO_2–Fe_2O_3–MoO_3$. (From H. Qiu, M. Kudo, and H. Sakata, *Mater. Chem. Phys.,* 51, 233–238, 1997.)

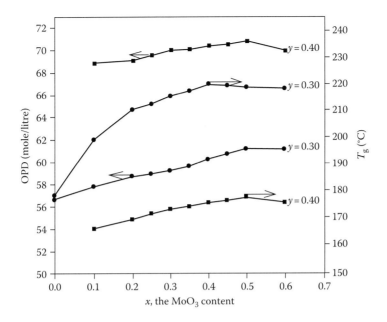

FIGURE 5.14 Glass transformation temperature of TeO_2–Ag_2O–MoO_3. (From B. Chowdari and P. Kumari, *J. Phys. Chem. Solids*, 58, 515–525, 1997. With permission.)

produced breaks in the network of the tellurium chains, which was confirmed from the calculated molar V of these binary glasses. As the T increased, there was increased chain alignment to reduce the internal energy of the system, and this self-alignment gave rise to nucleation and crystal growth. The values of T_m for binary transition-metal glasses obtained from the DTA curves were in the range of 993–1033 K, as in Table 5.1. The values of T_g/T_x were in the range of 0.6–0.63, satisfying the so-called "two-thirds rule." Komatsu et al. (1997a) analyzed the thermal stability of tellurite glasses of the form TeO_2–Nb_2O_5–K_2O as shown in Figure 5.15, where it can be seen that small values of $\Delta T = T_x - T_g$) are indicators of stable glass, and that values for super cooling ($[T_m - T_g]/T_m$) for these tellurite glasses have different behaviors. Figure 5.15 presents correlations between the degree of super cooling and thermal stability, and between T_m and T_g for TeO_2–Nb_2O_5–K_2O glass. It is also evident in Figure 5.15 that the tellurite glasses obey the two-thirds rule, as do silicate, borate, phosphate, and germinate glasses.

Also in 1995, El-Mallawany calculated the glass stability ($S = T_x - T_g$) and forming tendency (K_g), which were useful parameters for comparing the devitrification tendency of the glass as shown by Equation 5.7:

$$K_g = \frac{(T_c - T_g)}{(T_m - T_c)} \tag{5.7}$$

and had the values of 0.31 to 0.47, respectively, as seen in Table 5.1. Low values of K_g suggested high tendencies to devitrify. The α_{th} decreased from 15.5°C^{-1} for pure TeO_2 glass to 12.5°C^{-1} for binary samarium–tellurite glasses in the first series by El-Mallawany (1992). Table 5.1 collects the α_{th} data of sodium–tellurite glass by Mochida et al. (1978), tungsten–tellurite glass by Zeng (1981), and tungsten–lead–erbium-tellurite glasses by Heckroodt and Res (1976).

The correlations between T_g and the structure parameters—that is, stretching force constant and n_c, α_{th} and vibrational properties, and mean Gruniesen parameters—are explained in Sections 5.5 and 5.6.

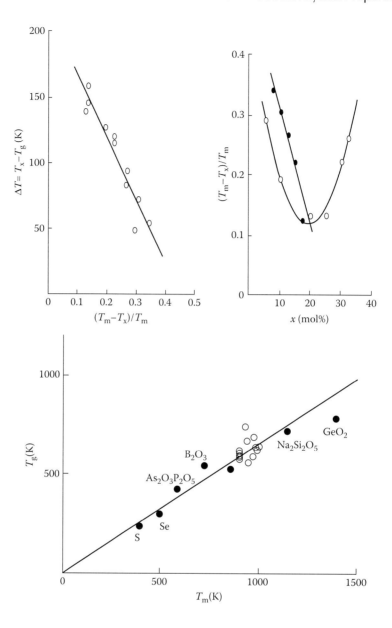

FIGURE 5.15 Thermal properties of tellurite glasses of the form $TeO_2–Nb_2O_5–K_2O$. (From T. Komatsu, H. Kim, and H. Oishi, *Inorganic Mater.*, 33, 1069, 1997. With permission.)

5.3.3 VISCOSITY AND FRAGILITY

In 1968, Yakakind et al. measured the viscosity of $TeO_2–V_2O_5–BaO$ glasses as a function of T as shown in Figure 5.16. As the percentage of TeO_2 oxide increased, the T of viscosity decreased, while entropy increased. For $TeO_2–Li_2O$ glasses, Tatsumisago et al. (1994) studied the local structure change with T in relation to m of liquid as shown in the 1st Ed. The viscosities of TeO_2-based glass were in the range $10^9–10^{13}$ Pa s$^{(-1)}$, as measured by the beam-bending technique developed by Hagy (1963) using a Mac-Science thermomechanical analyzer with a beam-bending attachment. The diagram of typical behavior of viscosity for strong and fragile liquids (E_η/T_g) is shown as a measure of the m. The viscosity data at Ts around the T_g for SiO_2, 20 mol% Na_2O–80 mol% SiO_2, and 20

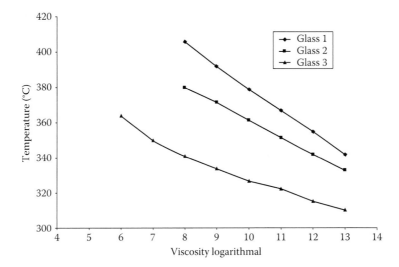

FIGURE 5.16 Viscosity of TeO$_2$–V$_2$O$_5$–BaO glasses as a function of temperature. Molecular percentages: glass 1, 20–35–45; glass 2, 30–60–10; glass 3, 80–10–10. (Original data from A. Yakakind, N. Ovcharenko, and D. Semenov., *Opt. Mech. Prom.*, 35, 34, 1968.)

mol% Li$_2$O–80 mol% TeO$_2$ glasses are plotted in Arrhenius form where the normalizing T is T_g as determined by DSC. The slope of the plots in this viscosity range (E_η/T_g) is a measure of the m of the liquid because the non-Arrhenius character of the T dependence of viscosity for glassy liquids corresponds to the slope E_η/T_g at around the T_g, assuming that the limit of viscosity at infinite T is equal to that of the glassy liquids. The composition dependence of the E_η/T_g values for 20 mol% Na$_2$O–80 mol% SiO$_2$ and 20 mol% Li$_2$O–80 mol% TeO$_2$ glasses has been shown in the 1st Ed. The values of the E_η/T_g range from 1.35 to 1.64 kJ/mol K^{-1}, and from 0.59 to 1.00 kJ/mol K^{-1} for the systems 20 mol% Li$_2$O–80 mol% TeO$_2$ and 20 mol% Na$_2$O–80 mol% SiO$_2$ glasses, respectively, indicating that the tellurite glass is much more fragile than the silicate glass. In both glasses, the values of E_η/T_g monotonically increase with an increase in the alkali–oxide content, indicating that both liquids become fragile with increasing amounts of network modifier content.

Komatsu et al. (1995) measured the glass T_g values, as shown in Figure 5.17, and the viscosity of $(20 - x)$ mol% Li$_2$O–x mol% Na$_2$O–80 mol% TeO$_2$ glass at Ts in the glass transition region by using a penetration method. The T dependence of viscosity at the glass transition region for Li$_2$O–Na$_2$O–TeO$_2$ glasses is shown in Figure 5.17. The T for viscosity measurements was about 50°C near the T_g. The viscosity range from 10^7 to 10$^{10.5}$ Pa s and the plot of log η against 1/T show a straight line because of the narrow T range. The E_η for each glass was calculated by using the so-called Andrade equation: η = A exp(E_η/RT), where R is the gas constant and A is a constant as shown in Figure 5.17. The activation energy E_η for viscous flow at the glass transition region decreases rapidly due to the mixing of Li$_2$O and Na$_2$O: E_η is 595 kJ mol^{-1} for 20 mol% Li$_2$O–80 mol% TeO$_2$, and E_η is 418 kJ mol^{-1} for 10 mol% Li$_2$O–10 mol% Na$_2$O–80 mol% TeO$_2$. The curve shows a negative deviation at an Li$_2$O/Na$_2$O ratio of 1. The dependence on Li$_2$O/Na$_2$O ratios for the 10^8, 10^9, and 10^{10} viscosity levels is shown in Figure 5.17. The m for the $(20 - x)$ mol% Li$_2$O–x mol% Na$_2$O–80 mol% TeO$_2$ glass is also shown in Figure 5.17. The values of the E_η at T_g for the estimation of m were obtained as indicated in Figure 5.17. Values of m ranged from 41 for 10 mol% Li$_2$O–10 mol% Na$_2$O–80 mol% TeO$_2$, to 57 for 20 mol% Li$_2$O–80 mol% TeO$_2$ glass. It is clear that mixed alkali–tellurite glasses were less fragile (i.e., stronger), than single alkali glasses such as Li$_2$O–TeO$_2$ glass. The values of m were 57 for silicate flint glass and 45 for 33 mol% Na$_2$O–67 mol% SiO$_2$ glass or 37 for 20 mol% Na$_2$O–80 mol% SiO$_2$ glass. Of course, it is clear that even the mixed alkali–tellurite glasses (e.g., 10 mol% Li$_2$O–10 mol% Na$_2$O–80 mol% TeO$_2$ glass with m of 41) are generally more fragile than most SiO$_2$ glasses with m of 20.

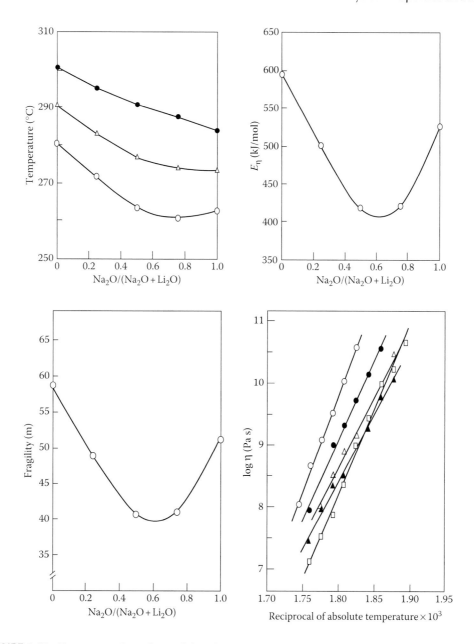

FIGURE 5.17   Temperature dependence of viscosity at the glass transition region for Li_2O–Na_2O–TeO_2 glasses. (T. Komatsu, R. Ike, R. Sato, and K. Matusita, *Phys. Chem. Glasses*, 36, 216–221, 1995. With permission.)

5.3.4   SPECIFIC HEAT CAPACITY

Zahra and Zahra (1995) studied the calorimetric properties of AgI-doped tellurite glasses. The heat capacity changes during the glass transition of tellurite glasses containing AgI or $AgI_{0.75}$–$TlI_{0.25}$ up to a mole fraction of 0.4 or 0.55 were measured by DSC. The T_gs of the binary compositions decrease with increasing $TlI_{0.5}$/TeO_2 ratios. With rising Ag^+ concentrations, the cohesion of the glass network is weakened, and the structural contributions to the relaxation phenomenon, as well as its activation enthalpies, diminish. There was no interaction between the iodide in crystalline form and in the host glass network. Thallium–tellurite glass was less stable in comparison with binary and ternary silver tellurite glasses, as shown in Table 5.2.

TABLE 5.2

Specific Heat Capacity of Tellurite Glasses

Glass	ΔC_p at T_g (J/g K)	C_p/C_{pg}	$\Delta C_p/C_p l$
TeO$_2$–0.66 AgO$_{0.5}$ (Zahra and Zahra 1995)	0.231	1.53	0.35
TeO$_2$–0.66 (AgO$_{0.5}$)$_{0.95}$ (AgI)$_{0.05}$	0.230	1.57	0.36
(TeO$_2$–0.86 AgO$_{0.5}$)$_{0.5}$	0.217	1.52	0.35
TeO$_2$–0.86 (AgO$_{0.5}$)$_{0.9}$ (AgI)$_{0.1}$	0.217	1.54	0.35
TeO$_2$–0.86 TlO$_{0.5}$	0.196	1.62	0.38
(TeO$_2$–0.86 TlO$_{0.5}$)$_{0.9}$ (AgI)$_{0.1}$	0.194	1.64	0.39
(TeO$_2$–0.86 TlO$_{0.5}$)$_{0.74}$ (AgI–TlI$_{0.25}$)$_{0.26}$	0.176	1.62	0.38
(TeO$_2$–0.86 TlO$_{0.5}$)$_{0.56}$ (AgI–TlI$_{0.25}$)$_{0.44}$	0.142	1.52	0.34
(TeO$_2$–0.86 TlO$_{0.5}$)$_{0.45}$ (AgI–TlI$_{0.25}$)$_{0.55}$	0.102	1.38	0.28

80 TeO$_2$–(20 – x)Li$_2$O–x Na$_2$O (Komatsu et al. 1997)	C_{pg} (200°C) (J/mol K)	C_{pe} (300°C) (J/mol K)	ΔC_p (J/mol K)	C_{pe}/C_{pg}
	≈75	≈120	≈45	1.6

TeO$_2$–V$_2$O$_5$ (mol%) (El-Mallawany 2000)	C_{pg} (J/g/K)	C_{pl} (J/g/K)	ΔC_p (J/g/K)	C_{pe}/C_{pg}
90–10	0.76	1.0	0.25	1.33
75–25	0.98	1.28	0.3	1.33
60–40	1.3	1.75	0.45	1.3
55–45	1.45	2.12	0.67	1.5

The specific heat capacity of the mixed alkali–tellurite glass 80 mol% TeO$_2$–(20 – x) mol% Li$_2$O–x mol% Na$_2$O has been measured by Komatsu et al. (1997b) as shown in the 1st Ed. They also studied the structural relaxation at the glass transition point. In another article in the same year, Komatsu and Noguchi (1997) studied the heat capacity at T_g in mixed alkali–tellurite glasses of the same composition, 80 mol% TeO$_2$–(20 – x) mol% Li$_2$O–x mol% Na$_2$O, and calculated the degree of *m* as shown in Figure 5.18.

Tanaka et al. (1997) studied the *T* dependence of the specific heat capacity of sodium–tellurite glasses of the form 85 mol% TeO$_2$–15 mol% Na$_2$O and 80 mol% TeO$_2$–20 mol% Na$_2$O as shown in Figure 5.19. Also, Kosuge et al. (1998) measured the thermal stability and heat capacity changes at the T_g in TeO$_2$–WO$_3$–K$_2$O glasses as shown in the 1st Ed.

Komatsu et al. (1997b) measured the specific heat and structural relaxation at T_g in mixed alkali–tellurite glasses. Komatsu et al. (1997b) also measured the specific heat capacity changes and enthalpy relaxation of the (20 – x) mol% Li$_2$O–x mol% Na$_2$O–80 mol% TeO$_2$ with different *m*s, as shown in Figure 5.18. An increase in the heat capacity, $\Delta C_p = 45$ J mol^{-1} K^{-1}, was observed, and that value was almost the same, irrespective of the Na$_2$O/(Li$_2$O + Na$_2$O) ratio. The ratio of the C_ps of the glassy state (C_{pg}) and super-cooled liquid state (C_{pe}) was $C_{pe}/C_{pg} = 1.6$, thus indicating that the above system should be included in the category of fragile liquids. The *E* for the relaxation process at the glass transition, ΔH, has been calculated as shown in the 1st Ed.

Although the structural units of TeO$_4$ and TeO$_3$ are connected weakly with each other, and thus the intermediate structure varies with increasing *T*, the rearrangement kinetic of TeO$_4$ and TeO$_3$ during heating is affected by alkali mixing. Komatsu and Noguchi (1997) measured and studied extensively the specific heat capacity of the tellurite glass system (20 – x) mol% Li$_2$O–x mol% Na$_2$O–80 mol% TeO$_2$. They noticed a large increase in the C_p value at the T_g in all compositions. The values

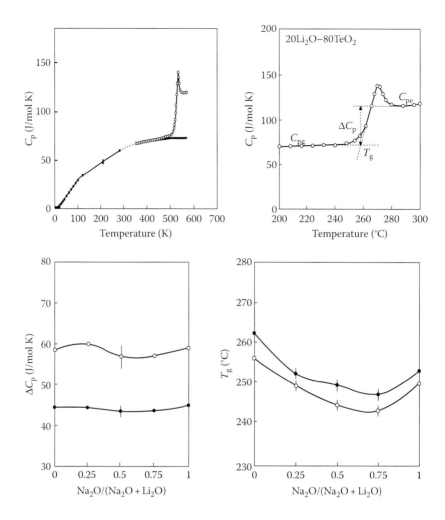

FIGURE 5.18 Heat capacity at the glass transition in mixed alkali–tellurite [80 mol% TeO_2–$(20 − x)$ mol% Li_2O–x mol% Na_2O]. (From T. Komatsu and T. Noguchi, *J. Am. Ceram. Soc.*, 80, 1327–1332, 1997.)

of C_{pg}, C_{pe}, and ΔC_p at T_g demonstrated that $(20 − x)$ mol% Li_2O–x mol% Na_2O–80 mol% TeO_2 are included in the category of fragile liquids. Komatsu and Noguchi (1997) compared the specific heat capacity of the annealed and crystallized glass 10 mol% Li_2O–10 mol% Na_2O–80 mol% TeO_2 with that of paratellurite TeO_2, as shown in Figure 5.18.

Tanaka et al. (1997) measured the T dependence of the specific heat capacity of the binary sodium–tellurite glasses 85 mol% TeO_2–15 mol% Na_2O and 80 mol% TeO_2–20 mol% Na_2O in the range 80–673 K to evaluate the characteristics and structural relaxation in the glass transition region as shown in Figure 5.19. The T variation of specific heat at 80 K to room T was analyzed in terms of a three-band model that describes the vibration density of the glass state in terms of three characteristic Ts, Θ_1, Θ_2, and Θ_E; which correspond to the Ts of interatomic vibration in the glass network, intermolecular type interaction in glass, and vibration of the network-modifying cation, respectively. The values of Θ_1 of sodium–tellurite glasses range from 820 to 840 K, which are less than those of silicate, borate, germinate, and phosphate glasses. This fact reflects the weak bond strength of Te–O. Specific heat data from room T to 673 K have been used to analyze glass transition. The activation enthalpy relevant to the cooling-rate dependence of T_g obtained with the Tool (1946) and Narayanaswamy (1971) models lies in the 690–1100 kJ/mol range, depending on glass composition. The Θ_1 values for SiO_2–, B_2O_3–, GeO_2–, and P_2O_5–TeO_2 glasses are 1550, 1920, 1250, 1500, and 840 K, respectively.

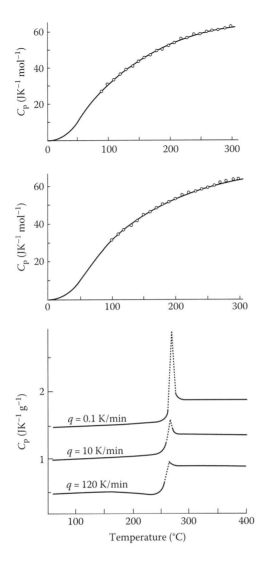

FIGURE 5.19 Specific heat capacity of sodium–tellurite glasses of the forms 85 mol% TeO_2–15 mol% Na_2O and 80 mol% TeO_2–20 mol% Na_2O. (From K. Tanaka, K. Hirao, K. Kashima, Y. Benino, and N. Soga, *Phys. Chem. Glasses*, 38, 87–91, 1997. With permission.)

In 1998, Kosuge et al. measured the thermal stability and heat capacity changes at T_g in K_2O–WO_3–TeO_2 glasses. Heat capacity changes of ΔC_p were 48–58 J mol^{-1} K^{-1}, and the range of the ratio C_p liquid (C_{pl})/C_{pg} was 1.6–1.8 for x mol% K_2O–x mol% WO_3–(100 – 2x) mol% TeO_2 glasses. The ΔC_p and ratio C_{pl}/C_{pg} increased with decreasing TeO_2 content, thus indicating an increase in thermodynamic m with decreasing TeO_2 content, as shown in the 1st Ed. However, the kinetic m estimated from the E_η for viscous flow was almost constant, irrespective of TeO_2 content. These behaviors have been analyzed using the configuration entropy model proposed by Adam and Gibbs. The results indicated that in K_2O–WO_3–TeO_2 glasses, Te–O–Te bonds are weak and bond-breaking occurs easily in the glass transition region, leading to large configuration entropy changes and thus large ΔC_p values.

El-Mallawany (2000b) measured the specific heat capacity of binary vanadium–tellurite glasses of the form (100 – x) mol% TeO_2–x mol% V_2O_5 as shown in Figure 5.20. El-Mallawany prepared a series of vanadium–tellurite glasses in this study by using a method described by Sidky et al. (1997). Batches of each glass were heated in an alumina crucible in an electric furnace, first at 400°C for

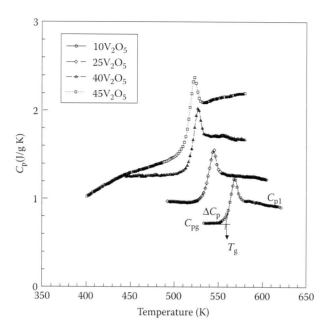

FIGURE 5.20 Specific heat capacity of binary vanadium-tellurite glasses of the form $(100 - x)$ mol% TeO$_2$-x mol% V$_2$O$_5$. (From R. El-Mallawany, *Phys. Stat. Solids*, 177, 439–443, 2000b.)

10 min to reduce volatilization, and then at 700°C–800°C for 30 min to melt, with periodic shaking to ensure homogeneity and attainment of thermal and chemical equilibrium. The melts were quenched at room T on polished stainless steel plates. Disks of glass were approximately 5 mm in thickness and 10 mm in diameter. Glasses were annealed at 200°C for 1 h in order to avoid the shattering of glass disks due to thermal stresses generated while quenching. C_p values were measured using a Perkin–Elmer DSC scanning calorimeter. Glass pieces weighing 40–50 mg were enclosed in alumina pans and heated at 10 K/min from room T to 650 K. The T_g was determined as the intersection of extrapolated straight lines around the glass transition in the C_p versus T plot. The ΔC_p of the glass and the super-cooled liquid C_{pg} and C_{pl} at T_g was also measured by extrapolating the respective C_p values from regions that were sufficiently away from T_g.

Figure 5.20 represents the plots of C_p data over a wide T range for a binary amorphous solid. Table 5.2 summarizes the experimental values of C_p for these glasses, together with the values of the T_g (the phenomena which cause the onset of flow or liquid) for all samples. The values of C_{pg}—the specific heat capacity just below the glass transition—are in the range 0.76–1.45 J/g K^{-1}. The values of C_p above the T_g, C_{pl}, are in the range 1.0–2.12 J/g K; that is, both C_{pg} and C_{pl} increase due to higher V$_2$O$_5$ content in the tellurite glasses. These values of the C_ps are higher than the values of 78 mol% TeO$_2$–22 mol% TlO$_{0.5}$ glass, and Zahra and Zahra (1995) found that C_{pg} and C_{pl} are 0.34 and 0.51 J/g K, respectively. Variations in the ΔC_ps of the glass and the super-cooled liquid were also measured, and the values of $\Delta C_p = C_{pl} - C_{pg}$ for these glasses increased from 0.24 to 0.67 J/g K^{-1}, all of which are higher than the ΔC_p of 78 mol% TeO$_2$–22 mol% TlO$_{0.5}$ glass (0.17 J/g K^{-1}), as reported by Zahra and Zahra (1995). The T_g decreased from 570 to 520 K due to higher V$_2$O$_5$ content. From Figure 5.27, it is clear that C_p increases as V$_2$O$_5$ content increases. The increase in C_p with T might be attributed to the anharmonicity of vanadium ions that are present in the samples, whereas the increase in C_p with V$_2$O$_5$ content in the sample might be attributed to transverse vibration due to the V$_2$O$_5$ content, which forms more holes in the structure.

Previously, Sidky et al. (1997) found the values of the densities and calculated molar Vs for every glass composition. The density of binary vanadium–tellurite glasses decreased from 5.2 to 4.1 g/cm^3 with the presence of 10 to 45 mol% V$_2$O$_5$ due to the replacement of tellurium atoms of atomic weight

127.6 with vanadium atoms of atomic weight 50.94. In addition, the molar V increased from 32 to 41 cm^3; that is, the resulting structures of these tellurite glasses had higher specific values. It is interesting to note that both density and molar V changed linearly with the percentage of vanadium oxide that was present in the glass. The molar V of these glasses should roughly correspond to the total contribution of oxygen, tellurium, and vanadium atoms; therefore, it is obvious that vanadium oxide–tellurite glasses have a more open structure resulting from corner sharing of the pentahedral and tetrahedral units connected through highly directional covalent bonds.

Measurements of the T variation of the C_p have been a rich source of information on the electronic, atomic, and molecular dynamics of solids. The C_p of a system of n interacting particles under an external constraint x is defined by the ratio of the added energy to the corresponding T rise of the system as in Equation 5.8:

$$C_x = \left(\frac{\partial E}{\partial T} \right)_x \tag{5.8}$$

The C_p of solids is generally measured at constant pressure, whereas statistical mechanics lead more naturally to formulas for the constant V quantity. The increase of specific C_p in the presence of V_2O_5 is caused by the substitution of V_2O_5 (7 particles) for TeO_2 (3 particles), which increases the number of oscillators. Because the glass also possesses frozen-configuration C_p, the low C_p values indicate the presence of strongly coordinated groups of atoms (rather than isolated atoms) that behave as independent oscillators. The addition of modifier oxides produces complimentary effects on the magnitude of C_p. Because modifier oxides degrade the network structure, the melt viscosity decreases. Glasses formed in such systems tend to become more "ideal" in the sense that they have a lower magnitude of entropy and, hence, of frozen-configuration C_p. At the same time, the degraded units produce defects in which the oxygen is more tightly bound to the central cation; therefore, the number of independent oscillators and the C_p are increased.

This large value of C_{pl}/C_{pg}, as reported by El-Mallawany (2000b), is 1.3–1.5 for higher V_2O_5 mol%, thus indicating that the binary vanadium–tellurite glasses (Table 5.2) should be included in the category of fragile liquids as defined by Angel (1985 and 1991). The so-called strong (kinetic) liquids have a tendency toward small changes in C_p values at T_g (i.e., $C_{pl}/C_{pg} \cong 1.1$), whereas the fragile (kinetic) glasses have large increases in C_p values at T_g. The C_{pl}/C_{pg} value of the glass system $(20 - x)$ mol% Li_2O–x mol% Na_2O–80 mol% TeO_2 was included in the category of fragile liquids as mentioned by Komatsu and Noguchi (1997). Also, Zahra and Zahra (1995) reported that TeO_2-based glasses containing $TlO_{0.5}$, $AgO_{0.5}$, or AgI have large values of C_{pl}/C_{pg} (1.43 to 1.66). From the high T Raman spectra of TeO_2-based glasses, such as 20 mol% Li_2O–80 mol% TeO_2 glass, Kowada et al. (1992) and Tatsumisago et al. (1994) reported that the TeO_2 trigonal bipyramid units that are present in the glass at room T change to TeO_3 trigonal pyramid units with NBO atoms as the T is increased above T_m, as shown in Chapter 10. Such a structural change above the glass transition causes a large change in viscous flow behaviors. Liquids with low ΔC_p values, such as SiO_2, have generally tetrahedrally coordinated network structures that are expected to experience relatively little disruption during heating; by contrast, liquids with large ΔC_p values have intermediate-range structures that are expected to change substantially as the T increases. Therefore, in this binary vanadium–tellurite glass with large values of ΔC_p and relatively low T_gs, the decrease in T_g for higher percentages of V_2O_5 confirms that increases in C_{pg}, C_{pl}, and ΔC_p of the structural units of TeO_4 are weakly connected to one another, and thus the intermediate structure varies easily as the T increases.

In conclusion, El-Mallawany (2000b) stated that the C_p changes at the glass transition in binary vanadium–tellurite glasses (with different percentages) have been measured to obtain more information on the structure of tellurite glasses. A large increase in C_p was clearly observed at the glass transition in whole-glass compositions. The values of the C_p of glass and super-cooled liquids differed between both of these C_ps (ΔC_ps), and the ratio of the C_p of the glassy state to that of the super-cooled liquid state (i.e., the C_{pl}/C_{pg}) increased for higher percentages of V_2O_5. The large value of the ratio C_{pl}/C_{pg} suggested that binary vanadium–tellurite glasses should be included in the category of fragile liquids. Avramov et al. (2000) measured the specific C_p of LiO_2–TeO_2 glasses in the T range 170°C–270°C. They found that $C_{pl}/C_{pg} \cong 0.62$ and Einstein T (θ_E) = 840 K, and they calculated the activation energy E (T_f) for structural relaxation.

The correlation between the specific C_p, Gruneisen parameter (γ) and the ultrasonic attenuation of tellurite glasses is through the so-called Woodruff–Ehrenreich relations. Izumitani and Msuda (1974) used the Woodruff–Ehrenreich equation $\alpha = (C_v T \gamma^2 / 3 \rho v)(\omega^2 \tau_l)$ for TeO_2–WO_3–Li_2O glass, where α is the ultrasonic attenuation coefficient, C_v is the specific C_p at constant V, ρ is density, v is ultrasonic wave velocity, ω is angular frequency of the ultrasonic waves, and τ_l is the relaxation time for lattice phonons. In this ternary tellurite glass, the τ_l of lattice phonons was calculated to be 5.68×10^{-11} s, using $\alpha = 3$ dB/cm at 100 MHz, $C_v = 0.34$ calories cm^3 deg, $\rho = 5.87$ g/cm^3, $v = 3.48 \times 10^5$ cm/s, and $\gamma = 1.65$. The calculated relaxation time of tellurite glass is roughly the same as those of α quartz and fused silica, whose acoustic loss is explained by phonon–phonon interaction. This suggests that the acoustic loss in tellurite glass is also due to phonon–phonon interaction.

The next three subsections focus on the structural effects of the above thermal properties on noncrystalline solids. Glass, like other substances, expands in proportion to any increase in T and contracts when cooled; the magnitude of a thermal property in glass varies according to the composition of the glass. The V increases with increasing T because of the increasing amplitude of vibration of atoms associated with changing interatomic distances and bond strengths. Therefore, the purpose of this section is to analyze thermal properties according to three types of guidelines:

1. Models of glass transformation and crystallization in tellurite glasses (Section 5.4)
2. Factors associated with composition and structure that affect the T_g; for example, the n'_c and F (Section 5.5)
3. Atomic vibration; that is, the correlation between the α_{th} and the average mode γ (Section 5.6)

5.4 GLASS TRANSFORMATION AND CRYSTALLIZATION ACTIVATION ENERGIES

5.4.1 GLASS TRANSFORMATION ACTIVATION ENERGIES

El-Mallawany (1995) calculated the E_c of pure tellurite glasses at various heating rates by using DTA as shown in Figure 5.21. That E_c was 115×10^{22} eV/mol. The E_η calculated by the method of Kissinger (1956) or Ozawa (1971) decreased for higher percentages of TeO_2. El-Mallawany et al. (1997) collected DSC data at different heating rates to gain some information about the thermal stability and calorimetric behavior of tricomponent tellurite–silver–vanadate glasses of the form 50 wt% TeO_2–(50 – x) wt% V_2O_5–x wt% Ag_2O, where x = 20, 22.5, 25, and 27.5 wt%. The T_g and T_c data are collected in Figure 5.22. From the variation in heating rate, the E_c and effective E_c (E_t) were calculated using different methods. In thermal measurements, two basic methods were used—isothermal and nonisothermal methods. In the isothermal method, the sample is brought quickly to a T_g, and the heat that evolves during the crystallization process at a constant T is recorded as a function of time. In the nonisothermal method, the sample is heated at a fixed heating rate (ϕ), and the heat that evolves is recorded as a function of T. Crystallization studies have been carried out under nonisothermal conditions, with samples heated at several uniform ϕs. For different ϕs, the T_g, and both the T_c and T_p, of the tricomponent tellurite glasses are shown in Figure 5.23: (a through d [20, 22.5, 25, and 27.5 wt% Ag_2O, respectively]).

It was clear from the obtained data for the T_g at $\phi = 2$ K/min that the T_g depends on the percentage of Ag_2O oxide; for example, tellurite glasses with higher percentages of Ag_2O oxide have lower values of T_g—that is, Ag_2O oxide creates a weaker tellurite glass, whereas V_2O_5 oxide creates a stronger glass. The T_g is discussed according to the models.

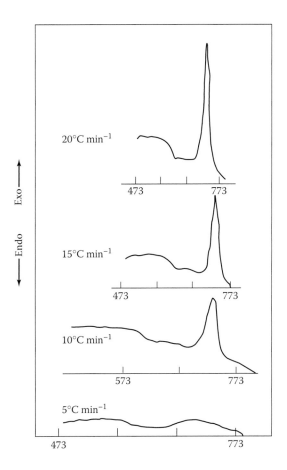

FIGURE 5.21   Differential thermal analysis of pure tellurite glasses at various heating rates to find the activation energy for crystallization. (From R. El-Mallawany, *J. Mater. Sci.*, 6, 1–3, 1995.)

5.4.1.1   Lasocka Formula

One approach to analyzing the dependence of T_g on ϕ in tellurite glasses is based on the empirical formula by Lasocka (1979):

$$T_g = A + B \ln(\phi) \tag{5.9}$$

where A and B are constants for a given glass composition. Figure 5.24 indicates the validity of this empirical formula for the tricomponent tellurite glasses.

5.4.1.2   Kissinger Formula

A second approach is the use of Kissinger's formula (1956), which originally applied to crystallization studies as given by Equation 5.14:

$$\ln\left(\frac{\phi}{T_c^2}\right) + \text{constant} = -\left(E_t/RT_c\right) \tag{5.10}$$

where E_t is the effective E_c and R is the gas constant. Although this model was originally derived for the crystallization process, Colemenero and Barandiaran (1978) suggested that it is also valid for glass transition. This equation has often been used to calculate E_t. Plots of $\ln(\phi/T_g^2)$ versus $1/T_g$

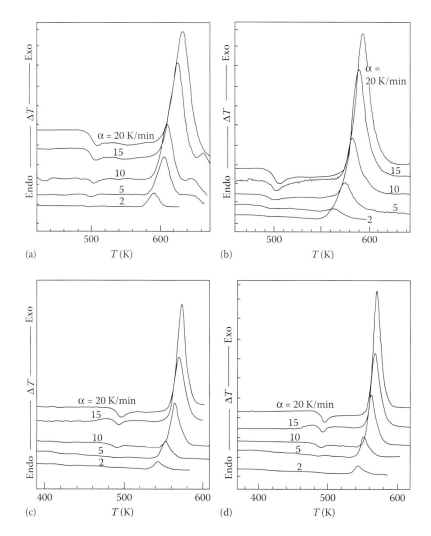

FIGURE 5.22   (a–d) Differential scanning calorimetric analyses at different heating rates for 50 wt% TeO$_2$–(50 – x) wt% V$_2$O$_5$–x wt% Ag$_2$O glasses, for x = 20, 22.5, 25, and 27.5. (From R. El-Mallawany, A. Abousehly, A. El-Rahamani, and E. Yousef, *Phys. Stat. Solids*, 163, 377–386, 1997. With permission.)

for tricomponent tellurite glasses indicate a linear relation as shown in Figure 5.23. The obtained values of E_t are listed in Table 5.3.

5.4.1.3   Moynihan et al. Formula

El-Mallawany et al. (1997) discussed a third approach to calculating glass E_t using an expression given by Moynihan et al. (1974):

$$\ln(\phi) = -\left(\frac{E_t}{RT_g}\right) + \text{constant} \qquad (5.11)$$

for the relations between ln(ϕ) versus T_g, ln(ϕ/T_g^2) versus 1000/T_g, and ln(ϕ) versus 1000/T_g for the ternary tellurite glass 50 mol% TeO$_2$–(50 – x) mol% V$_2$O$_5$–x mol% Ag$_2$O. The values of the glass E_t deduced from Equations 5.10 and 5.11 are listed in Table 5.3. It is clear from the data obtained for T_g at a ϕ of 2.0 K/min that T_g depended on the E_t of this glass, which decreased from 391.897 kJ/mol to 237.25 kJ/mol when Ag$_2$O increased from 20 to 27.5 mol% (using the Moynihan et al.

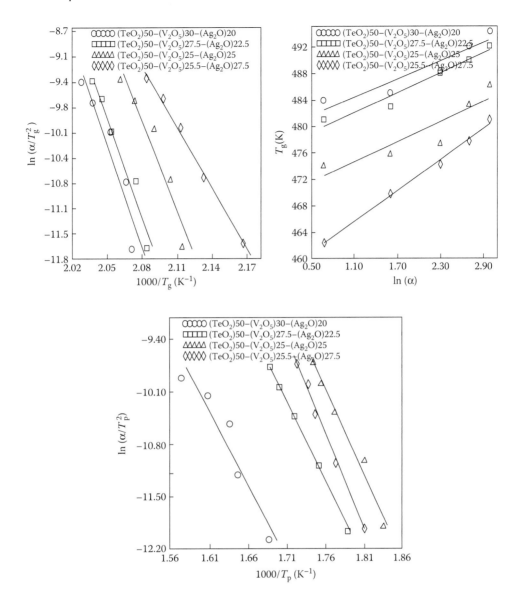

FIGURE 5.23 The glass transformation temperature (T_g) and both the onset temperature of crystallization (T_c) and the crystallization peak temperature (T_p) of the tricomponent tellurite glasses for different heating rates (ϕ). (From R. El-Mallawany, A. Abousehly, A. El-Rahamani, and E. Yousef, *Phys. Stat. Solids*, 163, 377–386, 1997. With permission.)

[1974] model), whereas E_t decreased from 383.3 kJ/mol to 229.34 kJ/mol for the same amount of Ag$_2$O, but using the Kissinger model (Kissinger 1956); therefore, these two models confirm each other. The decrease in the glass E_t due to the presence of Ag$_2$O in the network of the glass sample can be attributed to the creation of NBOs in this glass, which need less energy to break a bond in the network, as explained in Chapter 2.

5.4.2 CRYSTALLIZATION ACTIVATION ENERGY

Crystallization studies of tellurite glasses at various ϕs have been done by many authors, such as El-Mallawany (1995), as indicated in Table 5.3. The exothermic peak corresponds to the

TABLE 5.3

Glass Transition Temperature and Transition Activation Energy of Ternary Silver–Vanadate–Tellurite Glasses

Glass	Reference	E_t (kJ/mol)	E_t (kJ/mol)		
TeO_2–V_2O_5–Ag_2O	El-Mallawany et al. 1997				
50–30–20		383.3	391.89		
50–27.5–22.5		373.83	377.62		
50–25–25		340.87	348.93		
50–22.5–27.5		229.34	237.25		
		E_c (kJ/mol)	E_c (kJ/mol)	N	M
		174.6	164.47	2.62	2
		190.86	181.5	2.51	2
		200.69	192.02	2.85	2
		225.1	215.6	2.64	2
TeO_2	El-Mallawany 1995		E_c 115 (eV/ mol)		
TeO_2–P_2O_5	El-Kholy 1995	T_g	$N_b \times 10^{-22}$		
81.09–18.9		355	8.14		
71.82–28.18		312	6.89		

crystallization T_c. The E_c of pure tellurite glasses is calculated from the slope of the plot of $\ln\phi$ versus $1/T_c$, as shown in Figure 5.21. The ϕ and T_c fit the following relationship, as stated in Equation 5.15: $\ln\phi = -E/RT_c + $ constant. The value of the E_c of pure tellurite glasses is 115×10^{22} eV/mol. Later, El-Mallawany et al. (1997) studied the crystallization process in tellurite glasses containing TMO such as V_2O_5, and ionic oxides like Ag_2O. The T_p data for tricomponent tellurite glasses can be taken from Figure 5.24 and are tabulated in Table 5.3. The behavior of the T_p is typical of the T_g. The E_c values of these tricomponent tellurite glasses have been calculated using a method proposed by Ozawa (1965) to deduce the order of the crystallization reaction (n) from the variation of the V fraction (χ) of crystals precipitated in a glass at a given ϕ. The relationship between $\ln[-\ln(1-\chi)]$ and $\ln\phi$ at constant T is found by:

$$\frac{d\left\{\ln\left[-\ln\left(1-\chi\right)\right]\right\}}{d\left[\ln\left(\phi\right)\right]} = -n \tag{5.12}$$

where n is an integer constant depending on the morphology of the growth when the nuclei start to form during the DSC scanning and is equal to 1.0, 2.0, 3.0, or 4.0. On this basis, the value of n can be evaluated by plotting $\{\ln[\ln(1-\chi)]\}$ versus $\ln\phi$ at different ϕ (where χ is obtained from the crystallization exothermic peaks at the same T but at different ϕ). From the slopes of the plots of $\{\ln[-\ln(1-\chi)]\}$ versus $\ln\phi$ at different T, the value of n equals 2.6245, 2.655, 2.85, and 2.6488 (Table 5.3). The calculated value of n indicates that TeO_2–V_2O_5–Ag_2O glasses crystallize with two-dimensional growth, as stated by Chen (1978).

The Johnson–Mehl–Avrami formula (Equation 5.13), which is used to calculate kinetic transformation involving nucleation and growth under isothermal conditions, is generally used to analyze crystallization processes. Many attempts have been made to study E_c in the nonisothermal crystallization process. In nonisothermal crystallization, the χ of crystals precipitated in a glass heated at a uniform ϕ is shown to be related to E_c through the expression:

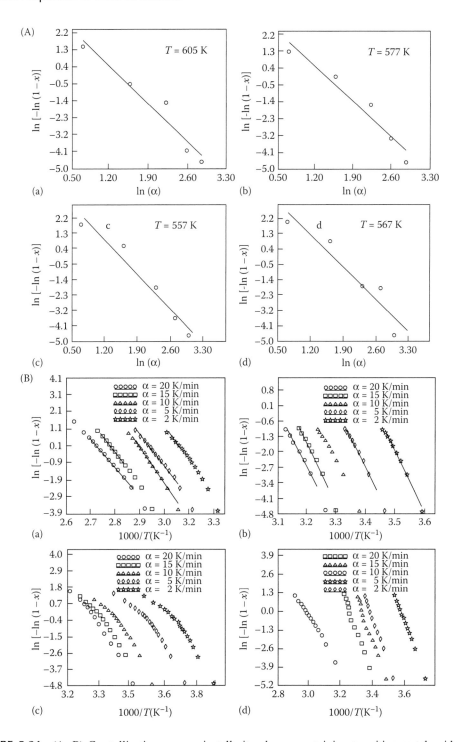

FIGURE 5.24 (A, B) Crystallization process in tellurite glasses containing transition metal oxide (e.g., V_2O_5) and ionic oxide (e.g., Ag_2O). (From R. El-Mallawany, A. Abousehly, A. El-Rahamani, and E. Yousef, *Phys. Stat. Solids*, 163, 377–386, 1997. With permission.)

$$\ln\left[-\ln\left(1-\chi\right)\right] = -n\ln\left(\alpha\right) - \left(mE_c/RT\right) + \text{constant} \qquad (5.13)$$

where m and n are constant integers depending on the morphology of growth. Figure 5.24 shows the plots of $\ln[-\ln(1-\chi)]$ versus $1/T$ at different ϕs. From this slope, the values of E_c were calculated for all values of ϕ. The average values of the E_c of the tricomponent tellurite glasses are given in Table 5.3. In the evaluation of the E_c from the variation of T_p with ϕ, the Kissinger (1956) equation was used:

$$\ln\left(\frac{\phi}{T_p^2}\right) = -\left(m/n\right)\left(\frac{E_c}{RT_p}\right) + \text{constant} \qquad (5.14)$$

Ozawa (1965) and Chen (1978) used another relationship between ϕ and T_p. However, Mahadevan et al. (1986) suggested an approximation of the Ozawa–Chen equation, written in the form:

$$\ln\left(\phi\right) = -\left(m/n\right)\left(\frac{E_c}{RT_p}\right) + \text{constant} \qquad (5.15)$$

The values of the E_c using the Kissinger and modified Ozawa–Chen equations as obtained for tellurite glasses are summarized in Table 5.3. The E_c increased from 174.628 kJ/mol to 225.14 kJ/mol due to higher values of Ag_2O in the glasses (from 20 to 27.5 mol%). The same behavior was shown by the Ozawa–Chen model. The increase in E_c can be attributed to the presence of separate unit cells (due, in turn, to the presence of NBO in the network of glass).

The conclusions achieved for both T_g and T_c were as follows: (1) both depend on the percentage of Ag_2O and also on the ϕ, and the results from Kissinger's and also Ozawa-Chen's models for calculating the Eg are very close (within 2%) for every tellurite glass, using Equations 5.14 and 5.15; and (2) both models agree to within 6% using Equations 5.14 and 5.15. Also, Elkholy (1995) studied E_g and E_c for the TeO_2–P_2O_5 glasses, as shown in Figure 5.25.

Elkholy (1995) also studied E_g and E_c for the TeO_2–P_2O_5 glasses, as shown in Figure 5.25.

5.5 CORRELATIONS BETWEEN GLASS TRANSFORMATION TEMPERATURE AND STRUCTURE PARAMETERS

El-Mallawany (1992) adapted the concepts of the n_c' and the average F of glass (see Chapter 2) to interpret the T_g of multicomponent tellurite glasses in the form x mol% $A_{n1}O_{m1}$–$(1-x)$ mol% $B_{n2}O_{m2}$, where x is the mole fraction and A–O and B–O are the bonds. The T_g is an increasing function of both the average crosslink density n_c' and the average stretching for constant F' that is,

$$T_g = f\left(n_c', F'\right) \qquad (5.16)$$

where

$$n_c' = \left(\frac{1}{\Psi}\right)\left[\left(n_c'\right)\left(N_c'\right)_1 + \left(n_c'\right)\left(N_c'\right)_2\right]$$

and ψ is the number of cations per glass formula unit (explained in detail in Chapter 2):

$$\Psi = \sum\left(N_c\right)_i = xn_1 + \left(1-x\right)n_2 \qquad (5.17)$$

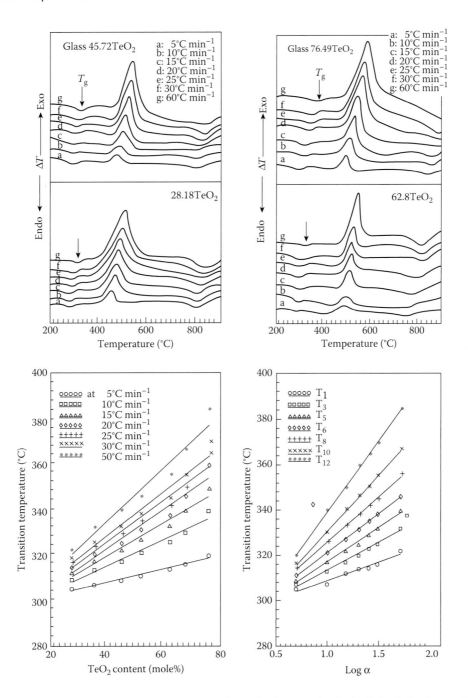

FIGURE 5.25 Glass transformation and crystallization activation energies for the TeO$_2$–P$_2$O$_5$ glasses. (From M. Elkholy, *J. Mater. Sci.*, 6, 404–408, 1995. With permission.)

and F' is the average force constant of the glass, calculated as:

$$F' = \frac{\left[F_1(n_1)(N_c)_1 + F_2(n_2)(N_c)_2 \right]}{\left[(n_1)(N_c)_1 + F_2(n_2)(N_c)_2 \right]}$$

(5.18)

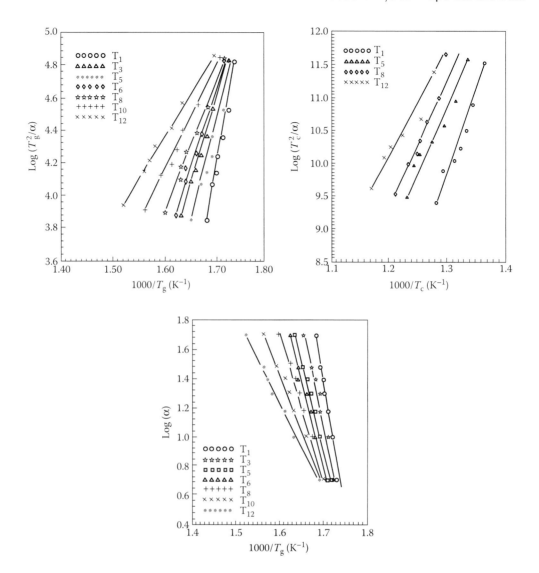

FIGURE 5.25 (Continued)

where F_1 is the first-order F calculated by the formula of Grody (1946) (used previously in Chapter 2):

$$F = 5.28N \left[\frac{(\chi_a \times \chi_b)}{r^2} \right]^{3/4} + 30(N/m) \tag{5.19}$$

where r is the bond length in nanometers, N is the bond order (i.e., the effective number of covalent or ionic bonds acting between the cation and anion), and χ_a and χ_b are the electronegativities of the two cations. Table 5.4 shows the data that have been adapted from crystal structures to calculate n_c and F of each oxide. The values of n_c, F', n'_c, and F' of tellurite glasses have been collected—for example, for RE tellurite glasses, in Table 5.4. It is clear from Table 5.4 that the n'_c increases from 2.0 for pure TeO_2 glass to 2.55, 2.4, and 2.54 for binary RE tellurite glasses when modified with La_2O_3, CeO_2, and Sm_2O_3, respectively. From the values of n'_c, it appears that the structure becomes more linked with the introduction of RE oxides. The average F of these glasses changes from 314

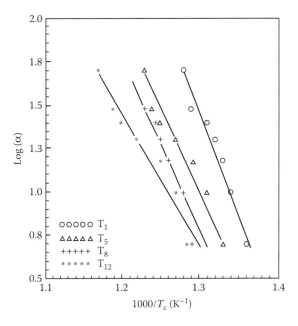

FIGURE 5.25 (Continued)

N/m for pure tellurite glass to 253, 273, and 257 N/m with the addition of La_2O_3, CeO_2, and Sm_2O_3, respectively. Conclusions from these quantitative analyses of binary RE tellurite glasses include the following: first, the structure of these glasses is weaker and more linked; second, the density data and molar V show that RE oxides act more as network formers than as network modifiers in tellurite glass, which is evident in the increase of the n_c of TeO_2.

These findings are also reflected in the elastic properties (Chapter 2) and infrared absorption spectra (Chapter 10) of these glasses. Quantitative analyses explain the high values of the T_g of these glasses and can be extended qualitatively to the low values of the α_{th} of these glasses.

TABLE 5.4

Bond Length, Electronegativity χ_c, Coordination Number C.N., and Calculated Stretching Force Constant and Cross-Link Density of RE Oxides n_c and Tellurite Glasses, χ_b, Oxygen = 3

Oxide	Reference	χ_a (Cation)	r (nm)	C.N.	F (N/m)	n_c
TeO_2	A.F. Walls	2.3	0.199	4	314.3	2
La_2O_3		1.1	0.253	7	144.05	5
CeO_2		1.1	0.248	8	147.5	6
Sm_2O_3		1.2	0.249	7	154.6	5
Glass						
TeO_2					314	2.0
$0.9\ TeO_2$–$0.1\ La_2O_3$					253.5	2.55
$0.9\ TeO_2$–$0.1\ CeO_2$					272.6	2.4
$0.9\ TeO_2$–$0.1\ Sm_2O_3$					257.5	2.54

5.6   CORRELATIONS BETWEEN THERMAL EXPANSION COEFFICIENT AND VIBRATIONAL PROPERTIES

In the same article as that discussed above (El-Mallawany 1992), in which the concepts of n_c' and F' of tellurite glasses are applied and data are collected, the relationship between α_{th} and γ of tellurite glasses is also explored. The linear α_{th} of any solid material depends strongly on the anharmonic nature of interatomic forces and is often discussed in terms of the mean γ. Included in Chapter 2 is a discussion of the relationship between anharmonic vibration and the thermal properties of tellurite glasses. These glasses are viewed as a set of harmonic oscillators whose frequencies are V dependent. Each oscillator has a mode γ, expressed as $\gamma_i = -[d(\ln \omega_i)/d(\ln V)]$, and a mean γ (γ^{th}) is obtained by summing across these modes: $\gamma^{th} = \Sigma C_i \gamma_i / \Sigma C_i$, where C_i is the Einstein C_p of the mode. Experimentally, γ^{th} is defined by $\gamma^{th} = 3\alpha_{th} V B^T / C_v = 3\alpha_{th} V B^S / C_p$, where C_v and C_p are the specific heat capacities, and B^T and B^S are the isothermal and adiabatic bulk moduli.

It has been argued that the secondary derivatives of the energy of any cation–anion pair ($U = ax^2 - bx^3 - cx^4$) should be considered functions of V ($F = -\partial U/\partial x = -2ax + 3bx^2 + cx^3$). On the basis of the potential energy of a diatomic chain, Bridge and Patel (1986) have shown that the α_{th} can be written in the form $\alpha = (bK_B/a^2 r_o) - (K_B/2a'r_o^2)$, where K_B is the Boltzman constant, r_o is the equilibrium atomic separation, a and a' are the first-order stretching and bending force constants, respectively, and b is the second-order F describing the asymmetry of the longitudinal interatomic force potential. In the quasiharmonic theory, small displacements of atoms from their equilibrium positions are assumed so that the potential energy expansion in terms of these displacements can be terminated at the quadratic term. The second derivatives of the energy are then considered to be functions of V. However, in the long-wavelength limit, the regime of ultrasonic waves, an amorphous solid appears isotropic and behaves as an elastic medium. These vibrational excitations are more heavily damped—that is, they have a shorter lifetime than those in a crystalline solid. Measurement of the pressure dependencies of the elastic stiffness can be used to quantify the anharmonicity of these low-frequency modes by determining the acoustic-mode γs (γ_{el}) as discussed in Chapter 2.

In general, the mean γ_{el} in the long-wavelength limit must be obtained at high enough temperatures, ($T > \Theta_D$), where Θ_D is the Debye temperature estimated from the room temperature ultrasonic wave velocity (as discussed in Chapter 2) to ensure that all vibrational modes are excited by the expression:

$$\gamma_{el} = \left(\frac{\gamma_L + 2\gamma_S}{3} \right) \tag{5.20}$$

This is a reasonable approximation at room-T for glasses. The fact that the long-wavelength acoustic modes do contribute substantially to collective summations over the modes in amorphous materials is indicated by the strong correlation between room-T thermal expansion and γ_{el}. Information on vibrational anharmonicity of acoustic modes at the long-wavelength limit for tellurite glasses has been obtained from the hydrostatic pressure dependencies of ultrasonic wave v values as measured by El-Mallawany and Saunders (1987, 1988). Both γ_L and γ_S are positive for tellurite glasses, as shown in Table 5.5. It is clear that the vibrational anharmonicity of the acoustic modes of these tellurite glasses does not exhibit the unusual features of silica glass, possibly as a result of their low coordination numbers and relative ease of bending vibrations. Vitreous silica has anomalously negative elastic constant pressure derivatives and hence a negative γ_{el}; the α_{th} is small because of the existence of individual γ_ls with both negative and positive signs. The γ values for all of the tellurite glasses, when modified by RE oxides or TMOs, are positioned in the range of glasses that show the more usual positive value of γ_{el}, coupled with moderate α_{th} close to

TABLE 5.5
Long-Wavelength Acoustic-Mode Gruneisen Parameters of Tellurite Glasses

Glass (mol% Composition)	Reference	γ_l	γ_s	γ_{el}	Born Exponent
TeO_2	Lambson et al. 1984	+2.14	+1.11	+1.45	4.7
85 TeO_2–15 WO_3	El-Mallawany and Saunders 1987	1.98	1.3	1.52	5.12
79 TeO_2–21 WO_3		2.28	1.09	1.49	4.94
67 TeO_2–33 WO_3		2.22	1.02	1.42	4.52
90 TeO_2–10 $ZnCl_2$		2.29	2.5	2.24	9.44
80 TeO_2–20 $ZnCl_2$		2.5	1.5	1.8	6.8
90 TeO_2–10 La_2O_3	El-Mallawany and Saunders 1988	2.15	1.03	1.4	4.4
90 TeO_2–10 CeO_2		1.88	1.07	1.34	4.04
90 TeO_2–10 Sm_2O_3		1.89	0.89	1.32	3.92
60 TeO_2–30 WO_3–10 Er_2O_3		2.83	1.09	1.67	6.02
70 TeO_2–20 WO_3–5 Sm_2O_3		1.83	1.26	1.45	4.7

that of the pure TeO_2 glass as shown in Table 5.5. Using the general Mie–Gruneisen relationship (Gruneisen 1926):

$$\gamma = \frac{(n+4)}{6} \tag{5.25}$$

where n denotes the Born repulsion exponent. Anderson and Anderson (1970) considered oxides in the n range 3.79–4.95. The value of n for pure TeO_2 glass is 4.7 and changes for binary RE and transition metal tellurite glasses as shown in Table 5.5. The Born repulsion exponents of tellurite glasses in their pure, binary, or ternary forms have been found to be greater than that of n ($SrTiO_3$) = 4.3 by Ledbetter and Lei (1990). Real progress has been made in understanding the elastic and anelastic properties of the new noncrystalline-solid tellurite glasses since the relations between the thermodynamic, compressibility, and elastic properties have been compiled (El-Mallawany, unpublished data). However, it will become much easier to unravel the dynamics of relaxation spectroscopy in the future because there are a variety of advanced applications in nonlinear optics and laser glass envisioned for tellurite glasses.

5.7 TELLURITE GLASS-CERAMICS

As shown in previous chapters, the structure of TeO_2-based glasses is of interest because the basic structural unit of glasses with high TeO_2 content is an asymmetrical TeO_2 trigonal bipyramid with a lone pair of electrons in the equatorial position. Tellurite glasses are also of great interest for their unique crystallization behavior; in some tellurite systems, crystallization occurs easily on heating the glasses, but it hardly occurs at all when the corresponding liquids are cooled. In 1979, McMillan's book "Glass Ceramics" started from a definition and goes through the whole process, including crystallization, devitrification, nucleation agents, properties of glass-ceramics, and applications. He stated that ceramics made by the controlled crystallization of glass have outstanding mechanical, thermal, and electrical properties and that it is useful to compare the properties of glass-ceramics with those of related materials such as glasses or ceramics made by conventional methods. In addition, whenever possible, it is of value to consider how the properties of a glass-ceramic are related to its chemical composition, its crystallographic constitution, and its microstructure. Kikuchi et al. (1990) studied the phases with fluorite superstructures in the pseudoternary system TeO_2–PbO–Bi_2O_3. Mixtures of oxides were heated in a quench furnace to 550°C–650°C for 24 h in air, followed by further heating at 600°C–800°C for 24 h in air. The specimens were examined by x-ray powder diffraction (XRPD) using a conventional diffractometer with nickel-filtered CuK_α radiation as shown in Figure 5.26. When the reactions were not completed at a particular T, the specimen was powdered and reheated at the same or a higher T. From Figure 5.26, it seems clear that an orthorhombic solid solution of TeO_2–x mol% PbO–$(4 - x)$ mol% Bi_2O_3, where $x = 0.75$–1.50, was formed at 700°C. The strong reflections of XRPD data were believed to be those derived from the fluorite subcell, which was slightly distorted to orthorhombic symmetry. The XRPD pattern of TeO_2–PbO–3 mol% Bi_2O_3 is shown in Figure 5.26. The XRPD pattern of a face-centered cubic phase of 10 mol% TeO_2–30 mol% PbO–60 mol% Bi_2O_3 is also shown for comparison in the same figure.

Also in 1990, Stavrakeva et al. established the real composition of crystalline phases in Te–Cu–O by electron microscopic analysis, as shown in Figure 5.27. Taking into account the possibility of oxide reduction processes owing to the complex phase equilibrium in the system, the specimens were produced using two methods:

1. Solid-state synthesis between the starting oxides (regimen A)
2. Melting and consequent crystallization (regimen B)

The synthesis of compositions was controlled by x-ray powder specimens. The results of the spectral analysis were treated, and the chemical formulae were evaluated by the oxygen method, as well as the data about the chemical composition of the crystalline phases. Scanning electron microscopy at 470× for the Te–Cu–O crystalline units was achieved as shown in Figure 5.34 and summarized in Table 5.6.

In 1991, Komatsu et al. studied the properties and crystallization behaviors of $(100 - x)$ mol% TeO_2–x mol% $LiNbO_3$ for values of x ranging from 10% to 60%. The glasses were prepared by using a conventional melt-quenching method, and some properties such as the thermal stability, density, refractive index, and crystallization behaviors of these glasses were measured as shown in the 1st Ed. For the values $x = 10$ and 20, α-TeO_2 (paratellurite) crystals were precipitated by heat treatment above 450°C. In glasses with $x \geq 40$, x mol% $LiNbO_3$ crystals were formed through the transformation of a metastable pyrochlore-type compound at Ts above 500°C. This study indicates that it is possible to fabricate transparent TeO_2-based glass-containing ferroelectric $LiNbO_3$ microcrystallites. Also, in 1992, Rojo et al. studied the relationship between microstructure and ionic conduction properties in oxyfluoride–tellurite glass-ceramics. Crystalline domains of LiF and α-TeO_2 were observed inside an amorphous matrix by transmission electron microscopy. The extent of the domains increased with the fluorine ratio.

Balaya and Sunandana (1993) studied the crystallization behavior of 70 mol% TeO_2–30 mol% Li_2O glass, as shown in the 1st Ed. The thermal stability of this glass, identified as the eutectic composition from a TeO_2–Li_2O binary-phase diagram, has been determined by measurement by DSC, x-ray diffraction (XRD), and AC electrical conductivity. The crystallization seems to be a three-step process for powdered glass and a two-step process for bulk glass. The nonisothermal kinetics of first-stage crystallization have been analyzed in terms of both the modified Ozawa and the modified Kissinger methods, and the crystallization products have been identified by XRD. Surface nucleation appears to be the dominant mechanism for crystallization, with an E barrier of 429 ± 10 kJ/mol—a value higher than that for silicate glasses, thus indicating the lower stability of the tellurite glasses. The products of crystallization—an intermediate unidentified phase, a metastable monoclinical phase, and a stable, final orthorhombic phase—are revealed by XRD. Variation in the crystallization percentage (χ) as a function of different heating rates is shown in the 1st Ed. The crystallized (orthorhombic) phase has a lower conductivity and higher E for conduction than the glass, as shown in Chapter 6. In 1995, Tatsumisago et al. continued the study of the crystallization

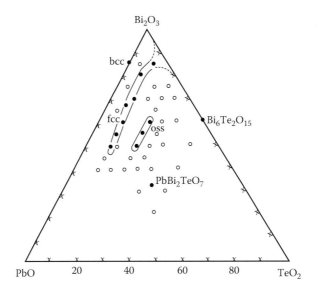

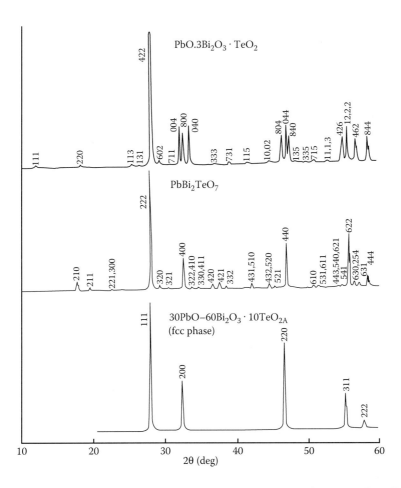

FIGURE 5.26 Results of x-ray analysis of the TeO$_2$–PbO–Bi$_2$O$_3$ glass-ceramic system. (From T. Kikuchi, T. Hatano, and S. Horiuchi, *J. Mater. Sci. Lett.*, 9, 580, 1990. With permission.)

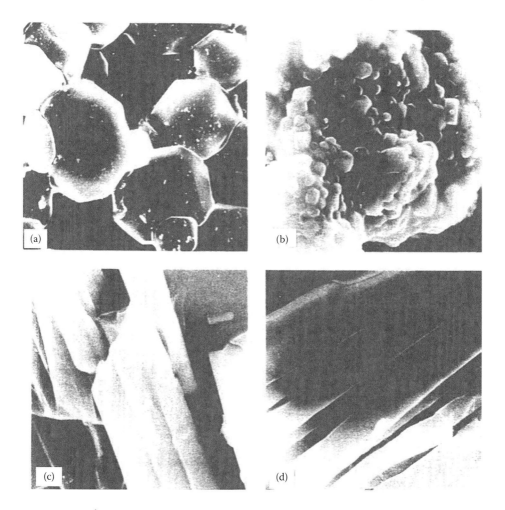

FIGURE 5.27   (a–d) Crystalline phases in Te–Cu–O by electron microscopy. (From D. Stavrakeva, Y. Ivanov, and Y. Pyrov., *J. Mater. Sci.*, 25, 2175, 1990. With permission.)

kinetics of TeO_2–Li_2O binary glass as shown in the 1st Ed. Unique crystallization behavior was observed in this tellurite glass system. This crystallization occurred easily on heating the glasses but hardly occurred at all when the corresponding liquids were cooled. The local structure and crystallization kinetics of the rapidly quenched glasses and liquids in this system were studied by high-T Raman spectroscopy and DSC. The TeO_2–Li_2O liquids exhibited much higher fractions of TeO_3 trigonal pyramidal units with NBOs than the corresponding glasses did. Whether crystallization from the liquids occurs easily depends on the similarity in local structure between the liquids and the main crystallization products. Crystallization kinetics reveal that the number of nuclei might possibly have become saturated during the quenching of the liquids in glass preparation, which would explain the unique crystallization of the TeO_2–Li_2O system as shown in the 1st Ed.

   In 1995, Shioya et al. prepared and studied the optical properties of transparent glass-ceramics of the form TeO_2–K_2O–Nb_2O_5, as shown in the 1st Ed. A transparent glass-ceramic material in a cubic crystalline phase with crystallites having diameters between 20 and 40 nm has been fabricated in compositions of 70 mol% TeO_2–15 mol% K_2O–15 mol% Nb_2O_5. The cubic structural phase is formed by postheat treatment at around 390°C for 1 h, and transforms into a stable phase at Ts above 450°C. This glass-ceramic crystalline phase is opaque, as shown in the 1st Ed. The transparency of glass-ceramics is attributed to the small particle size of the cubic crystalline phase. The optical and electrical properties are discussed Chapters 6 and 7.

TABLE 5.6

Composition of Crystal Phases in the Te–Cu–O System

Given Stoichiometric Composition	Composition of the Crystal Phases from X-ray	Regime of Crystallization
$2TeO_2$–CuO	$5TeO_2$–$2CuO$ (s.s.) from $7TeO_2$–$3CuO$ to $11Te_2$–$4CuO$	A at 600°C B from a melt at 680°C
TeO_2–CuO	From TeO_2–CuO to $6TeO_2$–$5CuO$ (s.s.)	A at 600°C
TeO_2–CuO	$2TeO_2$–$5CuO$	B from a melt at 1000°C overcooled and crystallization at 630°C
$2TeO_2$–$3CuO$	TeO_2–CuO + $2TeO_2$–$5CuO$	B from a 520°C melt
TeO_2–$3CuO$	$2TeO_2$–$5CuO$	A at 720°C
TeO_3–$3CuO$	$2TeO_2$–$5CuO$	A at 720°C
TeO_3–$3CuO$	$2TeO_2$–$5CuO$	B at 800°C
$2TeO_2$–$5CuO$	$2TeO_2$–$5CuO$	A at 720°C B at 800°C
$2TeO_2$–$5CuO$	$2TeO_2$–$5CuO$	A at 720°C B at 800°C

Source: D. Stavrakeva, Y. Ivanov, and Y. Pyrov, *J. Mater. Sci.*, 25, 2175, 1990.

[a] Regimen A, solid-state synthesis between the starting oxides (solid phase interaction at temperatures below 600°C); regime B, synthesis by melting and following crystallization above 600°C; s.s., solid state

Komatsu et al. (1996) studied the formation of an Rb-doped crystalline phase with second harmonic generation (SHG) in transparent TeO_2–K_2O–Nb_2O_5 glass-ceramics as in Figure 5.28. From DTA, they calculated the stability ($T_x - T_g$) of the TeO_2–K_2O–Nb_2O_5 glass and concluded that this glass is largely improved by a small substitution of Rb_2O for K_2O. The crystalline phase appears in the sample after heat treatment at 420°C and 430°C for 1 h, and it has a face-centered cubic structure. Figure 5.29 represents the lattice constants of the cubic phase as a function of the Rb_2O content. Also, in 1995, Saltout et al. studied the lattice structures, crystallization behavior, effects of heat treatment, thermal properties, and Raman spectroscopy of the $(100 - x)$ mol% TeO_2–x mol% WO_3 glasses with x values from 5 to 50 mol%, as shown in Figure 5.29. Thermal parameters such as the T_g, T at the onset of crystallization (T_o), and heat of crystallization δH were determined. Saltout et al. found the relation $\ln T_g = 1.6Z + 2.3$, where Z is some average coordination number of the constituent atoms, which is known for molecular and chalcogenide glasses. This relation was found to be satisfied by these oxide glasses up to a certain threshold (27.5 mol%) of WO_3.

Kim et al. (1996), Komatsu and Shioya (1997), Komatsu et al. (1997a), and Kim and Komatsu (1998) succeeded in fabricating transport TeO_2-based glass-ceramics and measured the SHG. This discovery is undoubtedly very important in glass science and technology. Kim et al. (1996) found that the crystalline phase of the system TeO_2–K_2O–Nb_2O_5 is basically a face-centered cubic structure. However, in 1997, Komatsu et al. (1997b) found differences in the crystalline phases between glass-ceramics and sintered ceramics in the TeO_2–K_2O–Nb_2O_5 system. Also, Komatsu et al. (1997c) measured the particle size of the TeO_2–K_2O–Nb_2O_5 crystalline phase, which was 10–20 nm in a face-centered cubic structure. The cubic phase is formed by annealing the original glass at around 390°C and transferring it to another stable phase at Ts above 450°C. Kim and Komatsu (1998) found the glass-forming region in the TeO_2–Na_2O–Nb_2O_5 system.

Investigations within the TeO_2–WO_3 system—including a phase equilibrium diagram and glass crystallization analysis—were conducted by Blanchandin et al. (1999a) (Figure 5.30). The TeO_2–WO_3 pseudobinary system was investigated by T-programmed XRD and DSC. The investigated

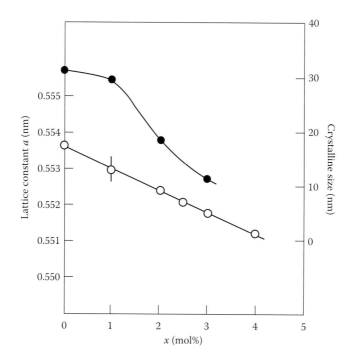

FIGURE 5.28 Formation of Rb-doped crystalline phase with SHG in transparent TeO$_2$-K$_2$O-Nb$_2$O$_5$ glass-ceramics. (From T. Komatsu, J. Onuma, H. Kim, and J. Kim, *J. Mater. Sci. Lett.*, 15, 2130, 1996. With permission.)

samples were prepared by the air-quenching of totally or partially melted mixes of TeO$_2$ and WO$_3$. Identification of the compounds appearing during glass crystallization revealed two new metastable compounds: the first one, which appears both for the low-WO$_3$-content binary glasses and pure TeO$_2$ glasses, was attributed unambiguously to a new TeO$_2$ polymorph called γ, which irreversibly transforms into the stable α-TeO$_2$ form at about 510°C. It crystallizes with orthorhombic symmetry and

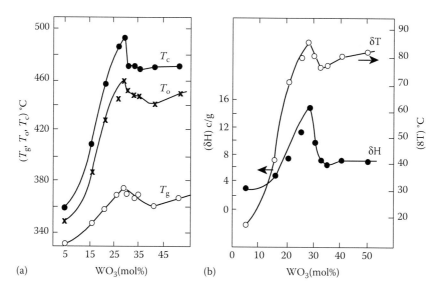

FIGURE 5.29 (a, b) Crystallization behavior, effects of heat treatment, and thermal properties of $(100 - x)$ mol% TeO$_2$–xWO$_3$ glasses with x = 5–50. (From I. Saltout, Y. Tang, R. Braunstein, and A. Abu-Elazm, *J. Phys. Chem. Solids*, 56, 141, 1995.)

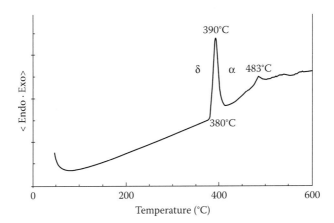

FIGURE 5.30 Thermal properties of the TeO$_2$-WO$_3$ system. (From S. Blanchandin, P. Marchet, P. Thomas, J. Ch-Mesjard, B. Frit, and A. Chagraoui, *J. Mater. Sci.*, 34, 4285, 1999a. With permission.)

unit cell parameters of a = 0.8453 nm, b = 0.4302 nm, c = 0.4302 nm, and Z = 4. The second compound was detected for samples containing about 5–10 mol% WO$_3$. It is cubic (F mode, a = 0.569 nm, Z = 4) and seems to have a fluorite-like structure. In the phase equilibrium diagram, this binary system appears to be a true binary eutectic one. Blanchandin et al. (1999b) studied the equilibrium and nonequilibrium phase diagrams for the TeO$_2$-rich part of the TeO$_2$–Nb$_2$O$_5$ glass system by *T*-programmed XRD and DSC. A large glass-forming domain was evidenced (0–25 mol% Nb$_2$O$_5$). The thermal behavior of these glasses, including T_g and T_c values, as well as the nature of their crystalline phases, have been determined.

In 2000, Taniguchi et al. reported the thermal behavior of 30 mol% Li$_2$O–70 mol% TeO$_2$ glass analyzed *in situ* by high-*T* XRD with a position-sensitive detector. The dynamic structural changes were also observed by DTA and DSC measurements. Differences and contradictions exist in previous studies related to this phase relation at high *T*, including the process of phase change. It was found that in the exothermic process after glass transition between 553 and 623 K, an unknown phase of lithium–tellurite (LT–*X*) crystallizes first. At almost the same *T*, a minor component of vitreous TeO$_2$ appears during decomposition of vitreous 30 mol% Li$_2$O–70 mol% TeO$_2$ of the initial phase. After that, the thermal changes of the LT–*X* and the TeO$_2$ components proceed independently. The LT–*X* phase converts to α-Li$_2$Te$_2$O$_5$, and the vitreous TeO$_2$ phase is followed by the crystallization of an α-TeO$_2$ phase. In the endothermic process between 673 and 793 K, the α-Li$_2$Te$_2$O$_5$ phase transforms into α-Li$_2$Te$_2$O$_5$ phase, and the minor crystalline α-TeO$_2$ phase melts. The α-Li$_2$Te$_2$O$_5$ phase then incongruently melts to form Li$_2$TeO$_3$ and a molten TeO$_2$ phase. Li$_2$TeO$_3$ continuously melts at around 763 K. The results show a possibility of reversible change between the α-Li$_2$Te$_2$O$_5$ and the α-Li$_2$Te$_2$O$_5$ phases. Both sides of this comparison are the same.

5.8   NONOXIDE-TELLURITE GLASSES

Deneuville et al. (1976) studied the effect of annealing on an amorphous (100 – *x*) mol% Te–*x* mol% Ge matrix with Te crystallite, as shown in the 1st Ed. Annealing shows very different behavior for Te crystallites in the amorphous matrix as a function of *x* = 0.1–0.5. For *x* < 0.2, annealing at increasing *T*s increases the number (size) of Te crystallites with subsequent GeTe +Te crystallization.

Tranovcova et al. (1987) measured the optical, mechanical, thermophysical, electrical, and dielectric properties of the systems Te–Se–Ge and Te–Se–Ge–Sb for use as short-distance power-transmitting fibers with CO and CO$_2$ lasers, as shown in the 1st Ed. They found optimization of multicomponent seleno-telluride glasses for optical fibers applied to power transmission at 10.6 μm. The optimum glass composition was 35 mol% Te–30 mol% Ge–35 mol% Se. The samples were prepared by using purified materials heated for 12 h in a high vacuum. The T_g and T_c were measured by using DTA in dry N_2 at a rate of 5°C/min. Heat capacity and thermal diffusivity were determined by a flash method at 20°C–150°C. Electrical and dielectric measurements at frequencies from 0 to 100 kHz and T levels up to 150°C under vacuum or dry Ar have been made with painted graphite electrodes, as shown in the 1st Ed., which summarizes all the physical properties for these glasses with density at room T by the Archimedes method and applied Vickers microhardness of 500 N for 10 s at room T.

In 1990, El-Fouly et al. measured the thermally induced transformations in the glassy chalcogenides $(60 - x)$ mol% Te–x mol% Si–30 mol% As–10 mol% Ge, as shown in Figure 5.31. Temperature-induced transformations are considered to be interesting characteristics of amorphous materials, including the $(60 - x)$ mol% Te–x mol% Si–30 mol% As–10 mol% Ge system, with x values of 5, 10, 12, and 20. Density and XRD and DTA data were used to characterize these compositions. DTA traces of each glass composition at different Φ from 5°C/min to 300°C/min were obtained and interpreted. Fast and slow cooling cycles were used to determine the rate of structure formation. Cycling studies of materials showed no memory effect but only ovonic switching action. The compositional dependence of crystallization activation was used to characterize E, and the K_g values were calculated. The thermal transition Ts and associated changes in specific heat have been examined as functions of the Te/Si ratio by DSC. It was found that ρ and E increase linearly with increasing tellurium content, whereas the C_p and K_g decrease with increasing tellurium content [$E = 154$ eV and $C_p = 0.246$ J g^{-1} K^{-1} for $x = 20$, whereas $E = 2.74$ eV and $C_p = 0.22$ J g^{-1} K^{-1} for $x = 5$].

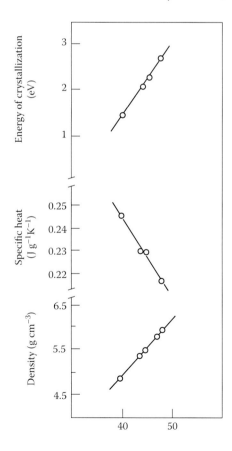

FIGURE 5.31 Thermally induced transformations in glassy chalcogenide $(60 - x)$ mol% Te–x mol% Si–30 mol% As–10 mol% Ge. (From M. El-Fouly, A. Maged, H. Amer, and M. Morsy. *J. Mater. Sci.*, 25, 2264, 1990. With permission.)

DSC data at different heating rates for 30 mol% Te–70 mol% Se chalcogenide glass were reported by Afify (1991), as shown in the 1st Ed. From the heating rate dependence of values of T_g, T_c, and T_p, the author derived the glass E_t and E_c. The crystallization data are interpreted in terms of recent analyses developed for nonisothermic crystallization, and also for the evaluation of E_c and the characterization of a crystallization mechanism. The results indicate that bulk crystallization with two-dimensional growth occurs for this glass. The calculated E_t and E_c were 143 ± 3 and 174 ± 8 kJ/mol, respectively. Li et al. (1991) measured the glass formation range and α_{th} of the ternary Te–Se–I or Br glasses as shown in the 1st Ed. In 1995, the novel chalcogenide glasses in the Te–As–Ge–Ag–Se–I system were prepared by Cheng et al. The glass-forming region of the ternary system TeGe–As$_2$Se$_3$–AgI was obtained. The thermal properties (i.e., T_g, T_c, and T_ms), along with the α_{th}, are summarized in Table 5.7. In 1996, Mahadevan and Giridhar studied silver as a dopant and as a constituent in Te–As–Ag glasses. Variations of the mean atomic V, T_g, and E_t have been reported. An analysis of the results indicates that for Ag content of 1 at%, Ag particles are uniformly dispersed in the network of the Te$_3$–As$_2$ glass without affecting either the short- or medium-range order of the parent glass. For Ag content of 3.5 at%, Ag forms Ag–Te bonds, and the observed variation of the properties can be explained by assuming that the resulting three-component glasses are homogeneous mixtures of Te$_3$–As$_2$ and Ag$_2$–Te units.

In 2000, Ramesh et al. studied glass formation in germanium–tellurite glasses containing metallic additives. DSC studies were carried out on germanium–telluride glasses containing Cu and Ag, as shown in the 1st Ed. Both x mol% Cu$_x$–15 mol% Ge–85 mol% Te ($2 < x < 10$) and x mol% Ag–15 mol% Ge–($85 - x$) mol% Te ($2.5 < x < 21.5$) glasses were found to exhibit single-glass transition

TABLE 5.7
Thermal Properties of Nonoxide–Tellurite Glasses

Glass	Reference	$T_g(°C)$	$T_c(°C)$	$T_m(°C)$
Te–Se–Ge	Sarrach et al. 1976			
50–50–0		57	90	
80–0–20		159	220	
33.3–33.3–33.3		292	382	
Te$_2$–Se$_7$–I	Hong et al. 1991	58		
Te$_2$–SeI–As$_6$		152		
Te$_5$–Se$_3$–I$_2$		65		
Te–Se$_3$I$_2$–As$_4$		101		
Te$_3$–Se$_6$–I		68		
Te$_{0.5}$–Se$_6$–I–As$_{0.5}$		95		
Te$_2$–Se$_7$–Br		65		
Te$_2$–Se$_2$–Br–As$_5$		132		
Te$_{35}$–As$_{35}$–Se$_{10}$–Ge$_{15}$–I	Klaska et al. 1993	194		
Te$_{40}$–As$_{35}$–Se$_5$–Ge$_{15}$–I$_5$		203		
Te$_3$–C$_{12}$	Lucas et al. 1990	82	189	242
Te–Cl		65	183	237
Te$_2$–Br		75	154	227
Te$_3$–Br$_2$		73	152	210
50 Te$_3$–As$_2$–50 As$_2$–Se$_3$	Zhenhua et al. 1993	146	245	306
40 Te$_3$As$_2$–40 As$_2$Se$_3$–20 PbCl$_2$		141	232	307
40 Te$_3$As$_2$–40 As$_2$Se$_3$–20 PbBr$_2$		134	229	294
40 Te$_3$As$_2$–40 As$_2$Se$_3$–20 PbI$_2$		135	235	301
70 TeGe–18 As$_2$Se$_3$–12 Ag$_2$	Cheng et al. 1995	249	373	$\alpha = 149 \times 10^{-7}$
Te$_3$Ge$_2$Se$_5$	Wang et al. 1995	171	$T_{c1} = 292$ $T_{c2} = 350$	410

and single crystallization. On the basis of the devitrification behavior of these glasses, one can conclude that the network connectivity of the parent Ge–Te matrix is not improved by the addition of Cu, whereas Ag improves the connectivity. Overconstraining of the structural network for $x > 5$ is rapid in Cu-added glasses and more gradual in Ag-added glasses. The difference in glass formation in the Cu–Ge–Te and Ag–Ge–Te systems is understood in light of the above differences in their structural networks.

5.9 THERMAL PROPERTIES OF NEW TYPES OF TELLURITE GLASS

As far as glass scientists are aware, the glass transition temperature is the most important property of the produced glass because it not only identifies the preparation temperature, but it also identifies how to deal with the produced glass—especially when it is subjected to heat. In 2011, Wang et al. prepared glass samples of $(1 - x)$[15Ge–10Ga–75Te]–x(CsI), where $x = 0$, 5, 10, and 15 wt %, using the traditional melt-quenching method. Glass compositions, density, ρ (g/cm^{-3}); thickness, cm; glass transition temperature, T_g; onset crystallization temperatures, T_x, $\Delta T = T_x - T_g$; and the weight stability criterion $H_w = (\Delta T/T_g)$ for the glass compositions 15Ge–10Ga–75Te and 85(15Ge–10Ga–75Te)–15CsI were equal to: 5.525–6.701 g/cm^{-3}, 0.28–0.16 cm, 175°C–133°C, 289°C–214°C, 114°C–81°C, and 0.65–0.61, respectively. The maximum T_g value of 175°C was obtained for the glass composition Ge15–Ga10–Te75, while the highest ΔT value was 130°C, which corresponds to 95(Ge15–Ga10–Te75)–5(CsI) as shown in Table 5.8. Their relatively good thermal stabilities, broad optical transmission region, and high optical band gaps make these glasses promising candidate materials for bio-optic sensors and far-IR transmission lasers. The Ge–Ga–Te–CsI glasses have an effective transmission window of between 1.7 and 25 μm, encompassing the region of interest for biosensing applications. The thermal properties of other chalcogenide glasses show low glass-transformation temperatures and high thermal expansion coefficients, such as: 60Te–40Cl glass, where $T_g = 82°C$ and $\alpha = 31.0 \times 10^{-6}$ °C^{-1}; and 10Te–70Se–20I glass, where $T_g = 53°C$ and $\alpha = 44.6$–31.0×10^{-6} °C^{-1} (Weber 2003).

In 2011, Chillcce et al. prepared and studied the physical properties of Er^{3+}-doped oxyfluoride–tellurite glasses in the form 7500 ppm Er$_2$O$_3$–(80 – x)TeO$_2$–xZnF$_2$–20ZnO, where $x = 5$, 10, 15, 20, 25, and 30 mol%. The authors observed that T_g decreases as a consequence of the ZnF$_2$ increase. Also, it was clearly observed that the glasses with ZnF$_2$ concentrations of 10, 20, and 30 mol% show two melting temperatures, while the glasses with ZnF$_2$ concentrations of 25 and 30 mol% show two onset crystallization temperatures. For the change of ZnF$_2$ mol% from 5% to 30%, the glass transition temperature T_g changed from 317°C to 302°C; the onset temperature T_x changed from 415°C to 428°C and 488°C; the melting temperature T_m changed from 629°C to 602°C and 627°C; and $(T_x - T_g)$ changed from 98°C to 126°C, as shown in Table 5.8. The ZnF$_2$ concentration increase in oxyfluoride–tellurite glasses was favorable to the thermal stability increase. The authors proved that it is possible to draw optical fibers from oxyfluoride–tellurite glasses without crystallization.

In addition, Moraes et al. (2010) investigated the relation among thermal, optical and thermo-optical properties and niobium concentration in tellurite glasses of the form 80TeO$_2$–(20 – x)Li$_2$O–xNb$_2$O$_5$ (where $x = 5$, 10, and 15 mol%). The authors used the thermal lens spectrometry, x-ray diffraction, Raman spectroscopy, modulated DSC, and the Brewster angle method. The influence of Nb$_2$O$_5$ in the glass structure was interpreted by the thermal diffusivity D, specific heat C_p, thermal conductivity K, temperature coefficient of the optical path length ds/dT, refractive index n, and band gap energy E_g. The C_p and K were almost constant for these glasses, at around 0.383 J/gK and 6.5×10^{-3} WK^{-1}cm^{-1}, respectively. It is clear that the thermal conductivity of tellurite glass agrees well with Weber (2003); that is, because of its disordered atomic structure, the thermal conductivity of glass is much lower than that for crystalline materials. The thermal conductivity of optical glasses ranges from about 0.5 to 1.5 W/m K, being high for silica and low for glasses containing large quantities of heavy elements such as lead, tantalum, barium, and lanthanum. The thermal conductivity of glass increases with temperature, but only slightly above 300 K. The temperature

TABLE 5.8

Thermal Properties of Recent Tellurite Glasses

Glass and Reference	T_g °C	T_x °C & T_m °C	$\Delta T = T_x - T_g$ °C	$H_w = \Delta T/T_g$
15Ge–10Ga–75Te	175	289	114	0.65
85(15Ge–10Ga–75Te)–15CsI	133	214	81	0.61
Wang et al. 2011				
7500 ppm Er_2O_3–(80 − x)TeO_2–$xZnF_2$–	$317 \geq T_g \geq 302$	$415 \geq T_x \geq 428, 488$	$98 \geq \Delta T \geq 126$	
20ZnO, x = 5, 10, 15, 20, 25 and 30 mol%		$629 \geq T_m \geq 602, 627$		
Chillcce et al. 2011				
$xPbO$–(100 − x)TeO_2				
x = 13, 15, 17, 19 and 21 mol%				
13 mol% of PbO	290	T_{c1} = 314, T_{c2} = 324, T_{c3} = 379 & T_{m1} = 516,		
19 and 21 mol% of PbO	310	T_{c1} = 665 & T_{m1} = 498, T_{m2} = 532, T_{m3} = 604,		
ZnO = 18 to 35 mol%	313 to 333	T_{c1} = 380, T_{c2} = 450, T_{c3} = 568 &		
Kaur et al. 2010		T_{m1} = 662, T_{m2} = 644		
(90 − x)TeO_2–10Bi_2O_3–rZnO		Viscosity Log 5–Log 8 Pa/s		
x = 15, 17.5, and 20				
80TeO_2–(20−y)Na_2O–yZnO				
y = 0, 5, and 10 at.%				
Tincher, et al. 2010				
75TeO_2–20ZnO–5Na_2O	572 K	Fragility, m(ΔH_η /2.302RT_g) = 78 ± 7		
Belwalkar et al. 2010				
TeO_2–WO_3–PbO	$280 < T_g < 410$	$30 < T_c − T_g < 140$	$13 < \alpha < 22 \times 10^{-6}(K^{-1})$	
Munoz–Martin, et al. 2009				
Bi_2O_3–CaO–TeO_2			T_{c1}-T_g = 20	
0–15–85	365	T_{c1} = 385, T_{c2} = 451, T_{c3} = 525		
Chagraoui et al. 2010				
TeO_2(100 − x)–$x$$La_2O_3$7.5 < x < 17.5	$624 < T_g < 705$(K)	$661 < T_c < 730, 923 < T_m < 953$(K)	0.21 < K_g < 0.25	
TeO_2(100 − x)–$x$$V_2O_5$10 < x <50	$563 < T_g < 511$(K)	$664 < T_c < 714, 993 < T_m < 1180$(K)	0.31 < K_g < 0.44	
El-Mallawany et al. 2010				
(90 − x)TeO_2–10GeO_2–$x$$WO_3$	$346 < T_g < 378$	$93 < \Delta T < 138$	12.6 > C_{pg} > 4.8	6.2 > ΔC_p > 1.3
x = 7.5, 15, 22.5, 30 mol%			18.8 > C_{pl} > 4.8	1.65 > $\Delta C_{pl}/C_{pg}$ > 1.27
Upendera et al. 2010				

(continued)

TABLE 5.8 (Continued)
Thermal Properties of Recent Tellurite Glasses

Glass and Reference	T_g °C	T_x°C & T_m°C	$\Delta T = T_x - T_g$ °C	$H_w = \Delta T/T_g$
$(100-x-y)TeO_2-xTl_2O-yZnO$, $5 < x < 30$, $10 < y < 40$, mol%	$125 < T_g < 318$	$200 < T_x < 388$	$55 < \Delta T < 159$	
Soulis et al. 2010				
$20Nb_2O_5-80TeO_2$	497	$T_c = 589$	$\Delta T = 77$	
$4ZnO-20Nb_2O_5-76TeO_2$	499	576	69	
$4MgO-20Nb_2O_5-76TeO_2$	506	609	86	
$4CaO-20Nb_2O_5-76TeO_2$	509	580	59	
$4SrO-20Nb_2O_5-76TeO_2$	511	591	67	
$4BaO-20Nb_2O_5-76TeO_2$	512	618	89	
Hayakawa et al. 2010a				
$5TiO_2-30TiO_{0.5}-65TeO_2$	209			
$10ZnO-30TiO_{0.5}-60TeO_2$	206			
$10GaO1.5-36TiO_{0.5}-54TeO_2$	175			
$5PbO-28.5TiO_{0.5}-66.5TeO_2$	197			
$5BiO_{1.5}-38TiO_{0.5}-57TeO_2$	161			
Hayakawa et al. 2010b				
$90TeO_2-5Bi_2O_3-5GeO_2$	361	$T_c = 470$ & $T_m = 741$	$S = 129$	$K_g = 0.514$
$60TeO_2-20Bi_2O_3-20GeO_2$	382	$T_c = 490$ & $T_m = 753$	$S = 88$	$K_g = 0.311$
Shivachev et al. 2009				
$90TeO_2-5Nb_2O_5-5Bi_2O_3$	387	$T_x = 526$		
$85TeO_2-10Nb_2O_5-5Bi_2O_3$	402	$T_x = 557$		
$75TeO_2-10Nb_2O_5-15Bi_2O_3$	439	$T_x = 502$		
Wang et al. 2009				
$60TeO_2-30WO_3-10PbO$	363	$\Delta C_p = 0.148$ & $T_{max\,cryst} = 545C$	$\Delta H_{cryst} = 2.56[Jg^{-1}]$	Crystallizing phases
$577TeO_2-30WO_3-10PbO-3La_2O_3$	391	0.477		TeO_2,WO_3
Golis et al. 2008				
$Bi_2O_3-TiO_2-TeO_2$				
0–5–95 mol%	313	$T_o = 361$, $T_x = 368$	$T_o-T_g = 48$	
5–15–80 mol%	362	$T_o = 441$, $T_x = 450$	$T_o-T_g = 79$	
25–10–65	354	$T_o = 387$, $T_x = 405$	$T_o-T_g = 33$	
Udovic et al. 2006				

Glass	Value	Property 1	Property 2	Property 3
$20Li_2O$–$80TeO_2$	259			
$10Li_2O$–$25WO_3$–$65TeO_2$	335			
$10K_2O$–$25WO_3$–$65TeO_2$	339			
Lim et al. 2004				
P_2O_5–TeO_2–ZnO				
11.9–30.9–57.2	432.6	$T_f = 465.3$	$\alpha = 111.3 \times 10^{-7}°C^{-1}$	T_x–$T_g = 161$
16.3–54.5–29.2	371.0	$T_f = 401.3$	$\alpha = 139.2 \times 10^{-7}°C^{-1}$	T_x–$T_g = 118$
Konishi et al. 2003				
$(90-x)TeO_2$–$xZnF_2$–$10Na_2O$				
60–30–10	235	$T_x = 396$	$T_m = 467$	
85–5–10	273	$T_x = 391$	$T_m = 494$	
Donnell 2003				
$76.5TeO_2$–$13.5Na_2O$–$10Al_2O_3$ + 2 CoO	320	$\alpha = 190 \times 10^{-7}°C^{-1}$, C_p(at 25°C) = 0.44(J/g/K)	$C_{pmax} = 0.68$(J/g/K)	
$72.3TeO_2$–$12.7Na_2O$–$15GeO_2$ + 2CoO	302	$\alpha = 201 \times 10^{-7}°C^{-1}$, C_p(at 25°C) = 0.54	$C_{pmax} = 0.80$	
$76.5TeO_2$–$13.5Na_2O$–$10TiO_2$ + 2 CoO	300	$\alpha = 201 \times 10^{-7}°C^{-1}$, C_p(at 25°C) = 0.55	$C_{pmax} = 0.74$	
$0.9TeO_2$–$0.1PbO$	289	$T_x = 331$–250	$T_p = 346$–256	
$0.5TeO_2$–$0.5PbO$	231			
Silva et al. 2001				
Vycor (Corning 7913)	890 K	$\alpha = 1.38 \times 10^{-7}°C^{-1}$	$C_p = 890$ J/gK	
Silica 96% SiO_2		Thermal conductivity = 0.75 W/mK		
Marvin J. Weber 2003				
Pyrex (Corning 7740) borosilicate	560 K	$\alpha = 1.13 \times 10^{-7}°C^{-1}$	$C_p = 890$ J/gK	
SiO_2–B_2O_3–Na_2O–Al_2O_3		Thermal conductivity = 1.05 W/mK		
Marvin J. Weber 2003				
GeO_2	749 K			
Marvin J. Weber 2003				
ZBLAN				
$56ZrF_4$–$14BaF_2$–$6La_2O_3$–$4AlF_3$–20NaF	543 K	$\alpha = 17.5 \times 10^{-7}°C^{-1}$	$C_p = 0.52$ J/gK	
Marvin J. Weber 2003		Thermal conductivity = 0.4 W/mK		
60Te–40Cl	82°C			
10Te–70Se–20I	53°C			
20Te–30Se–40As–10I	120°C			
Marvin J. Weber 2003				

coefficient of the optical path length is related to the thermo-optical coefficient (dn/dT) by the relation; $(ds/dT) = (n + 1)(1 + \nu)\alpha + (dn/dT)$, in which n is the refractive index, ν is Poisson's ratio, and α is the linear thermal expansion coefficient. The values of the temperature coefficient of the optical path length were 3 ± 0.3, 2.5 ± 0.2, and 2.1 ± 0.2 10^{-5} K^{-1} for 5, 10, and 15 mol% Nb$_2$O$_5$, respectively. The results indicated that changing Li$_2$O for Nb$_2$O$_5$ content in the tellurite glass, the thermal properties did not exhibit a significant difference. The same was not observed by the optical, thermo-optical, and spectroscopic parameters: by increasing the Nb$_2$O$_5$ concentration from 5% to 15% and decreasing Li$_2$O from 15% to 5%, (ds/dT) decreases ~30%, the refractive index increases ~6% and E_g decreases ~4.5%. By developing the calculation of the optical basicity, it was possible to predict the expected changes in the refractive index and band gap, with good agreement between both experimental and calculated refractive indices, as shown in Chapters 8 and 9. It has been indicated that the addition of Nb$_2$O$_5$ increases the polarizability, thus increasing the nonlinear optical properties of the glass. All the studied parameters combined indicated that the TeO$_2$–Li$_2$O–Nb$_2$O$_5$ tellurite system, with higher concentration of Nb$_2$O$_5$ than Li$_2$O, can be interesting for photonic devices.

N.B.: The thermal conductivity, K (Vycor, [Corning 7913]) = 0.75 W/mK; K{Pyrex (Corning 7740) borosilicate, SiO$_2$–B$_2$O$_3$–Na$_2$O–Al$_2$O$_3$} = 1.05 W/mK. ZBLAN glass (56ZrF$_4$–14BaF$_2$–6La$_2$O$_3$–4AlF$_3$–20NaF), T_g = 543 K, α = 17.5 × 10^{-7} °C^{-1}, C_p= 0.52 J/gK, and thermal conductivity = 0.4 W/mK (Weber 2003).

Because Er^{3+} and Tm^{3+}-doped tellurite glasses find wide applications that involve Er^{3+}-doped fiber amplifiers (EDFA) for C and L bands, and Tm^{3+}-doped fiber amplifiers (TDFA) for the S band (Balada 2009; Digonnet 1993), Ersundu et al. (2010) prepared tellurite glasses with the compositions of $(1 - x)$TeO$_2$–xWO$_3$, where x = 0.10, 0.15, and 0.20 in molar ratio and all three samples were doped with 0.5 and 1.0 mol% of Nd$_2$O$_3$, Er$_2$O$_3$, Tm$_2$O$_3$, and Yb$_2$O$_3$. Different melt-quenching techniques that were used in the experiments showed that different glass-preparing methods do not cause a significant change in the thermal behavior. However, the addition of rare-earth elements to undoped samples, and increasing their content from 0.5 to 1.0 mol%, changed the thermal behavior of the glasses by shifting the glass transition and exothermic reaction temperatures to higher values, and increasing the thermal stability. With the introduction of rare-earth dopants, the temperature values of the first endothermic peak corresponding to the eutectic reaction significantly decreased and liquid reaction temperatures were slightly shifted to lower temperature values. The addition of rare-earth elements with a higher atomic number (Er$_2$O$_3$, Tm$_2$O$_3$ and Yb$_2$O$_3$) caused the splitting of the first endothermic peak corresponding to the eutectic reaction of the system. An important study has also been conducted by Ersundu et al. (2010). The Ersundu et al. (2010) investigated thermal, phase and microstructural on the binary TeO$_2$–WO$_3$, TeO$_2$–CdO and ternary TeO$_2$–WO$_3$–CdO glasses samples in order to characterize the metastable δ-TeO$_2$ phase in the TeO$_2$-rich part. Therefore, as-cast samples were heat-treated above the first and second crystallization onset temperatures in order to identify the formation and transformation of the δ-TeO$_2$ phase. For the 90TeO$_2$–10WO$_3$ sample, both DTA and XRD results showed that the metastable δ-TeO$_2$ phase forms at 390°C and transforms into the stable α-TeO$_2$ form at 410°C, which is in agreement with the literature. For 90TeO$_2$–10CdO and 90TeO$_2$–5WO$_3$–5CdO samples, the formation of the metastable δ-TeO$_2$ phase was observed at 335°C and 365°C, respectively. The formation of α-TeO$_2$ phase was not observed for CdO containing samples in the studied temperature interval, showing that the transformation of the δ-TeO$_2$ into the α-TeO$_2$ phase was not completed. The δ-TeO$_2$ phase was detected for the first time in the literature for CdO containing binary and ternary TeO$_2$-based glasses. The addition of CdO into the tellurite glasses increased the stability range of the δ-TeO$_2$ phase, and in CdO-containing samples it was also proved that the formation of the δ-TeO$_2$ phase occurred in a wider compositional range. Scanning electron microscope images of the δ-TeO$_2$ phase were taken, revealing the formation of δ-TeO$_2$ phase on the surface as grain-like crystallites.

Moreover, Kaur et al. (2010) prepared and characterized glasses of two systems: xPbO–$(100 - x)$TeO$_2$ (where x = 13, 15, 17, 19, and 21 mol%) and yZnO–$(100 - y)$TeO$_2$ (where y = 18, 20, 22, 25, 30,

33, and 35 mol%) at two melt-cooling rates. Density increases from 6.132 ± 0.002 to 6.405 ± 0.004 g/cm^{-3} as PbO concentration is increased from 18 to 35 mol% in normally quenched lead tellurites, and there is a similar increase in density from 6.056 ± 0.007 to 6.389 ± 0.022 g/cm^{-3} in the same composition range. In the case of normally quenched zinc tellurites, density decreases slightly from 5.577 ± 0.003 to 5.506 ± 0.001 g/cm^{-3} as ZnO is increased from 18 to 35 mol%. In case of splat-quenched zinc tellurites, density decreases from 5.545 ± 0.032 to 5.488 ± 0.017 g/cm^{-3}. The first lead tellurite glass containing 13 mol% of PbO (sample 13PbTe-n) has glass transition at 290°C (midpoint value). The two slowly cooled samples contain 15 and 17 mol% of PbO exhibit a very weak and broad glass transition just before the exothermic crystallization peak, making it difficult to identify the correct glass transition temperature. Normally quenched samples containing 19 and 21 mol% of PbO (samples 19PbTe-n and 21PbTe-n, respectively) show a broad glass transition centered at 310°C clear and abrupt glass transition, which decreases systematically from 290°C to 279°C as PbO is increased from 13 to 21 mol%. For the second tellurite glass series, T_g increases significantly from 313°C to 333°C as ZnO content increases from 18 to 35 mol%. In both normal and splat-quenched glasses, T_g rises by 20°C, as ZnO is increased from 18 to 35 mol%. Normally quenched samples with 18 to 25 mol% ZnO show at least two crystallization peaks in their DSC patterns: first, a broad exothermic peak between 380°C and 418°C, and a second, sharper peak around 450°C. Glasses with a higher ZnO concentration of 30–35 mol% showed only one crystallization peak in the temperature range of 428°C–450°C, as shown in Table 5.8. The authors suggested that normally quenched samples containing 19 and 21 mol% of PbO (samples 19PbTe-n and 21PbTe-n, respectively) are not true glasses but are glass-ceramics, and the glass transition temperature of amorphous phase in the composite material dramatically enhances to 310°C, probably due to pressure exerted by coexisting crystals. All splat-quenched lead tellurite glasses showed at least two broad crystallization peaks, while normally quenched samples (except the first sample 13PbTe-n) showed only one crystallization peak. This was because splat-quenched samples have more amorphous content than normally quenched samples. The composition of lead tellurite glasses was such that it can produce $PbTeO_3$, TeO_2, and $Pb_2Te_3O_8$ phases in crystallization during DSC heating run by the following routes: $A-2(13PbO–87TeO_2) \rightarrow 13Pb_2Te_3O_8 + 135TeO_2$; $B-(13PbO–87TeO_2) \rightarrow 13PbTeO_3 + 74TeO_2$. The network connectivity increases for the Zno concentration. This result was consistent with molar volume data, which show that the network becomes more compact with increasing ZnO concentration. The glass transformation temperature is shown in Figure 5.32 with other tellurite glasses. Crystallization and liquid behavior of normal and splat-quenched zinc tellurite glasses of equal composition were very similar, though not identical. This indicated that higher quenching rates freeze the melt earlier (at a higher temperature), but the structure of two glasses may be the same, except for small differences in bond lengths.

As the glass transition temperature is an important parameter, so too is viscosity an important physical and technological property of a glass in the fabrication of performs and fibers. It is especially important for determining the processing conditions—such as shear rate and temperature—when using extrusion to make preforms prior to drawing fibers. Tincher et al. (2010) studied the viscosity behavior of glasses with the compositions $(90 – x) TeO_2–10Bi_2O_3–xZnO$ with $x = 15, 17.5,$ and 20 (TBZ glasses) and $80TeO_2–(20 – y) Na_2O–yZnO$ system with $y = 0, 5,$ and 10 (TNZ glasses) as a function of temperature using a beam-bending (BBV) and a parallel–plate viscometer (PPV). The viscosity of the investigated glasses in the softening range (Log 5–Log 8 Pa/s) is shown in Table 5.8. It has been shown that the fragility of the glasses increases as x, the concentration of ZnO, increased in the TBZ system but decreased as y, the concentration of ZnO, increased in the TNZ system. The increase of the fragility parameters with an increase of x has been related to a decrease of the TeO_2 content, which leads to the creation of a depolymerized network associated with a reduction in the $[TeO_4]$ units and an increase of the $[TeO_3]/[TeO_{3+1}]$ units. The decrease in the fragility parameter with an increase of y has been related to the increase of the $[TeO_4]$ units number that was induced by the reduction of Na_2O, which is known to depolymerize the tellurite network.

Belwalkar et al. (2010) measured the viscosity of sodium zinc–tellurite glass. Tellurite glass was prepared by mixing 75 mol% TeO_2, 20 mol% ZnO, and 5 mol% Na_2O powders, which were

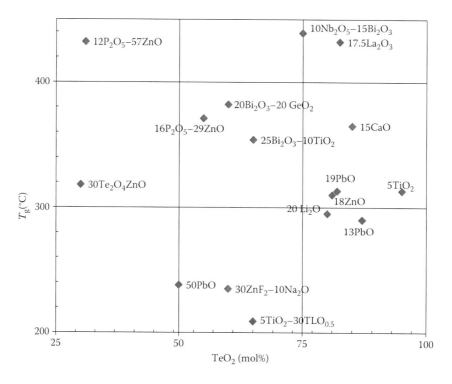

FIGURE 5.32 Reported glass transformation temperature (T_g) value vs. concentration of TeO_2 for tellurite glass with various compositions. (Exact composition and references are in Table 5.8.)

carefully ground and well mixed into a fine powder. The powder was dried at 473 K for 1.5 h and the temperature was then raised to a melting temperature of 1073 K in 1.5 h. The melt was held at 1073 K for another 3 h. The samples were prepared by pouring the melt into 2 mm thick ring-like steel molds, which were kept on a brass plate preheated to 505 K. The samples were then annealed at 572 K for 7 h and were finally allowed to cool down to room temperature. The samples were carefully polished to about 2 mm thickness using a precision polishing machine. Thermal analysis was performed on glass pieces weighing between 10 and 12 mg sealed in aluminum pans, using a DSC 2910 under flowing N_2 at a heating rate at 4.72 K/s. The steady-state shear viscosity was measured at multiple temperatures between T_g and T_x by applying a steady-state shear strain using an Advanced Rheometric Expansion System (ARES) from TA Instruments in a parallel plate configuration. The samples used in the experiment were 8 mm in diameter with thicknesses of between 1.5 and 2 mm. Measurements were made over a wide range of shear rates to study the shear rate dependence of the viscosity. The rheometer was used in the continuous rotation mode to apply strain. The viscosity was obtained from the following expression: $\eta_s = \varsigma/\gamma$, where η_s is the steady-state viscosity, ς is the shear stress, and γ the shear rate. Shear stresses were measured for logarithmically incremental shear rates, and the shear stress–to–shear rate ratio was calculated to yield the viscosity ηs. The viscosity curve for the TZN glass is significantly rounded when compared with the viscosity curves for SiO_2 and GeO_2; a characteristic of "soft" or "fragile" glasses as compared with "strong" ones.

The glass $75TeO_2$–$20ZnO$–$5Na_2O$ was to behave as a Newtonian fluid (elastically) up to a critical shear rate, γ_c, and a shear thin beyond (viscoelastic). In other words, full relaxation in the glass takes place before γ_c, but the relaxation is not reversible beyond γ_c; that is, the structure does not remain the same at all shear rates and in both directions. This was clearly revealed by the fact that at shear rates of up to γ_c, higher activation energy was obtained exhibiting a fully connected network structure, and rearrangement of the network structure took place without breaking the bonds, simply by changing the local configuration. A lower activation energy was obtained beyond γ_c exhibiting viscoelastic stretching and breaking of bonds leading to a permanent decrease in viscosity or shear thinning. The Cross model accurately describes the shear rate dependence of the steady-state viscosity in the temperature range between 609 and 663 K. In the same temperature range, an Arrhenius-type equation provides a good fit to the temperature dependence of both the viscosity and the relaxation time. The activation energy for flow and the fragility near T_g are about 200 ± 20 kcal/mol and 78 ± 7, respectively, as in Table 5.8, which makes TZN a fragile glass when compared to "strong" vitreous silica glass with a fragility of 20. Because of the large anisotropy of its elongated basic TeO_4 molecular unit, tellurite glasses may be compared to polymers, which also have a relatively high fragility index.

N.B. Vitreous silica and GeO_2: $T_g = 1476$ K and 49 K; fragility, $m(\Delta H_\eta / 2.302RT_g) = 20$ and 24, respectively.

A stable glass has been synthesized in Bi_2O_3–CaO–TeO_2 system at 800°C by Chagraoui et al. (2010). The vitreous crystallization of the samples rich in TeO_2 occurs for the γTeO_2 and αTeO_2 polymorphs. The γTeO_2 variety transforms completely to αTeO_2 up 500°C. Some physical properties of the glass have been measured. The high linear refractive index of glasses (≈ 2.14) indicates a high optical nonlinearity, and so they are promising materials for optical communication system. The densities and molar volume of the glasses decrease in CaO content. Some data of the glass transition and the crystallization temperatures have been provided in Table 5.8 and are represented in Figures 5.32 and 5.33, along with other values. The influence of a gradual addition of the modifier oxides on the coordination geometry of tellurium atoms has been elucidated. Based on IR absorption curves, the Raman spectra of glasses show systematic changes in structural units, from TeO_4 trigonal bipyramids (tbps) to TeO_3 trigonal pyramids (tps) via $[TeO_{3+1}]$ entities with increasing CaO content in the glass. A solid-state investigation by x-ray of the system allowed a synthesis of the stable new compounds Bi_2CaTeO_7 and $Bi_4Ca_3Te_5O_{19}$.

Because the glass transformation temperature needs quantitative interpretation, El-Mallawany et al. (2010) therefore investigated both volume and thermal properties of binary tellurite glass systems of the forms $TeO_2(100 - x)$–xA_nO_m, where $A_2O_m = La_2O_3$ or V_2O_5 and $x = 5, 7.5, 10, 12.5, 15, 17.5,$ and 20 mol% for La_2O_3 and 10, 20, 25, 30, 35, 40, 45, and 50 mol% for V_2O_5, as in Table 5.8. The authors measured the glass density and the calculated molar volume, number of bonds per unit volume, and the average stretching force constant for both glass series. The density was in the range 5.18–5.64 and 5.04–4.01 g/cm^{-3}, the molar volume 32.42–34.21 and 32.11–42.58 cm^3 mol^{-1}, the number of bonds N_b 8.07–9.58 and 7.69–6.36 ($\times 10^{28}$ m^{-3}), and the stretching force constant 196.7–160.5 and 234.8–264.8 N/m, respectively. Also, the glass transformation energy was

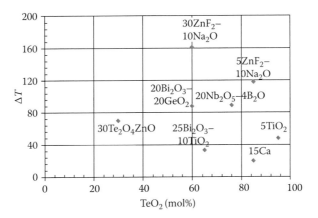

FIGURE 5.33 Reported glass stability range ΔT value vs. concentration of TeO_2 for tellurite glass with various compositions. (Exact composition and references are in Table 5.8.)

calculated using Chen's and Moynihan's formulas. Both models were very close for every glass series. The crystallization energies of these glasses were calculated using the Kissinger and Ozawa–Chen models. Upendera et al. (2010) prepared and measured the thermal properties of tellurite glass of the form: $(90 - x)$ TeO_2–$10GeO_2$–xWO_3 with $x = 7.5, 15, 22.5,$ and 30 mol%. The thermal properties; glass transition temperature T_g, onset crystallization T_o, thermal stability ΔT, glass transition width ΔT_g, heat capacities in the glassy and liquid states C_{pg} and C_{pl}, heat capacity change ΔC_p, and ratios C_{pl}/C_{pg} are shown in Table 5.8. The values of the ratio C_{pl}/C_{pg} ranged between 1.27 and 1.65, suggesting that these glasses forming melts are fragile liquids. DSC results indicate that the addition of germanium oxide into the tungsten–tellurite glass network improved the glass thermal stability. The $67.5TeO_2$–$10GeO_2$–$22.5WO_3$ glass showed a better thermal stability than a lead–tellurium–germinate glass, GTPC and tellurite-based glass, TZN = $75TeO_2$–$20ZnO$–$5Na_2O$, as reported previously by Wang (1994). This improved thermal stability may be a potentially useful candidate material for the drawing of optical fibers. Also, the present $67.5TeO_2$–$10GeO_2$–$22.5WO_3$ glass is compared with the $75TeO_2$–$20ZnO$–$5Na_2O$ tellurite glass (TZN) (Wang 1994) and the $33GeO_2$–$30TeO_2$–$27PbO$–$10CaO$ germano-tellurite glasses (GTPC) (Pan 1997), which have previously been reported as new candidates for fiber-drawing glasses. The working temperature ranges of TZN and GTPC glasses are 118°C and 150°C, respectively. On the other hand, the present $67.5TeO_2$–$10GeO_2$–$22.5WO_3$ glass has a working temperature of 154°C. Therefore, the thermal stability of this glass is superior to those of the TZN and GTPC glasses, and this glass is preferable for performing fabrication and crystal-free fiber drawing.

In 2010, Soulis et al. prepared new tellurite glasses with a large glass-forming domain elaborated within the TeO_2–Tl_2O–ZnO ternary system. The evolution of the glass transition, T_g and onset crystallization, T_x, temperatures for such tellurite glasses was studied—in particular, as a function of the Tl_2O addition, as shown in Table 5.8 and in Figures 5.32 and 5.33. A decrease of both T_g and T_o was observed; the former being more affected. Both T_g and T_x are reduced with the increase in the Tl_2O content; the former being more affected. As a consequence, the $(T_x - T_g)$ difference, which reflects the thermal stability of the glass, was increased. Such results constitute some interesting information in the case of potential fiber-drawing applications.

Following the above discussions, Hayakawa et al. (2010a) prepared TeO_2-based glasses containing metal oxides (Nb_2O_5, ZnO), and alkaline-earth oxides were prepared to obtain the high refractive and high nonlinear optical glass using commercially available chemicals ZnO, MgO, $CaCO_3$, $SrCO_3$, $BaCO_3$, Nb_2O_5 (<99%, Kishida Chemical Co.), and TeO_2 (<99.99%, Koujyundo Chemical Co.). In the Nb_2O_5–TeO_2 binary systems, 20 mol% of Nb_2O_5 contents was the maximum, and a refractive index of $20Nb_2O_5$–$80TeO_2$ glass was 2.06. The compositions studied were xNb_2O_5–$(100 - x)TeO_2$ $(x = 10, 20)$, $yZnO$–$20Nb_2O_5$–$(100 - y)TeO_2$ $(y = 2, 4, 6, 8, 15)$, and $4MO$–$20Nb_2O_5$–$76TeO_2$ with M = Mg, Ca, Sr, and Ba. The powders were weighed precisely, mixed thoroughly, and melted in alumina crucible at 900°C for 1 h in air. The melts were poured onto a carbon mold, annealed at 400°C for 4 h, and were then cooled to room temperature. Each of the planar surfaces were polished to optically flat for the optical measurements. The glass samples appeared slightly yellow. The thermal properties T_g, T_x, T_c, and ΔT are given in Table 5.8 and are represented in Figures 5.32 and 5.33. In order to prepare TeO_2-based glasses with higher nonlinear optical properties and better vitrification, the authors substituted ZnO with an alkaline-earth oxide so as to obtain $4MO$–$20Nb_2O_5$–$76TeO_2$ glasses (M = Mg, Ca, Sr, and Ba). The glass transition temperature, T_g, temperature of crystallization onset, T_x, crystallization temperature, T_c, and $\Delta T = T_x - T_g$ of the glasses studied have been selected and shown in Table 5.8. The values of T_g, T_x and T_c increased as the Nb_2O_5 content increased, resulting in an increase in the melting temperature of the glasses. The value of T_g of $4ZnO$–$20Nb_2O_5$–$76TeO_2$ glass was almost the same value as that of $20Nb_2O_5$–$80TeO_2$ glass, while T_g of $4MO$–$20Nb_2O_5$–$76TeO_2$ glasses increased. In the glasses that contained more ZnO content, T_g increased further (not shown here). It was also noted that ΔT of $4ZnO$–$20Nb_2O_5$–$76TeO_2$ glass was smaller than that of $20Nb_2O_5$–$80TeO_2$ glass. On the contrary, ΔT of $4MgO$ (or $4BaO$)–$20Nb_2O_5$–$76TeO_2$ glasses were larger than the others. In the series of alkaline-earth cations, except Mg, the increase in the atomic number gave rise to the increase in ΔT for $4MO–20Nb_2O_5–76TeO_2$ glasses. The authors concluded that $4MO–20Nb_2O_5–76TeO_2$ (M = Mg and Ba) glasses have better vitrification than $xNb_2O_5–(100 – x)$ TeO_2 (where $x = 10, 20$) and $4ZnO–20Nb_2O_5–76TeO_2$ glasses do. Also, Hayakawa et al. (2010b) studied the third-order nonlinear optical susceptibilities (χ^3) of thallium–tellurite ($Tl_2O–TeO_2$) glasses doped separately with various metal oxides, $MO_{Y/X}$, where M = 22Ti(IV), 30Zn(II), 31Ga(III), 82Pb(II), and 83Bi(III). These were investigated using a femtosecond Z-scan technique. The glass transition temperature was in the range from 209°C to 161°C, as shown in Table 5.8. The authors concluded that the tellurite glasses under study were found to consist of TeO_{3+1} units, thus representing intermediate systems between the two types of $TeO_2–TlO_{0.5}–MO_{y/x}$ glasses: those of island-type structures (M = Pb(II), Zn(II), Ga(III) and Bi(III)) and those of framework-type structures (M = Ti(IV)). Therefore, the incorporation of TiO_2 into $Tl_2O–TeO_2$ glasses was highly favorable for glass polymerization, resulting in an increase in their third-order hypersusceptibilities.

Munoz-Martin et al. (2009) prepared 22 glasses in the $TeO_2–WO_3–PbO$ system using conventional melting methods at temperatures ranging between 710°C and 750°C. The glass-forming area was determined for a wide region of the corresponding ternary diagram. From the thermal stability values for the glasses studied, the authors concluded that for a constant TeO_2 percentage, the thermal stability increases when PbO is substituted mole by mole by WO_3. The highest values are reached for glasses containing $[WO_3] \geq 10$ mol%. This behavior agreed with the fact that both WO_3 and TeO_2 share the role of network formers in $TeO_2–WO_3–PbO$ glasses for a wide composition range. Thermal expansion coefficients vary linearly with the molar percentage of each component for each glass series; that is, α behaves as an additive magnitude as shown in Table 5.8. With the aim of evaluating the contribution of each component to the α value, the corresponding additive molar factors have been calculated from a hyperabundant equations system:

$$f\ TeO_2 = (0.200 \pm 0.001) \times 10^{-6}$$
$$f\ WO_3 = (-0.005 \pm 0.004) \times 10^{-6}$$
$$f\ PbO = (0.275 \pm 0.006) \times 10^{-6}$$

According to these results, the WO_3 contribution to the thermal expansion coefficient of the glasses is zero or slightly negative. This agrees with the high strength of the W–O bond, as well as with the network-forming role of WO_3, attributed above, which yields W–O–W and W–O–Te bonds.

The effects of alkali metal oxides R_2O (R = Li, Na, K, Rb, and Cs) and network intermediate MO (M = Zn, Mg, Ba, and Pb) in tellurite glasses of the form of $R_2O–MO–TeO_2$ have been studied by Desirena et al. (2009). Low glass transition temperature, high viscosity, and low thermal expansion are desirable for a good fiber optic preparation. These parameters could be changed with the glass composition—increased or decreased—by introducing the right modifiers. The coefficient of thermal expansion was obtained from the slope in the 30°C–300°C range, while the glass transition temperature corresponds to the point where slope changes. These glasses were analyzed as a function of field strength calculated as $F_s = Z_c /(r_c – r_o)^2$, where Z_c is the valence of cation, and r_c and r_o are the ionic radii of cation and oxygen, respectively. These values were in the range 158–224×10^{-7}°C^{-1} showing a decreasing trend from Cs to Li through Rb, K, and Na for all glasses. The observed increase of the coefficient of thermal expansion CTE indicates a large ionic radius, and then a decrease in the strength of cationic field. Since Li has the highest cation field strength, it was expected that $Li_2O–MO–TeO_2$ glasses could provide the lowest thermal expansion. The obtained values of CTE reported in Desirena et al. (2009) were below 200×10^{-7} °C^{-1}, lower than values measured from 50 to 200×10^{-7} °C^{-1} reported in phosphate by other authors (e.g., Heo 1992). The coefficient of thermal expansion was also influenced by the presence of the network intermediate (Zn, Mg, Ba and Pb). In this case, an increment of CTE with an increase in the molecular weight or atomic number of intermediate (Pb, Ba, Zn, Mg) was observed. The glass transition temperature T_g showed an opposite trend to thermal expansion for all samples. A monotonic increment from Li to Cs in the ranges 277°C–306°C, 295°C–335°C, 311°C–347°C, and 352°C–373°C for the series Tl

(Pb), T2 (Zn), T3 (Ba), and T4 (Mg), respectively, was observed. The increment was produced by the very large ionic radii and small field strength of alkali metal modifiers. When Cs was replaced by Li through Rb, K, and Na, the ionic radii decreased and the field strength became greater. Thus, the Li–O bonds in the glass network are stronger and T_g increases. The relationship of T_g with intermediate ion was not completely clear; it apparently increases by reducing the molecular weight. However, there is a different behavior between Ba and Zn, thus suggesting the presence of an additional mechanism. Also, the authors measured the chemical durability as good chemical durability is a very important parameter in making high quality optical fiber. The dissolution rate (DR) of glasses was calculated using the relation DR = ΔW/(St), where ΔW is the difference on initial and final weight, S is the total surface area of the cube, and t is the immersion time of glasses in water. The DR for glasses decreases monotonically with the increment of field strength, from Cs to Li through Rb, K, and Na. However, a specific trend as a function of field strength was not observed for network intermediate. The dissolution rate decreases from Mg to Zn through Ba and Pb, but in all cases the chemical durability was improved when Cs was replaced by Li. The dissolution rate (DR) was 2.5, 3, 3.5, and 4 times bigger for R_2O–ZnO–TeO_2, R_2O–PbO–TeO_2, R_2O–BaO–TeO_2, and R_2O–MgO–TeO_2 glasses, respectively. Since Li has higher field strength and Li–O bonds are stronger, the chemical durability of Li_2O–MO–TeO_2 glass would be higher.

Based on the experimental results obtained from the systematic characterization of R_2O–MO–TeO_2 glasses, it was concluded that optical, chemical, and thermomechanical properties of tellurite glasses are strongly influenced by the introduction of alkali metals and network intermediate. The introduction of such ions modifies the glass network and its properties. Some of those properties are improved while others are reduced. It is then necessary to make a compromise of desired properties and to define the glass composition. Experimental results indicate that alkali metals with small ionic radii improve chemical (DR) and mechanical (α, T_g) properties. However, they also increase n and χ^3 (and then n_2). The former is good for the spectroscopic properties in laser and amplifier applications, but nonlinear response induces a deleterious effect. The trend of optical properties of tellurite glasses observed for the introduction of network intermediate is complicated. A lower thermal expansion coefficient was obtained for Zn and Mg, while better chemical durability corresponds to Pb and Zn. A lower refractive index was obtained for Mg and Ba. Although all intermediates under study present the same behavior, the minimum value was obtained for Ba. The R_2O–MO–TeO_2 glasses have a wide transmission range from 0.35 to 6.4 μm, and from these results it has been established that UV and IR transmission $\lambda_{\text{cut off}}$ seem less sensitive to alkali metal than network intermediate. Based on these results, it is suggested that R_2O–MO–TeO_2 glasses could be used in nonlinear optical devices because of their large third-order nonlinear optical susceptibility. Additionally, they provide other properties such as high refractive index, good chemical durability, wide transmission spectrum, and low coefficient of thermal expansion.

In 2009, Shivachev et al. successfully synthesized novel TeO_2–Bi_2O_3–GeO_2 glasses by using the melt-quenching technique. The density was in the range from 5.84 to 6.28 g/cm^{-3} and the refractive index was 2.13. The glasses are characterized by a higher T_g and a better thermal stability, as compared with other tellurite glasses, which makes TeO_2–Bi_2O_3–GeO_2 glasses suitable for applications over a wide temperature range, up to $\approx$290°C. Glass composition (mol%) and thermal properties are given in Table 5.8, and Figures 5.32 and 5.33. Wang et al. (2009) prepared the ternary tellurite glass of the form: TeO_2–Nb_2O_5–Bi_2O_3 by using the conventional melt-quenching method, and the formation range of the TeO_2–Nb_2O_5–Bi_2O_3 glass system was investigated. The TNB1, TNB2, TNB4, and TNB5 sample glasses have a high difference, T (between T_g and T_x) of 139°C, 130°C, 155°C, and 104°C, respectively, and are very stable toward crystallization. The lower values of ΔT observed in this study were 48°C and 63°C for compositions that contain 15 mol% Bi_2O_3 in TNB3 and TNB6 glasses, as selected and shown in Table 5.8 and Figure 5.32. The glasses have high refractive indices of more than 2.0. The largest refractive index value 2.1927 at 632.8 nm was obtained for $75TeO_2$–$10Nb_2O_5$–$15Bi_2O_3$ glass. These glasses have good thermal stability against crystallization for Bi_2O_3 < 10 mol%.

Because electronic ceramics is the fastest-growing field in ceramics today, it was very important to know topics such as theory and the properties of processing and testing of ceramics; that is, to know the principles necessary for understanding ceramic materials and their production. Golis et al. (2008) investigated tellurite glasses TeO_2–WO_3–PbO–La_2O_3 for optoelectronics devices. The effect of lanthanum oxide content on the tendency toward the crystallization of glassy matrix was investigated. The authors measured the glass transition temperature, T_g, the maximum of crystallization temperature, $T_{max.cryst.temp}$, the specific heat accompanying the glass transition, ΔC_p, and the enthalpy of crystallization, ΔH_{crys}, as shown in Table 5.8. The addition of La_2O_3 content in glasses from the system TeO_2–WO_3–PbO causes an increase of the transformation temperature to 391°C, and an increase of the specific heat ΔC_p accompanying the glass transition region, which may be evidence for an increased flexibility of the glass network. The obtained glasses have a good thermal stability. Tellurite glasses with 1.0, 2.0, and 3.0 La_2O_3 do not show a tendency toward crystallization. The authors concluded that the presence of lanthanum ions in the glass structure decreases the tendency toward crystallization.

El-Mallawany and Ahmed (2008) prepared the quaternary tellurite glass systems of the form: $80TeO_2$–$5TiO_2(15 - x)WO_3$–xA_nO_m, where A_nO_m is Nb_2O_5, Nd_2O_3, and Er_2O_3, and $x = 0.01, 1, 3$, and 5 mol% for Nb_2O_5 and $x = 0.01, 0.1, 1, 3, 5$, and 7 mol% for Nd_2O_3 and Er_2O_3 by using the melt-quenching method. The density and molar volumes were measured and calculated for every glass system. The thermal behavior of the glass series was studied by using the DTA DSC. The glass transition temperature T_g, crystallization temperature T_c, and the onset of crystallization temperature T_x were determined. The glass stability against crystallization $S \approx 100°C$ and glass-forming tendency $K_g \approx 0.3$ were also calculated. The specific heat capacity $C_p > 1.4$ J/g°C was measured from room temperature and above the T_g for every composition in each glass series. A quantitative analysis of the above thermal properties of these new tellurite glasses—with structure parameters such as average cross-link density $n_c \geq 2.4$, number of bonds per unit volume n_b ($\geq 8 \times 10^{28}$ cm^{-3}), and the average stretching force constant F—have been studied for every glass composition. The conclusions were:

- The decrease in molar volume indicates that Nb_2O_5 and Nd_2O_3 modifiers have been accommodated in the glass structure and have created a more compact glass network. By contrast, the molar volume increases as the content of the modifier Er_2O_3 increases. Although Er_2O_3 has the highest density glass among the studied modifiers, it is accommodated in the more open structure of the glassy state.
- These glass samples possess a high thermal stability of greater than 100°C, which is required by the thermal stability criteria. Hence, all the studied samples are promising candidates for rare-earth-doped optical fiber, with one glass sample exception (7 mol% Er_2O_3) due to its low thermal stability of 74°C, and Hurby's criterion of value 0.3.
- The recorded increase of the glass transition temperature with the content of the modifier is attributed to the increase in the cross-link density and the number of bonds per unit volume in the glass. Glass networks formed in the prepared samples are stronger than those of the pure host material TeO_2 alone. This can be inferred directly from the increasing values of T_g with modifier percentages and it finds direct support through the quantitative calculations of the number of bonds per unit volume, the cross-linked density, and the average stretching force constant.
- Theoretical calculations of the number of bonds per unit volume revealed the following: increases in the Nb_2O_5 concentration leads to an increase in the number of bonds per unit volume. The average stretching force constant was found to increase with the increasing mol% of the modifiers, which explain quantitatively the decrease in the glass transition temperature of the present tellurite glasses.
- Large values of specific heat capacity C_p have been observed compared with silicate glasses. Calculations of the ratios of the specific heat capacities of the super-cooled liquid to that at the glassy state revealed that these glass samples belong to the category of fragile liquids.

In 2007, Villegas et al. examined the glass-forming region in the TeO_2–TiO_2–Nb_2O_5 ternary system as shown in Figure 5.34 (a-Crystalline sample, b-Opal glass). The TiO_2 incorporation (and even more Nb_2O_5 incorporation) causes an improvement in the thermal stability of the glasses and a network reinforcement, which is pointed out by the T_g increasing and the α decreasing. The glass transformation temperature and thermal expansion coefficient were in the range from 340°C to 420°C and 10 to 17×10^{-6} K^{-1}.

Kabalci et al. (2006), investigated the $(1 - x)$ TeO_2–$xPbF_2$ binary glass system. Samples were prepared by melting the mixture of TeO_2 and PbF_2 in a platinum crucible at 800°C in air. Glass transition and the crystallization temperatures as functions of the glass composition were measured by DTA. An exothermic peak of the crystallization temperature was observed at about 340 ± 1°C for all three samples. A second peak of the crystallization temperature was observed at about 400°C for only the sample with 0.15 mol PbF_2. The mechanism and the activation energy for each crystallization peak were determined from the DTA curves measured with different heating rates between 5 and 20°C/min. The mechanism of the crystallization was found to be surface crystallization for the first exothermic peak (only the 0.15 mol PbF_2 sample); on the other hand, bulk crystallization was found for all samples. Corresponding activation energies are 814, 748, and 387 kJ/mol for the samples with 0.10, 0.15, and 0.25 mol PbF_2, respectively. The mechanism and the activation energy for the second exothermic peak observed in the sample with 0.15 mol of PbF_2 were found to be bulk crystallization and 415 kJ/mol.

In 2006, Jose et al. added WO_3 (up to 10 mol%) and P_2O_5 (up to 16 mol%) to the TeO_2–BaO–SrO–Nb_2O_5 (TBSN) glass system and studied the thermal and optical properties of the resultant glasses. The glass compositions studied were bound by $56 \leq TeO_2 \leq 78$; $2 \leq Nb_2O_5 \leq 8$; $0 \leq WO_3 \leq 10$; $0 \leq P_2O_5 \leq 16$ mol%. The dependences of the additive concentration on glass transition T_g and crystallization T_x temperatures were presented as follows:

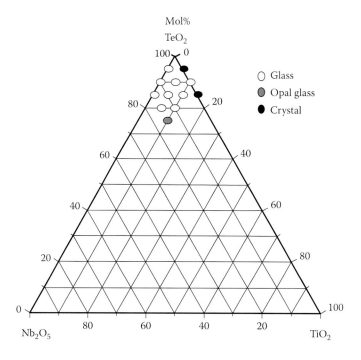

FIGURE 5.34 The glass-forming region in the TeO_2–TiO_2–Nb_2O_5 ternary system. (From Villegas et al. 2007, *J. Eur. Ceram. Soc.*, 27, 2715. With permission.)

1. $(70 - x)$ TeO_2–$3.5BaO$–$10.5SrO$–$8Nb_2O_5$–xWO_3–$8P_2O_5$ (mol%) primary series, where $x = 0$ to 10 mol% in steps of 2 mol%. The starting member was for $x = 0$ mol%, and the ending member was for $x = 10$ mol%.
2. DSC curves for $70TeO_2$–$3.5BaO$–$10.5SrO$–mNb_2O_5–$(12 - m)WO_3$–$4P_2O_5$, secondary series, where $m = 2$ to 8 mol% in steps of 2 mol%. The starting member was for $m = 2$ mol%, and the ending member was for $m = 8$ mol%.

The glass transformation temperature was 360°C to 460°C, and the crystallization temperature was in the range from 400°C to 700°C. The TBSN glass added with 4 mol% WO_3 and P_2O_5 showed high stability against crystallization. Both additives (i.e., WO_3 and P_2O_5) increased the T_g and T_x. The TBSN glass added with P4 mol% WO_3 and P_2O_5 showed high stability against crystallization. The increases in T_g and T_x are understood as being due to the formation of WO_4 and P_2O_5 tetrahedra. The $T_x - T_g$ was greater than 100°C for the glasses with the P_2O_5 content more than 4 mol%.

Udovic et al. (2006) investigated the Bi_2O_3–XO_2–TeO_2 (X = Ti, Zr) systems, a large glass-forming domain was found for X = Ti, but no glass formation was evidenced for X = Zr. Densities, glass transition T_g, crystallization T_c, and the Raman spectra of the relevant glasses were studied as functions of the composition, as in shown in Table 5.8. A glass-forming domain was evidenced in the $yBiO_{1.5}$–$xTiO_2$–$(100 - x - y)TeO_2$ system, but no glass formation was observed in the $yBiO_{1.5}$–$xZrO_2$–$(100 - x - y)TeO_2$ system. For $y = 0$, the evolution of the Raman spectra of the TiO_2–TeO_2 glasses at increasing TiO_2 content allows us to conclude that no tellurite anions are built up in that structure, and the Te–O–Te linkages are progressively substituted by Te–O–Ti. In contrast to this, the addition of Bi_2O_3 results in breaking off the initial glass framework—thus provoking its depolymerization—and in forming local complex tellurite anions interconnected via atoms of Bi. The increase of the T_g values observed in both the cases indicates a relative strength of the Ti–O–Te and Bi–O–Te bridges replacing the Te–O–Te ones.

Lim et al. (2004) measured the thermal analysis of (a) $10Li_2O$–xWO_3–$(90 - x)TeO_2$ and (b) xK_2O–$(10 - x)Li_2O$–$25WO_3$–$65TeO_2$ glasses. The glass transition temperature increased monotonically with increasing WO_3 concentration from 259°C to 339°C, as shown in Table 5.8 and Figure 5.32. The glass structure consists of a continuous network mainly of TeO_4 tbps at the low percent of WO_3. The increase of T_g was due to the formation of W–O–Te linkages with the increase in WO_3 content. The electronic structure of two alkali–tungsten–tellurite glass series has been investigated by x-ray photoelectron spectroscopy. In contrast to alkali–tellurite glasses, the core level spectra of the various elements appear unaffected when TeO_2 is replaced by WO_3, with the alkali oxide concentration remaining constant. The O1s spectra do not indicate a clear separation of the bridging and nonbridging oxygen contributions. Thus, the authors concluded that WO_3 behaves essentially as a network former in tellurite glasses. The authors also obtained the valence band spectra of the glasses and observed major changes with the addition of WO_3. In particular, the intensity corresponding to the lowest energy part of the spectra, attributed to the O2p bonding orbital, decreases significantly with increasing WO_3, thus indicating the appearance of a new bonding configuration. Weng et al. (2004) achieved controllable sol-gel processing of tellurite glasses through the use of Te (VI) precursors. The endothermic event from 130°C to 220°C in DSC curve was due to evaporation of the solvent entrapped in gel. Between 220°C and 250°C, a significant exothermic event in DSC curve occurred, corresponding to a large weight loss.

Konishi et al. (2003) investigated the glass formation and color properties in the P_2O_5–TeO_2–ZnO system as shown in Chapter 1. The glass transition temperature T_g and the softening point T_f increased with the increasing ZnO content. The compositional dependence in the thermal expansion coefficient showed the opposite tendency to that in the glass transition temperature, as shown in Table 5.8. Donnell et al. (2003) measured the thermal properties: glass transformation temperature T_g, onset crystallization temperature T_x and the liquids T_m, obtained from DTA, as shown in Table 5.8, using IR spectra and refractive indices of glasses in the ternary fluorotellurite system $(90 - x)TeO_2$–$xZnF_2$–$10Na_2O$ ($x = 5$, 10, 15, 20, 25, and 30 mol%) with a view to making fluorotellurite optical fibers. Glass stability $(T_x - T_g)$ is shown to increase, and T_g decreases, with the addition of ZnF_2. Also, Inoue et al. (2003) correlated the specific heat and the refractive index change formed by laser spot heating on tellurite glass surfaces. The ternary tellurite glasses of TeO_2–Na_2O–Al_2O_3, TeO_2–Na_2O–GeO_2, and TeO_2–Na_2O–TiO_2 doped with 2 mol% of CoO were irradiated by a green light beam spot (532 nm) from a second harmonic generator of a Q switch pulse YAG laser. The correlation between the specific heat and the refractive index change formed by laser spot heating on the tellurite glass surfaces was investigated. The addition of Al_2O_3 was found to decrease the specific heat of the tellurite glass systems, giving a large change of refractive index. The refractive index change of the irradiated parts correlated well with the specific heat on the high power with short irradiation cases, suggesting that the phenomenon derived from the heating and cooling above T_g of the glasses; in other words, the change of the fictive temperature. The maximum temperatures reached by the irradiated parts were estimated based on the simple two-stage (under T_g and over T_g) heating model. The estimated maximum temperatures correlated well with the refractive index change of less than 0.1, thus supporting the evidence that the mechanisms of the index change were driven by the fictive temperature change as shown in Table 5.8. Silva et al. (2001) prepared glasses in the $xPbO$–$(1 - x)TeO_2$ ($x = 0.1, 0.2, 0.3, 0.4,$ and 0.5) systems by using the conventional method of powder fusion and quenching. The characteristic temperatures T_g (glass transition), T_x (beginning of the crystallization), and T_p (maximum crystallization temperature) were obtained by DSC for the $0.9TeO_2 \pm 0.1PbO$, $0.7TeO_2 \pm 0.3PbO$, and $0.5TeO_2 \pm 0.5PbO$ samples as in Table 5.8. Tellurite-based glasses, as shown in the above data of thermal properties, are emerging as a promising material to compete for a lot of applications.

REFERENCES

Afify, N., *J. Non-Cryst. Solids*, 136, 67, 1991.
Anderson, D., and Anderson, O., *J. Geophys. Res.,* 75, 3494, 1970.
Angel, C., *J. Non-Cryst. Solids*, 131–133, 13–31, 1991.
Angel, C., *J. Non-Cryst. Solids*, 73, 1–17, 1985.
Avramov, I., Guinev, G., Rodrigues, A., *J. Non-Cryst. Solids*, 271, 12, 2000.
Balaya, P., and Sunandana, C., *J. Non-Cryst. Solids*, 162, 253, 1993.
Balda, R., Fernandez, J., and Fernandez-Navarro, J., *Opt. Express*, 17, 8781, 2009.
Baynton, P., Rawson, H., and Stanworth, J., *Proc. 4th Int. Glass Congr. Paris,* 52, 1956.
Belwalkar, A., Misiolek, W. Z., and Toulouse, J., *J. Non-Cryst. Solids*, 356, 1354, 2010.
Blanchandin, S., Marchet, P., Thomas, P., Ch-Mesjard, J., Frit, B., and Chagraoui, A., *J. Mater. Sci.*, 34, 4285, 1999a.
Blanchandin, S., Thomas, P., Marchet, P., Ch-Mesjard, J., and Frit, B., *J. Mater. Chem.*, 9, 1788, 1999b.
Bohmer, R., Ngai, K., Angel, C., and Plazek, D., *J. Phys. Chem.*, 99, 4201, 1993.
Bridge, B., and Patel, N., *Phys. Chem. Glasses*, 27, 239, 1986.
Burger, H., Vogel, W., and Kozhukharov, V., *Infrared Phys.*, 25, 395, 1985.
Burger, H., Vogel, W., Kozhokarov, V., and Marinov, M., *J. Mater. Sci.*, 19, 403, 1984.
Chagraoui, A., Tairi, A., Ajebli, K., Bensaid, H., and Moussaoui, A., *J. Alloys Compd.*, 495, 67, 2010.
Chen, H., *J. Non-Cryst. Solids,* 27, 257, 1978.
Cheng, J., Chen, W., and Ye, D., *J. Non-Cryst. Solids*, 184, 124, 1995.
Chillcce, E., Mazali, I., Alves, O., and Barbosa, L., *Opt. Mater.*, 33, 389, 2011.
Chowdari, B., and Kumari, P., *J. Non-Cryst. Solids*, 197, 31–40, 1996a.
Chowdari, B., and Kumari, P., *Solid State Ion.*, 86–88, 521–526, 1996b.
Chowdari, B., and Kumari, P., *J. Phys. Chem. Solids*, 58, 515–525, 1997.
Colememero, J., and Barandiaran, J., *Mater. Sci. Eng.*, 30, 263, 1978.
Cuevas, R., Barbosa, L., Paula, A., Lui, Y., Reynoso, V., Alves, O., Aranha, N., and Cesar, C., *J. Non-Cryst. Solids*, 191, 107–114, 1995.

Deneuville, A., Gerard, P., and Devenyi, J., *J. Non-Cryst. Solids*, 22, 77, 1976.

Desirena, H., Schulzgen, A., Sabet, S., Ramos-Ortiz, G., de la Rosa, E., and Peyghambarian, N., *Opt. Mater.*, 31, 784, 2009.

Digonnet, M., *Rare Earth Doped Fiber Lasers and Amplifiers,* Dekker, New York, 1993.

Donnell, M., Miller, C., Furniss, D., Tikhomirov, V., and Seddon, A., *J. Non-Cryst. Solids*, 331, 48, 2003.

El-Damrawi, G., *Phys. Stat. Sol.*, 177, 385, 2000.

El-Fouly, M., Maged, A., Amer, H., and Morsy, M., *J. Mater. Sci.*, 25, 2264, 1990.

El-Kholy, M., *J. Mater. Sci.*, 6, 404–408, 1995.

El-Mallawany, R., Abdel-Kader, A., El-Hawary, M., and El-Khoshkhany, N., *J. Mater. Sci.*, 45, 871, 2010.

El-Mallawany, R., Abousehly, A., El-Rahamani, A., and Yousef, E., *Phys. Stat. Sol.*, 163, 377–386, 1997.

El-Mallawany, R., and Ahmed, I., *J. Mater. Sci.*, 43, 5131, 2008.

El-Mallawany, R., *J. Mater. Res.*, 7, 224–228, 1992.

El-Mallawany, R., *J. Mater. Sci.*, 6, 1–3, 1995.

El-Mallawany, R., *Phys. Stat. Sol.*, 177, 439–443, 2000.

El-Mallawany, R., *Thermodynamics, Compressibility and Elastic Properties*, to be published.

Ersundu, A., Karaduman, G., Celikbilek, M., Solak, N., and Aydin, S., *J. Eur. Ceram. Soc.*, 30, 3087, 2010a.

Ersundu, A., Karaduman, G., Celikbilek, M., Solak, N., and Aydin, S., *J. Alloys Compd.*, 508, 266, 2010b.

Gent, A., *J. Appl. Phys.*, 11, 85, 1960.

Golis, E., Reren, M., Wasylak, J., and Filipecki, J., *Optica Applicata*, XXXVIII, 1, 163, 2008.

Grody, W., *J. Chem. Phys.*, 14, 315, 1946.

Gruneisen, E., In *Handbuch der Physik*, Vol. 10, Berlin, Springer-Verlag, 1, 1926.

Hagy, H., *J. Am. Ceram. Soc.*, 46, 93, 1963.

Hayakawa, T., Hayakawa, M., Nogami, M., and Thomas, Ph., *Opt. Mater.*, 32, 448, 2010a.

Hayakawa, T., Koduka, M., Nogami, M., Duclere, J., Mirgorodsky, A., and Thomas, Ph., *Scripta Materialia*, 62, 806, 2010b.

Heckroodt, R., and Res, M., *Phys. Chem. Glasses*, 17, 217, 1976.

Heo, J., Lam, D., Sigel, Jr., G., Mendoza, E., and Hensley, D., *J. Am. Ceram. Soc.*, 75, 277, 1992.

Hiroshima, H., Ide, M., and Youshida, T., *J. Non-Cryst. Solids*, 86, 327, 1986.

Hu, L., and Jiang, Z., *Phys. Chem. Glasses*, 37, 19–21, 1996.

Inoue, S., and Nukui, A., *Phys. Non-Cryst. Solids*, L. Pye, W. LaCourse, and H. Steven (eds), Taylor & Francis, London, 406, 1992.

Inoue, S., Nukui, A., Yamamoto, K., Yano, T., Shibata, S., and Yamane, M., *J. Non-Cryst. Solids,* 324, 113, 2003.

Jayasinghe, G., Dissanayake, M., Careem, M., and Souquet, J., *Solid State Ion.*, 93, 291–295, 1997.

Jose, R., Suzuki, T., and Ohishi, Y., *J. Non-Cryst. Solids*, 352, 5564, 2006.

Kabalci, I., Ozen, G., Ovecoglu, M., and Sennaroglu, A., *J. Alloys Compd.*, 419, 294, 2006.

Kaur, A., Khanna, A., Pesquera, C., Gonzalez, F., and Sathe, V., *J. Non-Cryst. Solids*, 356, 864, 2010.

Kikuchi, T., Hatano, T., and Horiuchi, S., *J. Mater. Sci. Lett.*, 9, 580, 1990.

Kim, H., and Komatsu, T., *J. Mater. Sci. Lett.*, 17, 1149, 1998.

Kim, H., Komatsu, T., Shioya, K., Matusita, K., Tanaka, K., and Hiroa, K., *J. Non-Cryst. Solids*, 208, 303, 1996.

Kissinger, H., *J. Res. Nat. Bur. Standards*, 57, 217, 1956.

Komatsu, T., and Mohri, H., *Phys. Chem. Glasses*, 40, 257, 1999.

Komatsu, T., and Noguchi, T., *J. Am. Ceram. Soc.*, 80, 1327–1332, 1997.

Komatsu, T., and Shioya, K., *J. Non-Cryst. Solids*, 209, 305, 1997.

Komatsu, T., Ike, R., Sato, R., and Matusita, K., *Phys. Chem. Glasses*, 36, 216–221, 1995.

Komatsu, T., *J. Non-Cryst. Solids*, 185, 199, 1995.

Komatsu, T., Kim, H., and Oishi, H., *Inorgan. Mater.*, 33, 1069, 1997a.

Komatsu, T., Noguchi, T., and Benino Y., *J. Non-Cryst. Solids*, 222, 206–211, 1997b.

Komatsu, T., Onuma, J., Kim, H., and Kim, J., *J. Mater. Sci. Lett.*, 15, 2130, 1996.

Komatsu, T., Shioya, K., and Kim, H., *Phys. Chem. Glasses*, 38, 188–192, 1997c.

Komatsu, T., Tawarayama, H., Mohri, H., and Matusita, K., *J. Non-Cryst. Solids*, 135, 105, 1991.

Konishi, T., Hondo, T., Araki, T., Nishio, K., Tsuchiya, T., Matsumoto, T., Suehara, S., Todoroki, S., and Inoue, S., *J. Non-Cryst. Solids*, 324, 58, 2003.

Kosuge, T., Benino, Y., Dimitrov, V., Sato, R., and Komatsu, T., *J. Non-Cryst. Solids*, 242, 154–164, 1998.

Kowada, Y., Habu, K., Adachi, H., Tatsumisago, M., and Minami, T., *Chem. Express*, 17, 965, 1992.

Lambson, E., Saunders, G., Bridge, B., and El-Mallawany, R., *J. Non-Cryst. Solids*, 69, 117, 1984.

Lasocka, M., *Mater. Sci. Eng.*, 23, 173, 1979.

Ledbetter, H., and Lei, M., *J. Mater. Res.* 5, 241, 1990.

Li, M., Hua, Z., and Laucas, J., *J. Non-Cryst. Solids*, 135, 49, 1991.

Lim, J., Jain, H., Toulouse, J., Marjanovic, S., Sanghera, J., Miklos, R., and Aggarwal, I., *J. Non-Cryst. Solids,* 349, 60, 2004.

Mahadervan, S., Giridhar, A., and Singh, A., *J. Non-Cryst. Solids*, 88, 11, 1986.

Mahadevan, S., and Giridhar, A., *J. Non-Cryst. Solids*, 197, 219, 1996.

Marvin J. Weber, *Handbook of Optical Materials,* CRC Press, Boca Raton, London, New York, Washington DC, pp. 246, 249, 254, 258, 2003.

McMillan, P., *Glass-Ceramics*, 2nd ed., London, UK, Academic Press, 1979.

Mochida, N., Sekiya, T., Ohtsuka, A., and Tonokawa, M., *Yogyo-Kyokai Shi*, 96, 973–979, 1988.

Mochida, N., Takashi, K., and Nakata, K., *Yogyo-Kyokai Shi*, 86, 317, 1978.

Monihan, C., Eastead, A., Wilderand, J., and Tucker, J., *J. Phys. Chem*, 78, 2673, 1974.

Moraes, J., Nardi, J., Sidel, S., Mantovani, B., Yukimitu, K., Reynoso, V., Malmonge, L., Ghofraniha, N., Ruocco, G., Andrade, L., and Lima, S., *J. Non-Cryst. Solids,* 356, 2146, 2010.

Munoz-Martín, D., Villegas, M., Gonzalo, J., and Fernández-Navarro, J., *J. Eur. Ceram Soc.*, 29, 2903, 2009.

Muruganandam, K., and Seshasayee, M., *J. Non-Cryst. Solids*, 222, 131–136, 1997.

Narayanswamy, O., *J. Ceram. Soc.*, 54, 491, 1971.

Nishida, T., Yamada, M., Ide, H., and Takashima, Y., *J. Mater. Sci.*, 25, 3546–3550, 1990.

Optical Glass, UDC 666, 113.546.24:666.11.01, 317

Ozawa, T., *Bull. Chem. Soc. Japan*, 38, 351, 1965.

Pan, Z., and Morgan, S., *J. Non-Cryst. Solids*, 210, 130, 1997.

Qui, H., Kudo, M., and Sakata, H., *Mat. Chem. Phys.*, 51, 233–238, 1997.

Ramesh, K., Asokan, S., Sangunni, K., and Gopal, E., *J. Non-Cryst. Solids*, 61, 95, 2000.

Rojo, J., Herrero, P., Sanz, J., Tanguy, B., Portier, J., and Reau, J., *J. Non-Cryst. Solids*, 146, 50, 1992.

Rossignol, S., Reau, J., Tanguy, B., Videau, J., Portier, J., Ghys, J., and Piriou, B., *J. Non-Cryst. Solids*, 162, 244, 1993.

Sahar, M., and Noordin, N., *J. Non-Cryst. Solids*, 184, 137–140, 1995.

Saltout, I., Tang, Y., Braunstein, R., and Abu-Elazm, A., *J. Phys. Chem. Solids*, 56, 141, 1995.

Sekiya, T., Mochida, N., and Ogawa, S., *J. Non-Cryst. Solids*, 185, 135–144, 1995.

Sekiya, T., Mochida, N., Ohtsuka, A., and Tonokawa, M., *J. Non-Cryst. Solids*, 144, 128–144, 1992.

Senapati, U., and Varshneya, A., *J. Non-Cryst. Solids*, 197, 210, 1996.

Shioy, K., Komatsu, T., Kim, H., Sato, R., and Matusita, K., *J. Non-Cryst. Solids*, 189, 16, 1995.

Shivachev, B., Petrov, T., Yoneda, H., Titorenkova, R., and Mihailova, B., *Scripta Materialia*, 61, 493, 2009.

Sidky, M., El-Mallawany, R., Nakhla, R., and Moneim, A., *J. Non-Cryst. Solids*, 215, 75, 1997.

Silva, M., Messaddeq, Y., Ribeiro, S., Poulain, M., Villain, F., and Briois, V., *J. Phys. Chem. Solids*, 62, 1055, 2001.

Soulis, M., Duclere, J. Hayakawa, T., Couderc, V., Dutreilh-Colas, M., and Thomas, P., *Mater. Res. Bull.,* 45, 551, 2010.

Stanworth, J., *J. Soc. Glass Technol.*, 36, 217, 1952.

Stanworth, J., *J. Soc. Glass Technol.*, 38, 425, 1954.

Stavrakeva, D., Ivanov, Y., and Pyrov, Y., *J. Mater. Sci.*, 25, 2175, 1990.

Tanaka, K., Hirao, K., Kashima, K., Benino, Y., and Soga, N., *Phys. Chem. Glasses*, 38, 87–91, 1997.

Taniguchi, T., Inoue, S., Mitsuhashi, T., and Nukui, A., *J. Appl. Crystallogr.*, 33, 64, 2000.

Tatsumisago, M., Kato, S., Minami, T., and Kowada, Y., *J. Non-Cryst. Solids*, 192/193, 478, 1995.

Tatsumisago, M., Lee, S., Minami, T., and Kowada, Y., *J. Non-Cryst. Solids*, 177, 154–163, 1994.

Tatsumisago, M., Minami, T., Kowada, Y., and Adachi, H., *Phys. Chem. Glasses*, 35, 89, 1994.

Tincher, B., Petit, J., and Richardson, K., *Mater. Res. Bull.*, 45, 1861, 2010.

Tool, A., *J. Ceram. Soc.*, 29, 2401, 1946.

Trnovcova, V., Pazurova, T., and Sramkova, T., *J. Non-Cryst. Solids*, 90, 561, 1987.

Udovic, M., Thomas, P., Mirgorodsky, A., Durand, O., Soulis, M., Masson, O., Merle-Mejean, T., and Champarnaud-Mesjard, J., *J. Solid State Chem.*, 179, 3252, 2006.

Upendera, G., Vardhania, C., Suresha, S., Awasthib, A., and Mouli, V., *Mater. Chem. Phys.,* 121, 335, 2010.

Villegas, M., and Navarro, J., *J. Eur. Ceram. Soc.*, 27, 2715, 2007.

Walls, A., *Structure of Inorganic Compounds*, 4th ed., Clarendon Press, London, 1975, p. 452, 455, 582.

Wang, G., Nie, Q., Barj, M., Wang, X., Dai, S., Shen, X., Xu, T., and Zhang, X., *J. Phys. Chem. Solids*, 72, 5, 2011.

Wang, J., Vogel, E., and Snitzer, E., *Opt. Mater.*, 3, 187, 1994.

Wang, Y., Dai, S., Chen, F., Xu, T., and Nie, Q., *Mater. Chem. Phys.*, 113, 407, 2009.

Weng, L., Hodgson, S., Bao, X., and Sagoe-Crentsil, K., *Mater. Sci. Eng. B*, 107, 89, 2004.

Woodruff, O., and Ehrenrich, H., *Phys. Rev.*, 123, 1553, 1961.

Yakakind, A., and Chebotarev, S., *Fiz. I Khim. Stekla*, 6, 164, 1980.

Yakakind, A., Ovcharenko, N., and Semenov, D., *Opt. Mech. Prom.,* 35, 34, 1968.

Yakhkind, A., and Martyshenko, N., *Izvest. Akad. Nauk. S.S.S.R.*, 6, 1459, 1970.

Zahra, C., and Zahra, A., *J. Non-Cryst. Solids*, 190, 251–257, 1995.

Zhang, X., Fonteneau, G., and Lucas, J., *Mater. Res. Bull.*, 23, 59–64, 1988.

Part III

Electrical Properties

Chapter 6: Electrical Conductivity of Tellurite Glasses

Chapter 7: Dielectric Properties of Tellurite Glasses

6 Electrical Conductivity of Tellurite Glasses

Chapter 6 examines the conduction mechanisms of pure tellurite glass; tellurite glasses containing transition metal, rare-earth, or alkaline components—oxide and nonoxide forms—and tellurite glass-ceramics. The effects of different temperature ranges, pressures, energy frequencies, and modifiers on AC and DC electrical conductivity in tellurite-based materials are summarized. Theoretical considerations and analyses of the electrical properties and conductivity of tellurite glasses and glass-ceramics based on their "hopping" mechanism are compared in high-, room, and low-temperature conditions. This chapter explores the dependence of the semiconducting behavior of tellurite glasses on the ratio of low- to high-valence states in their modifiers, on activation energy, and on electron-phonon coupling. Ionic properties of tellurite glasses are also summarized. Chapter 6 clearly shows that electrical-conduction parameters in tellurite-based materials are directly affected by temperature, frequency, and pressure, as well as by the kinds and percentages of modifiers. Most of the new data on the electrical properties of tellurite glasses are collected.

6.1 INTRODUCTION TO CURRENT–VOLTAGE DROP AND SEMICONDUCTING CHARACTERISTICS OF TELLURITE GLASSES

In 1952, Stanworth began preparing ternary tellurite glasses of the form $TeO_2–PbO–M$, where M, the third component, was one of the following compounds: BaO, Li_2O, Na_2O, B_2O_3, Cb_2O_3, P_2O_5, As_2O_3, MoO_3, WO_3, ZnF_2, MgO, TiO_2, GeO_2, or La_2O_3. The DC electrical resistivities ($\log_{10} \rho$) of these glass systems at 150°C were 9.8, 9.9, 13.2, 11.4, 8.8, 11.2, 10.5, 11.2, 10.4, 10.4, 10.3, 10.2, and 9.9 Ω.cm, respectively. Later, Stanworth (1954) continued his work by measuring the electrical resistivity of binary $TeO_2–WO_3$ glasses containing different weight percentages of WO_3, ranging from 16.7 to 34.5. The subsequent log DC resistivity in this group of glasses at 150°C ranged from 10.9 to 11.4 Ω.cm, respectively. Later, Ulrich (1964) prepared and measured the electrical resistivity of binary tellurite–bismuth glasses by varying the temperature range from −196°C to 27°C. The electrical resistivity ranged from 3.6×10^{11} to 3.3×10^{10} Ω.cm for the binary glass system 75 mol% TeO_2–25 mol% Bi_2O_3, although it was 2.6×10^{11}, 5.7×10^{10}, and 4.7×10^9 Ω.cm for the binary glass system 90 mol% TeO_2–10 mol% Bi_2O_3 at temperatures of −196°C, 27°C, and 127°C, respectively.

A number of new semiconductor glasses with interesting properties have been synthesized as a result of studies carried out over several years on multicomponent tellurite–vanadate glasses at the Department of Silicate Technology, Higher Institute of Chemical Technology, Sofia, Bulgaria, by Dimitriev et al. (1973) and Kozhouharov and Marinov (1974, 1975). Dimitriev et al. (1973) covered a large number of compositions of the two-component $TeO_2–V_2O_5$ system and of three-component $TeO_2–V_2O_5–M$ systems, with the participation of oxides from groups I, II, and III of the periodic table ($M = Ag_2O$, CuO, MgO, CaO, SrO, BaO, CdO, ZnO, B_2O_3, Al_2O_3, Ga_2O_3, In_2O_3, CeO_2, Pr_2O_3, or Nd_2O_3). Studies reported by Kozhouharov and Marinov (1974, 1975) focused on the electrical properties of the tellurite glass systems $TeO_2–V_2O_5$ and $TeO_2–V_2O_5–Fe_2O_3$, respectively. They found (Kozhouharov and Marinov 1974) that electrical conductivity (σ) in these systems is linear in the temperature range from 500°C to 1000°C, with V_2O_5 content ranging between 5% and 85%. The approximate values of the pre-exponential factor σ_0 range from −0.5 to 2.0 Ω^{-1} cm^{-1}, and the electrical activation energy (E) ranges from 0.258 to 0.3 eV. Flynn et al. (1977) measured σ as a function of both temperature and vanadium valence ratio. Above 200 K, the DC σ has a constant E of about 0.25–0.34 eV, depending on the composition, whereas below this temperature the σ decreases continuously to a lowest temperature of 77 K. Flynn et al. (1977) noticed that the σ of tellurite glasses is 2.5- to 3-fold greater for similar vanadium concentrations. From the low-temperature thermopower (S) data, the disorder energy (ΔE_D) is ~0.02 eV. We know that semiconductor glasses can be used in making thermistors, channel-type photoamplifiers, and other devices.

Gattef and Dimitriev (1979), and Dimitriev et al. (1981), discovered reversible monopolar switching phenomena in vanadium–tellurite glass that are useful for threshold devices, including bipolar threshold switching, memory switching, and current verus voltage drop (I–V)-characteristic curves. They found that monopolar threshold switching has a voltage lower than that for bipolar switching and has been observed in vanadium–tellurite glasses. By changing the external circuit conditions, a transition from a monopolar to a bipolar mode of operation and vice versa is feasible. This unusual mode of operation demonstrates memory ability related to the direction of the electric field. A monostable memory effect is a fully reversible and repeatable operation. These authors explained the mechanism of both monopolar and bipolar switching. This mechanism is based on the variation of subcritical-nucleus concentrations from a phase with metallic conduction under the action of an electric field. However, in 1981, Dimitriev et al. found that, by increasing the temperature, the threshold power decreases linearly and tends to zero at approximately the temperature of the metal–semiconductor phase transition.

Youshida et al. (1985) measured the memory switching caused by crystallization in TeO_2–V_2O_5–CuO glasses. They investigated crystallization and change in σ. The substitution of CuO for TeO_2 with the same V_2O_5 content increased σ. The σ of 30 mol% TeO_2–45 mol% V_2O_5–25 mol% CuO glass increased about two orders of magnitude from 10^{-3} to $10^{-0.5}$ Ω^{-1} cm^{-1} at 150°C by crystallization. The E for conduction of the crystallized glass was 0.1 to 0.2 eV, which was nearly the same as that of the developed crystal V_2O_5, which is known as a semiconductor. In 1986, Hirashima et al. measured the memory switching of a binary TeO_2–V_2O_5 glass film of 2 to 20 μm thickness, using a DC electric field $\leq 10^6$ V/cm. The switching characteristics and mechanisms of glass films of the system TeO_2–V_2O_5 were discussed by Hirashima et al. (1986) as follows:

- During memory switching, conductance (G) increases by at least two orders of magnitude.
- The DC σ of glass also increases after crystallization by heat treatment.
- Threshold voltage decreases with increasing temperature.
- In the range of thickness above 5 μm, the delay time of switching depends on the thickness of the sample to the power 2, according to the electrothermal switching theory, but when thickness is <5 μm, dependence of the delay time on applied voltage and thickness does not obey the electrothermal theory. In this region, the electronic process must play an important role.

Montani et al. (1998) confirmed that the switching observed in TeO_2–V_2O_5 glasses (disks of thickness between 0.3 and 0.5 mm and with mass of about 3×10^{-4} kg) results mainly from an electrothermal effect. Thus, they found that the appearance of that effect can be avoided by minimizing the amount of heat produced by Joule self-heating. Conversely, that effect might occur when the sample is subjected to current flow for prolonged durations.

Oxide glasses containing transition metal ions (TMI) were first reported to have semiconducting properties by Denton et al. (1954). Since then, most studies have been on systems based on phosphate, although semiconducting oxide glasses based on other glass network formers have also been made. In this chapter, I consider the theoretical and experimental relationship between tellurite glasses and various families of semiconducting oxide glasses, especially in light of the conclusions of Mott (1968) and Austen and Mott (1969). A general condition for semiconducting behavior is that the TMI should be present in more than one valence state, so that conduction can take place by transfer of electrons from low- to high-valence states. A second localization process occurs mainly in noncrystalline solids. This kind of localization was first discussed by Anderson (1958), who had shown that all the physical states are localized if the ratio of the mean disorder potential energy (W_D) between the ions to the total bandwidth (J) approaches some critical value.

In 1987, Hampton et al. measured the electrical σ properties of pure TeO_2 glass and binary tellurite glasses containing 33% WO_3. This impedance (Z) spectroscopy was the first determination of the electrical σ of tellurite glasses, indicating the nature of electrical mechanisms in such material. A sharp break in the slope of the log σ with reciprocal temperature divides σ data in these tellurite glasses into two distinct temperature regimes: σ mechanisms at low temperatures have much lower E than those at higher temperatures. These data are consistent with the charge carrier being a small polaron. Their σs have been analyzed using small-polaron theory to establish the thermal, disorder, and activation energies for the carriers in both glasses. Through the whole range of temperatures from 90 to 430 K, the σ of binary glass is several times larger than that of pure glass. Calculated activation energies are consistently smaller for binary glass than for pure vitreous TeO_2; the addition of WO_3 results in phonon-assisted hopping processes with substantially smaller barriers to surmount. The electrical conduction on TeO_2 single crystal has been measured by Jain and Nowick (1981), who found that below 400°C, for a single Arrhenius plot obtained both in parallel and in the normal direction of tetragonal symmetry, the activation energies are 0.54 and 0.42 eV, respectively. The σ of the paratellurite TeO_2 crystal in this temperature range is ionic, falling into the extrinsic dissociation range of behavior. The oxygen-ion vacancies could be responsible. At high temperature, σ increases sharply due to reduction of the crystal and the onset of electronic carriers.

Chalcogenide glassy semiconductors have been of great interest for materials researchers over the past 40 years, since the discovery of their switching properties by Ovshinsky back in 1968. Nevertheless, some aspects relating to their transport properties are still not well established. Among these is the problem of an electrical conduction mechanism, which is, in all probability, the aspect that has received the most attention. Several theoretical models have been proposed by Mott (1993) to explain the electrical properties of chalcogenide glasses, but some features of these glasses remain unresolved. The effect of the evolution of a glassy structure from an initial disordered lattice to a more orderly one, as in a forced-aging process, is one of the aspects typically studied.

As far as we know, if a glass-ceramic is to be used for electrical insulation, its resistivity should be as high as possible. In many cases, the insulating material is required to operate at nearly normal ambient temperatures; but in some devices the insulator is required to operate at elevated temperatures, and for these applications the variation of electrical resistivity with temperature is important. The volume conductivities of oxide glasses and ceramics generally result from the transport of mobile ions through these materials. There are glasses, ceramics, and glass-ceramics in which the transport process involves the transport of electrons or positive holes, and these are therefore classified as semiconductors; however, most glass-ceramics are ionic conductors, and it is this process that I mainly discuss in this chapter.

The importance of tellurite glasses has led to considerable international technological and fundamental interest as reflected in some more recent publications—that is, since the mid-1980s when pure tellurite glasses were first prepared and their physical properties ascertained. Reviewing this last approximately three decades since Lambson et al. (1984) reported their method to prepare pure TeO_2 glass, I have found that the level of performance in tellurite glasses in the 2000s was much higher than in the period during the 1980s and 1990s. The present chapter collects σ data for distinct tellurite glasses.

6.2   EXPERIMENTAL PROCEDURE TO MEASURE ELECTRICAL CONDUCTIVITY

The electrical σ of semiconducting oxide glasses is generally several orders of magnitude less than that of conventional crystalline semiconductors. This characteristic causes its measurement to be very difficult and inaccurate unless special care is taken. Sayer and Mansingh (1979) used ohmic contact between the bulk and metallic electrode. Sayer and Mansingh (1979) stated that a vapor-deposited gold film could be a good ohmic electrode for vanadium-containing glasses. However, the situation of electrodes in nonoxide glasses is different. For example, silver paste has been used for electrical contacts in the Te–Se–In system by Gadkari and Zope (1988). Jain (1993) stated in detail the factors that affect electrical σ measurement, including the following:

- Technique
- Sample
- Electrodes
- Temperature range
- Environment
- Measurement cell

6.2.1 Preparation of the Sample

For electrical measurements, all samples under investigation must have accurately parallel and highly polished surfaces. This is done in two steps:

1. Samples are sanded, using 600 grade silicon carbide paper and a special holder. A mobile light oil is used as a lubricant.
2. Samples are polished using, for example, Meta-Serve Universal Polisher with the aid of different grain size alumna powder (2–50 μm) and Hyperz polishing oil as a lubricant. The well-polished surfaces of the samples are coated with a thin film of silver paste from both sides to make electrodes of known areas with good ohmic contacts.

The glass sample should be either rectangular, a cylindrical bar, or a disk; free from any macroscopic defect such as bubbles, voids, or cracks; and of a uniform thickness. The size should be sufficient for handling and small enough to be held at a constant temperature, because a temperature gradient in the sample not only causes variation in σ, but also generates thermoelectric voltage. The electrodes and connections must make ohmic contact with the sample (with negligible potential drop at the electrode–glass interface), have negligible resistance (R), and be chemically inert to both sample and atmosphere. Therefore, an electronically conducting inert-metal electrode is often used. Up to a temperature of about 800°C, Ag, Au, or Pt may be used, but for higher temperatures Pt is suitable. Silver, being a relatively small ion, might diffuse into the glass during measurements at high temperatures and can thus introduce errors in the σ value. Platinum is fairly inert, but is a good catalyst and can promote undesirable reactions between the sample and the atmosphere. Colloidal graphite electrodes can be used under nonoxidizing conditions, but their connection with leads can be a problem. The sample should be placed in the homogeneous-temperature zone of the furnace; the temperature probe and the thermocouple should be placed close to the sample in accordance with the precautions described by the American Society for Testing and Materials (1981). The choice of thermocouple depends on the atmosphere around the sample and the temperature range of interest. The copper–constantan (type-T) thermocouple is suitable in temperatures ranging from −200°C to 370°C in oxidizing, reducing, inert, or moist atmospheres. The type-E Chromal–constantan thermocouple develops the highest electron motive force °C^{-1} and has a wider useful temperature range of −200°C to 900°C in oxidizing and inert atmospheres, but it is not suitable for reducing environments or vacuums at high temperatures. The measurement cell is described below.

6.2.2 DC Electrical Conductivity Measurements at Different Temperatures

Electrical σ is one of the essential properties for understanding these materials and their possible applications. The glasses under investigation might possess different mechanisms of conduction depending on the temperatures. For this reason, two experiments were set up to study σ of these materials in a wide range of temperatures (25°C–200°C) for a sample with cross-sectional area A and thickness d. The DC $\sigma(\sigma_{dc})$ is given by:

$$\sigma = \frac{1}{V}\frac{d}{A} \tag{6.1}$$

where I is the current passed through the sample and V is the voltage drop across it. The circuit arrangement is simple. The voltage across the sample is measured by a high-Z multimeter. A current meter is used to record the current through the sample.

There is no standard design for σ measurement cells; therefore, a cell should provide good temperature and environmental control, as well as the ability to change specimens. It is important to ensure that there are no short-circuiting paths and that the current flows only through the specimen. Common short-circuiting paths are the surface of the specimen, sample holder, and ambient gas. To minimize surface conduction, the sample and electrode geometry should be selected such that the surface conduction path is much greater than the volume conduction (thickness) path. The first method is called the "two-probe" method. Two same-sized electrodes can be deposited in the centers of large parallel surfaces of a thin sample. The electrode R is determined by using a bridge or from the ratio of V and I. The second method, called the "three-probe" method, is shown in Figure 6.1a. This method is designed to eliminate short-circuiting via the surface of the sample. A guard-ring electrode is placed around the central electrode on the top face, and any current between the guard and bottom electrodes is excluded from the measurement. The presence of a guard ring also reduces fringe effects, which make a nonuniform field between the central and bottom electrodes. Electrical σ is then given by:

$$\sigma = \left(\frac{d}{A_e}\right)\left(\frac{1}{R}\right) \tag{6.2}$$

where $A_e = \pi r^2$ is the effective electrode area, $r = r_1 + (g/2) - (2d/\pi) \ln \cosh (\pi g/4d)$, r_1 is the central electrode radius, and g is the gap width. The last term in the A_e is neglected.

The "four-probe" method has been used to eliminate the non-ohmic contacts between the electrode–glass interface, as shown in Figure 6.1b. A voltage drop in the sample is measured away from the electrode–glass interface by connecting a high-Z voltage-measuring system to the central leads. Two thin platinum wires are embedded in the sample, which act as voltage probes. Care should be taken not to cause any Joule heating of the sample by high currents. Measurements above room temperature are obtained using a specially designed sample holder and electronic temperature control equipment. The sample and two thin wire leads are situated on a long probe inserted inside an electric furnace. The furnace consists of a stainless steel tube with a 50 mm diameter and 300 mm length with heating wire wound around it. A thermocouple controller is put in back of the furnace. Measurements of current passing through the sample are performed at constant voltage and at constant values of the sample temperatures. The system can measure temperatures with an accuracy of ± 0.2°C. Measurements of electrical σ at low temperatures are made by using a cryogenic device that allows the measurement to be made under vacuum with helium as a working medium to room temperature 20 to 300 K; for example, Displex Closed-Cycle Refrigeration System CSA-202. The measuring chamber is evacuated, using a two-stage Edwards rotary pump, to 10 torr, as indicated, with an Edwards Pirani 1001 pressure gauge. Although DC σ reflects the steady-state flow of current, its magnitude is often modified by the presence of electrode polarization or contact R. Therefore, in the presence of electrode polarization effects, the recommended technique to measure the DC bulk σ is to perform AC measurements of the equivalent R and C as a function of the AC frequency using a C bridge, as shown in the next section.

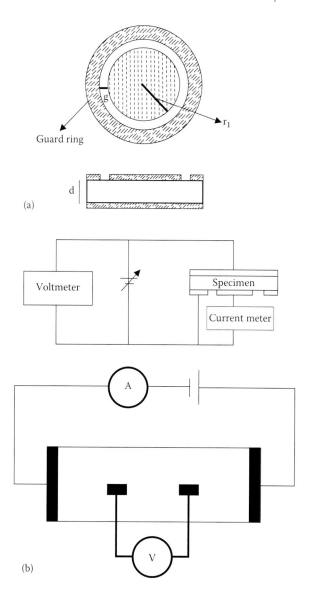

FIGURE 6.1   Methods for conductivity measurements. (a) Three-prop method; (b) four-prop method.

6.2.3   AC ELECTRICAL CONDUCTIVITY MEASUREMENTS AT DIFFERENT TEMPERATURES

The AC σ, dielectric constant, and dielectric loss of glass samples are measured using a C bridge to determine the frequency dependence of electrical σ and dielectric relaxation phenomena. The two-, three-, or four-probe methods described before can be used for AC measurements, but the leads should be shielded in order to minimize induction coupling and stray C. It is also important to apply a very low potential across the sample. The bridge, such as a Hewlett–Packard with a 4192A LF Z analyzer, is suitable for low- and high-frequency measurements. Recent AC bridges (LCR) are available with computer-controlled network analyzers; however, to understand and appreciate these instruments, it is first necessary to go through the details of calculating AC σ. The sample and electrodes are represented by an R in parallel to a C (e.g., R_s/C_s and R_e/C_e, respectively). From these parameters along with Z, angle (θ), and C, one can calculate the AC electrical σ, dielectric constant, and dielectric loss of a material at different frequencies and temperatures. The complex Z (Z^*) can be written in the form:

$$Z^* = Z' - jZ''$$

$$Z^* = \frac{R}{\left(1 + j\omega RC\right)}$$

$$Z^* = \frac{R\left(1 - j\omega RC\right)}{\left[1 + \left(\omega RC\right)^2\right]} \tag{6.3}$$

where $j = \sqrt{(-1)}$, Z' is the real part, and Z'' is the imaginary part. One then plots Z'' against Z' for each value of the ω, as shown in Figure 6.2. A semicircular arc is plotted where bulk σ is represented by the highest-frequency arc. Polarization effects, if any, are indicated by the lowest-frequency arc. The intersection (Z_o') of the high-frequency arc with the Z'-axis is used to determine the DC R (R_o). The meter can be set to treat the sample equivalent to either a series or parallel combination of L, C, and R elements. The choice between series and parallel arrangements depends on which arrangement best describes conduction through a glass/electrode assembly. Transformation from series to parallel can be done with negligible inductive effects. The G, C, and phase θ for each temperature and frequency can be changed to give the σ and dielectric behavior of the glass sample. The relation between real and imaginary parts of Z^* is as follows:

$$Z' = \frac{G}{\left(G^2 + \omega^2 C^2\right)}$$

and

$$Z'' = \frac{\omega C}{\left(G^2 + \omega^2 C^2\right)} \tag{6.4}$$

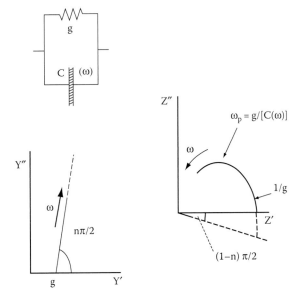

FIGURE 6.2   Impedance plane plot.

Once the true σ and dielectric constant values are obtained for a glass at various temperatures and frequencies, it is important to present this behavior in terms of the electrical E and a pre-exponential factor (σ_o) by fitting the experimental data to the Arrhenius equation $\sigma = \sigma_o \exp(-E/K_B T)$, where K_B and T are the Boltzman constant and absolute temperature, respectively. From the C data, the dielectric constant can be calculated as shown in detail in the next chapter.

6.2.4 ELECTRICAL CONDUCTIVITY MEASUREMENTS AT DIFFERENT PRESSURES

Sakata et al. (1996) stated the experimental procedure to measure the σ_{DC} of a glass under different pressures. The high-pressure experiments were carried out at room temperature with a cubic-anvil-type high-pressure apparatus (DIA-6, Kobe Steel Co.) with sintered tungsten carbide anvils. The schematic diagram of the apparatus has been shown in the 1st Ed. The upper and lower guide blocks are compressed uniaxially by a 250 ton hydraulic press. On compression, the four side blocks, together with the upper and lower blocks, are advanced toward the cube-shaped pressure-transmitting medium. The sample is embedded in the pressure-transmitting medium made of pyrophyllite [$Al_2Si_4O_{10}(OH)_2$], with a size of $8 \times 8 \times 8$ mm. The maximum pressure is 6 GPa. High-pressure experiments were performed using both powder and bulk glass samples.

Powder samples are used because, for this kind of sample, high-pressure effects on the refractive index of oxide glasses appear at relatively low pressures, as stated by Cohen and Roy (1961) and by Tashiro et al. (1964). The bulk specimen is used as a standard to compare the results with those for the powder. The powder glass sample is prepared by dry-crushing the glass in a mortar in air for 20 min, after which it is packed in a tube made of pyrophyllite (2.0 mm outer diameter by 1.2 mm inner diameter by 0.05 mm). The bulk glass sample is cubic (0.5 ± 0.05 mm per side), and Au wires are glued on each side using Au paste. The powder samples are directly embedded in the pyrophyllite pressure-transmitting medium, whereas the bulk glass sample is encapsulated in a Teflon capsule filled with a 4:1 mixture of methanol–ethanol and then embedded in the pyrophyllite. The schematic arrangement is shown in the 1.st Ed. The pressure of the sample is evaluated using a calibration curve obtained by another experiment with pressure-fixed points (2.55 and 2.70 GPa) of bismuth. In the experiment illustrated, two copper electrodes are put on each side of the packed sample for measuring electrical R. The upper and lower electrodes, drawn parallel to each other, intersect perpendicularly in practice, approximately in the directions of the load application. The rates of pressure increase and decrease are 75 and 94 MPa/min, respectively. Sample R is measured by the two-point probe method at each step of the pressure increase and decrease, applying a DC current of 1 μA.

6.2.5 THERMOELECTRIC POWER

Thermoelectric power (S) measurements were described previously by Ghosh (1989), who used disk-shaped samples with diameters of ~8 mm, and thicknesses of ~1 mm, which were cut and polished with very fine lapping papers. A schematic diagram of the sample holder used is shown in Figure 6.3. The sample is placed between two stainless steel electrodes, with the lower electrode supported by a spring to ensure proper electric contact. A temperature difference ($\approx$5–10 K) between the two surfaces of the sample is established with a heater and proper isolation around the upper electrode. The thermocouple (electron motive force) developed between the two surfaces is measured by a vibrating capacitor electrometer. The temperature difference is measured using two copper–constantan thermocouples insulated from the electrodes by mica sheets. For measurements at low temperatures, the sample holder is placed inside a cryogenic unit. Measurements are taken during heating as well as during the cooling run, and the two results should agree well within experimental error for S (μV/K).

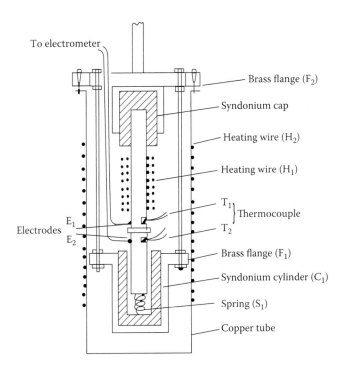

FIGURE 6.3 Schematic diagram of the sample holder for thermoelectric power measurements. (From A. Ghosh, *J. Appl. Phys.*, 65, 227, 1989. With permission.)

6.3 THEORETICAL CONSIDERATIONS IN THE ELECTRICAL PROPERTIES OF GLASSES

6.3.1 DC CONDUCTIVITY

6.3.1.1 DC Conductivity of Oxide Glasses at High, Room, and Low Temperatures

Oxide glasses containing TMI are of greater interest because of their switching properties. The semiconducting behavior of these glasses is due to the presence of TMI in multivalent states in the glass matrix. The importance of studying such materials is twofold: from an applications viewpoint, oxide glassy semiconductors (OGS) are promising materials for the production of thin-film switches and especially thin-film memory elements; from a scientific viewpoint, OGS are convenient model materials for studying polaron charge-transfer processes and for establishing the characteristics of polaron hopping σ in disordered solids. The most important parameters of the process of charge-transport by polarons and the polarizability of OGS based on TeO_2—when modified with TMI such as tungsten—are determined from the analysis of the temperature dependence of σ_{DC}:

- High-temperature E for conduction with this type of glass and for TMI doping dominates the magnitude of σ. This behavior at high temperature can be described as nonadiabatic, small-radius polaron (SRP) hopping between nearest neighbors. The hopping energy (ΔE_H) and binding energy of the polarons, and ΔE_D, have been calculated.
- At intermediate temperatures, a simple model is used in which excitation by optical phonons independently contributes to jump frequency. The electron-coupling coefficient has been calculated.
- Low-temperature electrical σ can be described by using Mott's variable-range hopping to calculate the density of states near the Fermi energy level.

6.3.1.1.1 High-Temperature Conductivity

The dependence of $\sigma (T)$ is described approximately by the expression $\sigma (T) = \sigma_o \exp(-E/KT)$, with a constant E at high temperature. In the literature, experimental data on the electrical properties of OGS are analyzed, as a rule, under the assumption that charge carriers (CC) are SRP whose densities do not depend on temperature. Under this condition, it is assumed that the E of σ corresponds to the E of the hopping σ of SRP. The predominant existence of only one reduced state suggests the transfer of CC by hopping from the lower state to the higher state and vice versa. In 1994, El-Mallawany implemented a computational procedure to determine the ΔE_H and binding polaronic E with an electron–phonon coupling coefficient to achieve agreement with σ data.

A general condition for semiconducting behavior is that conduction can take place by the transfer of electrons from low- to high-valence states. Drift mobility is low owing to the strong interaction between the network and electrons; the potential well that is produced by this interaction deforms this network and traps transport electrons. Under a conventional electric field, electrons are forced to move in one direction so that a net current is observed. A schematic description of our hopping model is shown in Figure 6.4. The initial state is represented by the dotted line, and an electron is trapped at Site 1. The depth of the potential well and the difference in potential energy between Sites 1 and 2 are given by E_p and E_D, respectively, and r_p is the radius of the polaron. At $T > (\Theta_D/2)$, where Θ_D is the Debye temperature, a coincident state appears by optical phonon assistance, as shown by the solid line. The trapped electron has a chance to move to Site 2, creating another state as shown by the chain line. The general equation for small-radius hopping conduction is expressed as:

$$\sigma(T) = \frac{v_{phonon} \, e^2 c (1-c)}{KTR} \Big[\exp(-2\alpha R)\Big]\Big[\exp(-\Delta E/kT)\Big] \tag{6.5}$$

where c is the ratio between the two valences of the transition metal cations, R is the hopping distance, α is the decay constant, and $\exp(-2\alpha R)$ is the tunneling term. R, c, and $\exp(-2\alpha R)$ are temperature-independent variables. In present conduction theory, crystallization or inhomogeneity alters the factors in Equation 6.1 such that R and E induce changes in the σ. The tunneling term $\exp(-2\alpha R)$ makes a large contribution. The conduction mechanisms are nonadiabatic if the plot of $\log \alpha T$ vs. E at high temperature yields values of temperature far from the real temperature, and vice versa. R is not only related to the tunneling and the pre-exponential terms but also to the E. Thus, the inclusion of any discussion of R into that of the E is appropriate. Assuming that the distribution of TMI in tellurite glasses is uniform, it is possible to evaluate the hopping distance R from the condition $R^3N = 1$, where N is the density of TMI, which is easily determined from the composition and density (ρ) of the glass. The number of ions per unit volume is:

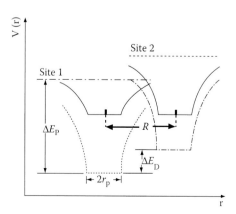

FIGURE 6.4 Hopping mechanism in semiconductors.

$$N = \frac{\rho P N_A}{(100 A W)} \tag{6.6}$$

where ρ is the density of the glass, P is the weight percent of the TMI in the glass matrix, N_A is Avogadro's number, and AW is the atomic weight of TMI. The polaron radius (r_p) can be calculated by the relation:

$$r = (0.5)\left(\frac{\Pi}{6N}\right)^{1/3} \tag{6.7}$$

The E at high temperature can be estimated based on the model in Figure 6.5. For $T > (\theta_D/2)$,

$$\Delta E = \Delta E_H + 0.5 \Delta E_D \tag{6.8}$$

where ΔE, ΔE_H, and ΔE_D are experimental, hopping, and disorder energies, respectively. The ΔE_H is shown as:

$$\Delta E_H = 0.5 \Delta E_D \tag{6.9}$$

where ΔE_P is the small-polaron binding energy. Two methods of calculating this binding energy have been suggested. The most general expression has been given by Holstein (1959) as:

$$\Delta E = (2N)^{-1} \Sigma_q /\gamma_q/^2 \hbar \omega_q \tag{6.10}$$

where $/\gamma_q/^2$ is the electron–phonon coupling constant and ω_q is the frequency of optical phonons of wave number q, and N is the site density. Austen and Mott (1969) derived a more direct expression for the small-polaron binding energy:

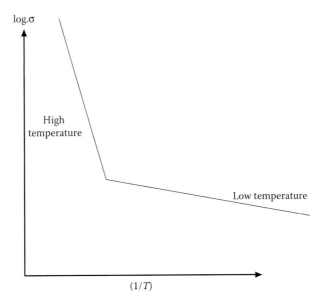

FIGURE 6.5 Change in the slope of conductivity vs. temperature.

$$\Delta E_{\mathrm{p}} = 0.5e2\varepsilon_{\mathrm{p}}r_{\mathrm{p}} \tag{6.11}$$

where ε_{p} is the effective dielectric constant given by $1/\varepsilon_{\mathrm{p}} = (1/\varepsilon_{\infty} - 1/\varepsilon_{\mathrm{s}})$, and $1/\varepsilon_{\infty}$ and ε_{s} are the high and static dielectric constants. The last equation is correct only when the hopping distance R is large, as given by Killas (1966). When the concentration of the sites is large enough that two polarization clouds overlap, E_{H} will depend on the jumping distance.

6.3.1.1.2 Room-Temperature Conductivity

The polaron model predicts that an appreciable departure from a linear log σ vs. $(1/T)$ plot should occur below $\Theta_{\mathrm{D}}/2$, where Θ_{D} is the Debye temperature, $hV = k\Theta_{\mathrm{D}}$, and V is the mean frequency of longitudinal optical phonons. This behavior is shown in Figure 6.5, and the value of the temperature Tt at which the slope of the σ behavior changes is determined from the intersection of the high- and low-temperature regions, assuming a strong electron–lattice interaction, the E ($2\Delta E_{\mathrm{H}} = \Delta E_{\mathrm{p}}$), and an energy difference (ΔE_{D}) between the initial and final sites due to local ion vibration. An estimate of the SRP coupling constant γ is calculated as $2\Delta E_{\mathrm{H}}/k\Theta_{\mathrm{D}}$. The value of γ for tellurite glasses has been calculated by El-Mallawany (1994) as shown below.

6.3.1.1.3 Low-Temperature Conductivity

As the temperature decreases, small-polaron conduction is difficult because of large energy differences between neighboring sites and a decrease in the thermal energy (kT) in a disordered structure. According to the present electronic conduction theory as discussed by Austen and Mott (1969), the contribution of variable-range hopping conduction can appear below $\Theta_{\mathrm{D}}/4$. The transport electron can hop to various neighboring sites determined by the distance and the energy difference between the two sites. The σ for variable-range hopping is given by:

$$\sigma = A\exp\left(-B/T^{1/4}\right)$$

$$T \leq \left(\Theta_{\mathrm{D}}/4\right) \tag{6.12}$$

where A and B are constants. The constant B is given by:

$$B = 2.1\left[\frac{\alpha^{3}}{kN(E_{\mathrm{F}})}\right]^{1/4}$$

$$B = 2.4\left[\Delta E_{\mathrm{D}}\left(\alpha R\right)^{3}/k\right]^{1/4} \tag{6.13}$$

where α is the wave function decay of the electron, ΔE_{D} is the energy difference between the two sites, and $N(E_{\mathrm{F}})$ is the density of states near the Fermi level. A transport electron can hop to various neighboring sites; the optimal path is determined by the distance between the two sites (which affects the α of the electron) and the ΔE_{D} between two sites.

6.3.1.2 DC Electrical Conductivity in Chalcogenide Glasses and Switching-Phenomenon Mechanisms

Mott and Davis (1979) devoted a complete chapter in their book to reviewing the properties of amorphous semiconductors containing one or more of the chalcogenide elements S, Se, or Te. They described σ_{DC}, S, the Hall effect, states in the gap (e.g., screening length and field effect), drift mobility, luminescence, photoconductivity, charge transport in strong fields, density of electron states in conduction, valence bands, optical properties, and switching properties. These materials obey the so-called "8-N bonding rule" previously proposed by Mott (1969), according to which all electrons are taken up in bonds so that large changes of σ with small changes of composition do not occur. The structures of As–Te glasses and films have been investigated and discussed by Cornet and Rossiter (1973). These authors proposed a breakdown in the 8-N valence rule in Te-rich alloys, in which some Te atoms become threefold coordinated, and some As atoms become sixfold coordinated. EXAFS studies are very important because they give an idea of the degree of chemical ordering that occurs at compositions near stoichiometry. The σ_{DC} values of most of the chalcogenide glasses that are near room temperature obey the relation $\sigma = C \exp(-E/kT)$ and vary from ~0.3 eV to >1 eV. Although the values of $2E$ lie close to the photon energy corresponding to the onset of strong optical absorption, intrinsic conduction must not be assumed, as discussed in Part IV of this book. Measurements of S show these glasses to be p-type. In general, the relation of log σ to $1/T$ is a fairly linear function, and variable-range hopping σ behavior even approximately equal to [$A \exp(-B/T^{1/4})$] is not generally observed.

6.3.1.2.1 Other Models of Electrical Conductivity of Chalcogenide Glasses

The Cohen–Fritzsche–Ovshinsky model (Cohen et al. 1969) assumes that the tail states extend across the gap in a structureless distribution. This gradual decrease of the localized states destroys the sharpness of conduction and valence band edges. In chalcogenide alloys, the disorder is sufficiently great that the tails of the conduction and valence bands overlap, leading to an appreciable density of states in the overlapping band so that a redistribution of the electrons must take place; thus redistribution forms filled states in the conduction band tail, which are negatively charged, but empty states in the valence band, which are positively charged. This places the Fermi energy somewhere near the gap center. The character of the wave function changes at critical energies E_C and E_V, which separate the extended and localized states. The electron and hole mobilities drop sharply from a low-mobility band transport with finite mobility at $T = 0$ K to a thermally activated hopping between localized gap states, which disappear at $T = 0$ K. The so-called mobility edges define a mobility gap ($E_C - E_V$), which contains only localized states. This model is believed to apply to alloy glasses, which contain compositional as well as positional disorder. It was originally intended to be applied to multicomponent chalcogenide glasses. It can be used to account for well-defined electrical E, the pinned Fermi energy, thermally stimulated currents, and some aspects of photoluminescence. Its major disadvantages are the sharp optical absorption, the absence of variable-range hopping near the Fermi energy, and the complete absence of any unpaired-spin density at all temperatures.

Davis and Mott (1970) suggested another model of electrical σ in chalcogenide glasses in which the tails of localized states are rather narrow and extend a fraction of an electron volt into the forbidden gap. This model distinguishes between:

- Localized states lying in the energy ranges ΔE_C (between E_C and E_A) and ΔE_V (between E_B and E_V), caused by the lack of long-range order
- Hypothetical "tails" caused by defects in the structures, in which the energies E_C and E_V separate the ranges of energy where states are localized and nonlocalized (extended), and the quantity $E_C - E_V$ defines a mobility gap

Marshall and Owen (1971) proposed a model showing bands of donors and acceptors in the upper and lower halves of the mobility gap. In this model, the concentrations of donors and acceptors adjust themselves by self-compensation to be almost equal, so that the Fermi level remains near the gap center. At a low temperature, it moves to one of the impurity bands because self-compensation is not likely to be complete.

The small-polaron model by Emin (1973) suggests that the charge carriers in some amorphous semiconductors enter a self-trapped (small-polaron) state as a result of the polarization of the surrounding atomic lattice. In support of this hypothesis, Emin (1973) argued that the presence of disorder in a noncrystalline solid tends to slow down a carrier. This slowing might lead to a localization of the carrier, and, if the carrier stays at an atomic site for a sufficient time for atomic rearrangements to take place, it may induce displacements of the atoms in its immediate vicinity, causing small-polaron formation. Field effect screening is accomplished by the redistribution of small polarons within the space charge layer. The small density of polarons can be calculated, their thermal E being the separation in energy of the polaron band from the Fermi level. The difference between this energy and σE is equal to the ΔE_H for polaron movement. The temperature dependence of the field effect is therefore useful for testing the small-polaron model in a range in which the field effect screening is controlled by polarons.

Licciardello (1980) proposed a modified Cohen–Fritzsche–Ovshinsky model by superimposing the suggestion of Anderson (1975), which states that disordered systems are characterized by sufficiently strong electron–phonon coupling to produce a negative effective correlation energy, which can account for the sharpness of the absorption edge and the absence of unpaired spins. This model has major difficulties for explaining the correlation between amorphous and crystalline solids, as well as the differences between chalcogenides and tetrahedral bonded amorphous semiconductors and their luminescence results.

Ovshinsky and Adler proposed a model in 1978 in which many types of chemical-bonding arrangements are discussed, which could characterize primarily covalent amorphous semiconductors. They showed how the removal of crystalline constraints inherent in all amorphous solids can allow a wealth of unusual chemical-bonding possibilities. Knowledge of these unusual chemical configurations is necessary in order to understand the recent experimental results on both doped and chemically modified amorphous semiconductors. This overall chemical approach provides a unified method for understanding the electronic structure of all amorphous semiconductors. It identifies normal structural bonds and their low-energy states and deviant electronic configurations, which characterize primarily amorphous solids in which atoms in the solids are coordinated differently from their normal structural bonding.

In real materials, the additional existence of significant densities of valence alternation pairs ordinarily compensates for any increase of the Fermi level, and the conversion of some of the positively charged centers with threefold coordinated chalcogen atoms to negatively charged centers with singly coordinated chalcogen atoms keeps the E_f pinned. The existence of a band of localized levels near the middle of the gap accounts for three basically different processes leading to conduction in amorphous semiconductors. Their contribution to the total σ changes markedly in different temperature ranges. The theoretical models and factors affecting switching and memory phenomena in chalcogenide glass are normally classified into "threshold" (or monostable) and "memory" (or bistable) devices. The first references to a switching in chalcogenide glasses occurred over 50 years ago (Ovshinsky 1959), but a later paper by the same author (Ovshinsky 1968) properly marked the point at which the subject became of serious interest in solid-state electronics. The switching process observed in amorphous semiconductors is characterized not only by the breakdown of the high-R state of the material, but also, very importantly, by the presence of a positive feedback mechanism that provides the high G state so that a breakdown of nondestructive and repetitive switching is possible. Generally, switching can be interpreted with any mechanism based on a shift of the Fermi level and band edge toward each other. The various models proposed to explain the switching process can be categorized into thermally induced crystallization, thermal switching mechanism, electronic mechanism, and electrothermal theories. It was shown earlier that amorphous semiconductors display memory and switching phenomena. The factors affecting these characteristics are temperature, composition, uniaxial pressure, and sample thickness.

6.3.1.3 DC Electrical Conductivity in Glassy Electrolytes

Rapid developments have occurred in electrochemical sensors, prototype batteries, and electrochromic devices in the last 20 years. For a current to flow in either a crystalline ceramic or an amorphous glass, charge must somehow be transported. Electrons can move through the lattice in one of two ways:

1. By ionic σ (i.e., moving with nuclei)
2. By electronic σ (i.e., becoming detached from one atom and moving to another)

Of course, both electronic- and ionic-conduction processes take place simultaneously. In the 1990s, Hench presented his principles of electronic ceramics. He stated that the complete expression of ionic G (Ω^{-1}), σ (Ω^{-1} cm^{-1}), and resistivity ρ (Ω/cm) involves the long-range migration of ionic-charge carriers through a material under the driving force of an applied electric field according to the equation:

$$\log G\left(ohm^{-1}\right) = \log\left[\frac{n\alpha Z^2 e^2 b^2}{2h}\right] - \left(\frac{\Delta F_{dc}}{RT}\right)$$

and

$$\log \rho = A + \left(\frac{B}{T}\right) \tag{6.14}$$

which is similar to the Rasch–Hinrichsen law of electrical resistivity of glass, where n is charge carriers per unit volume, α is the accommodation coefficient related to the irreversibility of the jump, Z is the valance of the ion, e is the electronic charge, b is the difference between two wells, h is Plank's constant, and ΔF_{DC} equals (N_A/V) the change in free energy for DC conduction in units of kilocalories/mol; where N_A is Avogadro's number (6.023×10^{23}), V is the height of the barrier, R is the gas constant, T is absolute temperature, kT/h is the vibrational frequency of an ion in the well, and k is Boltzmann's constant. When ions hop from one site to another, the relation between ordinary diffusion (D_i) and mobility (μ) is derived from the Boltzmann transport equation, which yielded the so-called Einstein relation: $\mu = (eD_i/kT)$. Hench (1990) reviewed the electrode polarization, glass composition and structure effects, molten silicate criteria for fast ion conduction, and grain boundary effects.

In 1980, Tuller et al. reported their discovery of fast ion transport in solids, which in recent years has stimulated much interest in the scientific community with regard to both improving our understanding of this phenomenon and applying such materials in advanced battery systems. More recently, the phenomenon of fast ion transport has also been observed in increasing numbers of amorphous systems. In 1980, Tuller et al. also reviewed transport data obtained in over 100 glasses that appear to exhibit exceptionally high Ag, Li, Na, and F ion conductivities at temperatures far from the melting point. They summarized the common characteristics of these glasses and compared them with the predictions of classical diffusion theory. Relatively low and composition-dependent values of the σ indicate poor agreement with a simple isolated ion diffusion model, although composition-dependent Es are related to structural changes. Some glasses share similar properties with their crystalline counterparts, while others do not. Tuller et al. (1980) discussed these similarities and differences in terms of the relative disorder already existing within the crystals. They also discussed the need for the improved characterization of glasses. Fast ion conductors (FICs)—which maintain high ionic conductivity over extended temperatures and chemical activities but remain electronic insulators—can be used instead of traditional liquid electrolytes in batteries and fuel cells. To achieve high-energy storage capabilities, highly active reaction couples, such as molten alkali metals and sulfur or chlorine, are required. Confinement of such reactants is greatly simplified by the use of newly developed solid electrolytes.

Tuller et al. (1980) mentioned >100 FICs that have been identified, more or less characterized, and listed in a number of articles (e.g., McGeehin and Hooper 1977, and Whitmore 1977). In general, these conductors are characterized by unusually high ionic conductivity. Tuller et al. stated their ionic-transport theory by assuming that conduction occurs predominantly for single-ionic species.

They obtained a more detailed expression for μ by invoking a diffusion model characterized by isolated jumps with random walks of the mobile species. Ions are visualized as vibrating within a potential well of energy (E_m) at a characteristic vibrational frequency (v_o). Assuming Boltzmann statistics, the diffusion constant (D) depends on the jump distance (d) and on a geometrical factor (α), as:

$$D = \alpha d^2 v_o \exp\left(\frac{-E_m}{kT}\right) \tag{6.15}$$

The Nernst–Einstein equation, $\mu KT = ZeD$, which relates μ to the D, may be used to obtain the σ:

$$\sigma = \left[\frac{n(Ze)^2 \alpha d^2 v_o}{kT}\right] \exp\left(\frac{-E_m}{kT}\right) \tag{6.16}$$

which shows the Arrhenius-like temperature dependence normally observed for ionic conductors.

In conventional ionic conductors, the concentration of charge carriers depends on the formation of point defects by either a Schottky (cation–anion vacancy pair) or Frenkel (cation or anion vacancy-interstitial pair) process. In either case, the defect concentration is given by:

$$n = n_o \exp\left(\frac{-E_d}{kT}\right) \tag{6.17}$$

where E_d is one-half of the defect-pair formation energy. The fraction of ions that are mobile (n/no) is often designated by the symbol β, as stated by Huggins (1975). Because E_d is generally large ($E_d = \chi 1 [eV]$ for NaCI), the fraction of defect sites (n/no) is usually quite small. Overall, the σ is given by:

$$\sigma = \left[\frac{n_o(Ze)^2 \alpha d^2 v_o}{kT}\right] \exp\left\{\frac{-(E_m + E_d)}{kT}\right\}$$

$$= \left[\frac{n_o(Ze)^2 \beta d^2 v_o}{kT}\right] \exp\left(\frac{-E_m}{kT}\right) \tag{6.18}$$

Modification of these relations must be made for heavily doped solids in which n is fixed by the concentration of alkali-valent additives, and care must be taken to include the degree of defect–impurity association. Due to the characteristic disorder of the mobile ion sublattice in FICs, high-defect concentrations are expected without the need for thermal generation. Thus, for FICs, β can be considered to be nearly independent of temperature and much larger than in normal ionic conductors, and the E for FICs should represent that of migration alone. The constants in Equation 6.18 are not known, and one customarily considers the more general form given by $\sigma = (\sigma_0/T) \exp(-E/kT)$, which is satisfied by the majority of both crystalline and glassy conductors. FICs appear in glasses composed of either group IA or IB cations, or group VIA or VIIA anions, which are mobile; for example, group IB—silver conductors; group IA—alkali conductors (e.g., alkali silicates, alkali phosphates, etc.); and group VIIA—conductors (fluorine ion conductors, e.g., fluorozirconate glasses).

In 1987, Ingram reviewed the ionic σ in glass. He was concerned with advances in new FICs, the mechanisms of ionic σ, the mixed mobile-ion effect, and AC relaxation. He also included the calculation of activation energies by classical methods (Anderson–Stuart theory), weak-electrolyte and defect models, universal aspects of the AC response, and the use of the decoupling index $R\tau$. Finally, Ingram (1987) suggested that ion mobility is most readily related to structural disorder in glass via some form of the amorphous cluster theory. However, Ingram and Robertson (1997) put forward some very important questions:

- What is the ion transport in glassy electrolytes?
- Is there a theory of glassy ionics?
- What are the steps to optimize the ionic σ?

The classical theory of ion transport by Anderson and Stuart (1954) has since been reinterpreted and extended by many authors. Anderson and Stuart (in their "A-S model") considered E_a as the energy that is required to overcome electrostatic forces plus the energy required to open "doorways" in a structure large enough for the ions to hop from one site to another and pass through a doorway that opens as an ion passes through (Figure 6.6). E_a is then the sum of two terms, E_b is the electrostatic binding energy to the original site, and E_s is the electrostatic strain energy as shown in the next equation:

$$E_A(\sigma) = E_b + E_s$$

$$= \left[\frac{\beta z z_o e^2}{\gamma(r+r_o)} \right] + 4\Pi G r_D (r - r_D)^2 \tag{6.19}$$

where z and z_o are the charges on the mobile ion and the fixed concentration (in this case oxygen), respectively, and r_1 and r_o are corresponding ionic radii, e is the electronic charge, and r_D is the effective radius of the unopened doorway. According to Martin and Angel (1986), both E_b and E_s are also related to the dissociation and migration energies, respectively, of Ravaine and Souquent's weak-electrolyte theory (1977).

The affected parameters in the A-S model are the elastic moduli (G); the Madelung constant (β), which depends on how far apart the ions are; and a covalency parameter (γ), which indicates the degree of charge neutralization between an ion and its immediate neighbors. The big success of the A-S model is the γ parameter, which pointed the way to the replacement of oxide ions by sulfide ions in glass, and to the further doping of such glasses with LiI. This way, anionic polarization and covalency are enhanced. This theory was used as a guide to the successful development of new superionic glass compositions. However, that advance was not enough. It might be more important in the long run to select glass compositions with good chemical stability and then to take steps to optimize the ionic σ''. How can this be achieved?

The A-S model has no information about the availability of empty cation sites in glass nor, indeed, about how far apart they are, nor is there any indication whether the dynamics of network relaxation become involved or whether they can influence ion transport. Hunter and Ingram (1994) examined why so many glasses in practice have similar conductivities at the glass transition temperature (T_g). Ingram and Robertson (1997) investigated the sites that the ions occupy in the glass structure and how they get there. The answers to such questions require a radically different theoretical model. This, at least, was the view adopted by Bunde et al. (1994), who arrived at a new dynamic structure model. Their model is based on the idea that mobile cations create well-defined sites and their own ion-specific pathways. The two processes of facile ion transport and site relaxation are illustrated in Figure 6.6. The A^+ cations occupy A- sites and then "leave these sites behind" as they move through the glass. The empty A- sites then act as "stepping stones" for the other A^+ ions to

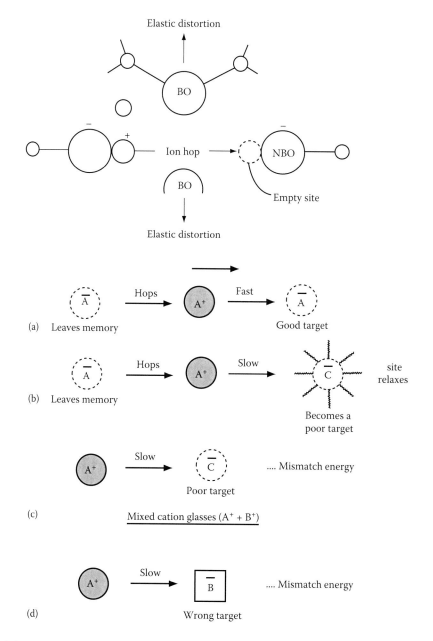

FIGURE 6.6 (a–d) Creation and loss of conducting pathways in solid electrolytes. (From D. Ingram and A. Robertson, *Solid State Ionics* 94, 49, 1997.)

move onto as they in turn move through the glass. The subsequent relaxation of the glass network can remove empty A- sites and replace them with less optimal C- sites. Also, Figure 6.6 shows how a cation pathway is reestablished. This process, which could in principle occur either in the melt or in the glass at temperatures not far below the T_g, allows A^+ ions to enter C- sites and reconvert them into A- sites. As the glass cools, these dynamically determined A- pathways become frozen into the structure. This simple model has been explored by Monte Carlo methods. It is necessary only to introduce a mismatch energy for processes involving ions moving into poor sites or into the "wrong" (B-) kind of site—that is, those left behind by the wrong kind of ion (B^+). The existence of empty A- and B- sites leads logically to the observed mixed-alkali effect.

6.3.1.4   Thermoelectric Power at High to Low Temperatures

Since the early 1960s, the S has been discussed by many authors, such as Heikes (1961).

6.3.1.4.1   Thermoelectric Power at High Temperatures ($T > 250$ K)

Heikes' (1961) model for the temperature-independent S was given by the relation:

$$S = (k/e)\left\{\ln\left[C/(1-C)\right] + \alpha'\right\} \tag{6.20}$$

where C is the ratio of the concentration of reduced TMI to the concentration of total TMI (usually measured experimentally by chemical analysis or by electron spin resonance [ESR]) and $\alpha'kT$ is the kinetic energy of a carrier. Ghosh (1993) stated that if $\alpha' = 0$ and, using values of C to calculate S, if the difference is not negligible, then α' is not negligible. However, Heikes (1961) showed that the relationship $\alpha' = \Delta S'/k$ does exist in the expression of the S in narrow-band semiconductors. Austen and Mott (1969) reported the relationship $\Delta S'/k \approx \Delta\omega_o/\omega_o$, where $\Delta\omega_o$ is the change in characteristic phonon frequency (ω_o) of the ion due to perturbation caused by the presence of an electron. The predicted values of α are between 0.1 and 0.2. Austen and Mott (1969) suggested that $\alpha > 2$ for a large polaron, whereas Mott (1968) suggested that $\alpha' \leq 1$ for a small polaron. Thus, Austen and Mott (1969) showed that for finite ΔE_D between the occupied and nonoccupied sites, $\alpha' = (1 - \theta) W_H/(1 + \theta)kT$, where W_H is the disorder ΔE_H, and θ is a constant correlated to the amount of disorder in the system. By knowing the experimental values of S and C, the α' can be calculated. By knowing α' and W_H, the electrical σ measurements, θ, can be calculated.

6.3.1.4.2   Thermoelectric Power at Low Temperatures ($T < 250$ K)

From the low-temperature S, it is possible to estimate the disorder energy W_D in two ways. Mott (1968) argued that for conduction in a material having an impurity bandwidth W_D, in this case $>kT$, regardless of whether transport is by hopping or not, the metallic-type S is calculated by:

$$S = \left(\Pi^2 kT/3e\right)\left\{kT\frac{d\left(\ln\sigma_o\right)}{dE} - \frac{dW_D}{dE}\right\} \tag{6.21}$$

Also, Emin (1973) regarded the ΔE_D as the energy required to produce an equilibrium number of carriers at low temperatures and calculated S as:

$$S = -(k/e)\left\{W_D/kT\right\}$$

$$S = -\left(W_D/eT\right) \tag{6.22}$$

The values of W_D can be estimated from the temperature dependence region by the least-squares-fit procedure.

6.3.2   AC Electrical Conductivity in Semiconducting and Electrolyte Glasses

There are three mechanisms that contribute to the σ_{AC}. For the first mechanism, the Drude formula applies:

$$\sigma(\omega) = \frac{\sigma_o}{\left(1 + \omega^2\tau^2\right)} \tag{6.23}$$

Deviations from this formula are expected if the density of states varies with energy over a range h/τ. For $\leq 10^7$ Hz, $\sigma(\omega)$ associated with carriers in extended states does not depend on ω. Second, it is difficult to derive the frequency dependence of $\sigma(\omega)$ for hopping under conditions of nondegenerate statistics. Therefore, we might expect the same result as derived under degenerate conditions; that is,

$$\sigma(\omega) \propto \omega \ln\left(v_{ph}/\omega\right)^4 \tag{6.24}$$

or $\sigma(\omega) \cong \omega^s$, where $S < 1$ for $\omega < v_{ph}$. For the third mechanism of electrical conductivity, Pollak (1971) stated that $\sigma(\omega)$ should increase with ω in a manner similar to that in the second case; $\sigma(\omega) \propto T$ if kT is smaller than the energy range over which $N(E_F)$ may be taken as a constant, independent of T, and kT is larger than the width of some well-defined defect band in which E_F lies. Austen and Mott (1969) gave the following equation:

$$\sigma(\omega) = \frac{\pi}{3} e^2 kT \left\{N(E_F)\right\}^2 \alpha^{-5} \omega \left\{\ln\left(\frac{v_{phonon}}{\omega}\right)\right\}^4 \tag{6.25}$$

The frequency dependence predicted from this equation can be written as $\sigma(\omega) = \omega^s$, where s is a weak function of ω, if $\omega \ll v_{ph}$ and is given by:

$$s = \frac{d\left[\ln\left(\omega \ln^4\left(v_{ph}/\omega\right)\right)\right]}{d\left(\ln\omega\right)} \tag{6.26}$$

These relations were discussed and compared with the experiment by Mott and Davis (1979). Elliot (1977, 1987) proposed a model for the mechanisms responsible for the σ_{AC} in chalcogenides. This model overcomes most of the problems associated with interpretation in terms of the Austen–Mott formula, which is given as:

$$\sigma(\omega) = \frac{\pi^2 N^2}{24} \varepsilon \left(\frac{8e^2}{\varepsilon W_M}\right)^6 \frac{\omega^s}{v_{ph}^\beta} \tag{6.27}$$

where $N/2$ is the number of pairs of sites, ε is the dielectric constant, and W_M is the barrier height separating distant pairs and can be approximately equated to the band gap. The exponent s is related to W_M by:

$$1 - S = \left(\frac{6K_B T}{W_M}\right) \tag{6.28}$$

When the σ_{AC} and σ_{DC} are caused by completely different processes, then the total σ (σ_t) measured at a particular frequency ω and temperature can be expressed as:

$$\sigma_t = \sigma_{DC} + \sigma_{AC}(\omega) \tag{6.29}$$

On the other hand, when AC and DC σs are caused by the same process and σ_{DC} is equal to σ_t in the limit $\omega \to 0$, then Equation 6.29 is not valid. The σ_{AC} values of a large group of amorphous semiconductors and glasses have been found to follow the relation described by Mott and Davis (1979):

$$\sigma(\omega) = A\omega^s \tag{6.30a}$$

where A is a constant weakly dependent on temperature, and s is the frequency exponent, generally less than unity. Different theories have been developed to explain the frequency-dependent σ_{AC} data of amorphous semiconductors. Two fundamental mechanisms developed to account for the sublinear frequency-dependent σ_{AC} are the quantum-mechanical tunneling (QMT) mechanism proposed by Pollak and Geballe (1961), and Austen and Mott (1969), and the hopping-over-barrier (HOB) mechanism proposed by Pike (1972), Elliot (1977), and Springett (1974). According to the premise of the QMT mechanism, Pollak (1965, 1971), Bottger and Bryksin (1979), and Efros (1981) found that expressions of the σ_{AC} can be written:

$$\sigma_{AC}(\omega) = \frac{ke^2 k_B T\left(N(E_F)^2 \omega R_\omega^4\right)}{\alpha} \tag{6.30b}$$

where $N(E_F)$ is the density of states at the Fermi level, the constant factor k varies slightly between various treatments, $R\omega$ is the characteristic tunneling distance given by:

$$R_\omega = \frac{1}{2\alpha}\ln\left(\frac{1}{\omega\tau_o}\right) \tag{6.31}$$

and the frequency exponent takes the form:

$$s = 1 - \frac{4}{\ln(1/\omega\tau_o)} \tag{6.32}$$

Thus, the QMT model predicts a linear temperature dependence of $\sigma_{AC}(\omega)$ and seems to be valid only in the low-temperature region ($T < 120$ K) for the present glass system, whereas at higher temperatures the dependence of σ_{AC} is much stronger. The correlated-barrier-hopping (CBH) model, which correlates the barrier W with the intersite separation R for single-electron hopping (Pike 1972), was extended subsequently by Elliot (1977, 1987) for two-electron hopping simultaneously between the sites. For neighboring sites at a separation R, the coulomb wells overlap, resulting in a lowering of the effective barrier height which, for the single-electron CBH model, is given by Elliott 1987:

$$W = W_M - \frac{4ne^2}{\varepsilon R} \tag{6.33}$$

where W_M is the maximum barrier height, ε_o is the dielectric constant of the free space, e is the electronic charge, and n is the number of electrons that hop. The corresponding $\sigma_{AC}(\omega)$ has the following expression:

$$\sigma_{AC}(\omega) = \frac{\pi^2}{24}N^2\varepsilon\varepsilon_o\left(\frac{8e^2}{\varepsilon W_M}\right)\left(\frac{s}{\tau_o^\beta}\right) \tag{6.34}$$

where N is the concentration of localized sites, and the frequency exponent s was evaluated by Elliot (1987) by:

$$\beta = \left(\frac{6K_B T}{W_M} \right) \tag{6.35}$$

and $S = 1 - \beta$.

To explain the σ_{AC} data, Long (1982) and Long et al. (1982) discussed a case in which an appreciable overlap of the polaron distortion clouds occurred, thereby reducing the value of the polaron ΔE_H. For such polarons, overlap of the potential wells of neighboring sites is possible because of the long-range nature of the dominant coulomb interaction—hence the name "large" or dielectric polaron. The potential extending over many interatomic distances has important consequences for AC loss. When the wells of the two sites overlap, the E associated with the particle transfer between them is reduced, which, according to Austen and Mott (1969), and Mott and Davis (1979), takes the form:

$$W_H = W_{Ho} \left(1 - \frac{r_p}{R} \right) \tag{6.36}$$

where r_p is the polaron radius, R is the mean distance between centers and is a random variable, and W_{Ho} is a constant for sites defined as:

$$W_{Ho} = \left(\frac{e^2}{4\varepsilon_p r_p} \right) \tag{6.37}$$

where e is the charge of an electron and

$$\varepsilon_p = \left(\frac{1}{\varepsilon_\infty} - \frac{1}{\varepsilon_s} \right)$$

where ε_s and ε_∞ are the static and high-frequency dielectric constants, respectively. The σ_{AC} in the OLPT model given by Long (1982) can be written:

$$\sigma_{AC}(\omega) = \frac{\pi^4}{12} e^2 \left(K_B/T \right)^2 N^2 \left(E_F \right) \left[\frac{\omega R_\omega^4}{2\alpha K_B T + \left(W_{Ho} r_p / R_\omega^2 \right)} \right] \tag{6.38}$$

where $N(E_F)$ is the density of states at the Fermi level, α is the wave function decay constant, K_B is Boltzmann's constant, and $R\omega$ is the optimum hopping length, which is calculated from the quadratic equation:

$$R_\omega'^2 + \left[\beta W_{Ho} + \ln \left(\omega \tau_o \right) \right] R_\omega' - \beta W_{Ho} r_p' = 0 \tag{6.39}$$

where

$$R_\omega' = 2\alpha R_\omega$$
$$r_p' = 2\alpha r_p$$

and

$$\beta = 1/K_B T$$

then the frequency exponent s is calculated as:

$$s = 1 - \left[\frac{8\alpha R_\omega + 6\beta W_{Ho} r_p / R_\omega}{\left(2\alpha R_\omega + \beta W_{Ho} r_p / R_\omega \right)^2} \right] \qquad (6.40)$$

Thus, according to the OLPT model, s should be both temperature and frequency dependent. It is also observed from Equation 6.40 that s decreases from the value unity with increasing temperature. For large values of r_p^1, s continues to decrease with increasing temperature, eventually tending to the value predicted by the QMT model of nonpolaron-forming carriers, whereas for small values of r_p^1, s exhibits a minimum at a certain temperature and subsequently increases with increasing temperature.

Ingram (1987) reviewed the AC conduction and relaxation behavior in glassy electrolytes. He mentioned several methods of data presentation for the σ data, including the following:

1. Complex admittance, $Y^* = (R_p)^{-1} + j\omega C_p$ and $Z^* = (Y^*)^{-1} = R_s - j/\omega C_s$
2. Complex permittivity, $\varepsilon^* = \varepsilon' - j\varepsilon''$
3. Complex modulus, $M^* = (M' + jM''') = (\varepsilon^*)^{-1} = j\omega(\varepsilon_o Z^*)$ (6.41)

where ω is the angular frequency $2\pi f$, and ε_o is the permittivity of free space ($8.854 \times 10^{-1} F$ cm^{-1}).

Ingram (1987) discussed two additional issues in his review: determining what is happening in different regions of the σ spectrum and seeing what can be learned from the corresponding relaxation spectrum.

Based on the AC behavior of glass and other electrolytes in light of Jonscher's general treatment (1981) that the $\sigma(\omega)$ can be expressed by $\sigma(\omega) = \sigma(\omega) + A\omega^n$, A is a temperature-dependent parameter, and n is allowed to take values between 0 and 1. The assumption relates to the σ_{DC} so that $\sigma(\omega) = K\omega_p$, where K is an empirical constant, which depends on the concentration of mobile ions. The σ expression is then:

$$\sigma(\omega) = K\omega_p + K\omega_p^{(1-n)\omega^n} \qquad (6.42)$$

The σ relaxation was discussed by Ngai and Jain (1986), who suggested instead that there is a single microscopic relaxation time (τ_o) thermally activated so that $\tau_o = \tau_o \exp(E_A/RT)$, which would be the relaxation time for a single ion hopping process in the absence of interference from other mobile ions. It is this interference that causes the "stretching out" of the relaxation process and the observed (macroscopic) time (τ_p) to be greater than τ_o. They showed that if the macroscopic relaxation time is also thermally activated,

$$\tau_p = \tau_o * \exp(E_A/RT) \qquad (6.43)$$

then $E_A^* = E_A/(1 - n)$, and n is independent of temperature.

6.4 DC ELECTRICAL-CONDUCTIVITY DATA OF TELLURITE GLASSES AT DIFFERENT TEMPERATURES

This section summarizes the electrical σ data and classifies these data into three main groups:

1. σ_{DC} data of semiconducting oxide–tellurite glasses containing transition metal oxides (TMOs) or rare-earth oxides (REOs)
2. σ_{DC} data of oxide–tellurite glasses containing alkaline oxides
3. σ_{DC} data of nonoxide–tellurite glasses

6.4.1   DC Electrical Conductivity Data of Oxide–Tellurite Glasses Containing Transition Metal Ions or Rare-Earth Oxides

Although literature about glass is full of references to σ, it is impossible to cover all of these data. Instead, the following data are representative of the majority of the physical properties and parameters—for example, temperature ranges of σ—providing a basis for the majority of electrical processes in these new noncrystalline materials in both thin-film and bulk forms. According to Flynn et al. (1977), the σ and S of binary vanadium–tellurite glasses of the form $(100 - x)$ mol% TeO_2–x mol% V_2O_5, where $x = 10$–50, have been studied as a function of temperature and vanadium-valence ratio. Above 200 K, the DC has a constant E of about 0.25–0.34 eV, depending on the composition, whereas below this temperature it decreases continuously with temperature to 77 K. Compared with phosphate glasses, the σ of tellurite glasses is 2.5 to 3 orders of magnitude greater for similar vanadium concentrations. The σ of tellurite glasses is very important because the variation of σ with composition in tellurite glasses is due to the variation in the pre-exponent constant rather than the E, as in phosphate glasses. Figure 6.7a represents the σ behavior of the binary TeO_2–V_2O_5 glasses for different $V^{4+}/V_{total} = 0.2$, and Figure 6.7b represents the variation of the electrical σ with the total vanadium that is present in tellurite or phosphate glasses. Figure 6.7c gives the S of binary vanadate glasses vs. inverse temperature for different compositions of binary vanadium– tellurite glasses. From Figure 6.7C, the slope of S vs. $1/T$ in the temperature-dependent region gives the E (W_D). The E is 0.02 eV in binary vanadium–tellurite glasses, the electrical σ is nonadiabatic hopping, and the tunneling parameter α is 0.97 Å^{-1}.

Hampton et al. (1987) measured the σ_{DC} of pure TeO_2 and binary 67 mol% TeO_2–33 mol% WO_3 tellurite glasses in the temperature range 90–430 K as shown in the 1.st Ed. The σ_{DC} is measured from the σ_{AC} at zero frequency. It is clear that the electrical σ of binary 67 mol% TeO_2–33 mol% WO_3 is several times larger than that of the parent glass. A sharp break in the slope of the log σ with reciprocal temperature divides the σ data in these glasses into two distinct temperature regimes; the σ mechanisms at low temperatures have a much lower E than those at higher temperatures. The σ data are consistent with a small-polaron charge carrier, as shown in Table 6.1. The conductivities have been analyzed using small-polaron theory to establish the thermal ΔE_D and E for carriers in both glasses, as collected in Table 6.2. In 1985, El Zaidia et al. measured the ratio of the lower-valence state to the total valence state that presents the TMO in binary 67 mol% TeO_2–33 mol% WO_3 glasses. Figure 6.8a represents the ESR for the same tellurite glass samples at room temperature. From these ESR measurements, the ratio of W^{5+}/W^{6+} has been deduced (Figure 6.10 bottom) to explain the σ data shown in Figure 6.8b. The ratio W^{5+}/W^{6+} is 13.03×10^{-6}. In 1994, El-Mallawany analyzed the electrical σ data of these binary glasses. That analysis was divided according to temperature; that is, high, room, and low temperatures, as shown in Table 6.2.

6.4.1.1   DC Electrical Conductivity Data of Oxide–Tellurite Glasses at High Temperatures

The predominant existence of only one reduced state (W^{5+}) suggests the transfer of a charge carrier by hopping from W^{5+} to state W^{5+}-O-W^{6+} would be W^{6+}-O-W^{5+}. We implemented a computational procedure to determine the ΔE_H and binding polaronic E with an electron–phonon coupling coefficient to achieve agreement with the σ data. Treatment of the electrical properties of semiconducting tellurite glasses is based mainly on the theories of Mott (1968) and Austen and Mott (1969). A general condition for semiconducting behavior is that conduction can take place by the transfer of electrons from low- to high-valence states. The drift mobility is low due to the strong interaction between the network and the electrons: the potential well produced by this interaction deforms this network and traps transport electrons. Under a conventional electric field, electrons are forced to move in one direction so that a net current is observed. The r_p, the R, and the tunneling term [$\exp(-2\alpha R)$] have been calculated for the binary tellurite glass system containing WO_3. Table 6.2 reports the values of r_p and R for binary TeO_2–WO_3 glasses. The r_p value is 2.19 Å, and the average W–W spacing value is 5.45 Å (i.e., $r_p < R$), which satisfies the criterion for SRP formation in the

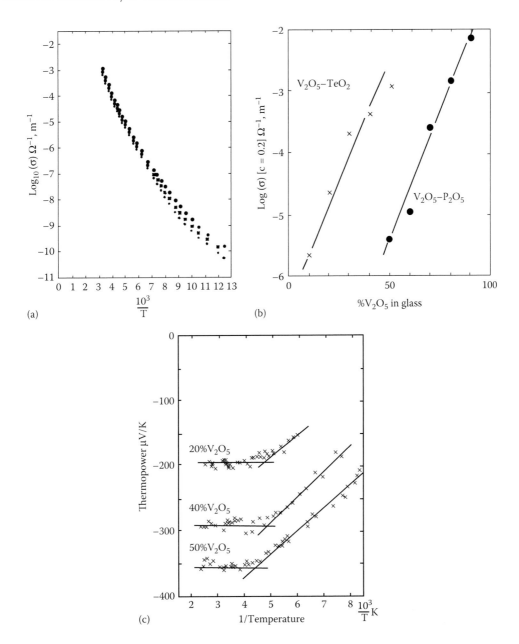

FIGURE 6.7 (a–c) Conductivity and thermopower of binary vanadium–tellurite glasses ([100 − x] mol% TeO₂–x mol% V₂O₅, where x = 10–50), as a function of temperature and vanadium-valence ratio. (From Flynn et al., *Proc. 7th Int. Conf. Amorphous and Liquid Semi.*, 678, 1977. With permission.)

present glass. The E at the high temperature is estimated based on the hopping model for $T > \theta_D/2$. The electron–phonon coupling constant is $|\gamma_q|^2$, ω_q is the frequency of optical phonon (ω) of wave number q, and the site density (N) is calculated and tabulated in Table 6.2. The polaron ΔE_H has been calculated by taking the values of r_p and R (Table 6.2). The polaron ΔE_H of TeO₂–WO₃ glasses is equal to 0.145 eV. The value of E is lower than that of TeO₂–V₂O₅ glass, as seen in Table 6.2 This lower ΔE_H is the main reason for the higher σ observed in these glasses. This lower ΔE_H of tungsten–tellurite glass also provides a lower thermal electrical E. Generally, E tends to decrease

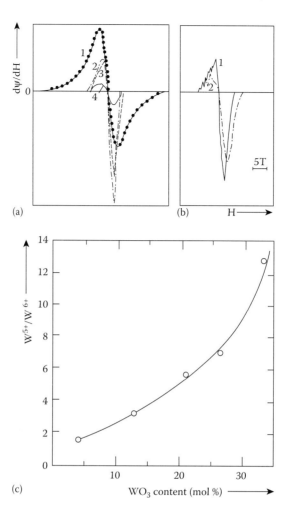

FIGURE 6.8 (a–c) Electron spin resonance analysis at room temperature for binary TeO_2–WO_3 glasses. (From M. El Zaidia, A. Amar, and R. El-Mallawany, *Phys. Stat. Solids*, 91, 637, 1985. With permission.)

with increasing TMO content in a disordered system, as predicted by theory. However, E is dependent on the kind of TMI and the glass composition. The E of tungsten–tellurite glass with 33 mol% WO_3 is comparable to that of 55 mol% V_2O_5 in the same host. Furthermore, the ΔE_D calculated is 0.23 eV. Usually, ΔE_D is around 0.1 eV, as measured by Sayer et al. (1971).

To complete the analysis, it is very important to compare the σ of tungsten oxide with that of other glass-forming oxides such as the phosphates. The electrical σ and E of P_2O_5–WO_3 glass, as determined by Mansingh et al. (1978) (>60 mol% WO_3), are $7 \times 10^{-4}\,\Omega^{-1}\,cm^{-1}$ and 0.35 eV, respectively. Thus, the electrical properties of the TMI are affected by this glass-forming oxide. For the system with higher WO_3 content, the σ is nearly the same, but the E is higher, meaning that the tungsten exists in a strongly defined local configuration and the electron transport is influenced by the network former. The glass-forming oxide is a dominating factor affecting the ΔE_H. The tungsten–phosphate glasses described by Mansingh et al. (1978) have a ΔE_D of 0.15–0.24 eV. As mentioned before, the tungsten ion has two stable coordination states, and both can be present in TeO_2–WO_3 glass, as indicated by El Zaidia et al. (1985). Therefore, the large ΔE_D is attributed to the distribution of tungsten ions into two coordination sites, and this agrees with the explanation of Mott (1968).

TABLE 6.1
Experimental Electrical Conductivity Data for Tellurite Glasses

Glass and mol% Composition (Reference)	Experimental Condition[a]	σ_o (Ω^{-1} cm^{-1})[a,b]	E (eV)[a,b]
TeO$_2$ (Hampton et al. 1987)	High temperature	4.9×10^{-3}	0.35
	Low temperature	2.6×10^{-8}	0.02
TeO$_2$–WO$_3$ (Hampton et al. 1987) 67-33	High temperature	7.2×10^{-4}	0.26
	Low temperature	1.2×10^{-7}	0.03
TeO$_2$–CuO–Lu$_2$O$_3$ (Malik and Hogarth 1990a)	Temperature = 444 K		
65–30–0		9×10^{-8}	0.69
65–31–4		1.7×10^{-7}	0.93
TeO$_2$–CuO–NiO (Malik and Hogarth 1990c)	Temperature = 454 K		
65–31–4		8.6×10^{-8}	0.72
TeO$_2$–CuO–CoO$_3$			
65–31–4		5.9×10^{-10}	0.83
TeO$_2$–V$_2$O$_5$ (Mansingh and Dahawan 1983)	Frequency = 1 kHz		
90–10		4.47×10^{-11}	4.20
50–50		1.78×10^{-10}	0.30
20–80		5.01×10^{-10}	0.23
Te–Si–As–Ge (El-Fouly et al. 1989)	Room temperature		
55–5–30–10		1.8×10^{-3}	$Ec - Ef = 0.7$
TeO$_2$–V$_2$O$_5$ (Ghosh 1990)			
63.26–36.74		9.98×10^{-2}	0.66
31.63–68.37		6.98×10^{-2}	0.51
TeO$_2$–P$_2$O$_5$ (El-Kholy 1994)			
28.18–71.82		$S = 0.91, \beta = 0.09$	$W = 2.15$
76.49–23.51		$S = 0.69, \beta = 0.31$	$W = 2.88$
TeO$_2$–MoO$_3$ (El-Kholy and El-Mallawany 1995)			
80–20		$S = 0.71, \beta = 0.29$	$N(E) = 3.48 \times 10^{15}$
70–30		$S = 66, \beta = 35$	$N(E) = 1.05 \times 10^{16}$
55–45		$S = 0.9, \beta = 0.9$	$N(E) = 7.66 \times 10^{18}$
TeO$_2$–CeO$_2$ (El-Mallawany et al. 1995)			
95–5	High temperature	2.0×10^{-7}	$W = 0.195$
	Low temperature	6.7×10^{-11}	$W = 0.005$
TeO$_2$–PbO–FeO$_3$ (El-Samadony 1995)	Temperature = 310 K	S (10^3 Hz) = 0.875	W(DC) = 0.475
80–10–10		S (10^4 Hz) = 1.138	W(AC) = 0.302
Te–Se (Gadkari and Zope 1988)	Room temperature		
30–70		7.0×10^{-11}	$W = 0.37$
23–70–7 In		4.0×10^{-10}	$W = 0.20$
TeO$_2$–V$_2$O$_5$ (Hirashima et al. 1986)			
60–40		Log σ_o = −3.85	$W = 0.40$
45–55		Log σ_o = −3.46	$W = 0.36$
xNa$_2$P$_2$O$_6$–(1 − x)Na$_2$Te$_2$O$_3$ (Jayasinghe et al. 1996)	Temperature = 190°C		
$x = 0$		Log σ_o = −7.5	$W = 0.80$
$x = 0.48$		Log σ_o = −6.3	$W = 0.70$
$x = 0.9$		Log σ_o = −5.5	$W = 0.73$
TeO$_2$–PbO–ZnF$_2$ (Ravikumar and Veeraiah 1998)			
68–12–20		$S = 0.92$	$W = 1.15$
46–9–45		$S = 0.88$	$W = 1.2$

(continued)

TABLE 6.1   (Continued)
Experimental Electrical Conductivity Data for Tellurite Glasses

Glass and mol% Composition (Reference)	Experimental Condition[a]	σ_o (Ω^{-1} cm^{-1})[a,b]	E (eV)[a,b]
0.5Ag$_2$O–0.5[xTeO$_2$–(1 – x)P$_2$O$_5$] (Chodari and Kumari 1996a)			
x = 0		Log σ_o = 3.54 ± 0.27	W = 0.68 ± 0.02
x = 0.5		Log σ_o = 2.34 ± 0.21	W = 0.48 ± 0.01
0.3Ag$_2$O–0.7[xB$_2$O$_3$–(1 – x)TeO$_2$] (Chodari and Kumari 1996b)			
x = 0		Log σ_o = 1.8 ± 0.06	W = 0.61 ± 0.02
x = 0.5		Log σ_o = 1.1 ± 0.04	W = 0.51 ± 0.02
yAg$_2$O–(1 – y)[xMoO$_3$–(1 – x)TeO$_2$] (Chodari and Kumari 1997)			
y = 0.3, x = 0		Log σ_o = 1.6 0 ± 12	W = 0.576 ± 0.04
y = 0.3, x = 0.42		Log σ_o = 1.32 ± 0.02	W = 0.546 ± 0.08
TeO$_2$–MoO$_3$–Fe$_2$O$_3$ (Qiu et al. 1997)	Temperature = 437 K		
80–10–10	μ = 2.71 × 10^{-5}	Log σ_o = –6.31	W = 0.67
60–20–20	(cm^2/VS)	Log σ_o = –4.88	W = 0.53
TeO$_2$–PbO–V$_2$O$_5$ (Sakata et al. 1996)	Temperature = 437 K		
50–10–40		Log σ_o = –4.84	W = 0.47
TeO$_2$–V$_2$O$_5$–NiO (Sega et al. 1998a)	Temperature = 416 K		
60–30–10	μ = 8.2 × 10^{-9} (cm^2/VS)	Log σ_o (Scm^{-1} K) = 2.4	W = 0.482
TeO$_2$–V$_2$O$_5$–MnO (Sega et al. 1998b)			
60–30–10		Log σ_o (Scm^{-1} K) = 3.38	W = 0.532
TeO$_2$–MoO$_3$ (Singh and Chakravarthi 1995)	Temperature = 364 K		
50–50		3.98 × 10^{-10}	W = 0.66
TeO$_2$–SnO–V$_2$O$_5$ (Mori et al. 1995)	Temperature = 437 K		
70–10–20		S (μV/K) = –53	W = 0.44
60–20–20		S (μV/K) = +5	W= 0.44
TeO$_2$–Na$_2$O (Mori et al. 1995)	Temperature = 200°C		
90–10		Log10 σ_o = –8.74	W = 1.06 ± 0.01
70–30		Log10 σ_o = –6.8	W = 0.93 ± 0.01

[a] Tσ_o, pre-exponent factor (DC and AC); s, frequency exponent; N(E$_f$), concentration of localized states; μ, mobility (cm^2/ VS); S, Sebeck Coefficient (μV/K).

[b] Data are units in headings unless otherwise indicated in the table itself.

6.4.1.2   DC Electrical Conductivity Data of Oxide–Tellurite Glasses at Room Temperature

From Figure 6.9 by Hampton et al. (1987), the value of T_t is determined from the intersection of the high- and low-temperature regions, 270 K. The value of θ_D is 540 K, which corresponds to the fundamental frequency of the vibrational spectrum of TeO$_2$–WO$_3$ glass—670 cm^{-1} as found by Al-Ani et al. (1985; explained in Chapter 10). This value of θ_D is the first determination of the optical-mode Debye temperature for TeO$_2$–WO$_3$ glass and is higher than that of P$_2$O$_5$–WO$_3$ glass. The Debye temperature is associated with the acoustic mode θ_D = 266 K, which has been measured by El-Mallawany and Saunders (1987) using ultrasonic velocities in this glass. The high value of θ_D for the present binary transition metal tellurite glass is caused by the replacement of tetrahedral Te atoms with octahedral W atoms, which increases cross-linking and in turn increases vibrating frequency. Generally, the Debye temperature depends on the number of vibrating atoms per unit volume, which changes from 5.77 × 10^{22} to 6.23 × 10^{22} cm^3 when TeO$_2$ is modified with

TABLE 6.2

Theoretical Values of the Conduction Process in Tellurite Glasses

Temperature Condition and Analytical Parameter	TeO$_2$ (El-Mallawany 1994)	67 TeO$_2$–33 WO$_3$ (El-Mallawany 1994)	45 TeO$_2$–55 V$_2$O$_5$ (Cheng and Mackenzie 1980)	40 TeO$_2$–60 MoO$_3$ (Ghosh 1990)	50 TeO$_2$–40 V$_2$O$_5$–10 Bi$_2$O$_3$ (Ghosh 1993)	40 TeO$_2$–60 MoO$_3$ (Singh and Chak 1995)
High Temperature						
N (TMI) (cm^{-3})		6.175×10^{21}	15.6×10^{21}	1.32×10^{22}	1.02×10^{22}	1.17×10^{22}
R (Å)		5.450	4.00	4.29	4.61	4.4
r_p (Å)		2.196	1.60	1.71	1.81, 1.89	1.77
E_H (eV)		0.145	0.28	0.31		0.59
E (eV)		0.115	0.09	0.20		
Room Temperature						
Θ_D (K)		588				480
$(N/V) \times 10^{22}$ (cm^{-3})	5.77	6.23				
Electron–Phonon Coupling γ		7				11.1
Low Temperature						
E_D (eV)		0.03		10.01	0.13	0.09
$N(E_F)$ (eV^{-1} cm^{-3})		1.2×10^{23}		1×10^{20}	7.7×10^{19}	
Decay constant α (Å-1)		1.14		0.58	0.81	

33 mol% WO$_3$. The number of vibrating atoms in TeO$_2$–WO$_3$ glass is lower than that of the system P$_2$O$_5$–WO$_3$ (60 mol%) as shown in Table 6.2. As a result, the optical-mode Debye temperature of TeO$_2$–WO$_3$ is lower than that of P$_2$O$_5$–WO$_3$ glass, as obtained by Mansingh et al. (1978). Assuming a strong electron–lattice interaction, the E (ΔE) is the result of polaron formation with a binding energy $\Delta E = 2\Delta E_H$ and an energy difference ΔE_D between the initial and final sites due to local ion vibrations. An estimate of the SRP coupling constant (γ) can be made in which $\gamma = \Delta E_P/\hbar\omega_0$. If θ_D and ΔE_H are known, it is possible to evaluate γ. The γ value equals 7 for binary tungsten–tellurite glasses, indicating a strong electron–phonon interaction that justifies the applicability of SRP theory in binary transition metal tellurite glasses.

6.4.1.3 DC Electrical Conductivity Data of Oxide–Tellurite Glasses at Low Temperatures

A plot of log σ against $T^{1/4}$ for the TeO$_2$–WO$_3$ glass is shown in Figure 6.9, which suggests that the variable-range hopping mechanism may be valid in this system below 185 K. The high-temperature slope, however, leads to an unacceptable value of α. From Figure 6.9, the low-temperature E $\Delta E = 0.03$ eV (Table 6.2). The values for α and $N(E_F)$ are 1.14 Å^{-1} and 1.2×10^{23} eV^{-1} cm^{-3}, respectively. These data are reasonable for the localized states as stated by Mott (1979). The reason for using the low-temperature E is that at low temperatures, when the polaron binding energy or ΔE_H values are small, the static ΔE_D of the system dominates the conduction mechanism. A transport electron can hop to various neighboring sites; the optimal path is determined by the distance between the two sites (which affects the wave function decay of the electron) and the energy difference between the two sites. The estimated values of α for the present TeO$_2$–WO$_3$ glass are lower than that of P$_2$O$_5$–WO$_3$ glass, as stated by Mansingh et al. (1978). Finally, as seen in Table 6.2, the requirements for applying polaron theory, namely that $\alpha < r_p < R$, are fulfilled in TeO$_2$–WO$_3$ glass.

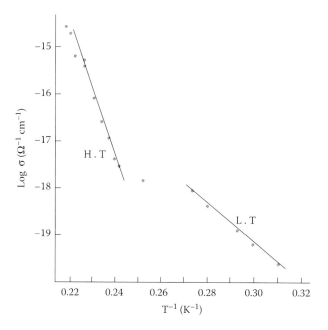

FIGURE 6.9   Representation of log σ vs. $T^{-1/4}$. (From R. El-Mallawany, *Materials Chem. and Phys.*, 37, 376, 1994. With permission.)

Also in 1987, Hirashima and Youshida studied the effect of additive oxide on the DC electrical σ of tellurite glasses of the form TeO_2–V_2O_5–MO, where M is Ca, Ba, Sr, Zn, or Pb. The glass composition contained 50 to 60 mol% of V_2O_5, and the measurements were in the temperature range 100°C–200°C as shown in the 1.st Ed. The temperature dependence of the σ was exponential, $σ = (σ_o/T) \exp(−W/KT)$, whereas the E of W increased with increasing concentrations of MO. The pre-exponential factor ($σ_o$) decreased more than one order of magnitude when BaO or SrO content increased, whereas it decreased little with increasing content of another oxide, as shown in the 1.st Ed. Hirashima and Youshida (1987) concluded that, when the basicity of MO—that is, cationic field strength of M^{2+}—and the amount of ionic character of the M–O bond are high, the hopping probability is reduced by the addition of more MO, and the hopping regime changes from adiabatic to nonadiabatic. The pre-exponential factor ($σ_o$) is a function of the ion spacing as shown by:

$$a_v = V − V; c_v = f(V − V) \tag{6.44}$$

and the transition probability ($ν_o \exp[−2αa]$) is also a function of the vanadium ion spacing.

Chopa et al. (1990) measured the electrical and structural properties of amorphous $100 − x$ mol% TeO_2–x mol% V_2O_5 blown films, where $x = 10$–50 mol%. The thickness was in the range 1–7 μm. The σ measurements were carried out in the temperature range from room temperature to 430°C. The obtained values of the E for different percentages of V_2O_5 confirmed that no marked dependence of E on temperature was observed in this temperature region. The values of the activation energies of these films decreased linearly in accordance with the increase in the percentage of the V_2O_5. Chopa et al. (1990) suggested that the conduction mechanism of blown films is similar to that reported for bulk glasses. The DC E is considerably less than the optical band gap, which has been found optically.

Hogarth and his coworkers (Hassan and Hogarth 1988; Malik and Hogarth 1990a, 1990b, 1990c, and 1990d; Malik et al. 1990) published a series of articles on the electrical properties of tellurite glasses in forms including TeO_2–CuO, 65 mol% TeO_2–$(35 − x)$ mol% CuO–x mol% $CuCl_2$, and 65 mol% TeO_2–$(35 − x)$ mol% CuO–x mol% TMO. where TMO = NiO, CoO, or Lu_2O_3; 65 mol%

TeO2–(35 − x) mol% CuO–x mol% *MO*, where *MO* = NiO or CoO and 65 mol% TeO$_2$–(35 − x) mol% CuO–x mol% Lu$_2$O$_3$. Figure 6.10 represents some of the σ and *E* data of tellurite glasses. The increase or decrease of the electrical σ of glasses containing either <2mol% or >2mol% CuCl$_2$ is interpreted in terms of the electronic transitions between the orbits of tellurium three-dimensional electrons. Ghosh (1990) measured the electrical transport properties of molybdenum–tellurite glassy semiconductors in the temperature range 100–500 K, as shown in Figure 6.11, and concluded that polaronic conduction in the high-temperature region is most appropriate, variable-range hopping dominates in the intermediate temperature range, and the glass-forming oxide rather than the TMO greatly affects the conduction mechanism. Tanaka et al. (1990) measured the electronic conduction in TeO$_2$–Fe$_2$O$_3$–P$_2$O$_5$ glasses as shown in the 1st Ed. The *E* for electrical conduction of the TeO$_2$-rich glass is much lower than that of the P$_2$O$_5$-rich glass, and its variation with the composition corresponds to the variation of the electrical σ.

Abdel-Kader et al. (1991) measured the σ$_{DC}$ of TeO$_2$–P$_2$O$_5$ glasses and later (1992) of TeO$_2$–Bi$_2$O$_3$–P$_2$O$_5$ glasses. They measured the effect of γ-radiation on the σ$_{DC}$ as shown in Figure 6.12. The σ$_{DC}$ suffers from an increase in its value due to the radiation, which is attributed to an increase in charge carriers produced by γ-rays. Abdel-Kader et al. (1993) measured the σ$_{DC}$ (50°C–300°C) of 0.95(81 mol% TeO$_2$–19 mol% P$_2$O$_5$)–0.55 mol% R_nO_m, where *R* = La, Ce, Pr, Nd, Sm, and Yb, as shown in Figure 6.13. The REOs lowered the value of the electrical σ, whereas the value of the electrical *E* was dependent on the kind of rare-earth ion. Abdel-Kader et al. (1993) found the variation of log (σ*T*) vs. the *E* (W) at different temperatures and concluded that the σ of rare-earth tellurite–phosphate glass is a function of the *E*.

Ghosh (1993) measured the σ$_{DC}$ of semiconducting vanadium–tellurite glasses containing bismuth oxide in the temperature range 80–450 K. The experimental results were interpreted in terms of polaron-hopping theories. The results suggested that the addition of bismuth oxide to the binary vanadium–tellurite glasses changes the hopping mechanism from nonadiabatic to adiabatic, as shown in Figure 6.14.

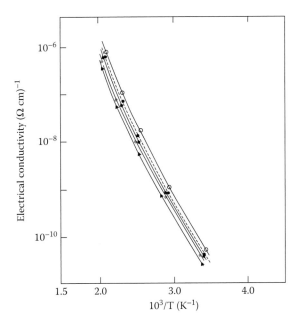

FIGURE 6.10 Conductivity and activation energy data of the tellurite glasses 65 mol% TeO$_2$–(35 − x) mol% CuO–x mol% *MO*, where MO = NiO or CoO. (From M. Malik and C. Hogarth, *J. Mater. Sci.*, 25, 2585, 1990. With permission.)

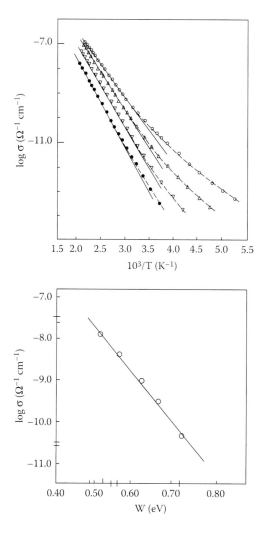

FIGURE 6.11 Conductivity and activation energy data of molybdenum–tellurite glasses. (From A. Ghosh, *Phil. Mag. B*, 61, 87, 1990. With permission.)

In 1994, Mori and Sakata measured the Seebeck coefficient of TeO_2–V_2O_5–SnO glasses. The thermoelectric power of semiconducting $(90, 80 - x)$ mol% TeO_2–x mol% V_2O_5–10, 20 mol% SnO glasses was measured from 373 to 473 K. The glasses were found to be n-type for $x = 30$–50, whereas 60 mol% TeO_2–20 mol% V_2O_5–20 mol% SnO glasses were p-type. The Seebeck coefficient, Q, was determined to be from -188 to $+5$ µV/K at 473 K, whereas for n-type semiconducting glasses, Q satisfied Heikes' relation, giving evidence for small-polaron hopping conduction. The Seebeck coefficients for $(80 - x)$ mol% TeO_2–x mol% V_2O_5–20 mol% SnO glasses, where $x = 20$–60, and for $(90 - x)$ mol% TeO_2–x mol% V_2O_5–10 mol% SnO glasses, where $x = 20$–50, increased with decreasing V_2O_5 content. It was concluded that the dominant factor affecting the Seebeck coefficient is the fraction of reduced ion concentration $C_V = (V^{4+}/V_{total})$ as shown in Table 6.1. Also in 1994, Mori et al. measured the σ_{DC} of TeO_2–Sb_2O_3–V_2O_5 glasses. Conduction was confirmed to be due to small-polaron hopping between vanadium ions and was adiabatic for $V_2O_5 > 50$ mol%, and nonadiabatic for $V_2O_5 < 50$ mol%. The polaron bandwidth ranged from $J = 0.02$ to 0.11 eV, which depended on the V–V spacing. The carrier concentration was ~10^{21} cm^{-3}. The estimated carrier mobility was from 10^{-7} to 10^{-5} cm^2 V^{-1} S^{-1} and varied significantly with V_2O_5 content. The dominant factor determining σ was carrier mobility in these glasses.

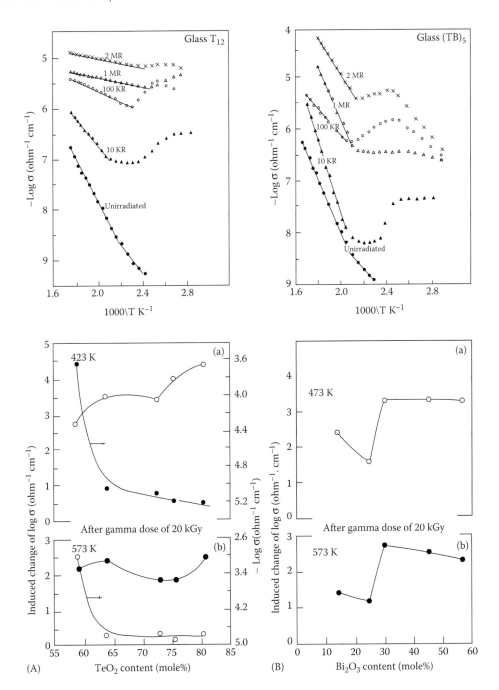

FIGURE 6.12 (A, B) Effect of γ on the DC electrical conductivity of tellurite glasses TeO$_2$–P$_2$O$_5$ and TeO$_2$–P$_2$O$_5$–Bi$_2$O$_5$. (From A. Abdel-Kader, A. Higazy, R. El-Mallawany, and M. El-Kholy, *Rad. Effects Defects Solids*, 124, 401, 1992. With permission.)

Singh and Chakravarthi (1995) measured the σ_{DC} of molybdenum–tellurite glasses of the form TeO$_2$–MoO$_3$ in the temperature range 100–400 K. Mott's model of thermally activated small-polaron hopping between the nearest neighbors was found to be consistent with the data in the high-temperature range as shown in Figure 6.15. According to the detailed theoretical model of the temperature dependence of σ given by Schnakenberg (1968), the σ parameters are calculated as shown in Table 6.2.

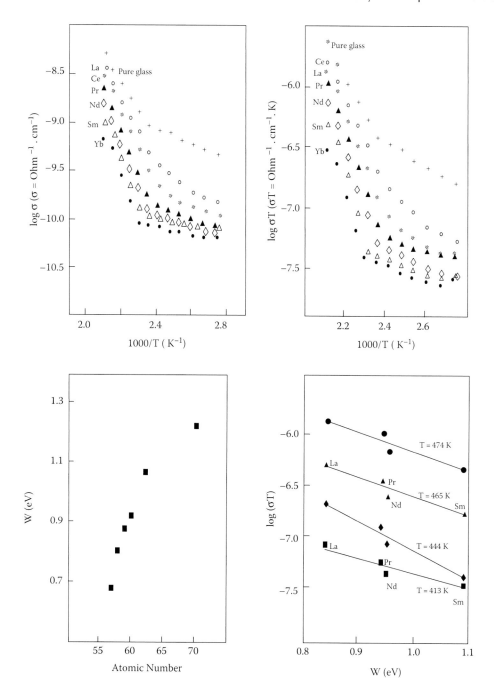

FIGURE 6.13 Effect of rare-earth ions on the electrical conductivity of tellurite glasses. (From A. Abdel-Kader, R. El-Mallawany, and M. El-Kholy, *J. Appl. Phys.*, 73, 75, 1993. With permission.)

The small-polaron coupling $\gamma = (W_H/h\nu)$ is between 11.1 and 15.5. According to Austen and Mott (1969), a value of $\gamma > 4$ usually indicates strong electron–phonon interaction in solids; El-Mallawany (1994) determined a value of 7 for binary TeO_2–WO_3 glasses, which means that the electron–phonon coupling in binary TeO_2–MoO_3 glasses is smaller than in binary TeO_2–WO_3 glasses.

El-Mallawany et al. (1995) measured the σ_{DC} of the rare-earth TeO_2–CeO_2 tellurite glasses and their σ after determining the ratio of their valence states at different temperatures. The σ values

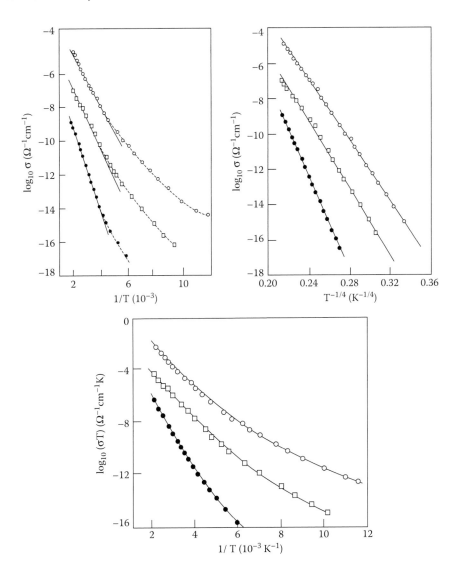

FIGURE 6.14   DC electrical conductivity of semiconducting vanadium–tellurite glasses containing bismuth oxide, in the temperature range 80–450 K. (From A. Ghosh, *J. Phys. Condens. Matter*, 5, 8749, 1993. With permission.)

of binary rare-earth tellurite glasses were lower than those of the pure and binary transition metal tellurite glasses as shown in Figures 6.16a and b. The ESR spectra at high temperatures are characterized by two overlapping lines, a broad line with g = 2.00, and a narrow line with g = 1.967. At temperatures lower than 258 K, the narrow line disappears and the broad line remains, as shown in Figure 6.16b. This suggests that carriers become localized at $T < 258$ K. At temperatures >273 K, σ increases as carriers hop between cerium sites. They found that the reduced-valence ratio also increases with temperature.

Peteanu et al. (1996) measured the Cu^{2+} contained in 70 mol% TeO_2–25 mol% B_2O_3–5 mol% PbO glasses by using EPR over a wide range of contamination. Mori and Sakata (1996) measured the σ_{DC} of the 33.3 mol% TeO_2–4.3 mol% Sb_2O_3–62.3 mol% V_2O_5 glasses. These glasses were prepared by press-quenching in Ar and O_2 gas atmospheres at temperatures between 303 and 473 K. The glass was an *n*-type semiconductor. The σ was lower in O_2 and higher in Ar than in air. The explanation for this was that V^{4+} ions in the glass are oxidized by the O_2 gas that has diffused into the glass, resulting

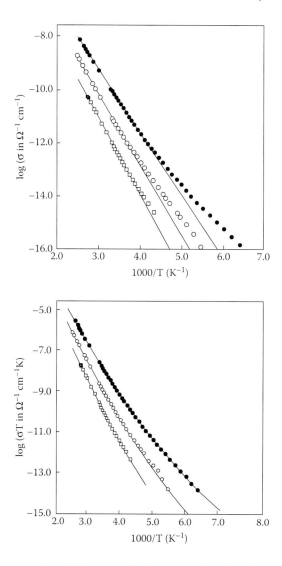

FIGURE 6.15 DC conductivity of molybdenum–tellurite glasses of the form TeO_2–MoO_3 in the temperature range 100–400 K. (From R. Singh and J. Chakravarthi, *J. Phys. Rev. B*, 51, 16396, 1995.)

in an increase in V^{5+} with time. The experimental relationship between the σ and the oxygen partial pressure (pO_2) agrees quantitatively with the theoretical relation ($\sigma \alpha [pO_2]^{-1/4}$). Variations in σ on switching the atmospheres between O_2 and Ar gasses were found to be reproducible, and the O_2 gas sensitivity S at 473 K was 1.2. These dynamic changes have been explained quantitatively with the oxygen diffusion model. From the data on the gas-sensing behavior, the 33.3 mol% TeO_2–4.3 mol% Sb_2O_3–62.3 mol% V_2O_5 glass is potentially applicable as an O_2 gas sensor. Kowada et al. (1996) studied the electronic states of binary tellurite glasses by the DV-Xα cluster method. Their results with various polymerization degrees and with various second-component ions—such as glass-forming ions, intermediate ions, and modifier ions—are shown in the 1st Ed. The bonding nature of each bond in the clusters of binary tellurite glasses was discussed and compared with that of silicate glasses. In the TeO_2 trigonal bipyramid unit, bond order between the Te- ion and the equatorial oxygen ion (O_{eq}) was larger than that between the Te- ion and the axial oxygen ion (O_{ax}). In the cluster for binary tellurite glasses, the bond order of the clusters with the modifier ions is almost zero. The bond order with the glass-forming ions has values larger than those with the modifier ions. For the intermediate ions,

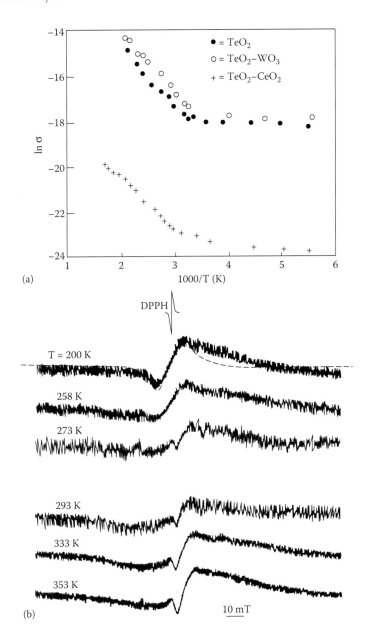

FIGURE 6.16 (a, b) DC electrical conductivity of the rare-earth tellurite glass system TeO_2–CeO_2 and the ESR spectra. (From R. El-Mallawany, A. El-Sayed, and M. El-Gawad, *Mater. Chem. Phys.*, 41, 87, 1995. With permission.)

the bond order ranges from −0.3 to 0.1. This change of the bond order of M–$O_{Te–O–M}$ with second-component ions had a similar tendency to that in the binary silicate glasses.

Qiu et al. (1997) measured the σ_{DC} of TeO_2–Fe_2O_3–MoO_3 glasses and found that these glasses are *n*-type semiconductors with a polaron band width J ranging from 0.09 to 0.18 eV, depending on the Fe–Fe spacing. In 1998, Sega et al. measured the σ_{DC} of TeO_2–NiO–V_2O_5 glasses. These glasses are also *n*-type semiconductors, and their charge densities range from 5.3×10^{18} to 7.3×10^{20} cm^{-3} for 35 mol% TeO_2–10 mol% NiO–55 mol% V_2O_5 and 60 mol% TeO_2–10 mol% NiO–30 mol% V_2O_5. Sega et al. (1998a, 1998b) measured the σ_{DC} of TeO_2–MnO–V_2O_5 glasses in the temperature range 375–475 K. These glasses are *n*-type semiconductors, and the electron–phonon interaction coefficients (γ)

are in the range from 21 to 26 for 30 mol% TeO$_2$–10 mol% MnO–60 mol% V$_2$O$_5$ and 60 mol% TeO$_2$–10 mol% MnO–30 mol% V$_2$O$_5$ glasses. In 1998, Montani et al. studied the behavior of the I–V-characteristic curves on TeO$_2$–V$_2$O$_5$ glasses. Their experimental evidence supports an electrothermal origin for the non-ohmic behavior of TeO$_2$–V$_2$O$_5$-based glasses with regard to a previously proposed model for the so-called "switching effect." Montani et al. (1998) supported the assumption that the switching effect observed in TeO$_2$–V$_2$O$_5$ glasses resulted mainly from an electrothermal effect and that it can be avoided by minimizing the heat amount produced by Joule heating.

El-Damrawi (2000) studied the electrical σ of the tellurite glass system TeO$_2$–PbO–PbF$_2$ as shown in Figure 6.17. The addition of PbF$_2$ to the lead–tellurite network increases the σ by increasing both carrier concentration and mobility. The E for conduction decreases with increasing PbF$_2$ contents. El-Damrawi (2000) concluded that this decrease in the E by increasing PbF$_2$ content provides a basis for modifying the channel network by expanding, opening, and weakening the bonds constituting the skeleton of the glass. The open structure of the lead–tellurite network was found to have a direct effect on the ionic conduction of the glass. The σ is enhanced by more than seven orders of magnitude upon addition of 50 mol% PbF$_2$. The random site model has been proposed to explain the fluorine ion transport mechanism in these glasses.

6.4.2 DC ELECTRICAL CONDUCTIVITY DATA OF OXIDE–TELLURITE GLASSES CONTAINING ALKALIS

In 1986, Sunandana and Kumaraswami reported measurements on fast-ion-conducting tellurite plate glass in the temperature range 300–670 K. In 1988, Tanaka et al. started measuring both σ_{DC} and σ_{AC} of the TeO$_2$–Li$_2$O–LiCl glasses in the temperature range 0°C–300°C as shown in the 1st Ed. The electrical σ of the glasses with high concentrations of Li$^+$ ion reached 10 S cm^{-1} at 25°C. The composition dependence of the electrical σ was discussed in relation to the glass structure, especially in terms of the bonding characteristics of Te–Cl and Te–O. The electrical σ and E are mainly dependent on the LiCl content due to the weaker Li–Cl bond than the Li–O bond:

$$\{(Te_{eq} - O_{ax}) + (LiCl)\} \rightarrow \{Te_{eq} - Cl \ldots Li - O_{ax} - Te\},$$

while Li$_2$O causes the next one:

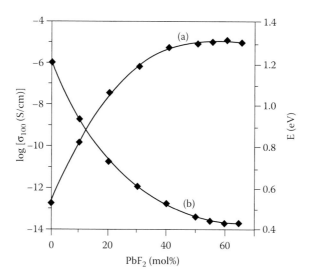

FIGURE 6.17 (a, b) Electrical conductivity of the tellurite glass system TeO$_2$–PbO–PbF$_2$. (From G. El-Damrawi, *Phys. Stat. Sol.*, 177, 385, 2000. With permission.)

$$Te_{eq}-O_{ax}-Te + Li_2O \Rightarrow Te - O_{eq} \overset{\diagup \text{Li} \diagdown}{\underset{\diagdown \text{Li} \diagup}{}} O_{ax} - Te.$$

Moreover, formation of the nonbridging oxygen by the addition of LiCl may assist the creation of the diffusion path, which leads to the high electrical σ and low E on the LiCl-rich side. Balaya and Sunandana (1994) studied the mixed-alkali effect in tellurite glasses of the form 70 mol% TeO_2–30 mol% [$(1 - x)$ mol% Li_2O–x mol% Na_2O], where $x = 0.0, 0.2, 0.4, 0.6, 0.8,$ and 1.0 as shown in the 1st Ed. The σ and E for the conduction process show nonlinear behavior after the substitution of one alkali ion for another, at $Na/(Na + Li) \approx 0.6$. Balaya and Sunandana (1994) also found the same ratio as found in borate glasses. The experimental results have been discussed in light of current mechanisms (interchange transport process) along with the relevant structural aspects, which show that the interchange transport mechanism operates in both single and high-alkali–tellurite glasses. In 1995, Komatsu et al. (1995) studied the mixed-alkali effect in tellurite glasses and their change in fragility. Negative deviations from additivity in the electrical σ at 200°C and 1 kHz indicated the mobility of Li^+ and Na^+ in alkali–tellurite glasses.

In 1996 and 1997, in three articles, Chowdari and Kumari reported their studies of the silver ion-conducting tellurite glasses, in which they used the same experimental technique but different kinds of tellurite glasses: first, 0.5 mol% Ag_2O–0.5{x mol% TeO_2–[$(1 - x)$ mol% P_2O_5–x mol% Na_2O]}; second, 0.3 mol% Ag_2O–0.28 mol% TeO_2–0.42 mol% B_2O_3; and third, y mol% Ag_2O–$(1 - y)$[$(1 - x)$ mol% TeO_2–x mol% MoO_3]. To obtain their results, Chowdari and Kumari (1996) first synthesized by progressive replacement a 50 mol% Ag_2O–0.05{x mol% TeO_2–[$(100 - x)$ mol% P_2O_5–x mol% Na_2O]} in the frequency range 10–1 MHz. The ionic σ was found to increase by about two orders of magnitude with increases in TeO_2 content. Although σ at 25°C $= (1.93 \pm 0.43) \times 10^{-6}$ S cm^{-1}, the electronic σ is $(3.11 \pm 0.14) \times 10^{-10}$ S cm^{-1} for the glass composition 50 mol% Ag_2O–30 mol% TeO_2–20 mol% P_2O_5, as shown in the 1st Ed. The XPS studies of these glasses reveal information about the structural modifications leading to the creation of bridging oxygen and nonbridging oxygen. A detailed analysis of the Te-3d (XPS) spectrum of the binary and ternary systems shows the existence of different Te-based structural units like TeO_4, TeO_{3-}, TeO_3^{2-}, and Ag–Te bond. The results from the ternary system suggest the unique role played by P_2O_5 in the conversion of TeO_4 units leading to the creation of Ag–Te bond and TeO_3^{2-} units. Also, in 1996, Chowdari and Kumari (1996a, 1996b) examined the synthesis and ionic σ of 30 mol% Ag_2O–0.70[$(100 - x)$ mol% TeO_2–(x mol% B_2O_3)]. The mixed-former effect has also been observed in these glasses. A maximum value of the ionic σ ($\sigma = 2.82 \times 10^{-6}$ S cm^{-1}) at 100°C was observed for the composition 30 mol% Ag_2O–28 mol% TeO_2–42 mol% B_2O_3. In 1997, Chowdari and Kumari studied the effect of mixed glass formers in the system: y mol% Ag_2O–$(100 - y)$[$(100 - x)$ mol% TeO_2–x mol% MoO_3]. The σ at 25°C showed two maxima for $y = 0.3$ at $x = 0.2$ and 0.4, and for $y = 0.4$ at $x = 0.3$ and 0.5.

From recent progress on the σ of tellurite glasses containing alkaline oxides, tellurite glasses can be considered to be glassy electrolytes with a number of advantages:

1. Ionic σ is isotropic and does not involve any gain boundary effects like those in polycrystalline solids.
2. The electronic contribution to the total σ is weak, which is a consequence of the periodic potential fluctuations imposed by the disordered structure.
3. Metal impurities are insufficient to enhance the electronic σ because they can exist in their own distinct local environment in the glass structure.
4. Electronic leakage is then unlikely to occur in electrochemical devices that use a glass membrane as an ionic separator.
5. Good contacts can be obtained between electrode materials and ionic conductive glasses of low T_g.

6. Tellurite glasses can be electrochemically active materials (by changing their composition).
7. Bulk glass samples can be obtained with ionic transport properties at one end and electronic transport at the other end of the sample.

These advantages open the way to make a compact battery with improved performance through the delocalization of electrode/electrolyte interfaces. Tellurite glasses are now considered as possible candidates for positive electrode materials in lithium batteries, which opens the way for making monolithic vitreous batteries. Within the next decade, all integrated electronic circuits will certainly be powered by implanted vitreous microbatteries. Semiconductor nanotechnology, novel materials and devices for electronics, photonics, and renewable energy applications have now reached an advanced stage in their development, as stated by Goodnick et al. 2010, who go on to state that cost-effective improvements to current technological approaches have made great progress, although certain challenges remain.

Regarding electrode materials for lithium ion or RC batteries, electrochemical properties of innovative vanadium–tellurite glasses in a lithium battery have been studied by Amarilla et al. (1997). The shape of the discharge curves, the Li insertion degree (x), and the capacity retention during cycling strongly depend on glass composition. For the binary V_2O_5–TeO_2 glass, x diminishes on cycling. The addition of Li_2O and/or NiO increases (x) and the reversibility of the Li insertion process. Regarding the specific capacity of the 15 wt% NiO-doped V_2O_5–TeO_2 glass, at $x = 2.04$, the Li^+ was the highest value reported in the literature for vanadium-based electrodes. These properties, as well as the low reduction potential (E » 2V), make this glass a very well-suited material for negative electrodes in RC lithium batteries.

6.4.3 DC OF NONOXIDE–TELLURITE GLASSES

During the 1980s and 1990s, a number of authors reported their measurements of the electrical properties of amorphous Te combined with other elements at near room temperature. Concerning the electrical properties of chalcogenide glasses, Zope and Zope (1985) studied the effect of γ-irradiation on the nonlinear behavior of both I–V and thermoelectric power of Te–As–Se glasses in the range 283–350 K. The sample was held between two copper rods of identical size and shaped with indium contacts. The nonlinear I–V behavior and the p-type nature of nonirradiated and γ-irradiated Te_2As_2Se glasses were interpreted on the basis of charged defect states existing and also created after γ-irradiation in the material. Zope and Zope discussed the thermoelectric power for nonirradiated and γ-irradiated Te_2As_2Se on the basis of a quantity related to σ. Kastner and Hudgens (1978) proposed that in amorphous semiconductors, the lower energy defect states are positively charged, threefold coordinated C_3^+ and negatively charged, onefold coordinated C_1^- chalcogen atoms. The p-type nature of Te_2As_2Se amorphous materials has been explained on the basis of the charged defect states C_3^+ and C_1^- existing in the materials. During thermoelectric measurements of Te_2As_2Se, the thermal gradient establishes changes in the density of charged defect states by capturing electrons and changes in the density of holes by C_3^+ and C_1^- centers, respectively.

In 1986, Kahnt and Feltz published two articles summarizing their examination of the influence of the preparation method on σ in amorphous chalcogenides of the form x mol% Te–(60 − x) mol% Se–40 mol% Ge. The first of these articles (Kahnt and Feltz 1986a) compared the bulk $TeSe_5Ge_4$ with thin amorphous films of identical composition prepared by vacuum evaporation at a relatively low temperature and by flash evaporation, respectively. For σ measurements, the bulk samples were polished to a thickness of ~0.5 mm and a cross section of ~50 mm² and were put in contact with aluminum in a sandwich arrangement. Differences in σ of up to three orders of magnitude and variations in the field dependence were discussed in terms of Poole–Frenkel active single- and screened-charge centers of the intrinsic valence alternation pair type. The second article (Kahnt and Feltz 1986b) reported the σ of amorphous chalcogenides in the series x mol% Te–(60 − x) mol% Se–40 mol% Ge prepared at a relatively low temperature and by the flash evaporation technique. The σ values differed by about three orders of magnitude for the same chemical composition. These differences are caused by a lower band gap in the films and the influence of annealing on photocurrent spectra and field-dependent σ of the x mol% Te–$(60 - x)$ mol% Se–40 mol% Ge films prepared by the flash evaporation technique. Annealing precess yields an approach of the films to the bulk glasses. The σ differences are discussed in terms of chemical bonding and the resulting electronic states in amorphous lone-pair semiconductors. Shortly after these reports by Kahnt and Feltz, Tanovcova et al. (1987) found the optimal composition and preparation method to make multicomponent seleno-tellurides applicable for optical fibers applied to power transmission at 10.6 (μm); for this work the authors used the systems x mol% Te–$(100 - x - y)$ mol% Se–y mol% Ge, where $x = 0$–50 and $y = 20, 25, 30,$ and 33; and x mol% Te–$(80 - x - y)$ mol% Se–y mol% Sb–20 mol% Ge, where $x = 0$–5 and $y = 15$–20 glasses. The glass samples were sandwiched with painted graphite electrodes. The electrical σ increased monotonously due to the sharp increase in the pre-exponential factor.

Gadkari and Zope (1988) measured the σ_{DC} and thermoelectric power of bulk chalcogenide glasses of the forms 30 mol% Te–70 mol% Se and $(30 - x)$ mol% Te–70 mol% Se–x mol% In in the temperature range 270–373 K, as shown in Table 6.1. The σ measurements were accomplished by using pellets covered with silver paste as electrical contacts. The E values were obtained from the σ and thermoelectric power measurements. The addition of indium to the samples increased their electrical σ. The following year, El-Fouly et al. (1989) studied the effect of the Te/Si ratio on the electrical characteristics of the amorphous chalcogenide film switches $(60 - x)$ mol% Te–30 mol% Ae–10 mol% Ge–x mol% Si, where $x = 5, 10, 12,$ and 20, in the temperature range from liquid nitrogen to below T_g, as shown in Figure 6.18. The samples used for σ were prepared by evaporation of chalcogenide materials on substrates of mica. El-Fouly et al. (1989) applied the three mechanisms

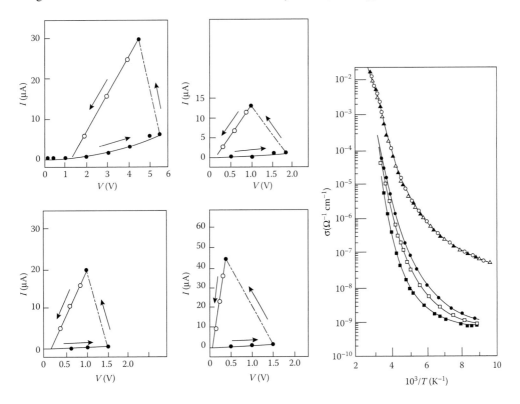

FIGURE 6.18 Electrical characteristics of the amorphous chalcogenide film switches $(60 - x)$ mol% Te-30 mol% Ae-10 mol% Ge-x mol% Si, where $x = 5, 10, 12,$ and 20 and the temperature ranges from liquid nitrogen to below T_g. (From M. El-Fouly, A. Morsy, A. Ammar, A. Maged, and H. Amer, *J. Mater. Sci.*, 24, 2444, 1989. With permission.)

of conduction defined by Mott et al. (1975) to find the density of states at the Fermi level $N(E_F)$. They found that the density of states at the Fermi level increases with decreases in silicon content. In 1999, Hegazy measured the electrical I–V curves of glasses of form $(100 - x)(22$ mol% Ge–14 mol% Se–64 mol% Te)–x mol% I, where $x = 0, 4, 6, 8$, and 10, at different ambient temperatures from room temperature up to 343 K. Measurements of the electrical σ at constant current value to estimate the temperature of the electrical path have also been achieved.

6.5   AC ELECTRICAL CONDUCTIVITY DATA OF TELLURITE GLASSES

For nonoxide–tellurite glasses, Rockstad (1970) measured the electrical σ of As_2Te_3, Te_2AsSi, and 47.7 mol% Te–29.9 mol% As–12.6 mol% Si–9.8 mol% Ge. The results were similar for the three compositions. The σ had been considered to consist of two components: (1) a frequency-independent component σ_0 and (2) a frequency-dependent σ_1.

The frequency-dependent component varied as ω^s with s usually ~0.8–0.9, and it attained a magnitude comparable with the σ_0 at 10^5–10^6 Hz, except for point contact with light contact pressures. The frequency component was attributed to the hopping mechanism, and the independent component was attributed to intrinsic band conduction.

For oxide–tellurite glasses, Mansingh and Dahwan (1983) measured the σ_{AC} of the semiconducting vanadium–tellurite glasses $(100 - x)$ mol% TeO_2–x mol% V_2O_5, where $x = 10$–80, the frequency range is 0.1–100 kHz, and the temperature range is 77–400 K, as shown in Figure 6.19. The frequency dependence of the σ_{AC} satisfies the relation $\sigma(\omega) = A\omega^s$, and the values of exponent s are 0.95–0.80. Mansingh and Dahwan (1983) analyzed the σ_{AC} in light of the different theoretical models based on QMT and HOB. The analysis showed that neither QMT nor HOB models explain the data quantitatively. A linear dependence of the σ and a temperature dependence s at low temperatures with a nonlinear increase of σ at higher temperatures for higher frequencies agree, respectively, with QMT and HOB models. Hampton et al. (1987) measured the first complex admittance profiles of pure (89.5 mol%) TeO_2 glass and binary 67 mol% TeO_2–33 mol% WO_3 glass in the temperature range 90–430 K and frequency range 10–5 kHz as shown in the 1st Ed, by using the complex admittance techniques and the small-polaron model to estimate the polaron E.

In 1990, Rojo et al. studied the influence of ion distribution on ionic σ of the lithium–tellurite glasses $(100 - x)$ mol% TeO_2–x mol% Li_2O, where x is 18%–50%, and $(100 - x)$ mol% TeO_2–x mol% LiF, where $x = 10\%$–47%, in the frequency range 10^{-2}–10^4 Hz and in the temperature range from 100 K to T_g as shown in the 1st Ed. A parallel study by ^{7}Li- and ^{19}F-NMR spectroscopy allowed identification of Li^+ ions as the carriers responsible for the ionic σ of these solids. In 1990, Malik and Hogarth studied the effect of cobalt oxide on the σ_{AC} mechanism of copper–tellurite glasses of the form 65 mol% TeO_2–35 mol% CuO and 65 mol% TeO_2–34 mol% CuO–1 mol% CoO, at temperatures between 293 and 458 K in the frequency range $\leq 10^6$ Hz. The measured σ showed a frequency dependence obeying the relation $\sigma(\omega) = A\omega^s$, with $s < 1$ but with different values at different temperatures. After that, Rossignol et al. (1993) studied the ionic σ and structure of thallium–tellurite glasses containing AgI for glasses of the form $[(100 - x)$ mol% $(TeO_2:RTlO_{0.5})] - (x$ mol% AgI) and compared with $[(100 - x)$ mol% $(TeO_2:RAgO_{0.5})] - (x$ mol% AgI), where $R = (AgO_{0.5}/TeO_2)$ by using the Z^* in the frequency range 10^{-2}–10^6 Hz and temperatures between 20°C above and 20°C below T_g. Rossignol et al. (1993) concluded that from the σ variation with composition, the ionic σ properties are bound essentially to the mobility of the Ag^+ ions located in an iodide environment. Ghosh (1993) found the frequency and temperature dependence of the real and imaginary parts of the σ_{AC} of tellurium composite glassy semiconductors in the frequency range 10^2–10^5 Hz and in the temperature range 80–350 K. Ghosh (1993) analyzed experimental data on the QMT of electrons and on the classical hopping of electrons over the same barrier separating the localized states. This model quantitatively predicts the temperature dependence of the σ_{AC} and its frequency

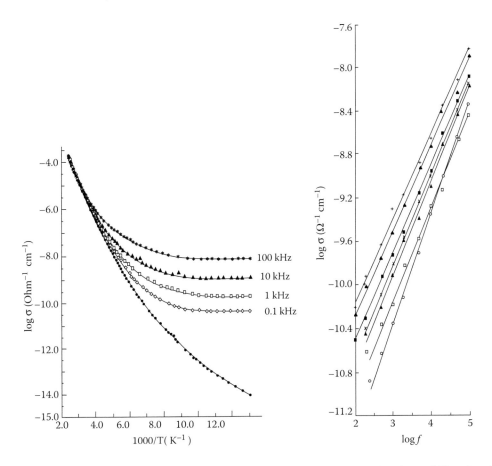

FIGURE 6.19   AC electrical conductivity of the semiconducting vanadium-tellurite glasses $(100 - x)$ mol% TeO_2-x mol% V_2O_5, where $x = 10$–80, the frequency range is 0.1–100 kHz, and the temperature range is 7–400 K. (From A. Mansingh and V. Dahwan, *J. Phys. C: Solid State Phys.*, 16, 1675, 1983.)

exponent for the measured temperature range. In 1994, El-Kholy measured the σ_{AC} of TeO_2–P_2O_5 glasses in the frequency range 100–10 kHz and temperature between 300 and 573 K. The density of states was calculated using the Elliott model. El-Kholy (1994) showed that the CBH model was the most appropriate mechanism for conduction in the TeO_2–P_2O_5 glass system. For the binary transition metal tellurite glasses of the form $(100 - x)$ mol% TeO_2–x mol% MoO_3, El-Kholy and El-Mallawany (1995) measured the σ_{AC} in the frequency 100–10 kHz and temperature range 300–573 K as shown in Figure 6.20. The σ data show a strong dependence on the percentage of TMI in the glass. The density of localized states has been calculated using the QMT model and found to be compositionally dependent.

Jayasinghe et al. (1996) studied the mixed-former effect on the σ_{AC} of sodium phosphor–tellurite glasses of the form x mol% $Na_2P_2O_6$–$(100 - x)$ mol% $Na_2Te_2O_5$, where $0 < x < 1$, in the temperature range 100°C–250°C and the frequency range 5–13 MHz, as in Table 6.1. The σT product obeys an Arrhenius relationship. Isothermal σ curves with x and shows a maximum in the ionic σ. This mixed-network-former effect might be justified by assuming an endothermic mixture between sodium–phosphate and sodium–tellurite. However, in 1997, Jayasinghe et al. found the electronic-to-ionic σ of glasses in the $2TeO_2$–xNa_2O–$(1 - x)$ V_2O_5 glasses, the frequency range 500–10 MHz, and temperature range 25°C–250°C. These experimental data suggest that a polaron hopping mechanism operates in the electronic-conduction domain of $0 < x < 0.5$, and that an interstitial-pair

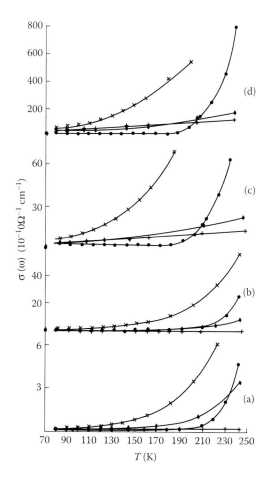

FIGURE 6.19 (Continued)

mechanism operates in the ironically conducting domain of $0.5 < x < 1$. In 1999, Jayasinghe et al. measured and discussed the electronic-to-ionic σ of the glasses $3TeO_2–xLi_2O–(1 − x)V_2O_5$, where $0 < x < 1$, as shown in Figure 6.21, under the same conditions as described for their previous article. The experimental values suggested that an electronic σ would operate between vanadium cations and would prevail in the first region. In the alkali–oxide region, ionic σ would operate by an interstitial pair mechanism between nonbridging oxygens.

In 1998, Ravikumar and Veeraiahi measured the σ_{AC} of $(1 − x − y)TeO_2–xPbO–yZnF_2$, in the frequency range $5 \times 10^2–10^5$ Hz and temperature range 30°C–200°C. The E for σ_{AC} in the high-temperature region is decreased by increasing the PbO content, which has been related to the increase in the degree of depolymerization of the glass network with increase in the modifier (PbO) concentration. Pan and Ghosh (1999) measured the E and σ relaxation of sodium–tellurite glasses of the form $(100 − x)$ mol% $TeO_2–x$ mol% Na_2O in the frequency range 10–2 MHz, the temperature range 373–533 K, as shown in Figure 6.22, and with the electric modulus as shown in Chapter 7. The Anderson–Stuart (A-S) model was used to explain the DC E. The relaxation behavior of these glasses was analyzed in light of the modulus formalism. The variation of the stretching exponential parameter with composition was explained in terms of cation–cation distance correlation. The structural model of tellurite glasses was used to explain the composition dependence of the σ relaxation time.

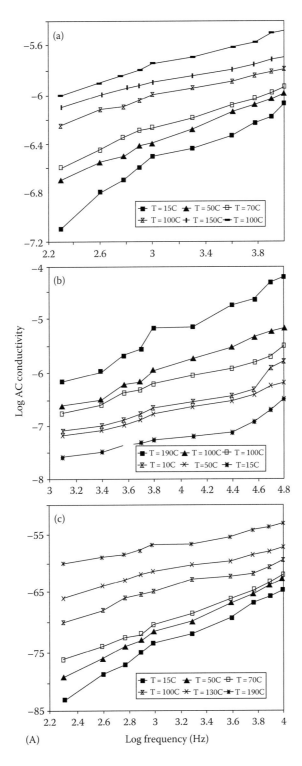

FIGURE 6.20 (A–C) AC electrical conductivity of $(100 - x)$ mol% TeO_2–x mol% MoO_3, in frequencies from 100 to 10 kHz and temperatures from 300 to 573 K. (From M. El-Kholy and R. El-Mallawany, *Mater. Chem. Phys.*, 40, 163, 1995. With permission.)

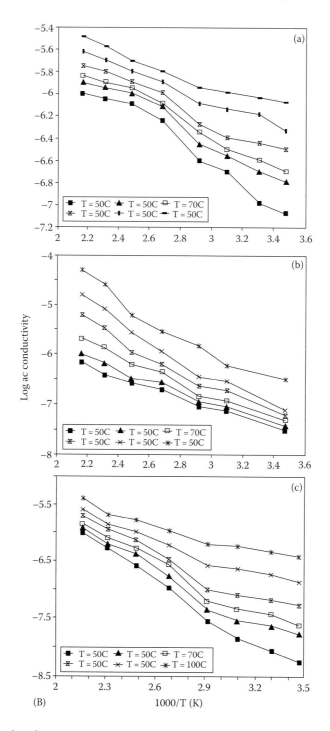

FIGURE 6.20　(Continued)

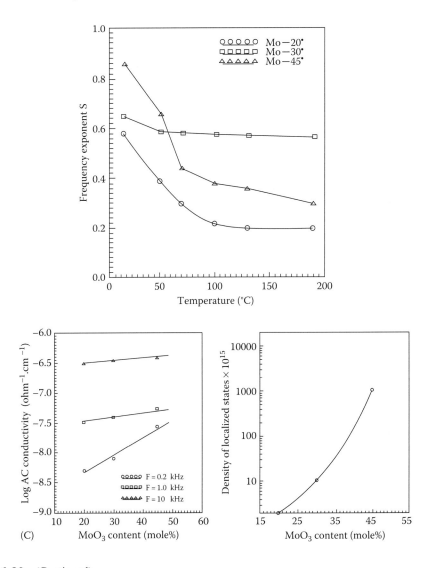

FIGURE 6.20 (Continued)

6.6 ELECTRICAL CONDUCTIVITY DATA OF TELLURITE GLASS-CERAMICS

Deneuville et al. (1976) examined the effect of annealing on the crystallization process of amorphous $(1 - x)$Te–xGe, where $x = 0.1, \ldots 0.5$. Crystallites of either Te or Ge–Te have extrinsic effects that add to the intrinsic ones. E decreases (increases) and G increases as the number of defects increases (decreases). In all cases, the final metallic state is obtained only when both Ge–Te and Te crystallites are present. The memory switching caused by crystallization in TeO$_2$–V$_2$O$_5$–CuO glasses was considered by Yoshida et al. (1985). Four kinds of glasses consisting of 30 mol% TeO$_2$ plus various CuO and V$_2$O$_5$ contents were crystallized by heat treatment from 300°C to 460°C, stepwise. The σ of 30 mol% TeO$_2$–45 mol% V$_2$O$_5$–25 mol% CuO glasses is increased about two orders of magnitude from 10^{-3} to $10^{-0.5}$ Ω^{-1} cm^{-1} at 150°C by the crystallization. The E for conduction of the crystallized glass is 0.1–0.2 eV, which is nearly the same as that of the developed crystal V$_2$O$_5$, known as a semiconductor. In 1986, Hirashima et al. observed a memory switching of TeO$_2$–V$_2$O$_5$ glass film. The switching phenomena are observed at a DC electric field of $\leq 10^5$ V/cm and for film of

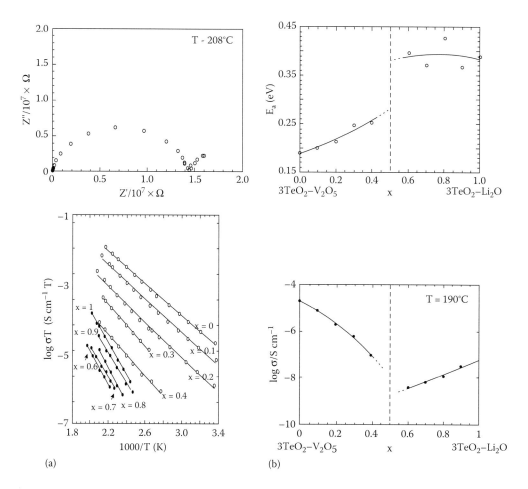

FIGURE 6.21 (a, b) Electrical conductivity of the glasses $3TeO_2–xLi_2O–(1 − x)V_2O_5$, where $0 < x < 1$. (From G. Jayasinghe, M. Dissanayake, P. Bandaranayake, J. Souquet, and D. Foscallo, *Solid State Ionics*, 121, 19, 1999. With permission.)

thickness ~2–20 μm. Crystallization of these glasses by heat treatment causes an increase in the σ of at least two orders of magnitude, as shown in the 1st Ed. Rojo et al. (1992) measured the relationship between microstructure and ionic conduction properties in oxyfluoride–tellurite glass-ceramics. The ionic σ of 50 mol% TeO_2–(50 − x) mol% $LiO_{0.5}$–x mol% LiF glass-ceramic, where $0 < x < 50$, has been measured as a function of temperature and chemical composition by the Z^* method in the temperature range 100°C–240°C. Crystalline domains of LiF and α-TeO_2 have been observed inside an amorphous matrix by transmission electron microscopy. The extent of the domains increases with the fluorine ratio. In samples with higher fluorine content (F/Te > 0.5), formation of the crystalline domains progresses and a decrease in ionic σ is observed. The association of Li+ and F- ions induces the formation of α-TeO_2 domains, which promotes phase separation and devitrification. The composition dependence of electrical σ is often discussed in terms of the cluster-bypass model. In 1993, Balaya and Sunandana measured the bulk ionic conductivities of 70 mol% TeO_2–30 mol% Li_2O glass and crystallized samples as a function of temperature. The polycrystallized samples had a lower σ and higher E compared with those of glassy samples. Whereas the E in these glasses can be attributed to the migration of Li+ ions, in polycrystalline samples it includes the energy necessary

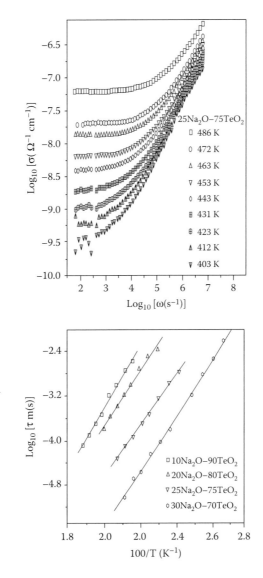

FIGURE 6.22 Electrical conductivity relaxation of sodium–tellurite glasses $(100 - x)$ mol% TeO_2–x mol% Na_2O at 10–2 MHz and in the temperature range 373–533 K. (From A. Pan and A. Ghosh, *Phys. Rev. B*, 59, 899, 1999. With permission.)

both for the migration of Li^+ ions and for the creation of defects and existence of grain boundaries in the orthorhombic crystalline phase.

Mandouh (1995) studied the transport properties in Se–Te–Ge by measuring the resistivities of their thin films from room temperature to 300°C. Crystallization of that sample was achieved at 573 K. The addition of germanium to the Se–Te alloy causes structural changes that modify the band structure and hence the electrical properties of the Se–Te alloy. The amorphous Se–Te–Ge system has been considered a good photovoltaic material. Rojas et al. (1996) studied the effect of annealing on changes in the electrical σ of amorphous semiconductors of the Te–Se–Ge system (Figure 6.23). This study included the determination of I–V characteristics, the electrical σ, and the relationship of σ to temperature and the aging of samples started by annealing and by thermal switching during Joules self-heating.

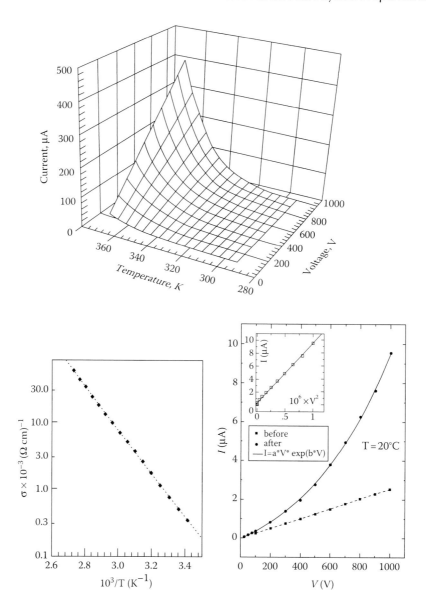

FIGURE 6.23 Electrical conductivity of amorphous semiconductors in the Te–Se–Ge system. (J. Roja, M. Dominguez, P. Villares, and R. Garay, *Mat. Chem. Phys.*, 45, 75, 1996. With permission.)

6.7 ELECTRICAL CONDUCTIVITY DATA OF NEW TELLURITE GLASSES

Electronic, photonic information technology and renewable energy, such as solar energy, fuel cells, and batteries have now reached an advanced stage in their development. The cost-effective improvements to current technological approaches have made great progress, but certain challenges remain. Superionic glasses are of current interest because of their application in different technical devices. Significant effort is currently being focused on the development of "all-solid-state batteries" that may be used in the future in high-technology applications, as reported by Weppner in 2003, and also by Schwenzel et al. in the same year. Superionic glasses with high silver ion mobility are important solid electrolytes, finding commercial applications in low-current, long-life batteries. Glasses containing silver halide, particularly silver iodide, can provide superionic behavior at room temperature. The electrical properties of glasses have been studied extensively for a number of years due to their potential use in solid-state devices. A large number of binary and ternary tellurite glassy systems have been synthesized using tellurium oxide as a network former and other different oxides as network modifiers, as mentioned in the previous sections. It is well known that the physical and chemical properties of glasses depend on the glass composition. Thus, it is important to know the chemical elements that should be added into the host in order to design glasses with desirable optical properties and high performance. The right selection of glass composition depends on the application, and it is often a compromise among many factors. Desirena et al. (2009) emphasized the importance of the glass composition used in the design of optical amplifiers, lasers, and other applications to achieve the best properties.

Moawad and his coworkers (Moawad, Jain and El-Mallawany, 2002, 2009, and 2010) published a series of articles on the electrical properties of tellurite glasses in form $0.5[xAg_2O–(1 − x)\ V_2O_5]–0.5[TeO_2]$. In 2002, Moawad et al. investigated the mixed electronic–ionic conduction in $0.5[xAg_2O–(1 − x)V_2O_5]–0.5TeO_2$ glasses, where the Ag_2O amount varies over a range of 5–40 mol%. The AC conductivity was measured at frequencies from 10 to 100 kHz and temperatures of 300–425 K. The data indicate that the conduction mechanism changed from being predominantly electronic to ionic for Ag_2O contents of $x >27.5$ mol%, as in Table 6.3. This transition is due to the change in glass structure, which affects both electronic- and ionic-transport properties. The electronic DC conductivity results were analyzed in terms of a small-polaron hopping model. The XPS spectra show that the addition of Ag_2O disrupts the Te–O–Te, Te–O–V, and V–O–V bonds, and concurrently forms nonbridging groups, which apparently restrict the ease of intervalence electron transfer, and thus reduce electron mobility. To understand the variation of d.c. activation energy, $E_{d.c.}$, one must recognize the changes in the glass structure. In the $yV_2O_5–(1 − y)TeO_2$ binary system, the basic structural units are VO_5, TeO_4, and TeO_3 polyhedra that share corners according to the Zachariasen rules. Pure TeO_2 glass consists of TeO_4 trigonal bipyramids, in which one equatorial site is occupied by a lone pair of electrons and the other two equatorial and axial sites are occupied by O atoms. The Te atoms are connected at vertices by $TeO–_{eq}O_{ax}–OTe$ linkages. $Te–_{eq}O_{ax}–V$ linkages that form from the $Te–_{eq}O_{ax}–Te$ and V–O–V species may connect the VO_5 and TeO_4 or TeO_3 polyhedra. Thus, the present $0.5[xAg_2O–(1 − x)V_2O_5]–0.5TeO_2$ glass system can be considered as a $V_2O_5–TeO_2$ glass modified by Ag_2O according to the following reactions:

$$Ag^+$$
$$/\quad \backslash$$
$$Ag_2O + Te–_{eq}O_{ax}–Te \rightarrow Te–_{eq}O^-\ O–Te$$
$$\backslash\quad /$$
$$Ag^+$$

and

$$Ag_2O + Te–_{eq}O_{ax}–V \rightarrow Te–_{eq}O^-Ag^+ \text{ (or } Te–_{ax}O^-Ag^+) + V–O^-Ag^+$$

Hence, oxygen species such as $Te–_{eq}O_{ax}–Te$, V–O–V, V=O, Te=O, Te–O⁻V, V–O⁻Ag⁺, $Te–_{eq}O^-$ Ag⁺, and $Te–_{ax}O^-Ag^+$ may be present in $0.5[xAg_2O–(1 − x)V_2O_5]–0.5[TeO_2]$ glasses. The Ag⁺ ion has an ionic radius of 1.26 Å and an electronegativity of 1.9 eV, as mentioned by Weast and Astle (1980); therefore, it would occupy more space than the V ion (ionic radius of 0.59 Å and electronegativity of 1.63 eV) near the NBO site, which would cause an overlap of wave function of silver with tellurium (ionic radius of 0.7 Å and electronegativity of 2.1 eV; Chowdari and Kumari 1996).

In 2009, Moawad, Jain and El-Mallawany, continued their effort by investigating the mechanism of DC conductivity changes in $0.5[xAg_2O–(1 − x)V_2O_5]–0.5TeO_2$ glasses with $x = 0.1–0.8$

TABLE 6.3

Experimental Data of Electrical Conductivity

Glass	Reference	DC Conductivity at Room Temperature (Ω^{-1} cm^{-1})	DC Activation Energy (eV)	AC Conductivity (Ω^{-1} cm^{-1})	AC Activation Energy (eV)	S&β
yV_2O_5–$(1 - y)$Te O_2 mole%	Moawad et al. 2002					
10–90		3.42×10^{-9}	0.439			
60–40		1.96×10^{-9}	0.340			
TeO_2–V_2O_5–Ag_2O						
50:45:5		3.35×10^{-6}	0.349			
50:22.5:27.5		1.40×10^{-9}	0.645			
50:10:40		3.45×10^{-7}	0.525			
$0.5[xAg_2O$–$(1 - x)$ $V_2O_5]$–$0.5[TeO_2]$	Moawad et al. 2009					
45 mol% V_2O_5		3.35×10^{-6}	0.349			
30 mol % V_2O_5		3.361×10^{-8}	0.412			
V_2O_5–TeO_2–BaO	Szu et al. 2005					
40–40–20			0.57			
40–30–30			0.62			
70–15–15			0.33			
$80TeO_2 + 20ZnO$	Shabann et al. 2006			1.4×10^{-10}	2.57	0.94 & 0.06
$80TeO_2 + 20ZnO +$ $0.01g$ Ho_2O_3				1.22×10^{-10}	1.717	0.91 & 0.09
$80TeO_2 + 20ZnO +$ $0.1g$ Ho_2O_3				7.44×10^{-10}	0.907	0.83 & 0.17
$0.4 V_2O_5$–$x Li_2O$– $(0.6 - x)$ TeO_2	Kumar 2008a					
$X = 0.1$		2.58×10^{-3} at 450 K	0.418			
$X = 0.5$		1.56×10^{-4} at 450 K	0.826			
$0.4(V_2O_5)$–$(TeO_2)0.6$	Kumar 2008b	3.47×10^{-3} at 450 K				
0.4 (V_2O_5)–$0.3(CoO)$– $0.3(TeO_2)0.3$		6.99×10^{-4} at 450 K				
$(Ag_2O)x$ $(Bi_2O_3)30$ $(B_2O_3)60 - x$ $(TeO_2)10$	Ali and Shabann 2008					
$X = 0$				2.71×10^{-11}	1.973	0.922
$X = 20$				7.94×10^{-11}	7.698	0.980
$(100 - x)TeO_2$–$xCuO$	Sandhya Rani and Singh 2010					
$X = 10$		3.14×10^{-14}	0.72			

TABLE 6.3 (Continued)
Experimental Data of Electrical Conductivity

Glass	Reference	DC Conductivity at Room Temperature (Ω–1 cm–1)	DC Activation Energy (eV)	AC Conductivity (Ω^{-1} cm^{-1})	AC Activation Energy (eV)	S&β
X = 50		7.38×10^{-10}	0.61			
60B$_2$O$_3$–10TeO$_2$–5TiO$_2$–25Li$_2$O	Suresh et al. 2010				0.99	
60B$_2$O$_3$–10TeO$_2$–5TiO$_2$–25Na$_2$O					1.16	
60B$_2$O$_3$–10TeO$_2$–5TiO$_2$–25K$_2$O,					1.11	

over a wide temperature range (70–425 K). The mechanism of DC conductivity changes from predominantly electronic to ionic within the $30 \leq$ mol% Ag$_2$O ≤ 40 range. The temperature dependence of electronic conductivity has been analyzed quantitatively to determine the applicability of various models of conduction in amorphous semiconducting glasses. At high temperature, $T > \theta_D/2$ (where θ_D is the Debye temperature) the electronic DC conductivity is due to nonadiabatic small-polaron hopping of electrons for $0.1 \leq x \leq 0.5$. The density of states at Fermi level is estimated to be $N(E_F) \approx 10^{19} – 10^{20}$ eV^{-1}cm^{-3}. The carrier density was of the order of 10^{19}cm^{-3}, with mobility $\approx 2.3 \times 10^{-7} – 8.6 \times 10^{-9}$ cm^{-2}V^{-1}s^{-1} at 300 K. The electronic DC conductivity within the whole range of temperature is best described in terms of the Triberis–Friedman percolation model. For $0.6 \leq x \leq 0.8$, the predominantly ionic DC conductivity was described well by the A-S model. All of the experimental conductivity data are in Table 6.3, while the theoretical conductivity parameters like the Debye temperature, θ_D, optical phonon frequency, $\nu_o = k\,\theta_D/h$, polaron hopping energy, W_H, the polaron band width, J, density of states at Fermi level, $N(E_F)$, polaron coupling constant, $\gamma_p = 2W_H/h\nu_o$, mobility, μ, carrier density, N_c, according to different conductivity models are in Table 6.4. Also, in 2010, the same authors identified the origin of electrical relaxation in tellurite glasses of same tellurite glass systems as explained in Chapter 7.

In 2005, Sungping Szu and Fu-Shyang Chang did an impedance study of V$_2$O$_5$–TeO$_2$–BaO glasses of different compositions as in Table 6.3. All the results of the impedance plots in this study were semicircle only, which indicates that the conduction was mainly electronic. The conductivity of sample 50–30–30 is two orders of magnitude smaller than that of sample 40–40–20, although they have the same amount of V$_2$O$_5$ contents. It is known that barium ions act as glass modifiers. It will create nonbridging oxygen (NBO) and break the glass network. In 1997, Jayasinghe et al. have suggested the existence of two kinds of independent migrating paths in sodium vanadium–tellurate glasses. The authors mentioned that one kind of path consists of an electronic transfer polaron, and the other kind of path, which is made by the regular position of NBO along the network former chains, allows ion displacement. When adding sodium ions, the electronic paths are progressively blocked, causing a decrease of electronic conductivity. The blocking of electronic paths by barium ions causes lower conductivity for sample 50–30–30. The activation energy E_r corresponds to an activation energy for a long-distance charge transport. The activation energy E_s corresponds to an activation energy for a short-distance charge transport. In this study, E_r is almost the same as E_s. The same value of E_r and E_s indicates that the DC conduction and its relaxation process have the same thermal-activated hopping processes as reported by Dutta and Ghosh (2005), and also by Bhattacharya and Ghosh (2003). Sungping Szu and Fu-Shyang (2005) explained that the decrease of conductivity for glass having more barium ions was due to the blocking of the polaron movement by NBOs created by barium ions.

TABLE 6.4

Theoretical Data of Electrical Conductivity According to Polaronic Hopping and Other Models

Glass	Reference	θ_D K	v_o s⁻¹	W_H eV	J eV	$N(E_F)$ eV⁻¹cm⁻³	γ_p	μ cm²V¹s⁻¹ at 300 K	N_c cm⁻³ at 300 K
$0.5[xAg_2O–(1-x)$ $V_2O5]–0.5[TeO_2]$	Moawad et al. 2009								
45 V_2O_5		300	6.26×10^{12}	0.093	0.012	1.0×10^{20}	10.8	2.3×10^{-7}	9.2×10^{19}
Schnakenberg's model			9.1×10^{12}	0.37					
Greaves' variable-range hopping						8.2×10^{22}			
$0.4\ V_2O_5–x\ Li_2$ $–(0.6-x)\ TeO_2$	Kumar 2008a								
X = 0.1		757.6	1.578×10^{13}	0.250	0.0160	4.00×10^{21}	7.65		
X = 0.5		809.7	1.686×10^{13}	0.493	0.0162	2.39×10^{21}	14.14		
$(V_2O_5)x\ (TeO_2)1–x$ X = 0.4	Kumar 2008b	704	1.47×10^{13}	0.238	0.608	6.31×10^{21}	7.85		
$0.4\ (V_2O_5)–0.3$ $(CoO)–0.3$ (TeO_2)		750	1.56×10^{13}	0.270	0.132	15.1×10^{21}	8.36		

In 2005, Balaya and Goyal examined the non-Debye conductivity relaxation in a mixed glass former system. The electrical conductivity and conductivity relaxation of different network glasses $(Na_2O)0.2–(GeO_2)0.8$, $(Na_2O)0.2–(TeO_2)0.8$, and $(Na_2O)0.2–(TeO_2)0.4–(GeO_2)0.4$ have been investigated as a function of frequency and temperature. Conductivity relaxation in these glasses is found to be non-Debye as the value of stretched exponent is smaller than 1. The microscopic activation energy E_a is determined using the coupling model. The activation energy $E\sigma$ for DC conduction consists of two contributions; the correlation between energy barrier E_{2a} in which ion resides, and the energy required to overcome the coupling between the ions and β expressed the degree of coupling/correlation between microscopic processes. The values of E_a for $(Na_2O)0.2–(TeO_2)0.8$, $(Na_2O)0.2–(TeO_2)0.4–(GeO_2)0.4$, and $(Na_2O)0.2–(GeO_2)0.8$ glasses were 0.85, 0.65, and 0.45 eV, respectively. The authors concluded that the change in glass structure results in a change in the potential depth in which ions reside and this, in turn, affects the correlation among mobile ions and their cooperative motion and the relaxation behavior.

Montani and Frechero (2006) measured the electrical properties of mixed-ion-polaron conducting vanadium–tellurite glasses of the form $xLi_2O\ (1-x)[0.5V_2O_5–0.5MoO_3]2TeO_2$. The obtained results confirm the existence of a transition from a typically electronic (polaronic) conductive regime when the molar fraction (x) of Li_2O is equal to 0, to an ionic conductive regime when X tends to 1. This transition is characterized by a deep minimum in the electrical conductivity of about three orders of magnitude. The correlated behavior between conductivity and the mean distance between lithium ions and between vanadium ions reinforces the key idea of two independent migrating paths for both electrons and ions, respectively. Shaaban et al (2006) studied the AC conductivity of zinc-tellurite glasses of composition $80TeO_2–20\ ZnO$ (mol%) doped with Ho^{3+} ions (0.01, 0.05, and 0.1 g of Ho ions for 10 g of blank glass) in the temperature range 298–478 K and in the frequency range 0.16–100 kHz. The AC conductivity was found to increase with increasing frequency and to follow the power low $\sigma_{ac}(\omega) = A\omega^s$. The frequency exponent s was found to decrease with increasing temperature. The AC conductivity was also found to increase with increasing temperature. The correlation barrier hopping (CBH) model was found to apply to the AC conductivity data as in Table 6.3.

In 2006, Nascimento and Watanabe presented and discussed the ionic conductivity s data in the binary alkali–tellurite system, including 47 glasses that extend the ionic conductivity range by more than 14 orders of magnitude in a wide compositional range. A "universal" behavior is obtained, using log σ or log σT vs. $E_A/k_B T$, where E_A is the activation enthalpy for conduction, k_B is the Boltzmann constant, and T is the absolute temperature. This finding further indicates the importance of a scaling factor F, recently proposed, which is correlated to the free volume of glass composition. For a given value of $E_A/k_B T$, the difference between large and small values of s is only one order of magnitude in 87% of these glass systems. The influence of alkali content and temperature was minor on the pre-exponential terms, considering both expressions log10s and log10sT. Indeed, the pre-exponential term s0 varies around an average value of 50 $\Omega^{-1}cm^{-1}$ considering different compositions in this system. The fact that s lies on these single "universal" curves for so many ion-conducting binary tellurite glasses means that s is governed mainly by E_A. The composition dependence of the activation enthalpy is explained in the context of the A-S theory. Krins et al. (2006) investigated the structural and electrical properties of $xLi_2O–(1 − x)[0.3V_2O_5–0.7\ TeO_2]$ with $0 \leq x \leq 0.4$. At low Li_2O content, the electronic conductivity was predominant (0.18 eV $\leq E_A \leq$ 0.28 eV), whereas at high Li_2O content the Li^+ transport contribution becomes the most important one (0.35 eV $\geq E_A \geq$ 0.32 eV). The relationship between the structural characteristics of the $xLi_2O–(1 − x)(0.3V_2O_5–0.7TeO_2)$ system and its electrical behavior was studied by Raman spectroscopy, static ^{51}V NMR, and impedance spectroscopy measurements. The Li-free sample ($x = 0$) presents mainly electronic conductivity resulting from electron hopping between the aliovalent sites of V (V^{4+} to V^{5+}). When increasing x, the coordination of vanadium progressively changes from 5 to 4. The creation of NBOs leads to the opening of the glassy network and results in an increase of the distance between V hopping sites.

Kumar and Sankarappa (2008a) investigated the DC electrical conductivity of lithium–vanadium–tellurite glasses in the composition $(V_2O_5)0.4–(Li_2O)x–(TeO_2)0.6$ where $0.10 \leq x \leq 0.50$, in the temperature range 300 to 525 K. The density increased with increasing Li_2O up to 40 mol% and then decreased for further doping of Li_2O. Mixed-electronic-ionic conduction was observed in this set of glasses. The changeover of conduction mechanism from predominantly electronic to ionic was observed to occur at $x = 0.4$. The high temperature ($T > \theta_D/2$) conductivity data obeyed Mott's small-polaron hopping model (SPH) and the low temperature ($T > \theta_D/2$) conductivity has been analyzed in the light of Mott's and Greaves' variable-range hopping (VRH) conduction models. Various physical parameters associated with theoretical models are estimated. Both the experimental values of the electrical conductivity and the electrical parameters are in Table 6.3 and Table 6.4. Kumar et al. (2008b) continued their work in measuring the electrical properties of tellurite glasses of two sets of tellurium-based glasses doped with vanadium and vanadium–cobalt oxides in the forms:

1. $(V_2O_5)0.4–(Li_2O)x–(TeO_2)0.6$; and $x = 0.1, 0.2, 0.3$ and 0.4 mol%,
2. $(V_2O_5)0.4–(CoO)x–(TeO_2)0.6–x$; and $x = 0.3, 0.4$ and 0.5 mol%.

The authors measured the DC electrical conductivity in the temperature range 300–500 K. The obtained experimental and theoretical data are given in Tables 6.3 and 6.4. Ali and Sabann (2008) measured the AC electrical conductivity in $(Ag_2O)x\ (Bi_2O_3)30–(B_2O_3)60–x\ (TeO_2)10$ glass ($x = 0$, 2, 4, 5, 10, 15 and 20) at different temperatures and frequencies. The results obtained indicated that glasses containing silver 0.5 mol% have values that are approximately equal to AC electrical conductivity. A slight decrease was observed with increasing Ag_2O concentration up to 4 mol%. However, the AC electrical conductivity values increase with increasing silver content (i.e. ≥ 5 mol%). The AC electrical conductivity values increased with increasing frequency and follow the power law, $\sigma_{ac}(\omega) = A\omega^s$. The frequency exponent s was found to be dependent on frequency and temperature. The s values tended to increase to unity as the temperature decreased. Such results suggest that the CBH model is appropriate for explaining the AC electrical conductivity in these glasses. A pronounced increase in the dielectric loss values was observed with increasing silver content. These reflect the effect of Ag^+ ions charge carriers on the electrical conductivity of such glasses as in Table 6.3.

In 2009, Kubliha et al. studied the influence of crucibles (Au or Pt) on the structure, electrical, dielectric, and optical properties of $70TeO_2$–$30PbCl_2$ glasses doped with Pr^{3+} added as a metal, chloride, or oxide, in concentrations of 500–1500 wt-ppm. The DC conductivity of "pure" glasses prepared in Au crucibles is two orders of magnitude higher than that of those prepared in Pt crucibles. Upon doping, the DC conductivity of glasses prepared in Pt and Au crucibles increases or decreases, respectively. In 2010, Sandhya Rani and Singh measured the DC electrical conductivity of tellurite glasses of composition $(100 - x)TeO_2$–$xCuO$ ($x = 10, 20, 30, 40, 50$ mol%). The value of the electrical conductivity increased from 3.14×10^{-14} to 7.38×10^{-10} Ω^{-1} cm^{-1}, and the electrical activation energy decreased from 0.72 to 0.61 eV due to the increase in the CuO from 10 mol% to 50 mol%. The electrical conduction in these glasses is due to a polaron hopping mechanism in the adiabatic regime. The ESR spectra of the $x = 10$ glass consists of a broad symmetrical line with no hyperfine splitting and is characteristic of Cu^{2+} clusters. The line width of the ESR signal increases with an increase in CuO content. The signal intensity decreases drastically with an increase in CuO content. No ESR could be observed for the glass samples with $x = 40$ and 50. The loss of the ESR signal may be due to antiferromagnetic coupling between the Cu^{2+} clusters.

Suresh et al. (2010) carried out AC conductivity and impedance measurements for an alkali boro-tellurite glass system with the composition $60B_2O_3$–$10TeO_2$–$5TiO_2$–$25R_2O$ (where R = Li, Na, and K). The impedance plots ($Z''[\omega]$ vs. $Z'[\omega]$) for all glass samples were recorded and found to exhibit good, single, and well-shaped semicircles over the studied temperature range. The frequency dependencies of the imaginary part of impedance $Z''(\omega)$ and the imaginary part of modulus (M'') for all glass samples at different temperatures were also investigated. The conductivity isotherms showed a transition from an independent DC region to a dispersive region where the conductivity continuously increased with increasing frequency. For the low-frequency region, the conductivity values (σ_{ac}) at RT are relatively low in the orders of $10-8$ and $10-9$ (Ohm-cm)$^{-1}$. However, there is an increase in conductivity with temperature for all glass samples showing conductivities at 300°C (σ_{300}) in the ranges $10-6$ and $10-5.5$ (Ohm-cm)$^{-1}$. In the high-frequency region, conductivity was enhanced by one order of magnitude ($10-5$ [Ohm-cm]$^{-1}$) as temperature is increased. The power law parameter (s) was determined for all the glass samples at various temperatures, and lowest "s" values are found to be for lithium-containing glasses. It is found that AC conductivity is more for lithium followed by potassium and sodium, but this amount of increase between sodium and potassium is negiotable. The lowest "s" values are found to be for lithium-containing glasses. Usually in alkali-borate glasses, the ionic conductivity increases in the order $\sigma_{Li} < \sigma_{Na} < \sigma_K$ but this order of conductivity is found to be different in the present glass series BTTR (borate rich) where, the conductivity of lithium is followed by potassium and sodium, though the amount of increase between sodium and potassium is negotiable. The activation energy calculated from conductivity measurements (from Z''), E_τ (eV) were, 0.99, 1.16, and 1.11 eV for the three glass systems $60B_2O_3$–$10TeO_2$–$5TiO_2$–$25Li_2O$, $60B_2O_3$–$10TeO_2$–$5TiO_2$–$25Na_2O$, and $60B_2O_3$–$10TeO_2$–$5TiO_2$–$25K_2O$, respectively. The authors understood that the conductivity of lithium was due to its high mobility. Variation of conductivity in these glasses might be attributed to the alkali effect, and the highest conductivity of lithium content glasses might form a highly disordered matrix with more opened channels for the easy migration of Li^+ ions. An excellent time–temperature superposition in the imaginary part of the dielectric modulus (M'') confirms that the dynamical process is temperature independent and was also indicative of a common ion transport mechanism as shown in Chapter 7.

A better understanding of the physical and chemical phenomena involved in tellurite glasses was attained through the comparative study of the electron paramagnetic resonance (EPR). Prohaska et al. (1995) described the existence of three paramagnetic centers in $5Na_2O–25ZnO–85TeO_2$ mol% glass induced by excimer laser light, and further progress on the subject was published. In 2010, Giehl et al. observed intrinsic paramagnetic responses in the $60TeO_2–25ZnO–15Na_2O$ and $85TeO_2–15Na_2O$ mol% glasses, after γ-irradiation at room temperature:

1. A shoulder at $g_1 = g_\parallel = 2.02 \pm 0.01$ and an estimated $g_\perp \sim 2.0$ attributed to tellurium–oxygen hole center (TeOHC)
2. A narrow resonance at $g_2 = 1.9960 \pm 0.0005$ related to the modifiers
3. A resolved resonance at $g_3 = 1.9700 \pm 0.0005$ ascribed to a tellurium electron center (TeEC) of an electron trapped at an oxygen vacancy (V^+_O) in a tellurium oxide structural center. It is suggested that the creation of ($NBO–V^+_O$) pair follows a mechanism where the modifier oxide molecule actuates as a catalyst. An additional model for the NBO radiolysis produced by the γ-irradiation is proposed on the basis of the evolution of the g_1, g_2, and g_3 intensities with increasing dose.

7 Dielectric Properties of Tellurite Glasses

In Chapter 7, the electric properties of tellurite glasses are explained, and the dielectric constant (ε) and loss factor data are summarized for both oxide and nonoxide–tellurite glasses. These values vary inversely with frequency (f) and directly with temperature (T). The rates of change of ε with f and T, complex dielectric constants, and polarizability depend on the types and percentages of modifiers that are present in tellurite glasses. Data on the electric modulus and relaxation behavior of tellurite glasses are also reviewed according to their stretching exponents. The pressure dependence of the ε is also examined. Quantitative analysis of the ε is discussed in terms of the number of polarizable atoms per unit volume, and data on the polarization of these atoms are related to the electrical properties of tellurite glasses. Recent data of the dielectric and magnetic properties of tellurite glasses and tellurite ceramics are collected.

7.1 INTRODUCTION

Dielectric properties are of special interest when glasses or ceramics are used either as capacitative elements in electronic applications or as insulation. The dielectric constant (ε), dielectric loss factor (ε″), and dielectric strength usually determine the suitability of a particular material for such applications. Variations in dielectric properties with frequency (f), field strength (E), and other circuit variables influence performance. Environmental effects such as temperature (T), humidity, and radiation levels also influence dielectric applications. Glasses and ceramics have definite advantages over plastics—their major competitors—as materials for capacitative elements and electronic insulation.

When a material is inserted between two parallel conducting plates, capacitance (C) is increased. The dielectric constant (ε) is defined as the ratio of the C of a condenser or capacitor to the C of two plates with a vacuum between them (C_o) or the ratio of the ε between the plates to ε_o (ε with a vacuum between the plates): $\{\varepsilon = C/C_o = [(\varepsilon A/d)/(\varepsilon_o A/d)] = (\varepsilon/\varepsilon_o)\}$, where ε is the permittivity of the dielectric material and ε_o is the permittivity of the vacuum, measured both in square Coloumbs per square meter and Farads per meter. When an electrical field is established in a dielectric substance, electricity is stored within the material, and, on removal of the field, this energy might be wholly recoverable, but is usually only partially recoverable. Energy is lost in the form of heat. In an alternating field, therefore, the rate of power loss depends on the effectiveness of insulating materials. The ratio between irrecoverable and recoverable parts of this electrical energy is expressed as the tangent of the power factor (tan δ). The dielectric losses are also dependent on the value of ε, because they depend on the product of ε and tan δ. Both ε and tan δ should not increase markedly with increases in T. Nonlinear ε values characterize an important class of crystalline ceramics that exhibit very large ε values (>1000). The large ε values accompany the spontaneous alignment or polarization (α) of electric dipoles.

There are four primary mechanisms of polarization (α) in glasses and ceramics. Each mechanism involves a short-range motion of charge and contributes to the total α of the material. The α mechanisms include electronic α, atomic α, dipole α, and interfacial α; for example, electronic α is the shift of the valence electron cloud of ions within a material with respect to the positive nucleus. The dielectric behavior of amorphous solids at low T values differs completely from that of crystalline solids. For example, acoustic absorption and dielectric absorption are strongly enhanced in amorphous solids compared with those values of crystals, and a large absorption peak is found near the T of liquid nitrogen in many glasses. More than 50 years ago, Stanworth (1952) examined the ε of binary 77 mol% TeO_2–23 mol% WO_3 glass. Since then, numerous review articles and excellent chapters in solid-state and materials science books have been published on the electrical properties of glasses, many of which contribute to the following description.

7.2   EXPERIMENTAL MEASUREMENT OF DIELECTRIC CONSTANTS

Experimental techniques for measuring low-f dielectric behavior in glasses can be separated into two groups: (1) those using a C bridge, and (2) those using charging–discharging current as a function of time. Both types of measurements can be performed at various T and f levels and, as expected, usually give identical results. Particularly in the C bridge methods, as in the previous chapter, the problems of electrodes and the choice of appropriate electrodes should be considered.

7.2.1   Capacitance Bridge Methods

In the f range 10 Hz–100 kHz, C bridge measurements are most frequently used to obtain the ε and dielectric loss (or AC conductivity). These methods are based on the balance of the resistive and capacitive components of a sample compared with the known variables in standard components (Figure 7.1). Electrodes have finite areas, which, when evaporated on both opposite polished surfaces of a sample, provide well-defined electrode sites and facilitate the connection of wire leads to the sample. Other wires are attached to the electrodes themselves with a paste (of the same material). The other ends of these wires are passed through holes in silica rods and connected to the "bridge." Figure 7.2 illustrates a standard four-terminal method on a Hewlett–Packard 4192A LF impedance analyzer, which can be used for high-T measurements of the permittivity of a solid material, as mentioned by Thorp et al. (1986); however, for low-T values, the heater should be replaced by a cryogenic instrument for cooling. Thin, parallel-faced glass samples are prepared. The ε values are obtained using the equation:

$$\varepsilon = \left(C_m - C_s\right)\left(\frac{d}{A\varepsilon_o}\right) \tag{7.1}$$

where d, A, C_m, and C_s are the sample thickness, the sample area, the measured C, and the C of the sample holder, respectively. The Hewlett–Packard bridge method calculates the value of C_s for all f values used and outputs the quantity $C_m - C_s$ automatically, thus continually correcting for lead C.

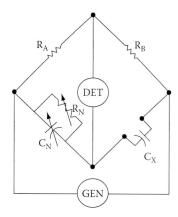

FIGURE 7.1   Capacitive and resistance bridges.

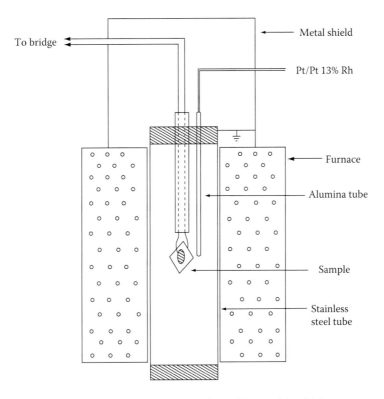

FIGURE 7.2 Apparatus for permittivity measurement of a solid material at high temperatures as designed by Thorp, J., Rad, N., Evans, D., and Williams, C., *J. Mater. Sci.*, 21, 3091, 1986. (From J. Thorp, N. Rad, D. Evans, and C. Williams, *J. Mater. Sci.*, 21, 3091, 1986. With permission.)

T values between room and low T are attained in a conventional glass double-dewier system. The final stage of the electrical connections to sample electrodes includes a short run of coaxial cable, with the measurement method consequently reverting to a two-terminal technique. A gold–iron/chrome thermocouple mounted next to the sample ensures that measurements of the sample T are highly accurate (± 1 K). The sample dimensions are measured with a digital micrometer.

7.2.2 ESTABLISHING THE EQUIVALENT CIRCUIT

When admittance data are plotted as an f dispersion at low-T values, they fall on straight lines inclined at an angle of $n\pi/2$ to the real admittance axis. Unequal scales have been used to establish more easily the inclination of these profiles and the very small zero f (y') intercept. When plotted on equal scales, these profiles are nearly vertical. Although the f range of the Hewlett–Packard impedance analyzer is 5 Hz–13 MHz, the impedance of the sample must be within the impedance bandwidth of the instrument, which limits the f range to 1–500 kHz. The profiles produced are independent of the oscillation and bias voltages within the ranges 5 mV–1 V and −35–+35 V, respectively. It is crucial to verify the absence of boundary effects. Inversion of admittance data into the complex impedance plane results in curved arcs, which conveniently enable the small DC conductance values, which are not easily identified from admittance plots alone, to be evaluated. The forms found in these two representations suggest that the electrical properties of the glass can be represented by an equivalent circuit as shown in the 1st Ed., which consists of a conductance component and a "universal" C component connected in parallel. The admittance and impedance profiles produced by such an equivalent circuit are shown schematically in Figure 7.3. Such schematic profiles are used to analyze "admittance/impedance" data to extract the values of the equivalent circuit components.

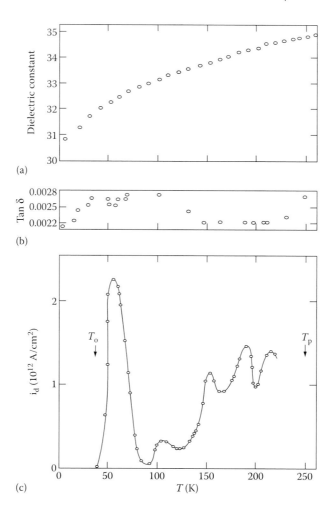

FIGURE 7.3   (a–c) The dielectric constant (a) and loss factor (b) of 77 mol% TeO_2–23 mol% WO_3 glass in the temperature range 4.2–300 K. (From R. Braunstein, I. Lefkowitz, and J. Snare, *J. Solid State Commun.*, 28, 843, 1978.)

7.2.3   LOW-FREQUENCY DIELECTRIC CONSTANTS

An admittance representation is extremely useful in ascertaining the resistive components of a circuit but not necessarily the reactive components; it is normal practice to use an alternative representation of the data (i.e., an "impedance profile") to extract the values of these latter components. This introduces the problem of how to extract the C values of the circuit (the f at the vertex of the profile cannot be found). However, due to the simple nature of the equivalent circuit, the C can be assessed from an analysis of the imaginary part of the complex admittance ($y'' = j\omega C$).

7.2.4   MEASUREMENT OF DIELECTRIC CONSTANTS UNDER HYDROSTATIC
PRESSURE AND DIFFERENT TEMPERATURES

Both hydrostatic pressure and T effects have been studied by Hampton et al. (1989), who used the Bridgman opposed-anvil apparatus for high-pressure measurements. With this apparatus, the sample is sandwiched between two epoxy resin disks in the center of a washer manufactured from MgO-loaded epoxy resin. To measure the sample T, a copper–constantan thermocouple is included between the epoxy disks, as shown in the 1st Ed. The cell is inserted between the Bridgman anvils. Above a critical load, the epoxy resin is able to exert hydrostatic pressure on the sample. The pressure within the cell is determined from the load applied to the anvils, using a prior calibration of the system from the resistance discontinuities exhibited at pressure-induced phase transitions in bismuth, as mentioned by Bundy (1958), Yomo et al. (1972), and Decker et al. (1975). Attainable pressures are in the range 25 to 70 kbar (1 kbar is ~0.1 GPa).

The T variation within the cell is achieved by direct control of the anvil T. For high-T values, an electrical heating element around the top anvil is used. The electrical power to the heater is controlled by a T controller with the copper–constantan thermocouple within the pressure cell as a sensor. Using this system, the sample T can be maintained at a level that remains constant within a 10°C range. Low Ts are attained by placing a copper cooling collar around the sample area and passing liquid nitrogen through the collar onto the anvil surfaces. The rate of cooling is controlled by adjustment of the overpressure (exerted from a cylinder of gaseous nitrogen) inside the nitrogen storage dewier. It is not necessary to correct for the effect of the applied hydrostatic pressure on the electronmotive force of the thermocouple; the copper–constantan thermocouple has a very small pressure coefficient.

7.3 DIELECTRIC CONSTANT MODELS

Knigry et al. (1976) state in Introduction to Ceramics that the ε of a single crystal or glass sample results from electronic, ionic, and dipole orientations and space-charge contributions to α, as follows:

1. Electronic α is a shift in the center of gravity of the negative electron cloud in relation to that of the positive atom nucleus in an electric field. This kind of α is common in all materials.
2. Ionic α arises from the displacement of ions of the opposite sign from their regular lattice sites under the influence of an applied electric field and also from the deformation of the electronic shells resulting from the relative displacement of these ions.
3. Dipole α (or ion jump α) arises in glasses when there are multiple sites available to a modifier ion that cannot contribute to observed DC conductivity. It is noteworthy that the jump fs for ion motion in glasses at room T are slow even when compared with low-f-dielectric-property measurements. As a result, the static $\varepsilon(\varepsilon_s)$ may be considerably larger than that measured for f values as low as 100 Hz.

The effectiveness of charge carriers in giving an increased ε for single crystals and glasses depends critically on the electrode materials, α effects at the electrodes, and the resulting space charges, which cause space-charge α at low f. When a DC voltage is applied on two parallel conducting plates separated by a narrow gap or vacuum, the charge instantly builds up on the plate (the ε of a vacuum [ε_o] = 8.8554 × 10^{-12} F/m). However, when a dielectric material such as glass is placed between the plates, an additional charge builds up on the plate and is time-dependent and controlled by the rate of α of the glass. The electric flux associated with total charge on the surface of the plate is called the "electric displacement" (D) and is numerically equal to the total charge density on the plate. The D is related to the DC electric field by $D = \varepsilon_s E$. When a low-f periodic field $E = E_o \exp(i\omega t)$, where ω is the angular f and is applied across the dielectric material, the charge on the plate varies in a periodic way, but the produced charge on the plates lags behind the applied field. D can be expressed by the following set of calculations:

$$D = D_o \text{exo}\left(i\omega t - \delta\right)$$

$$= \left(D_o \cos\delta \cos\omega t\right) + \left(D_o \sin\delta \sin\omega t\right)$$

$$= \left\{\left[\left(D_o \cos\delta\right)\left(E_o \cos\omega t\right)\right]/E_o\right\} + \left\{\left[\left(D_o \sin\delta\right)\left(E_o \sin\omega t\right)\right]/E_o\right\}$$

$$= \left(\varepsilon' - i\varepsilon''\right)E_o \exp\left(i\omega t\right)$$

$$= \varepsilon^* E_o \exp\left(i\omega t\right) \tag{7.2}$$

The phase δ is expressed by:

$$\tan\delta = \varepsilon''/\varepsilon' \tag{7.3}$$

In general, D and E are vector quantities but they are scalars only in isotropic materials like glass. ε^* is the complex dielectric constant, and ε' and ε'' are the real and imaginary parts of ε, respectively.

7.3.1 Dielectric Losses in Glass

Dielectric energy losses result from the following processes (Mackenzie 1974):

1. Conduction losses
2. Dipole relaxation losses
3. Deformation losses
4. Vibration losses

7.3.1.1 Conduction Losses

Conduction losses occur at low f and are attributed to the migration of alkali ions over large distances. They are obviously related to DC conductivity. If a condenser containing a glass dielectric component is represented by a simple circuit containing a resistance (R) in parallel with a capacity (C), then in an AC field of angular f (ω),

$$\tan\delta = \frac{1}{\omega RC} \tag{7.4}$$

Thus, tan δ is inversely proportional to f. Except at very low f or at high T, conduction losses are small. At f levels in excess of 100 Hz, this type of loss is generally negligible.

7.3.1.2 Dipole Relaxation Losses

Dipole relaxation losses are attributed to the motion of ions over short distances. If a spectrum of energy barriers exists in a random glassy network, then an ion might be able to move over relatively short distances and be stopped at a high-energy barrier. In the AC field, most ions can oscillate only between two high-energy barriers separated by a short distance. It is, of course, assumed that these barriers exist between the two limiting ones. In general, ion jump relaxation between two equiv-alention positions is responsible for the largest part of the dielectric loss factor (ε'') for glasses at moderate f. If the relaxation time for an atom jump is τ, then the maximum energy loss occurs for f values equal to the jump f (i.e., when $\omega = 1/\tau$). When the applied alternating field f is much smaller than the jump f, atoms follow the field and the energy loss is small.

7.3.1.3 Deformation and Vibrational Losses

Similarly, if the applied f is much larger than the jump f, the atoms do not have an opportunity to jump at all, and losses are small. The dielectric loss is equivalent to an "σ_{AC} conductivity" given by:

$$\sigma(AC) = \omega\varepsilon'' \tag{7.5}$$

In fused silica, for instance, deformation losses can be attributed to the lateral oscillation of an oxygen ion between two silicons. On the other hand, vibrational losses are attributed to the vibration of the ions in a glass about their equilibrium positions at some f determined by their mass and the potential wells they are in. These two sources of loss are not important for values of f below 10^{10} Hz.

7.3.2 Dielectric Relaxation Phenomena

Insertion of a dielectric material between the plates of a condenser reduces the voltage between the two plates by an amount equal to the ε_s. The charge surface density (σ) is also reduced by an amount:

$$p = \left(\frac{\varepsilon_s - 1}{\varepsilon_s}\right)\sigma \tag{7.6}$$

where p is called the polarization of the dielectric. A displacement field (D) is introduced in terms of the original charges on the condenser where:

$$D = 4\pi Q \tag{7.7}$$

The electric field strength is shown as:

$$E = 4\pi\sigma/\varepsilon_s \tag{7.8}$$

Therefore,

$$D = \varepsilon_s E \tag{7.9}$$

From Equations 7.6, 7.7, and 7.9:

$$D = E + 4\pi P \tag{7.10}$$

In an alternating field, there is generally a phase difference between D and E. Using the complex notation:

$$E^* = E_o e^{j\omega\tau}$$

and

$$D^* = D_o e^{-j(\omega\tau - \delta)} \tag{7.11}$$

where δ represents the difference in phase, and ω is the angular f of the applied field.
From Equations 7.9 through 7.11:

$$D^* = \varepsilon^* E^* \tag{7.12}$$

$$\varepsilon^* = \varepsilon_s^* e^{-j\delta}$$

hence,

$$= \varepsilon_s \left(\cos\delta - j\sin\delta\right) \tag{7.13}$$

ε^* is in the complex form:

$$\varepsilon^* = \varepsilon' - j\varepsilon'' \tag{7.14}$$

Comparing Equations 7.13 and 7.14 gives:

$$\varepsilon' = \varepsilon_s \cos\delta \text{ and } \varepsilon'' = \varepsilon_s \sin\delta$$

and

$$\tan \delta = \frac{\varepsilon''}{\varepsilon'} \qquad (7.15)$$

where ε'' is the dielectric loss factor and $\tan \delta$ is the loss tangent or power factor. However, in time-dependent fields, generally the displacement of charge is related to the polarization of the dielectric medium. The P' does not reach its static value immediately so that its value can be given by:

$$P^* = \left[\left(\frac{\varepsilon_\infty - 1}{4\pi} + \frac{1}{4\pi}\right)\left(\frac{\varepsilon_s - \varepsilon_\infty}{1 + j\omega\tau}\right)\right]E_o e^{j\omega\tau} \qquad (7.16)$$

where τ is a constant having the dimension of time and is a measure of the time lag τ, which is called the τ and ε_∞ is the instantaneous ε. From the definition of electric displacement we have:

$$D^* = E^* + 4\pi p^*$$

$$= \left[\varepsilon_\infty + \frac{\varepsilon_s + \varepsilon_\infty}{1 + j\omega\tau}\right]E_o e^{j\omega\tau} \qquad (7.17)$$

and hence the ε^*:

$$\varepsilon' = \varepsilon_\infty + \frac{\varepsilon_s - \varepsilon_\infty}{1 + \omega^2\tau^2} \qquad (7.18)$$

$$\varepsilon'' = \frac{(\varepsilon_s - \varepsilon_\infty)\omega\tau}{1 + \omega^2\tau^2} \qquad (7.19)$$

$$\tan \delta = \frac{\varepsilon''}{\varepsilon'} = \frac{(\varepsilon_s - \varepsilon)\omega\tau}{(\varepsilon_s + \varepsilon)\omega^2\tau^2} \qquad (7.20)$$

It is obvious from Equations 7.18, 7.19, and 7.20, which are commonly called the Debye equations, that ε' and ε'' are f dependent. It has been found (McDowell 1929) that $\tan \delta$ can be represented by a simple empirical equation: $\tan \delta = BF^{-n}$, where B and n are constants. Also, Macedo et al. (1972) have provided the following set of equations for the electric modulus [$M^*(\omega)$]:

$$M^*(\omega) = \frac{1}{\varepsilon^*(\omega)}$$

$$= M'(\omega) + iM''(\omega)$$

$$= M\left\{1 - \int_0^\infty e^{-i\omega t}\left[d\phi t/dt\right]dt\right\}$$

and

$$\phi(t) = \exp\left[-(t/\tau_m)^\beta\right], 0 < \beta < 1 \qquad (7.21)$$

where $\phi(t)$ is the function that gives the time evolution of the electric field within the dielectrics, τ_m is the most probable τ, and β is the stretching exponent parameter.

7.3.3 Theory of Polarization and Relaxation Process

Consider a dielectric for which the total polarization P_s (in Debye units = 10^{-24} cm³) in a static field is determined by three contributions:

$$P_s = P_e + P_a + P_d \tag{7.22}$$

Subscripts e, a, and d refer, respectively, to the electronic, atomic, and dipolar P.

In general, when such a substance is suddenly exposed to an external static field, a certain length of time is required for P to build to its final value. It can be assumed that values of P_e and P_a are attained instantaneously. The time required for P_a to reach its static value can vary from days to 10^{-12} s, depending on T, chemical constitution of the material, and its physical state. The phenomenological description of transient effects is based on the assumption that a τ can be defined. Consider the case of an alternating field. Let P_{ds} denote the saturation value of P_d obtained after a static field E is applied for a long time. Then the field is switched on, which is given by:

$$P_d(t) = P_{ds}\left(1 - e^{-t/\tau}\right) \tag{7.23}$$

hence

$$\left(dP_d/dt\right) = \left(\frac{1}{\tau}\right)\left[P_{ds}(t) - P_d\right] \tag{7.24}$$

Accounting for the decay that occurs after the field is switched off leads to a well-known proportionality with $e^{-t/\tau}$. For an alternating field $E = E_0 e^{i\omega t}$, Equation 7.24 can be used if P_{ds} is replaced by a function of time $P_{ds}(t)$ representing the saturation value that is obtained in a static field equal to the instantaneous value $E(t)$. Hence, for alternating fields, we can use the differential equation:

$$\left(\frac{dP_d}{dt}\right) = \left(\frac{1}{\tau}\right)\left[P_{ds}(t) - P_d\right] \tag{7.25}$$

To then express the real and imaginary parts of the ε in terms of the $f(\omega)$ and τ, we first define the "instantaneous" ε (ε_{ea}) by:

$$P_e + P_a = \frac{\varepsilon_{ea} - 1}{4\pi} E \tag{7.26}$$

We can then write:

$$P_{ds} = P_s - (P_e + P_a)$$

$$= \frac{\varepsilon_s - \varepsilon_{ea}}{4\pi} E \tag{7.27}$$

Substitution of P_{ds} yields:

$$\frac{dP_d}{dt} = \frac{1}{\tau}\left(\frac{\varepsilon_s - \varepsilon_{ea}}{4\pi} E_0 e^{i\omega t} - P_d\right) \tag{7.28}$$

Solving this equation, we obtain:

$$P_d(t) = Ce^{-t/\tau} + \frac{1}{4\pi} \frac{\varepsilon_s - \varepsilon_{ea}}{1 + i\omega\tau} E_o e^{i\omega t} \tag{7.29}$$

The first term is a transient in which we are not interested here. The total P is also a function of time, and is given by $P_e + P_a + P_d(t)$.

7.3.4 Dielectric Dependence on Temperature and Composition

To analyze the composition dependence of ε, it is important to study the polarization factor α. Two sources are identified from the macroscopic Clausius–Mossotti equation:

$$\frac{\varepsilon - 1}{\varepsilon + 2} = \frac{4\pi\alpha}{3}\left(\frac{N}{V}\right) \tag{7.30}$$

The number of polarizable (α) atoms per unit volume (N/V) and their polarizability α are factors that explain, for instance, how vanadium lowers the ε of tellurite glass.

Quantitative analysis of the temperature dependence of dielectric constant is based on an equation by Bosman and Havinga (1963) for an isotropic material at constant pressure:

$$\left(\frac{1}{(\varepsilon-1)(\varepsilon+2)}\right)\left(\frac{\partial\varepsilon}{\partial T}\right) = \left[\frac{-1}{3V}\left(\frac{\partial V}{\partial T}\right)\right] + \left[\frac{1}{3\alpha}\left(\frac{\partial\alpha}{\partial V_T}\right)\left(\frac{\partial V}{\partial V_p}\right)\right] + \left[\frac{1}{3\alpha}\left(\frac{\partial\alpha}{\partial T_v}\right)\right]$$

$$= A + B + C \tag{7.31}$$

where V is volume. The three constants (A, B, and C) have the following significance. Factor A represents the decrease in number of α particles per unit V as T increases, which has a direct effect on V expansion. Factor B results from an increase in the α of a constant number of particles as their available V increases. Factor C reflects the change in α due to T changes at constant V. It may be noted that A is inversely related to ε (but both B and C are directly related to ε). The sum A + B is always positive, and hence it contributes to an increase of ε with increasing T, as stated by Hampton et al. (1988). Furthermore, it contributes to an increase of ε with increasing T.

7.3.5 Dielectric Constant Dependence on Pressure Models

Hampton et al. (1989), using the method of Gibbs and Hill (1963), proved that:

$$\left(\frac{\partial \ln \varepsilon}{\partial P}\right) = \frac{(\varepsilon+2)(\varepsilon-1)}{3\varepsilon}\left[\frac{\partial \ln \alpha}{\partial P} - \frac{\partial \ln V}{\partial P}\right] \tag{7.32}$$

whereas Bosman and Havinga (1963) gave:

$$\frac{1}{(\varepsilon-1)(\varepsilon+2)}\left(\frac{\partial\varepsilon}{\partial P}\right) = (A+B)\left[\left(\frac{\partial V/\partial P}{\partial V/\partial T}\right)\right] \tag{7.33}$$

In general, Equations 7.32 and 7.33 have similar forms. The difference between the two approaches is purely the manner in which terms are collected after differentiation. However, this is not as straightforward as it might first appear, because each analysis places a different emphasis on the pertinent terms and hence on the effects of cumulative errors. To describe the T dependence of ε, a similar approach was used in the Bosman and Havinga (1963) equations.

A useful quantity that can be extracted from both analyses is $\partial \ln \alpha / \partial V$:

$$\left(\frac{\partial \ln \alpha}{\partial V} \right) = \left(-\frac{B}{A} \right)$$

$$= \left[\frac{\left(\dfrac{\partial \ln \alpha}{\partial P} \right)}{\left(\dfrac{\partial \ln V}{\partial P} \right)} \right] \tag{7.34}$$

which enables the assessment of the overall agreement between the two different methods of analysis of the same data.

7.4 DIELECTRIC CONSTANT DATA OF OXIDE–TELLURITE GLASSES

7.4.1 DEPENDENCE OF DIELECTRIC CONSTANT ON FREQUENCY, TEMPERATURE, AND COMPOSITION

Stanworth (1952) measured the ε at room T with $f = 50$ cycles/s, using 3 mm thick disks. The forms of tellurite glass that were used and the values of both density (ρ) and ε values were:

1. 80.4 mol% TeO_2–13.5 mol% PbO–6.1 mol% BaO ($\rho = 5.93$, $\varepsilon = 29.0$)
2. 88.2 mol% TeO_2–9.8 mol% PbO–6.1 mol% Li_2O ($\rho = 5.56$, $\varepsilon = 27.0$)
3. 42.5 mol% TeO_2–29.6 mol% PbO–27.9 mol% B_2O_3 ($\rho = 4.41$, $\varepsilon = 13.5$)
4. 44.1 mol% TeO_2–42.5 mol% PbO–13.4P_2O_5 ($\rho = 5.99$, $\varepsilon = 29.0$)
5. 63.5 mol% TeO_2–22.2 mol% PbO–14.3MoO_3 ($\rho = 5.93$, $\varepsilon = 29.5$)
6. 41.1 mol% TeO_2–19.1 mol% PbO–39.8 mol% WO_3 ($\rho = 6.76$, $\varepsilon = 32.0$)
7. 66.3 mol% TeO_2–23.2 mol% PbO–10.5 mol% ZnF_2 ($\rho = 6.08$, $\varepsilon = 24.5$)
8. 85.9 mol% TeO_2–12.0 mol% PbO–2.1 mol% MgO ($\rho = 5.77$, $\varepsilon = 27.0$)
9. 73.25 mol% TeO_2–22.7 mol% PbO–4.05 mol% TiO_2 ($\rho = 6.03$, $\varepsilon = 30.5$)
10. 72.3 mol% TeO_2–22.4 mol% PbO–5.3 mol% WO_3 ($\rho = 6.07$, $\varepsilon = 26.0$)

Stanworth (1952) explained his data as follows:

- Tellurite glasses containing a high percentage of B_2O_3 have not only a lower ρ than the other tellurite glasses, but a correspondingly lower ε.
- The high ε of tellurite glasses occurs when tellurite glasses are modified with WO_3.
- No significant change in ε occurs over the f range 50 Hz/s–1.2 MHz/s; the loss angle is small (0.0035–0.0024°) and approximately constant over the whole range of f studied.

Ulrich (1964) measured the ε of 75 mol% TeO_2–25 mol% Bi_2O_3 and 90 mol% TeO_2–10 mol% Bi_2O_3 glasses at T levels of –196°C and 27°C. For the first glass, ε was 23 at –196°C and 25.8 at 27°C, whereas for the second glass, ε was 21 at –196°C and 26.1 at 27°C; that is, Ulrich concluded that the ε of binary bismuth–tellurite glasses decreases with T.

Braunstein et al. (1978) measured the C and thermally stimulated depolarization current on 77 mol% TeO_2–23 mol% WO_3 glass in the T range 4.2–300 K. Their results (Figure 7.3) showed that ε increases from about 31 to 35 with increased T, whereas the thermally stimulated depolarization current has a polarization $T(T_p)$ of 250 K, polarization time t_p of 8.5 min, linear heating rate of 3.4 K/min, polarization field E_p of 7.1 kV/cm, and initial T (T_o) of 40 K. The data indicate that the local WO_6^- octahedral determines the dielectric properties of this glass, and that dipole–dipole correlations contribute to the ferroelectric-like character of this amorphous system.

A detailed investigation of $(100 - x)$ mol% TeO_2–x mol% V_2O_5 glasses, where $x = 10$, 20, 30, 40, 50, 60, 70, and 80 has been done by Mansingh and Dahwan (1983), as shown in Figure 7.4. An Au–Ge (alloy) electrode was put on the samples by vacuum evaporation, and by using a bridge in the f range 0.1–100 kHz at different T levels in the range 77–400 K, both AC conductivity and ε were measured within 2% accuracy in different runs. The $\varepsilon(\varepsilon')$ at four fixed f values as a function of T is reported in Figure 7.4 for the binary glass system 80 mol% TeO_2–20 mol% V_2O_5. A nonlinear variation of ε with concentration is observed as in Figure 7.4. The dielectric constant ε decreases with increasing f and shows little dependence on T, with a sharp increase at 100 Hz and ~120 K. Mansingh and Dahwan (1983) attributed this increase to dielectric relaxation, and they ignored it. They also measured the variation of ε' for different compositions of binary vanadium–tellurite glasses with logarithm f. After that they calculated the P_s present in these glasses from the values of ε' (at 100 kHz). The P was found to increase with the percentage of TeO_2. The abnormally high ε (30) of the low concentration of V_2O_5 glasses was attributed to the characteristic role played by the

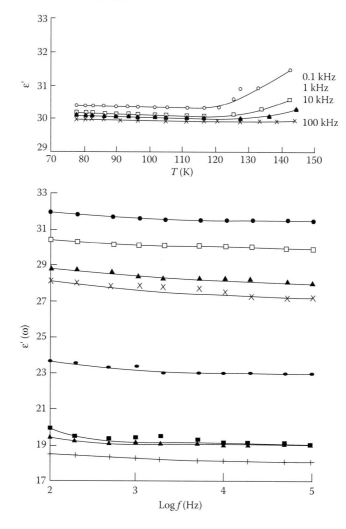

FIGURE 7.4   The dielectric constant of $(100 - x)$ mol% TeO_2–x mol% V_2O_5 glasses for $x = 10$, 20, 30, 40, 50, 60, 70, and 80. (From A. Mansingh and V. Dahwan, *J. Phys. C Solid State Phys.*, 16, 1675, 1983.)

glass former TeO_2. However, it was difficult to estimate the absolute value of the characteristic τ of the system or to arrive at a definite conclusion about the conduction mechanism with the Butcher and Morys (1973) model, owing to the experimental uncertainties in the measured values of $\varepsilon'(\omega)$ and the lack of knowledge of ε_∞.

The dielectric data of pure tellurite glass and binary 67 mol% TeO_2–33 mol% WO_3 and 80 mol% TeO_2–20 mol% $ZnCl_2$ were measured in 1988 by Hampton et al. at Ts of 93, 179, and 292 K, and f values of 1–500 kHz as in Figure 7.5.

7.4.1.1 Establishing the Equivalent Circuit

Schematic profiles have been used to analyze the admittance/impedance data to extract the values of the equivalent circuit components, as shown in Chapter 6. It is clear that at high Ts the admittance profiles according to the relation $y'' = I\omega C$ show deviations from linearity, which can be attributed to a perturbation of the equivalent circuit, and the numbers beside experimental points in the figure correspond to the measurement fs in kilohertz. This perturbation takes the form of additional C, which is due to the dielectric response at high fs in circuits connected in parallel

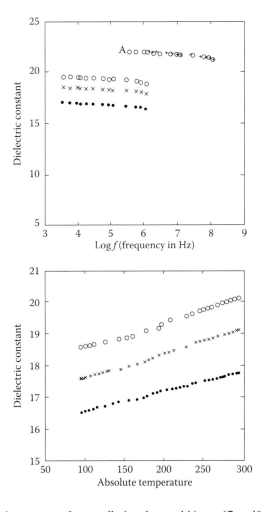

FIGURE 7.5 The dielectric constant of pure tellurite glass and binary 67 mol% TeO_2–33 mol% WO_3 and 80 mol% TeO_2–20 mol% $ZnCl_2$ glasses as measured by Hampton et al. (1988) at temperatures of 93, 179, and 292 K, and frequencies from 1 to 500 kHz. (Reproduced from R. Hampton, W. Hong, G. Saunders, and R. El-Mallawany, *Phys. Chem. Glasses*, 29, 100, 1988. With permission of the Society of Glass Technology.)

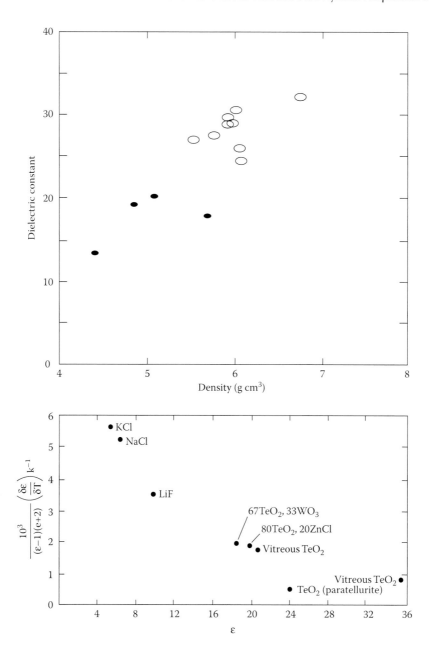

FIGURE 7.5   (Continued)

with the existing equivalent circuit. Estimates of the magnitudes of refractive indices (Table 7.1) required to produce the observed deviations from linearity are consistent with refractive indices of other tellurite glasses.

7.4.1.2   Low-Frequency Dielectric Constants

The data show an almost linear portion at low fs for each type of glass; $d\varepsilon/df$ for vitreous TeO_2 is -6×10^{-7} Hz^{-1} (Table 7.1). However, at higher fs ($f > 500$ kHz), data are affected by the final, two-terminal stages of the electrical connections. Figure 7.5 represents the low-f (hertz) dependence of

TABLE 7.1

Dielectric Constant, Frequency, Temperature, Derivatives, and Polarizability of Tellurite Glasses and Comparison with Other Crystalline Solids and Plastics

Glass or Material	ε_s (~300 K)	$d\varepsilon/df$ (Hz^{-1})	$d\varepsilon/dT$ (K^{-1})	$(d\varepsilon/dT)(\varepsilon-1)$ $(\varepsilon+2)$ (K^{-1})	Static Polarizability $(\times 10^{-6})$
		Glasses			
TeO$_2$–23 mol% W$_3$O$_2$	35.0		12×10^{-3}	9.5×10^{-6}	
TeO$_2$–24 mol% BaO	27.5				
TeO$_2$–ZnF$_2$	24.5				
TeO$_2$–B$_3$O$_3$	13.5				
TeO$_2$–18 PbO	27.5				
TeO$_2$–50 PbO	25.0				
TeO$_2$	20.1	-6.2×10^{-7}	8.2×10^{-3}	1.9×10^{-5}	2.2
TeO$_2$–33 mol% WO$_3$	17.8	-5.7×10^{-7}	6.3×10^{-3}	2.1×10^{-5}	
TeO$_2$–20 mol% ZnCl$_2$	19.2	-5.2×10^{-7}	7.9×10^{-3}	2.0×10^{-5}	
TeO$_2$ crystal	23.4		3.9×10^{-3}	0.7×10^{-5}	2.8
LiF	9.3			3.7×10^{-5}	
KCl	4.7			5.8×10^{-5}	
NaCl	5.6			5.4×10^{-5}	
BaTiO$_3$	>1000.0			-0.6×10^{-5}	
SiO$_2$	3.8				
95 mol% P$_2$O$_5$–5 mol% Sm$_2$O$_3$	5.4				
SiO$_2$	3.8				
		Crystalline Solids			
NaCl	5.6				
MgO	9.8				
Diamond	5.6				
		Plastics			
Epoxy resin	3.6				
Polyethylene	2.3				
Polystyrene	2.6				
Ice I	99.0				
Ice V	193.0				

Source: From R. Hampton, W. Hong, G. Saunders, and R. El-Mallawany, *Phys. Chem. Glasses*, 29, 100, 1988.

the ε (at 292 K) for pure TeO$_2$, 67 mol% TeO$_2$–33 mol% WO$_3$ and 80 mol% TeO$_2$–20 mol% ZnCl$_2$ glasses, which is consistent with the universal capacitor description of the equivalent circuit (used to account for the inclination of the admittance profiles).

The f-dependent $C(\omega)$ has the form

$$C(\omega) = B\left[\sin\frac{(n\pi)}{2} - j\cos\frac{(n\pi)}{2}\right]\omega^{n-1} \qquad (7.35)$$

Thus, on the basis of this description, the gradient of a plot of log ε against log ω gives the value $n-1$, which can be compared with the value n calculated from the gradient of the admittance profiles. In vitreous TeO$_2$ at 292 K, the level of agreement between the values derived by the two routes is reasonable; $n - 1 = -6 \times 10^{-4}$ for log ε against log ω, and $n - 1 = -2 \times 10^{-4}$ for y' against y''. The T dependence of the admittance profile inclination $[d(n - 1)/dT]$ is in fact small and positive (8.5×10^{-6} K^{-1}), indicating that the "universal" C becomes more ideal (i.e., more f independent) as the measured T is decreased and the f-dependent processes are frozen out. Thus, the slight f dependence of the ε that has been observed supports the introduction of a universal C.

The ε values of the tellurite glasses measured by Hampton et al. (1988) were somewhat lower than those reported by other workers for binary and ternary tellurite glasses, but they were consistent with other results as shown in Table 7.1. A large ε of ~35, measured for 77 mol% TeO$_2$–23 mol% WO$_3$ glass, was considered by Braunstein et al. (1978) to be evidence for a possible ferroelectric-like character of the material, resulting from dipole–dipole correlations between WO$_6$ octahedra. However, they were unable to ascertain definitively whether this large value was due to TeO$_2$ units, WO$_3$ units, or a complex interplay between both. It was suggested that if the correlation distances for ferroelectric or antiferroelectric clustering occur, the contribution of WO$_3$ should be considered the prime factor. However, the polarizability of the TeO$_2$ matrix also contributes to the observed ε. No attempt was made by Braunstein et al. (1978) to examine the separate contributions of TeO$_2$ and WO$_3$ to the ε by separating these components, because they did not think it possible to form a glass from pure TeO$_2$ or pure WO$_3$. However, the later measurements of ε by Hampton et al. (1988) went some way toward resolving the earlier suggestions of Braunstein et al. The ε of vitreous TeO$_2$ was actually greater than that of the 80 mol% TeO$_2$–20 mol% ZnCL$_2$ glass. It is instructive at this stage to compare the ρ and ε values of vitreous TeO$_2$ and those of the crystalline modification paratellurite (Table 7.1); the ratios of these parameters are essentially identical. Hence, the fact that the ε of the glass is smaller than that of the crystal correlates extremely well with the reduced number of TeO$_2$ units that can be accommodated in the more open structure of the vitreous state. Hampton et al. (1988) compared the static α values derived from the macroscopic Clansins–Mossotti equation (Equation 7.30) for the crystalline and vitreous states and found them in very good agreement as given below:

- Paratellurite: $\rho = 5.99$, $\varepsilon = 23.44$, static $\alpha = 2.8 \times 10^{-6}$
- Vitreous TeO$_2$: $\rho = 5.10$, $\varepsilon = 20.1$, static $\alpha = 2.2 \times 10^{-6}$
- Ratios: $(\rho_{para-tell.}/\rho_{glass}) = 1.18$ $(\varepsilon_{para-tell.}/\varepsilon_{glass}) = 1.17$ $(\alpha_{para-tell.}/\alpha_{glass}) = 1.27$

This agreement between the two material states indicates the important role of the TeO$_2$ unit in determining the ε of binary tellurite glasses. The ε_s values of selected tellurite glasses as a function of ρ, with previous data for binary and ternary tellurite glasses, are collected in Figure 7.7C. Stanworth (1952) examined the static constants of a number of tellurite glasses as a function of ρ; his results are included in Figure 7.5C, together with data for the binary glasses 77 mol% TeO$_2$–23 mol% WO$_3$, 67 mol% TeO$_2$–33 mol% WO$_3$, and 80 mol% TeO$_2$–20 mol% ZnCl$_2$, and the parent tellurite glass. A number of glasses that have large ε values do not have modifiers that exhibit a ferroelectric character in the crystalline state.

The inclined straight lines of the complex admittance plots clearly identify the resistive element of the glasses, although it is not so clear that plots of $C^* = y^*/j\omega$ will also produce straight-line plots, which give nonzero intercepts with the real C axis. Figure 7.5B identifies an f dependence of the ε (calculated from the C_m), which does not show a flattening at low f values. The low-f plateau region can be used to obtain the value of the ε_s, but in the example shown, this approach is inappropriate. However, a C^* profile does make it possible to identify the low-f C. A similar approach, illustrated in Figure 7.5, is to use the results from complex profile analysis to determine ε_s values by extrapolation of the data to low fs (extrapolation to 10 Hz is usually sufficient for agreement with C^* profiles). The zero-$f\varepsilon_s$ values of these materials are significantly lower than those reported previously (Table 7.1). Figure 7.7 shows the T dependence of the ε_s values for each glass, measured below room T, and values of $d\varepsilon/dT$ are given in Table 7.1.

There are a number of possible reasons for the finding that TeO_2–WO_3 glass has a smaller ε than that measured previously by Braunstein et al. (1978). One explanation for the large ε of the ternary glasses is that there is significant segregation of the component phases, which alters the environment of the components. Another explanation is that if electrode effects change the C_m as found by Branstein et al. (1978), such effects could show minimal dispersion. The critical test for electrode effects, namely a dependence of C on voltage, has been exhaustively applied in recent work, and the observed independence of voltage has established that only minimal electrode effects occur in the f range of these studies. The change in slope of the ε with T was cited by Braunstein et al. (1978) as further evidence for a dominant contribution from the WO_6 component in the glass. However, the T dependence of the ε values of TeO_2 and the binary glasses in Figure 7.7 are all similar, so this property cannot be intrinsically caused by dipole–dipole correlations of WO_6 octahedra in TeO–WO_3 glass. The T dependence of the ε of solid insulators has been shown to depend on three factors, as stated by Jonsher (1983), which are related to the polarization and thermal expansion of the material as stated by Equation 7.31.

The significance of each term has been adequately discussed before, so only a brief description need be given here. The term

$$A = \left[\frac{-1}{3V} \left(\frac{\partial V}{\partial T} \right) \right]$$

corresponds to the decrease in the number density of the polarizable particles,

$$B = \left[\frac{1}{3\alpha} \left(\frac{\partial \alpha}{\partial V_T} \right) \left(\frac{\partial V}{\partial T_p} \right) \right]$$

is the increase in the α of a constant number of particles, and C is the change in α with temperature

$$C = \left[\frac{1}{3\alpha} \left(\frac{\partial \alpha}{\partial T_V} \right) \right]$$

The experimental values of

$$\frac{1}{(\varepsilon - 1)(\varepsilon + 2)} \left(\frac{\partial \varepsilon}{\partial T} \right)$$

obtained for the TeO_2 glasses are given in Table 7.2, in comparison with data for selected cubic materials as mentioned by Havinga (1961).

The normal trend is for the T dependence of ε to decrease with increasing values of ε as stated by Bosman and Havinga (1963); $\partial \varepsilon / \partial T$ becomes negative for materials having ε greater than ~20. The positions of tellurite glasses and crystalline paratellurite in this scheme are illustrated in Figure 7.5D. The tellurite glasses have rather more positive values of $\partial \varepsilon / \partial T$ than the values found by Bosman and Havinga (1963) for several other oxide materials having the same magnitude of ε. The comparatively small but positive T dependence of the ε for TeO_2 glasses measured as shown in Figure 7.7 are consistent with the pattern $\varepsilon < 20$ and $(\partial \varepsilon / \partial T) > 0$. However, the ε obtained by Braunstein et al. (1978) for 77 mol% TeO_2–23 mol% WO_3 was 35, so that a negative value of $\partial \varepsilon / \partial T$ might have been expected, but they measured a positive T coefficient.

TABLE 7.2

Pressure and Temperature Dependencies of Dielectric Constants of Tellurite Glasses Compared with Other Glasses and Materials

Analysis, Parameter (Units)	TeO₂	TeO₂–33 mol% WO₃	P₂O₅–5 mol% Sm₂O₃
$(\partial \ln \varepsilon / \partial P)$ (10^{11} Pa^{-1})	4.40	1.70	3.50
Havinga Analysis (1969)			
$(\partial \varepsilon / \partial T)//[(\varepsilon - 1)(\varepsilon + 2)]$ (10^{-5} K^{-1})	1.75	1.95	5.08
$(\partial \varepsilon / \partial P)/[(\varepsilon - 1)(\varepsilon + 2)]$ (10^{12} Pa^{-1})	2.85	2.75	5.87
$(\partial \ln V / \partial T)$ (10^{-5} K^{-1})	4.65	3.75	1.00
$(\partial \ln V / \partial P)$ (10^{-11} Pa^{-1})	−3.08	−2.59	−2.60
A + B (10^{-6} K^{-1})	−4.30	−3.98	−2.26
A (10^{-5} K^{-1})	−1.55	−1.23	−0.33
B (10^{-5} K^{-1})	1.12	0.8	−0.10
C (10^{-5} K^{-1})	2.18	2.35	5.30
$(\partial \ln \alpha / \partial V)$ (no unit)	0.73	−0.68	0.32
Gibbs Analysis (1963)			
$(\partial \ln \varepsilon / \partial P)$ (10^{11} Pa^{-1})	4.41	1.68	3.54
$[(\varepsilon + 2)(\varepsilon - 1)]/3\varepsilon$	7.00	6.23	2.01
$(\partial \ln V / \partial P)$ (10^{-11} Pa^{-1})	−3.08	−2.59	−2.60
$(\partial \ln \alpha / \partial P)$ (10^{-11} Pa^{-1})	−2.45	−2.32	−0.84
$(\partial \ln \alpha / \partial V)$	0.79	0.89	0.32
Polystyrene NaCl			
$(1/\varepsilon)(\partial \varepsilon / \partial P)$ (10^{11} Pa^{-1})	35.20	−9.80	
$\{[(\varepsilon + 2)(\varepsilon - 1)]/3\varepsilon\}(\partial \ln V / \partial P)$ (10^{11} Pa^{-1})	32.80	8.70	
$(1/\alpha)(\partial \ln \alpha / \partial P)$ (10^{11} Pa^{-1})	1.60	−8.90	
$(1/V)(\partial \ln V / \partial P)$ (10^{11} Pa^{-1})	−34.70	−4.20	
$(\partial \varepsilon / \partial T)$ at $P = 0$ kbar (K^{-1})	0.007	0.007	0.005
$(\partial \varepsilon / \partial T)$ at $P = 25$ kbar (K^{-1})	0.0140		0.026
$(\partial \varepsilon / \partial T)$ at $P = 30$ kbar (K^{-1})		0.067	
$(\partial \ln \varepsilon / \partial P)$ (10^{11} Pa^{-1})	4.400	1.700	3.500

Source: R. Hampton, I. Collier, H. Sidek, and G. Saunders, *J. Non-Cryst. Solids*, 94, 307, 1987.
For NaCl = −9.8, Diamond = −0.1, Polystyrene = 32.2, Ice I = 1.5, Ice V = 13.7.

Hampton et al. (1988) used complex admittance techniques to measure the ε_s of TeO₂ glass and the value was 20.1. In addition, the ε values of binary glasses were measured by the same method, giving $\varepsilon = 19.2$ for 80 mol% TeO₂–2O mol% ZnCl₂ and $\varepsilon = 17.8$ for 67 mol% TeO₂–33 mol% WO₃ glasses. Although large, these values for the binary glasses are rather smaller than those previously reported. The ε values and their T dependence for pure and binary glasses are similar to each other and to those of crystalline TeO₂, as are the molar static α values. The ε values of the glasses studied here seem to be derived mainly from the α of the TeO₂ unit, even at relatively high concentrations of the second component.

El-Mallawany (1994) discussed the dielectric together with the electrical properties of tellurite glasses. The N/V and the α of these atoms are factors in how tungsten lowers the electric activation energy of tellurite glass. The N/V in our material has been increased from 5.77 to 6.23×10^{22} cm^{-3} by modification with 33 mol% WO₃ using the Clausius–Mossoti equation. El-Mallawany (1994) obtained the value of α[TeO₂] (=112×10^{-24} cm^{-3}) and α[TeO₂–WO₃]

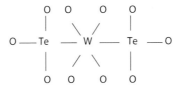

FIGURE 7.6 Two-dimensional representation of binary TeO$_2$–WO$_3$ glass. (Reprinted from *Mater. Chem. Phys.*, 37, R. El-Mallawany, 376, 1994, with permission from Elsevier.)

(=105 × 10^{24} cm^{-3}), respectively. This slight decrease in atomic polarization can be attributed to a hopping process that occurs in this glass system, but the high ε is a property of the TeO$_2$ glass matrix. The polarization of binary P$_2$O$_5$–WO$_3$ is higher than that of TeO$_2$–WO$_3$, which is due to the lower hopping process as discussed above. The decrease in the ε of TeO$_2$–WO$_3$ glass can be explained qualitatively if one assumes an increase in energies of Te–O and W–O bonds. Because the total α is the sum of electronic and ionic α values, the electronic α of oxygen diminishes the effects of its lower ionic α. In TeO$_2$–WO$_3$ glass, the relative displacements of oxygen are greater than those of either tungsten or tellurium ions (Figure 7.6). The total polarization of the glass is expected to be mainly due to the relative displacement of oxygen in W–O and Te–O bonds. The net polarization thus depends on the number of W–O and Te–O bonds per unit V and their strengths.

El-Mallawany (1994) correlated the conductivity and dielectric properties of these glasses with T as shown in Figure 7.7. Both types of properties decrease with decreasing T below room T. The values of $\partial \varepsilon / \partial T$ are 8.2 × 10^{-3} and 6.3 × 10^{-3} for TeO$_2$ and TeO$_2$–WO$_3$ glasses, respectively. To understand a fundamental property of a glass such as the ε, its behavior with T, and its structure, it is important to use the differentiated version of the Clausius–Mossotti equation, also called the

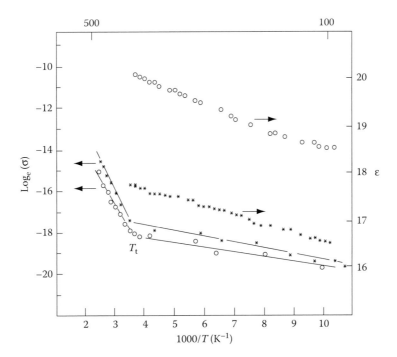

FIGURE 7.7 Variation of both conductivity and the dielectric properties of TeO$_2$–WO$_3$ glasses with temperature. (Reprinted from *Mater. Chem. Phys.*, 37, R. El-Mallawany, 376, 1994, with permission from Elsevier.)

Bosman–Havinga equation. Using the analysis of Bosman and Havinga (1963), the decrease in the $\partial\varepsilon/\partial T$ for the presence of WO_3 oxide in tellurite glass is due to the following factors:

1. The greater decrease in N/V in TeO_2 glass, which is a direct result of the greater thermal expansion of TeO_2 glass with lower N/V. The presence of WO_3 (octahedra) decreases the thermal expansion and increases N/V. As the T increases, the glass expands less than TeO_2 glass, and therefore A (TeO_2) > A (TeO_2–WO_3).
2. The greater increase in α with increase in available V for TeO_2 glass than for the TeO_2–WO_3 glass as the T is raised; that is, $\partial\alpha/\partial V_T$ decreases with increasing N/V; that is, B(TeO_2) > B(TeO_2–WO_3).
3. The apparently small change in α with T for a constant V corresponding to process C.

In 1996, El-Mallawany et al. measured the dielectric behavior of the $(10 - x)$ mol% TeO_2–x mol% MoO_3 glasses, for $x = 0.2$, 0.3, and 0.45, in the f range 0.1–10 kHz and T range 300–573 K, as shown in Figure 7.8. Both static- and high-f ε for these binary tellurite glasses decreased with the increase in MoO_3 content. The T dependence of the ε of these glasses was positive. The f dependence of the ε identifies an f dependence that does not show a flattening at low f. The room $T\varepsilon_s$ was discussed in terms of the MoO_3 concentration as shown in Figure 7.9. The T dependence of the ε has been analyzed in terms of the T changes of both V, α, and also V change of the α.

Kosuge et al. (1998) stated the f dependence of the ε at room T (ε_r) for x mol% K_2O–x mol% WO_3–$(100 - 2x)$ TeO_2 glasses as shown in the 1st Ed. The values decrease gradually with decreasing TeO_2 content, but the glass with $x = 25$ still had a large value of $\varepsilon_r = 19$, implying again that the polarization of WO_3 is large. In 1999, Shankar and Varma measured the ε of $(100 - x)$ mol% TeO_2–x mol% $LiNbO_3$ glasses with the precipitated $LiNbO_3$ microcrystals on the surface of 50 mol% TeO_2–50 mol% $LiNbO_3$ glass, by a single-step heat treatment as shown in the 1st Ed. A scanning electron micrograph of the surface-crystallized 50 mol% TeO_2–50 mol% $LiNbO_3$ glass sample is in the 1st Ed. The compositional dependencies of ρ, ε at 100 Hz, and f for glass samples with x values of 50, 30, and 10 were illustrated. Shankar and Varma (1999) measured the dependence of the pyroelectric coefficient of the surface-crystallized 50 mol% TeO_2–50 mol% $LiNbO_3$ glass sample and the variation of the static α of glass samples. The values of the pyroelectric coefficients of these kinds of tellurite glasses at T levels of 200–650 K range from 200 to 1,400 $\mu C/m^2$ K^{-1}. The pyroelectric coefficient of these binary tellurite glasses is nearly fourfold higher than that of roller-quenched $LiNbO_3$ (at 300 K) by Glass et al. (1977). Shankar and Varma (1999) concluded that tellurite glass containing 50% lithium niobate yields a surface layer of microcrystalline lithium niobate on heat treatment at 200°C for 12 h. Surface-crystallized glasses exhibit optical nonlinearity (as determined by χ^2-based analysis) and polar characteristics as seen in Part 4 of this book.

In 1999, Pan and Ghosh measured the electric moduli (M' and M'') for binary tellurite glasses of the form 75 mol% TeO_2–25 mol% Na_2O at Ts of 486, 472, 463, and 453 K (Figure 7.10). The most probable τ (τ_m) was calculated from the peak of M'' (ω) spectra. The values of the $\log_{10}\tau_m$ at 200°C were −5.04, −4.1, −3.2, 2.26, and 2.76 ± 0.01 for 30, 25, 20, 15, and 10 mol% Na_2O content, respectively. The values of τ_m decreased as Na_2O content increased.

7.4.2 DIELECTRIC CONSTANT DATA UNDER HYDROSTATIC PRESSURE AND DIFFERENT TEMPERATURES

Hampton et al. (1989) studied the effect of hydrostatic pressure on the ε and the T dependence of both tellurite and phosphate glasses. The dependence of the ε of tellurite glasses (pure TeO_2 and 67 mol% TeO_2–33 mol% WO_3) on applied hydrostatic pressure at selected Ts is shown in the 1st Ed. The pressure was in the range 0–70 kbar, and the T was in the range 77–380 K. The ε was determined from a low-f complex analysis of the glass disks. For these glasses, the effect of increasing

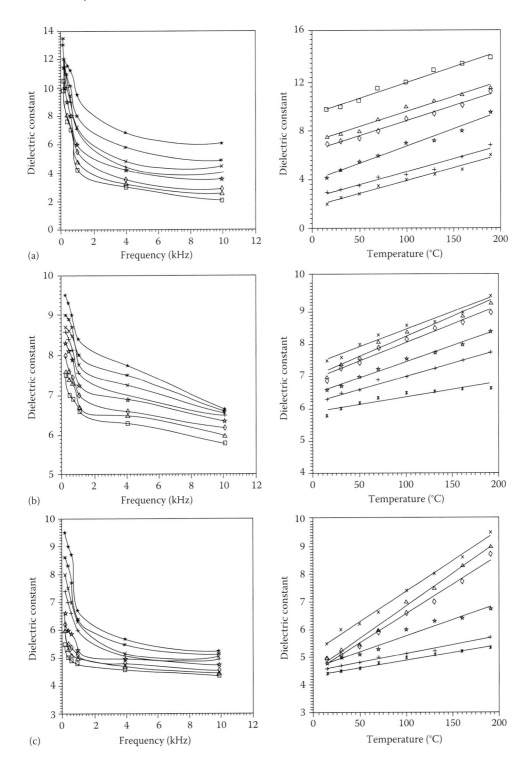

FIGURE 7.8 (a-c) The dielectric constant of the $(100 - x)$ mol% TeO_2–x mol% MoO_3 glasses, for $x = 0.2, 0.3$, and 0.45, in the frequency range 0.1–10 kHz and the temperature range 300–573 K. (From R. El-Mallawany, L. El-Deen, and M. El-Kholy, *J. Mater. Sci.*, 31, 6339, 1996. With permission.)

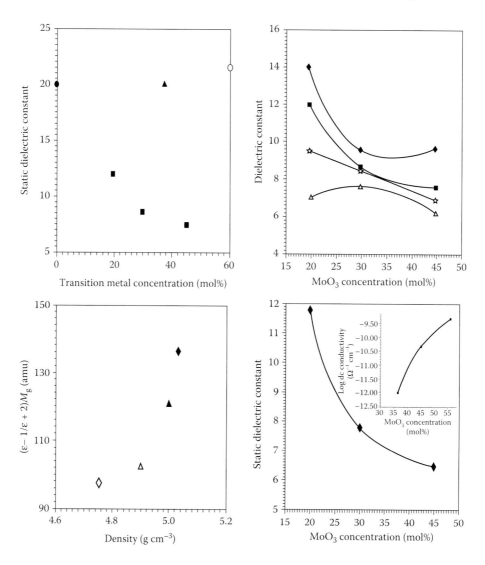

FIGURE 7.9   Room temperature, static dielectric constant as a function of MoO_3 concentration. (From R. El-Mallawany, L. El-Deen, and M. El-Kholy, *J. Mater. Sci.*, 31, 6339, 1996. With permission.)

either the pressure or T was to increase the ε. At pressures up to 80 kbar and Ts up to 370 K, the pressure coefficient of the ε was positive.

7.4.2.1   Combined Effects of Pressure and Temperature on the Dielectric Constant

The main effect of increasing T on the ε is little change in the pressure coefficient. Results obtained at elevated Ts show that the effect of pressure is to increase $\partial\varepsilon/\partial T$ as given in Table 7.2. The curve $[(\varepsilon-1)(\varepsilon+2)]^{-1}$ decreases with increasing electrical activation energy, and thus Hampton et al. (1989) concluded that the observed increase of $[(\varepsilon-1)(\varepsilon+2)]\,(\partial\varepsilon/\partial P)$ with T is a real effect caused by increases in $\partial\varepsilon/\partial P$.

The rather simplistic approach adopted here is from Hampton et al. (1989), who assumed that the effect of pressure on the α $(\partial\ln\alpha/\partial P)$ is much smaller than its effect on compressibility (i.e., $[\partial^2\ln\alpha/\partial P^2] \ll [\partial^2\ln V/\partial P^2]$). In other words, the prime effect of pressure is to macroscopically reorient the glass matrix rather than to alter the α on a microscopic level. In this approximation, the factor

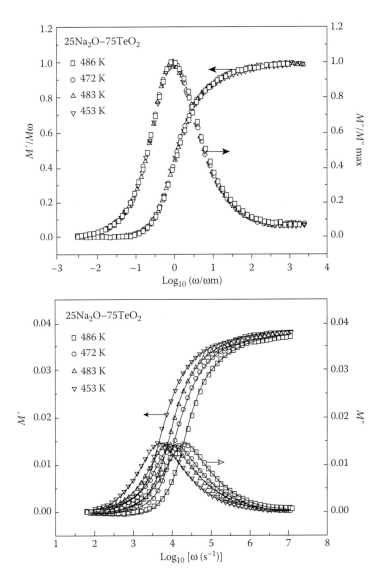

FIGURE 7.10 The electric moduli (M' and M'') for binary tellurite glasses of the form 75 mol% TeO_2–25 mol% Na_2O at temperatures of 486, 472, 463, and 453 K. (From A. Pan and A. Ghosh, *Phys. Rev. B*, 59, 899, 1999. With permission.)

B (at a pressure P) is inversely proportional to $[\ln V/P]_{(P)}$. A similar argument suggests that the C is independent of pressure.

The compressibility of these glasses at elevated pressures can be estimated using the Murgnahan equation, as stated by El-Mallawany and Saunders (1987), and Lambson et al. (1984), and as explained in detail in Chapter 2. The calculation of the high-pressure term (term B in Equation 7.31) is shown in Table 7.2 for the 67 mol% TeO_2–33 mol% WO_3 and in comparison with samarium–phosphate glasses. The data for the tellurite glass show that $\partial \varepsilon / \partial T$ is in fact increased by pressure due to its influence on the B term (i.e., the change in the α of a constant number of particles). However, the anomalous decrease of the bulk modulus with pressure found for the samarium–phosphate glasses as measured by Mierzejewski et al. (1988) causes the B term to decrease with pressure; but $\partial \varepsilon / \partial P$ does not decrease with pressure, and therefore, the effects of pressure on the other terms are sufficient to counteract this effect. Hampton et al. (1989) concluded the following:

1. The ε increases with applied pressure for all glasses studied (up to a T level of 370 K). The pressure dependence of the ε is controlled by the competing effects of compressibility and pressure on the α. For glasses, it appears that the compressibility term is the dominant factor, which contrasts with the situation for crystalline insulators in which the effect of pressure on the α is dominant and consequently $\partial\varepsilon/\partial P$ is negative. The primacy of compressibility in determining the effect of pressure on the ε appears to hold even for samarium–phosphate glasses, which have anomalous elastic behavior under pressure—as these glasses are compressed, they become easier to compress. This also seems to be the case for silica glass, and it would be useful to confirm this by measurements of the ε under greater pressures.

2. The positive-pressure coefficient of ε is a characteristic feature of glasses under pressure. However, there is strong evidence for universality in the increase of $\partial\ln\varepsilon/\partial P$ to a less negative value in the glassy state, compared with its crystalline polymorph. It then follows that for some glasses this increase would not be enough to force $\partial\varepsilon/\partial P$ to become positive.

3. It has been found that analyses of $\partial\varepsilon/\partial P$ using the techniques of Bosman and Havinga (1963), and usually applied to crystalline insulators, are appropriate and afford a useful tool in studies of the dielectric properties of glasses.

4. The effect of pressure is to increase $\partial\varepsilon/\partial T$. A simple analysis estimating the effect of pressure on the A, B, and C terms of the Bosman and Havinga analysis supports the observed increase in $\partial\varepsilon/\partial T$ for tellurite glasses but shows a decrease for samarium–phosphate glasses. This calculated decrease in $\partial\varepsilon/\partial P$ is perhaps not unexpected, considering the problems in determining the A term, which plays a large part in the analysis and anomalous elastic properties of these particular phosphate glasses.

In 2000, El-Mallawany studied the change in V with compressibility of pure and binary tellurite glasses according to the Mukherjee, Ghosh, and Basu models (1992), as discussed in Chapter 1.

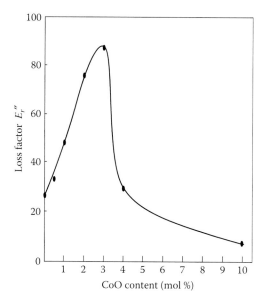

FIGURE 7.11 Dielectric properties of 65 mol% TeO_2–35 mol% CuO and 65 mol% TeO_2–$(35 - x)$ mol% CuO–x mol% CoO glasses between 293 and 458 K and at the general frequency of 10^{-6} Hz. (From S. Malik and C. Hogarth, *J. Mater. Sci.* 25, 3208, 1990. With permission.)

7.4.3 DIELECTRIC CONSTANT AND LOSS DATA IN TELLURITE GLASSES

Malik and Hogarth (1990) measured the C of 65 mol% TeO_2–35 mol% CuO and 65 mol% TeO_2–(35 − x) mol% CuO–x mol% CoO glasses at between 293 and 458 K in an f range up to 10^{-6} Hz. The solid glass was pressed between two electrodes and, by using the guard-ring method, was found to have a C equivalent to that with R in parallel; the relative permittivity ($\varepsilon_r = \varepsilon_r' - i\varepsilon_r''$, where ε_r' is the real part and ε_r'' is known as the dielectric loss factor) or ε was measured given that tan $\delta = \varepsilon_r''/\varepsilon_r'$. The C of these glasses shows an increase in value with the increase in T at low fs. The tan δ and ε_r'' are found to increase with the increase in T and decrease in f, but the usual Debye loss peaks are absent even at high Ts, as shown in Figure 7.11 for 65 mol% TeO_2–35 mol% CuO and Figure 7.11 for 65 mol% TeO_2–34 mol% CuO–1 mol% CoO glasses. The ε_r is unexpectedly high, ~10^3, and decreases very slightly with increasing f. Overall the effect of a small amount of CoO (1 mol%) has a dominant effect on the dielectric properties of copper–tellurite glasses, as shown in Chapter 6.

In 1995, Komatsu et al. studied the mixed-alkaline effect on dielectric properties of tellurite glasses of the form 80 mol% TeO_2–(20 − x) mol% Li_2O–x mol% Na_2O at room T and with an f of 1 kHz as shown in Figure 7.17. The values of the ε_r and tan δ at room T are given in Figure 7.12. Although

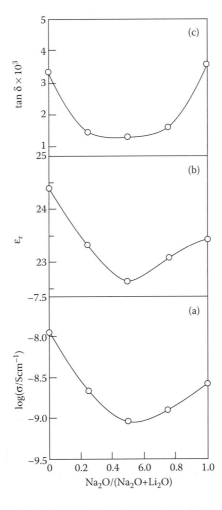

FIGURE 7.12 (a–c) The mixed-alkali effect on dielectric properties of tellurite glasses of the form 80 mol% TeO_2–(20 − x) mol% Li_2O–x mol% Na_2O at room temperature and a frequency of 1 kHz. (From T. Komatsu, R. Ike, R. Sato, and K. Matusita, *Phys. Chem. Glasses*, 36, 216, 1995. With permission of the Society of Glass Technology.)

the changes in these properties caused by mixing Li_2O and Na_2O are small, the negative deviations from additivity are clearly observed. The values of ε_r from 23 to 25 for TeO_2–Li_2O–Na_2O glasses are almost the same as those for other TeO_2-based glasses, as reported by Stanworth (1952) and Hampton et al. (1988), indicating that mixed alkali–tellurite glasses also have large ε values. The values of tan $\delta = 1.5 \times 10^{-3}$ were almost the same as those of other tellurite glasses. Komatsu et al. (1995) pointed out that the mixed alkali–tellurite glasses have the characteristic properties of tellurite glasses of high ε and small ε''. (Noted in passing, silicate glasses show a mixed-alkali effect, as stated by Isard [1969].)

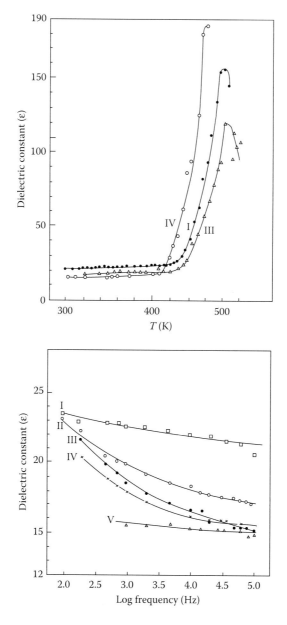

FIGURE 7.13 Dielectric properties of 80 mol% TeO_2–10 mol% Fe_2O_3–10 mol% PbO, 80 mol% TeO_2–10 mol% Fe_2O_3–10 mol% B_2O_3, 70 mol% TeO_2–10 mol% Fe_2O_3–10 mol% B_2O_3–10 mol% PbO, and 70 mol% TeO_2–10 mol% Fe_2O_3–10 mol% PbO–10 mol% SiO_2 glasses in the frequency range 10^2–10^5 Hz and the T range 300–500 K. (From M. El-Samadony, *J. Mater. Sci.*, 30, 3919, 1995. With permission.)

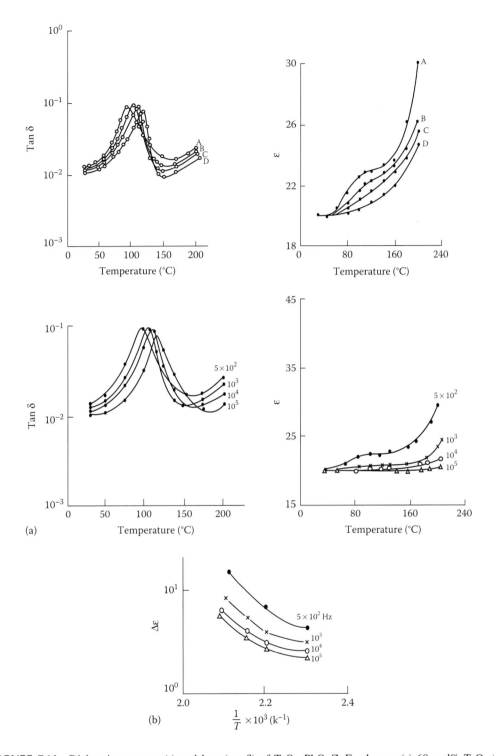

FIGURE 7.14 Dielectric constant (ε) and loss (tan δ) of TeO_2–PbO–ZnF_2 glasses: (a) 68 mol% TeO_2–20 mol% ZnF_2–12 mol% PbO; (b) 59 mol% TeO_2–30 mol% ZnF_2–11 mol% PbO; C, 50 mol% TeO_2–40 mol% ZnF_2–10 mol% PbO; D, 46 mol% TeO_2–45 mol% ZnF_2–9 mol% PbO. (From V. Kumar and N. Veeraiah, *J. Phys. Chem. Solids*, 59, 91, 1998.)

El-Samadony (1995) measured the electrical properties of TeO_2–Fe_2O_3 glasses doped with PbO, B_2O_3, and SiO_2 in the f range 10^2–10^5 Hz and the T range 300–500 K by using a lock-in amplifier. The contacts were of a silver paste. The compositions of the samples were 80 mol% TeO_2–10 mol% Fe_2O_3–10 mol% PbO, 80 mol% TeO_2–10 mol% Fe_2O_3–10 mol% B_2O_3, 70 mol% TeO_2–10 mol% Fe_2O_3–10 mol% B_2O_3–10 mol% PbO, and 70 mol% TeO_2–10 mol% Fe_2O_3–10 mol% PbO–10 mol% SiO_2. The ε for these tellurite glasses increases with increasing f, whereas the tan δ decreases with log f, with peaks as shown in Figure 7.13. In 1998, Kumar and Veeraiah measured the ε and tan δ of TeO_2–PbO–ZnF_2 glasses with successive decreases in PbO and TeO_2 concentrations, as shown in Figure 7.14. The f range was 5×10^2–10^5 Hz and the T range was 30°C–200°C. The variation of dielectric loss with T exhibits dielectric relaxation effects in these conditions. The rate of increase of ε and tan δ with T beyond the relaxation region increases with an increase in the PbO concentration, whereas the dielectric breakdown strength and activation energy for dipoles in the high-T region decrease with an increase in the concentration of PbO.

7.5 DIELECTRIC CONSTANT DATA OF NONOXIDE-TELLURITE GLASSES

In 1978, Panwar et al. measured the ε of $(90 - x)$ mol% Te–x mol% As_2–10 mol% Ge_2 glasses in the T range 77–383 K and an f range from DC to 5 MHz. The DC or AC measurements were made on pellets held between steel electrodes. Variations of ε with T at 0.5, 1.0, 2.6, and 5.0 MHz for the glasses 75 mol% Te–15 mol% As_2–10 mol% Ge_2 and 85 mol% Te–5 mol% As_2–10 mol% Ge_2 are shown in the 1st Ed. The dielectric constant ε values change very little with T at these f values up to about 253 K, which seems to be common behavior for these chalcogenide glasses. After a T of 253 K, the ε becomes activated and proportional to log (ω). Panwar et al. (1978) used the relation $\omega\varepsilon_o\varepsilon'' = \sigma(\omega) = \sigma_{tot} - \sigma_{DC}$ to measure the ε''. These data are consistent with a system of multiple τ values. The various parameters, including distribution parameter α_H, macroscopic τ (ε_s), and ε_∞, have been calculated from the Cole and Cole (1941) plot as follows: $\alpha_H = 0.597, 0.561, \tau = 1.91 \times 10^{-7}$, 2.307×10^{-7} (S), and $\varepsilon_s = 30.9, 33.3, \varepsilon_\infty = 6.0, 6.8$. Panwar et al. (1978) correlated the dielectric behavior with the existence of states in the gap. Since the defect states are paired so that half of the charge carriers (dangling bonds) are positively charged and the other half are negatively charged, Panwar et al. (1978) assumed that they may act as dipoles. The very little change in ε from 83 to 253 K reveals that these dipoles are frozen at lower T levels and obtain complete freedom at high Ts in random environments.

In 1987, Tanovcova et al. measured the dielectric properties of x mol% Te–$(100 - x - y)$ mol% Se–y mol% Ge for x = 0–50 and y = 20, 25, 30, or 33; of x mol% Te–20 mol% Ge–$(80 - x - y)$ mol% Se–y mol% Sb for y = 15–20 and x = 0–5; and of 15 mol% Te–85 mol% Se to optimize the multicomponent seleno-telluride glasses for optimal applied power transmission at 10.6 μm. The dielectric measurements were performed at f values of 0–100 kHz and at a T of 150°C, in a vacuum. Samples were sandwiched with painted graphite electrodes. The dielectric constant (ε) was believed to increase due to increases in x mol% Te.

7.6 DIELECTRIC AND MAGNETIC DATA OF NEW TELLURITE GLASSES AND CERAMICS

Not too much scientific research has been done on the dielectric properties of tellurite glass in comparison to the work done on the conductivity. The growing need of the microelectronics industry for small, lightweight, multifunctional devices that use less battery power and have low processing costs has been the driving force behind the search for optimized materials. Thick films of materials with improved properties that are capable of fabrication at low temperatures to replace the currently used bulk dielectric components are presently of paramount importance. The requirement for a low sintering temperature is of particular relevance for applications in which materials have to be co-fired, as in the case of low-temperature co-fired ceramics (LTCCs) and base-metal electrode multilayer ceramic capacitors (BME-MLCCs). Because these technologies require the co-firing of the inner metallic electrode and the ceramic layer, the sintering temperature must be lower than the melting point of the metal. Furthermore, if low-cost metals are to be used, the dielectric material has to densify well below 950°C.

Durga and Veeraiah (2002) measured the dielectric parameters of tellurite glasses of the form $40ZnF_2-(10-x)As_2O_3-xCr_2O_3-50TeO_2$ with $x = 0, 0.1, 0.2, 0.3, 0.4, 0.5$ and 0.6, all by wt%; namely, ε, tan δ and σ_{ac}. The dielectric constant ε at room temperature of pure glass at 10^5 Hz was 15.4 and was found to be dependent on frequency with slightly larger values at lower frequencies. The dielectric constant ε, dielectric loss factor, tan δ, and AC electrical conductivity σ_{ac} were found to increase, whereas the dielectric breakdown strength and activation energy for AC conduction were found to decrease with the increase in Cr_2O_3 concentration up to 0.3 wt%, thus indicating an increase in the concentration of Cr^{3+} ions (in the glasses 0.1 wt% Cr_2O_3 and 0.3 wt% Cr_2O_3) that act as modifiers. However, when Cr_2O_3 is present in higher concentrations (greater than 0.3 wt%). The authors observed that the values of the dielectric parameters decrease with an increase in Cr_2O_3 concentration; such changes may be understood due to increase in the concentration of chromium ions that take part in the tetrahedral positions with CrO_4^{2-} structural units causing a decrease in the degree of depolymerisation of the glass network. Also, the dielectric breakdown strength and the activation energy are found to increase when Cr_2O_3 concentration is greater than 0.3 wt%, indicating a gradual conversion of chromium ions from Cr^{3+} state to Cr^{6+} state (which take part in network-forming positions) in these glasses. The dielectric breakdown strength for glass with zero Cr_2O_3 was equal to 18.5 kV/cm. With the introduction of Cr_2O_3 (0.3 wt%), the value of breakdown strength is lowered to 14.2 kV/cm, and for further increase of Cr_2O_3 the breakdown strengths increased. Breakdown strengths (kV/cm) were 18.5, 16.32, 14.58, 14.21, 18.7, 19.2, and 21.5 kV/cm. The magnetic moments evaluated from susceptibility measurements of the glasses showed a decreasing trend for glasses from $Cr_2O_3 = 0.4-0.6$ wt%, indicating a gradual conversion of Cr^{3+} ions into Cr^{6+} ions, with increase in the concentration of Cr_2O_3. Data on magnetic susceptibility, χ (10^{-6} emu) and moment μ_{eff} (μB) of $ZnF-As_2O_3-TeO_2-Cr_2O_3$ glasses were: 0.463, 0.932, 1.415, 1.552, 1.81, and 1.92 (10^{-6} emu), and 3.82, 3.82, 3.822, 3.44, 3.32, and 3.1 (μB), respectively.

Turky and Dawy (2002) measured the electrical properties of ternary $[65TeO_2-(35-x)V_2O_5-xSm_2O_3]$ glasses where $x = 1, 2, 3, 4$, and 5 mol%. The results obtained showed that glasses with 4 and 5 mol% Sm_2O_3 have higher ε and ε due to the weakness of the bonds. The values of both ε and ε at 5 kHz and room temperature were $\varepsilon < 50$ and $\varepsilon \leq 50$ and increased with temperature to be $\varepsilon \leq$ 200 and $\varepsilon \leq 500$ at 200°C. Also, the conductivity decreases until about 3 mol% and then increases with the increase of Sm_2O_3 due to the introduction of Sm ions in the network structure of the glass. The dielectric constant of glasses depends primarily upon the electric and ionic polarization, while the observed dielectric loss commonly depends on the relaxation of the latter. The dielectric constant tends to increase with temperature as it increases the amplitude of the ionic displacement, and it decreases very slowly with increasing frequency over a wide range. The loss is commonly small and subjected to only a small change over the range of frequency. As the temperature increases above 80°C, the enhancement in DC conductivity causes a strong frequency dependence and rapid increase in the dielectric loss and the apparent dielectric constant. Glasses with 4 and 5 mol% of Sm_2O_3 have higher ε' and ε''. This can be attributed to the weakness of the bonds and increase in the nonbridging oxygen.

Kumar and Sankarappa (2008a) investigated the DC electrical conductivity of lithium–vanadium–tellurite glasses in the composition $(V_2O_5)0.4-(Li_2O)x-(TeO_2)0.6$ and x; $0.10 \leq x \leq 0.50$, in the temperature range 300–525 K. Kumar et al. (2008b) continued their work in measuring the electrical properties of tellurite glasses of two sets of tellurium-based glasses doped with vanadium and vanadium–cobalt oxides in the forms:

1. $(V_2O_5)x (TeO_2)1-x$; and $x = 0.1, 0.2, 0.3$, and 0.4 mol%
2. $(V_2O_5)0.4 (CoO)x (TeO_2)0.6-x$; and $x = 0.3, 0.4$, and 0.5 mol%

Kumar and Sankarappa (2008a) and Kumar et al. (2008b) calculated the effective dielectric constant, ε_p, which can be determined from the relation $\varepsilon_p = e^2/4Wr_p$ and the small-polaron radius $r_p = (1/2)(p/6N)1/3$. Errors in W_H and r_p were ± 0.01 eV and ± 0.004 nm, respectively. The estimated values of ε_p and r_p are listed in Table 7.3.

Superionic glasses are of current interest because of their application in different technical devices. Dutta and Ghosh (2009) measured the capacitance and conductance of the samples in the frequency range 10–2 MHz and in the temperature range 128–303 K by using an LCR meter. The DC conductivity for different temperatures was computed from the complex impedance plots. Dutta and Ghosh (2009) correlated the ion dynamics of AgI-doped Ag_2O–TeO_2 glasses, and Ag_2S-doped glass nanocomposites have been studied using impedance spectroscopy and have been correlated with the structures investigated by using Fourier transform infrared spectroscopy. The electric modulus was fitted to a one-sided Fourier transform. Dutta and Ghosh (2009), in their study, used the Kohlrausch–Williams–Watts function to measure the evolution of the electric field within the materials, and obtained the values of the stretched exponential parameter β from the fitting and observed that for different compositions is almost independent of temperature. The activation energy E_τ of the relaxation frequency ω_τ is shown in Table 7.3, and is the same as the activation energy E_σ of the DC conductivity. These results indicate that the ions face the same barrier during conduction and relaxation. Dutta and Ghosh (2009) concluded that the conductivity for the tellurite glasses is higher than that for the borate and phosphate glasses. A structural change occurs in the AgI-doped Ag_2O–TeO_2 glass at $x = 0.2$. The variation of the crossover frequency for compositions below and above $x = 0.2$ was explained on the basis of the variation of the relative population density of TeO_4 and TeO_3 structural units. The scaling of the conductivity spectra indicates that the relaxation dynamics in these glasses is independent of temperature but is dependent on AgI content. The variation of the stretched exponential parameter indicates more cooperative ion motion with an increase in the AgI content as compared with the Ag_2S-doped silver tellurite nanocomposites, where the cooperation saturates with the increase in the Ag_2S content. Also, Moawad et al. (2009) calculated the effective dielectric constant, ε_p, for $0.5[xAg_2O–(1-x)V_2O_5]–0.5TeO_2$ glasses. The values of the effective dielectric constant, ε_p, for $0.5[xAg_2O–(1-x)V_2O_5]–0.5TeO_2$ glasses decreased from 4.49 to 3.92 due to the increase of the percentage of V_2O_5 in tellurite glasses as shown in Table 7.3.

Kumer et al. (2009) measured the dielectric properties of the two sets of tellurite glasses over a wide range of frequencies and temperatures. The two sets of tellurite glasses were:

1. $(V_2O_5)x–(TeO_2)1-x$, with $x = 0.10, 0.20, 0.30$, and 0.40,
2. $(V_2O_5)0.4–(CoO)x–(TeO_2)0.6-x$, with $x = 0.3, 0.4$, and 0.5 mol%.

The frequency-dependent measurements of capacitance, C, and dissipation factor, tan d, were obtained using a computer-controlled LCR HiTester for different frequencies in the range 50 Hz–5 MHz and the temperature range 300–500 K. The dielectric constant (ε), dielectric loss factor (tan δ), and AC conductivity (σ_{ac}) were determined. The decrease in dielectric constant and dielectric loss with increase in frequency was attributed to the decrease in electronic contribution and increase in dipolar contribution to the total polarizability. Kumar et al. (2009) used Hunt's model to determine the activation energies for dielectric relaxation processes in both sets of glasses and these activation energies were found to be in close agreement with DC activation energies as shown in Table 7.3. Both dielectric components ε' and ε'' were given in Table 7.3. Therefore, the authors concluded that in both tellurite glasses, the relaxation process has a local character implying hops of polarons between multivalent transition metal ion sites.

Gandhi et al. (2009) measured the spectroscopic and dielectric properties of ZnF_2–As_2O_3–TeO_2 glasses mixed with different concentrations of V_2O_5 (ranging from 0 to 0.6 mol%) as in the form of the ZnF_2–As_2O_3–TeO_2 glass system, a particular composition $20ZnF_2$–$30As_2O_3$–$(50-x)$ $TeO_2:xV_2O_5$ (with x ranging from 0 to 0.6). A variety of properties—such as optical absorption, photoluminescence, infrared, ESR spectra, and dielectric properties (e.g., constant ε', loss tan δ, AC conductivity

TABLE 7.3
Dielectric Properties of New Tellurite Glasses

Glass	Reference	ε	E	$W_{(eV)}$	Electrical Parameters
$(V_2O_5)0.4–(Li_2O)$ $x–(TeO_2)0.6$	Kumar et al. 2008a				
$X = 0.1$		$\varepsilon_p = 2.84 \pm 0.02$			
$X = 0.5$		$\varepsilon_p = 1.52 \pm 0.02$			
$(V_2O_5)x–(TeO_2)1 – x$	Kumar et al. 2008b				
$X = 0.1$		$\varepsilon_p = 1.80 \pm 0.02$			
$X = 0.4$		$\varepsilon_p = 3.15 \pm 0.02$			
$(V_2O_5)\ 0.4–(CoO)$ $x–(TeO_2)0.6 – x$					
$X = 0.3$		$\varepsilon_p = 3.32 \pm 0.02$			
$X = 0.5$		$\varepsilon_p = 4.20 \pm 0.02$			
$0.5[xAg_2O–(1 – x)$ $V_2O_5]–0.5TeO_2$	Moawad et al. 2009				
45 mol% V_2O_5		$\varepsilon_p = 4.49 \pm 0.03$			
40 mol% V_2O_5		$\varepsilon_p = 4.45 \pm 0.03$			
35 mol% V_2O_5		$\varepsilon_p = 4.41 \pm 0.03$			
30 mol% V_2O_5		$\varepsilon_p = 4.00 \pm 0.03$			
25 mol% V_2O_5		$\varepsilon_p = 3.92 \pm 0.03$			
	Kumar et al. 2009				
$(V_2O_5)x–(TeO_2)1 – x$					
$X = 0.1$ mol%	$\varepsilon' \approx 200$	$\varepsilon'' \approx 1500$			$W = 0.449$
$X = 0.4$ mol%	$\varepsilon' \approx 3750$	$\varepsilon'' \approx 38000$			$W = 0.391$
$(V_2O_5)0.4–(CoO)$ $x–(TeO_2)0.6 – x$					
$X = 0.3$ mol%	$\varepsilon' \approx 100$	$\varepsilon'' \approx 550$			$W = 0.469$
$X = 0.5$ mol%	$\varepsilon' \approx 1150$	$\varepsilon'' \approx 2800$			$W = 0.428$
$xAgI–(1 – x)$ $(0.30Ag_2O–0.70TeO_2)$	Dutta and Ghosh 2009	E_r (eV)	E_c (eV)	E_s (eV)	n
$x = 0.1$		0.39	0.41	0.41	0.65
0.2		0.38	0.40	0.40	0.65
0.4		0.31	0.26	0.32	0.66
	Suresh et al. 2010	Temperature °C	C (pF)	Resistance (Ω)	FWHM (Hz)
$60B_2O_3–10TeO_2–$ $5TiO_2–25Li_2O$		200 and 300	39.8 and 30.4	8×105 and 4.9×103	5006 and 181350
$60B_2O_3–10TeO_2–$ $5TiO_2–25Na_2O$		200 and 300	48.5 and 34.7	5×106 and 1.6×104	2108 and 66463
$60B_2O_3–10TeO_2–$ $5TiO_2–25K_2O$		200 and 300	30.4 and 27.3	7.9×105 and 3×103	1363 and 41877
	Gandi et al. (2009)	Relaxation temp. region °C		Dipols activation energy eV	
$20ZnF_2–30As_2O_3–$ $50TeO_2$	$(\tan\delta_{max})_{avg} = 0.026$	130–147		2.52	
$20ZnF_2–30As_2O_3–$ $49.7TeO_2: 0.3V_2O_5$	$(\tan\delta_{max})_{avg} = 0.031$	103–126		2.18	
$20ZnF_2–30As_2O_3–$ $49.4TeO_2: 0.6V_2O_5$	$(\tan\delta_{max})_{avg} = 0.039$	70–97		2.27	

σ_{ac} over a wide range of frequency and temperature)—of these glasses have been explored. The optical absorption, together with the electron spin resonance studies, indicated vanadium ions coexist in V^{4+} with V^{5+} state in these samples. The IR spectra of these samples have exhibited bands due to v-TeO$_{2ax}$ and AsO$_3$ structural groups; these results indicated the most structural disorder in the network by increasing the concentration of V_2O_5. The luminescent emission spectra recorded at room temperature of glasses excited at 640 nm have exhibited a broad emission band in the spectral wavelength range of 750–850 nm. The luminescence efficiency is found to be the highest for the sample doped with 0.6 mol%. The dielectric constants $\varepsilon'(\omega, T)$ and $\varepsilon''(\omega, T)$ of tellurite glass with high concentration of vanadium of the form: 20ZnF$_2$–30As$_2$O$_3$–49.4TeO$_2$: 0.6V$_2$O$_5$ was about $16 < \varepsilon' < 17$ and $0.3 < \varepsilon'' < 0.4$ at room temperature, as shown in Table 7.3. Gandhi et al. (2009) pointed out that there is a gradual increase in the concentration of V^{4+} ions that act as modifiers in the glass network, as shown in Figure 7.15. The analysis of dielectric loss studies indicated that these glasses exhibit dipolar effects. The AC conduction could be explained both due to classical activation energy and to the tunneling phenomena. The dielectric parameters—namely, ε', loss tan δ, and σ_{ac}—are found to increase, and the activation energy for AC conduction is found to decrease with the increase in the concentration of V_2O_5 up to 0.6 mol%, reflecting an increase in the concentration of V^{4+} ions that take part-modifying positions in the glass network. The AC conduction in these glasses could satisfactorily be explained by both classical activation energy and the tunneling phenomena.

Gandhia et al. 2010, studied the structure of Nd^{3+} ions in the TeO$_2$ glass network with tungsten ions in tetrahedral and octahedral positions. Gandhia et al. 2010, found that more connectivity is related to the higher coordination number of the modifiers exists in the tetrahedral tellurite glasses of the form: $(50 − x)$ZnF$_2$–xWO$_3$–49TeO$_2$:1Ln$_2$O$_3$ (with x ranging from 5 to 20) and where Ln = Nd, Sm, and Eu. Glasses of the composition $(45 − x)$ZnF$_2$–xWO$_3$–49TeO$_2$:1.0Nd$_2$O$_3$/Sm$_2$O$_3$/Eu$_2$O$_3$, with x varying from 5 to 20 mol%, have been prepared. Differential scanning calorimetry, IR, Raman, ESR, optical absorption and photoluminescence studies have been carried out. Electron spin resonance (ESR) studies have indicated that the tungsten ions exist in the W^{5+} state, in addition to the W^{6+} state. IR and Raman spectral studies have suggested that there is a growing degree of disorder in the glass network with an increase in the concentration of WO$_3$ up to 15 mol%, as shown in Figure 7.16

Suresh et al. (2010) measured the AC conductivity and impedance measurements in alkali borotellurite glasses of the form: 60B$_2$O$_3$–10TeO$_2$–5TiO$_2$–25Li$_2$O, 60B$_2$O$_3$–10TeO$_2$–5TiO$_2$–25Na$_2$O, and 60B$_2$O$_3$–10TeO$_2$–5TiO$_2$–25K$_2$O. The authors measured the frequency dependence of the imaginary part of impedance ($Z''(\omega)$) and the imaginary part of modulus (M'') for a glass sample (BTTL) at different temperatures because a complex modulus spectrum indicates electrical phenomenon, with the smallest capacitance occurring in a dielectric medium. The capacitance, resistance, and FWHM values from impedance plots (Z'' vs. Z', and Z'' vs. frequency) at different

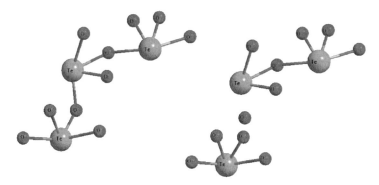

FIGURE 7.15 (**See color insert**) A mechanism for structural changes in TeO glass network from TeO$_4$ (tbp) to TeO$_{3+1}$ polyhedra induced by modifiers. (Reprinted from Y. Gandhi, N. Venkatramaiah, V. Ravi Kumar, and N. Veeraiah, *Physica B: Condensed Matter*, 1450, 2009. With permission.)

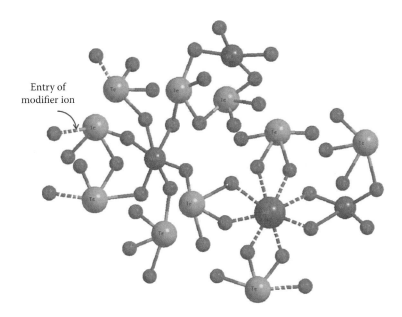

FIGURE 7.16 **(See color insert)** An illustration of a Nd^{3+} ion embedded in the TeO_2 glass network with tungsten ions in tetrahedral and octahedral positions. (Reprinted from Y. Gandhi, I.V. Kityk, M.G. Brik, P. Raghava Rao, and N. Veeraiah, *J. Alloys and Comp.*, 278, 2010. With permission.)

temperatures for all glass series are given in Table 7.3. It is clear from the table that FWHM values increase with an increase in temperature. The AC conductivity and relaxation behavior of alkali boro-tellurite glasses have been examined. It was found that AC conductivity is more for lithium followed by potassium and sodium, but this amount of increase between sodium and potassium is negotiable. Lowest "s" values were found for the lithium-containing glasses. Scaling analysis of dielectric modulus shows excellent time–temperature superposition, indicating common ion transport mechanism.

Tellurium glasses exhibit a high refractive index and good optical transmission in the IR region. They also have interesting semiconducting properties and can be prepared at low temperatures. In contrast to glasses, tellurium-based oxide ceramics have not yet been seriously considered for use in electronics. Most tellurium-based oxide materials can be synthesized and sintered at temperatures below 900°C, which makes them potential candidates for use in low-temperature co-fired ceramics (LTCC) technology. Other requirements for LTCC materials, in terms of their electrical properties, are high relative permittivity (k'), high Q-value, and a temperature coefficient of resonant frequency (τ_f) tunable around zero. For ceramic processing, the most important oxidation states of tellurium are Te^{+4} and Te^{+6}, where Te^{+6} possesses a lone electron pair. In some cases, when the thermodynamic conditions are fulfilled, an oxidation from Te^{+4} to Te^{+6} occurs in air. However, pure TeO_2 never oxidizes in air to the +6 oxidation state (Bayer 1969). Udovic and Suvorov (2001) prepared the ceramic sample from reagent grade TeO_2 and TiO_2 via a solid-state reaction method. The powders were weighed in different molar ratios, mixed in alcohol, dried, and calcined for 80 h at 620°C–700°C with intermediate crushing until phase equilibrium was achieved by x-ray diffraction analysis (XRD). After the calcinations, the samples were milled for 1 h in a planetary ball mill at 200 rpm using 3 mm f zirconia balls. An analysis revealed a mean particle size of below 1 mm measured by granulometer. Powders were isostatically pressed into pellets and sintered in air. For the investigation of possible solid solubilities, unit-cell parameters were determined from XRD patterns using LCLSQ11 software. Microstructural analyses were performed on a scanning electron microscope (SEM). Electrical properties were measured by a reflection resonant-cavity method at ≈5GHz with a HP 8719C network analyzer. The temperature dependence of resonant frequency was determined in the temperature range 20°C–60°C. A cell-parameter determination of the 0.6 TiO_2–0.4 TeO_2 sample indicates no changes in the unit-cell parameters, neither for the TiO_2 nor the $TiTe_3O_8$ phase, which again indicates a negligible solid solubility. The $TiTe_3O_8$ ceramics exhibit 5% porosity and a grain size of 2–5 m. The microwave–dielectric properties of these ceramics, measured at ≈5 GHz, are a $Q \times f$ value of 30,600 GHz, relative permittivity of 50, and a temperature coefficient of resonant frequency +133 ppm/°C. The $Q \times f$ value and the relative permittivity of these ceramics are much higher than previously reported due to a significant reduction in the porosity.

Su et al. (2009) prepared the thick films of $TiTe_3O_8$, which were fabricated by electrophoretic deposition EPD on Pt-coated Si substrates, using acetone as suspension media with triethanolamine additions. Monophasic $TiTe_3O_8$ films are well sintered at 700°C for 5 h when compared with $TiTe_3O_8$ ceramics sintered at 740°C for 5 h. The relative permittivity (ε_r) and dielectric loss (tan δ) of these films were ≈54 and 0.009, measured at 100 kHz. $TiTe_3O_8$ films display dielectric losses of lower than 0.01 and good temperature stability, thus making them suitable for integrated capacitor and microwave device applications. The temperature coefficients of permittivity of $TiTe_3O_8$ films and ceramics between 35°C and 200°C were +78 and –100 ppm °C^{-1}, respectively. In particular, Su X et al. mentioned that the ability to sinter $TiTe_3O_8$ films below 800°C means that these materials can be considered as potential candidates for low-temperature co-fired ceramics (LTCCs) and base-metal electrode multilayer ceramic capacitors (BME-MLCCs).

REFERENCES

Abdel-Kader, A., El-Mallawany, R., and El-Kholy, M., *J. Appl. Phys.*, 73, 75, 1993.
Abdel-Kader, A., Higazy, A., El-Kholy, M., and El-Bahnasawy, R., *J. Mater. Sci.*, 26, 4298, 1991.
Abdel-Kader, A., Higazy, A., El-Mallawany, R., and El-Kholy, M., *Radiat. Eff. Defect. Solids*, 124, 401, 1992.
Abelard, P., and Baumard, J., *Phys. Rev. B*, 26, 1005, 1982.
Al-Ani, S., Hogarth, C., and El-Mallawany, R., *J. Mater. Sci.*, 20, 661, 1985.
Ali, A., and Shaaban, M., *Physica B*, 403, 2461, 2008.
Amarilla, J., Colomer, M., Acosta, J., Morales, E., and Jurado, J., *Proc. Symp. Batteries for Portable and Electrical Vehicle Appl.*, 1997.
Anderson, O., and Bommel, H., *J. Am. Ceram. Soc.*, 38, 125, 1955.
Anderson, O., and Stuart, D., *J. Am. Ceram. Soc.*, 37, 573, 1954.
Anderson, P. W., *Phys. Rev. Lett.*, 34, 954, 1975.
Anderson, P., *Fizika Dielektrikov*, Ed. I. Skanvig, Moscow, Akad. Nauk. S.S.S.R., 1960.
Anderson, P., *Phys. Rev.*, 109, 1492, 1958.
Austen, I., and Mott, N., *Adv. Phys.*, 18, 41, 1969.
Balaya, P., and Goyal, P., *J. Non-Cryst. Solids*, 351, 1573, 2005.
Balaya, P., and Sunandana, C., *J. Non-Cryst. Solids*, 162, 253, 1993.
Balaya, P., and Sunandana, C., *J. Non-Cryst. Solids*, 175, 51, 1994.
Bayer, G., *Zur Kristallchemie des Tellurs. Fortschr. Miner.*, 46, 42, 1969.
Bhattacharya, S., and Ghosh, A., *Solid State Ion.*, 161, 61, 2003.
Bogomolov, V., Kudinov, E., and Frisov, Y., *Sov. Phys. Sol. Stat.*, 9, 3175, 1967.
Bosman, A., and Havinga, E., *Phys. Rev.*, 129, 1593, 1963.
Bottger, H., and Bryksin, V., *Phys. Stat. Sol.*, 78, 415, 1979.
Braunstein, R., Lefkowitz, I., and Snare, J., *Sol. Stat. Commun.*, 28, 843, 1978.
Bunde, A., Ingram, M., and Maass, P., *J. Non-Cryst. Solids*, 172, 378, 1994.
Bundy, F., *Phys. Rev.*, 110, 314, 1958.
Butcher, P., and Morys, P., *J. Phys. C*, 6, 2147, 1973.
Chopra, N., Mansingh, A., and Chadha, G., *J. Non-Cryst. Solids*, 126, 194, 1990.
Chowdari, B., and Kumari, P., *J. Non-Cryst. Solids*, 197, 31, 1996.
Chowdari, B., and Kumari, P., *J. Phys. Chem. Solids*, 58, 515, 1997.
Chowdari, B., and Kumari, P., *Sol. Stat. Ion.*, 86, 521, 1996.
Chung, C., and Mackenzie, J., *J. Non-Cryst. Solids*, 42, 1980.
Cockran, W., *Adv. Phys.*, 9, 387, 1960.
Cohen, H., and Roy, R., *J. Am. Ceram. Soc.*, 44, 523, 1961.
Cohen, M., Fritzsche, H., and Ovshinsky, S. R., *Phys. Rev. Lett.*, 22, 1065, 1969.

Cole, K., and Cole, R., *J. Chem. Phys.*, 9, 341, 1941.

Cornet, J., and Rossiter, D., *J. Non-Cryst. Solids*, 12, 85, 1973.

Davis, E. A., and Mott, N. F., *Philos. Mag.*, 22, 903, 1970.

Decker, D., Jorgensen, J., and Young, R., *High Temp. High Press*, 7, 331, 1975.

Dekker, A., *Solid State Physics*, Madras, India, Macmillan India Ltd., 1995.

Deneuville, A., Gerard, P., and Deveny, I., *J. Non-Cryst. Solids*, 22, 77, 1976.

Denton, E., Rawson, H., and Stanworth, J., *Nature* 173, 1030, 1954.

Desirena, H., Schulzgen, A., Sabet, S., Ramos-Ortiz, G., de la Rosa, E., and Peyghambarian, G., *Opt. Mater.*, 31, 784–789, 2009.

Dimitrive, Y., Gattef, E., and Eneva, A., *Int. J. Electron.*, 50, 385, 1981.

Dimitrive, Y., Marinov, M., and Gatev, E., *C. R. Acad. Bulgar. Sci.*, 26, 675, 1973.

Durga, D., and Veeraiah, N., *Physica B*, 324, 127, 2002.

Dutta, A., and Ghosh, A., *J. Non-Cryst. Solids*, 351, 203, 2005.

Dutta, D., and Ghosh, A., *J. Non-Cryst. Solids*, 355, 1930, 2009.

El-Mallawany, R., and Saunders, G., *J. Mater. Sci Lett.*, 6, 443, 1987.

Efros, A., *Philos. Mag B.*, 43, 829, 1981.

El Zaidia, M., Amar, A., and El-Mallawany, R., *Phys. Stat. Solids*, 91, 637, 1985.

El-Damrawi, G., *Phys. Stat. Sol.* 177, 385, 2000.

El-Fouly, M., Morsy, A., Ammar, A., Maged, A., and Amer, H., *J. Mater. Sci.*, 24, 2444, 1989.

El-Kholy, M., and El-Mallawany, R., *Mater. Chem. Phys.*, 40, 163, 1995.

El-Kholy, M., *J. Mater. Sci.*, 5, 157, 1994.

Elliot, S. R., *Adv. Phys.*, 36, 135, 1987.

Elliot, S. R., *Philos. Mag.*, 36, 1291, 1977.

El-Mallawany, R., and Saunders, G., *J. Mater. Sci. Lett.*, 6, 443, 1987.

El-Mallawany, R., El-Deen, L., and El-Kholy, M., *J. Mater. Sci.*, 31, 6339, 1996.

El-Mallawany, R., El-Sayed, A., and Abd El-Gawad, M., *Mater. Chem. Phys.*, 41, 87, 1995.

El-Mallawany, R., *Mater. Chem. Phys.*, 37, 376, 1994.

El-Samadony, M., *J. Mater. Sci.*, 30, 3919, 1995.

Emin, D., *Electrical and Structural Properties of Amorphous Semi*, P. LeComber and N. Mott (eds), London, Academic Press, 201, 1973.

Flynn, B., Owen, A., and Robertson, J., *Proc. 7th Int. Conf. Amorphous and Liquid Semiconductors*, UK, 678, 1977.

Gadkari, A., and Zope, J., *J. Non-Cryst. Solids*, 103, 295, 1988.

Gandhi, Y., Kityk, I., Brik, M., Rao, P., and Veeraiah, N., *J. Alloys Compd*, 508, 278, 2010.

Gandhi, Y., Venkatramaiah, N., Ravi Kumar, V., and Veeraiah, N., *Physica B.*, 404, 1450, 2009.

Gattef, E., and Dimitriev, Y., *Philos. Mag. B.*, 40, 233, 1979.

Ghosh, A., *J. Appl. Phys.*, 65, 227, 1989.

Ghosh, A., *J. Phys. Condens. Matt.* 5, 8749, 1993.

Ghosh, A., *Philos. Mag. B*, 61, 87, 1990.

Ghosh, A., *Phys. Rev. B.*, 47, 15537, 1993.

Gibbs, D., and Hill, G., *Philos. Mag.*, 367, 1963.

Giehl, J., Pontuschka, W., Barbosa, L., and Ludwig, Z., *J. Non-Cryst. Solids*, 356, 1762, 2010.

Glass, A., Lines, M., Nassau, K., and Shiever, J., *Appl. Phys. Lett.*, 31, 249, 1977.

Goodnick, S., Korkin, A., Krstic, P., Mascher, P., Preston, J., and Zaslavsky, A., *Nanotechnol*, 21, 130201, 2010.

Hampton, R., Collier, I., Sidek, H., and Saunders, G., *J. Non-Cryst. Solids*, 110, 213, 1989.

Hampton, R., Hong, W., Saunders, G., and El-Mallawany, R., *J. Non-Cryst. Solids*, 94, 307, 1987.

Hampton, R., Hong, W., Saunders, G., and El-Mallawany, R., *Phys. Chem. Glasses*, 29, 100, 1988.

Hassan, M., and Hogarth, C., *J. Mater. Sci.*, 23, 2500, 1988.

Havinga, E., *J. Phys. Chem. Solids*, 18/23, 253, 1961.

Hegazy, H., M.S. Thesis, Faculty of Science, Menofia University, Menofia, Egypt, 1999.

Heikes, R., and Use, R., *Thermoelectricity*, New York, NY, Interscience, 81, 1961.

Hench, L., *Principals of Electronic Ceramics*, New York, NY, Wiley-Interscience, 1990.

Hirashima, H., Ide, M., and Youshida, T., *J. Non-Cryst. Solids*, 86, 327, 1986.

Hirashima, H., and Yoshida, T., *J. Non-Cryst. Solids*, 95/96, 817, 1987.

Holstein, T., *Ann. Phys.*, 8, 325, 1959.

Huggins, R., *Diffusion in Solids, Recent Development*, A. Nowick and J. Burton (eds), New York, NY, Academic Press, 445, 1975.

Hunter, C., and Ingram, M., *Sol. Stat. Ion.*, 14, 31, 1994.

Ingram, D., and Robertson, A., *Sol. Stat. Ion.*, 94, 49, 1997.

Ingram, M., *Phys. Chem. Glasses*, 38, 215, 1987.

Isard, J., *J. Non-Cryst. Solids*, 1, 235, 1969.

Jain, H., and Nowick, A., *Phys. Stat. Solids*, 67, 701, 1981.

Jain, H., *Experimental Technology* II changed "Tach." in original to "Technology", correct? of Glass Science, Eds. C. Simmons and O. El-Baoumi, Westerville, OH, Am. Ceram. Soc., Ohio, 1993.

Jayasinghe, G., Bandaranayake, P., and Souquet, J., *Sol. Stat. Ion.*, 86, 447, 1996.

Jayasinghe, G., Dissanayake, M., Bandaranayake, P., Souquet, J., and Foscallo, D., *Sol. Stat. Ion.*, 121, 19, 1999.

Jayasinghe, G., Dissanayake, M., Careem, M., and Souquet, J., *Sol. Stat. Ion.*, 93, 291, 1997.

Jonsher, A., *Dielectric Relaxation in Solids*, London, Chelsea Dielectric Press, 1983.

Jonsher, A., *J. Mater. Sci.,* 16, 2037, 1981.

Kahnt, H., and Feltz, A., *J. Non-Cryst. Solids*, 86, 33, 1986.

Kahnt, H., and Feltz, A., *J. Non-Cryst. Solids*, 86, 41, 1986.

Kastner, M., and Hudgens, S., *Philos. Mag. B.,* 37, 665, 1978.

Killias, H., *Phys. Lett.*, 20, 5, 1966.

Knigry, W., Brwen, H., and Uhlman, D., *Introduction to Ceramics*, 2nd Ed., New York, NY, Wiley-Interscience, 1976.

Komatsu, T., Ike, R., Sato, R., and Matusita, K., *Phys. Chem. Glasses*, 36, 216, 1995.

Kosuge, T., Benino, Y., Dimitriov, V., Sato, R., and Komatsu, T., *J. Non-Cryst. Solids*, 242, 154, 1998.

Kowada, Y., Morimoto, K., Adachi, H., Tatsumisago, M., and Minami, T., *J. Non-Cryst. Solids*, 196, 204, 1996.

Kozhokarov, V., and Marinov, N., *Akad. Bulgar. Sci.*, 27, 1557, 1974.

Kozhokarov, V., and Marinov, N., *Akad. Bulgar. Sci.*, 28, 505, 1975.

Krins, N., Rulmont, A., Grandjean, J., Gilbert, B., Lepot, L., Cloots, R., and Vertruyen, B., *Sol. Stat. Ion.*, 177, 3147, 2006.

Kubliha, M., Trnovcova, V., Furar, I., Kadlec ikova, M., Pedlikova, J., and Greguš, J., *J. Non-Cryst. Solids*, 355, 2035, 2009.

Kumar, M., and Sankarappa, T., *Sol. Stat. Ion.*, 178, 1719, 2008a.

Kumar, M., Sankarappa, T., and Awasthi, A., *Physica B*, 403, 4088, 2008b.

Kumar, M., Sankarappa, T., Kumar, B., and Nagaraja, N., *Solid State Sci.*, 11, 214, 2009.

Kumar, V., and Veeraiah, N., *J. Phys. Chem. Solids*, 59, 91, 1998.

Lambson, E., Saunders, S., Bridge, B., and El-Mallawany, R., *J. Non-Cryst. Solids*, 69, 117, 1984.

Licciardello, D. C., *Sol. Cells,* 2, 191, 1980.

Long, A., *Adv. Phys.*, 31, 553, 1982.

Long, A., Balkan, N., Hogg, W., and Ferrier, R., *Philos. Mag. B.*, 45, 497, 1982.

Macedo, P., Moynihan, C., and Bose, R., *Phys. Chem. Glasses*, 13, 171, 1972.

Mackenzie, J., *Electrical Conductivity in Ceramics and Glass, Part B*, N. M. Tallan (ed), New York, NY, Marcel Dekker, 586, 1974.

Malik, M., and Hogarth, C., *J. Mater. Sci.*, 25, 1913, 1990.

Malik, M., and Hogarth, C., *J. Mater. Sci.*, 25, 2493, 1990.

Malik, M., and Hogarth, C., *J. Mater. Sci.*, 25, 2585, 1990.

Malik, M., and Hogarth, C., and Lott, K., *J. Mater. Sci.*, 25, 1909, 1990.

Malik, M., and Hogarth, C., *J. Mater. Sci.*, 25, 3208, 1990.

Mandouh, Z., *J. Appl. Phys.*, 78, 7158, 1995.

Mansingh, A., and Dahwan, V., *J. Phys. C*, 16, 1675, 1983.

Mansingh, A., Dhawan, A., Tandon, R., and Vaid, L., *J. Non-Cryst. Solids*, 27, 309, 1978.

Marshall, J. M., and Owen, A. E., *Philos. Mag.*, 24, 1281, 1971.

Martin, S., and Angel, C., *Sol. Stat. Ion.*, 23, 185, 1986.

McGeehin, P., and Hooper, A., *J. Mater. Sci.*, 12, 1, 1977.

McMillan, P., *Glass-Ceramics*, 2nd Ed., New York, Academic Press, 1979.

Mierzejewski, A., Saunders, G., Sidek, H., and Bridge, B., *J. Non-Cryst. Solids*, 104, 333, 1988.

Moawad, H., Jain, H., and ElMallawany, R., Ramadan, T., and El-Sharbiny, M., *J. Am. Ceram. Soc.*, 85, 2655, 2002.

Moawad, H., Jain, H., and El-Mallawany, R., *J. Phys. Chem. Solids*, 70, 224, 2009.

Moawad, H., Jain, H., and El-Mallawany, R., *Sol, Stat. Ion.*, 181, 1103, 2010.

Montani, R., Robledo, A., and Bazan, J., *Mater. Chem. Phys.*, 53, 80, 1998.

Montani, R.A., and Frechero, M.A., *Sol. Stat. Ion.,* 177, 2911, 2006.

Mori, H., and Sakata, H., *J. Ceram. Soc. Japan*, 102, 562, 1994.

Mori, H., and Sakata, H., *Mater. Chem. Phys.*, 45, 211, 1996.

Mori, H., Kimami, T., and Sakata, H., *J. Non-Cryst. Solids*, 168, 157, 1994.

Mott, N., and Davis, E., *Electronic Processes in Non-Crystalline Materials*, 2nd ed., Oxford, UK, Clarendon, 1979.

Mott, N., *Conduction in Non-Crystalline Materials*, 2nd ed., Oxford, UK, Clarendon, 99, 1993.

Mott, N., *J. Non-Cryst. Solids*, 1, 1, 1968.

Mukherjee, S., Ghosh, U., and Basu, C., *J. Mater. Sci.*, 11, 985, 1992.

Murawski, L., *J. Mater. Sci.*, 17, 2155, 1982.

Nascimento, M., and Watanabe, S., *Brazilian J. Phys.*, 36, 795, 2006.

Ngai, K., and Jain, H., *Sol. Stat. Ion.*, 18/19, 362, 1986.

Ovshinsky, S. R., and Adler, D., *Contemp. Phys.*, 1.19, 109, 1978.

Ovshinsky, S. R., *Electronics*, 32, 76, 1959.

Ovshinsky, S., *Phys. Rev. Lett.*, 21, 1450, 1968.

Pan, A., and Ghosh, A., *Phys. Rev. B.*, 59, 899, 1999.

Peteanu, M., Ardelean, I., Filip, S., and Ciorcas, F., *J. Mater. Sci.* 7, 165, 1996.

Pike, G., *Phys. Rev. B,* 6, 1672, 1972.

Pollak, M., and Geballe, T., *Phys. Rev.*, 122, 1742, 1961.

Pollak, M., *Philos. Mag.*, 23, 519, 1971.

Pollak, M., *Phys. Rev.*, 138, 1822, 1965.

Prohaska, J., Li, J., Wang, J., and Bartram, R., *Appl. Phys. Lett.*, 67, 1841, 1995.

Qiu, H., Kudo, M., and Sakata, H., *Mater. Chem. Phys.*, 51, 233, 1997.

Ravaine, D., and Souquent, J., *Phys. Chem. Glasses*, 18, 27, 1977.

Ravikumar, V., and Veeraiah, N., *J. Phys. Chem. Solids*, 59, 91, 1998.

Rockstad, H., *J. Non-Cryst. Solids*, 2, 192, 1970.

Rojas, J., Dominguez, M., Villares, P., and Garay, R., *Mater. Chem. Phys.*, 45, 75, 1996.

Rojo, J., Herrero, P., Sanz, J., Tanguy, B., Portier, J., and Reau, J., *J. Non-Cryst. Solids*, 146, 50, 1992.

Rojo, J., Sanz, J., Reau, J., and Tanguy, B., *J. Non-Cryst. Solids*, 116, 167, 1990.

Rossignol, S., Reau, J., Tanguy, B.,Videau, J., Porties, J., Dexpert, J., and Piriou, B., *J. Non-Cryst. Solids,* 162, 244, 1993.

Sakata, H., Amano, M., and Yagi, T., *J. Non-Cryst. Solids*, 194, 198, 1996.

Sandhya Rani, P., and Singh, R., *J. Mater Sci.,* 45, 2868, 2010.

Sayer, M., and Mansingh, A., *J. Non-Cryst. Solids*, 31, 385, 1979.

Sayer, M., Mansingh, A., Reyes, J., and Rosenblatt, G., *J. Appl. Phys.*, 42, 2857, 1971.

Schwenzel, J., Thangadurai, V., and Weppner, W., *Ionics*, 9, 348, 2003.

Sega, K., Kasai, H., and Sakata, H., *Mater. Chem. Phys.*, 53, 28, 1998.

Sega, K., Kuroda, Y., and Sakata, H., *J. Mater. Sci.*, 33, 1303, 1998.

Shaaban M., Ali, A., and El-Nimr, M., *Mater. Chem. and Phys.*, 96, 433, 2006.

Shankar, M., and Varma, K., *J. Non-Cryst. Solids*, 243, 192, 1999.

Singer, G., and Tomozawa, M., *Phys. Chem. Glasses*, 30, 86, 1989.

Singh, R., and Chakravarthi, J., *Phys. Rev. B*, 51, 16396, 1995.

Springett, B., *J. Non-Cryst. Solids*, 15, 179, 1974.

Stanworth, J., *J. Soc. Glass Technol.*, 36, 217, 1952.

Stanworth, J., *J. Soc. Glass Technol.*, 38, 425, 1954.

Su, X., Wu, A., Vilarinho, P., Scripta Materialia, 61, 536, 2009.

Sunandana, C., and Kumaraswami, T., *J. Non-Cryst. Solids*, 85, 247, 1986.

Suresh, S., Prasad, M., and Mouli, V., *J. Non-Cryst. Solids*, 356, 1599, 2010.

Suresh, S., Prasad, M., Mouli, V., *J. Non-Cryst. Solids*, 356, 1599, 2010.

Szu, S., and Chang, Fu.-Sh., *Sol. Stat. Ion.*, 176, 2695, 2005.

Tanaka, K., Yoko, T., Nakano, M., and Nakamura, M., *J. Non-Cryst. Solids*, 125, 264, 1990.

Tanaka, K., Yoko, T., Yamada, H., and Kamiya, K., *J. Non-Cryst. Solids*, 103, 250, 1988.

Tanovcova, V., Pazurova, T., and Sramkova, T., *J. Non-Cryst. Solids*, 90, 561, 1987.

Tashiro, M., Sakka, S., and Yamamoto, T., *Yogyo-Kyokai-Shi*, 72, 108, 1964.

Thorp, J., Rad, N., Evans, D., and Williams, C., *J. Mater. Sci.*, 21, 3091, 1986.

Tomashpolski, Y., Sevostianov, M., Pentagova, M., Sorokina, L., and Venevstsev, Y., *Ferroelectrics*, 7, 257, 1974.

Tomozawa, M., *Treatise on Materials Science,* Vol. 12, 1979.

Tuller, H., Button, D., and Uhlman, D., *J. Non-Cryst. Solids*, 40, 93, 1980.

Turky, G., and Dawy, M., *Mater. Chem. and Phys.*, 77, 48, 2002.

Udovic, M., Valant, M., and Suvorov, D., *J. Eur. Ceram. Soc.*, 21, 1735, 2001.

Ulrich, D., *J. Am. Ceram. Soc.*, 47, 595, 1964.

Varshneya, A., *Fundamentals of Inorganic Glasses*, San Diego, Academic Press, 1994.

Weast, R., and Astle, M., (eds.), *CRC Handbook of Chemistry and Physics*, 60th Ed., pp. B30–B31, CRC Press, Boca Raton, FL, 1980.

Weppner, W., *Ionics*, 9, 444, 2003.

Whitmore, D., *J. Cryst. Growth*, 39, 160, 1977.

Xinming, S., Aiying, W., and Paula, M. V., *Scripta Materialia*, 61, 536, 2009.

Yomo, S., Mori, N., and Tadayasu, M., *J. Phys. Soc. Japan*, 32, 667, 1972.

Yoshida, T., Hirashima, H., and Kato, M., *Yogyo-Kyokai-Shi*, 93, 244, 1985.

Zope, M., and Zope, J., *J. Non-Cryst. Solids*, 74, 47, 1985.

Part IV

Optical Properties

Chapter 8: Linear and Nonlinear Optical Properties
            of Tellurite Glasses in the Visible Region

Chapter 9: Optical Properties of Tellurite
            Glasses in the Ultraviolet Region

Chapter 10: Infrared and Raman Spectra
             of Tellurite Glasses

8 Linear and Nonlinear Optical Properties of Tellurite Glasses in the Visible Region

Linear and nonlinear refractive indices for oxide and nonoxide–tellurite glasses and glass-ceramics (bulk and thin-film) are discussed in Chapter 8, along with experimental measurements of optical constants in these glasses. The refractive indices and dispersion values at different frequencies and temperatures are summarized for most tellurite glasses, together with new data. Reflection, absorption, and scattering are related to dielectric theory. The relationship between refractive indices and numbers of ions/unit volume (N/V) is explored, along with the values of polarizability. The reduced N/V of polarizable ions is primarily responsible for reductions in both the dielectric constant and refractive indices, although reductions in electronic polarization also affect optical properties. Data on reductions in refractive indices, densities, and dielectric constants that occur with halogen substitution are also discussed, along with thermal luminescence, fluorescence, phonon sideband spectra, and optical applications of tellurite glasses.

8.1 INTRODUCTION TO OPTICAL CONSTANTS

Although glass was already in widespread use more than 4500 years ago, the first examples of transparent glasses occurred much later. The earliest glasses were opaque and were frequently colored. Dilich et al. (1989) cite an ancient Assyrian hieroglyphic text, translated to read "take 60 parts of sand, and 5 parts of chalk, and you get glass." The Egyptians did produce transparent glass. Glass beads dating to 2500 BC have been found in Egypt and other parts of the Near East. The transparency of some glasses was first explained only in the mid-1970s. The process of making tellurite glass also has a past, a present, and a future. The past can be defined as the 40-year period from 1952 to 1992, the present as the period from 1992 to 2010. In the future the glass will gain greater use for fiber applications in communications and medicine.

Ikushima (1994) reported other recent developments in optoelectronic-material research in Japan, where Nippon Telegraph and Telephone Public Corp. has announced plans for its so-called "Fiber-to-the-Home" program. Optical switches with picosecond-order switching times will be devoted to third-order, nonlinear optical materials.

Quanta of light (i.e., photons) can interact with solids such as crystalline ceramics or amorphous glasses in a number of ways. The types of such photon interactions depend primarily on the composition of the materials and the nature and types of phases and interfaces present within materials and their ambient media. The incident radiant flux of photons (Φ) is split into beams of reflected (φ_r), transmitted (φ_t), scattered (φ_s), and absorbed (φ_a) radiation—that is,

$$\Phi = \varphi_r + \varphi_t + \varphi_s + \varphi_a$$

and

$$\rho + \tau + \sigma + \alpha = 1 \tag{8.1}$$

where ρ is the refractivity, defined as $[(n - n')/(n + n')]^2$, n is the refractive index of the incident medium, n' is the refractive index of the material, and

$$n(\lambda) = \sin\Theta_1/\sin\Theta_2 \tag{8.2}$$

is equal to (velocity in vacuum)/(velocity in material). λ is the given wavelength, and Θ_1 and Θ_2 are the incident and refracted angles, respectively. Refraction (R) (i.e., the bending of a light beam) and dispersion (υ) (i.e., the splitting of white light into colors) occur because different light frequencies have different velocities. We know that light travels faster in a vacuum than through any collection of atoms, so the n of any substance is >1. Dielectric materials interact differently than free space because they contain electrical charges that can be displaced, as discussed in detail in Chapter 7. The change in wave velocity (v) and intensity of a sinusoidal electromagnetic wave, as indicated by the complex coefficient of R, $n^*(\omega)$, is shown by the following expression: $n^*(\omega) = n(\omega) + ik(\omega)$, where n is the index of R, $k(\omega)$ is the index of absorption, and (ω) is angular frequency. According to solutions derived from Maxwell's equations, the dielectric constant and the n of condensed media with nonvanishing conductivity are the complex quantities:

$$n^*(\omega) = n(\omega) + ik(\omega)$$

and

$$\varepsilon^*(\omega) = \varepsilon'(\omega) + i\varepsilon''(\omega) \tag{8.3}$$

where $\varepsilon'(\omega) = (n - k)$ and $\varepsilon'' = 2nk$. The response of the medium to an electromagnetic field can be fully described on a macroscopic level in terms of a pair of optical constants, namely, n and k, or ε' and ε''.

The n is a function of the frequency of light, normally decreasing as the λ increases. This change with λ is known as dispersion (υ) of the n and has been defined by the relation:

$$\upsilon = (dn/d\lambda) \tag{8.4}$$

υ can be determined by a comparison of the n to the λ. However, most practical measurements are made by using the n at fixed λ values rather than by determining the complete υ curve. Values are most commonly reported as reciprocal relative υ, for:

$$\upsilon = \frac{(n_D - 1)}{(n_F - n_C)} \tag{8.5}$$

where n_D, n_C, and n_F refer to the values of n obtained by measurements with sodium D-line, the hydrogen line, and the hydrogen C line (5,893 Å, 4,861 Å, and 6,563 Å, respectively). The n' values of glasses can be classified into three major categories: dense-flint, light-flint, and borosilicate (e.g., Pyrex) glasses. Dense-flint glasses have n' values from 1.6 to 1.7 µm, light-flint glasses have n' values from 1.55 to 1.62 µm, and the n' values of borosilicate glasses range from 1.5–1.55 to 0.3–1.0 µm. Pure tellurite glasses have n' values in the range 2.24–2.14 µm at λs from 500 to 1,000 nm and are considered the highest-optical-density flint glasses (discussed below). The constituents of most practical glasses have the greatest effect in raising the index of the fraction and decreasing its υ value.

In 1993, Yasui and Utsuno announced that it is possible to construct a universal material design model for any compositional system. They investigated fluorides and chalcogenides as these represent two extremes of the relationship between composition and properties, which is called "additivity." For fluoride glass, a model using additivity principles should reliably reproduce properties derived from glass compositions, and a prototype system has been constructed. For chalcogenide glass, additivity is obvious for some properties, but a model glass is needed in this system for other properties.

Some light is refracted at an angle equal to the incident angle, which is called "specular reflection." The normal incidence of the fraction of light reflected in this way is given by Fresnel's formula:

$$R = \frac{(n-1)^2 + k^2}{(n+1)^2 + k^2} \tag{8.6}$$

where k is the absorptive factor. Within the optical region of the spectrum, the $k(\omega)$ is much less than the n'; thus, Equation 8.6 reduces to

$$R = \frac{(n-1)^2}{(n+1)^2} \tag{8.7}$$

The absorption coefficient (β) is a function of λ and is defined by $\beta = 4\pi k/\lambda$. The transmission coefficient (τ) is the fraction of light transmitted as given by the β and sample thickness:

$$\tau = I/I_o$$
$$= \exp(-\beta x) \tag{8.8}$$

where I_o is the initial intensity, I is the transmitted intensity, and x is the optical path length. Scattering is the υ of part of a light beam and its de-intensification in a heterogeneous medium such as a transparent medium containing small particles. For a single particle with a projected area πr^2, the fraction of beam intensity lost is given by

$$(\Delta I/I) = -K(\pi r^2/A) \tag{8.9}$$

where r is the particle radius, A is the beam area, and K is the scattering factor, which varies between 0 and 0.4.

The transparency of glass represents a major dilemma in theoretical physics because it is not known why a given glass absorbs radiation at some frequencies while transmitting others and why different frequencies travel at different speeds.

The transparency of a crystalline insulator has been shown to be a consequence of the forbidden gap between the valence and conduction bands. The transparency of a crystalline insulator occurs when photon energies in the visible range (VIS) are insufficient to excite an electron across the forbidden energy region. However, glass is also transparent. It must therefore have a full valence band and an empty conduction band, and there must be a gap lying between them. In Part III, I examine the work of Mott (1969), who realized that electron traps must be a natural attribute of the noncrystalline state. The most common sources of color in glass are the electronic processes associated with transition elements. These have inner electrons that are not involved in covalent bonding and that are therefore available for absorbing and scattering light radiation. Moreover, many of their transitions fall within the VIS range of the spectrum. In the periodic table, the row of transition elements that starts with scandium contains the first elements encountered with electrons in the d orbital. These are special in that their electron probability density (ρ) distributions have a multiplicity of lobes extending in various directions. This characteristic, which distinguishes these materials from those with the more symmetrical lower-energy states, leads to enhanced sensitivity to the local atomic environment. In an isolated atom of a transition element, the d-type energy levels are degenerate; they all have the same energy. In condensed matter, on the other hand, the relatively large spatial excursions made by the d electrons periodically bring them into close proximity with the surrounding ions. It is generally recognized that the n and ρ of many common glasses can be varied by changing the base glass composition, changing the sample temperature, subjecting the sample to pressure, and inducing a new transformation range history, such as by reannealing.

Coloration of glass is caused by the presence of impurity atoms. The most common sources of color in glass are the electronic processes associated with transition elements. One important way to deliberately color both glass and opaque ceramics is to produce in them small percentages of impurity atoms. Radiation energy is converted to kinetic energy of electron motion through the conduction band, and subsequent electron–atom collisions lead to a further conversion to heat energy. If the transfer is sufficiently large, the material is opaque to radiation at the frequency in question. In the absence of interband transitions the material is transparent, with the atoms both absorbing and re-emitting light energy.

Before the optical properties of tellurite glasses are discussed, note that in the VIS, light rays are generally referred to by λ in nanometers (SI units) or in angstroms. In the infrared (IR) region, rays are referred to by their wave number ($v = [1/\lambda]$) and measured in centimeters. In the ultraviolet (UV) region, the v or the energy E (eV) of the photons is used. For conversion, 1 eV = 1,239.8/λ, where λ is in nanometers. The different techniques to measure optical constants in the VIS are reviewed below. Current data on these optical constants—including the causes and effects of composition, frequency, and temperature dependence—are presented. Optical spectroscopy in both the IR and UV regions is covered in Chapters 9 and 10, respectively.

Interaction between an electromagnetic wave and bulk material is governed by the n^* of the material in question. This, in turn, can be split into resistive and reactive components, and the electric and magnetic fields can be seen to have two components: a regular sine wave with a modified period, and a damping term with exponential decay. The former term is responsible for the reflection characteristics of the medium, but the latter is responsible for its absorption. It has been mentioned that the n of a solid material varies with the λ. In crystalline solids, the n values depend on the direction of the wave vector and differ in various directions. The transmission characteristics of the material change depending on the direction of the wave vector, and the material becomes birefringent and the light beam becomes polarized. In highly electric fields, most materials exhibit optical nonlinearity.

8.1.1 What Is Optical Nonlinearity?

A material consists of nuclei surrounded by electrons. In dielectric materials, these particles are bound and the bonds have a degree of elasticity derived from the interaction between charged particles. When an electric field (E) is applied across the material, the positive charges (ions) are attracted and the negative charges (electrons) are repelled slightly, which results in the formation of a collection of induced dipole moments and hence polarization (P). The E can be in the form of voltage across the material, a light passing through the material, or a combination of both. A lightwave comprises electric and magnetic fields, which vary sinusoidally at "optical" frequencies (10^{13}–10^{17} Hz). The motion of charged particles in a dielectric medium when a light beam is passed through it is therefore oscillatory and the particles form oscillating dipoles. Because the effect of the E is much greater than the effect of the magnetic field, the latter can, for all intents and purposes, be ignored. Also, because the electrons have a lower mass than the ions, it is the motion of the former that is significant at high optical frequencies (UV and VIS regions of the spectrum). The electric dipoles oscillate at the same frequency as the incident E and modify the way that the wave propagates through the medium. The electric displacement (D) is shown by the following equation:

$$D = \varepsilon_o \left(E + P \right) \tag{8.10}$$

where ε_o is the free-space permittivity and the P can be expressed in terms of the E and the susceptibility (χ):

$$P = \varepsilon_o \left[\chi^{(1)} E + \chi^{(2)} E + \chi^{(3)} E + ... \right] \tag{8.11}$$

where $\chi^{(1)}$ is the linear electric χ, $\chi^{(2)}$ is the quadratic nonlinear χ, and $\chi^{(3)}$ is the cubic nonlinear χ, and the field-dependent n can be written as:

$$n^2 = 1 + \chi^{(1)} + \chi^{(2)} + ... \tag{8.12}$$

Oscillation of the electromagnetic wave results in oscillation of the induced dipole moment, which in turn results in the generation of light with different frequencies from that of the incoming lightwave. It should be noted that the applied E used in the above equation is the total field and can be made up of a number of contributions (i.e., both optical and electric). Good reviews of "optically nonlinear materials" have been published by Wood (1993). Simmons et al. (1992) reported on nonlinear optical processes in glasses and glass-based composites. Applications of $\chi^{(1)}$ have been examined by R in lenses, optical fibers, and physical optics; whereas applications of $\chi^{(2)}$ have been accomplished by frequency doubling, in the second harmonic generation (SHG), and applications of $\chi^{(3)}$ are performed by frequency tripling in the third harmonic generation (THG).

8.1.2 ACOUSTO-OPTICAL MATERIALS

Acoustic vibration is induced in an acousto-optical material when a frequency is imposed via a piezoelectric transducer bonded to the side of a block of the material, which induces a standing wave. These acoustic waves result in changes in n with the periodicity of the waves, such that a beam of light passing through the glass is diffracted. The induced P or perturbation varies spatially with a periodic length equal to the acoustic λ. Normal technology limits the choice of acousto-optic material to $LiNbO_3$–TeO_2, quartz, or extra-dense flint glasses to achieve relatively high diffraction efficiencies. The cells themselves are fairly simple, usually comprising the cube of acousto-optic material, the transducer bonded onto one side (e.g., lead zircotantalate, lithium niobate, strontium barium niobate, or a ferroelectric compound) and an acoustic damping layer bonded to the other side of the cube to reduce reflections and complicated ripple interference patterns. Triangular cells with transducers on each of the side faces have been constructed to achieve ultra-fast switching with minimal reflections for laser Q-switch operations. Acousto-optic devices can be used in three configurations:

1. Amplitude modulation is produced by using the first-order diffracted output beam and varying the radiofrequency (RF; acoustic wave) power. This can also be used as a switch.
2. Beam deflection is effected by using the diffracted output beam and varying the RF.
3. A laser beam with frequency sidebands can be produced by recombining the undiffracted and the first-order beams.

The switching speed of the effect depends on the transit time of the acoustic wave across the optical beam and can therefore be increased by focusing the laser beam down to a small spot size. Commonly available devices have switching speeds ranging from 10^{-5} to 10^{-8} s. The ability to rapidly modulate the amplitude and direction of a laser beam by simply applying an RF field to a block of suitable material has resulted in the widespread use of acousto-optic technology throughout the fields of reprographics, imaging, laser Q-switching, and the laser entertainment industry. A true two-dimensional deflection can be gained by simply using two thin, large-aperture acousto-optical cells orthogonally to each other. By generating two optical beams (input and first-order) and redirecting them, it is possible to produce an interference pattern. This fringe pattern is not stationary because there is an RF shift; therefore, the pattern moves in the direction perpendicular to the fringes at the RF. This phenomenon has been used to measure particle flow in a medium sited at the point of beam recombination. Stationary particles generate a pulsed scatter signal at the RF, and a moving particle generates a signal that is Doppler-shifted to higher or lower frequencies.

Glasses are also widely used as host materials for lasers, optical fibers, and wave guides. Relative to crystal lasers, they offer improved flexibility in size and shape and are readily obtained as large, homogeneous, isotropic bodies with high optical quality. The n values of glass hosts can be varied between ~1.5 and 2.0, and both the temperature coefficient of the n and the strain optic coefficient can be adjusted by changes in the composition. The principal differences between the behaviors of glass and crystal lasers are associated with greater variation in the environments of lasing ions in the glasses, which leads to a broadening of fluorescence levels in the glasses. As an example of this broadening, the width of the Nd^{3-} emission in $Y_3Al_5O_{12}$ glass (YAG) is ~10 Å; that in oxide glasses is typically ~300 Å. The broadened fluorescence lines in glasses make it more difficult to obtain continuous-wave laser operation, compared with the same lasing ions in a crystalline host.

The broadened fluorescence lines in glasses are used to advantage in the so-called "Q-switched operation." In this case, the complete reflecting mirror of the previous example is replaced by a reflector of low reflectivity during pumping. After an inversion in the population at fluorescence and lower levels, the reflector is rapidly increased to a high-reflectivity value, and the energy stored by the excited ions is rapidly converted to a pulse of light whose peak power can be orders of magnitude greater than power levels reached in continuous or long-pulsed operation. The duration of such Q-switched pulses is typically in the range of $10-50 \times 10^{-9}$ s. A number of ions have been made to lase in glass, including Nd, Yb, Er, Ho, and Tm. The most important is neodymium, a four-level system, because it can be operated at room temperature with high efficiency. The erbium-glass laser, a three-level system with emission at about 1.54 μm, provides greatly improved eye safety relative to neodymium lasers. Efficient erbium-glass has been obtained by doping ytterbium.

Because of its total internal reflection, a glass rod can transmit light around corners. An image incident on one end of the rod can be seen at the other end as an area of approximately uniform intensity. The development of laser devices is based on using light as the carrier wave in communications systems. For single-mode operation, core diameters in the range of a few micrometers are required. Because most fiber handling requires outer cladding diameters in the range of at least 100 μm, single-mode fibers would have cladding–core diameter ratios in the range of 20 or 30 to 1. Such fibers represent a most critical test of fabrication techniques and the ability to control form.

8.2   EXPERIMENTAL MEASUREMENTS OF OPTICAL CONSTANTS

8.2.1   Experimental Measurements of Linear Refractive Index and Dispersion of Glass

A series of books have explained the experimental techniques for measuring optical constants (e.g., Varshneya 1994). The n of a thin film was explained by Swanepoel (1983). "Experimental Techniques of Glass Science," edited by Simmons and El-Bayoumi (1993), collected and discussed various methods, including

- n and υ measurements: single-surface methods, Abbe refractometer measurements, n measurement using Brewster's angle (prism methods that allow for minimum deviation), autocollimator techniques, V-block measurement techniques, index-mating methods, immersion methods, and interferometric-index measurement
- Thermal dependence of optical characterization

- Reflectance measurements
- Attenuation measurements
- Absorption measurements
- Scattering measurements
- Fiber-loss measurements

The dependence of n on temperature has also been discussed by Scholze (1990) in his book *Glass: Nature, Structure, and Properties*. He reported that the n temperature dependence is based on the specific R relation to give

$$dn/dT = (\partial n/\partial T)_\rho + (\partial n/\partial \rho)_T \, d\rho/dT$$

$$= (\partial n/\partial T)_\rho - \beta\rho(\partial n/\partial \rho)_T \tag{8.13}$$

β is the cubic coefficient of expansion, $(\partial n/\partial T)_\rho$ represents the temperature dependence of the n at constant ρ and is thus dependent only on the R—that is, α.

8.2.2 EXPERIMENTAL MEASUREMENTS OF NONLINEAR REFRACTIVE INDEX OF BULK GLASS

To obtain the $\chi^{(3)}$ values, the THG $I_{(3\omega)}$ was measured by the THG method using a nonlinear optical measurement apparatus as shown schematically in Figure 8.1 (by Kim et al. 1993). The

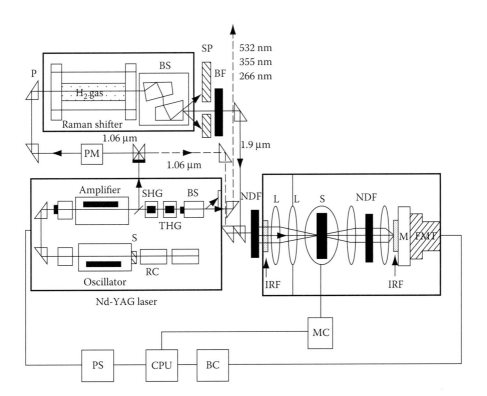

FIGURE 8.1 THG method, using a nonlinear optical measurement apparatus. (From S. Kim, T. Yoko, and S. Sakka, *J. Am. Ceram. Soc.,* 76, 2486, 1993. With permission.)

Q-switched Nd-YAG laser was operated at the fundamental λ of 1.064 μm with a pump pulse duration of 10 ns and peak power ρ of about 200 MW/cm². The laser frequency at 1.9 μm was obtained directly by stimulated Raman scattering in a high-pressure hydrogen cell, which was excited by the 1.064-μm Nd-YAG line. The conversion efficiently was 10%. To avoid laser damage to a glass sample, the 1.9 μm beam is appropriately attenuated by means of neutral-ρ filters. The third-harmonic-wave output at 0.633 μm is detected by a photomultiplier (PM) after being isolated by a monochromatic. The sample is mounted on a goniometer and rotated from +30° to –30° by a computer-controlled stepper motor with respect to the axis perpendicular to the 1.9 μm incident beam.

Sabadel et al. (1997) used the Z-Scan technique (single-beam method) to measure nonlinear optical characterizations. As shown in Figure 8.2, the sample is moved along the Z direction and very small effects are recorded. For this reason, very high sample quality is required. The transmittance (D_1/D_2) of a tightly focused Gaussian beam through a finite aperture S (linear transmission) in the far field is measured as a function of the sample position (Z) with respect to the focal plane. At each position, the sample experiences a different incident field (amplitude and phase). Nonlinear R in the sample causes beam broadening or narrowing in the far field and thus modifies the R of light passing through the aperture as the sample position is changed. For a material with a positive nonlinear n (n_2), the beam broadening (or narrowing) in the far field occurs when the sample is situated before (or after) the focus. A decrease in transmittance followed by an increase in transmittance (valley-peak) for increasing Z then denotes a positive nonlinear R. In a material with a negative n_2, a peak-valley configuration characterization has been obtained.

The n_2 is obtained in terms of the difference in transmittance between the peak and the valley ($\Delta T_{p,v}$) as stated by Sheik-Bahae et al. (1990):

$$\Delta T_{P,V} = 0.406\left(1-S\right)^{0.25}\left(2\pi\lambda\right)\left(n_2/\sqrt{2}\right)\left(I_o\right)\left(L_{eff}\right)$$

and

$$L_{eff} = \left(1-e^{-L\alpha_o}\right)\Big/\alpha_o \tag{8.14}$$

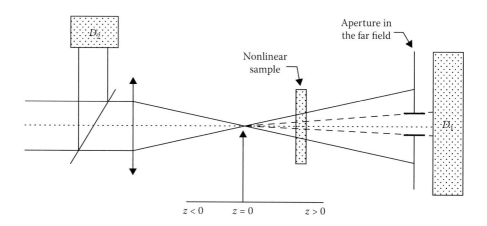

FIGURE 8.2 Z-Scan technique (single-beam method) to measure the nonlinear optical characterizations. (From J. Sabadel, P. Armand, D. Herreillat, P. Baldeck, O. Doclot, A. Ibanez, and E. Philippot, *J. Solid State Chem.*, 132, 411, 1997.) With Permission.

where n_2 is expressed in square meters/watt and $n = n_o + n_2I$, where I_o is the peak axis irradiance at focus, expressed in watts per square meter inside the sample, λ is in meters, L_{eff} is the effective length, α_o is the linear absorption of the sample, and $(1/(\sqrt{2}))$ is the time average factor used for temporally Gaussian pulses having pulse widths longer than the nonlinearity response time. Removing the aperture in the far field enables nonlinear absorption measurements because a Z scan with a fully opened aperture ($S = 1$) is insensitive to nonlinear R. Such Z-scan traces are expected to be symmetric with respect to the focus ($z = 0$), where they exhibit a minimum transmittance for nonlinear absorption (e.g., multiphonon absorption). In this case, the normalized energy transmittance is given by Sheik-Bahae et al. (1990):

$$T(Z) = \frac{1}{2\pi} \left\{ \frac{1}{q_o(z,0)} \int \ln\left[1 + q_o(z,0)\exp\left(-t^2\right)\right] dt \right\} \tag{8.15}$$

where $q(z, t)$ was expressed by

$$q_o(z,t) = \frac{\beta L_{\mathrm{eff}} I_o(t)}{1 + (z/z_o)}$$

and β is the nonlinear β expressed in meters per watt or, more commonly, in centimeters per gigawatt with $\alpha = \alpha + \beta I_o$ representing the sample absorption expressed in centimeters.

Sabadel et al. (1997) reported measurements performed at 532 nm with a frequency-doubled single pulse extracted from a mode-locked and Q-switched Nd-YAG laser by a Pockels cell. The laser is operated in the TEM_{00} mode at a 60 Hz repetition rate. The temporal width of the pulses was 250 ps (full width at half maximum). The laser parameters (waist, I_o) were calibrated to a 1 mm cell filled with CS_2 with a known n_2.

It is also very important to measure changes in n_2 under mechanical stress or changes in the electric or magnetic field. Cherukuri et al. (1993) explained and collected various experiments, including:

- Photoelastic effects (stress-optical effect, strain-optical effect)
- Kerr effect (quadratic electro-optical effect)
- Magnetic effects (linear magneto-optical effect [Faraday effect], quadratic magneto-optic effect [Cotton-Mouton effect])

8.2.3 Experimental Measurements of Fluorescence and Thermal Luminescence

The fluorescence spectra can be measured by fluorescence spectrophotometry, with a dye laser as the up-conversion excitation source (Figure 8.3a and b). Good-optical-quality glasses a few millimeters thick and tens of square millimeters in area should be chosen for absorption measurements. Samples should have the same dimensions and be well polished in order to minimize scattering loss on sample surfaces. The diode laser is used as an excitation source for the up-conversion spectra. The spectra are measured with a fluorescence spectrophotometer. The fluorescence of the glasses excited by UV radiation is also measured and is called "normal fluorescence." The main pump used for fluorescence and lifetime measurements is the coherent argon ion pump. The fluorescence spectra and lifetime measurements are obtained at 90° to the incident pump beam so that there is minimum interference by the pump beam with the measurements. Absorption spectra are obtained using a UV/VIS/near-IR double-beam spectrophotometer. Also, the photon sideband spectra (explained in Chapter 1) of a material can be measured by the excitation of light with a certain λ and then monitoring $^5D_0 \rightarrow {}^7F_2$ emission.

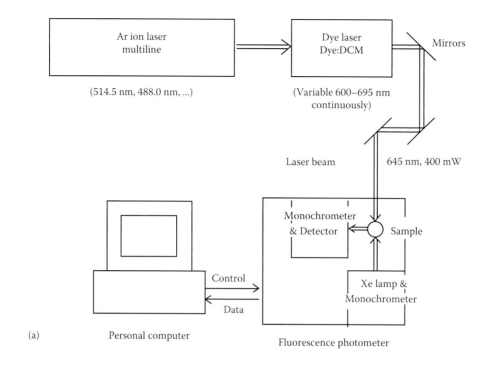

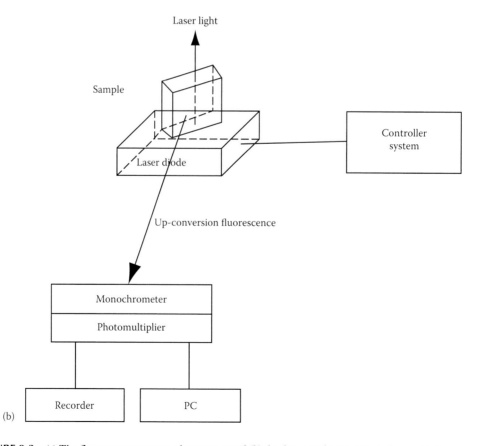

FIGURE 8.3 (a) The fluorescence spectrophotometer and (b) dye laser as the up-conversion excitation source.

The photoluminescence spectra of a material can be measured by using a spectrophotometer. Both excitation and emission spectra are carried out on the spectrofluorimeter. The thermal luminescence (TL) intensity can be measured using a photomultiplier (PM). The TL intensity of the resulting glow curve is directly proportional to the output signal of the PM tube. The TL intensity is recorded as a function of temperature by using a TL analyzer containing a temperature–time control and a special PM tube. The basic function of the detector is to heat the material using a reproducible controlled temperature cycle and to detect the light emitted from the sample by means of a low-noise, high-gain PM that converts the emission from TL material into a current signal that is amplified, integrated, and displayed.

8.3 THEORETICAL ANALYSIS OF OPTICAL CONSTANTS

8.3.1 QUANTITATIVE ANALYSIS OF THE LINEAR REFRACTIVE INDEX

To understand the effects of R, it is helpful to use the model of a forced harmonic oscillator. El-Mallawany (1992) used classic dielectric theory to provide a number of relationships between the n and the ρ of solids. For a given glass composition, the mathematical relationships between n and ρ that describe these changes are of importance because of their association with structural chemistry and also because of the need to elucidate more clearly the effect of atomic interactions on atomic αs.

8.3.1.1 Refractive Index and Polarization

At the optical frequencies where n is measured, the dielectric constant arises almost entirely from electronic P. Electronic α is expressed by the following:

$$\alpha = \frac{e^2}{4\pi m} \sum_i \frac{f_i}{v_i^2 - v^2} \tag{8.16}$$

where e is the electronic charge, m is the electronic mass, f represents the oscillator strength, v_1 is the electronic frequency, and v is the frequency of the incident photons. According to Maxwell's theory, the relationship $n^2 (\lambda \to \infty) = \varepsilon$ holds for infinity long waves. The Clausius–Mossotti equation (explained in detail in Chapter 7):

$$\frac{4\pi N\alpha}{3M} = \frac{\varepsilon - 1}{\varepsilon + 2}$$

$$= \frac{n^2 - 1}{n^2 + 2} \tag{8.17}$$

$$= N \frac{e^2}{12\varepsilon_0 \pi^2 m} \sum_i \frac{f}{v_i^2 - v^2}$$

applies, where N is the number of electrons in the solid. As indicated in the above equation, the n depends on the composition of an optical material; the more polarizable the outer electrons, the higher the n. When the Clausius–Mossotti equation is expressed in terms of the specific mass or ρ, it reduces to:

$$\left\{ \frac{n^2 - 1}{n^2 + 2} \right\} \left(\frac{1}{\rho} \right) = R \tag{8.18}$$

and describes the R of the material. The R_M is:

$$\left\{\frac{n^2-1}{n^2+2}\right\}\left(\frac{M}{\rho}\right) = R_M \tag{8.19}$$

where M is the molecular weight of the material and M/ρ equals the molar volume (V_m). Equations 8.18 and 8.19 are the well-known "Lorentz-Lorenz" equations. The R_M and n depend on the α of the material, which in turn is determined, approximately, by the sum of R_I. The R_M is proportional to α; that is,

$$R_M = 4\pi N_A \alpha / 3 \tag{8.20}$$

where N_A is Avogadro's number.

Consequently, the R_M and n depend on the α of the material, which in turn is determined, approximately, by the sum of R_I. The value of R_I in crystal compounds depends on at least the following factors:

- The electronic α of the ion, which increases as the number of electrons in the ion increases
- The coordination number of the ion
- The α of the first neighbor ions coordinated with it
- The field intensity (z/a^2), in which z is the valence of the ion and a is the distance of separation

El-Mallawany (1992) used the values of R_I, ionic α, and coordination number for some cations like Li$^+$, Na$^+$, K$^+$, Rb$^+$, Mg$^+$, and Sr$^+$, and some anions such as O^{2-}, Cl$^-$, and F$^-$, as shown in Table 8.1. El-Mallawany (1992) also summarized some data on the n of tellurite glasses by previous authors including Vogal et al. (1974), as given in Tables 8.2 through 8.4. These properties data (ρ and n from Tables 8.2 through 8.4) relate to the lower and upper limits of the glass formation range in each system. If halides are melted with TeO$_2$, evaporation is more or less inevitable; thus, ternaries of the volatile component result, such as TeO$_2$–Me$_2$O–MeCl$_2$. The optical data listed show that

TABLE 8.1

Values of Ionic Refraction, Ionic Polarizability, and Coordination Number for Some Cations and Anions

Species	Coordination Numbers	R_i(cm^3)	($\alpha \times 10^{-24}$ cm^3)	Radius of Ion Coordination (nm)
Li$^+$	6, 6	0.08	0.029	0.068
Na$^+$	6, 8	0.47	0.179	0.098
K$^+$	6, 10, 12	2.24	0.830	0.133
Rb$^+$	10, 12	3.75	1.400	0.149
Mg^{2+}	4, 6	0.26	0.094	0.074
Al^{3+}	4, 6	0.14	0.052	0.570
Ti^{4+}	4, 6	0.6	0.185	0.064
O^{2-}		6.95	3.880	0.136
F$^-$		2.44	1.040	1.330

TABLE 8.2

Area of Glass Formation (Second Component), Density, Linear Refractive Index, and Abbe Number of Binary Oxide–Tellurite Glasses

Second Component	Glass Formation Range (mol%)		Density Range (g/cm³)	Refractive Index Range	Abbe Number Range
	Lower Limit	Upper Limit			
$LiO_{1/2}$	12.2	34.9	5.27–4.511	2.108–1.931	18.4–20.1
$NaO_{1/2}$	5.5	37.8	5.43–4.143	2.148–1.801	17.0–22.2
$KO_{1/2}$	6.5	19.5	5.27–4.607	2.122–1.955	17.2–18.8
Rb_2O	5.6	21.0	5.357–4.725	2.110–1.893	18.4–20.1
Tl_2O	13.0	38.4	6.262–7.203	2.190–2.280	15.9–11.3
BeO	16.5	22.7	5.357–5.246	2.110–2.075	18.4–19.1
MgO	13.5	23.1	5.347–5.182	2.107–2.048	
SrO	9.2	13.1	5.564–5.514	2.128–2.117	17.8–17.9
BaO	8.0	23.1	5.626–5.533	2.156–2.052	17.6–19.2
ZnO	17.4	37.2	5.544–5.466	2.125–2.038	18.4–20.7
PbO	13.6	21.8	6.054–6.303	2.202–2.206	16.5–16.4
La_2O_3	4.0	9.9	5.662–5.707	2.160–2.108	17.6–19.5
Al_2O_3	7.8	16.8	5.287–4.85	2.082–1.947	19.2–23.1
TiO_2	7.8	18.9	5.580–5.242	2.202–2.220	16.0–14.9
WO_3	11.5	30.8	35.782–5.953	2.192–2.191	16.7–16.3
Nb_2O_5	3.6	34.4	5.558–5.159	2.192–2.211	16.7–15.7
Ta_2O_5	3.8	33.4	5.659–6.048	2.190–2.156	17.0–18.0
ThO_2	8.4	17.1	5.762–5.981	2.165–2.133	17.7–19.0
GeO_2	10.2	30.2	5.437–5.057	2.126–2.011	22.1–18.4
P_2O_5	2.2	15.6	5.433–4.785	2.149–1.993	17.8–21.9
B_2O_3	11.8	26.4	5.205–4.648	2.082–1.964	19.2–21.7
1 SiO_2:5.2 base glass ($Al_2O_3TeO_3$)	1.2	17.9	4.854–4.484	1.959–1.882	22.2–24.3

Source: W. Vogel, H. Burger, B. Muller, G. Zerge, W. Muller, and K. Forkel, *Silikattechnel*, 25, 6, 1974.

a valuable family of optical glasses has been discovered. To understand how tellurite glasses have high n values, the number of polarizable atoms/unit volume (N/V), the electronic α of the ion, and α_{ionic} must all be considered. The values of n for TeO_2 vary from that of glass (2.2) to that of crystal (2.37), which is due to the variation of ρ from that of glass (5.1 g/cm³) to that of crystal (5.99 g/cm³). The N/V decreases from 6.78×10^{22} cm⁻³ for TeO_2 crystal to 5.77×10^{22} cm⁻³. By knowing the R_M of both TeO_2 crystal and glass, and by applying Equation 8.22, the α is decreased from 7.37×10^{-24} cm³ to 6.96×10^{-24} cm³ (i.e., $\alpha_{cryst}/\alpha_{glass} = 1.1$). Previous conclusions confirm that the dielectric constant decreases from 23.44 to 20.1 and the static α decreases from 2.8×10^{-6} to 2.2×10^{-6} cm³. The ratios of the ρ, dielectric constant, and static α of TeO_2 crystal to TeO_2 glass vary between 1.18 and 1.27. Thus, tellurite glass has a lower n from low electronic α and higher specific V_m. This indicates the important role of the TeO_2 unit in determining the n of binary and ternary tellurite glasses.

El-Mallawany (1992) used the chemical composition of the glass and its ρ from Tables 8.2 through 8.4 to calculate the N/V, which can be shown to depend on the modifier as in Table 8.5. However, these changes in N/V are not sufficient to account for the observed decrease in the n. With the substitution of alkali oxide M_2O for TeO_2 in alkali–tellurite glasses, bridging Te–O–Te bonds are broken, and nonbridging Te–O–M^+ bonds are formed. The nonbridging oxygen (NBO) bonds have a much greater ionic character and much lower bond energies. Consequently, the NBO bonds have higher αs and cation Rs, and the n of TeO_2–M_2O glass depends on the ionic R of M as listed in Tables 8.2 through 8.4. The n of each TeO_2–M_2O system at higher compositions is

TABLE 8.3

Halide-Tellurite Glass Systems, Areas of Glass Formation, and Properties

System and Reference	Halide (Range [mol%])	Metal Oxide (Range [mol%])	Density (Range [g/cm³])	Refractive Index (Range)	Abbe Number (Range)
Vogel et al. (1974)					
TeO₂–LiF					
TeO₂–LiCl	11.9–33.2	4.3–11.6	5.197–4.318	2.138–1.982	17.2–19.2
TeO₂–LiBr	5.0–6.3	6.5–7.9	5.363–5.294	2.150–2.137	16.5–16.8
TeO₂–NaF	19.0–32.0		5.207–4.901	2.058–1.947	18.2–19.9
TeO₂–NaCl	16.6–22.0	0.8–1.1	5.001–4.802	2.096–2.059	17.2–17.3
TeO₂–NaBr	8.6–26.4	4.6–2.9	5.196–4.724	2.121–2.035	16.1–16.3
TeO₂–KF	16.1–42.0		5.103–4.241	2.052–1.796	18.2–22.8
TeO₂–KCl	15.2–20.0	0.1–0.1	4.956–4.731	2.088–2.050	17.0–17.3
TeO₂–KBr	10.8–23.8	2.7–2.5	5.056–4.594	2.096–2.008	16.5–16.5
TeO₂–RbF	12.2–23.9	2.5–1.4	5.199–4.952	2.002–1.870	18.3–19.7
TeO₂–RbCl	21.9–27.4	0.1–0.3	4.777–4.589	2.019–1.965	17.5–18.1
TeO₂–RbBr	9.8–18.0	2.1–0.5	5.156–5.000	2.099–2.062	16.5–16.2
TeO₂–BeF₂					
TeO₂–BeCl₂					
TeO₂–BeBr₂					
TeO₂–MgF₂					
TeO₂–MgCl₂	5.0–15.7	1.2–9.0	5.255–4.543	2.123–1.981	17.6–19.7
TeO₂–MgBr₂	3.0–8.4	3.4–1.7	5.473–5.209	2.147–2.107	16.8–17.2
TeO₂–CaF₂					
TeO₂–CaCl₂	6.6–15.5	0.8–4.5	5.221–4.801	2.140–2.063	17.3–17.9
TeO₂–CaCl₂	3.2–4.8	2.9–5.2	5.383–5.323	2.165–2.149	16.6–16.7
TeO₂–SrF₂					
TeO₂–SrCl₂	7.2–46.7	1.4–9.1	5.271–3.966	2.132–1.833	17.3–26.5
TeO₂–SrBr₂	4.5–11.1	2.7–10.1	5.420–5.155	2.154–2.063	16.5–17.1

Glass	Composition (mol%)	x Value (mol%)		Refractive Index	Abbe No.
TeO2–BaF2	2.0–15.4		5.573–5.494	2.116–2.036	18.3–19.8
TeO2–BaCl2	7.2–19.7		5.334–4.988	2.135–2.509	17.3–18.4
TeO2–BaBr2	6.3–37.3		5.463–4.973	2.147–1.910	16.4–22.9
TeO2–LaF2					
TeO2–LaCl2	3.8–7.8	3.6–10.1	5.382–5.280	2.136–2.065	18.0–20.5
TeO2–LaBr2	2.0–2.0	3.8–14.0	5.537–5.632	2.159–2.072	17.1–20.7
TeO2–LaF2	13.7–26.0		5.993–6.295	2.158–2.119	17.8–18.0
TeO2–LaCl2	14.8–49.1	1.3–5.4	5.600–5.625	2.180–2.162	16.2–15.9
TeO2–LaBr2	5.6–39.4	9.6–12.7	5.925–6.207	2.209–2.295	15.7–12.0
TeO2–CdCl2	2.0–4.8	10.8–7.1	5.751–5.498	2.109–2.123	18.5–18.0
TeO2–CdBr2	5.1–3.2	18.1–14.3	5.813–5.619	2.115–2.134	17.5–16.6
TeO2–PbF2	13.7–26.0		5.993–6.295	2.158–2.119	17.8–18.0
TeO2–PbF2	14.8–49.1	1.3–5.4	5.600–5.625	2.180–2.162	16.2–15.9
TeO2–PbF2	5.6–39.4	9.6–12.7	5.925–6.207	2.209–2.295	15.7–12.0
TeO2–ZnF2	4.3–36.0	16.2–28.8	5.538–5.308	2.088–1.793	19.2–31.4
TeO2–ZnF2	2.5–33.6	9.3–14.2	5.202–4.316	2.110–1.893	17.5–23.8
TeO2–ZnF2	0.3–29.2	9.9–39.9	5.343–4.694	2.168–1.894	16.6–22.9

Yakhkind and Chebotarev (1980)	Composition (mol%)	x Value (mol%)	Refractive Index	Abbe No.
TeO2–WO3–ZnCl2	(80–20)–x	0.0	2.194	16.4
TeO2–BaO–ZnCl2	(85.7–14.3)–x	42.9	2.128	17.5
		0.0	2.098	18.7
		36.0	1.924	15.6
TeO2–Na2O–ZnCl2	(83.3–16.7)–x	0	2.040	18.0
		12.0	1.984	17.0

Source: W. Vogel, H. Burger, B. Muller, G. Zerge, W. Muller, and K. Forkel, *Sillikattechnel*, 25, 6, 1974; A. Yakhkind and S. Chebotarev, *Fiz. I. Kjim. Stekla*, 6, 485, 1980.

TABLE 8.4

Tellurite–Sulfate Glasses, Halide–Tellurite Glass Systems, Areas of Glass Formation, and Properties

System	Percentage of Halide (Range [mol%])	Metal Oxide (Range [mol%])	Density (Range [g/cm³])	Refractive Index (Range)	Abbe Number (Range)
TeO_2–Li_2–SO_4					
TeO_2–Na_2–SO_4					
TeO_2–K_2–SO_4	3.4–6.8	0.6–1.1	5.283–4.931	2.148–2.931	17.0–17.5
TeO_2–Rb_2–SO_4	3.3–4.9	0.9–1.3	5.292–5.118	2.115–2.081	17.4–17.7
TeO_2–Be–SO_4	3.8–17.2	3.7–6.8	5.303–4.487	2.101–1.900	18.7–23.7
TeO_2–Mg–SO_4	6.7–43.5	4.1–17.2	5.051–3.432	2.025–1.642	19.0–43.1
TeO_2–Ca–SO_4					
TeO_2–Sr–SO_4					
TeO_2–Ba–SO_4	2.6–8.0	0.3–4.3	5.437–5.194	2.152–2.032	17.1–18.1
TeO_2–La_2–$(SO_4)_3$	4.0–0.7	7.4–2.1	5.476–5.591	2.053–2.162	20.8–17.7
TeO_2–Pb–SO_4	5.0–12.4	13.6–23.0	6.020–6.271	2.165–2.116	17.2–18.0

Source: W. Vogel, H. Burger, B. Muller, G. Zerge, W. Muller, and K. Forkel, *Sillikattechnel*, 25, 6, 1974.

lower than those at lower concentrations of the oxide, which can be explained by the decrease in N/V for each binary glass system according to the molecular weight percent portion of the metal oxide. The R_M decreases for higher concentrations of alkali–metal oxide. The difference between the αs of binary systems can be attributed to the valence of the cation (monovalence or divulgence), the coordination number, and the field intensity (Z/a^2), where Z is the valence of the ion and a is the distance of separation.

Multicomponent glass generally behaves in a similar manner; for example, n depends on the M_2O and MCl_2 content in TeO_2–M_2O–MCl_2 glasses. There is also a general trend for n to depend on the value of the R_1 of the cations added. The low αs and ionic refractivities of no-oxygen anions, such as the halides F- and Cl- (Table 8.1), lower the ns. Using the Clausius–Mossotti equation, the effects of lower αs due to the substitution of 2 moles of halogen for 1 mole of oxygen can be seen, and the effect in reducing the N/V can also be distinguished. The results of these calculations are summarized in Table 8.5. It is clear that the difference in N/V is primarily responsible for the observed decrease in the n for each halide–tellurite glass system at higher halogen content. The lower αs of halogens seems to offer a satisfactory explanation for the reduction in n of these halide glasses. Figure 8.4a presents the Lorentz–Lorenz plot for the binary oxide–tellurite glasses when modified with alkaline metal oxide, and Figure 8.4b provides the plot for the ternary oxyhalide–tellurite glasses. The relationship between both n and ρ of each glass system is linear, and its position in the figure depends on the M. The increase of alkaline content alters the α of the tellurite glass as shown in Table 8.5 and calculated by the Clausius–Mossotti equation. Because the total α is the sum of the electronic and ionic αs, the ionic α of fluorine, for example, diminishes or counteracts the effect of its lower electronic α. When 1 mole of oxygen is replaced by 2 moles of fluorine, the total α is reduced as the electronic component is reduced though this is offset by a slight increase in the ionic polarizability. Ionic polarizability results from the displacement of ions of opposite signs from their regular sites due to the applied E. As nonbridging fluorine ions replace bridging-oxygen ions in the glass structure, the glass network is weakened, thus making ion displacements easier. The weakening of the glass network with fluoride substitution is clear from the decrease in the elastic modulus as found by El-Mallawany (1993).

TABLE 8.5
Refractive Index, Number of Atoms per Unit Volume (N/V), and Polarizability of Tellurite Glasses

Glass	Area of Glass Formation (Range [mol%])		Refractive Index (Range)	$(N/V) \times 10^{22}$ (Range [cm³])	Polarizability $(10-24)$ (Range [cm³])
	Metal	Halide			
TeO₂ (crystal)			2.37	6.78	7.37
TeO₂ (glass)			2.20	5.77	6.07
TeO₂–LiO₂	12.1–34.9		2.11–1.93	6.62–7.13	5.78–4.78
TeO₂–Na₂O	5.4–4.1		2.15–1.80	6.34–6.10	7.26–5.03
TeO₂–K₂O	6.5–19.5		2.12–1.95	6.13–5.67	6.31–6.12
TeO₂–Rb₂O	5.6–21.0		2.11–1.89	6.00–5.17	6.39–6.43
TeO₂–MgO	13.5–23.1		2.11–2.05	6.43–6.54	5.71–5.23
TeO₂–SrO	9.2–13.1		2.13–2.12	6.31–6.26	5.95–5.87
TeO₂–Li₂O–LiCl	11.0–33.2	4.2–11.6	2.14–1.98	6.43–6.57	5.81–4.79
TeO₂–Li₂O–LiBr	5.0–6.3	6.5–7.9	2.15–2.14	2.18–2.20	5.98–5.89
TeO₂–Li₂O–NaF	0.0–0.0	19.0–32.0	2.06–1.95	2.28–2.42	5.42–4.76
TeO₂–Li₂O–NaCl	16.6–22.0	0.8–1.1	2.10–2.06	2.12–2.06	5.98–6.1
TeO₂–Na₂O–NaBr	8.6–26.4	4.6–2.9	2.12–2.04	2.08–2.00	6.18–6.09
TeO₂–K₂O–KF	0.0–0.0	16.1–42.0	2.05–1.80	2.03–2.18	5.76–4.67
TeO₂–K₂O–KCl	0.1–0.1	15.2–20.0	2.09–2.05	2.03–1.99	6.20–6.16
TeO₂–K₂O–KBr	2.7–2.5	10.8–23.8	2.10–2.01	1.98–1.94	6.39–6.44
TeO₂–Rb₂O–RbF	2.5–1.4	12.2–23.9	2.02–1.87	2.04–2.03	5.79–5.34
TeO₂–Rb₂O–RbCl	0.1–0.3	21.9–27.4	2.02–1.97	5.52–5.05	6.35–5.94
TeO₂–Rb₂O–RbBr	2.1–0.5	9.8–18.0	2.10–2.06	5.61–6.53	6.56–5.37
TeO₂–MgO–MgCl₂	1.2–0.9	5.0–15.7	2.12–1.98	6.11–4.73	6.30–7.25
TeO₂–MgO–MgBr₂	3.4–1.7	3.2–8.4	2.15–2.11	6.26–5.86	6.19–6.50
TeO₂–SrO–SrCl₂	1.4–9.1	7.2–46.7	2.13–1.83	5.97–2.86	6.47–10.07
TeO₂–SrO–SrBr₂	2.7–10.1	4.5–11.1	2.15–2.06	5.99–5.50	6.50–6.55

Source: R. El-Mallawany, *J. Appl. Phys.*, 72, 1774, 1992. With permission.

8.4 LINEAR REFRACTIVE INDEX DATA OF TELLURITE GLASSES

8.4.1 TELLURIUM OXIDE BULK GLASSES AND GLASS-CERAMICS

In 1952, Stanworth observed that very little work had previously been done on tellurite glasses, and there initially appeared to be no mention at all of such glasses in the technical literature before that time. However, Stanworth was surprised to find that Berzelius (1834) had mentioned such glasses. Only very brief mention was made of alkali or barium–tetratellurite glasses, and no measurements of physical properties were recorded. Stanworth (1952) stated a formula to estimate the n as shown in Equation 8.21:

$$(n-1)V_o = \sum \alpha_M N_M \tag{8.21}$$

where V_o is the volume of the glass containing 1 g of oxygen, and N_M is the number of atoms of M in the glass per atom of oxygen. The cation radius of Te^{4+} is 0.84 Å in tellurium dioxide. The partial ρ characteristic of a component M_mO_n can be found from the above equation and by using the ρ of the glass as that of the crystalline α-TeO_2 (5.67 g/cm³), V_o about 14, and $[(m/n)a_M]$ equal to 13 by plotting $[(m/n)a_M]$

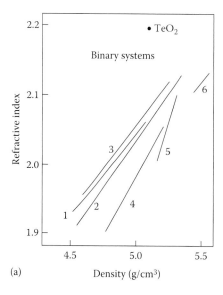

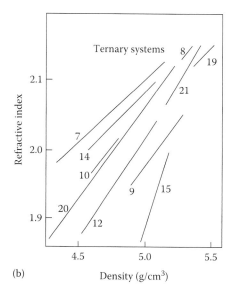

(a) Density (g/cm³)   (b) Density (g/cm³)

FIGURE 8.4 (a) Lorentz–Lorenz plot for the binary oxide–tellurite glasses when modified with alkaline metal oxide. (b) Lorentz–Lorenz plot for the ternary oxyhalide–tellurite glasses. Numbers on the lines are for glass composition as in Table 8.5. (From El-Mallawany, R., *J App Phys*, 72, 1774, 1992.)

against the cation radius (a_M). The value of n is 1.93 for the major proportion of TeO_2, which confirms the rough expectation that these glasses might have unusually high values for the n. Stanworth (1952) measured the n of tellurite glasses at 5780 Å, as determined by a 45° prism (Table 8.6).

Later, Weissenberg et al. (1956) prepared an optical glass that was essentially of tellurium oxide and aluminum oxide, and they developed a process for its production. The optical glass was of the form TeO_2–Al_2O_3–SiO_2–Na_2O_2 and a fifth component, such as La_2O_3, ZnO, In_2O_3, TiO_2, GeO_2, Bi_2O_3, WO_3, Ta_2O_5, Tl_2O, LiF, CaF_2, SrF_2, or BaF_2. The n and υ of these glasses were as shown in Table 8.6 by Weissenberg et al. (1956). It is clear that the presence of the alkali–oxide ionic bond causes the n to decrease and the υ value to increase. In 1966, Yakhkind announced the boundaries of regions of glass-forming compositions in binary tellurite glasses containing alkali, alkaline-earth,

TABLE 8.6
Values of Ionic Refraction, Ionic Polarizability, and Coordination Number for Some Cations and Anions

Species	Coordination Numbers	R_i(cm³)	$\alpha \times 10^{-24}$ cm³	Radius of Ion Coordination (nm)
Li^+	4, 6	0.08	0.029	0.068
Na^+	6, 8	0.47	0.179	0.098
K^+	6, 10, 12	2.24	0.830	0.133
Rb^+	10, 12	3.75	1.400	0.149
Mg^{2+}	4, 6	0.26	0.094	0.074
Al^{3+}	4, 6	0.14	0.052	0.570
Ti^{4+}	4, 6	0.6	0.185	0.064
O^{2-}		6.95	3.880	0.136
F^-		2.44	1.040	1.330

and heavy metal oxides in binary and ternary TeO_2–WO_3–(Ta_2O_5, BaO, Bi_2O_3, or Tl_2O) as shown in Table 8.6. Yakhkind (1966) investigated the crystallizability, ρ, optical constants, spectral-transmission characteristics, and chemical durability of these tellurite glasses as functions of their chemical compositions. The optical-constant data led to the development of a new flint-type optical glass, with $n = 2.1608$ and $\nu = 17.4$, possessing a high chemical durability and resistance to devitrification. The spectral properties and transmission regions of these binary and ternary tellurite glasses have been found at 390–410 nm, and at 450–500 nm the transmission reaches a maximum value of 70%–74%. An increase in tungsten oxide content caused displacement of the short-λ transmission boundary (390–410 nm) to longer λs. Losses due to reflection from each surface in the visible region of the spectrum were, on average, 13%–15%. Yakhkind and Loffe (1966) prepared prisms of the extra-dense flint-tellurite glasses ($n = 2.16$) with high chemical stability for commercial and laboratory use. The composition was 59–60 wt% TeO_2–10 wt% WO_3–12 wt% Nb_2O_5–2 wt% Na_2O–4.5 wt% BaO–19 wt% La_2O–3–4 wt% B_2O_3 and ≤ 2 wt% (CaO + ZnO). The n was 1.9858, and the υ coefficient was 23.2.

Redman and Chen (1967) prepared binary 65 mol% TeO_2–35 mol% ZnO glasses. The n was 2.036 (at 5,892 Å and 23°C). Ovcharenko and Yakhkind (1968) reported another family of tellurite glasses containing tungsten, bismuth, and lead oxides. These glass forms were TeO_2–(4–10 mol%) Nb_2O_5, TeO_2–WO_3–Bi_2O_3, TeO_2–WO_3–PbO, TeO_2–WO_3–TlO_2, TeO_2–Bi_2O_3–Nb_2O_5, and TeO_2–Tl_2O–Nb_2O_5. The n values were in the range 2.2–2.31 and the Abbe numbers ranged from 17 to 14. In 1972, Ovcharenko and Yakhkind measured the n of tellurite glasses containing scandium, yttrium, lanthanum, gadolinium, gallium, arsenic, zinc, and cadmium.

Izumitani and Namiki (1972) prepared high-n TeO_2–WO_3 tellurite glasses of no lower than 2.0 and with Abbe numbers of ≤ 21 with greatly improved stability with respect to both devitrification and chemical durability, by inclusion of Li_2O and at least one other oxide selected from a group consisting of K_2O, MgO, BaO, ZnO, CdO, TiO_2, PbO, PbO, La_2O_3, B_2O_3, Nb_2O_5, and Bi_2O_3. However, in 1976, Heckroodt and Res prepared erbium–tellurite glasses in the form 60 mol% TeO_2–20 mol% WO_3–20 mol% Er_2O_3 and measured the refractive index at 589.6 nm. The value of the refractive index was 2.134. In 1978, Mochida et al. published their article about the glass formation ranges of the binary tellurite systems TeO_2–$MO_{1/2}$ (M: Li, Na, K, Rb, Cs, Ag, and Tl) and TeO_2–MO (M: Be, Mg, Ca, Sr, Ba, Zn, Cd, and Pb) as shown in Table 8.6.

Yakhkind and Chebotarev (1980) determined the optical constants of tellurite–halide glasses. In 1981, Blair et al. reported that the high n of the tellurite glass 68 mol% TeO_2–7 mol% ZnO–13 mol% Li_2O–9 mol% PbO–3 mol% BaO leads to a figure of merit of 22×10^{-18} sec³/g. For this reason, tellurite glasses are used for acousto-optic devices. In 1985, Burger et al. measured the n of TeO_2–R_nX_m/R_nO_m as shown in Figure 8.5. They also positioned tellurite glasses in the diagram with other optical glasses.

Al-Ani and Hogarth (1987) correlated the n and ρ of tellurite glasses of the binary form TeO_2–WO_3 and TeO_2–Ta_2O_5 and ternary TeO_2–WO_3–Ta_2O_5 as $n = (C_{1\rho} + 1)^{1/2}$, where C_1 is a constant and is the Newton–Drude refractivity. Some of the data were tabulated, as in Table 8.7.

In 1991, Komatsu et al. prepared $(100 - x)$ mol% TeO_2–x mol% $LiNbO_3$ glasses by conventional melt-quenching methods and measured the ρ and n for an λ of 632.8 nm (He–Ne laser) at room temperature by using an ellipsometer (Mizojiri Optical Co., DVA-36L model). The ρ of the glasses decreased monotonically with increasing $LiNbO_3$ content from 5.5 g/cm³ ($x = 10$) to 4.58 g/cm³ ($x = 60$). The values of n of these glasses were 2.04–2.11 and showed an anomalous compositional dependence. Komatsu et al. (1991) explained that these glasses are composed of structural units of trigonal bipyramid TeO_4, trigonal pyramid TeO_3, and octahedral NbO_6. Burger et al. (1992) prepared tellurite glasses of the binary $(100 - x)$ mol% TeO_2–x mol% ZnO ($x = 17.4$, 19.9, 24.6, 29.6, 33.2, and 36.4; standard error, ± 0.1 mol%). Burger et al. (1992) found that these glasses were characterized by a high n, which increases with TeO_2 content, as shown in the 1st Ed. They also measured the υ, relative υ, and the Abbe numbers. The Abbe numbers were 18.3, 18.7, 19.1, 19.7, 20.0, and 20.6, respectively, for each glass.

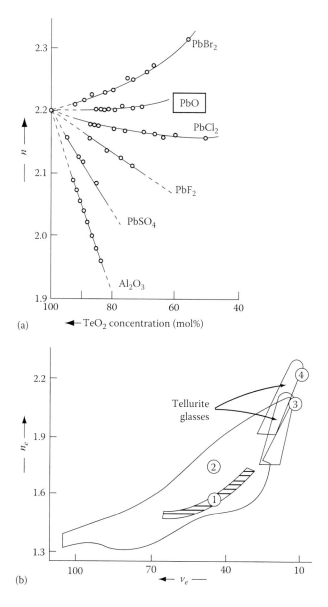

FIGURE 8.5   (a, b) Refractive indices of TeO_2–R_nX_m/R_nO_m. (From H. Burger, W. Vogal, and V. Kozhukarov, *Infrared Phys.*, 25, 395, 1985. With permission.)

Kim et al. (1993a) measured both the linear and nonlinear optical properties of TeO_2 glass by making a thin plate (0.22 mm). The glass sample was obtained by a rapid-quenching method. The n_2 was measured as a function of λ from 486.1 to 1,000 nm by using Mizojiri Optical Co., model DVW-35VW ellipsometer. The n decreased from 2.239 to 2.14 with increasing λ, as shown in Figure 8.6. Komatsu et al. (1993) prepared and measured the optical properties of TeO_2–BaO–TiO_2 glasses, and the crystalline phases were obtained by heat treatment to fabricate transparent TeO_2-based glasses by counting ferroelectric $BaTiO_3$ crystals. The n values of glasses and heat-treated samples were measured at a 632.8 nm λ (He–Ne laser) at room temperature by an ellipsometer. The small difference in the ns, between the TeO_2-based glasses ($n = 2.1$) as matrix and $BaTiO_3$ crystals ($n = 2.39$) leads to transparency, even though the particle size is much larger than the λ of visible light.

TABLE 8.7

Relation between Density and Refractive Index for Binary and Ternary Tellurite Glasses

Glass (mol%)	Density (g/cm³)	Refractive Index	C_1 (cm³/g)	C_2 (cm³/g)	C_3 (cm³/g)
TeO$_2$–WO$_3$					
88.76–11.24 65.00–35.00	5.806	2.178	0.644	0.202	0.112
	6.185	2.150	0.585	0.185	0.104
TeO$_2$–Ta$_2$O$_5$					
96.8–3.2	5.747	2.164	0.640	0.202	0.112
92.0–8.0	5.923	2.157	0.616	0.195	0.109
TeO$_2$–WO$_3$–Ta$_2$O$_5$					
80.9–16.1–2.91	5.927	2.184	0.626	0.199	0.110
85.13–6.45–8.42	5.695	2.168	0.650	0.205	0.114
TeO$_2$–WO$_3$–BO					
72.12–24.02–3.85	5.976	2.072	0.551	0.179	0.104
79.36–1.54–9.09	5.854	2.178	0.639	0.201	0.111

$n = (C_1\rho + 1)^{1/2}$

Source: S. Al-Ani and C. Hogarth, *J. Mater. Sci. Lett.*, 6, 519, 1987. With permission.

Takabe et al. (1994) measured the n of binary tellurite glasses in the region from 0.4 to 1.71 μm by using the minimum-deviation method. The binary glasses were in the form $(100 - x)$ mol% TeO$_2$–x mol% A_nO_m, where x mol% A_nO_m was 20 mol% Na$_2$O, 10 mol% Al$_2$O$_3$, 25 mol% ZnO, 20 mol% BaO, 15 mol% BaO, 10 mol% Sb$_2$O$_3$, 20 mol% MoO$_3$, 20 mol% WO$_3$, 10 mol% Nb$_2$O$_5$, and 20 mol% Tl$_2$O, as shown in Figure 8.7. An empirical relation based on the single-oscillator "Drude-Voigt" υ relation was:

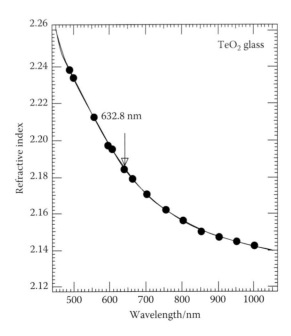

FIGURE 8.6 Linear refractive index measured as a function of wavelength from 486.1 to 1,000 nm. (From S. Kim, T. Yoko, and S. Sakka, *J. Am. Ceram. Soc.*, 76, 2486, 1993a. With permission.)

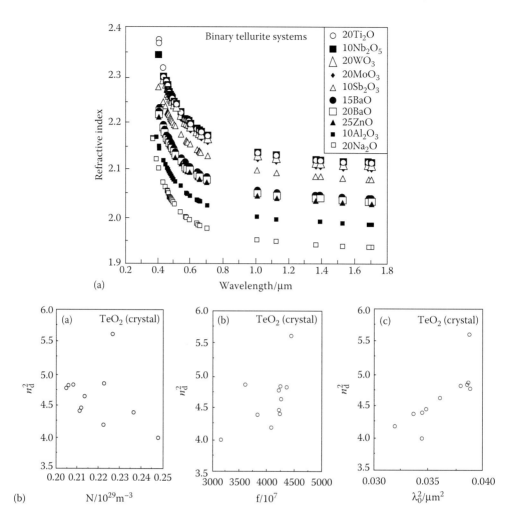

FIGURE 8.7 (a, b) Refractive indices of binary tellurite glasses of the form $(100 - x)$ mol% TeO_2–x mol% A_nO_m, where x mol% A_nO_m is 20 mol% Na_2O, 10 mol% Al_2O_3, 25 mol% ZnO, 20 mol% BaO, 15 mol% BaO, 10 mol% Sb_2O_3, 20 mol% MoO_3, 20 mol% WO_3, 10 mol% Nb_2O_5, and 20 mol% Tl_2O in the region from 0.4 to 1.71 μm, by the minimum-deviation method. (From H. Takabe, S. Fujinio, and K. Morinaga, *J. Am. Ceram. Soc.*, 77, 2455, 1994. With permission.)

$$n_d^2 = AfN\lambda_o^2 + B \tag{8.22}$$

where n_d is the n at 0.5876 μm, N is the number of molecules per unit V, f is the average oscillator strength, λ_o is the average resonance λ, and A and B are constants. The n_d of tellurite glasses is substantially determined by the resonance λ at the UV region, which is affected by the main constituent, TeO_2 oxide. Takabe et al. (1994) found that each tellurite glass had a λ_o of ≤ 0.2 μm, and the n at 0.5876 μm represents the variation of n_d^2 versus $Nf\lambda_o^2$ for binary tellurite glasses. Finally, these authors also discussed the parameters that affect the n_d of the present binary tellurite glasses, noting that n_d:

- Is independent of the N of molecules per unit V.
- Is independent of the dimensionless oscillator strength (f).
- Increases monotonically with increasing λ_o, which is affected by TeO_2.

In 1995, Kim et al. prepared and extruded a tellurite glass fiber with a diameter of 14 μm. The tellurite glass fiber was of the form 80 mol% TeO_2–10 mol% Li_2O–10 mol% Na_2O and was produced using the conventional melt-quenching method. The 80 mol% TeO_2–10 mol% Li_2O–10 mol% Na_2O glass and sintered $LiNbO_3$ ceramics were pulverized, and these powders were well mixed (6 mass%). The mixtures were heated in a platinum crucible at 980°C. After sintering with a silica glass rod and holding for a few minutes, the glass fibers were drawn by pulling out viscous melt using a silica glass rod. The $LiNbO_3$-doped bulk samples were also prepared by pouring the re-melted samples onto an iron plate. The presence of $LiNbO_3$ by using the incorporation method in the $LiNbO_3$-doped tellurite glass was confirmed by XRD analysis at room temperature. The n of the matrix glass at 632.8 nm λ (He–Ne laser) was measured at room temperature using an ellipsometer. This technique is applicable for fabrications of nonlinear optical glass fibers containing various nonlinear optical crystals. The n was 1.98 and the ρ was 4.73 g/cm³.

Also in 1995, Shioya et al. prepared the transparent glass-ceramic 70 mol% TeO_2–15 mol% K_2O–15 mol% Nb_2O_5, consisting of a cubic crystalline phase. This glass-ceramic in its stable crystalline phase is opaque. The transparency of glass-ceramics is attributed to the small particle size (10–20 nm) of cubic crystalline phase. The transparent glass-ceramics were obtained by heat treatment at 425°C for 1 h. The change of the n and ρ of binary heat-treated tellurite glasses at different temperatures is illustrated as shown in the 1st Ed., together with a photograph of the heat-treated samples. Based on the analysis made by El-Mallawany (1992), Shioya et al. (1995) calculated the αs of 70 mol% TeO_2–15 mol% K_2O–15 mol% Nb_2O_5 glasses at temperatures of 380°C, 390°C, 400°C, 410°C, 425°C, 435°C, 450°C, and 475°C: 7.18, 7.09, 7.10, 7.12, 7.12, 7.07, 7.10, and 7.06 × 10^{-24} cm³ ± 0.01. El-Mallawany (1992) calculated the αs of TeO_2 and 93.5 mol% TeO_2–6.5 mol% K_2O, which were 6.965 and 6.31 × 10^{-24} cm³, respectively (Table 8.5). Shioya et al. (1995) concluded that 70 mol% TeO_2–15 mol% K_2O–15 mol% Nb_2O_5 glass is useful for practical applications.

Ghosh (1995) calculated the Sellmeier coefficients and chromatic υs for some tellurite glasses. The Sellmeier coefficients are calculated by using the relation:

$$n = A + \frac{B}{\left(1 - C/\lambda^2\right)} + \frac{D}{\left(1 - E/\lambda^2\right)} \qquad (8.23)$$

and they are necessary in order to optimize the design parameters of optical devices. Although the λ dependence of the n is by the chromatic υs V, where $V = -[(\lambda/c)d^2 n(\lambda)/d\lambda^2]$, and c is the speed of light. Ghosh (1995) represented the variation of the n of binary tellurite glasses of the form $(100 - x)$ mol% TeO_2–x mol% M_mO_n, where x mol% M_mO_n is 20 mol% Na_2O, 10 mol% Al_2O_3, 25 mol% ZnO, 10 mol% Sb_2O_3, 20 mol% MoO_3, 20 mol% Tl_2O, 10 mol% Nb_2O_5, 20 mol% WO_3, 15 mol% BaO, and 20 mol% BaO, as shown in Figure 8.8.

Kim et al. (1996a) found that small differences in n values between the TeO_2-based glasses ($n = 2.07$) and the incorporated $LiNbO_3$ crystals ($n = 2.296$) were a significant reason for the transparency. This glass was of the form 80 mol% TeO_2–$(20 - x)$ mol% Li_2O–x mol% Nb_2O_5 with $x = 5$, 8, or 10 mol%. The ρ of these glasses was 5.06, 5.11, and 5.16 g/cm³, respectively, whereas the n is 2.07, 2.08, and 2.09, respectively. The transparency of the samples has been examined for the re-melted mixture of the powders 80 mol% TeO_2–15 mol% Li_2O–5 mol% Nb_2O_5 at 950°C at different periods of time as shown in Figure 8.9. The transparency was compared against the thickness in micrometers. Figure 8.9 shows that the higher-percentage $LiNbO_3$ gives higher transmittance for the thicker glass samples.

Komatsu et al. (1997a) prepared $(100 - x)$ mol% TeO_2–$0.5x$ mol% K_2O–0.5 mol% Nb_2O_5, for $x = 5$–32 and 70 mol% TeO_2–$(30 - x)$ mol% K_2O–x mol% Nb_2O_5, for $x = 5$–20. A linear correlation between the n (1.83–2.13) and ρ (4.12–5.18) was observed as in Figure 8.10. Nb_2O_5 acts to maintain

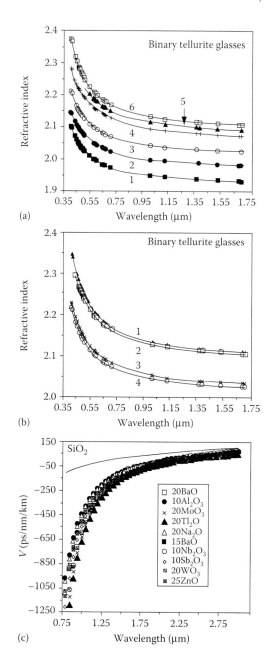

FIGURE 8.8 (a–c) Variation of the refractive indices and the chromatic dispersions, V, of binary tellurite glasses of the form $(100 - x)$ mol% TeO_2–x mol% M_mO_n, where x is 20 mol% Na_2O, 10 mol% Al_2O_3, 25 mol% ZnO, 10 mol% Sb_2O_3, 20 mol% MoO_3, 20 mol% Tl_2O, 10 mol% Nb_2O_5, 20 mol% WO_3, 15 mol% BaO, and 20 mol% BaO (solid lines were computed). (From G. Ghosh, *J. Am. Ceram. Soc.*, 78, 2828, 1995. With permission.)

relatively large values of n, R_M, and α of the glasses, but K_2O breaks Te–O–Te bonds, leading to low glass transition and low melting temperatures as shown in Chapter 5.

Also in 1997, Vithal et al. measured the n and R_M of 30 mol% TeO_2–70 mol% PbO glass. The n was 1.849 and the R_M was 13.44 cm^3 for the binary lead–tellurite glasses of the form 30 mol% TeO_2–70 mol% PbO. Vithal et al. (1997) calculated both the optical basicity (Λ [0.464]) and α (1.778×10^{-24} cm^3) of this tellurite glass system according to the relation:

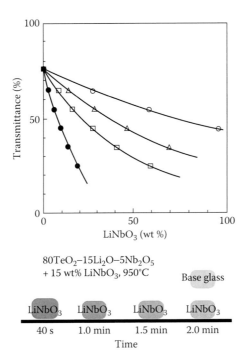

FIGURE 8.9 Transparency of samples examined for remelted mixtures of the powders 80 mol% TeO$_2$–15 mol% Li$_2$O–5 mol% Nb$_2$O$_5$ at 950°C and at different periods. Experiments showed that the higher-percentage LiNbO$_3$ gives higher transmittance for the thicker glass samples. (From H. Kim, T. Komatsu, *J. Mater. Sci.*, 31, 2159, 1996a. With permission.)

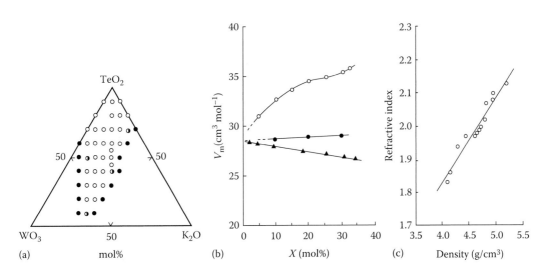

FIGURE 8.10 (a–c) Linear correlation between refractive index and density of $(100 - x)$ mol% TeO$_2$–0.5x mol% K$_2$O–0.5 mol% Nb$_2$O$_5$, for $x = 5$–32, and 70 mol% TeO$_2$–$(30 - x)$ mol% K$_2$O–x mol% Nb$_2$O$_5$, for $x = 5$–20. (From T. Komatsu, K. Shioya, and H. Kim, *Phys. Chem. Glasses*, 38, 188, 1997a.)

$$\Lambda = 1 - \sum \left(\frac{z_i r_i}{2} \right)\left(1 - \frac{1}{\gamma_i} \right) \tag{8.24}$$

where z_i, r_i, and γ_i are the valence of the ith cation, the ratio of the number of ith cations to that of oxide ions, and the basicity moderating parameter given by $\gamma_i = 1.36\ (\chi_i^{-0.26})$, respectively, where χ is the Pauling's electronegativity of the ith atom. In Chapter 1, however, Himi (1997) calculated the Λ, according to the relation:

$$\Lambda_{cal} = \sum \frac{z_i r_i}{2\gamma_i} \tag{8.25}$$

and its relation to O1st binding energy for tellurite glasses. Dimitrov and Sakka (1996) investigated the optical properties of single-component oxides and the optical basicity of oxides according to Duffy (1989) stated in the relation:

$$\Lambda = 1.67\left(1 - \frac{1}{\alpha_o^{2-}} \right) \tag{8.26}$$

This was estimated on the basis of average electronic oxide α calculated from the n and energy gap. The simple oxides have been separated into three groups according to the values of their oxide α, as shown in Figure 8.11.

1. Group one includes SiO_2, B_2O_3, Al_2O_3, GeO_2, and Ga_2O_3, which are characterized by a low α of oxide ions in the range 1–2 Å^3.
2. Group two consists mainly of d-transition metal oxides with an oxide α between 2 and 3 Å^3, which is attributed to the empty d states of the corresponding cations.
3. Group three oxides include Cd^{2+}, Pb^{2+}, Ba^{2+}, Sb^{3+}, and Bi^{3+}, with average oxide α of >3 Å^3.

Dimitrov and Sakka (1996) concluded that Pb^{2+}, Sb^{3+}, Bi^{3+}, and Te^{2+} ions possess lone pairs in the valence shell, and that the optical constants of tellurium oxides are:

- Linear n (n_o) = 2.279
- Cation α_i = 1.595 Å^3
- Oxide $\alpha_o^2 - n_o$ = 2.444 Å^3
- Λ (n) = 0.99

Kosuge et al. (1998) measured the n at a λ of 632.8 nm for ternary tellurite glass of the form $(100 - x - y)$ mol% TeO_2–x mol% K_2O–y mol% WO_3, as shown in Table 8.6. Also in 1998, Kim and Komatsu measured the n at room temperature and with λ of 632.8 nm for ternary tellurite glasses of the form $(100 - x - y)$ mol% TeO_2–x mol% Na_2O–y mol% Nb_2O_5, as shown in Figure 8.14.

8.4.2 Linear Refractive Index Data of Tellurium Oxide Thin-Film Glasses

In 1990, Chopra et al. prepared binary amorphous films of tellurite glass of the form $(100 - x)$ mol% TeO_2–x mol% V_2O_5, for $x = 10$–50, by a blowing technique. The thickness of these films was in the range 1–7 µm. The n values of the films were evaluated as shown in the 1st Ed. (similar behavior was observed in films of other compositions). The variation of the n is in accordance with the typical shape of a normal υ curve near electronic transition. Cuevas et al. (1995) prepared $(100 - x)$ mol% TeO_2–x mol% Li_2O–5 mol% TiO_2 thin films where $x = 5$, 10, 15, 20, and 25. The n_o has also been

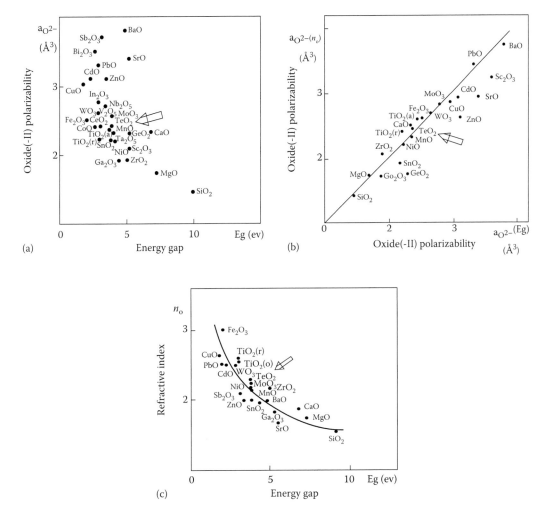

FIGURE 8.11 Optical properties of single-component oxides and optical basicity. Arrows indicate TeO$_2$. (a) Relationship of polarizability and energy gap. (b) Polarizability. (c) Relationship of energy gap and refractive index. (From V. Dimitrov and S. Sakka, *J. Appl. Phys.*, 79, 1736, 1996. With permission.)

measured by the Swanepoel (1983) method. In these tellurite glasses, the n_o decreases with increasing λ in the range 200–600 nm as in the 1st Ed., which showed that at $\lambda = 589.3$ nm, the linear n_o varies between 1.66 and 2.28, which is caused by an increase in the percentage of the x mol% Li2O.

Weng et al. (1999) prepared binary TeO$_2$–TiO$_2$ thin films by the sol-gel process, as the melt-quenching technique was usually adopted in the preparation of bulk tellurite glasses but could not be used to produce high-quality thin films with potential applications in optical and electronic devices. The n of the thin film TeO$_2$–TiO$_2$, estimated from interference fringes in the transmission spectrum according to Swanepoel's method, was about 2.19 at 633 nm, close to that of 90 mol% TeO$_2$–10 mol% TiO$_2$ bulk glass reported by Yamamoto et al. (1994).

8.4.3 LINEAR REFRACTIVE INDEX DATA OF TELLURIUM NONOXIDE BULK AND THIN-FILM GLASSES

In 1970, Hilton stated that oxide-based optical glasses do not transmit beyond 3 to 5 μm because of the strong absorption of chemical bonds formed with the light chalcogen element. Development of useful optical materials based on Te and other chalcogen elements began in the late 1950s as an outgrowth of exploratory programs searching for new semiconductors for use in thermal electrical

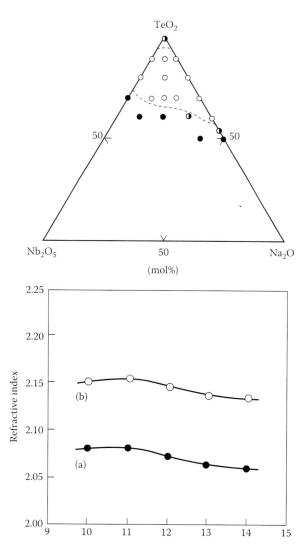

FIGURE 8.12 The refractive index at room temperature and wavelength of 632.8 nm for ternary tellurite glasses of the form $(100 - x - y)$ mol% TeO_2–x mol% Na_2O–y mol% Nb_2O_5. (From H. Kim and T. Komatsu, *J. Mater. Sci. Lett.*, 17, 1149, 1998. With permission.)

applications. Optical-materials investigations began in early 1960s. The temperature dependence of the n was measured and the data for the nonoxide–tellurite glasses are collected in Table 8.8. The major factor in the thermal change of n for IR materials was the thermal expansion coefficient. From Table 8.8, it is clear that the larger the value of thermal expansion coefficient, the more likely the $(\Delta n/\Delta T)$ is to be large and negative.

8.5 NONLINEAR REFRACTIVE INDEX DATA OF TELLURITE GLASSES

8.5.1 NONLINEAR REFRACTIVE INDICES OF TELLURIUM OXIDE BULK, THIN-FILM GLASSES, AND GLASS-CERAMICS

Kim et al. (1993a) examined the nonlinear optical properties of TeO_2-based glasses. For this purpose, it was important to determine the most significant optical parameters, which dominate the χ^3

TABLE 8.8

Glasses from Nonoxide Chalcogenide Systems

Chemical Composition (mol%)	Transmission Range (µm)	Refractive Index at 5 µm N	Thermal Change in Refractive Index (5 µm) $(\Delta N/\Delta T_c^\circ) \times 10^6$	Softening Point (°C)
25 Si–25 As–50 Te	2–9	2.93	+110	317
10 Ge–20 As–70 Te	2–20	3.55		178
15 Si–10 Ge–25 As–50 Te	2–12.5	3.06	+169	320
50 As–20 S–20 Se–10 Te	1–13	2.51		195
35 As–10 S–35 Se–20 Te	1–12	2.70		176

Source: A. Hilton, *J. Non-Cryst. Solids*, 2, 28, 1970.

values of tellurite glasses as shown in Table 8.9. Kim et al. (1993b) correlated the χ^3 of the TeO_2–La_2O_3 glasses with the following optical parameters:

- n_o
- n_2
- Abbe number (ν_d)
- Oscillation strength (energy band gap [E_d])
- Average excitation energy (E_o)
- Polarizability (α_m)

The TeO_2–La_2O_3 glasses were chosen because, in general, La_2O_3-containing silicate or borate glasses are characterized by high n and low υ. Kim et al. (1993b) prepared bulk tellurite glasses using the melt-quenching technique. The samples were $10 \times 10 \times 1.5$ mm in size. Both faces were optically polished to eliminate light scattering at the surfaces. The glasses used for the THG measurements were 1.0 ± 0.1 mm in thickness. The THG values were measured by using a nonlinear optical measurement apparatus. The n_o decreases due to the increase in λ and percentage of La_2O_3 oxide. The dependence of the n_o versus the photon energy (E) indicates a linear relationship. The molar V increases with the percentage of the La_2O_3. Kim et al. (1993b) identified variations in α, molar V, and R_M with the La_2O_3 content in tellurite glass and the relative intensity of the THG of binary La_2O_3 tellurite glasses in comparison with SiO_2 glass. The χ^3 of the binary lanthanum–tellurite glasses increases directly with both the linear and n_2. From the same study, it is clear that the χ^3 of the binary lanthanum–tellurite glasses decreases indirectly with the ν_d. The χ^3 increases linearly with E_d/E_o^2, which implies that a small E_o and large υ yield a high χ^3 value. Kim et al. (1993b) correlated the χ^3 of the binary lanthanum–tellurite glasses with the α_m and showed that the χ^3 increases with the α, but not linearly, which means that the relation is not simple. Thus, Kim et al. (1993b) correlated χ^3 with $[(n^2 + 2)^3(n^2 - 1)]_m$, which yielded a straight line. This means that both the n_o and α are important parameters that dominate the χ^3 in TeO_2–La_2O_3 glasses.

Previously, Kim et al. (1993a) reported their measurements of the n_o data of tellurite glasses, and in the same article they measured the nonlinear optical properties of TeO_2 glass. This glass was prepared using the melt-quenching technique. The pure tellurite glass sample was $5.0 \times 4.0 \times 0.25$ mm. χ^3, as determined by the THG method, was as high as 1.4×10^{-12} electrostatic units (esu), about 50-fold larger than that of SiO_2 glass. Kim et al. (1993a) interpreted the data with the Lines (1990) model, in which an influence of the cationic empty d-orbitals on the nonlinear properties was taken into account.

TABLE 8.9
Nonlinear Optical Properties of Tellurite Glasses

Glass (Composition [mol%])	Reference	Nonlinear Refractive Index (n_2) [esu]	Third-Order Nonlinear Optical Susceptibilities (χ^3)	Second-Order Nonlinear Coefficient (d_{33}) [pm/V]
TeO_2	Kim et al. (1993a)		14.0×10^{-13} esu	
SiO_2			0.28×10^{-13} esu	
$TeO_2–La_2O_3$	Kim et al. (1993b)			
90–10		10.48	1.19×10^{-13} esu	
80–20		6.84	1.04×10^{-13} esu	
$TeO_2–WO_3$	Tanaka et al. (1994)			
85–15	Poled at (3 kV,			0.05
75–25	280°C)			0.07
70 TeO_2–30 ZnO	Tanaka et al. (1996b)			
70 TeO_2–15 ZnO–15 MgO				0.22
$TeO_2–B_2O_3$				0.13
80–20				0.11
75–20				0.08
TeO_2	Yamamoto (1994)			
$TeO_2–Li_2O$			13.0×10^{-13} esu	
80–20			$(24.0 \pm 0.3) \times 10^{-13}$ esu	
75–25			$(38.0 \pm 0.3) \times 10^{-13}$ esu	
95 TeO_2–5 Sc_2O_3	Kim and Yoko (1995)		8.2×10^{-13} esu	
85 TeO_2–15 Ti_2O_3			16.6×10^{-13} esu	
90 TeO_2–10 V_2O_3			36.0×10^{-13} esu	
70 TeO_2–30 Nb_2O_3			16.9×10^{-13} esu	
95 TeO_2–5 MoO_3			6.9×10^{-13} esu	
95 TeO_2–5 Ta_2O_5			12.2×10^{-13} esu	
95 TeO_2–5 WO_3			15.9×10^{-13} esu	
70 TeO_2–15 Li_2O–15 Nb_2O_5	Tanaka et al. (1995) Poled at (3.2 kV, 250°C)		I (relative SHI) = 1.00 a.u.	
85 TeO_2–15 Nb_2O_5			I = 0.5 a.u.	
80 TeO_2–20 Li_2O			I = 0.45 a.u.	
70 TeO_2–15 K_2O–15 Nb_2O_5 (glass-ceramic)	Komatu et al. (1997)		3.3×10^{-13} esu	
80 TeO_2–20 Al_2O_3	Fragin et al. (1996)			$(10.9 \pm 10\%)$ 10^{-19} m²/W
90 TeO_2–10 Al_2O_3				$(17.4 \pm 10\%)$ 10^{-19} m²/W
70 TeO_2–30 Na_2O	Narazaki et al. (1998)			0.082
70 TeO_2–30 ZnO				0.45
70 TeO_2–10 Na_2–20 ZnO				0.23
90 TeO_2–10 Al_2O_3	Jeansannetas et al. (1999)		3.81×10^{-21} esu	5.38×10^{-19} m²/W
90 TeO_2–10 Nb_2O_5			5.63×10^{-21} V²/m²	6.93×10^{-19} m²/W
80 TeO_2–20 Nb_2O_5			4.93×10^{-21} V²/m²	5.94×10^{-19} m²/W
79 TeO_2–21 Tl_2O			6.9×10^{-21} V²/m²	8.9×10^{-19} m²/W

TABLE 8.9 (Continued)
Nonlinear Optical Properties of Tellurite Glasses

Glass (Composition [mol%])	Reference	Nonlinear Refractive Index (n_2) [esu])	Third-Order Nonlinear Optical Susceptibilities (χ^3)	Second-Order Nonlinear Coefficient (d_{33}) [pm/V])
85 TeO$_2$–13.5 Tl$_2$O–1.5 Bi$_2$O$_3$			7.48×10^{-21} V^2/m^2	9.1×10^{-19} m^2/W
CdeTe	Simmons et al. (1992)			0.2×10^{-9} esu at 1,700 nm

In 1994, Tanaka et al. measured the SHG in several electrically poled TeO$_2$-based glasses to discuss the relationship between second harmonic intensity and glass structure, as shown in Figure 8.13 and Table 8.9. The glasses were TeO$_2$–ZnO, TeO$_2$–BaO, TeO$_2$–WO$_3$, TeO$_2$–ZnO–BaO. The SHG was observed in TeO$_2$–ZnO and TeO$_2$–WO$_3$ systems. The second harmonic intensity increases with an increase in the concentration of ZnO in the binary TeO$_2$–ZnO systems. Therefore, the addition of ZnO to TeO$_2$ glass converts trigonal bipyramid TeO$_4$ to trigonal bipyramid TeO$_3$ and leads to the creation of NBOs. Tellurite glasses are oriented to the direction of an external DC

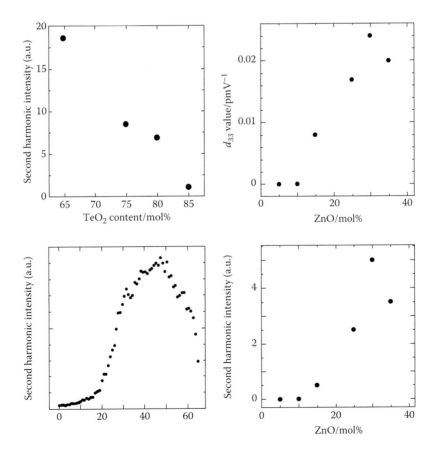

FIGURE 8.13 The second harmonic generation in several electrically poled TeO$_2$-based glasses. (From K. Tanaka, K. Kashima, K. Kajihara, K. Hirao, N. Soga, A. Mito, and H. Nasu, *SPIE*, 2289, 167, 1994.)

E to produce more orientations. Orientations take place easily in the ZnO-rich glasses. Tanaka et al. (1994) concluded that orientations of structural units scarcely occur in the TeO_2–BaO glasses because the external field mainly causes the electron P of Ba^{2+} ions.

Yamamoto et al. (1994) measured the high-third-order optical nonlinear tellurite glasses of the form TeO_2–TiO_3–PbO. The n values were measured using an ellipsometer equipped with an He–Ne laser (λ = 632.8 nm), whereas the χ^3 was determined by the THG method using the marker fringe pattern method as in Table 8.9. The pump light has a λ of 1.97 μm, which is used to avoid reabsorption of third harmonic light in the glass samples. For the SiO_2 glass, which is used as a reference, χ^3 = 2.8 × 10^{-14} esu. Yamamoto et al. (1994) collected the values of n, χ^3, and the coordination number of the ternary tellurite glasses. Both n and χ^3 increased with the molecular percentage of PbO, whereas the coordination number decreased, as shown in the 1st Ed. The effect of increases in the PbO content or replacement of a Te atom by PbO is that TeO_2 glass has a coordination number of 4 and χ^3 of 1.3 × 10^{-12} esu. Another subject to be considered was the role of TiO_2, because TiO_2 glass has high optical nonlinearity. The Ti atoms, which play an intermediate role similar to that of Pb in glass forming, can be present in four and six coordination.

Kim and Yoko (1995) measured the nonlinear optical properties of the TeO_2–MO_x (M = Sc, Ti, V, Nb, Mo, Ta, or W) binary glasses. The addition of TiO_2, Nb_2O_5, or WO_3 to TeO_2 glass increases the χ^3 value as well as the n, whereas other oxides decrease both values, as shown in Figure 8.19 and Table 8.9. The positive effect of TiO_2, Nb_2O_5, or WO_3 on χ^3 is interpreted in terms of the cationic empty d-orbit contribution. There is an almost linear relationship between χ^3 and the term $[(n^2 + 2)^3 (n^2 - 1) (E_d/E_0^2)]$, containing three measured parameters only, irrespective of the kinds of MO, which is derived based on the bond orbital theory developed by Lines (1991). The largest χ^3 value obtained is 1.69 × 10^{-12} esu for 30 mol% Nb_2O_5–70 mol% TeO_2 glass, about 60-fold larger than that of pure fused silica glass.

Also in 1995, Tanaka et al. measured the SHG in electrically poled TeO_2–Nb_2O_5–Li_2O glasses. The electrical poling was performed at 250°C for 30 min under an applied DC potential of 3–4 kV. The SHG was measured for the binary TeO_2–Nb_2O_5, TeO_2–Li_2O, and ternary 80 mol% TeO_2–10 mol% Nb_2O_5–10 mol% Li_2O glasses as shown in the 1st Ed. and in Table 8.9. The SHG in tellurite glasses that do not contain Li^+ ions leads to speculation that SHG is not attributed to the Li^+ ion migration but is presumably ascribed to a long-range orientation of electric dipoles due to asymmetrical structural units (i.e., trigonal bipyramid Te_4 and trigonal bipyramid TeO_3).

Kim et al. (1996b) prepared transparent tellurite-based glass-ceramics by the SHG method. These transparent glass-ceramics, in which the crystalline phase is basically a face-centered cubic structure, have been prepared for the system TeO_2–Nb_2O_5–K_2O, and SHG has been observed. The SHG in the glass-ceramics is comparable to that in electrically poled TeO_2-based glasses. Kim et al. (1996b) prepared a transparent glass-ceramic consisting of a cubic crystalline phase with TeO_2–Nb_2O_5–Li_2O or Na_2O, but SHG was not observed. Komatsu et al. (1996) investigated the SHG in transparent glass-ceramics of the form 70 mol% TeO_2–15 mol% Nb_2O_5–(15 – x) mol% K_2O–x mol% Rb_2O (x = 1, 2, 3, 4, or 5). SHG was observed in the heat-treated samples consisting of a cubic crystalline phase similar to the previous studies by Shioya et al. (1995).

Also in 1996, Tanaka et al. (1996a, 1996b) measured the optical SHG in poled TeO_2–ZnO–MgO and TeO_2–B_2O_3 glasses. The effect of poling temperature on the SHG for TeO_2–ZnO–MgO glass is shown in Figure 8.15, according to Tanaka et al. (1996a). Tanaka et al. (1996b) measured the values of the n at 532 nm, which were 2.02, 1.99, 1.98, 2.06, and 2.00 for the glass systems 70 mol% TeO_2–30 mol% ZnO, 70 mol% TeO_2–20 mol% ZnO–10 mol% Mg, 70 mol% TeO_2–15 mol% ZnO–15 mol% Mg, 80 mol% TeO_2–20 mol% B_2O_3, and 75 mol% TeO_2–25 mol% B_2O_3, respectively. The optical second-order nonlinear coefficient d_{33} (picomoles per unit V) of poled tellurite glasses is in the range 0.22–0.13 for TeO_2–ZnO and TeO_2–ZnO–Mg glasses, and from 0.11 to 0.08 for binary TeO_2–B_2O_3, as in Table 8.9. Tanaka et al. (1996b) concluded that the SHG of TeO_2–ZnO–MgO is greater than that of binary tellurite–borate glasses.

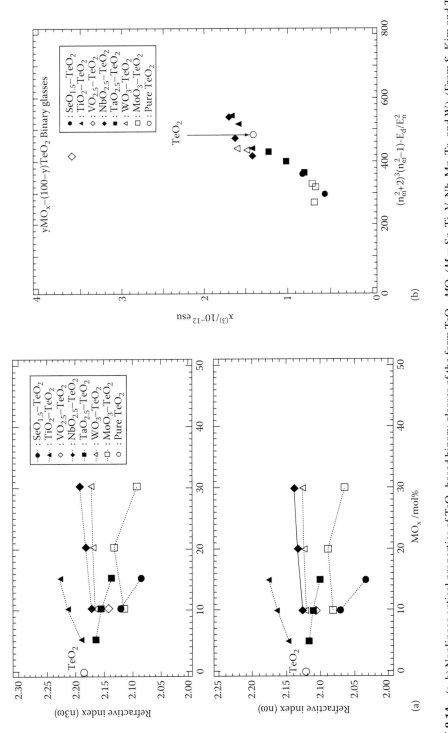

FIGURE 8.14 (a, b) Nonlinear optical properties of TeO_2-based binary glasses of the form TeO_2-MO_x (M = Sc, Ti, V, Nb, Mo, Ta, and W). (From S. Kim and T. Yoko, *J. Am. Ceram. Soc.,* 78, 1061, 1995. With permission.)

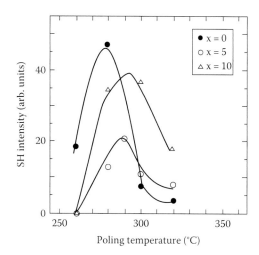

FIGURE 8.15 Effect of poling temperature on the second harmonic intensity for TeO_2–ZnO–MgO glass. (From K. Tanaka, A. Narazaki, K. Hirao, and N. Soga, *J. Appl. Phys.*, 79, 3798, 1996a. With permission.)

Fargin et al. (1996) stated the origin of the nonlinear response in two different types of oxide glasses containing transition metal ions with a *d*-shell such as Ti^{4+} or ions with lone pairs of electrons like Te^{4+}. The n_o and n_2 values have been obtained by interferometric measurements. Correlations between the local structure of polarizable entities within glasses and their optical activities have been tentatively established through *ab initio* calculation. The experimental linear indices, non-linear indices, and relative figure of merit for tellurite glasses $(100 - x)$ mol% TeO_2–x mol% Al_2O_3 ($x = 20$, 15, and 10) are 1.93, 1.96, and 2.04 (± 0.02); 10.9, 11.8, and 17.4×10^{-19} m²/W ($\pm 10\%$); and 1.6, 1.5, and 4.8 ($\pm 20\%$), respectively, as in Table 8.9.

Komatsu et al. (1997b) measured the SHG of the new transparent nonlinear tellurite-based glass-ceramics of the form 70 mol% TeO_2–15 mol% Nb_2O_5–15 mol% K_2O, heated-treated at 375°C for 5 h and then 425°C for 5 h, as shown in Figure 8.16 and in Table 8.9. Sabadel et al. (1997) measured the nonlinear optical process in TeO_2–BaO–TiO_2 glasses by using the Z-scan technique. The highest nonlinear absorption value, 15.7 cm GW⁻¹, was obtained for titanium–tellurite glasses of the form 90 mol% TeO_2–10 mol% TiO_2, and the highest *n* value was $(6.6 \pm 0.5) \times 10^{-18}$ m²/W for the 85 mol% TeO_2–15 mol% TiO_2 glass. This exceptional nonlinear coefficient was about 10-fold larger than that of TiO_2-containing phosphate or borophosphate glasses as stated by Boiteux et al. (1997).

Narazaki et al. (1998) measured the poling temperature dependence of the optical second harmonic intensity of TeO_2–Na_2O–ZnO glasses as shown in Figure 8.17a through d, and in Table 8.9. Narazaki et al. (1999) measured the induction and relaxation of optical second-order nonlinearity in tellurite glasses of the form 70 mol% TeO_2–30 mol% ZnO. Based on the linear relationship between the optimum poling temperature and glass transition temperature, the irreversible process is deduced to consist of some oxidation reactions such as a migration of NBO ions to, and subsequent evaporation of, oxygen gas at the anode side. Decay of the second harmonic intensity for 70 mol% TeO_2–30 mol% Na_2O glass, as well as 70 mol% TeO_2–30 mol% ZnO glass, has been examined at room temperature. Whereas the 70 mol% TeO_2–30 mol% ZnO glass does not show decay, the second harmonic intensity of the 70 mol% TeO_2–30 mol% Na_2O glass decays rapidly with an average relaxation time of 10 h. This relaxation behavior is explained in terms of the difference in mobility between Zn^{2+} and Na^+ ions. Jeansannetas et al. (1999) prepared the 85 mol% TeO_2–13.5 mol% Tl_2O–1.5 mol% Bi_2O_3 glasses. By solid-state chemistry, the highest n_2 measured 1.5 μm and was 9.1×10^{-19} m²/W, as given in Table 8.9.

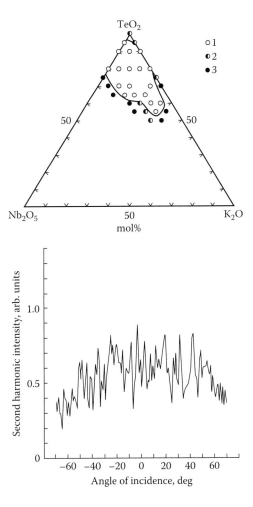

FIGURE 8.16 SHG of the new transparent nonlinear tellurite-based glass-ceramics of the form 70 mol% TeO_2–15 mol% Nb_2O_5–15 mol% K_2O. (From T. Komatsu, K. Shioya, and H. Kim, *Phys. Chem. Glasses*, 38, 188, 1997a. With permission.)

Dimitrov et al. (1999) deduced a relationship between oxide ion α, cation α, and Λ, which was used as a measure of acid–base properties of oxides and glasses and O1s-binding energy for simple oxides. This relationship is that O1s-binding energy decreases with increasing oxide ion α, cation α, and Λ. Dimitrov et al. (1999) also concluded that increasing oxide ion α means stronger electron donor ability of the oxide ions and vice versa, which is why O1s-binding energy can be used for the constriction of the common basicity scale of different amorphous and crystalline materials. On this basis, Dimitrov et al. (1999) separated simple oxides into three groups according to the values of their oxide ion α, Λ, and O1s-binding energy as shown in Figure 8.18; and Dimitrov and Komatsu (1999) correlated the average electronic α of the oxide ions of numerous binary oxide glasses (tellurite, phosphate, borate, silicate, germinate, and titanate) with Λ and nonlinear optical properties, as shown in the 1st Ed. They used the molar R (R_m), V_m, n_o, and the theory of metallization of condensed matter (by Herzfeld 1927) for the condition $R_m/V_m = 1$ in Lorentz–Lorenz, and the values of n become infinite, which corresponds to the metallization of covalent solid materials. The necessary and sufficient conditions to predict the nonmetallic or metallic nature of solids are $(Rm/Vm) < 1$ (nonmetal) and $(R_m/V_m) > 1$ (metal). The difference from 1 is $M = 1 − (R_m/V_m)$ and is called the "metallization criterion", as stated by Dimitrov and Sakka (1996). Conventional oxide glasses with low n values possess low α, low Λ, large metallization criterion, and small χ^3. Tellurite and titanate

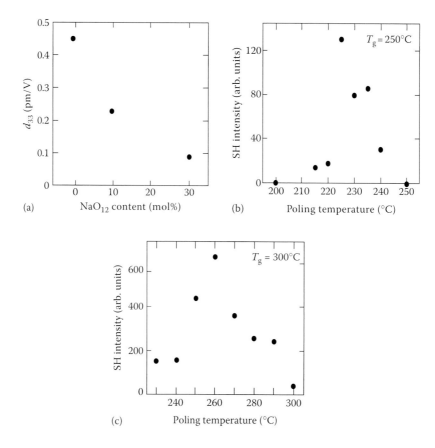

FIGURE 8.17 (a–c) Poling temperature dependence of optical second harmonic intensity of TeO_2–Na_2O–ZnO glasses. (From A. Narazaki, K. Tanaka, K. Hirao, and N. Soga, *J. Appl. Phys.*, 83, 3986, 1998. With permission.)

glasses, as well as borate glasses containing a large amount of Sb_2O_3 and Bi_2O_3 with high n values above 2.0, show a high α, high Λ, small metallization criterion, and large χ^3.

In 2000, Tanaka et al. examined the second-order nonlinear optical properties of thermally/electrically poled WO_3–TeO_2 glasses. Glass samples with different thicknesses (0.60 and 0.86 mm) have been poled at various temperatures to explore the effects of external E strength and poling temperature on second-order nonlinearity. The dependence of second harmonic intensity on the poling temperature manifests a maximum at a certain poling temperature. Second-order nonlinear χ of 2.1 pm/V is attained for 0.60 mm thick glass poled at 250°C. This value is fairly large compared with poled silica and tellurite glasses reported thus far. It is speculated that large third-order nonlinear χ of WO_3–TeO_2 glasses gives rise to the large second-order nonlinearly via $\chi^2 = 3\chi^3$.

Optical fiber amplifiers using tellurite fiber modules containing Er, Pr, Tm, and Nd as rare-earth (RE) ions (1,000 ppm by weight) have been produced commercially in Japan by NEL as shown in the 1st Ed. The fiber structure has a core diameter of ≈2.6 μm, low reflection, outer dimensions of $120 \times 120 \times 9$ mm, and a scattering loss of <0.1 dB/m. Fluoride- and tellurite-based fiber amplifiers can provide levels of amplification unobtainable with silica-based amplifiers. Praseodymium-doped fluoride amplifiers can operate in the 1.3 μm band, which is beyond the scope of practical optical amplifiers. Erbium-doped fluoride fiber amplifiers exhibit a flat spectrum gain without the need for a gain equalizer and offer advantages when used in wave division multiplexing transmission systems. Erbium-doped tellurite fiber amplifiers have an ultra-broad gain bandwidth of 80 nm

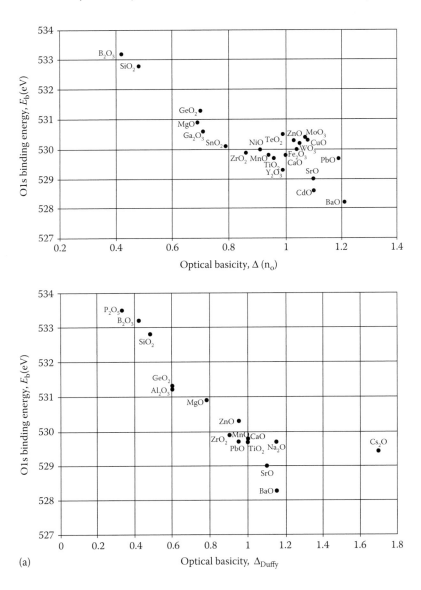

FIGURE 8.18   (a, b) Relationships among ion polarizability, optical basicity, and O1s-binding energy of indicated oxides. (From V. Dimitrov, T. Komatsu, and R. Sato, *J. Ceram. Soc. Japan*, 107, 21, 1999. With permission.)

and the potential for allowing the channel numbers of wave division multiplexing systems to be increased.

8.5.2   Nonlinear Refraction Index of Tellurium Nonoxide Glasses

Simons et al. (1992) reviewed the major sources of nonlinear optical behavior based on the third-order optical susceptibilities in glasses alone and glasses containing semiconductor clusters. They discussed the mechanisms of optical nonlinearity and the relative changes in n and β associated with each process. These authors compared the n values of semiconductor Cd–Te glasses with those of silica-based glasses in the range from UV to far IR (FIR) (2000 nm) on the basis of the

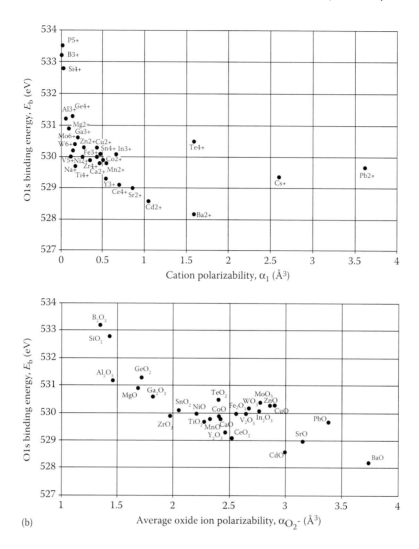

FIGURE 8.18 (Continued)

presence of free carrier resonance, 2-photon absorption, and Kerr effects. The n_2 of Cd–Te glass was 0.2×10^{-9} esu at 1700 nm (Table 8.9). Simons et al. concluded that:

- Cd–Te glasses have a wide range of free carrier resonance from UV to after the red region in the VIS, a wide range of 2-phonon absorption in the range from red to after the IR region, and then a narrow range in the FIR region for the Kerr effect.
- Silica has a very narrow range of free carrier resonance in the UV region, a small range of 2-phonon absorption in the UV region, and then a very wide range for the Kerr effect, which starts before violet and continues to the end of the FIR region.

8.6 OPTICAL APPLICATIONS OF OXIDE–TELLURITE GLASS AND GLASS-CERAMICS (THERMAL LUMINESCENCE FLUORESCENCE SPECTRA)

Starting in 1976, Reisfeld et al. measured the intensity parameters, radiative transition, and non-radiative relaxation of Ho^{3+} in various tellurite glasses. The glasses were 80 mol% TeO_2–20 mol% Na_2O, 85 mol% TeO_2–15 mol% BaO, and 65 mol% TeO_2–35 mol% ZnO. The set of "Judd-Ofelt" intensity parameters was discussed by Reisfeld (1973). From these parameters and the matrix elements of Ho^{3+}, transition probabilities and branching ratios from the excited states $(^5F_4, ^5S_2)$, $^5F_5$, $^5I_4$, $^5I_5$, and $^5I_6$ were calculated. Quantum efficiencies for the $(^5F_4, ^5S_2)$ state were obtained, along with multiphonon transition rates, which were calculated at room temperature. Nonradiative multiphonon relaxation was obtained from the $(^5F_4, ^5S_2)$ level. This relaxation increased in the order zinc–barium–sodium, and was assumed to be dependent on the local site symmetry of Ho^{3+} in glass. Weber et al. (1981) measured the optical absorption, emission spectra, and fluorescence lifetimes for Nd^{3+} in tellurite glasses containing various alkali and higher-valence-state cations, and in a series of new phospho-tellurite glasses as shown in Figure 8.19. The glasses were of the 79 mol% TeO_2–20 mol% K_2O–1 mol% Nd_2O_3 and 77 mol% TeO_2–22 mol% P_2O_5–1 mol% Nd_2O_3 forms. Judd–Ofelt intensity parameters were also determined and used to calculate radiative lifetimes and stimulated emission cross sections for $^4F_{4/2} \rightarrow ^4I_{11/2}$ and $^4F_{3/2} \rightarrow ^4I_{13/2}$ transitions. The cross sections for several

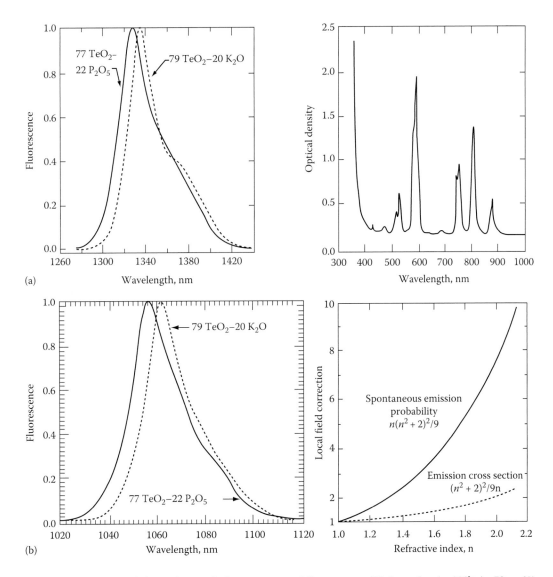

FIGURE 8.19 Optical-absorption, emission spectra, and fluorescence lifetimes for the Nd^{3+} in 79 mol% TeO_2–20 mol% K_2O–1 mol% Nd_2O_3 and 77 mol% TeO_2–22 mol% P_2O_5–1 mol% Nd_2O_3 tellurite glasses. (From M. Weber, J. Myers, and D. Blackburn, *J. Appl. Phys.*, 52, 2944, 1981. With permission.)

of the glasses were the largest obtained for any pure oxide glasses. The dependence of the spectroscopic properties on composition and the application of tellurite glasses for lasers were discussed.

In 1981, Blair et al. used tellurite glasses in acousto-optic devices and stated their compositions. An improved acousto-optic device and method featuring an acousto-optic element having an alkali metal oxide–tellurite glass composition ($\geq$92 wt% tellurium dioxide and one or more alkali metal oxides selected from the group consisting of sodium oxide and potassium oxide) were obtained. A process for decreasing the acoustic attenuation of the aforementioned glass composition by annealing is also disclosed. The figure of merit:

$$M_e = n^6 P^2 / \rho v^3 \tag{8.27}$$

where P is the photoelastic constant and v is the sound velocity, respectively. The values of the figure of merit of ternary tellurite TeO_2–WO_3–Li_2O glasses were measured by Izumitani and Masuda (1974) (see Chapter 3), and the values were in the range 15–18 $\times$ 10^{-18} (sec^3/g) as measured by Blair et al. (1981).

Romanowski et al. (1988) studied the influence of temperature and acceptor concentration on the energy transfer from Nd^{3+} to Yb^{3+} and from Yb^{3+} to Er^{3+} in tellurite glasses. The multicompnent tellurite glasses were of the form 85.6 wt% TeO_2–8.4 wt% BaO–4 wt% Na_2O–1 wt% MgO or ZnO–1 wt% (Nd_2O_3, Yb_2O_3, and Er_2O_3) as shown in the 1st Ed. Results of the measurements were interpreted on the basis of a kinetic model that describes the interactions between RE ions in terms of parameters characterizing the possibilities of energy migration and luminescence quenching. The probability of Nd^{3+}-to-Yb^{3+} energy transfer was found to be high and independent on the temperature in the range 13–300 K, due partly to the lack of back transfer from ytterbium to neodymium. Also, observed excitation energy transfer from Yb^{3+} to Er^{3+} was considerably less efficient because the small rate of multiphonon relaxation from the $^3I_{11/2}$ level Er^{3+} favors an efficient back transfer from erbium to ytterbium.

In the 1990s, considerable research work began on tellurite glasses. Borrelli et al. (1991) measured the electric-field-induced birefringence properties of the high-n tellurite glasses exhibiting large Kerr effects (B) (cm/V^2) $\times$ 10^{-6} as shown in Figure 8.20 for 88.5 mol% TeO_2–3.98 mol% ZnO–7.5 mol% BaO and 67.1 mol% TeO_2–11.4 mol% ZnO–21.5 mol% BaO glasses. The electro-optic Kerr effect and its λ υ had been measured for the glasses. The measured Kerr effect was found to be large for glasses having large n values. Comparison of the third-order nonlinear susceptibilities was performed from the measured electro-optic data.

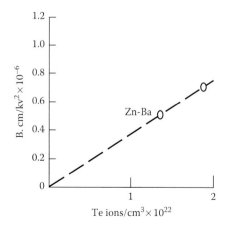

FIGURE 8.20   Kerr effect (B) ([cm/V^2] $\times$ 10^{-6}) of 858.5 mol% TeO_2–3.98 mol% ZnO–7.5 mol% BaO and 67.1 mol% TeO_2–11.4 mol% ZnO–21.5 mol% BaO glasses. (From N. Borrelli, B. Aitken, M. Newhous, and D. Hall, *J. Appl. Phys.*, 70, 2774, 1991. With permission.)

Hirao et al. (1992) observed an up-conversion fluorescence for the first time in tellurite glasses doped with Ho^{3+} at room temperature without using any sensitizing ions such as Yb^{3+} as shown in the 1st Ed. The tellurite glasses were in the forms:

- $(100 - x - y)$ mol% TeO_2–x mol% MgO–y mol% ZnO–0.25 mol% Ho_2O_3 ($x = 15$, $y = 0$; $x = 10, 20, y = 10$)
- $(100 - x - y)$ mol% TeO_2–x mol% SrO–y mol% ZnO–0.25 mol% Ho_2O_3 ($x = 10$, $y = 0$; $x = 10, y = 10$)
- $(100 - x - y)$ mol% TeO_2–x mol% BaO–y mol% ZnO–0.25 mol% Ho_2O_3 ($x = 15, 20, y = 0$; $x = 37, y = 0$; and $x = 10, 20, 30, y = 10$)
- $(100 - x)$ mol% TeO_2–x mol% (0.2 mol% PbO–0.8 mol% $PbCl_2$)–0.25 mol% Ho_2O_3 ($x = 10$, $20, 30, 40, 50$, and 60)
- 50 mol% AlF_3–20 mol% BaF_2–30 mol% CaF_2–0.5 mol% HoF_3, and 33.3 mol% BaF_2–6.1 mol% LaF_3–60 mol% ZrF_4–0.5 mol% HoF_3

The 645 nm laser beam from an Ar ion laser-pumped dye laser has been used as a pumping source for stepwise excitation as shown in Figure 8.29. The two-step excitation of up-conversion fluorescence was confirmed by the quadratic changes of up-conversion fluorescence intensity with the excitation power. The intensity of the up-conversion fluorescence was relatively high; it reached about half the value for AlF_3–BaF_2–CaF_2 glass containing the same amount of Ho^{3+}, indicating that these tellurite glasses are promising material for up-conversion lasers.

In 1992, Congshan et al. measured the up-conversion fluorescence near 525, 550, and 660 nm observed in TeO_2–PbO-based tellurite glasses containing Er_2O_3 at ambient temperatures as shown in the 1st Ed. A diode laser operating at 804 nm was used as the excitation source. The dependence of the fluorescence intensity on the pumping power fits the quadratic profile, which indicates the character of the up-conversion process. The influence of PbO contents in the glasses on the relative intensity of the up-conversion fluorescence has been estimated.

Abdel-Kader et al. (1993, 1994) measured the TL with a dosimeter in the microgray range of neodymium-doped tellurite–phosphate glass in the first article and La_2O_3, CeO_2, Sm_2O_3, and Yb_2O_3 in the second article. The optimum concentration of the RE was 0.001 wt%. The tellurite–phosphate glass system was 81 mol% TeO_2–19 mol% P_2O_5 as shown in Figure 8.21. For all samples, at each dose level, the recorded glow curve peaked at 1498–513 K for neodymium-doped tellurite–phosphate glass and 485–535 K for La_2O_3-, CeO_2-, Sm_2O_3-, and Yb_2O_3-doped tellurite–phosphate glasses, depending on both the type of the RE oxide and the gamma-ray dose. The peak height, as well as the area under the TL glow curves, was linear in the dose range of 0–200 μGy. An examination of TL emission spectra by Abdel-Kader et al. (1994) indicated that, in the temperature range from ambient to >300°C, the spectrum was that of forced forbidden 4f-transitions of the trivalent RE ion (RE^{3+}). Based on this observation, potential TL processes occurring under gamma-ray irradiation can be modeled using the reaction:

$$RE^{3+} + \gamma\text{-ray} \rightarrow RE^{2+} + h^+,$$

$$RE^{3+} + \gamma\text{-ray} \rightarrow RE^{2+} + h^+ + e^-$$

where h^+ and e^- represent holes and electrons, respectively. In the first process a hole trapped at a lattice site during irradiation recombines with RE^{2+}, producing TL as the material is heated. In the second process, both electrons and holes are trapped in the lattice during irradiation; the electron and hole then recombine at the RE^{3+} site as the material is heated during TL. Two principal mechanisms were suggested by Abdel-Kader et al. (1994) for the TL dosimeter in 81 mol% TeO_2–19 mol% P_2O_5 glass doped with different RE oxides.

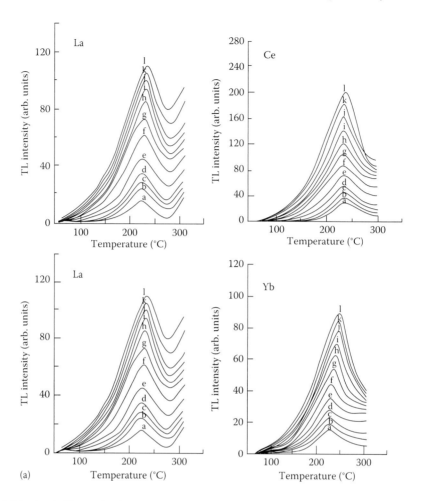

FIGURE 8.21 (a–c) Thermoluminescence (TL) of neodymium-doped tellurite–phosphate glass of the form 81 mol% TeO_2–19 mol% P_2O_5, doped by La_2O_3, CeO_2, Sm_2O_3, and Yb_2O_3. (From A. Abdel-Kader, R. El-Mallawany, M. El-Kholy, and H. Farag, *Mater. Chem. Phys.*, 36, 365, 1994. With permission.)

8.6.1 MECHANISM 1

Either TeO_4 or a PO_{4-} radical is split on irradiation into TeO_{3-} or PO_{3-}, respectively, and an interstitial oxygen ion O_{i-}. During thermal excitation, the O_{i-} diffuses back to the PO_3 or TeO_3 radical where it joins to form an excited phosphate or telluride ion. This excitation is then resonance transferred to the nearby RE ion, which absorbs and then emits the energy as TL. The above mechanism is written as:

$$\left(TeO_{3-} \text{ or } PO_{3-}\right) + RE^{3+} \rightarrow RE^{3+} \rightarrow \left(RE^{3+}\right)^* \rightarrow RE^{3+} + h\nu$$

$$+ \left(PO_4^{3-}\right)^* PO_4^{3-} PO_4^{3-}$$

or

$$O_i^- \left(TeO_4^{2-}\right)^* TeO_4^{2-} TeO_4^{2-}$$

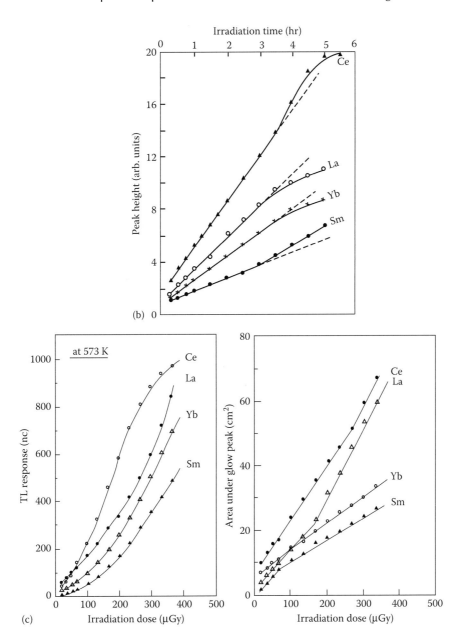

FIGURE 8.21   (Continued)

8.6.2 Mechanism 2

In the second mechanism, an RE^{3+} ion attacks the local O_{i-} charge compensation and captures an electron from either a TeO_4^{2-} or PO_4^{3-} site, which is ionized during irradiation, forming RE^{2+}. As the material is heated for TL, the RE^{2+} releases an electron, which then recombines with the TeO_4^- or PO_4^{2-} radical. The excitation is either transferred to an RE^{3+} ion, where TL is emitted, or it causes the excited TeO_4^{2-} or PO_4^{3-} to dissociate and, in so doing, forms PO_4^{2-} and O_i^- ions. The above mechanism is written as:

$$RE^{3+} \Rightarrow RE^{3+} + c \quad RE^{3+} \quad \Rightarrow RE^{3+} \Rightarrow RE^{3+} + hv$$

$$PO_4^{2-} \quad PO_4^{2-} \left[\begin{array}{c} \left(PO_4^{3-}\right)^* \Rightarrow PO_4^{3-} \quad PO_4^{2-} \\ RE^{3+} \qquad RE^{3+} \end{array} \right.$$

or

$$TeO_4^- \quad TeO_4^- \quad \left(PO_4^{3-}\right)^* \quad \left[PO_4^{2-} + O^- i\right]$$

In 1994, Wang et al. measured the 1.3 μm emission of neodymium and praseodymium in tellurite-based glasses of the forms 75 mol% TeO_2–20 mol% ZnO–5 mol% Li_2O, 75 mol% TeO_2–20 mol% ZnO–5 mol% Na_2O, 75 mol% TeO_2–20 mol% ZnO–5 mol% K_2O, 75 mol% TeO_2–20 mol% ZnO–5 mol% Rb_2O, and 75 mol% TeO_2–20 mol% ZnO–5 mol% Cs_2O as shown in the 1st Ed. The glass formation, wave-guide fabrication, optical properties, and suitability for fiber drawing of sodium, zinc, tellurite-oxide glasses doped with RE ions are described in Figure 8.32. Because their characteristic phonon energies are lower than those of silica glasses, praseodymium- and neodymium-doped tellurite glasses are considered for 1.3 μm fiber amplifiers.

Tanabe and Hanada (1994) studied the effect of the ligand field on the branching ratio of UV and blue up-conversions of Tm^{3+} ions in halide and oxide glasses as shown in Figure 8.22. The Tm^{3+}-doped zirconium, barium, lanthanum, chloride, germinate, and tellurite glasses were prepared and their up-conversion luminescence and excitation spectra were measured to investigate the effect of the host matrix glass on the branching ratio of 0.36 and 0.45 μm luminescence. These glasses were in the forms 60 mol% TeO_2–25 mol% BaO–10 mol% ZnO–4.5 mol% $YO_{3/2}$–1 mol% $TmO_{3/2}$, 80 mol% GeO_2–16 mol% $NaO_{1/2}$–3 mol% $YO_{3/2}$–1 mol% $TmO_{3/2}$, and 65 mol% ZrF_4–(32 – y) mol% BaF_2–y

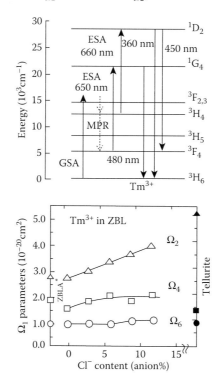

FIGURE 8.22   Judd–Ofelt intensity parameter Ω_t (t = 2, 4, 6) of Tm^{3+} ions in halide and oxide glasses. (From S. Tanabe and T. Hanada, *J. Appl. Phys.*, 76, 3730, 1994. With permission.)

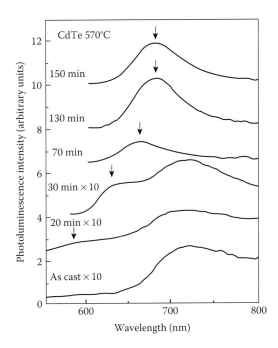

FIGURE 8.23 Photoluminescence spectra of the glass matrix 47.66 wt% SiO$_2$–16.55 wt% BO–37.57 wt% NaO–5.22 wt% ZnO + extra 3 wt% CdO–4 wt% Te-metal. (From Y. Liu, V. Reynoso, L. Barbosa, R. Rojas, H. Frangnito, C. Cesar, and O. Alves, *J. Mater. Sci. Lett.*, 14, 635, 1995. With permission.)

mol% BaCle$_2$–2.5 mol% LaF$_3$–0.5 mol% TmF$_2$ (y = 0, 5, 10, 15, and 20). The "Anion Fraction" (x = [Cl$^-$]/{[F$^-$] + [Cl$^-$]}) of the glass were 0, 3, 6, 9, and 12%, respectively, and these samples were called "ZBLx." The branching ratio β of 0.45 μm luminescence in zirconium–barium–lanthanum chlorine–fluoride glasses increased slightly with increasing Cl$^-$ doping, but the ratios were much smaller than those in tellurite and germinate glasses.

In 1995, Liu et al. studied trap elimination in Cd–Te quantum dots in glasses. The glass matrix was 47.66 wt% SiO$_2$–16.55 wt% BO–37.57 wt% NaO–5.22 wt% ZnO + extra {3% CdO–4% Te-metal}. The Te-metal was incorporated as the source of semiconductor elements. The photoluminescence spectra of the samples are in Figure 8.23. The transitions marked 1 and 2 are the direct absorption and direct recombination of band-to-band transitions, respectively. Transition 3 corresponds with the lower energy peak on the photoluminescence spectrum.

In 1996, McDougall et al. reported the spectroscopic characteristics of Er^{3+} in fluorozirconate, germinate, tellurite, and phosphate glasses. The three intensity parameters: Ω_2, Ω_4, and Ω_6 (cm^2), which had been introduced by the Judd (1962) and Ofelt (1962) theories, were calculated for Er^{3+}-fluorozirconate, germinate, tellurite, and phosphate glasses. McDougall et al. (1996) prepared tellurite glass of the form 88.92 mol% TeO$_2$–10.04 mol% Na$_2$O–1.04 mol% Er$_2$O$_3$. The lifetime 1.55 μm emission of Er^{3+} was reported. The values of Ω_2, Ω_4, and Ω_6 were 4.65, 1.21, and 0.87 × 10^{20} cm^2, respectively. McDougall et al. (1996) deduced a relationship between the product of the molar α and Ω_2 with the oscillator strengths of the hypersensitive transitions.

In 1997, Hussain et al. measured the spectra of Nd^{3+}-doped borotellurite glasses. The energy level structure, bonding, oscillator strength, and radiative lifetimes of the borotellurite glasses containing Li, Na, and K were computed. Also in 1997, Mori et al. fabricated Er^{3+}-doped tellurite single-mode fiber (Figure 8.24). Signal amplification and laser oscillation were demonstrated for the first time. A small signal gain of 16 dB at 1560 nm was obtained for a pump power of 130 mW at 978 nm. A laser oscillation was observed with a threshold pump power of 120 mW at 978 nm and a slope efficiency of 0.65% using this fiber.

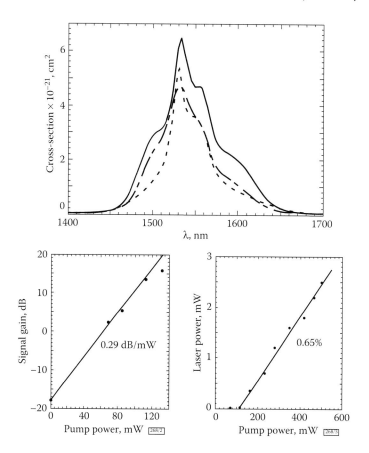

FIGURE 8.24 Characteristics of Er^{3+}-doped tellurite single-mode fiber. (From A. Mori, Y. Ohishi, and S. Sudo, *Electron. Lett.*, 33, 863, 1997. With permission.)

Sidebottom et al. (1997) measured the structure and optical properties of RE-doped zinc oxy-halide–tellurite glasses as shown in 1st Ed. They reported measurements of the emission lifetime, absorption, and vibrational density of states of the glass system $(100 - x - y)$ mol% TeO_2–x mol% ZnO–y mol% ZnF_2 doped (0.1 mol%) with a series of REs. The RE oxides were Eu_3O_3, Nd_2O_3, Ho_2O_3, Tm_2O_3, and Er_2O_3. Phonon sideband spectroscopy has been successfully used to prove vibrational structure in the immediate vicinity of RE ions. They also observed a significant increase in the emission lifetime (from $\approx$150 to 250 μs) of Nd^{3+} with increase in the fluorine substitution.

In 1998, Reddu et al. measured the absorption and photoluminescence spectra of some RE-doped TeO_2–B_2O_3–BaO–R_2O–REF_3 glasses, in which R = Li, Na, or Li + Na and RE = Sm^{3+}, Dy^{3+}, or Eu^+. The characteristic luminescence parameters have been evaluated for these tellurite glasses. Under a UV source, it was observed that Sm^{3+}-doped glasses are luminescent in greenish-yellow, Dy^{3+}-doped glasses are luminescent in the yellowish-blue, and Eu^+ glasses are luminescent in the red colors. These optical glasses are IR-transmitting (up to 4.5 μm) with good transparency and sufficient hardness.

Murata et al. (1998) studied the compositional dependence of IR-to-VIS up-conversion in Yb^{3+}- and Er^{3+}-codoped germanate, gallate, and tellurite glasses by up-conversion fluorescence analysis of the green $^4S_{3/2} \rightarrow {}^4I_{15/2}$ and red $^4F_{9/2} \rightarrow {}^4I_{15/2}$ transitions of the Er^{3+} ion, also studied for Yb^{3+}- and Er^{3+}-codoped sodium germinate, potassium tantalum gallate, and barium tellurite glasses by In–Ga–As laser-diode pumping. The phonon energies of the host glasses are determined by IR reflection measurements. Compositional effects of the Judd–Ofelt parameters for the Er^{3+} ion, the spontaneous emission probability of the $^2F_{5/2} \rightarrow {}^2F_{7/2}$ transition for the Yb^{3+} ion, and the phonon energy of the glass network are discussed below in terms of glass structure. The factors that affect the up-conversion fluorescence intensities of the Er^{3+} ion are discussed, using the phonon energy of the host glass and the spontaneous emission probability for the Yb^{3+} ion in the germinate, gallate, and tellurite glasses. Oishi et al. (1999) reported the optical properties of transparent-tellurite-based glass-ceramics doped by Er^+ and Eu^+. Both Er^+ and Eu^+ are effective for the formation of a crystalline phase showing an SHG, as discussed above. Fluorescence spectra of Er^+ and Eu^+ in glass-ceramics are similar to those in precursor glasses. The intensity of frequency up-conversion fluorescence at around 550 nm due to the $^4S_{3/2} \rightarrow {}^4I_{15/2}$ transition of Er^+ in the glass-ceramics is strong compared with that in the precursor glass. From the phonon sideband spectra associated with $^5D_2 \rightarrow {}^7F_o$ transition of Eu^+, it has been found that the phonon energy for transparent glass-ceramics is smaller than in the glasses, indicating that the multiphonon relaxation rate of Er^+ and Eu^+ is present in TeO_2-based glasses. Whether Er^+ and Eu^+ are incorporated into the crystalline phase is unclear. This study provides a new possibility for optical applications of transparent TeO_2-based glass-ceramics.

8.7 OPTICAL PROPERTIES OF NEW TELLURIUM GLASS

As far as we know that optical glasses have important parameters such as: refractive index and its dispersion in the optical material, relative and absolute refractive index, Brewster's angle, total internal reflection, dispersion formulas, differential modifications of the refractive index, influence of the annealing rate, variation of the refractive index with the temperature, photoelastic effects, variation of the refractive index with electric field strength, Kerr effect, variation of the refractive index with magnetic field strength, Faraday rotation, nonlinear refractive index, laser damage, reflectance, transmittance, and colored glasses.

Back in 2001, Ovcharenko and Sairnova prepared and measured the linear refractive indices of TeO_2–WO_3–Bi_2O_3 and TeO_2–WO_3–PbO glasses. The values were in the range 2.2263–2.2506 (TeO_2 content was 76.19–61.54 mol%) and 2.2337–2.3072 (TeO_2 content was 75.0–50.0 mol%) for both glass series, respectively. The authors also measured values of the magneto-optical parameter of these glasses, which were: 0.088–0.090 and 0.098–0.085 (min/Oe cm), respectively. Shen et al. (2002) studied the Er^{3+}-doped tungsten–tellurite glass as a host for broadband erbium-doped fiber amplifiers (EDFA). The tungsten–tellurite glass was in the form: $40WO_3$–$40TeO_2$–$20Li_2O$, $80TeO_2$–$10ZnO$–$10Na_2O$. The refractive index of the Er^{3+}-doped glasses tungsten–tellurite, zinc–sodium–tellurite and Al/silica were 2.1, 2.0, and 1.45, respectively. Inoue et al. (2003) found a correlation between specific heat and change of refractive index formed by the laser spot heating of tellurite glass surfaces. The ternary tellurite glasses were in the forms TeO_2–Na_2O–Al_2O_3, TeO_2–Na_2O–GeO_2 and TeO_2–Na_2O–TiO_2 doped with 2 mol% of CoO and irradiated by a green light beam spot (532 nm) from a second harmonic generator of a Q switch pulse YAG laser. The refractive index was equal to 2.165 for the glass 85 TeO_2–$15Na_2O$–$2CoO$, and changed to 2.192 for the tellurite glass system of the form $76.5TeO_2$–$13.5Na_2O$–$10TiO_2$–$2CoO$. Rolli et al. (2003) studied the erbium-doped tellurite glasses with high quantum efficiency and broadband-stimulated emission cross section at 1.5 µm. The refractive indices were 2.057 and 2.034 at 543.5 nm and 632.8 nm of the tellurite glasses of molar composition $75TeO_2$–$12ZnO$–$10Na_2O$–$2PbO$–$1Er_2O_3$, and 2.034 and 2.018 for the $75TeO_2$–$12ZnO$–$10Na_2O$–$2GeO_2$–$1Er_2O_3$, respectively. Prasad et al. (2003) prepared a highly transparent and yellow colored $20TeO_2$–$75B_2O_3$–$5Li_2O$ glass, containing 1 mol% of Ho^{3+} glass. The authors prepared this glass in order to study its optical properties from the measured UV–VIS–NIR absorption, emission, and up-conversion emission spectra. The values of the refractive indices were: $n_C(656.3$ nm$) = 1.471$, $n_D(589.3$ nm$) = 1.474$, $n_F(486.1$ nm$) = 1.480$, and Abbe number $(\nu_d) = 53$. In 2004, Wang et al. studied a series of new glasses of $70TeO_2$–$(20 - x)ZnO$–xWO_3–$5La_2O_3$–$2.5K_2O$–$2.5Na_2O$ (mol%) doped with Yb^{3+}. The authors used Cauchy's equation, $n(\lambda) = A + B/\lambda_2$, to determine the refractive index at the mean wavelength of Yb^{3+}:$2F_{7/2} \rightarrow 2F_{5/2}$ transition, and the value of the refractive index was 2.089 for $x = 20$ mol%.

In 2006, Xu et al. studied the spectral properties and thermal stability of Er^{3+}/Yb^{3+}-codoped tungsten–tellurite. The authors prepared tellurite glasses and measured the refractive index on a prism by the minimum deviation method, together with the density of the glass by using Archimedes' principle. The tellurite glasses were in the form: $70TeO_2$–$20WO_3$–$10BaO$ with density = 6.17 g/cm³ and refractive index, $n = 2.06$; $65TeO_2$–$25WO_3$–$10La_2O_3$ with density = 6.30 g/cm³ and refractive index $n = 2.05$; and $60TeO_2$–$30WO_3$–$10Bi_2O_3$ with density = 6.70 and $n = 2.32$. The densities of the glasses are higher than 6.0 g/cm³ and the refractive indices are all around 2. The Er^{3+} and Yb^{3+} ions concentrations (N_E and N_Y) in the glasses are calculated from the density values of the glass. Sakida et al. (2006) studied the planar waveguide of $12Na_2O$–$35WO_3$–$53TeO_2$–$1Er_2O_3$ glass (in mol%) by Ag^+–Na^+ ion exchange at 330°C for 5 h. The glass was prepared according to the following procedure: 20 g batch of well-mixed reagents was melted in a gold crucible at 800°C for 30 min. The melt was then poured onto a brass plate and immediately pressed by a stainless steel plate. The prepared glass was annealed near the glass-transition temperature (374.2°C) for 1 h. After annealing, the glass was cut into a plate of $50 \times 15 \times 2$ mm in size, and all faces mirror-polished for optical measurements and waveguide fabrication. The $12Na_2O$–$35WO_3$–$53TeO_2$ glass without Er^{3+} ions was also prepared by using the same procedures to examine the effect on optical properties of the addition of 1 mol% Er_2O_3. Ion exchange was performed by immersing the glass samples in $1.0AgNO_3$–$49.5NaNO_3$–$49.5KNO_3$ (mol%) molten salt at 330°C for 5 h for waveguide fabrication. The refractive indices of the substrate glasses and the effective mode indices and propagation losses of the waveguides at wavelengths of 473, 632.8, 983.1, and 1548 nm for TE and TM modes were measured by means of a prism coupler technique (Metricon Model 2010 Prism Coupler). Refractive indices (n) of $12Na_2O$–$35WO_3$–$53TeO_2$–$1Er_2O_3$ glass at the wavelengths of 473, 632.8, 983.1, and 1548 nm for TE and TM modes were:

Mode	n(473 nm)	n(632.8 nm)	n(983.1 nm)	n(1548 nm)
TE	2.1362	2.0720	2.0298	2.0103
TM	2.1360	2.0709	2.0300	2.0101

The glass has high refractive indices of more than 2.0. The refractive indices for the TE mode are almost the same as those for the TM mode in the same wavelength, indicating that the glass is optically isotropic. Also, in 2006, Eraiah prepared glasses with the composition, $(Sm_2O_3)x(ZnO)$ $(40 - x)(TeO_2)(60)$, by using the conventional melt-quenching method, and measured the optical energy band gap. The refractive index, molar refraction, and polarizability of oxide ion have been calculated using Lorentz–Lorentz relations. The calculated refractive index was equal to 2.397 and 2.646 for $x = 0.0$ and 0.5 mol%, respectively. Rivera et al. (2006) reported the preparation of planar waveguides by $Ag^+ \rightarrow Na^+$ ion exchange in Er^{3+}-doped tellurite glass with a composition of $75TeO_2$–$2GeO_2$–$10Na_2O$–$12ZnO$–$1Er_2O_3$ (mol%). In order to measure the refractive index of the glass (n_g), and the waveguides (n_w), a prism coupling method was used. The measurements were carried out at three different wavelengths. The results of these measurements were as follows:

Wavelength	n_g	$\Delta n = (n_g - n_w)$	Thickness of the refractive index increment
632.8 nm	2.0169	0.1063	0.5872 μm
1305 nm	1.9705	0.0697	1.1125 μm
1536 nm	1.9658	0.1154	1.5791 μm

Also, in 2006, Jose et al. studied the addition of WO_3 (up to 10 mol%) and P_2O_5 (up to 16 mol%) in the form of TeO_2–BaO–SrO–Nb_2O_5 (TBSN). The primary series of glasses takes the general formula $(78 - x - y)TeO_2$–$3.5BaO$–$10.5SrO$–$8Nb_2O5$–xWO_3–yP_2O_5, ($x = 0, 2, 4, 6, 8, 10$; $y = 0, 4, 8, 12, 16$). The first composition in the primary series was $78TeO_2$–$3.5BaO$–$10.5SrO$–$8Nb_2O_5$ and the second was $76TeO_2$–$3.5BaO$–$10.5SrO$–$8Nb_2O_5$–$2WO_3$. The seventh composition in this series was $74TeO_2$–$3.5BaO$–$10.5SrO$–$8Nb_2O_5$–$4P_2O_5$, the eighth was $72TeO_2$–$3.5BaO$–$10.5SrO$–$8Nb_2O_5$–$2WO_3$–$4P_2O_5$, and so on. The thermal and optical properties of the resultant glasses were as follows:

1. The refractive index of all the samples increased in the primary series monotonously with increasing WO_3 content. In the case of no phosphorous, refractive index varied from 2.114 to 2.118 when the WO_3 content varied from 0 to 10 mol%. Corresponding increments for P_2O_5 contents of 4, 8, 12, and 16 mol% were from 2.078 to 2.094; 2.038 to 2.066; 2.000 to 2.005; and 1.971 to 1.987, respectively.
2. The refractive index decreased linearly with an increase in the P_2O_5 content.
3. No observable variation in refractive indices was noted for the secondary series.

Eyzaguirre et al. (2007) used CsCl to act like a network modifier in glass systems, weakening the network by forming Te–Cl bonds. Tellurite glasses are important as a host of Er^{3+} ions because of their good solubility and because they present broadband optical gain compared with Er^{3+}-doped silica. They also have the potential to increase the bandwidth of communication systems. The refractive index of the cladding and core were 1.9407 and 1.9430, respectively.

El-Mallawany et al. (2008) measured the refractive index of the quaternary tellurite glass system of the form $80TeO_2$–$5TiO_2$–$(15 - x)WO_3$–xA_nO_m where A_nO_m is Nb_2O_5, Nd_2O_3, and Er_2O_3, $x = 0.01$, 1, 3, and 5 mol% for Nb_2O_5 and $x = 0.01$, 0.1, 1, 3, 5, and 7 mol% for Nd_2O_3 and Er_2O_3. The refractive index has been measured at room temperature and at wavelengths of 486.13, 587.56, 589, and 656.27 nm. The average dispersion ($n_F - n_C$) and Abbe number of this glass were estimated. In the range of visible spectrum, the linear refractive indices (n_λ) for the studied three glass series were found to be among the highest values of known glass. It has been found to increase from 2.10 to 2.15, with the increase of the Nb_2O_5 content from 0.01 to 5 mol%. It increases from 1.94 to 2.22 and from 1.96 to 2.21 with the increase of the modifier Nd_2O_3 and Er_2O_3 concentration from 0.01% to 7 mol%, respectively. High refractive indices of these glass samples were attributed to the high polarization of the host material, TeO_2. Moreover, the incorporation of the transition metal cation Nb^{5+} and rare-earth cations Nd^{3+} and Er^{3+} at the expense of the W^{4+} cation leads to further higher polarization glass and, consequently, an increase in the refractive index. As a measure of dispersion, the Abbe numbers were calculated for the three glass series and found to decrease from 23.69 to 16.02 as the concentration of Nb_2O_5 increased from (0.01 to 5 mol%). It was found to decrease from 25.16 to 16.02 and from 22.78 to 18.05 as the concentration of Nd_2O_3 and Er_2O_3 increased from 0.01 to 7 mol%, respectively. Also, Chen et al. (2008) studied both the linear and nonlinear optical properties of Bi_2O_3–WO_3–TeO_2 glasses. The values of the densities and refractive indices at 632.8 nm, 1310 nm, and 1532 nm of glass samples were as follows:

Glass	Density (g/cm³)	n(632.8 nm)	n(1310 nm)	n(1532 nm)
$10Bi_2O_3$–$20WO_3$–$70TeO_2$	6.370	2.1687	2.0626	2.0260
$10Bi_2O_3$–$30WO_3$–$60TeO_2$	6.510	2.1806	2.1002	2.0628

Both densities and refractive indices increase with increasing WO_3 content. Again, the linear refractive indices, n_o, decrease in the direction of the longer wavelength. It is well known that there are empirical relationships between linear and nonlinear optical susceptibility in solids. Accordingly, the mentioned relationships are used for a prediction that Bi_2O_3–WO_3–TeO_2 glass samples with large linear refractive index n_o possess large nonlinear refractive index n_2.

Oermann et al. (2009) studied the index matching between passive and active tellurite glasses for use in microstructured fiber lasers: erbium-doped lanthanum–tellurite glass. For undoped base glass the density $\rho = 5.34$ g/cm³, the refractive index $n = 2.0515$ (at 633 nm); for base glass doped with −1Er, $\rho = 5.37$ g/cm³, $n = 2.0505$; and for 10La-glass, $\rho = 5.37$ g/cm³ and $n = 2.0360$. Desirena et al. (2009) studied the effect of alkali metal oxides R_2O (R = Li, Na, K, Rb, and Cs) and network intermediate MO (M = Zn, Mg, Ba, and Pb) in tellurite glasses. The refractive index is a critical parameter in the control of mode profile and also affects the performance of optical fiber amplifiers. For this reason, it is necessary to know the effect of network modifiers' introduction into tellurite glasses. The refractive index at 632.8 nm and 1550 nm increased from Cs to Li for the four batches in ranges from 1.9705 to 2.0626, 1.8817 to 1.9728, 1.8603 to 1.9375, and 1.8598 to 1.9379 for $R_2O–PbO–TeO_2$, $R_2O–ZnO–TeO_2$, $R_2O–BaO–TeO_2$, and $R_2O–MgO–TeO_2$ glasses, respectively. The results showed that the highest values of refractive index were obtained when lithium was incorporated into the tellurite glasses. The observed trend suggested an increment of refractive index with a decrement of the ionic radii or an increment of field strength. Notice that such an effect depends on the intermediate ion being more pronounced for PbO and ZnO, but diminished for BaO and MgO. It was clear that by modifying the glass composition it is possible to increase or decrease the refractive index. Such dependence indicates the importance of the right glass composition of both modifiers and intermediate. The difference of refractive indices could be very useful as core and cladding glasses for optical fiber fabrication.

Liao et al. (2009) prepared and characterized new fluorotellurite glasses for photonics application. Glasses based on the $(85 – x)TeO_2–xZnF_2–12PbO–3Nb_2O_5$ ($x = 0–40$) system have been studied for the first time for fabricating mid-IR optical fiber lasers. Regarding glass densities, its decrease with increasing zinc fluoride content was not surprising as:

1. Substitution of ZnF_2 for TeO_2 results in a decrease of mass/unit volume of the glass due to different molar mass
2. Substitution of ZnF_2 for TeO_2 depolymerizes the glass network, resulting in a less dense structure

With respect to the refractive index of the glass, as the polarizability of ZnF_2 is lower than that of TeO_2, a decrease in refractive index is expected with an increase in ZnF_2 content. In addition, since the refractive index of a glass is also related to its density, the lowering of the glass density induced by the introduction of ZnF_2 concentration will also contribute to the decrease in refractive index: this result was in good agreement with other similar studies performed on silica-based glasses (by Kakiuchida et al. 2007). Moreover, Gao et al. (2009) measured the refractive index of: $70TeO_2–20ZnO–2.5Na_2O–2.5K_2O–5La_2O_3$ (TZNK). The refractive index of the TZNK sample was collected as a function of wavelength from ($n < 2.02$) 400 to 1300 nm ($n > 1.92$). It followed the trend of typical dispersion curves of tellurite glasses. The refractive index of TZNK glass was relatively high, and n_d (589.3 nm) was about 1.96. The design of glasses with high electro-optical sensitivity is important for modern optics. Wang et al. (2007) mentioned that a lack of clear understanding of the origin of Kerr sensitivity of glassy materials, except well studied family of alkaline–silica–niobate glasses. Duclère et al. (2009) found that the highest value of Kerr coefficient, $B \approx 190 \times 10^{-16}$ mV^{-2} was registered for $0.6TeO_2–0.3TlO0.5–0.1ZnO$ glass.

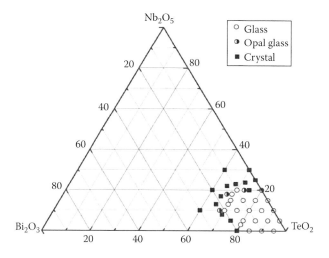

FIGURE 8.25 Glass formation of the ternary $TeO_2–Nb_2O_5–Bi_2O_3$ (Y. Wang, S. Dai, F. Chen, T. Xu, Q. Nie, *Mater. Chem. Phys.*, 113, 407, 2009. With permission.)

Wang et al. (2009) prepared ternary TeO_2–Nb_2O_5–Bi_2O_3 glass by using the conventional melt-quenching method and investigated the glass formation range of TeO_2–Nb_2O_5–Bi_2O_3 glass system as shown in Figure 8.25. The refractive indices decreased monotonously with increasing wavelength in the range from 632.8 to 1550 nm. For every wavelength, the refractive indices increased with increasing Bi_2O_3 content. At the same wavelength (632.8 nm), the glasses had high refractive indices, which were more than 2.0. Linear refractive indices, with the addition of Bi_2O_3 content, increased from 2.0857 to 2.1743 for 5 mol% Nb_2O_5, and increased from 2.0949 to 2.1927 for 10 mol% Nb_2O_5. It was concluded that the Bi_2O_3 content contributes to greater increase of the refractive index than Nb_2O_5 content. The obtained data on the refractive indices were associated with high concentration of Bi_2O_3, because the Bi^{3+} ion has cation polarizability equal to 1.508 A^3 (Dimitrov and Komatsu 1999) and a lone pair in the valence shell, which was also strongly polarizable. The addition of Nb_2O_5 content also leads to an increase in the refractive index, although the extent is not as large as with Bi_2O_3. The contribution of the cation polarizability of three oxides to the increase of the linear refractive index is in the order, $Te^{4+}(1.595A^3) > Bi^{3+}(1.508A^3) > Nb^{5+}(1.035A^3)$ (Dimitrov and Komatsu 1999).

In 2010, Chen et al. measured the linear and nonlinear refractive index of $(100 - x - y)TeO_2$–xBi_2O_3–$yTiO_2$, where $x = 0, 5, 10, 15$, and $y = 5, 10$. The linear refractive indexes n_o measured at 632.8 nm of the samples: TeO_2–Bi_2O_3–TiO_2 was equal to $n = 1.9757$, for 90-0-10%, $n = 2.1489$ for 85–5–10%, $n = 2.1152$ for 85–10–5%, $n = 2.1808$ for 80–10–10% and $n = 2.1870$ for 80–15–5%. It was notable that all samples, except for the first composition, exhibit large n_o (>2.1) while keeping low optical absorption at measured wavelengths, which indicates their good durability against intense light. On the other hand, the n_o value showed increasing tendency with decrease of optical gap E_{opt}. This behavior implies that linear optical absorption controls linear refractive index at wavelengths approaching the fundamental absorption edge. Afterward, glass samples with maximum Bi_2O_3 and TiO_2 content exhibit the largest n_o, thus implying that this sample could potentially have the largest third-order nonlinear TONL, which originates from nonresonant effect (Sugimoto et al. 1999). Also, in 2010, Mizuno et al. studied the distributed strain measurement using a tellurite glass fiber with Brillouin optical correlation-domain reflectometry (BOCDR). The measurement range (distance between the correlation peaks) is given by $d_m = c/2nf_m$, where c is the velocity of light in vacuum, n the refractive index, and f_m the modulation frequency of the light source. Mizuno et al. (2010) compared between silica fiber, bismuth oxide fiber, chalcogenide fiber, and tellurite fiber as follows: n(silica) = 1.46, n(bismuth oxide) = 2.22, n(chalcogenide) = 2.8, and n(tellurite) = 2.03, the Brillouin frequency shift $(BFS)_{Silica} = 10.86$, $(BFS)_{bismuth oxide} = 8.83$, $(BFS)_{chalcogenide} = 7.97$, and $(BFS)_{tellurite} = 7.95$. The challenge remains for tellurite glasses that their optical nonlinearity is more than 1 order smaller to compare with chalcogenides, although tellurite glasses are more stable chemically and structurally than chalcogenides.

As far as we know, optical fiber is a thin, flexible, transport fiber that acts as a waveguide to transmit light between two ends of the fiber. Optical fibers are widely used in fiber optic communications, which permits transmission over long distance and at higher bandwidths than other forms of communication. Fibers are used instead of metal wires because signals travel along with less loss and are also immune to electromagnetic interference. Fibers are also used for illumination, and are wrapped in bundles so they can be used to carry images. Optical fiber consists of a transparent core surrounded by a transparent cladding material with a lower index of refraction. Light is kept in the core by total internal reflection, which causes the fiber to act as a waveguide. Fibers that support many propagation paths or transverse modes are called "multimode fibers" (MMF), while those that can only support a single mode are called "single-mode fibers" (SMF). As mentioned before that tellurite glass fiber are under focused, Massera et al. (2010) processed and characterized of core–clad tellurite glass performs and fibers fabricated by rotational casting of high-purity core (72.5 mol% TeO_2–10mol% Bi_2O_3–17.5 mol% ZnO) and clad (70 mol% TeO_2–10 mol% Bi_2O_3–20 mol% ZnO). The resulting fiber was found to have an index step of 0.009 ± 0.002 between the fiber core and clad composition at 632 nm, and propagation losses of 3.2 ± 0.1 dB/m at 632 nm and 2.1 ± 0.1 dB/m at 1.5 μm. The primary source of loss in the near-IR (NIR) is associated with residual hydroxyl (OH$^-$) groups in the bulk preform, which remain in the glass fiber. The core–clad fiber has been drawn from a core–clad preform prepared via rotational casting. The step index (Δn) of the core–clad fiber was estimated (± 0.002) by measuring the numerical aperture NA of the fiber using the far-field output pattern at 632 nm. From these measurements, and using NA $= (n_1^2 - n_2^2)^{1/2}$, the numerical aperture (NA) was determined: where n_1 is the refractive index of the core glass and n_2 is the refractive index of the clad glass. The refractive index (n_{clad}) of cladding glass of composition 70TeO$_2$–10Bi$_2$O$_3$–20ZnO: (n_{clad}) = 2.1492 (at 630 nm), (n_{clad}) = 2.1165 (at 825 nm), and (n_{clad}) = 2.0847 (1533 nm). Also, the refractive index (n_{core}) of core glass of composition 72.5TeO$_2$–10Bi$_2$O$_3$–17.5ZnO: (n_{core}) = 2.1583 (at 630 nm), (n_{core}) = 2.1261 (at 825 nm), and (n_{core}) = 2.0925 (at 1533 nm). Then $\Delta n = n_{core} - n_{clad}$ (± 0.0002) = 0.0091, 0.0096, and 0.0078, respectively.

In linear optics, when two optical waves interact, they do not exchange energy. In nonlinear optics, two optical waves can interact and the medium in which they propagate (glass, crystal, plasma, gas) can undergo slight changes. Tellurite glasses have been widely studied from bulk materials to structured devices, with the emphasis on the development of nonlinear optical fibers to demonstrate the functionalities of supercontinuum generation, erbium-doped fiber and Raman amplifiers, and so on. Beyond all this progression, we do not forget that optical nonlinear property of tellurite glasses $\chi^3 \approx 50$ times bigger than of fused silica. Tellurite glasses are known as serious candidates for nonlinear optical and fast all-optical switching applications (Lines 1991). Superior laser damage resistance—even higher in comparison with chalcogenide glasses—allows the use of high-intensity laser pumping to reach high nonlinear optical effects (Donnell et al. 2007).

Hayakawa et al. (2010) studied the third-order nonlinear optical susceptibility χ^3 of tellurite (TeO$_2$)-based ternary glasses of MO–Nb$_2$O$_5$–TeO$_2$ (M = Zn, Mg, Ca, Sr, Ba) by Z-scan measurement using Ti–Sapphire femtosecond laser pulses. The compositions studied were xNb$_2$O$_5$–(100 – x) TeO$_2$ (x = 10, 20), yZnO–20Nb$_2$O$_5$–(100 – y)TeO$_2$ (y = 2, 4, 6, 8, 15), and 4MO–20Nb$_2$O$_5$–76TeO$_2$ (M = Mg, Ca, Sr, Ba). The third-order nonlinear optical susceptibility, χ^3, increased as the Nb$_2$O$_5$ content increased, and it also increased with simultaneous doping of the appropriate amount of ZnO content. The values of χ3 (4ZnO–20Nb$_2$O$_5$–76TeO$_2$) glass were equal to 2.71×10^{-12} esu. On the other hand, χ^3 of 4MO–20Nb$_2$O$_5$–76TeO$_2$ glasses (M = Mg, Ca, Sr, Ba) were almost the same as that of 4ZnO–20Nb$_2$O$_5$–76TeO$_2$ glass, although the values of linear refractive indexes, n_o, at 632.8 nm of 4MO–20Nb$_2$O$_5$–76TeO$_2$ glasses were smaller than those of 4ZnO–20Nb$_2$O$_5$–76TeO$_2$ glass. According to the relationship between χ^3 and the glass structures, it was revealed that χ^3 increased as the stretching band of Te–O$_{ax}$ in TeO$_4$ (tbp) unit increased while the stretching band of Te–O in TeO$_3$ (tp) unit and that of Te–O–Te decreased. This indicated that the amount of TeO$_4$ (tbp) units was deeply related to the value of χ^3. These authors discussed the role of niobium oxide in zinc or alkaline earth-tellurite glasses and proposed the mechanism for oxygen coordinated Nb to stabilize TeO$_4$ units despite the cleavage of Te$_{eq}$–O$_{ax}$–Te connection by the action of divalent ions. Also, in 2010, Ozdanova et al. prepared the Nb$_2$O$_5$–TeO$_2$, PbO–Nb$_2$O$_5$–TeO$_2$, and PbO–TeO$_2$ glasses from very pure oxides. The room temperature values of the experimental nondirect optical gap ($E_{g,non}$), the coefficient of the temperature dependence of the optical gap ($\gamma \times 10^{-4}$ eV/K), the refractive index (n at 800 nm), and the calculated values of the nonlinear refractive index ($n_2 \times 10^{-11}$ esu) were found as follows:

PbO–Nb$_2$O$_5$–TeO$_2$	$E_{g,non}$ (eV)	($\gamma \times 10^{-4}$ eV/K)	n (800 nm)	$n_2 \times 10^{-11}$ esu
0–0.1–0.9	3.42	4.7	2.16	0.92
0–0.2–0.8	3.45	4.3	2.27	0.89
0.1–0.1–0.8	3.45	4.3	2.14	0.88
0.1–0.2–0.7	3.39	4.7	2.27	0.95
0.2–0.1–0.7	3.51	4.7	2.15	0.83
0.2–0–0.8	3.61	6.8	2.19	0.74

In 2009 Gebavi et al., fabricated glass samples based on the following notation: $(100 - x)$ $(75TeO_2-20ZnO-5Na_2O)$ where $x = 0.36, 0.72, 1.08, 2.14, 3, 4, 5, 6, 7$, and 10 mol% of Tm^{3+}. The refractive index was measured for all samples at five different wavelengths (533, 825, 1061, 1312, and 1533 nm) by using the prism-coupling technique (Metricon, Model 2010). The resolution of the instrument was of ± 0.0001. Refractive index decreases with increasing Tm^{3+} concentration and increasing wavelength, as given by the least square fit: n ($\lambda = 633$ nm) $= -(395 \pm 7)10^{-5}c + (20.510 \pm 4)$ 10^{-4}. Also, the nonlinear refractive index of $(100 - x - y)$ $TeO_2-xBi_2O_3-yTiO_2$, where $x = 0, 5, 10, 15$, and $y = 5, 10$ has been studied by Chen et al. (2008). The third-order optical nonlinearity increases with decreasing optical band gap E_{opt}, since an increase of WO_3 content can provide the nonbridging oxygen ion content. Values of the nonlinear refractive indices were $n_2 = 4.92$ to 7.87 $(10^{-12}$ esu) for $10Bi_2O_3-20WO_3-70TeO_2$ and $10Bi_2O_3-30WO_3-60TeO_2$, respectively. Results showed that glass in the composition of $80TeO_2-10Bi_2O_3-10TiO_2$ (in mol%) was the best candidate for all-optical switching, while glass with the largest optical band gap E_{opg} had the greatest potential for optical limiters.

Desirena et al. (2009) found that an increase liner refractive n and χ^3 (and then n_2) when alkali metal oxides R_2O (R = Li, Na, K, Rb and Cs) and network intermediate MO (M = Zn, Mg, Ba and Pb) added in tellurite glasses. The former is good for the spectroscopic properties in laser and amplifier applications, but nonlinear response induces a deleterious effect. The trend observed for the introduction of intermediate is more complicated. Also, Lasbrugnas et al. (2009) measured the SHG of thermally poled tungsten–tellurite glass of the form $85TeO_2-15WO_3$ (mol%). Thermally poled glasses with $85TeO_2-15WO_3$ (mol%) composition have shown improved SHG performance with respect to silica glasses (generally <1 pm/V) and most of tellurium oxide-based glasses. After optimization of the thermal poling conditions, a high second harmonic signal ($\chi^2 = 1.5$ pm/V) was obtained. The SH active layer was thin, between 12 and 17 μm, and obtained at the anode side. Lasbrugnas et al. (2009) proposed two hypotheses to explain the origin of the second order nonlinearity of tellurite glasses: i- a reorientation of the TeO_4 glass structural entities could be envisaged under electric field, ii- the formation of an anodic depletion region of sodium ions. This generated the formation of an anodic depletion region of sodium ions, which can be considered as the main origin of the second-order nonlinearity property of this tellurite glass.

Shivachev et al. (2009) successfully synthesized, using the melt-quenching method, the novel $TeO_2-Bi_2O_3-GeO_2$ glasses possessing nonlinear optical properties. The linear refractive index was 2.13 for the three glasses: $90TeO_2-5Bi_2O_3-5GeO_2$, $80TeO_2-10Bi_2O_3-10GeO_2$, and $60TeO_2-20Bi_2O_3-20GeO_2$. To induce a nonlinear response in the $TeO_2-Bi_2O_3-GeO_2$ glass matrix, electrothermal poling was applied. A parallel plate of TBG-10 glass sample 1.6 mm thick, with surfaces polished to optical quality, was placed between two borosilicate glass slides and sandwiched with copper electrodes. The whole setup was placed inside a quartz vacuum chamber and kept in thermal contact with the chamber walls. The chamber and sample were slowly heated until the sample temperature stabilized at 230°C; the vacuum was about 5×10^{-6} torr. A 5 kV DC voltage was applied for 1.5 h, after which the heating was switched off. The DC voltage and vacuum were maintained until the sample reached ambient temperature. Measurements of SHG were conducted with a typical experimental setup for the Maker-fringe technique, using a picosecond fiber laser FIANIUM FP 1060-1 with a primary beam wavelength of 1064 nm, a repetition rate of 80 MHz, a pulse duration of $\Delta s = 5ps$, and an average output power of 800 mW. The base and the second harmonic beams were separated by a quartz prism in front of the detectors. The SHG intensity was monitored by a CCD camera as well as by a photodiode detector to check the angular dependence. The variation of the SHG intensity as a function of the incidence angle is similar to that observed in the other tellurite poled glasses. While for poled samples, SHG was anticipated, it was quite surprising to observe SHG signals in nonpoled, as-synthesized samples. Noguera and Suehara (2008) studied the high nonlinear optical properties in TeO_2-based glasses according to the modifier's influence from the localized hyperpolarizability approach. The substitution of the hydrogen atoms by alkali atoms (A = Li or Na) has much larger effects on second hyperpolarizabilities than on linear polarizabilities. The most influent electron pair is the Te 1p, but Te–O bp and O 1p have nonnegligible effects on these properties. However, the calculations were not able to reproduce qualitatively the experimental facts. Several explanations can be advanced:

- Usually, the decrease of these properties were attributed to the depolymerization of the network. The calculations should then include at least the break of one Te–O–Te bridge.
- The calculations of disrupted bonds were terminated by hydrogen atoms. The substitution of hydrogen atoms by alkali atoms constitutes a modification of the disrupted bond ending.
- These properties are sensitive to the modification of the structural unit.

Lin et al. (2004) measured the structure and nonlinear optical performance of TeO_2–Nb_2O_5–ZnO glasses. The addition of ZnO content from 0 to 15 wt% in a TeO_2–Nb_2O_5 base glass has two effects on the network structures:

1. It converts few Nb^{5+} ions from network modifiers to network formers and increases BO ions.
2. It changes the Te^{4+} bond from TeO_4 tbp to TeO_3 tp and increases NBO ions.

The former effect causes the optical nonlinearity of these tellurite glasses to increase, while the latter effect causes it to decrease. These two opposing effects of ZnO on the nonlinear optical response produce a maxima in the third-order nonlinear optical susceptibility χ^3 when about 2.5 wt% ZnO is added in place of TeO_2. The addition of ZnO > 2.5 wt% causes χ^3 to decrease. The average value of χ^3 for the $2.5ZnO$–$20N_2O_5$–$77.5TeO_2$ (wt%) glass is about 8.2×10^{-13} esu, which is the highest among all the nonlinear optical oxide glasses known up to this time. The nonlinear optical response time for these glasses containing from 0 to 15 wt% ZnO varied from 450 to 500 fs. The authors believed that the nonlinear optical response in these glasses arises primarily from the deformation of electron clouds. Such an ultra-fast nonlinear optical response makes these ZnO–Nb_2O_5–TeO_2 glasses highly suitable for all optic switching applications. Ferreira et al. (2003) measured the linear and nonlinear optical properties of tellurite glass with $70TeO_2$–$25Pb(PO_3)_2$–$5Sb_2O_3$ and found that: $n_o(532\ nm) = 1.986$, $n_o(800\ nm) = 1.97$ and $n_o(1064\ nm) = 1.957$, respectively. The third-order susceptibility of the same glass was $\chi^3 = 2.44 \times 10^{-21}(m^2/V^2)$ while χ^3 for silice infrasil = $0.11 \times 10^{-21}(m^2/V^2)$.

NB: χ^3 (Nb_2O_5–TiO_2–Na_2O–SiO_2) = 0.7–0.58×10^{-14} esu, χ^3 (Germanate glasses) = 4.8–8.0×10^{-14} esu and χ^3 ($2.5ZnO$–$20Nb_2O_5$–$77.5TeO_2$) = 82×10^{-14} esu.

Hagar et al., in 2011, prepared new tellurite glass series of the form $(70 - x)TeO_2$–$20WO_3$–$10Li_2O$–xLn_2O_3 where x = 0, 1, 3, and 5 mol%, and Ln are Nd, Sm, and Er. The luminescence spectra of the prepared glasses were measured at room temperature using a micro-Raman spectrometer. The obtained luminescence intensity ratio was correlated with the rare-earth ion concentrations, the short distance between the identical rare-earth ions r (Ln – Ln), density and number of bonds per unit volume, n_b, refractive index, molar refractivity, and optical polarizability. The dependence of` the luminescence intensity ratio is also correlated with the optical polarizability of the glasses. The authors calculated the luminescence ratio ρ ($\rho = I_L/I_R$, where I_L and I_R, are the intensity of luminescence line and the intensity of Raman line) with the optical polarizability α_o. The ρ (or the luminescence intensity) depends on the glass polarization, and also on the correlation between α_o and the short distance between the identical rare-earth ions r (Ln – Ln). This means that when the glasses become rigid and compact (i.e., short distance or high strength of r [Ln – Ln] with increase the rare-earth contents) α_o increases, which is the case of the studied glasses and vice versa. Saleem et al. (2011) studied the luminescent of Dy^{3+} ion in alkali–lead tellurofluoro–borate glasses of the form: $26LiF$–$20PbO$–$10TeO_2$–$43H_3BO_3$–$1Dy_2O_3$, $26NaF$–$20PbO$–$10TeO_2$–$43H_3BO_3$–$1Dy_2O_3$, and $26KF$–$20PbO$–$10TeO_2$–$43H_3BO_3$–$1Dy_2O_3$. The luminescence spectra show two intense bands at

483 and 575 nm, which were attributed to $^4F_{9/2}$-$^6H_{15/2}$ (blue) and $^4F_{9/2}$-$^6H_{13/2}$ (yellow) transitions, respectively.

Also, Babu et al. (2011) prepared and studied the spectroscopic and photoluminescence properties of Dy^{3+}-doped lead tungsten–tellurite glasses for laser materials. The glasses were in the form $15PbF_2$–$25WO_3$–$(60 - x)TeO_2$–xDy_2O_3 with $x = 0.1$, 0.5, 1.0, and 2.0 mol%, and were prepared using the melt-quenching technique. The color coordinates were evaluated for the appropriate combinations of yellow to blue emissions. The authors concluded that $15PbF_2$–$25WO_3$–$(60 - x)$ TeO_2–xDy_2O_3 glasses doped with 1.25 mol% of Dy^{3+} ions could be useful for potential laser material as well as to generate white light in color-displaying devices.

9 Optical Properties of Tellurite Glasses in the Ultraviolet Region

Experimental procedures to measure ultraviolet (UV) absorption and transmission spectra in bulk and thin-film forms of tellurite glasses are summarized in Chapter 9, along with theoretical concepts related to absorption spectra, optical energy gaps, and energy band tail width. The UV-spectrum data discussed here have been collected in the wavelength range from 200 to 600 nm at room temperature. Data are provided for the UV properties of oxide–tellurite glasses (bulk and thin film), oxide–tellurite glass-ceramics (bulk and thin film), halide–tellurite glasses, and nonoxide–tellurite glasses. From these experimental absorption spectra, the energy gap and band tail data are also summarized for these glasses. Analysis of these optical parameters is based on the Urbach rule, which is also explained.

9.1 INTRODUCTION—ABSORPTION, TRANSMISSION, AND REFLECTANCE

In recent years, descriptions of the optical properties of glasses have increasingly been more in electronic than in chemical terms. This is probably because improved understanding of amorphous and glassy states (e.g., Mott and Davies 1979) and, in particular, the relationship between structures and properties has led, for example, to increased applications of glasses and glassy complexes in electronic systems. Optical absorption in solids and liquids occurs by various mechanisms, all of which involve the absorption of photon energy ($\hbar\omega$) by either lattice structures or electrons, where this transferred energy can be conserved. Lattice (or phonon) absorption data can give information on the atomic vibrations involved in these properties, and the absorption of radiation normally occurs in the infrared (IR) region of the spectrum. At higher-energy levels, parts of the spectrum—particularly those associated with interband electronic transition—provide information about the electron states of a material. Electrons move by being excited from a filled to an empty band by photon absorption; as a result, a marked increase occurs in the frequency dependence (ω) of the absorption coefficient (α). The onset of this rapid change in $\alpha\omega$ against ω is called "the fundamental absorption edge," and the corresponding energy is defined as "the optical energy gap" (E_{opt}). A photon with a certain range of energy can be absorbed by modifying ions in an oxide glass via either of two processes: (1) internal transitions between the *d*-shell electrons, or (2) transfer of an electron from a neighboring atom to the modifier ion (transition metal ion) and vice versa.

The study of optical absorption, and particularly the absorption edge, is a useful method to investigate optically induced transitions and to obtain information about the band structures and energy gap (E_g) of both crystalline and noncrystalline materials. The principle behind this technique is that a photon with energies greater than the band gap energy will be absorbed. There are two kinds of optical transitions at the fundamental absorption edge of crystalline and noncrystalline semiconductors: direct and indirect transitions, both of which involve interaction of an electromagnetic wave with an electron in the valence band (VB), by which the electron is then raised across the fundamental gap to the CB. For direct optical transition from the VB to the CB, it is essential that the wave vector for an electron remains unchanged. For indirect transition, the interaction with lattice vibrations (or phonons) takes place; thus, the wave vector of the electron can change in the optical transition and the momentum of the change is taken or given up by a phonon. In other words, if the minimum of the CB lies in different parts of k space from the maximum of the VB, a direct optical transition from the top of the VB to the bottom of the CB is "forbidden."

9.2 EXPERIMENTAL PROCEDURE TO MEASURE UV ABSORPTION AND TRANSMISSION SPECTRA

The most commonly used apparatus for measuring the absorption of bulk samples is a spectrophotometer. This instrument compares the transmission intensities of a standard and an unknown with the response in the material being analyzed, as a measure of transmission or optical density (OD). The apparatus itself consists of a light source, chopper, light-directing optics, monochrometer, and detector. Typically, spectrophotometers are manufactured to measure certain transmission windows. Commercially available equipment generally measures responses in the energy range 0.1–2.0 μm on one apparatus and 2.0–50 μm on another. This wavelength specificity is a result of detector and monochrometer limitations. Many chapters have been published concerning the optical properties of glasses; for example, Fleming's work on "Optical properties and characterization," in a 1992 book edited by Simmons et al.

For light that is normally incident on a glass surface, which is the typical orientation for attenuation measurement, the first component of attenuation is surface reflection. The fraction of light absorbed by the material is determined by comparing the intensity (I) of the light actually injected into a specimen x, I_0, to the I detected after a given length of travel through x, which can be shown as:

$$\alpha = (1/x)\log(I/I_0) \tag{9.1}$$

where α is generally reported in units per centimeter.

Sample preparation includes optically polishing two parallel faces of the sample through which the beam will pass, which is done to remove surface irregularities that might scatter the incident light and to remove other impediments to transmission, because these are usually of no interest to an investigator. Thickness is an important parameter in calculating attenuation; it should be measured accurately. Thickness can be increased or decreased to optimize the measured signal. When one is observing very strong absorption, it is possible to dilute the sample by mixing it in powder form in a medium that transmits well in the wavelength region of interest. An example of such an appropriate immersion material is a low-melting-temperature salt, such as KBr.

9.2.1 SPECTROPHOTOMETERS

Both single-beam and double-beam spectrophotometers are commercially available. The single-beam instrument uses a single optical path for measurement of a sample. Quantification and standardization are accomplished through a standard or baseline measurement. To overcome the problem of sample surface reflection, two samples of different lengths with similarly polished faces are required. The sensitivity of this method is a function of the magnitude of the difference in the lengths of the samples. These measurements should be made consecutively. Double-beam spectrophotometers measure the attenuation of two glass samples simultaneously and provide a comparison of the two as the final instrument response, which can be either an AC or DC response, depending on the nature of the source and arrangement of optics. Double-beam methods are easier to use and generally more accurate.

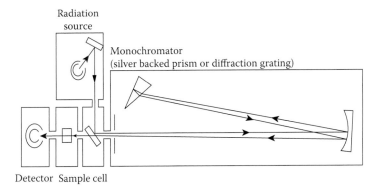

FIGURE 9.1 UV spectrophotometer. (From R. Denney and R. Sinclair, *Visible and UV Spectra*, John Wiley & Sons: New York, 1987. With permission.)

Several techniques have been developed to measure the attenuation of very-low-attenuation bulk glass specimens. One uses an apparatus called a "two-beam optical bridge" (Figure 9.1). Light from a laser beam is split by a beam splitter, and the two beams are then "chopped," by which synchronized pulses are created so that one is passed when the other is blocked. The beams next pass through polarizers and onto detectors. The sample is placed in one optical path between a polarizer and detector. The faces of the sample through which the beam passes must be optically polished and should also be parallel. To eliminate the complication of surface reflection, the sample faces are oriented at Brewster's angle with respect to the incident beam, as shown in Figure 9.1. Using the polarizer in this light path, surface reflectivity (R) can be eliminated as an attenuation factor. Another polarizer is used to adjust the attenuation on the reference beam so that the two beams are detected with equal I when no sample is in the light path. Thus, with the detectors connected in parallel, the electrical signal is a constant DC value. When a sample is placed in the light path, a small AC signal proportional to the loss in the sample is placed on the DC detector. This action can be monitored selectively with very high sensitivity, thus enabling very accurate measurements of attenuation. The optical bridge measures the total sample attenuation, which includes bulk absorption, bulk scattering, and surface scattering. By measuring different thicknesses of the same glass sample, one can isolate the bulk loss effects. The most critical elements in this measurement are the detectors. Most detectors do not have a uniform response to a given light signal across their surfaces. It is therefore important to orient the detector of the sample beam so that the beam falls on the same spot with and without the sample in place. For this purpose, the means for precision movement of the sample beam detector are incorporated into the apparatus.

Spectrometers can record absorption as either percent transmission [$100 \times (I/I_o)$] or as OD, which is defined as:

$$OD = \log_{10}\left(I_o/I\right) \tag{9.2}$$

For example, 10% transmission corresponds to 1-OD absorbency and 1% transmission to 2-OD absorbency. The α is related to the OD by the expression $\alpha = [(1/x) \ln(I_o/I)] = [(2.303/x)(OD)]$. The use of the decibel (dB) as a unit of optical loss is more common today because of the influence of optical-waveguide technology. (10 dB = 1 OD; fiber optic losses are usually quoted in units of decibels per kilometer.)

Reflective losses at surfaces are also encountered in practice. In a double-beam system, these are easily eliminated by the use of two specimens of different thickness in the sample and reference beams. In a single-beam system, samples of multiple thickness can also be used to separate the constant surface-reflective-loss contribution from thickness-dependent bulk absorption. In regions of extremely intense absorption, the reflection itself is used to determine α. Strong reflectance bands correspond to regions of strong absorption. This can be concluded by inspection of the expression for normal incidence R and optical absorption in dielectric media with complex refractive index $n^* = (n - ik)$, where n is the real component (index of refraction) and ik is the imaginary component. The attenuation of a light beam in such a medium is related to Equation 9.3 by the expression:

$$I = I_o \exp(-4\pi k/\lambda) x$$

$$= I_o \exp(-\alpha x) \tag{9.3}$$

The R at normal incidence at the first surface is given by the expression:

$$R = \frac{(n-1)^2 + k^2}{(n+1)^2 + k^2} \tag{9.4}$$

as stated in Chapter 8. The optical properties of thin film are characterized by n and the extinction coefficient k. Both parameters are combined to give the complex refractive index $n^* = (n - ik)$. The coefficient n is related to the magnitude of light absorption, and the coefficient k is related to the frequency of absorption. The absorption index (k) and the n of a thin absorbing film on a nonabsorbing substrate can be determined by measuring the transmittance (T) of the film substrate system alone, taking into account the multiple incoherent reflections at the interface of the thin film with the substrate, as mentioned by Amirtharaj (1991) and by Baars and Sorger (1972):

$$T = \frac{Me^{-\alpha d}}{P - Qe^{-2\alpha d}} \tag{9.5}$$

where d is the thickness of the films, $M = (1 - R_1)(1 - R_2)(1 - R_3)$, $P = (1 - R_2 R_3)$, and $Q = R_1 R_2 + R_1 R_3 - 2R_1 R_2 R_3$. The letters R_1, R_2, and R_3 are the reflection coefficients of the air–film, film–substrate, and substrate–air interfaces, respectively. These coefficients are used in their most general forms:

$$R_1 = \frac{(n-1)^2 + k^2}{(n-1)^2 + k^2}$$

and

$$R_2 = \frac{(n-n_s)^2 + k^2}{(n+n_s)^2 + k^2}$$

and

$$R_3 = \frac{(n_s - 1)^2}{(n_s + 1)^2} \tag{9.6}$$

where $n = \lambda \alpha/4\pi$, n_s is the n of the glass substrate, and λ is the wavelength of the incident beam.

The procedure to calculate optical transmission properties of glasses, including E_{opt} and the optical energy tail (E_{tail}), involve using the spectrophotometer at room temperature as follows:

- α is derived from the formula:

$$\alpha(\omega) = (1/d) \ln I_o \, \omega / I_t \, \omega$$
$$= (1/d) \log_{10} e \log_{10} \left[I_o \, \omega / I_t \, \omega \right] \tag{9.7}$$

where d is the sample thickness, and I_o and I_t are the intensities of incident and transmitted light, respectively.
- The energy of the incident photons is transferred to electrons by the relation:

$$E(eV) = \hbar\omega$$
$$= 1.24 / \lambda \, \mu m \tag{9.8}$$

- The inverse of the slope of the linear part of the representation of $\ln\alpha$ against ω produces E_e (in electron volts).
- The intersect of the slope of the linear part of the representation of $\alpha\hbar\omega^{1/2}$ on the y-axis against $\hbar\omega$ on the x-axis gives the values of the E_{opt} (in electron volts), and if it does not give a straight line, the representation of the y-axis is changed to $\alpha\hbar\omega^{1/3}$ to get the intersect of the linear part with x-axis.

9.3 THEORETICAL ABSORPTION SPECTRA, OPTICAL ENERGY GAP, AND TAIL WIDTH

In 1970, Davis and Mott measured and discussed in detail the conduction and optical absorption edge and photoconductivity of amorphous semiconductors. Kurik (1971) explained the Urbach rule, which was by Urbach in 1953. In 1979, Mott and Davis explained the optical properties of amorphous materials in detail. Two types of optical transitions occur at the fundamental edge in crystalline semiconductors; namely direct and indirect transitions. Both involve interaction of an electromagnetic wave with an electron in the VB, which is raised across the fundamental gap to the CB as in Figure 9.2. Indirect transitions also involve simultaneous interactions with lattice vibrations; thus, the wave vector of the electron can change in the optical transitions, with the momentum change being taken or given up by phonons. If the excitation formation or electron–hole interaction is neglected, the form of the α as a function of $\hbar\omega$ depends on the type of energy bands containing the initial and final states. In many crystalline and noncrystalline semiconductors, the $\alpha\omega$ depends exponentially on the $\hbar\omega$. This exponential dependence, known as the Urbach rule, can be written in the form:

$$\alpha(\omega) = B \exp\left(\frac{\hbar\omega}{\Delta E} \right) \tag{9.9}$$

where B is a constant and ΔE is the width of the band tails of the localized states.

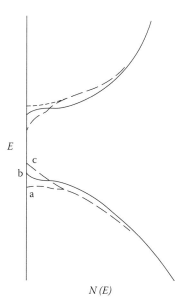

E

c

b

a

$N(E)$

FIGURE 9.2   Proposed density of states in amorphous solid material.

In general, both direct and indirect transitions can occur in a crystalline semiconductor. The smallest gap leads to direct transition. The indirect transition is associated with a smaller α. Mott and Davis (1979) suggested the following expression for direct transitions:

$$\alpha(\omega) = B(\hbar\omega - E_{\text{opt}})^n / \hbar\omega$$

(9.10)

where $n = 1/2$ or $3/2$, depending on whether the transition is allowed or forbidden, and B is a constant. Evidence for optical transitions from the VB state (or localized gap state) to unoccupied localized states above the Fermi level can be sought at energies near and below E_{opt}.

The origin of the exponential dependence of the α on $\hbar\omega$ is not clearly known. Dow and Redfield (1972) suggested that the tailing might arise from random fluctuations of the internal disorder in many amorphous solids. One possible reason, suggested by Davis and Mott (1970), is that the slopes of the observed exponential edges obtained from Equation 9.12 are very much the same as in many semiconductors, and the value of ΔE for a range of amorphous semiconductors lies between ~0.045 and 0.67 eV. The optical α of most amorphous semiconductors is also found to increase exponentially with $\hbar\omega$ in the range $1 < \alpha < 10^4$ cm^{-1}. At α values of $>10^4$ cm^{-1}, the absorption curves begin to level off, where transitions can take place between two bands of states. Mott and Davis (1979) assumed that the densities of states at the band edges E_C and E_V are linear functions of the energy so that:

$$N(E_C) = N(E_V)$$

and

$$(E_C - E_V) = \Delta E$$

(9.11)

where ΔE is the width of the tail of localized states in the band gap and the transition in which both the initial and final states are localized can be neglected. Hogarth and Kashani (1983) developed an equation for glassy materials of the form:

$$\alpha(\omega) = \left(\frac{4\pi\sigma_o}{cn}\right)\left(\hbar\omega - E_{opt}\right)^2 / \hbar\omega\Delta E \tag{9.12}$$

where $A = (4\pi\sigma_o/nc\Delta E)$, σ_o is the conductivity at $1/T = 0$, and c is the velocity of light.

As cited in Chapter 8, Dimitrov and Sakka (1996) showed that for simple oxides the average electronic oxide polarizability (α_m) calculated on the basis of two different properties (i.e., linear n (n_o) and E_{opt}) shows remarkable correlation. Vithal et al. (1997) combined the molar refractivity (R_m), n_o, molar volume (V_m), and optical basicity (Λ) of glass with E_{opt} data as discussed earlier; based on the optical band gap in amorphous systems, they concluded that E_{opt} is closely related to the E_g between the VB and CB. In glasses, although the CB is influenced by the glass-forming anions, the cations play an indirect but significant role. El-Mallawany (1992) assumed that R_m and α_m are additive quantities, but Dimitrov and Sakka (1996) identified the relation:

$$R_m = \left(pR_i + qR_{O2-}\right)$$
$$= 2.52\left(p\alpha_i + q\alpha_{O2-}\right) \tag{9.13}$$

where p and q denote the number of cations and oxide ions, respectively, and R_i is the ionic refraction of the action, R_{O2-} is the ionic refraction of the oxide ion, and α_i and α_{O2-} are the α_m of the cation and oxide ion, respectively, in the chemical compound A_pO_q. Vithal et al. (1997) deduced that:

$$\alpha_{o2} - n_o = \left[\left(\frac{V_m}{2.52}\right)\frac{n_o^2 - 1}{n_o^2 + 2} - \sum p\alpha_i\right]q^{-1} \tag{9.14}$$

using the Lorentz–Lorentz equation from Chapter 8. Dimitrov and Sakka (1966) used the following linear relationship:

$$\sqrt{E_{opt}} = (20)\left[1 - \left(R_m/V_m\right)\right] \tag{9.15}$$

whereas Vithal et al. (1997) studied a large number of glass systems as well, leading to another modified relationship; that is, the relationship between $E_{opt}^{1/2}$ and $[1 - (R_m/V_m)]$ of the form:

$$\sqrt{E_{opt}} = (1.23)\left(1 - \frac{R_M}{V_M}\right) + 0.98 \tag{9.16}$$

and the oxide ion α_m for glass may be written in the form:

$$\alpha_{o2} - E_{op} = \left[\left(\frac{V_m}{2.52}\right)\left\{1 - \frac{\left(E_{op}^{1/2} - 1.14\right)}{0.98}\right\} - \sum_i p_i\alpha_i\right]q^{-1} \tag{9.17}$$

where α_{o2} is the α_m of the cation and oxide ion, and p and q are the numbers of cations and oxide ions, respectively, in the chemical formula A_pO_q.

The transparency of glass is based on the size of the scattering particles. When the size of the scattering particles is small in comparison with the wavelength of the light, the angular distribution of scattered light (i.e., Rayleigh scattering) was expressed by Beall and Duke (1969) as follows:

$$I(\theta) = \left(\frac{1 + \cos^2 \theta}{r^2} \right) \left(\frac{8\pi^4}{\lambda^4} \right) d^6 \left| \frac{M^2 - 1}{M^2 + 2} \right| \left| (I_o) \right| \tag{9.18}$$

where $I(\theta)$ is the specific I, θ is the scattering angle, r (the radius) is the distance from the scattering center, λ is the wavelength of the incident light, d is the radius of the particle, and M is the ratio of the n of the particle to that of the surrounding medium.

9.4 UV PROPERTIES OF TELLURITE GLASSES (ABSORPTION, TRANSMISSION, AND SPECTRA)

9.4.1 UV-Properties of Oxide–Tellurite Glasses (Bulk and Thin Film)

Table 9.1 summarizes E_{opt} and E_{tail} data for selected tellurite glasses.

TABLE 9.1

Calculated Optical Energy Gaps and Energy Tails of Tellurite Glasses

Glass Composition (mol%)	E_{opt} (eV)	E_{tail} (eV)	Glass Composition (mol%)	E_{opt} (eV)	E_{tail} (eV)
TeO_2–CaO–WO_3			TeO_2–CuO		
65–0–35	2.82	1.22	95–5	3.55	
65–2.5–32.5	2.93	1.26	90–10	3.30	
65–10–25	3.15	1.34	85–15	3.31	
TeO_2	3.79	0.07	TeO_2–GeO_2–Fe_2O_3		
TeO_2–WO_3			90–5–5	2.095	0.17
95–5	3.49	0.11	85–10–5	2.025	0.285
90–10	3.43	0.13	75–20–5	1.90	0.195
85–15	3.40	0.11	65–30–5	1.73	0.342
80–20	3.38	0.14	95–0–5	2.1	0.269
75–25	3.38	0.11	95–5–0	2.7	0.168
70–30	3.32	0.12	90–10–0	2.236	0.275
TeO_2–CuO–Lu_2O_3			65 TeO_2–(35 − x)		
			CuO–x $CuCl_2$		
65–35–0	2.01	0.5	$x = 1$	2.08	0.57
65–34–1	2.04	0.74	$x = 2$	2.15	0.51
65–33–2	2.08	0.77	$x = 3$	2.23	0.53
65–32–3	2.13	0.76	$x = 4$	2.35	0.58
65–31–4	2.18	0.60	$x = 5$	2.43	0.66
TeO_2–BaO			TeO_2–P_2O_5		
90–10	3.85	0.134	36.15–73.85	2.17	0.67
80–20	3.98	0.090	59.42–40.58	2.39	0.42
70–30	4.12	0.178	73.51–26.49	2.83	0.28
90–10 thin film	3.48	0.555	81.09–18.91	2.97	0.33
80–20	3.60	0.500	Bi_2O_3–TeO_2–P_2O_5		
70–30	3.61	0.474	14.15–50.14–35.71	2.63	0.21
TeO_2–Ti_2O	2.30		26.50–43.45–30.05	2.30	0.42
TeO_2–ZnO	2.52		46.11–31.19–22.70	2.37	0.26
TeO_2–La_2O_3	2.90		TeO_2–$TeCl_4$	3.15	0.14
TeO_2–P_2O_5–La_2O_3	3.03	0.45	TeO_2–WCl_6	2.70	0.24
TeO_2–P_2O_5–CeO_2	3.04	0.30			
TeO_2–P_2O_5–Pe_2O_3	3.05	0.28			
TeO_2–P_2O_5–Nd_2O_3	3.07	0.38			
TeO_2–P_2O_5–Sm_2O_3	3.74	0.25			
TeO_2–P_2O_5–Yb_2O_3	3.04	0.27			

Hogarth and Kashani (1983) measured the E_{opt} of ternary tellurite–tungsten–calcium glasses. Their ternary tellurite glass formula was 65 mol% TeO_2–$(35 - x)$ mol% CaO–x mol% WO_3, where $35 > x > 0$. Thin films were prepared by a blowing technique using a silica rod. The thicknesses of the films were measured with a Sigma comparator, and the thicknesses ranged from 1.5 to 2.0 µm. The optical transmission properties of the glasses were analyzed using a spectrophotometer at room temperature. The optical α is derived from Equations 9.1 through 9.3, as shown in Figure 9.3A, B, C, D. The E_{opt} is in the range 3.15–2.82 eV, somewhat lower than for many oxide-based glasses. In ternary tellurite glasses, the E_{opt} decreases from 3.15 to 2.82 eV with increases in the percentage of WO_3 $(35 > x > 0)$, whereas the constant $A = 4\pi\sigma/nc\Delta E$ increases with increases in the WO_3. There are several possible explanations:

- The increase in the percentage of WO_3 in ternary tellurite glasses decreases the optical gap because WO_3 facilitates polaron hopping, and conductivity is increased by hopping, as mentioned in Chapter 6.
- The value of electrical activation energy is less than half of the optical gap on the same glass, which suggests that electronic activation is not across the mobility gap but is possibly from one or more trapping levels to the CB, from the bonding states to the CB, or from the bonding states to the trapping level.

In 1984, Ahmed et al. studied the effect of other glass formers, like $(100 - x)$ mol% GeO_2–x mol% TeO_2, for $x = 10, 15, 20, 25,$ and 30, on the UV optical properties as shown in Figure 9.4. The optical gap is on the order of 2.74 eV. They measured the electrical properties of these glasses and correlated both the activation energy and E_{opt}. They found that both quantities decrease with increases in the percentage of TeO_2. Burger et al. (1985) measured the optical properties of new families of tellurite glasses. The optical properties were in the UV, visible (VIS), and IR regions, as shown in Figure 9.5. The VIS optical properties are collected in Chapter 8, and the UV optical properties are presented in this chapter, whereas optical properties in the IR region are presented in Chapter 10. The new families of tellurite glasses are in the form TeO_2–R_nO_m or R_nX_m, $R_n(SO_4)_m$, $R_n(PO_3)_m$, and B_2O_3, and X_m is F, Cl, or Br. The UV properties of these glasses are represented in Figure 9.5, and the cutoff curves in the direction of the short-wavelength region of some selected glasses.

In 1985, Al-Ani et al. measured the UV optical absorption of pure tellurite glass and binary tungsten–tellurite thin-film glasses as shown in Figure 9.6A, B, C, D, and E. The E_{opt} decreases from 3.79 to 3.32 eV with increases in the molecular percentage of WO_3, whereas the tail increases from 0.07 to 0.12 eV for the same percentages of the transition metal oxide. A reduction in E_{opt} by introducing WO_3 means that reduction is caused when the energy levels of WO_3 are introduced between the E_{opt} of the TeO_2 glass. This reduction in E_{opt} is the reason that binary tellurite–tungsten glass is a semiconductor, and also why electrical conductivity increases with increased WO_3.

Ahmed et al. (1985) measured the absorption spectra of binary sodium–borate glasses (70 mol% B_2O_3–30 mol% Na_2O) doped with TeO_2, in the wavelength range 300–700 nm, as shown in the 1st Ed. Depending on the alkali content, the glasses obtained are either colorless, rose, or gray. The rose color is attributed to the formation of the polytellurite state, whereas the gray tint is attributed to the formation of elemental tellurium particles (this conclusion is supported by x-ray diffraction analysis). The sequence of formation of the different states of tellurium with increasing alkali content was in accordance with that predicted from a consideration of the acidity–basicity concept. The identification of such species in glasses studied does not exclude the probability of formation of other unidentified tellurium states that absorb only in regions outside the range of wavelengths reported in this study.

In 1988, Hassan and Hogarth measured optical absorbency as a function of wavelength in three samples of copper–tellurite glass of the form $(100 - x)$ mol% TeO_2–x mol% CuO, for which $x = 25,$ 35, and 50. These authors emphasized several points:

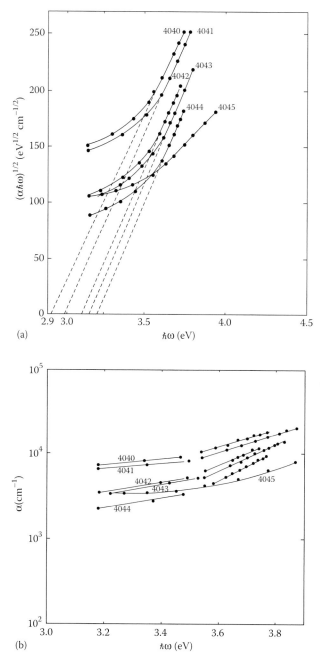

FIGURE 9.3   (a–d) UV spectra of ternary tellurite glasses of the form 65 mol% TeO_2–(35 − x) mol% CaO–x mol% WO_3, for which 35 > x > 0. Samples: 4040, 65 mol% TeO_2–0 mol% CaO–35 mol% WO_3; 4041, 65 mol% TeO_2–2.5 mol% CaO–32.5 mol% WO_3; 4042, 65 mol% TeO_2–5.0 mol% CaO–30.0 mol% WO_3; 4043, 65 mol% TeO_2–7.5 mol% CaO–27.5 mol% WO_3; 4044, 65 mol% TeO_2–10.0 mol% CaO–25 mol% WO_3; 4045, 65 mol% TeO_2–15.0 mol% CaO–20.0 mol% WO_3; 4046, 65 mol% TeO_2–20.0 mol% CaO–15 mol% WO_3. (From C. Hogarth and E. Kashani, *J. Mater. Sci.*, 18, 1255, 1983. With permission.)

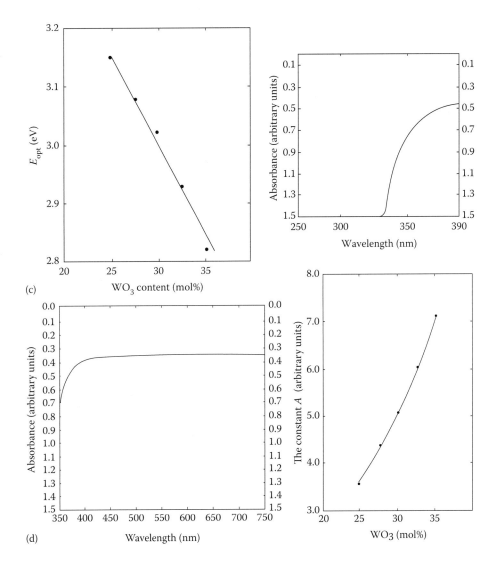

FIGURE 9.3 (Continued)

- The shift of the UV absorption band to longer λ with increasing CuO content
- The variation of $E_{opt.}$ with electrical activation energy
- The kind of conduction mechanism that is likely to be involved in these glasses

In 1989, Hassan et al. compared the UV optical properties of binary $(100 - x)$ mol% TeO_2–x mol% BaO glasses by blowing and by evaporation, for which $x = 10, 20$, and 30, as shown in the 1st Ed. The values of the optical gap for blown films are greater than those of thin evaporated films, but the ΔEs are greater for the thin evaporated films, which are expected to be more disordered than the thin blown-glass films. Malik and Hogarth (1989b) measured the UV optical properties of $(65 - x)$ mol% TeO_2–$(35 - x)$ mol% CuO–x mol% Lu_2O_3 glasses. They found that the optical gap increases with increasing percentages of the Lu_2O_3 as in Figure 9.7.

Since the 1990s, little new research work has been published on the UV-range optical properties of tellurite glasses. Binary TeO_2–V_2O_5 amorphous films with compositions in the range 10–50 mol% V_2O_5 and thicknesses in the range 1–7 μm were prepared by Chopa et al. (1990), using the

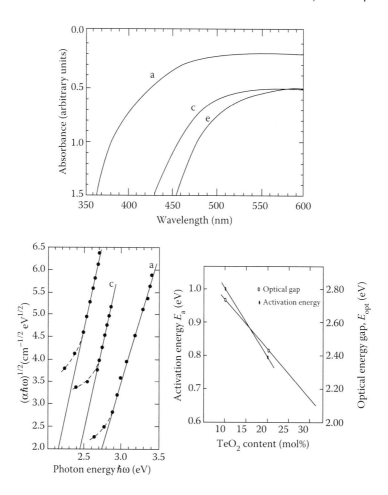

FIGURE 9.4 UV optical properties of $(100 - x)$ mol% GeO_2–x mol% TeO_2, for which $x = 10$, 15, 20, 25, and 30. (From A. Ahmed, R. Harini, and C. Hogarth, *J. Mater. Sci. Lett.*, 3, 1055, 1984. With permission.)

blowing technique. They also measured the relationship between the n of the binary vanadium films at different wavelengths, as shown in Chapter 8. The optical transmission spectra of these films were measured in the range 190–900 nm and in the E_g and E_{tail} of localized states, as shown in the 1st Ed. A sharp decrease in the value of the optical band E_g, with a small addition of V_2O_5 (10 mol%) in TeO_2 was observed. The decrease in E_g was considerably less than at higher concentrations of V_2O_5. Khan (1990) studied the effects of CuO on the structure and optical absorption of tellurite glasses. Khan (1990) found that the fundamental absorption edge is a function of the glass composition and that the optical absorption in this binary tellurite glass can be analyzed in terms of indirect transitions across E_{opt} as shown in Figure 9.8. The optical absorption of binary tellurite–phosphate and ternary tellurite–phosphate–bismuth glasses was recorded by Abdel-Kader et al. (1991a). The optical absorption spectra are in the spectral range 300–800 nm, and it was found that the fundamental absorption spectra of these glasses are dependent on the glass composition as shown in Figure 9.9. The E_{opt} values of binary glasses increase with increasing TeO_2 content, but the addition of Bi_2O_3 to TeO_2–P_2O_5 decreases the E_{opt}. The absorption edges of these glasses arise from direct forbidden transitions and occur at photon energies in the range 2.17–2.97 eV for TeO_2–P_2O_5 glasses and 2.63–2.32 eV for TeO_2–P_2O_5–Bi_2O_3 glasses, depending on their composition.

Also in 1991, El-Samadony et al. measured the optical absorption of tellurite glasses by counting ferric and germanium oxide of the form $(95 - x)$ mol% TeO_2–x mol% GeO_2–5 mol% F_2O_3, for

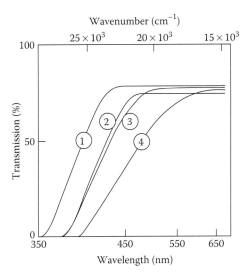

FIGURE 9.5 Optical properties in the UV spectra of $TeO_2-R_nO_m$ or R_nX_m, $R_n(SO_4)_m$, $R_n(PO_3)_m$, and B_2O_3, for which X_m = F, Cl, or Br. Curves: 1, 83.2 mol% $TeO_2-16.8$ mol% Al_2O_3; 2, 75 mol% TeO_2-5 mol% A_2O_3-20 mol% PbO; 3, 75 mol% TeO_2-20 mol% A_2O_3-5 mol% PbF_2; 4, 75 mol% TeO_2-5 mol% A_2O_3-20 mol% $PbCl_2$; 5, 75 mol% TeO_2-5 mol% A_2O_3-20 mol% $PbBr_2$, 6, 75 mol% TeO_2-5 mol% A_2O_3-20 mol% $PbSO_4$, 7, 85 mol% TeO_2-15 mol% P_2O_5. (From H. Burger, W. Vogal, and V. Kozhukarov, *Infrared Phys.*, 25, 395, 1985. With permission.)

which $5 < x < 35$, and $(90 - x)$ mol% TeO_2-x mol% GeO_2-10 mol% F_2O_3, for which $5 < x < 30$. The spectra are measured at room temperature in the wavelength range 350–900 nm, and the results show that the fundamental absorption edge is a function of composition (Figure 9.10), with optical absorption caused by indirect transitions. The optical band gap decreases with increasing GeO_2 content, and the Urbach rule is used. In 1992, El-Mallawany measured the optical properties of binary 90 mol% TeO_2-10 mol% R_nO_m, for which R is Ti, Zn, or La as shown in Figure 9.11. The UV spectra of these binary vitreous systems were recorded in the range 200–600 nm at room temperature. The experimental spectra and the optical absorption edge have been analyzed by their indirect, allowed transitions. The E_{opt} values were 2.3, 2.52, and 2.9 eV, respectively. The band tails were calculated by the Urbach equation, and values of 0.21, 0.18, and 0.4 eV have been obtained for the same binary tellurite glass samples. Explanation for the larger value of the optical gap in these binary rare-earth tellurite glasses compared with that of binary transition metal tellurite glasses is that rare-earth ions are larger in size, so distance between the top of the VB and the CB is large.

In 1992, Burger et al. measured the spectral UV T curves of zinc–tellurite glasses of thickness 10 mm in the region 300–500 nm as shown in Figure 9.15. The 50% T in the UV region is at 399, 393, 385, and 380 nm for 19.9, 24.6, 29.6, and 33.2 mol% ZnO, respectively. The UV T of ZnO edge shifts to shorter wavelengths with increasing ZnO content in binary tellurite glasses, whereas the increase in TeO_2 leads to a shift of the UV cutoff to longer wavelengths as shown in the 1st Ed. These two factors operate oppositely. The second parameter is dominant; thus, with increasing TeO_2 and decreasing nonbridging oxygen, a long-wavelength shift occurs.

The UV absorption bands were measured by Abdel-kader et al. in 1993; they identified the optical α of tellurite–phosphate glasses of the form $TeO_2-P_2O_5-B_nO_m$, in which B_nO_m is one of the rare-earth oxides La_2O_3, CeO_2, Pr_2O_3, Nd_2O_3, Sm_2O_3, or Yb_2O_3. Figure 9.12 shows the absorption I in arbitrary units as a function of wavelength for glasses in the group 81 mol% TeO_2-19 mol% P_2O_5, doped with 5 mol% rare-earth oxides of the type La_2O_3, CeO_2, Pr_3O_3, Nd_2O_3, Sm_2O_3, and Yb_2O_3. It is clear from this figure that the absorption edge of rare-earth-doped glasses occurs in the near-UV region. Two facts about these rare-earth-doped glasses and absorption regions are of particular interest:

1. The fundamental optical absorption edge becomes sharper as the rare-earth elements are introduced into the glass.
2. Sharp absorption peaks are observed, especially in Pr_2O_3- and Nd_2O_3-doped glasses.

These sharp bands are caused by forbidden transitions involving the $4f$ levels, and these $4f$ orbits are very effectively shielded from interaction with external fields of the hosts by $5s^2$ and $5p^6$ shells. Hence, the states arising from the various $4f^n$ configurations are only slightly affected by the surrounding ions and remain practically invariant for a given ion in various compounds. Reisfeld (1973) proposed that the rare-earth ions in these glasses occupy the center of a distorted cube, which is made of a four tetrahedral of phosphate, silicate, borate, and germinate glasses. Each tetrahedron contributes two oxygen atoms to the coordination of the rare-earth ions. The overall coordination number is eight, the most common coordination number of the rare-earth oxides. The concentration of rare-earth oxides in the glass under investigation is kept rather low to prevent a dominating energy transfer among the trivalent rare-earth ions. From the relatively high photon absorption energies in glasses, multiphonon relaxation from higher-energy excited levels to lower-energy levels has

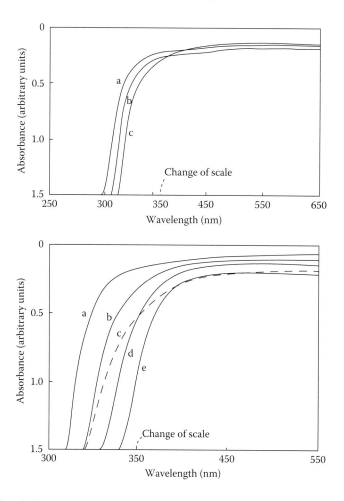

FIGURE 9.6 UV optical absorption of pure tellurite glass and binary tungsten–tellurite thin-film glasses. (From S. Al-Ani, C. Hogarth, and R. El-Mallawany, *J. Mater. Sci.*, 20, 661, 1985. With permission.)

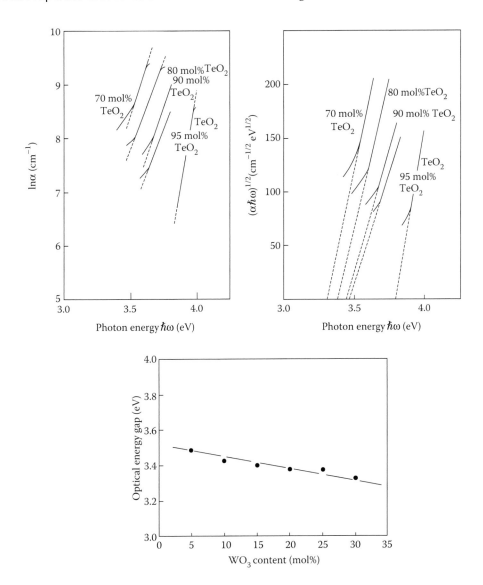

FIGURE 9.6 (Continued)

a high probability. Therefore, only some relatively broad bands are observed in the optical absorption spectra of these glasses.

From inspection of the absorption spectra of the studied glasses doped with rare-earth oxides, it can easily be seen that the absorption spectra of the rare-earth ions in the glasses differ from those of rare-earth ions in crystals. The broadening of these absorption bands is caused by the multiplicity of rare-earth sites in the glass. Smith and Cohen (1963), and Kutub et al. (1986), assigned the absorption bands observed at 442, 447, 454, and 585 nm for Pr_2O_3-doped glass to the transitions $^3H_4 \rightarrow {}^3P_2$, $^3H_4 \rightarrow {}^3P_1$, $^3H_4 \rightarrow {}^3P_0$, and $^3H_4 \rightarrow {}^1D_2$, respectively. These absorption bands are also reported for praseodymium–phosphate glasses by Ahmed et al. (1984) and Harani et al. (1984), and the same sharp peaks have been observed by Oshishia et al. (1983) in ZnF_2 glasses doped with Pr^{3+} ions. In Nd_2O_3-doped glasses, similar absorption bands are observed at 464, 522, 576, 742, 801, and 871 nm. These absorption bands are respectively caused by $^4I_{9/2} \rightarrow G_{11/2}$, $^4I_{9/2} \rightarrow {}^2(K_{3/2}\text{-}G_{7/2})$, $^4I_{9/2} \rightarrow ({}^4G_{5/2}, {}^4G_{7/2})$, $^4I_{9/2} \rightarrow ({}^4S_{3/2}, {}^4F_{7/2})$, $^4I_{9/2} \rightarrow ({}^2H_{9/2}, {}^4F_{5/2})$, and $^4I_{9/2} \rightarrow {}^4F_{3/2}$ transitions. The detected

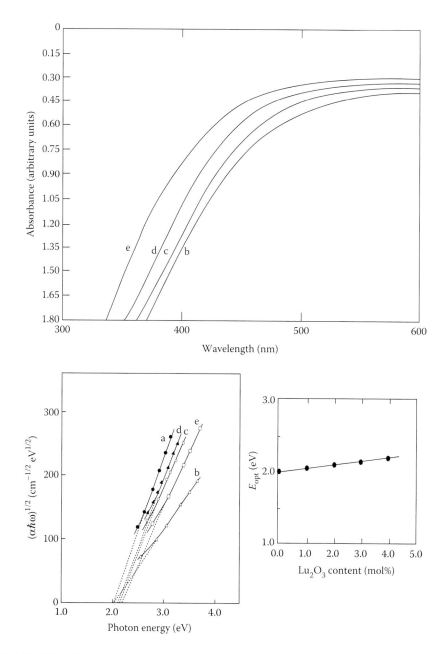

FIGURE 9.7   UV optical properties of $(65 - x)$ mol% TeO_2–$(35 - x)$ mol% CuO–x mol% Lu_2O_3 glasses. (From M. Malik and C. Hogarth, *J. Mater. Sci. Lett.*, 8, 655, 198b. With permission.)

absorption bands in Nd^{3+}-doped TeO_2–P_2O_5 glass are nearly identical to those of Nd^{3-}-doped glass, as stated by Hirayama and Lewis (1964) and by Lipson et al. (1975). The optical absorption spectra of Sm_2O_3-doped glass display three weak absorption bands at 341, 375, and 393 nm, which are attributed to $^6H_{5/2} \rightarrow {}^6F_{9/2}$, $^6H_{5/2} \rightarrow {}^6H_{7/2}$, and $^6H_{5/2} \rightarrow {}^6F_{5/2}$ transitions, respectively. Unfortunately, not all the rare-earth elements in glasses possess analytically recognized absorption bands in the VIS region.

The $\alpha(\omega)$ has been derived, and the results can be displayed in a number of ways as a function of $\hbar\omega$. The most satisfactory results are obtained by plotting the quantity $\alpha\hbar\omega^{1/2}$ as a function of $\hbar\omega$,

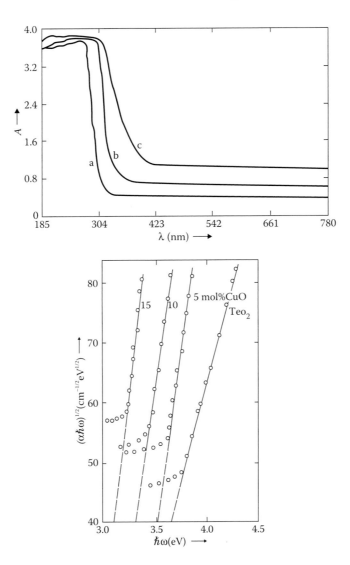

FIGURE 9.8 UV optical properties of TeO$_2$–CuO. (From M. Khan, *Phys. Stat. Solids*, 117, 593, 1990. With permission.)

as suggested by Davis and Mott (1970), for materials in which the optical transitions are indirect. The values of E_{opt} calculated by this method for different rare-earth-doped TeO$_2$–P$_2$O$_5$ glasses are as follows:

- E_{opt}: La, 3.03 eV; Ce, 3.04 eV; Pr, 3.05 eV; Nd, 3.07 eV; Sm, 3.74 eV; Yb, 3.04 eV
- ΔE: La, 0.45 eV; Ce, 0.3 eV; Pr, 0.28 eV; Nd, 0.38 eV; Sm, 0.25 eV; Yb, 0.27 eV

It is clear that the values of E_{opt} increase with an increase in the atomic number of the rare-earths. The above behavior can be attributed to the compactness of the glass, which increases with the addition of the rare-earth oxides. The width of the tails of the localized states within the gap is calculated for these glasses using the Urbach rule. The values of E are calculated from the slopes of the straight lines of the log ($\alpha\hbar\omega$) plot; ΔE is found to lie between 0.25 and 0.45 eV for these glasses.

Kim et al. (1993a and b) measured the UV optical properties of pure TeO$_2$ glass and found transmission spectra in the wavelength range from 200 nm (6.2 eV) to 3000 nm (0.41 eV) for a glass

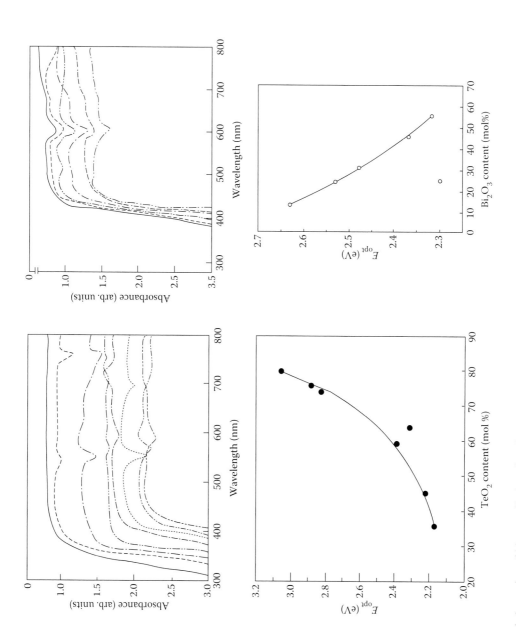

FIGURE 9.9 Optical absorption of binary tellurite–phosphate and ternary tellurite–phosphate–bismuth glasses. (From A. Abdel-Kader, A. Higazy, and M. Elkholy, *J. Mater. Sci.*, 2, 204, 1991b. With permission.)

sample of 0.222 mm thickness. The cutoff of the UV spectrum is >300 nm. The plot of $\alpha\hbar\omega$ vs. $(\hbar v)$ produces the E_{opt} value 6.85 eV, whereas that of SiO_2 glass was 13.38 eV, as stated by Kim et al. (1993a and b).

In 1994, Wang et al. measured the UV edge (in nanometers) and band gap (in electon volts) for tellurite glasses doped with praseodymium and neodymium of the forms:

- 75 mol% TeO_2–20 mol% ZnO–5 mol% Li_2O and 75 mol% TeO_2–20 mol% ZnO–5 mol% Na_2O
- 75 mol% TeO_2–20 mol% ZnO–5 mol% K_2O and 75 mol% TeO_2–20 mol% ZnO–5 mol% Rb_2O
- 75 mol% TeO_2–20 mol% ZnO–5 mol% Cs_2O

The UV edges at $\alpha = 5$ cm^{-1} are 375, 370, 365, 371, and 375 ± 1, whereas the band gaps are $E_{opt} = 3.12, 3.23, 3.26, 3.22$, and 2.002 ± 0.05 eV, respectively. The energy level diagram of Nd^{3+} and Pr^{3+} is shown in Chapter 8.

Cuevas et al. (1995) prepared and measured the UV-VIS cutoff (in nanometers), $E^{1/2}$-E (in electron volts) and E_{opt} (in electron volts) of the ternary tellurite glasses of the form $(100 - x)$ mol%

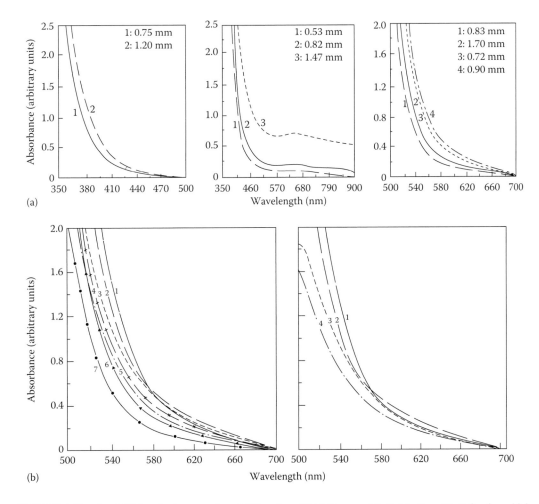

(a)

(b)

FIGURE 9.10 (a–e) UV optical properties of $(95 - x)$ mol% TeO_2–x mol% GeO_2–5 mol% F_2O_3, for which $5 < x < 35$, and $(90 - x)$ mol% TeO_2–x mol% GeO_2–10 mol% F_2O_3, for which $5 < x < 30$. The spectra were measured at room temperature in the wavelength 350–900 nm. (From M. El-Samadony, A. Sabry, E. Shaisha, and A. Bahgat, *Phys. Chem. Glasses*, 32, 115, 1991. With permission.)

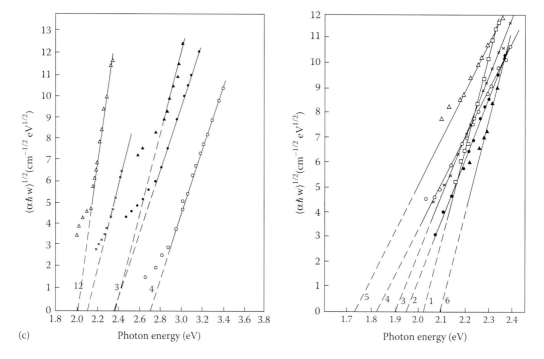

FIGURE 9.10   (Continued)

TeO$_2$–x mol% Li$_2$O–5 mol% TiO$_2$, for which x = 5, 10, 15, 20, and 25 as shown in the 1st Ed. Sabry and El-Samadony (1995) prepared and studied the optical properties of tellurite glasses of the form (100 – x) mol% TeO$_2$–x mol% B$_2$O$_3$, for which x = 5, 10, 20, 25, and 30. The optical spectra have been measured in the range 350–450 nm at room temperature, and the results show that the fundamental absorption edge was a function of composition and indirect transitions. The E_{opt} (in electron volts) increased with the percentage of the B$_2$O$_3$ as follows:

- 95 mol% TeO$_2$–5 mol% B$_2$O$_3$ (E_{opt} = 2.93 eV)
- 90 mol% TeO$_2$–10 mol% B$_2$O$_3$ (E_{opt} = 2.97 eV)
- 80 mol% TeO$_2$–20 mol% B$_2$O$_3$ (E_{opt} = 3.01 eV)
- 75 mol% TeO$_2$–25 mol% B$_2$O$_3$ (E_{op} = 3.08 eV)
- 70 mol% TeO$_2$–30 mol% B$_2$O$_3$ (E_{op} = 3.22 eV)

Kim et al. (1995) prepared the transparent mixed alkali–tellurite glass fiber 80 mol% TeO$_2$–10 mol% Li$_2$O–10 mol% Na$_2$O doped with LiNbO$_3$ crystals as a function of the remelt time as shown in Figure 9.13. Also in 1995, Inoue et al. measured the transmission spectra in the range from UV to VIS regions at various temperatures below the glass transition temperatures of tellurite glasses containing transition metal oxides of the form shown in the 1st Ed. The cutoff wavelength in the optical spectra was shifted toward the "red" wavelength at rates of 8 × 10^{-4} (eV/K) with increasing temperature. The shift of the transmission edge with temperature took place reversibly. The shift rates were compared with those of lead–silicate and lead–borate glasses containing transition metal oxides. Komatsu et al. (1997b) measured the 70 mol% TeO$_2$–(30 – x) mol% K$_2$O–x mol% Nb$_2$O$_3$ glasses as shown in Figure 9.14, in which Curve A represents this glass with x = 5; B is with x = 10; C is with x = 12.5, D is with x = 15, and E is with x = 17.5.

Dimitrov and Sakka (1996) calculated the optical constants for simple oxides. The E_{opt} for TeO$_2$ oxide is 3.79 eV, and the V_m is 28.15 cm^3/mol. Vithal et al. (1997) measured the UV spectra of TeO$_2$–PbO glass as shown in Figure 9.15. A modified linear relationship between the $E_{opt}^{1/2}$ and [1 – (R_m/V_m)]

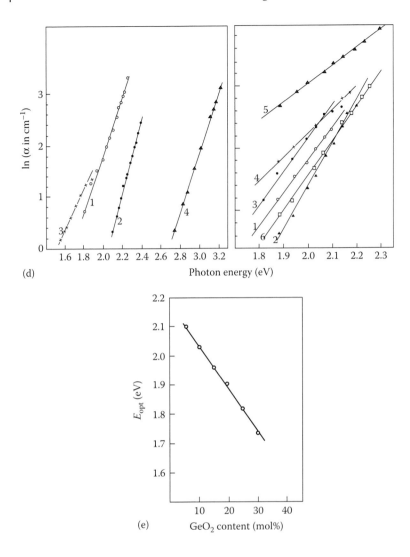

(d)

(e)

FIGURE 9.10 (Continued)

holds good for the glass. The oxide ion α_m deduced from two different experimental quantities—namely, n and E_{opt}—agree well for binary lead–tellurite and other glasses, including:

1. GeGaS
2. 85 mol% PbO–15 mol% TiO_2
3. 30 mol% TeO_2–70 mol% PbO
4. 70 mol% PbO–30 mol% CdO
5. 30 mol% Li_2O–70 mol% B_2O_3
6. 30 mol% CaO–70 mol% B_2O_3
7. 30 mol% BaO–70 mol% B_2O_3
8. 50 mol% PbO–50 mol% KF
9. 70 mol% PbO–30 mol% PbF_2
10. 70 mol% PbO–30 mol% CdF_2
11. 60 mol% PbO–40 mol% CdF_2
12. 50 mol% PbO–50 mol% CdF_2

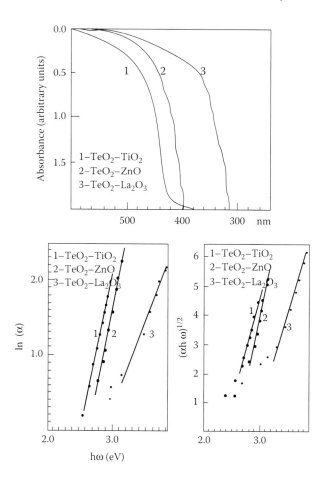

FIGURE 9.11 UV optical properties of 90 mol% TeO_2–10 mol% R_nO_m, where R = Ti, Zn, and La. (From R. El-Mallawany, *J. Appl. Phys.*, 72, 1774, 1992. With permission.)

The other values of the α_{O2-} for binary lead–tellurite glasses in the system 30 mol% TeO_2–70 mol% PbO by Vithal et al. (1997) are as follows: density, 6.79 ± 0.04 g/cm³; V_m, 30.1 ± 0.2 cm³; n_o, 1.849 ± 0.001; R_m, 13.44; Λ, 0.464; $\alpha_{O2-} - n_o$, 1.778 (10^{-24} cm³), $\alpha_{O2-} - E_{opt}$, 1.636 (10^{-24} cm³); E_{opt}, 2.82 ± 0.03 eV, and $\Delta E = 0.17 \pm 0.001$.

In late 1999, Weng et al. prepared thin films of 90 mol% TeO_2–10 mol% TiO_2 by the sol-gel process. The transmission spectrum of the binary 90 mol% TeO_2–10 mol% TiO_2 thin film is shown in Figure 9.16. Weng et al. expected that 90 mol% TeO_2–10 mol% TiO_2 thin film possesses high nonlinear optical properties, although further study on this aspect is needed.

9.4.2 Data of the UV Properties of Oxide–Tellurite Glass Ceramics

Shioya et al. (1995) measured the UV absorption curve of the ternary transparent tellurite glassceramics system of the form 70 mol% TeO_2–15 mol% Li_2O–15 mol% Nb_2O_5 for quenched and heat-treated samples as shown in the 1st Ed. The optical energy of the transmission cutoff wavelength E_{opt} is 2.84 eV (463 nm) for glass-ceramics obtained by heat treatment at 425°C for 1 h, which was less than the E_{opt} for quenched glass, 3.18 eV (390 nm). In their 1995 study, Shioya et al. found that the particle d of the cubic crystalline phase is 10–20 nm and the n of the matrix and cubic crystalline phase are 2.02 ± 0.02 and 2.11 ± 0.02, respectively, as shown in Chapter 8. Shioya et al. (1995) measured the transparency of these glasses. These values mean that both (d^6) and (λ^4) and $[(M^2 - 1)/(M^2 + 2)]^2$ in

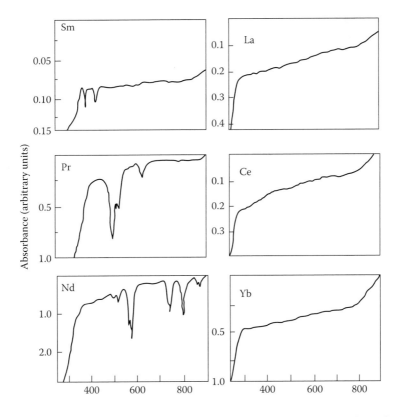

FIGURE 9.12   UV absorption bands of tellurite–phosphate glasses of the form TeO_2–P_2O_5–B_nO_m, where B_nO_m is rare-earth oxides La_2O_3, CeO_2, Pr_2O_3, Nd_2O_3, Sm_2O_3, and Yb_2O_3. (From A. Abdel-Kader, R. El-Mallawany, and M. Elkholy, *J. Appl. Phys.* 73, 71, 1993. With permission.)

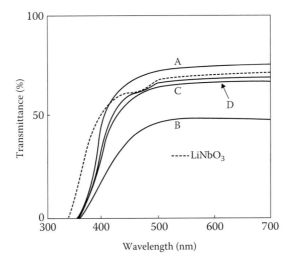

FIGURE 9.13   UV properties of the transparent mixed alkali–tellurite glass fiber 80 mol% TeO_2–10 mol% Li_2O–10 mol% Na_2O, doped with $LiNbO_3$ crystals as a function of the remelt time. (From H. Kim, T. Komatsu, R. Sato, and K. Matusita, *J. Ceram. Soc. Japan*, 103, 1073, 1995. With permission.)

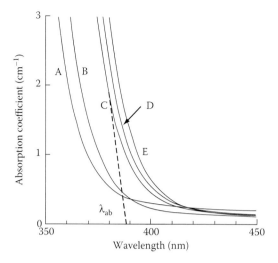

FIGURE 9.14 UV properties of the 70 mol% TeO_2–$(30 - x)$ mol% K_2O–x mol% Nb_2O_3. Curves: A, $x = 5$; B, $x = 10$, C, $x = 12.5$; D, $x = 15$; E, $x = 17.5$. (From T. Komatsu, H. Kim, and H. Oishi, *Inorgan. Mater.*, 33, 1069, 1997b.)

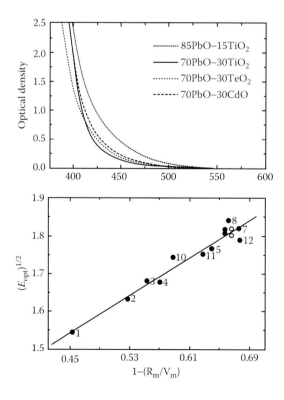

FIGURE 9.15 UV spectra of TeO_2–PbO glass, A modified linear relationship between $E_{opt}^{1/2}$ and $[1 - (R_m/V_m)]$ holds good for this glass. The oxide ion α_m values are deduced from two different experimental quantities; namely, n and E_{opt}. (From M. Vithal, P. Nachimuthu, T. Banu, and R. Jagannathan, *J. Appl. Phys.*, 81, 7922, 1997. With permission.)

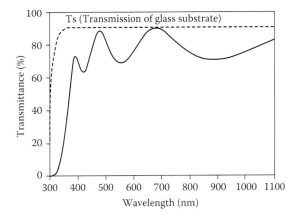

FIGURE 9.16 Optical T spectra of thin-film 90 mol% TeO_2–10 mol% TiO_2 by the sol-gel process. (From L. Weng, S. Hodgson, and J. Ma, *J. Mater. Sci. Lett.*, 18, 2037, 1999. With permission.)

Equation 9.18 are small (i.e., 4.9×10^{-4}), with $d = 20$ nm and $\lambda = 600$ nm and 8.7×10^{-4}, respectively. Therefore, as for other transparent glass-ceramics, the origin of the transparency of the glass-ceramics obtained in the 70 mol% TeO_2–15 mol% Li_2O–15 mol% Nb_2O_5 glass-ceramics is attributed to a particle size smaller than the wavelength of VIS light. The T of the 70 mol% TeO_2–15 mol% Li_2O–15 mol% Nb_2O_5 glass and glass-ceramics at wavelengths of 450 and 800 nm obtained by heat treatment at 450°C for 1 h is low compared with that of the quenched glass. The maximum particle diameter of cubic crystalline estimated from the full width at half maximum of the x-ray diffraction peak and from transmission electron microscopic observation is 40 nm, but coarser crystallites might be present in the glass-ceramics, which could be one of the reasons for the lowering of transparence. More controlled heat-treatment processes, such as two-stage heat treatment, give smaller crystallites, as mentioned by Beall and Duke (1969). In Chapter 8, the samples are transmitted in the VIS region because the sample heat treated at 390°C is not transparent but translucent, and the phase separation might occur before the formation of the cubic crystalline phase.

Komatsu et al. (1996a and b) prepared Rb-doped crystalline phase in transparent 70 mol% TeO_2–$(15 - x)$ mol% K_2O–15 mol% Nb_2O_5–x mol% Rb_2O glass-ceramics as shown in Figure 9.24. The tellurite glass-ceramics consisting of the cubic crystalline phase were optically transparent as shown in Figure 9.17 for $x = 1$, for a sample heat treatment at 389°C for 1 h and 430°C for 1 h, with a

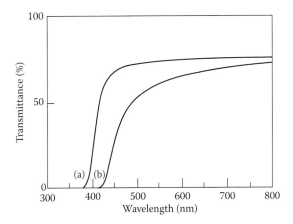

FIGURE 9.17 UV properties of prepared Rb-doped crystalline phase in transparent 70 mol% TeO_2–$(15 - x)$ mol% K_2O–15 mol% Nb_2O_5–x mol% Rb_2O glass-ceramics. (From T. Komatsu, H. Kim, and H. Mohri, *J. Mater. Sci. Lett.*, 15, 2026, 1996. With permission.)

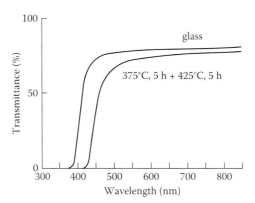

FIGURE 9.18 UV properties of transparent tellurite glass-ceramics of the form 70 mol% TeO₂–15 mol% Nb₂O₅–15 mol% K₂O, as shown for a thickness of 1 mm. (From T. Komatsu, H. Kim, and H. Oishi, *Inorgan. Mat.*, 33, 1069, 1997b. With permission.)

thickness of 1 mm. The two-step heat treatment exhibits a good transparency, close to that of the original glass.

Komatsu et al. (1997a) prepared new transparent tellurite glass-ceramics of the form 70 mol% TeO₂–15 mol% Nb₂O₅–15 mol% K₂O as shown in Figure 9.18, at a thickness of 1 mm. The glass-ceramics they obtained by two-step heat treatment for 5 h at 375°C plus 5 h at 425°C exhibit good transparency, close to that of the original glass. The first heat treatment is for nucleation, and the second is for crystal growth. The transparency of this ternary tellurite glass system is attributed to the fact that particle sizes are smaller than the wavelength of VIS light. The cubic crystal structure is another important factor for transparency. The optical and dielectric properties of 70 mol% TeO₂–15 mol% Nb₂O₅–15 mol% K₂O transparent glass-ceramics obtained by heat treatment at 425°C for 1 h are as follows: n, 2.11; relative permittivity, 1 kHz at 300 K; ε_R, 44; and third-order nonlinear optical susceptibility (χ^3) = 3.3 × 10⁻¹³ electrostatic units. These values are larger than those for the original glass.

Kim and Komatsu (1998) fabricated transparent glass-ceramics of the form TeO₂–Nb₂O₅–K₂O as shown in Figure 9.19. To decrease the particle size further and to fabricate better optical transparency, a two-step heat treatment, 370°C for 1 h and then 400°C for 1 h, was carried out for 72 mol% TeO₂–14 mol% Nb₂O₅–14 mol% K₂O glass. The optical absorption spectra for the original glass and the heat-treated sample are shown in Figure 9.19. The glass-ceramic obtained by a two-step heat treatment exhibits a good transparency, close to that of the original glass.

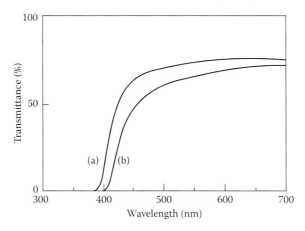

FIGURE 9.19 UV properties of transparent glass-ceramics of the form TeO₂–Nb₂O₅–K₂O. (From H. Kim and T. Komatsu, *J. Mater. Sci. Lett.*, 17, 1149, 1998. With permission.)

Komatsu and Mohira (1999) succeeded in preparing TeO_2–PbO–CdO glass and glass-ceramics as shown in Figure 9.20a and b. These glasses were heat treated at 320°C (below the crystallization temperature, 346°C) in nitrogen or 93% Ar–7% H_2 atmosphere, and they became shiny-black while keeping a good transparency. It was found that an unknown crystalline phase (neither CdTe nor $CdPbO_4$ semiconductor) formed at the surface. The optical absorption spectra at room temperature of some of these glasses are represented in Figure 9.20a: Curve A, 70 mol% TeO_2–20 mol% PbO–10 mol% CdO; B, 60 mol% TeO_2–20 mol% PbO–20 mol% CdO; C, 50 mol% TeO_2–20 mol% PbO–30 mol% CdO. Figure 9.20b charts the optical absorption spectra at room temperature of the heat-treated glasses as follows: Curve A, 60 mol% TeO_2–20 mol% PbO–15 mol% CdO–5 mol% BaO (original glass); B, 320°C, 3 h, in N_2; C, 320°C, 4 h, in 93% Ar–7% H_2; D, 320°C, 24 h, in 93% Ar–7% H_2. Shanker and Varma (1999) also measured the optical properties of the crystallized surface of TeO_2–$LiNbO_3$ glasses as shown in the 1st Ed.

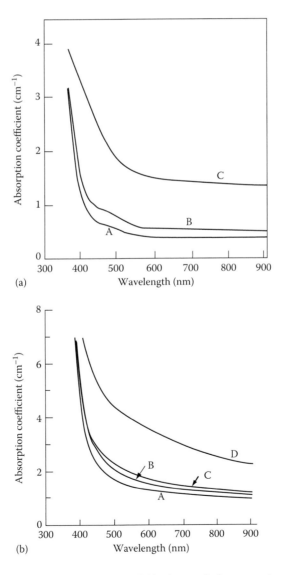

FIGURE 9.20 (a, b) UV properties of TeO_2–PbO–CdO glass and glass-ceramics. (From T. Komatsu and H. Mohira, *Phys. Chem. Glasses*, 40, 257, 1999. With permission.)

9.4.3 UV Properties of Halide–Tellurite Glasses

In 1990, the effect of chloride ions on the optical properties of TeO_2–CuO–$CuCl_2$ glasses was studied by Malik and Hogarth (1990) as shown in Figure 9.21. The compositions of these glasses were 65 mol% TeO_2–(35 – x) mol% CuO–x mol% $CuCl_2$, for which x was 0, 1, 2, 3, 4, or 5, and the glasses were prepared by the melt-quenching technique. The thin films were made by blowing and varied in thickness. The E_{opt} was studied and its variation with composition was discussed in terms of the effective role played by chloride ions, which reduce the nonbridging oxygen ions and modify the structure of the network. El-Mallawany (1991) measured the UV characterization of binary oxy-halide–tellurite glasses of the form TeO_2–$TeCl_2$, as shown in Figure 9.22. Stable bulk glasses were obtained by quenching the melt at room temperature. These new kinds of tellurite glasses have good optical properties; E_{opt} = 2.7 eV, and energy band tail = 0.14 eV.

In 1995, Sahar and Noordin measured the UV-VIS spectra of (100 – x) mol% TeO_2–x mol% $ZnCl_2$–y mol% ZnO, for which x = 10, 20, 30, 40, or 60 when y = 0; x = 30, 30, 30, 10, 20, and 0 when y = 10, 20, 30, 30, 30, and 50, respectively, as shown in the 1st Ed. For this suite of glasses, the

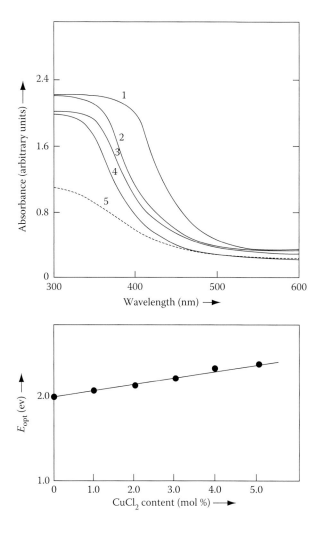

FIGURE 9.21   Effect of chloride ions on the optical properties of TeO_2–CuO–$CuCl_2$ glasses. (M. Malik and C. Hogarth, *J. Mater. Sci.*, 25, 116, 1990. With permission.)

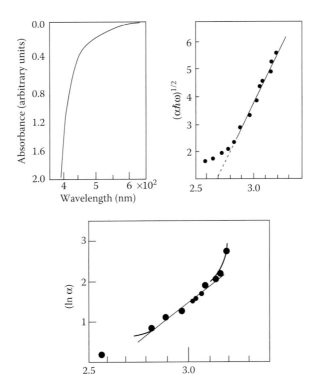

FIGURE 9.22 UV characterization of binary oxyhalide–tellurite glasses of the form TeO$_2$–TeCl$_2$. (From R. El-Mallawany, *Mater. Sci. Forum*, 67-68, 149, 1991.)

values of E_{opt} are in the range 2.0–2.5 eV, and Sahar and Noordin (1995) stated that the glass could provide some encouragement for long-wavelength applications.

9.4.4 UV PROPERTIES OF NONOXIDE–TELLURITE GLASSES

Since the 1970s, optical properties of amorphous chalcogenide films or bulk materials have been under intensive focus. Fagen and Fritzsche (1970) prepared two amorphous glasses: Type A, 40 mol% Te–35 mol% As–11 mol% Ge–11 mol% Si–3 mol% P; and Type B, 28 mol% Te–21 mol% S–35 mol% As–16 mol% Ge.

The optical absorption constants of both of the above glasses in both thin film and bulk states are represented in the 1st Ed. These constants increase exponentially with photon energy in the range 1.0–1.2 eV and do not exhibit a well-defined edge at 300 K. Fagen and Fritzsche (1970) searched for electro-absorption effects in thin film of type-B material. The modulation of optical transmission by transverse electric fields was performed at 77 K, and a small shift of 3×10^{-15} eV/cm^2/V^2 was recorded. The temperature dependence of absorption in the exponential region was ~3×10^{-15} eV/cm^2/V^2 at 77 K.

Rockstad (1970) measured the optical properties of amorphous chalcogenide films of the forms Te$_2$AsSe, TeAsSiGe, Te$_3$As$_2$, and Te$_{1.5}$As$_2$Se$_{1.5}$, as shown in the 1st Ed. The films were between 2,000 and 10,000 Å thick. Optical absorption measurements were made with a spectrophotometer. Photoconductivity measurements were made with a double-gating monochromatic and phase detection system. Although the optical-absorption-edge data in Figure 9.23 are not sufficiently precise for accurate calculations because of sample fluctuations, the differences in optical gaps are consistent with changes of conductivity composition (as shown in Chapter. 6). For example, the difference in absorption edges for Te$_3$As$_2$ and TeAsSiGe indicates a conductivity ratio of approximately 200, which is in good agreement with the observed ratio of average conductivity values. Trnovcova

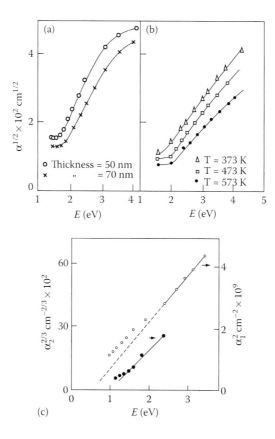

FIGURE 9.23   (a–c) Optical properties of Te-Se alloy film with germanium added as measured at different thicknesses and temperatures. (From Z. Mandouh, *J. Appl. Phys.*, 78, 7158, 1995. With permission.)

et al. (1987) measured the optimization of multicomponent seleno-telluride glasses for optical fibers applied to power transmission at 10.6 μm. The glass compositions were x mol% Te–(75 – x) mol% Se–25 mol% Ge, x mol% Te–(70 – x) mol% Se–30 mol% Ge, and 41.5 mol% Te–8.5 mol% Ge–30 mol% Se–20 mol% Ge. The ratio of the electrical activation energy/optical gap increases linearly with the percentage of x. The ratio of electrical activation energy/optical gap was constant and equal to 0.43 ± 0.01.

In 1995, Mandouh reported that the addition of germanium to Te–Se alloy causes structural changes that minify the band structure and, hence, the optical and electrical properties of Te–Se alloy. The optical α of the alloy film varies at different thicknesses. The film is deposited and heat treated. Figure 9.23 plots the α against E at two sample thicknesses (i.e., 50 and 70 nm) (Figure 9.23a) and at different temperatures (373, 473, and 573 K for the 70 nm thick sample [Figure 9.23b]). The curves are identical in character to the interband absorption curves of elemental amorphous semiconductors. The E_{opt} of the glass system 35 mol% Te–60 mol% Se–5 mol% Ge is 1.18 eV. The α decreases with thickness. This low value of the α of thin film is interpreted to be caused by the low transition probability of the carriers across the correspondingly large gap between the localized states. Mandouh (1995) also observed that the process of heat treatment gradually anneals out unsaturated defects, producing a large number of saturated bonds. This reduction in the number of unsaturated defects decreases the density of localized states in the band structure, consequently increasing the optical gap. Heat treatment at 573 K further reduces the disorder present in Te–Ge–Se films caused by the transition from amorphous to crystalline states, as is evident from structural studies. The optical gap increases as a consequence of the reduction of disorder.

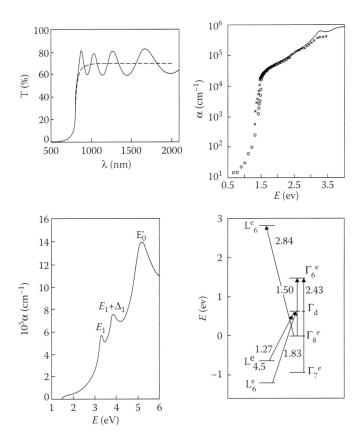

FIGURE 9.24 Optical absorption coefficient for electro-deposited films of CdTe and optical dispersions measured in the photon energy range $E < 1.5$ eV. (From A. Rakhshani, *J. Appl. Phys.*, 81, 7988, 1997. With permission.)

Rakhshani (1997) prepared electro-deposited films of CdTe and measured the optical dispersions in the $\hbar\omega$ range of $E < 1.5$ eV, which is in excellent agreement with measured dispersions in the single crystal. The optical α was determined to be in the $E < 3.5$ eV range and was compared with that for the single crystal as shown in Figure 9.24. Rakhshani also measured the experimental value of the substrate–air interface (R_3), which was 0.034 over the wavelength range 400–200 nm. Temperature measurements were performed with a glass slide placed in the reference beam compartment of the spectrometer; that is, the factor (M) in Equation 9.5 was used at $(1 - R_1)(1 - R_2)$. Film dispersion was measured from the positions of the maxima and minima of the interference pattern in the T spectrum using the relation:

$$4dn(\lambda_m) = m\lambda_m, m = 1, 2, 3,\ldots \tag{9.19}$$

where λ_m is the wavelength at which the peak (even m) or the valley (odd m) or order (m) occurs. The dispersion can be given in the modified Sellmeier form, as shown in Chapter 8. The results of Rakhshani (1997) revealed two directly allowed transitions at 1.5 eV (Γ_8 VB→Γ_6 CB) and 2.43 eV (Γ_7 VB→Γ_6 CB), and three indirectly allowed transitions at 1.27 eV ($L_{4,5}$ VB→ Γ_d), 1.83 eV (L_6 VB→Γ_d), and 2.84 eV (Γ_8 VB→Γ_6 CB). The transitions at 1.27 and 1.83 eV, which have not been reported previously nor detected in single-crystal data, were attributed to transitions to a grain-boundary-related defect energy band, Γ_6 0.65 eV, above Γ_8 VB. The indirect transitions at 1.83 and 2.84 eV are assisted by photons with energies of 80 and 84 meV, respectively. Rakhshani (1997) used Equation 9.20 for a single crystal of CdTe, as shown in Figure 9.24, in the $\lambda > 840$ nm region. The dispersion is given by the modified Sellmeier form as stated by Amirtharaj (1991) and by Baars and Sorger (1972):

$$n^2 = 5.304 + \frac{1.899}{1-\left(0.5713/\lambda\right)^2} - 4.188 \times 10^{-4}\lambda - 2.391 \times 10^{-7}\lambda^3 \qquad (9.20)$$

where λ is in nanometers. In the $\lambda < 820$ nm range, the n values and extinction coefficients were obtained with an ellipsometer. The optical α shown in Figure 9.24 is a function of $\hbar\omega$ for a thick (0.23 μm) CdTe thin film, deposited initially on molybdenum at 580 mV, and the solid line for a single-crystal CdTe. Figure 9.24 also provides a schematic energy diagram for optical transitions below 3 eV in electrodeposited CdTe.

9.5 UV PROPERTIES OF NEW TELLURITE GLASSES

As it has been stated in chapter 8, that Prasad et al. (2003) prepared a highly transparent and little yellow-colored $20TeO_2–75B_2O_3–5Li_2O$ containing 1 mol% of Ho^{3+} glass. The authors prepared these glasses to study its optical properties from the measured UV–VIS–NIR absorption, emission, and up-conversion emission spectra. The authors observed a prominent and bright red color when this optical glass was brought under a UV lamp. The recorded up-conversion emissions revealed a strong green color along with blue and violet colors upon excitation with $\lambda = 643$ nm, and the mechanism was explained by an energy level diagram as a three-photon absorption process. The optical band gaps (E_{opt}) for both indirect and direct transitions of $20TeO_2–75B_2O_3–5Li_2O$ glass were given and from this, the optical band gaps (E_{opt}) for indirect and direct transitions were equal to 2.7 and 3.3 eV, respectively. Also, a prominent and strong red emission at 643 nm ($^5F_5 \rightarrow {}^5I_8$) has been observed from it with an excitation at 399 nm. The up-conversion emission spectrum has revealed three up-conversion emissions at the wavelength of 425, 472 and 525 nm, respectively, upon excitation with 643 nm. The up-conversion emissions have been explained in terms of a three-photon absorption process through an energy level diagram. The authors mentioned that the holmium glass has a good potential application as a red luminescent optical glass and also as a green up-converting luminescent material.

Eraiah (2006) prepared glasses with the composition $(xSm_2O_3)–(40 – x)(ZnO)–60(TeO_2)$ and $x = 0.0, 0.1, 0.2, 0.3, 0.4,$ and 0.5 mol%. The author measured the optical energy band gap E_{opt}. The E_{opt} increases linearly from 3.1 to 3.4 eV with increase of Sm_2O_3 concentration up to 0.4 mol%, and then drops suddenly at 0.5 mol% of Sm_2O_3 to 2.1 eV. The sudden drop of E_g may be due to the variation of density as well as the variation of nonbridging oxygens. Another reason could be that at high dopant concentrations, the broadening of the impurity band and the formation of band tails on the edges of the conduction and valence bands would lead to a reduction in E_g as in semiconductors. Also, Jose et al. (2006) studied the addition of WO_3 (up to 10 mol%) and P_2O_5 (up to 16 mol%) in form of $TeO_2–BaO–SrO–Nb_2O_5$ and called TBSN glass. The optical band gaps of WO_3 and P_2O_5 added to the TBSN glasses were calculated from the experimental absorption spectrum using the Tauc equation. The value of E_g was determined from absorption onset curves. The optical band gap energy of the TBSN glass was 2.99 eV, which was decreased to 2.75 eV on 10 mol% WO_3 addition and increased to 3.18 eV on 16 mol% P_2O_5 addition. The observed shifts due to the WO_3 and P_2O_5 additions could be understood in terms of the variation in nonbridging oxygen (NBO) ion concentrations. In metal oxides, the valence band maximum (VBM) mainly consists of O(2p) orbital and the conduction band minimum (CBM) mainly consists of M(ns) orbital. The NBO ions contribute to the VBM. When a metal–oxygen bond is broken, the bond energy is released. The nonbonding orbitals have higher energies than bonding orbitals. Increase in concentration of the NBO ions results in the shifting of the VBM to higher energies and reduces the band gap. Thus, the lowering of band gap energy due to an increase in the WO_3 content suggests that nonbridging oxygen ion concentration increases with an increase in the WO_3 content, which lowers the band gap energy. On the other hand, an increase in the P_2O_5 content causes a decrease in the non bridging oxygens concentration which increases the band gap energy.

El-Mallawany et al. (2008) measured the refractive index of the quaternary tellurite glass system of the form $80TeO_2$–$5TiO_2$–$(15 - x)WO_3$–xA_nO_m where A_nO_m was Nb_2O_5, Nd_2O_3, and Er_2O_3, with $x = 0.01, 1, 3,$ and 5 mol% for Nb_2O_5, and $x = 0.01, 0.1, 1, 3, 5,$ and 7 mol% for Nd_2O_3 and Er_2O_3. In the range of ultraviolet, the position of the fundamental absorption edge shifts to a higher wavelength with an increase in Nb_2O_5, and Nd_2O_3, but it shifts to a lower wavelength with an increase in Er_2O_3 content. The shifts of the absorption edge were most likely related to structural rearrangements of the glass network and modifier. The cutoff wavelength for the Nb_2O_5 series has been found to increase from 418 to 442 nm as the content increased from 0.01 to 5 mol%, and it increased from 421 to 440 nm as the concentration of Nd_2O_3 increased from 0.01 to 7 mol%. By contrast, the cutoff wavelength has been found to decrease from 445 to 420 nm as the content of Er_2O_3 increased from 0.01 to 7 mol%. The optical energy gap (E_{opt}) was found to increase from 2.42 to 2.53 eV as the contents of the Nb_2O_5 modifier increased from 0.01 to 5 mol%. For the Nd_2O_3 and Er_2O_3 modifier series, it was found to increase from 3.51 to 3.59 eV and from 2.43 to 2.71 eV, respectively, as their concentration increased from 0.01 to 7 mol%. In addition, Sooraj et al. (2008) prepared and measured the optical characterization of Sm^{3+} and Dy^{3+}-doped lithium–boro–tellurite glasses in the following chemical compositions, $20TeO_2$–$74B_2O_3$–$5Li_2O$–$1RE_2O_3$ (where RE = Sm^{3+} and Dy^{3+}). UV-visible NIR absorption spectra (400–1600 nm) of the samples were measured by Shimadzu UV-3101 PC spectrophotometer. The excitation spectra were obtained on a SPEX Fluorolog-2 Fluorimeter with Datamax software and a Xe lamp (450 W) as the excitation source. The emission spectra were obtained in a Horiba Jobin Yvon excitation–emission spectrofluorometer equipped with a CCD detector and a Xe lamp excitation source. A bright orange ($4G5/2 \rightarrow 6H7/2$), along with a red ($^4G_{5/2} \rightarrow ^6H_{9/2}$) and a yellow ($^4G_{5/2} \rightarrow ^6H_{5/2}$) emission were observed from the Sm^{3+}-doped lithium–boro–tellurite glass. Both blue ($^4F_{9/2} \rightarrow ^6H_{15/2}$) and yellow ($^4F_{9/2} \rightarrow ^6H_{13/2}$) emissions have been observed for the Dy^{3+} glass, and both the emissions in this glass were in equal proportion in their emission intensities. The decay curves along with the evaluated stimulated emission cross sections of these emission bands of Sm^{3+} and Dy^{3+} glasses were found. Since the lifetime of the emission transitions were in milliseconds, such glasses could be labeled as phosphorescent optical materials. For Sm^{3+}, the stimulated emission cross sections (σ_p^E) of the measured orange emission transition have the minimum cross section value 0.38×10^{-20} cm^2 while the stimulated emission cross sections for blue emission band in the Dy^{3+} have 0.66×10^{-20} cm^2. Therefore, based on the spectral properties of these glasses, both of these optical glasses are promising as novel optical materials for use in luminescent systems.

Chen et al. (2008) studied both linear and nonlinear optical properties of $10Bi_2O_3$–xWO_3–$(100 - x)TeO_2$ glasses where $x = 20, 25,$ and 30 mol%. The optical gap (E_{opt}) and the band tail (E_e) behaved oppositely. The values (E_{opt}) were found to be 2.47, 2.53, and 2.70 eV, which correspond to the tungsten content of 30 mol%, 25 mol%, and 20 mol%, respectively. The values of E_e were equal to 0.14 eV, 0.21 eV, and 0.22 eV, respectively. As previously mentioned, it is known that the shift to lower energy or a change in the absorption band is related to the formation of nonbridging oxygen (NBO) contents which binds excited electrons less tightly than bridging oxygen (BO) content. Consequently, it could be suggested that the NBO content increases with increasing WO_3 content, shifting the band edge to lower energies and leading to a decrease in the values of E_{opt}. One can also recognize that tungsten oxide is one of the transition metal oxides (TM) having two different valence states; namely, W^{5+} and W^{6+}. Both W^{5+} and W^{6+} possess the relatively high polarizability of the oxide ions, which can be attributed to the empty d-orbitals of the corresponding cations, their high coordination number toward oxide ions. This affects the higher amount of NBO, allowing less tight oxygen anions. The less tightly the valence electrons, the larger optical nonlinearity is obtained. Such an assumption led to the implication that high WO_3 content can lead to the higher third-order optical nonlinearity, thus decreasing the optical band gap energies (E_{opt}). While exponential band tailing can explain the observed Urbach tail of the absorption coefficient versus photon energy, it is also possible to attribute the absorption tail behavior to strong internal fields arising, for example, from ionized dopants or defects. These internal electric fields lead to an exponential broadening of the energy states. Accordingly, an increase in the degree of the localization of electrons goes with the increase of NBO, thereby increasing the number of donor centers. A large concentration of donor centers will effectively lower the band gap and shift the absorption to the lower wavelength through the formation of a band gap.

Liao et al. (2009) prepared and characterized new fluorotellurite glasses based on $(85 - x)$ TeO$_2$–xZnF$_2$–12PbO–3Nb$_2$O$_5$ glass, where $x = 0$, 10, 20, 30, and 40 mol%. These glasses have been studied for the first time for fabricating mid-infrared optical fiber lasers. The absorption of glass in UV and Vis spectral ranges is caused by electron transitions from unexcited to excited states. The absorption edge in oxide glasses corresponds to the transition of an electron belonging to an oxygen ion to an excited state. The more weakly these electrons are bound, the more easily the absorption occurs at a longer wavelength. Values of the UV cutoff wavelength were 415, 405, 385, 342, and 358 nm (data has been taken from figure) respectively. The radii of cations are, respectively, 0.97 Å for Te^{4+}, 1.19 Å for Pb^{2+}, 0.74 Å for Zn^{2+}, and 0.64 Å for Nb^{5+}, and the radius of F is 1.33 Å (Greenwood and Earnshaw 1997). The calculated field strengths were 0.756 Å^{-2} for Te^{4+}, 0.315 Å^{-2} for Pb^{2+}, 0.467 Å^{-2} for Zn^{2+}, and 1.288 Å^{-2} for Nb^{5+}, respectively. Thus, the probability of the binding of F ions with these positive ions should be Nb^{5+} > Te^{4+} > Zn^{2+} > Pb^{2+}. Since the batching amount of Nb$_2$O$_5$ in these glasses was little when compared with that of TeO$_2$, ZnF$_2$, and PbO, the influence of Nb–F bond on the structure of the fabricated glass was very small and was consequently ignored. Therefore, it could be deduced that F ions, introduced by ZnF$_2$ into TeO$_2$–PbO–Nb$_2$O$_5$ system glass, would preferentially bind Te ions in TeO$_n$ units by replacing O ions in the TeO$_n$ units, and would consequently form Te(O, F)$_n$, such as Te(O, F)$_{3+1}$ and Te(O, F)$_3$. Liao et al. (2009) has reasonably speculated that the electrons of O$_{ax}$ ion in Te(O, F)$_{3+1}$, Te (O, F)$_3$ groups were bound more tightly than those in TeO$_{3+1}$, TeO$_3$ groups, and, consequently, the electron transition from unexcited to excited states would require more energy, which lead to the shift of the optical absorption edge from lower to higher energy with the increase of ZnF$_2$ content and Te (O, F)$_{3+1}$ and/or Te (O, F)$_3$ units in the glass structure.

Furthermore, Wang et al. (2009) prepared ternary TeO$_2$–Nb$_2$O$_5$–Bi$_2$O$_3$ glass in the forms 90–5–5, 85–5–10, 80–5–15, 85–10–5, 80–10–10, and 75–10–15 mol% and measured the UV spectra. The authors found that the direct allowed an optical band gap, and the indirect allowed an optical band gap decrease as Nb$_2$O$_5$ and Bi$_2$O$_3$ increase. The direct optical band gap and the indirect optical band gap were in the range 2.83, 273, 2.63, 2.78, 2.68, 2.54 eV, and 2.95, 2.92, 2.78, 2.95, 2.73eV, respectively. The lowering of the optical band gap due to an increase in the Nb$_2$O$_5$ and Bi$_2$O$_3$ content suggests that nonbridging oxygen ion concentration increases with an increase in the Nb$_2$O$_5$ and Bi$_2$O$_3$ content, which lowers the optical band gap. The metallization values of the glasses were 0.4724, 0.4586, 0.4459, 0.4486, and 0.4406, respectively. The small metallization (0.4406) obtained for the 75TeO$_2$–10Nb$_2$O$_5$–15Bi$_2$O$_3$ glass mean that the width of both valence and conduction bands become large, resulting in a narrow band gap (2.54 eV). The molar refraction R_m and molar volume V_m characteristics of the glass influence the results of the energy band gap, which leads to the energy band gap being larger than the optical band gap (Vithal 1997). The authors calculated the energy band gap E_g for these glasses by using the Lorenz–Lorenz equation. The values of the energy gap were 4.46, 4.21, 3.98, 4.41, 4.02, and 3.88 eV respectively. It was also found that the nonlinear optical susceptibility χ^3 increases with a decreasing optical band gap of glasses (Abdel Baki et al. 2006). It has been established experimentally that tellurite glasses containing Nb$_2$O$_5$ and Bi$_2$O$_3$ possess a very high linear refractive index and small optical band gap. Thus, Wang et al. (2009) established that there is a trend by which the energy gap increases with decreasing refractive index and a decreasing of the polarizability of the oxide ions, which indicates that the glasses in the TeO$_2$–Nb$_2$O$_5$–Bi$_2$O$_3$ system are probably useful for optical applications.

Ozdanova et al. (2010) prepared the $x(PbO)–y(Nb_2O_5)–(1 − x − y)(TeO_2)$ where $x = 0, 0.1, 0.2,$ and $y = 0, 0.1, 0.2$ mol% glasses from very pure oxides and measured the experimental nondirect optical gap ($E_{g.non}$). The room temperature values of the experimental nondirect optical gap ($E_{g.non}$) were equal to 3.42, 3.45, 3.45, 3.39, 3.51, and 3.6 eV for the 0–0.1–0.9, 0–0.2–0.8, 0.1–0.1–0.8, 0.1–0.2–0.7, 0.2–0.1–0.7, and 0.2–0–0.8 molar fractions, respectively. The optical band gap values of $0.2(PbO)–0.8(TeO_2)$ and $0.2(PbO)–0.1(Nb_2O_5)–0.7(TeO_2)$ glasses were 3.61 and 3.51 eV, respectively—higher than the optical band gap values for the other glasses. The values of reflectance were R $\approx$ 0.14, 0.135, 0.15, 0.13, 0.15, and 0.135 for $x, y = 0.2, 0; 0, 0.1; 0, 0.2; 0.1, 0.1; 0.1, 0.2;$ and 0.2, 0.1, respectively. Absorption coefficient K has been measured within a narrow spectral region of the short wavelength absorption edge (SWAE). The refractive index values at $\lambda = 800$ nm were calculated from transmission measurements. The values of the refractive index were: $n_{(\lambda=800)} = 2.16,$ 2.27, 2.14, 2.27, 2.15, and 2.19 for the compositions 0–0.1–0.9, 0–0.2–0.8, 0.1–0.1–0.8, 0.1–0.2–0.7, 0.2–0.1–0.7, and 0.2–0–0.8, respectively. Assuming the concept of local gaps additivity in simple alloys (Elliott 1991) the authors estimated the optical gap values according to the expression, $E_{g\text{-calc}} = aE_g (Nb_2O_5) + bE_g (PbO) + cE_g (TeO_2)$, where, a, b and c are the formal volume fractions of Nb_2O_5, PbO, and TeO_2, respectively. The average values of the optical band gap were as follows: $E_g(Nb_2O_5) = 3.37$ eV (Serenyi et al. 2008), $E_g(PbO) = 2.7$ eV (Strehlow 1973), and $E_g (TeO_2) = 3.5$ eV (Nayah et al. 2003), respectively. The calculated values of $E_{g\text{-calc}}$ were 3.48, 3.46, 3.41, 3.35, 3.41, and 3.36 eV for the compositions, 0–0.1–0.9, 0–0.2–0.8, 0.1–0.1–0.8, 0.1–0.2–0.7, 0.2–0.1–0.7, and 0.2–0–0.8, respectively. There was an agreement between the experimental and calculated $E_{g\text{-calc}}$ values, except the glasses $0.2(PbO)–0.8(TeO_2)$ and $0.2(PbO)–0.1(Nb_2O_5)–0.7(TeO_2)$. The values of the coefficient of the temperature dependence of the optical band gap are in the region $4.1 \times 10^{-4} – 6.8 \times 10^{-4}$ (eV/ K), and are close to c values for other tellurite glasses (Inoue et al. 1995). It should be noted that the optical band gap is an important quantity also in relation to the nonlinear refractive index (n_2). Fairly good correspondence between the experimental and calculated n_2 values indicates that, in tellurite glasses, the electronic contribution to n_2 could be important, as it was found at around 80% using accurate experiments by Montant et al. (2008). The authors used the simple correlation between the optical band gap and n_2, which was: $n_2 \approx B/(Eg)^4$ (by Ticha et al. 2002). The value of B for the above glasses was B $= 1.26 \times 10^{-9}$ esu(eV)^4.

Hayakawa et al. (2010) studied the UV absorption spectra of tellurite (TeO_2)-based ternary glasses of $MO–Nb_2O_5–TeO_2$ (M = Zn, Mg, Ca, Sr, Ba) in the forms of $xNb_2O_5–(100 − x)TeO_2$ ($x = 10, 20$) glasses and $yZnO–20Nb_2O_5–(100 − y)TeO_2$ ($y = 2, 4, 6, 8, 15$) glasses. Values of the absorption-edge wavelength of $10Nb_2O_5–90TeO_2$ glass were 386 nm, and the absorption-edge wavelengths of the other glasses were almost the same as to be approximately 390 nm: 386, 391, 393, 392, 397, 393, and 394 nm for the glasses, respectively. Also, the values of the absorption coefficient, α, at 800 nm wavelength for these glasses were 4.85, 3.23, 3.81, 1.66, 3.12, 3.65, and 3.92 cm^{-1}. The differences of α were in the range of 2–5 cm^{-1}. The values of linear refractive indexes, n_o (632.8 nm) were equal to 2.045, 2.060, 2.057, 2.150, 2.143, 2.148, and 2.073, respectively. In the binary $Nb_2O_5–TeO_2$ systems, the n_o value slightly increased as the Nb_2O_5 content increased. In the ternary $ZnO–Nb_2O_5–TeO_2$ systems, n_o increased at first as the ZnO content increased, and reached a maximum of 2.150 in $y = 4$. The authors stated that TeO_2-based glasses with high nonlinear optical properties are good vitrification glasses.

Chen et al. (2010) measured the UV spectra of $(100 − x − y)TeO_{opg}–xBi_2O_3–yTiO_2$, where $x = 0,$ 5, 10, and 15, and $y = 5$ and 10; and they studied the relationships between linear optics represented by the linear refractive index n_o and the optical band gap E_{opg}, which were calculated by extrapolating the linear portion of the curves to zero absorption. The strong absorption bands at the UV reign were attributed to internal electronic transition absorption. The optical band gaps E_{opg} of the glasses with compositions 90–0–10, 85–5–10, 85–10–5, 80–10–10, and 80–15–5 were: 2.876, 2.698, 2.715, 2.650, and 2.641 eV, respectively. It can be seen that the E_{opg} values of the glasses studied ranged from 2.6 to 2.9 eV, which are very close to those reported in $TeO_2–Bi_2O_3–WO_3$ (Chen et al. 2008) ternary glasses. In general, the optical band gap E_{opg} is used to present E_g (Chen et al. 2008), since E_{opg}, estimated from absorption edge, strongly reflects electronic transition from valence to conduction bands as well as absorption from excitation and defects. In tellurite glasses, the electronic transition from nonbonding O2pπ orbital to the empty nonbonding Te5d orbital dominates the absorption in ultraviolet range. With the incorporation of modified cations (namely, Bi^{3+} and Ti^{4+}), the absorption edges shift, leading to the variation of E_{opg} values. As a consequence, the two-photon absorption TPA coefficient β shows a dependence on E_{opg}, and increases when E_{opg} is close to 2E (3.1 eV). The deviation that occurs in the fifth composition can be explained in that the Bi^{3+} cations, as network formers, give a positive influence on the TPA cross section. Therefore, in order to modify TPA, certain values of E_{opg} needed to be achieved at a given light wavelength. In that case, the variation of E_{opg} values will be accomplished by the modification of glass components. The figure of merit was used to assess the suitability of a nonlinear material for all-optical switching (AOS) devices for the present glass at 800 nm [Marchese et al. 1998; and Yu et al. 1997), W = $\{\gamma I_o/\alpha\lambda\}$ > 1 and T = $\{\beta\lambda/\gamma\}$ < 1 and I_o is laser power density (3.0 GW/cm^2), α is linear absorption coefficient 120.90, 124.11, 125.59, 149.02 and 144.08 (m^{-1}), k was wavelength]. The W and T values define the consideration of influences of linear and nonlinear absorption on AOS, respectively. The criterion showes in the above expressions for W and T is required for 2p phase shift. For first criterion, first glass sample, W = 0.57, and for the third glass sample, W = 0.38 do not satisfy it since their c values at 800 nm are small. Secondly, all the T values of samples are unfortunately larger than 1 (6.46, 1.41, 8.1, 1.14 and 1.65, respectively), indicating the nonlinear absorption which is two-photon absorption in the present case seriously affects the performance of AOS at 800 nm. Notably, the fourth glass sample, with a T value of 1.14, almost satisfies the second criterion and also possesses a relatively large W value, thus indicating that it has the best third-order nonlinear TNOL properties among the present glasses, as well as the reference glasses that were measured at 800 nm as follows: γ (10^{-18} m^2/W) = 1.85, 4.211, 1.263, 4.196, and 5.108; and β (10^{-11}m/W) = 1.483, 0.742, 1.359, 0.599, and 1.053 respectively. Furthermore, Maciel et al. (2001) proposed the following criteria for materials being considered as optical limiters (OL):

1. High optical transmittance
2. Large (β/α) value

For TeO$_2$–Bi$_2$O$_3$–TiO$_2$ glasses, the first criterion can be easily satisfied since all the samples show excellent transparence in the visible region. Secondly, the largest β/α value was obtained in the first sample, thus indicating its greatest potential for OL considerations. The authors showed that glass with the composition of 80TeO$_2$–10Bi$_2$O$_3$–10TiO$_2$ (mol%) was the best candidate for all-optical switching, while glass with the largest optical band gap E_{opg} had the greatest potential for optical limiters.

10 Infrared and Raman Spectra of Tellurite Glasses

This chapter describes infrared (IR) and Raman spectroscopy, two complementary, nondestructive characterization techniques, both of which provide extensive information about the structure and vibrational properties of tellurite glasses. The description begins with brief background information on, and experimental procedures for, both methods. Collection of these data for tellurite glasses in their pure, binary, and ternary forms is nearly complete. IR spectral data for oxyhalide, chalcogenide, and chalcogenide–halide glasses are now available. The basis for quantitative interpretation of absorption bands in the IR spectra is provided, using values of the stretching force constants and the reduced mass of vibrating cations–anions. Such interpretation shows that coordination numbers determine the primary forms of these spectra. Raman spectral data of tellurite glasses and glass-ceramics are also collected and summarized. Suggestions for physical correlations are made.

10.1 INTRODUCTION

Glass and other solid materials possess vibrational modes characteristic of their composition and the arrangement of their structural bonds. Collective vibrations of molecules, atoms, and ions in an oxide glass network determine the optical absorption properties of such materials in the near and mid-infrared (IR) regions. Several approaches have been used to interpret both the complementary IR and Raman spectra of glasses. The IR spectra of tellurite glasses are less well studied than those of other glasses. IR and Raman spectroscopy are nondestructive techniques that provide extensive information about the optical, dielectric, vibrational, chemical, and structural properties of glasses. In Raman spectroscopy, monochromatic light is coupled to vibrational modes through the nonlinear polarizability associated with those modes. "Raman scattering" is a process in which vibrational excitations are either created (i.e., the Stokes process) or downshifted to a scattered-light frequency (i.e., "annihilated," the anti-Stokes process). The term "Raman scattering" is usually reserved for scattering at high frequency (i.e., via IR vibrations); elastic scattering is termed "Rayleigh scattering," and low-frequency (acoustic-wave) scattering is termed "Brillouin scattering."

Good reviews of these subjects have been produced by Sigel (1977), Bendow (1993), Efimov (1995), and El-Mallawany (1989). The present chapter summarizes the theoretical background of these measurements and the experimental techniques used for both IR and Raman spectroscopy; the primary goal is to explain how data collected by IR and Raman spectra clarify the properties of tellurite glasses. A new direction for research work is mentioned at the end of this chapter, intended to confirm previously compiled data and ultimately to complete them and promote their application via these strategically important solid materials.

10.2 EXPERIMENTAL PROCEDURE TO IDENTIFY INFRARED AND RAMAN SPECTRA OF TELLURITE GLASSES

A number of very interesting books describing the experimental techniques for measuring IR spectra already exist, such as that by McIntyre (1987). The IR absorption spectra of glass materials or pure oxides are improved when these materials are mixed with KBr in a nearly fixed ratio. This

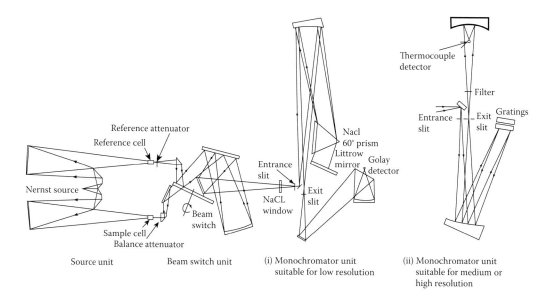

FIGURE 10.1 Spectrophotometer for IR measurements. (From G. Mclntyre, *Infrared Spectroscopy*, New York: John Wiley & Sons, 1987. With permission.)

mixture is shaken by a special machine to obtain a well-mixed powder. The powder is pressed under vacuum under ~10–15 tons of pressure for several minutes to create a uniform pellet. Characteristic IR absorption spectra for this material are then detected by putting the pellet in an IR spectrophotometer operated in the double-beam mode as shown in Figure 10.1, a spectrophotometer for IR measurements described by Mclntyre (1987). For IR transmission measurements, glasses are cut, ground, and polished by a standard method in which ethanol or water is applied as a liquid component. Flat glass must have parallel surfaces of suitable thickness, from 1 to 10 mm in some cases (depending on the darkness of the sample). The spectrometer is set in the range 700–4000 cm^{-1} (15–2.2 μm) to measure the IR transmission through the sample.

Raman spectrometers are available commercially, combining discrete or continuously tunable laser sources with dispersive elements and detection devices for scattered radiation. The first element is the source of monochromatic light, usually a visible laser. The laser beam is focused into a narrow volume of the sample, positioned in a chamber designed to facilitate the collection of scattered light at desired angles (often 90°) from the scattering volume. For high-resolution spectroscopy, single-mode lasers with very narrow spectral bandwidths are advantageous. Filters are inserted in the optical train to suppress unwanted signals. The collection of scattered light to the dispersion element is usually based on a diffraction grating. High-resolution spectroscopy might require enhancements such as multiple monochromators; for the same purpose, Fourier transform (interferometric) spectrometers for Raman systems have recently been introduced. The output from the monochromator is provided to a detector, usually a photomultiplier, which converts the optical signal to a train of electrical pulses. The photomultiplier's output is subsequently processed with "photon-counting" electronic apparatus consisting of a preamplifier, pulse-height analyzer, and count rate meter. The schematic diagram of a furnace for high-temperature Raman measurements, by Tatsumisgo et al. (1994), has been shown in the 1st Ed.

10.3 THEORETICAL CONSIDERATIONS FOR INFRARED AND RAMAN SPECTRA OF GLASSES

The quantitative justification for absorption bands in the IR spectra—that is, the wave number of vibrational modes in these spectra—is determined by the mass of a substance's atoms; the interatomic force within groups of atoms composing a solid network; and the chemical arrangement of units in its matrix. The effect of heavier atoms on absorption bands can be seen from Equation 10.1:

$$\upsilon_o = (1/\lambda) = \frac{1}{2\pi c}\sqrt{f/\mu}$$

(10.1)

which is valid for the fundamental stretching-vibration mode of a diatomic molecule, where υ_o is the wave number per centimeter, c is the speed of light, f is the force constant of the bond, and μ is the reduced mass of the cation–anion molecule, given by:

$$\mu = \frac{M_R M_o}{M_R + M_o}$$

(10.2)

where M_R and M_O are the atomic weights in kilograms of the cation cation R and anion O, respectively. The stretching or bending f can be calculated according to an empirical formula (as explained in Chapter 2) by the relation:

$$f = 1.67N\left[X_a X_b/r^2\right]^{3/4} + 0.3\left(md/\mathring{A}\right)$$

(10.3)

where N is the bond order, X_a and X_b are electronegativities, and r is the bond length. El-Mallawany (1989) calculated the theoretical wave number of the binary rare-earth tellurite glasses TeO_2–La_2O_3, TeO_2–Sm_2O_3, and TeO_2–CeO_2, and compared these values with the experimental values shown in Table 10.1. From Table 10.1, and using Equation 10.1, it can be seen that the calculated values of υ_o agree reasonably well with experimental values for all the assumed stretching vibrations. The experimental band wave number has been found to be higher than the theoretical one for Te–O_{eq}, Te–O_{ax}, and all La–O bonds, suggesting that the attribution of bonds to this group might be band, combining a stretching motion with the harmonics of a bending motion. However, for Ce–O and Sm–O, it has been found that the experimental band wave number is less than the calculated value obtained from Equation 10.1, assuming stretching f and suggesting that the vibration might involve mixed bending and stretching characteristics. For molecules, counting N atoms indicates that they possess $(3N-6)$ independent vibration modes. In addition to the vibrational modes associated with the translation of atoms, there are also rotational modes. From the vibrations of a linear diatomic chain composed of two different atoms (in a crystal form), this crystal has three types of modes and represents two acoustic branches, and one optical branch with finite frequency at $k = 0$, where k is the continuous-wave number. In glasses (amorphous solids), the r and bond angles are not fixed and distributed. Distribution is usually narrow, with mean values close to those in crystal.

In Chapter 8, the refractive index $n(\omega)$ and the absorption coefficient $[\alpha(\omega) = 4\pi k(\lambda)/\lambda]$ are shown to be directly and separately measurable by well-known experimental methods. The propagation of electromagnetic waves in refracting and absorbing media is governed by the relationship between the frequency (ω)-dependent complex dielectric constant $\varepsilon(\omega)$ and its square root, the complex n,

$$\varepsilon^{1/2} = n + ik$$

and

$$\varepsilon(\omega) = \varepsilon_1 + i\varepsilon_2$$

$$= \left(n^2 - k^2\right) + 2ink$$

(10.4)

TABLE 10.1

Theoretical IR Band Positions of Tellurium, Lanthanum, Cerium, and Samarium Oxides

Cation	Atomic Weight	Resting Mass of Cation (10–27 Kg/U)	Reduced Mass of Cation–O (10–27 Kg/U)	Bond Length (nm)	Stretching Force Constant (N/m)	Theoretical Wave Number (cm⁻¹)	Experimental Wave Number (cm⁻¹)
Te	127.61	211.887	2.361	0.190 eq.	248	544	720
				0.208 ax.	189	474	660
La	189.91	230.650	2.382	0.238	126	386	415
				0.245	115	369	
				0.272	84	315	
Ce	58.00	96.305	2.082	0.230	136	434	425
				0.266	90	349	
Sm	62.00	102.947	2.112	0.238	124	407	370
				0.271	84	325	

Source: R. El-Mallawany, *Infrared Phys.*, 29, 781, 1989. With permission.

The reflectivity (R) is described in Chapter 8 and expressed by Equation 8.5:

$$R = \frac{(n-1)^2 + k^2}{(n+1)^2 + k^2}$$

$\varepsilon(\omega)$ is expressed by the relation:

$$\varepsilon(\omega) = \varepsilon_\infty + \frac{\varepsilon_0 - \varepsilon_\infty}{1 - (\omega/\omega_o)^2 - i(\gamma/\omega_o)(\omega/\omega_o)}$$

where γ is the damping factor and ω_o is the oscillator frequency. The reflectivity $R(\omega)$ is a function of both n and the absorption index, thus varying with the frequency in the IR in a complex way. Efimov (1997) used IR reflection spectra of mixed glasses in a broad frequency range, based on a specific analytical model for the complex $\varepsilon(\omega)$ of the glasses, to obtain numerical data on optical constants, band frequencies, and band intensities. For the simplest case of nearly normal incident light, $R(\omega)$ is expressed by the optical constants:

$$R(\omega) = \frac{\varepsilon'(\omega) - 1}{\varepsilon'(\omega) + 1}$$

$$= \frac{[n(\omega) - 1]^2 + k^2(\omega)}{[n(\omega) + 1]^2 + k^2(\omega)}$$

(10.5)

Efimov (1997) proved that the complex $\varepsilon(\omega)$ of glasses could be given by the expression:

$$\varepsilon'(\omega) = \varepsilon_\infty + \sum_{j=i} \frac{S_j}{\sqrt{2\pi\sigma}} \int_{-\infty}^{+\infty} \frac{\exp\left[-(X-\omega)^2/2\sigma_j^2\right]}{x^2 - \omega^2 - i\gamma\omega} dx$$

(10.6)

where x is the variable inherent frequency of an oscillator, ω_j is the central frequency for the jth oscillator distribution, σ_j is the standard deviation for this distribution, and S_j is the oscillator strength. The complex amplitude of refraction is $r^\wedge(\omega) = r(\omega)\exp[i\phi(\omega)]$, where $r(\omega) = \sqrt{R(\omega)}$ and $\phi(\omega)$ is the phase angle of the reflected beam. The magnitude of the error function (Q) serving as the proof of computer fit to an experimental spectrum is obtained as:

$$Q = \sqrt{\frac{1}{b-a_a} \int_a^b \left[R_{mod}(\omega) - R_{exp}(\omega)\right]^2 d\omega}$$

(10.7)

where a and b are the limits of the frequency range studied and $R_{mod}(\omega)$ and $R_{exp}(\omega)$ are the computed and experimental reflectivities, respectively. Efimov (1995) used the calculation in Equation 10.7 to examine silicate, borate, germinate, and tellurite glasses in the binary forms, and all glasses contain nearly the same amount of Na_2O. Figure 10.2a represents the computer fit to the experimental IR reflection spectra with the dispersion analysis based on Equation 10.6 for 65 mol% TeO_2–35 mol% Na_2O. In Figure 10.3a, Line 1 is the experimental spectrum, and Line 2 is the mathematical fit. Figure 10.3b represents a comparison of the $k(\omega)$ spectra of two sodium–tellurite glasses with

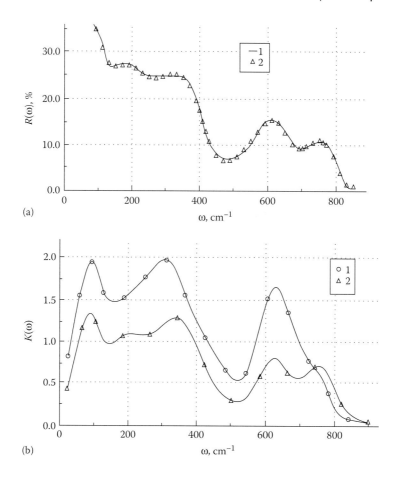

FIGURE 10.2 (a) Computer fit to the experimental IR reflection spectra with the dispersion analysis based on Equation 10.6 for 65 mol% TeO_2–35 mol% Na_2O. 1, experimental spectrum; 2, the fit. (b) Comparison of the k spectra of two sodium–tellurite glasses with maximum differences in Na_2O content. (From A. Efimov, *Optical Constants of Inorganic Glasses*, CRC Press: Boca Raton, FL, 1995.)

maximum difference in Na_2O content. These spectra demonstrate that an increase in intensity of the shoulder at about 750 cm^{-1} is accompanied by a considerable decrease in the peak absorption index k at ~600–620 cm^{-1} and an appreciable decrease in overall absorption throughout the 20–450 cm^{-1} interval of the spectrum. Efimov (1995) wrote that the IR spectra of tellurite glasses are less well studied than those of other glasses such as silicate, borate, germinate, and phosphate, and that there is also a lack of data on the crystal structure and IR spectra of alkali tetratellurites and ditellurites.

Bendaow (1993) discussed Raman spectra by the Stokes (frequency-downshifted) process, which originates from the ground state, or by the anti-Stokes process, which requires phonons to be in excited states and usually to be small—characteristics governed by the Bose–Einstein distribution. Raman spectrum displays peak as a function of the frequency shift. The Raman peaks are caused by discrete vibrational modes or features of continuous modes, depending on the selection rules and coupling strength for Raman processes in a particular geometry. Raman scattering reflects the dependence of electronic polarizability on atomic positions, whether for molecules, crystal, or amorphous structures, as well as the dipole moments accompanying changes in polarizability. Raman scattering involves a three-step process:

1. Changes in polarizability occur, which result in transitions from ground to "virtual" excited electronic states.

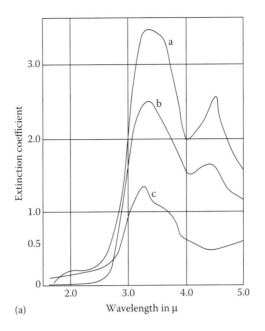

(a)

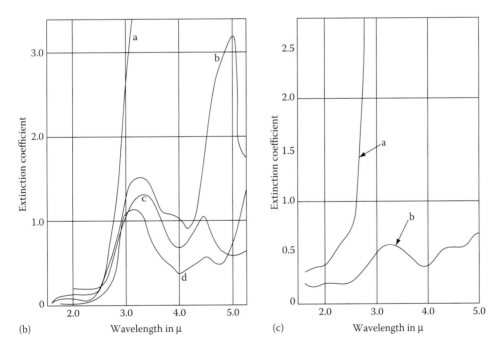

(b)   (c)

FIGURE 10.3 IR spectra of (a) $0.4PbO–0.6Na_2O–5.0TeO_2$, $0.4PbO–0.6Li_2O–5.0TeO_2$, $MoO_3–PbO–4.0TeO_2$; (b) $P_2O_5–3.0PbO–4.0TeO_2$, $0.25SO_3–PbO–4.75TeO_2$, $0.4BaO–0.6PbO–5.0TeO_2$, $2.0WO_3–PbO–3.0TeO_2$; (c) $3.0B_2O_3–PbO–2.0TeO_2$ and $ZnF_2–PbO–4.0TeO_2$ glasses. (From J. Stanworth, *J. Soc. Glass Technol.*, 36, 217, 1952. With permission.)

2. "Real" phonons are created via electron–phonon interactions.
3. The material returns to the electronic ground state.

Based on the definitions of polarizability in Chapter 8, the polarizability tensor (α_{lm}) is expressed in terms of normal mode displacement (U_j) as:

$$\alpha_{lm} = \alpha_{lm}^{(0)} + \alpha_{lm}^{(1)}U + \alpha_{lm}^{(2)}U^2 + \cdots$$

$$U_j = U_{jo}^{i\omega(j)t}$$

and

$$\alpha_{k1}^{(1)} = \alpha_{klj} = \left(\frac{\partial \alpha_{kl}}{\partial U_j}\right)_{U=0} \tag{10.8}$$

The associated electronic moment (M) is expressed by the relation:

$$M = \alpha E$$

So,

$$= \alpha^{(0)}E_o e^{i\omega t} + \alpha^{(1)}U_{j0}e^{i(\omega \pm \omega_j)t} + \cdots \tag{10.9}$$

This expression of electronic moment has several characteristics:

- The first term leads to elastic or Rayleigh scattering.
- The second term, involving the first derivation of polarizability, leads to the creation and annihilation of phonons.
- When the phonons are of low-frequency acoustic modes, typically 0.1–1.0 cm^{-1}, the process is referred to as "Brillouin scattering."
- When phonons at IR frequency (10^2–10^3 cm^{-1}) are involved, whether acoustic or optical, then the process is called Raman scattering.

The theory of Raman scattering in glasses was formulated in the work of Balkaniski (1971), based on the short correlation lengths in glasses relative to crystals. Assuming a simplified correlation function, Raman scattering theory yields the following expression for Stokes scattering intensity as a function of frequency shift:

$$I(\omega) = \omega^{-1}\left[1 + n(\omega)\right]\sum_b c_b g_b(\omega) \tag{10.10}$$

where b is the vibrational band index, $g_b(\omega)$ is the density of states for band b, and c_b is the factor that depends on correlation length associated with the modes in band b.

10.4   INFRARED SPECTRA OF TELLURITE GLASSES

10.4.1   Infrared Transmission Spectra of Tellurite Glasses and Glass-Ceramics

In 1952, Stanworth measured the IR spectra of the following tellurite-based materials:

1. 0.4 PbO–0.6 Na$_2$O–5.0 TeO$_2$; 0.4 PbO–0.6 Li$_2$O–5.0 TeO$_2$; and MoO$_3$–PbO–4.0 TeO$_2$
2. P$_2$O$_5$–3.0 PbO–4.0 TeO$_2$; 0.25 SO$_3$–PbO–4.75 TeO$_2$; 0.4 BaO–0.6 PbO–5.0 TeO$_2$; and 2.0 WO$_3$–PbO–3.0 TeO$_2$
3. 3.0 B$_2$O$_3$–PbO–2.0 TeO$_2$ and ZnF$_2$–PbO–4.0 TeO$_2$, as shown in Figure 10.3a through c

Some of these glasses have remarkably good IR transmission, at least to 5.5 μm. Measurements have been made of the IR transmission properties of disks prepared from several of these tellurite glass melts. The results were mainly in the wavelength range of 2–5 μm and are plotted as curves of the extinction coefficient (*k*) against wavelength (Figure 10.3). *k* is calculated from the relation:

$$I = I_o 10^{-kd} \tag{10.11}$$

where *I* is the intensity of radiation reaching the second face of the disk sample, I_o is the intensity of radiation leaving the first face, and *k* is the sample thickness; the ratio of the reflected light to light incident on each face was taken as $(n - 1)^2/(n + 1)^2$, where *n* is the refractive index. Tellurite glass containing molybdenum oxide has good transmission at least as far as 5 μm, with an absorption band in the region of ~3.2 μm. This particular band is much more marked in the glasses containing Na_2O or Li_2O, as shown also in Figure 10.3. These glasses also have an absorption band at ~4.5 μm, although tellurite glass containing P_2O_5 shows a sharp cutoff at ~3 μm, which is typical of ordinary phosphate glasses not containing TeO_2. The glass containing WO_3 has an absorption band at a wavelength somewhat beyond 5 μm; it also shows a band at about 3.1 μm and a smaller band at 4.5 μm observed with other tellurite glasses, but its absorption is low at wavelengths up to 5 μm. Lead–barium–tellurite glass also has good transmission at wavelengths ≥5 μm, and it again shows absorption bands at ~3.2 and 4.5 μm. Some interesting results have been obtained with a glass containing SO_3. This glass, at a calculated composition of 75.8 wt% TeO_2–22.2 wt% PbO–2.0 wt% SO_3 (corresponding to a molecular composition of 4.75 TeO_2, PbO, 0.25 SO_3), was melted in a zirconia crucible at about 950°C using lead sulfate as the source of SO_3. A clear glass was obtained with a density of 6.02 g/cm³, a thermal expansion coefficient in the temperature range 50°C–200°C of 185×10^{-6} °C⁻¹, and a deformation temperature of 305°C. A determination of the tellurium and lead contents of the glass also indicates, by subtraction, that SO_3 content is approximately 2%, although there is no physical proof that SO_3 is actually present. The IR transmission property of the glass shows the usual absorption at ~3.2 μm but, in addition, a very sharp absorption band with a peak at 5 μm. This result suggests that the SO_3 content of some tellurite glasses might well be determined by the intensity of the absorption band at 5 μm. The glass containing B_2O_3 was shown to have a sharp cutoff at ~2.8 μm, which agrees approximately with the absorption properties of borate and borosilicate glasses not containing TeO_2. Glass containing ZnF_2 has excellent IR transmission at least to 5 μm with only small absorption bands at 3.2 and 4.5 μm. This glass is the best tellurite glass sample prepared so far (Stanworth 1952).

Stanworth (1954) measured the IR transmission spectra of tellurite glasses containing 10 wt% V_2O_5, 15 wt% B_2O_3, 15–50 wt% MoO_3, or 16.7–36.7 wt% WO_3, as shown in Figure 10.4, for glass samples with thicknesses of 1.8 and 3.1 mm. These glasses, which absorb visible radiation very strongly, can transmit efficiently in the IR—in some cases at least to 5 μm. Adams (1961) measured the IR transmission spectra of a thin sample of 15.17 mol% ZnF_2–9.13 mol% PbO–75.7 mol% TeO_2 glass and crystalline TeO_2 in the range 7–16 μm as shown in Figure 10.5. He found that the crystal is built up of distorted TeO_6 octahedrals. Thus, the IR results agreed with the results of Barady (1957), who came to the conclusion from x-ray diffraction (XRD) studies on lithium glass that TeO_2 is six-coordinated in the vitreous state, as shown in Chapter 1.

In 1962, Cheremisinov and Zlomanov measured the IR absorption spectra of tellurium dioxide in crystalline (tetragonal) and glassy states. An examination of vibrational spectra showed that the structure of tellurium oxide glass is similar to that of the tetragonal dioxide crystal. This result conflicts with those based on x-ray analysis of tellurium dioxide glass. It has been demonstrated by analyzing the vibrational spectrum of crystalline tellurium dioxide and comparing it with these x-ray data that the spectrographic data concur most satisfactorily with a lattice structure whose elementary cells each contain four TeO_2 groups and belong to symmetry group D_4. An interpretation is given for the frequencies of the vibrational spectra of tellurium dioxide in tetragonal modification.

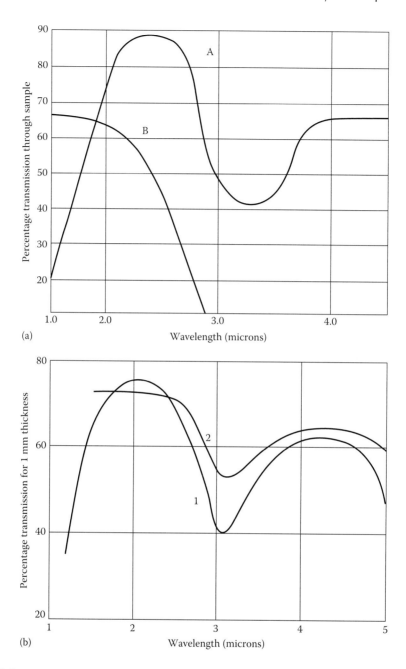

FIGURE 10.4 (a, b) IR transmission spectra of tellurite glasses containing 10 wt% V_2O_5 (Glass A), 15 wt% B_2O_3 (Glass B), 15–50 wt% MoO_3 (Glass 1), or 16.7–36.7 wt% WO_3. (From J. Stanworth, *J. Soc. Glass Technol.*, 38, 425, 1954. With permission.)

In 1964, Ulrich measured the IR transmission spectrum of 90 mol% TeO_2–10 mol% B_2O_3. Yakhkind et al. (1968) measured the transmission curves of barium–tellurite glasses with vanadium, and oxides of the form TeO_2–V_2O_5–BaO were investigated. Tantarintsev and Yakhkind (1972, 1975) studied the effect of water on the IR transmission of highly refracting tellurite glasses of the forms 80 mol% TeO_2–20 mol% WO_3 and 80 mol% TeO_2–20 mol% Na_2O.

Mochida et al. (1978) measured the IR absorption spectra of binary tellurite glasses containing monovalent and divalent cations such as Li, Na, K, Rb, Cs, Ag, Tl, Be, Mg, Ca, Sr, Ba, Cd, and Pb

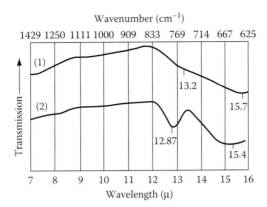

FIGURE 10.5 IR transmission spectra of a thin sample of 15.17 mol% ZnF_2–9.13 mol% PbO–75.7 mol% TeO_2 glass and crystalline TeO_2 in the range 7–16 μm. (From R. Adams, *Phys. Chem. Glasses*, 2, 101, 1961. With permission.)

as shown in Figure 10.6. The IR spectra of the binary glasses with high TeO_2 content resembled that of paratellurite but showed remarkable downward shifts of the peak positions. Mochida et al. concluded that imaginary TeO_2 glass, which means the network structure of high-percentage TeO_2 glasses, is classified into a group including B_2O_3, P_2O_5, and S_2O_3 glasses according to its expansion coefficient, which is estimated by extrapolation of the α-composition curve to 100% TeO_2. These results, and consideration of the coordination polyhedra around Te^{4+} of the tellurite crystals, suggest that the glasses with high TeO_2 content comprise a layered network of the distorted tetragonal-pyramidal TeO_4, described as a distorted trigonal bipyramid (tbp) in which one of the equatorial sits is occupied by a lone electron pair of Te. On the basis of the assignments of IR absorption peaks of the crystalline compounds, it has been revealed that the bands assigned to the Te–O stretching vibration of the trigonal pyramidal (tp) TeO_3 appear at 720 cm^{-1} in the $LiO_{1/2}$ glasses, 700 cm^{-1} in the $NaO_{1/2}^-$, $KO_{1/2}^-$, and BaO glasses, 660 cm^{-1} in the $TlO_{1/2}$ glasses, 675 cm^{-1} in ZnO glasses, and 695 cm^{-1} in the MgO glasses. Dimitriev et al. (1979) measured the IR spectra of crystalline phases and related glasses in the form 2 TeO_2–V_2O_5–2 Me_2O (where Me is Li, Na, K, Cs, or Ag) as shown in Figure 10.7. The absorption bands in the 970–880 cm^{-1} range were assigned to the stretching modes of the VO_2 isolated groups. A trend has been observed toward a shift of the high-frequency band by the replacement of an alkaline ion with another in the order Ag^+, Cu^+, Li^+, Na^+, K^+, Rb^+, and Cs^+, which is explained by their different polarizing abilities. With the aid of XRD, Dimitriev et al. (1979) showed that the basic structure units in the glasses studied were in the TeO_4 and VO_4 groups.

A structural model of tellurite glasses in the binary form modified by WO_3 and using IR spectral data has been done by Zeng (1981), who concluded that the distorted TeO_6 octahedral and WO_4 tetrahedral network are entangled. Hogarth et al. (1983) measured the IR spectra of $(100 - x - y)$ mol% TeO_2–x mol% CaO–y mol% WO_3 glasses, where x is 0, 2.5, 5.0, 7.5, 10, 15, or 20, and the corresponding values for y are 35, 32.5, 30, 27.5, 25, 20, and 15, respectively. Hogarth et al. (1983) concluded that the most important absorption bands in glasses are the same as for TeO_2. The shift of band position to lower frequency when Te–O stretching frequencies are important is caused by the creation of single bonds between bridging oxygen ions and a tungsten ion to form a Te–O–W unit, and the TeO_2 tetrahedra dominate this structure. Dimitrov et al. (1987) compared the IR spectra of tellurite glasses and their crystal products containing from 5 to 45 mol% WO_3, and indicated that the modifier does not change the coordination of tellurium as shown in Figures 10.8 and 10.9. Both figures represent the IR spectra of glasses containing small amounts of WO_3 and show a band at 925 cm^{-1}, which shifts to 950 cm^{-1} with an increase in the tungsten concentration. The effect is specific for the vitreous state and is explained by the change in coordination of tungsten. The tellurium is present in deformed TeO_4 groups (a band at 635 cm^{-1}), and when WO_3 increases from

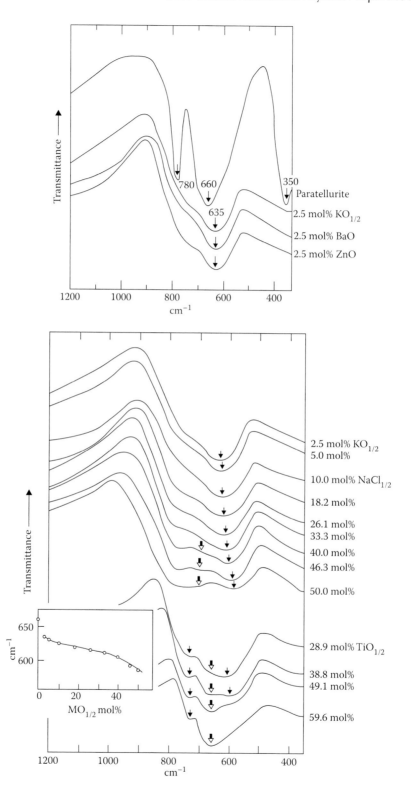

FIGURE 10.6   IR absorption spectra of binary tellurite glasses containing monovalent and divalent cations. (From N. Mochida, K. Takahshi, K. Nakata, *Yogyo-Kyokai-Shi*, 86, 317, 1978. With permission.)

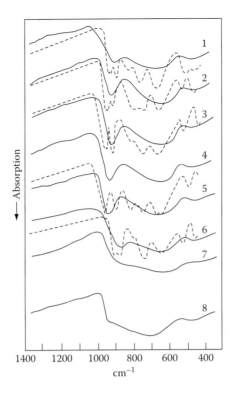

FIGURE 10.7 IR spectra of crystalline phases and related glasses in $2TeO_2–V_2O_5–2M_2O$ for which M = Li, Na, K, Cs, and Ag. (From Y. Dimitriev, V. Dimitrov, and M. Arnaudov, *J. Mater. Sci.*, 14, 723, 1979. With permission.)

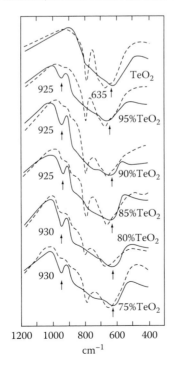

FIGURE 10.8 IR spectra of tellurite glasses in the range ≤ 75 mol% TeO_2. (From V. Dimitrov, *J. Sol. Chem.*, 66, 256, 1987. With permission.)

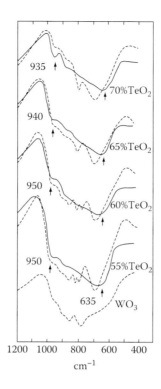

FIGURE 10.9   IR spectra of tellurite glasses in the range from pure 70 mol% TeO_2 to 55 mol% TeO_2. (From V. Dimitrov, *J. Sol. Chem.*, 66, 256, 1987. With permission.)

small amounts, tungsten participation changes from WO_4 (a band at 925 cm^{-1}) to WO_6 (a band at 950 cm^{-1}). Ahmed et al. (1984) measured the IR optical absorption of TeO_2–GeO_2 glasses as in Figure 10.10. Most of the sharp absorption bands characteristic of the basic materials TeO_2 and GeO_2 are modified with the formation of broad and strong absorption bands in the process of going from the crystalline state to the amorphous state.

Al-Ani et al. (1985) measured the IR absorption spectra of a series of binary $(100 - x)$ mol% TeO_2–x mol% WO_3 glasses, where x is 0, 2, 5, 7, 10, 15, 20, 25, and 33 mol%. The IR absorption spectra of TeO_2 and WO_3 oxides were also measured. The intensity and the band positions are indicated in Figure 10.11. A band of WO_3 was not detected in these glasses. The intensity and band positions show some chemical interaction between the two oxides rather than positions characteristic of a simple oxide mixture. Burger et al. (1985) measured the IR transmission spectra of TeO_2–R_nO_m, $-R_nX_m$, $-R_n(SO)_m$, $-R_n(PO_3)_m$, and B_2O_3 glasses as shown in Figure 10.12. It has been proven that Te–O stretching vibrations have a strong influence on multiphonon absorption, but in spite of this, they are close to some halide nonoxide medium-IR-transmitting glasses, exhibiting a similar transmission.

In 1986 and 1987, the Bulgarian group published a series of articles on IR spectral investigations of water in tellurite glasses (Arnaudov et al. 1986), effect of mode formation on the structure of tellurite glasses of the form 2 TeO_2–V_2O_5 (Dimitriev et al. 1987), and IR spectral investigations of 2 TeO_2–V_2O_5–Li_2O–V_2O_5–2 TeO_2 glasses by Dimitrov et al. (1987). Arnaudov et al. (1986) presented a study of the OH-stretching region for TeO_2–BaO and TeO_2–Nb_2O_5 glasses as shown in Figure 10.13. The lowering of OH stretching in the spectra of barium-tellurite glasses as the amount of modifier increases was explained by the formation of a ($O_5TeOH....O$–TeO_3) hydrogen bond, where oxygen atoms and electrons were strongly polarized and become practically nonbridging. The higher value in the IR spectra of TeO_2–Nb_2O_5 glasses is caused by the existence of a hydrogen bond between O_5TeOH and bridging oxygen atoms, because Nb is introduced directly

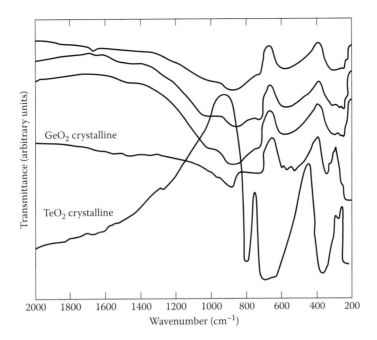

FIGURE 10.10 IR optical absorption of TeO_2–GeO_2 glasses. (From M. Ahmed, C. Hogarth, and M. Khan, *J. Mater. Sci.*, 19, 4040, 1984. With permission.)

through bridging Te–O–Nb bonds. Based on the IR spectra of $2TeO_2$–V_2O_5 glasses at different temperatures, Dimitriev et al. (1987) concluded that significant changes in structure are established in vitreous $2TeO_2$–V_2O_5 by a rise in the melting temperature. Dimitrov (1987) measured the IR absorption spectra of $2TeO_2$–V_2O_5–Li_2O–V_2O_5 glasses. From these spectra, the corresponding crystallization products and data of known crystal structures, a model of the short-range order in these glasses have been proposed.

In 1987, Bahgat et al. measured the IR spectra of tellurite glasses of the form $[100 – (2x + 5)]$ mol% TeO_2–x mol% Fe_2O_3–$(x + 5)$ mol% Ln_2O_3, where x is 0 and 5 and Ln is lanthanum, neodymium, samarium, europium, or gadolinium. From the figure, Bahgat et al. (1987) concluded that fractions of the Fe_2O_3 and Ln_2O_3 are incorporated into this network and act as a network intermediates.

Mochida et al. (1988) studied the structure of TeO_2–P_2O_5 glasses. The IR spectra of these glasses show that condensation of PO_4 tetrahedra occurs in compositions over 9.3 mol% P_2O_5 in spite of the fact that all PO_4 tetrahedra are isolated in the crystalline Te_4O_5 $(PO_4)_2$ (33.3 mol% P_2O_5). In 1988, Sabry et al. identified the IR spectra of binary tellurite glasses of the form TeO_2–NiO. They measured the integrated area of the IR absorption band at 460 cm^{-1} for the glass with composition 90 mol% TeO_2–10 mol% NiO for different periods at 400°C. These authors also compared the IR absorption spectra of heat-treated and untreated samples with a composition of 65 mol% TeO_2–35 mol% NiO for 112 h at 400°C. Later, Malik and Hogarth (1989a) studied some of the effects of substituted cobalt and nickel oxides on the IR spectra of copper–tellurite glasses.

Pankova et al. (1989) measured the IR spectra of binary TeO_2–Ag_2O, TeO_2–BaO, TeO_2–CuO, TeO_2–PbO, TeO_2–ZnO, and TeO_2–Bi_2O_3. The influence of the different modifiers on Te–O stretching vibrations in TeO_4 and TeO_3 groups and on the $TeO_4 \rightarrow TeO_3$ transition has been investigated. It has been established that the ion modifiers Ag^+ and Ba^{2+} create nonbridging Te–O bands that act as defects and destroy the three-dimensional glass network. The cations with partial covalent bonds with the oxygen (Cu^{2+}, Pb^{2+}, Zn^{2+}, and Bi^{3+}) stimulate the formation of Te_2O_5 and Te_3O_8 complexes. The structural $TeO_4 \rightarrow TeO_3$ transition begins at lower concentrations of these cations than with typical ion modifiers.

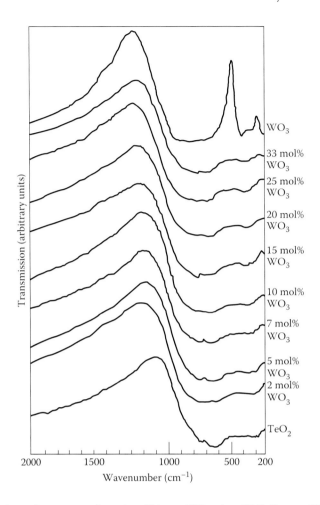

FIGURE 10.11   IR absorption spectra of a series of binary $(100 - x)$ mol% TeO_2–x mol% WO_3 glasses, where x is 0, 2, 5, 7, 10, 15, 20, 25, and 33. (From S. Al-Ani, C. Hogarth, and R. El-Mallawany, *J. Mater. Sci.*, 20, 661, 1985. With permission.)

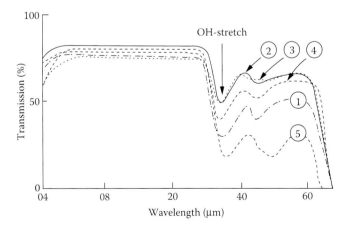

FIGURE 10.12   IR transmission spectra of TeO_2 –R_nO_m, –R_nX_m, –$R_n(SO)_m$ glasses. (From H. Burger, W. Vogel, and V. Kozhukarov, *Infrared Phys.*, 25, 395, 1985. With permission.)

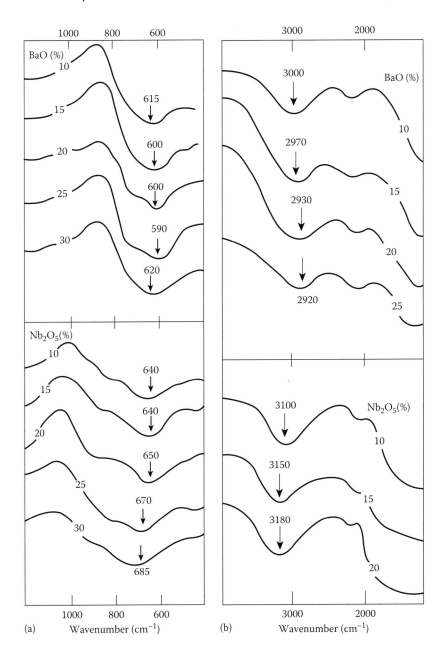

FIGURE 10.13 (a, b) IR spectra of the OH stretching region for TeO$_2$–BaO and TeO$_2$–Nb$_2$O$_5$ glasses. (From M. Arnaudov, Y. Dimitriev, V. Dimitrov, and M. Pankov, *Phys. Chem. Glasses*, 27, 48, 1986. With permission.)

In 1989, El-Mallawany measured and calculated both experimental and theoretical main absorption frequencies of the IR absorption spectra for rare-earth tellurite glasses as shown in Figure 10.14. The rare-earth oxides were samarium, cerium, and lanthanum oxides. The main shift in these glasses was found to be sensitive to the glass structure. The rare-earth element–oxygen bond vibration in these glasses has been calculated. The results are interpreted on the basis of stretching f of each oxide present in the glass. Also in 1989, Malik and Hogarth (1989a and 1989b) measured the IR spectra of TeO$_2$–CuO–Lu$_2$O$_3$ glasses. The bands were more resistant to stretching than to bending, but in spite of this fact, the presence of Lu$_2$O$_3$ (a rare-earth oxide) overcomes the stiffness of bonds and shows more broadening and less bending, which is a characteristic of the glassy state.

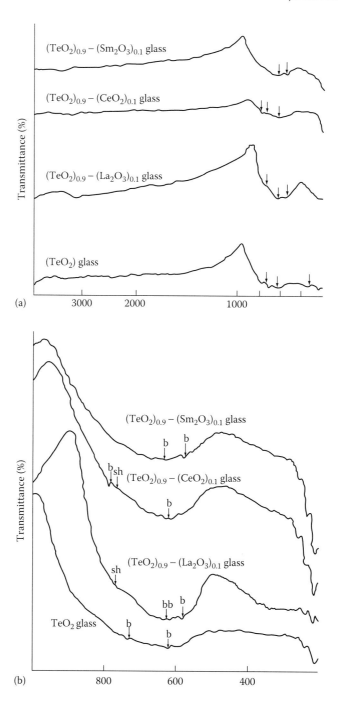

FIGURE 10.14   (a, b) Measured and calculated main absorption frequencies of the IR absorption spectra for rare-earth tellurite glasses. (R. El-Mallawany, *Infrared Phys.*, 29, 781, 1989. With permission.)

Chopa et al. (1990) measured the IR spectra of pure TeO_2 and binary TeO_2–V_2O_5 blown-glass films with modifier compositions in the range 10–50 mol%. The absorption peaks in the IR spectra of these films were not characteristic of mixtures with more oxide; thus, Chopa et al. (1990) concluded that this result indicates a chemical interaction between the two oxide materials. Khan (1990) studied the effect of CuO on the structure of tellurite glasses.

Abdel-Kader et al. (1991a) identified the compositional dependence of IR absorption spectra for TeO_2–P_2O_5 and TeO_2–P_2O_5–Bi_2O_3 glasses as shown in Figure 10.15. They found that the mid-band wave number and absorption intensity for the attributed bands are strongly and systematically dependent on glass composition. Quantitative analysis was also done in order to justify the attribution of the observed bands.

In 1992, Mizuno et al. measured the IR spectra while they used tellurite glasses of the following form as bonding glasses to magnetic heads: $(85 - x - y)$ wt% TeO_2–x wt% PbO–y wt% B_2O_3–5 wt% ZnO–10 wt% CdO. Mizuno et al. (1992) concluded that replacing PbO with TeO_2 strengthened the glass network with the change in the coordination number in the B–O bond. Burger et al. (1992) measured the IR transmittance of α–TeO_2, $Zn_2Te_3O_8$, and $ZnTeO_3$.

In 1993, Abdel-Kader et al. measured the IR absorption spectra of tellurite–phosphate glasses doped with different rare-earth oxides in the form 81 mol% TeO_2–19 mol% P_2O_5 as shown in Figure 10.16. The rare-earth oxides were La_2O_3, CeO_2, Pr_2O_3, Nd_2O_3, Sm_2O_3, and Yb_2O_3. The IR spectra of these glasses indicated that the rare-earth oxides were connected to the chains of TeO_4. Dimitriev (1994) summarized the structure of tellurite glasses by using IR spectroscopy for TeO_2–GeO_2 glasses with fast and slow cooling. Sabry and El-Samadony (1995) measured the IR spectra of binary TeO_2–B_2O_3 glasses and proved the distribution of the TeO_4 polyhedra, which determines the network and the basic oscillations of the building units in the tellurite glasses. The IR results also proved the distribution of the boroxal group.

Hu and Jian (1996) measured the IR spectra of TeO_2-based glasses containing ferroelectric components, including $KNbO_3$–TeO_2, $PbTiO_3$–TeO_2, $PbLa$–TiO_3–TeO_2, 5 mol% $KNbO_3$–5 mol% $LiTaO_3$–90 mol% TeO_2, and 10 mol% $KNbO_3$–5 mol% $LiTaO_3$–85 mol% TeO_2 glasses. Hu and Jian (1996) found sharp absorption peaks at 668 and 659 cm^{-1} in $PbTiO_3$- and $PbLaTiO_3$-containing TeO_2-based glasses, respectively, and they found broad peaks at around 680, 675, and 694 cm^{-1} in 10 mol% $KNbO_3$–90 mol% TeO_2, 5 mol% $KNbO_3$–5 mol% $LiTaO_3$–90 mol% TeO_2, 10 mol% $KNbO_3$–5 mol% $LiTaO_3$–85 mol% TeO_2 glasses, respectively. Small peaks were also observed at around 1090 cm^{-1} in $LaTiO_3$- and $KNbO_3$-containing TeO_2-based glasses.

Weng et al. (1999) measured, for the first time, the Fourier transform IR (FTIR) spectra of TeO_2–TiO_2 thin films prepared by the sol-gel process, as shown in Figure 10.17. In the FTIR spectrum reported by this group, a large, broadband at around 3320 cm^{-1} attributed to vibrations of the OH group in ethylene glycol was weakened greatly, which indicated that the H atom in the OH group of ethylene was replaced by a Te atom during the reaction of $Te(Oet)_4$ with ethylene glycol (as shown in Figure 10.18) structure of the sol-gel material.

10.4.2 IR Spectral Data of Oxyhalide–Tellurite Glasses

In 1980, Yakhkind and Chebotarev studied the IR transmission of ternary tellurite–halide systems of the forms TeO_2–WO_3–$ZnCl_2$, TeO_2–BaO–$ZnCl_2$, TeO_2–Na_3O–$ZnCl_2$. The concentration of "water" in the glass was calculated using the relation:

$$C = \frac{\lg(\tau_0/\tau)}{d\varepsilon}$$ (10.12)

where τ and τ_0 are the transmission coefficients of wet and virtually dry glasses, respectively, at a frequency of 2900 cm^{-1} (corresponding to the most absorption maximum of the OH groups), d is the thickness of the specimen, measured in centimeters, and ε is the water absorption coefficient of tellurite glass, previously defined by Tatarintsev and Yakhkind (1972). The concentrations (C) of water in the tellurite–halide glasses are as follows:

* 72 mol% TeO_2–19 mol% WO_3–9 mol% $ZnCl_2$; C = 0.0063
* 50 mol% TeO_2–5 mol% WO_3–45 mol% $ZnCl_2$; C = 0.0274

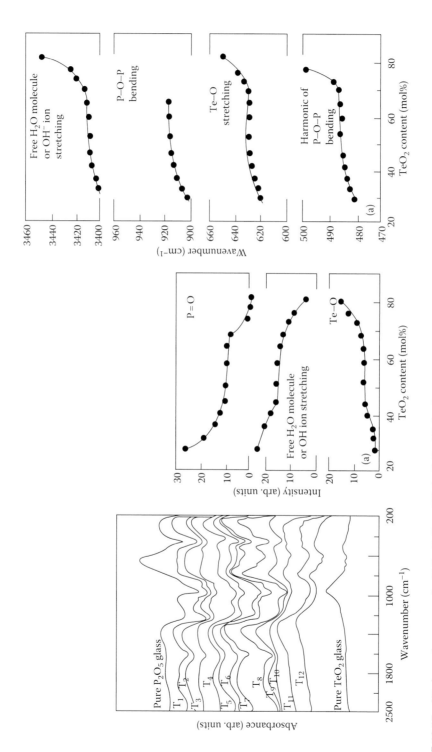

FIGURE 10.15 IR absorption spectra for TeO_2–P_2O_5 and TeO_2–P_2O_5–Bi_2O_3 glasses. (From A. Abdel-Kader, A. Higazy, and M. Elkholy, *J. Mater. Sci.*, 2, 157, 1991a. With permission.)

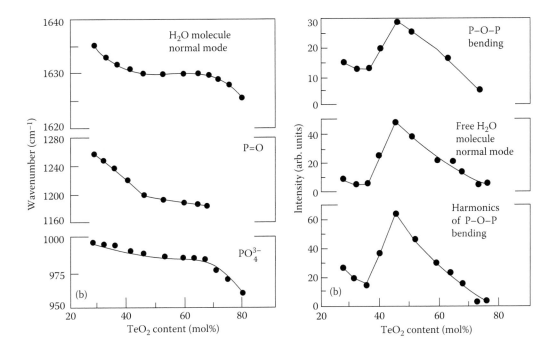

FIGURE 10.15   (Continued)

- 83.3 mol% TeO_2–16.7 mol% Na_2O–x mol% $ZnCl_2$; $x = 3$, C = 0.0092; $x = 12$, C = 0.0234
- 85.7 mol% TeO_2–14.3 mol% Na_2O–x mol% $ZnCl_2$; $x = 6$, C = 0.0074; $x = 36$, C = 0.0167
- 78 mol% TeO_2–22 mol% Na_2F–x mol% $NaCl$; $x = 3$, C = 0.0103; $x = 6$, C = 0.0161
- 73 mol% TeO_2–27 mol% $NaBr$–x mol% $NaCl$; $x = 3$, C = 0.0073; $x = 6$, C = 0.0231

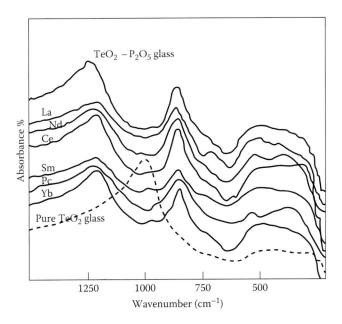

FIGURE 10.16   The IR absorption spectra of tellurite–phosphate glasses doped with different rare-earth oxides in the form 81 mol% TeO_2–19 mol% P_2O_5. (From A. Abdel-Kader, R. El-Mallawany, and M. Elkholy, *J. Appl. Phys.*, 73, 71, 1993. With permission.)

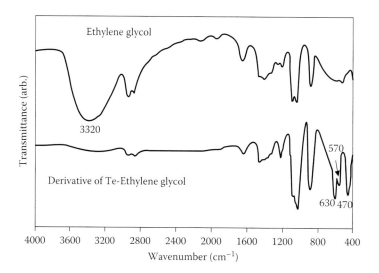

FIGURE 10.17 FTIR spectrum of TeO$_2$–TiO$_2$ thin films prepared by the sol-gel process. (L. Weng, S. Hodgson, and J. Ma, *J. Mater. Sci. Lett.*, 18, 2037, 1999. With permission.)

As the Br in the glass is replaced by Cl and F, the maximum of the IR absorption of the OH groups is shifted toward lower frequencies, and this is in good agreement with the formation process of the hydrogen bond between hydroxyl groups in the glass structure and the halide atoms. Burger et al. (1985) measured the IR transmission of TeO$_2$–R_nX_m, for X = F, Cl, or Br. The entire transmission range of these halide glasses is 0.4–7 μm, and they do not have OH vibration absorption bands at 3.2 and 4.4 μm. Burger et al. (1985) proved that heavy ions influence the absorption ability of glasses and shift the IR cutoff toward longer wavelengths.

Tanaka et al. (1988) measured the absorption spectra of ternary tellurite glasses of the form TeO$_2$–LiCl$_2$–Li$_2$O. They observed that the Te–O$_{ax}$ bond, where *ax* is the axial position of the TeO$_4$ tbp, becomes weaker with increasing LiCl content in binary LiCl–TeO$_2$ glasses, thus indicating that LiCl works as a network modifier. On the other hand, a gradual increase in wave number of the peak due to the Te–O$_{ax}$ bond was observed when Li$_2$O was replaced with LiCl. Malik and Hogarth (1990) measured the IR spectra of 65 mol% TeO$_2$–(35 – x) mol% CuO–x mol% CuCl$_2$, for *x* is 0, 1, 2, 3, 4, and 5, at room temperature in the frequency range 200–2400 cm^{-1}. Also in 1990, Ivanova measured the IR spectral transmission of TeO$_2$–PbCl$_2$–PbO–KCl–NaCl and TeO$_2$–BaCl$_2$–PbO–KCl–NaCl. El-Mallawany (1991) measured the IR absorption spectra of two forms of the oxyhalide–tellurite glasses: Glass A (TeO$_2$–TeCl$_4$) and Glass B (TeO$_2$–WCl$_6$), as shown in Figure 10.18. Sahar and Noordin (1995) measured the IR spectra of TeO$_2$–ZnO–ZnCl$_2$ glasses. The IR cutoff edge up to 6.5 μm and spectra were dominated by the presence of Te–O stretching vibration from TeO$_4$ units (for lower Zn [O, Cl$_2$] content) and TeO$_3$ or Te(O, Cl)$_3$ units (for higher Zn [O, Cl$_2$] content).

10.4.3 IR SPECTRA OF HALIDE–TELLURIDE GLASSES

Zhang et al. (1988) prepared a new class of tellurite glasses in the binary system Te–Br. They reported that the limits of the vitreous state are Te$_2$Br and TeBr, whereas the most stable composition toward crystallization is Te$_3$Br$_2$. The optical transmission range lies between 1.9 and 20 μm for Te$_3$Br$_2$ and Te$_3$Cl$_2$, as shown in Figure 10.19. Lucas et al. (1988) prepared a new group of IR-transmitting glasses based on tellurium halides, finding that these heavy TeX glasses have an IR edge located in the 20 μm region. Chiaruttini et al. (1989) investigated IR-transmitting TeX glasses belonging to the system Te–Se–Br.

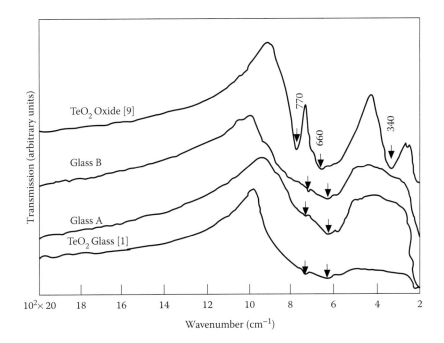

FIGURE 10.18 IR absorption spectra of the following oxyhalide–tellurite glasses: A, TeO_2–$TeCl_4$; B, TeO_2–WCl_6. (From R. El-Mallawany, *Mater. Sci. Forum*, 67/68, 149, 1991. With permission.)

Lucas and Zhang (1990) prepared new IR-transmitting tellurium–halogen glasses. The halogens in these glasses were Cl, Br, and I. The IR cutoff of the Te–Br glasses after the addition of S or Se was measured. Hong et al. (1991) substituted As for Se or Te in several tellurium–halide glasses, which led to a significant increase in the glass transition temperature, as shown in Chapter 5. Zhang et al. (1993) also investigated TeX glasses for X as Cl, Br, or I. Rossignol et al. (1993) measured the IR spectra of ternary $(TeO_2$–$RTlO_{0.5})_{(1-x)}$–$(AgI)_x$ glasses, which showed small modifications of the tellurite network. Rossignol et al. concluded that AgI had been introduced into the binary glasses. Klaska et al. (1993) measured the IR transmission of the TeGeAsI and TeGeAsSeI glasses, which are characterized by a broad optical window lying between 2 and 15 μm without any peak due to impurities. Neindre et al. (1995) prepared and measured the IR transmission spectra for TeX glass fiber in the range 3–13 μm.

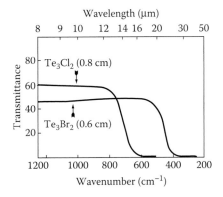

FIGURE 10.19 Optical transmission of Te–Br glass. (From X. Zhang, G. Fonteneau, and J. Lucas, *Mater. Res. Bull.*, 23, 59, 1988. With permission.)

10.4.4 IR Spectra of Chalcogenide Glasses

Hilton (1970) stated that Te, S, and Se were then under study as IR optical materials. Chalcogenide glasses become transparent on the long-wavelength side of an absorption edge, whereas their long-wavelength cutoff is determined by their lattice-type absorption. In the transport region, impurity absorption was found. Also, chalcogenide glasses were used as optical materials and compared favorably with conventional IR optical materials. In 1987, Lezal et al. reported a technique for the preparation of 30 mol% Ge–35 mol% Se–35 mol% Te glass, as well as their IR transmission spectra for a sample of thickness 40 mm. In 1993, Zhenhua and Frischat prepared chalcohalide glasses based on heavy metal halides of the system M_nX_m–As_2Se_3–As_2Te_3 as well as chalcogenide glasses of the system As(Ge)–Se–Te as shown in the 1st Ed. Wang et al. (1995) studied the effect of Sn and Bi addition on IR transmission and far-FTIR spectra of Ge–Se–Te chalcogenide glass. Xu et al. (1995) measured the IR spectra of Ge–As–Se basic glasses that had been affected by the addition of Te and I, both separately and together. Novel chalcohalide glasses in the As–Ge–Ag–Se–Te–I glass system were prepared and investigated by Cheng et al. (1995).

10.5 RAMAN SPECTRA OF TELLURITE GLASSES

10.5.1 Raman Spectra of Oxide–Tellurite Glasses and Glass-Ceramics

Starting in 1962, Bobovich and Yakhkind (1963) reported the Raman spectra of a series of tellurite glasses belonging to the systems TeO_2–Na_2O, TeO_2–BaO, TeO_2–WO_3, TeO_2–BeO_3, and crystal-line TeO_2 and $NaTeO_3$. They observed that the spectrum of crystalline TeO_2 consists of several narrow lines. The breakdown of part of the edge bonds between TeO_6 tetrahedra, which probably accompanies the formation of a tellurite glass, leads to a very marked broadening of these bands, which is also observed in all of the other glasses. Raman spectroscopic studies were performed on TeO_2–B_2O_3 and TeO_2–B_2O_3–K_2O glasses by Kenipp et al. (1984). In the binary system TeO_2–B_2O_3, a partial coordination change of the boron–oxygen coordination from three to four appeared, as indicated by a Raman frequency at ~760 cm^{-1}, which was found to correspond to that at 808 cm^{-1} in glasses with high B_2O_3 content. In the ternary glass system TeO_2–B_2O_3–K_2O, the insertion of TeO_2 leads to a significant change in the structure, contrary to its effect in glasses of type R_2O–B_2O_3 or R_2O–B_2O_3–SiO_2. The addition of TeO_2 does not lead to a continuous decrease of boroxol groups for the benefit of BO_3–BO_4 structure units with increasing R, as already known from the system R_2O–B_2O_3. This result is manifested in the Raman spectra by a relative maximum of the intensity ratio (I [808 cm^{-1}]/I [772 cm^{-1}]) independence on R contrary to a continuous regression of this ratio in other R_2O–B_2O_3 or R_2O–B_2O_3–SiO_2 glasses.

Mochida et al. (1988) measured the Raman spectra of binary TeO_2–P_2O_5 glasses. The Raman peak assigned to P=O stretching vibrations, which usually appears in the wave number region 1330–1390 cm^{-1}, was absent over the composition range of these glasses. The Raman spectra of paratellurite, tellurite, and pure TeO_2 glass were measured by Sekiya et al. (1989). The spectrum of pure TeO_2 glass was deconvoluted into symmetric Gaussian functions. The normal vibrations of paratellurite were described as combined movements of oxygen atoms in Te–$_{eq}O_{ax}$–Te linked with vibrations of TeO_4 tbps. When the resolved Raman peaks of pure TeO_2 glass are compared with normal vibrations of paratellurite, all Raman peaks from 420 to 880 cm^{-1} are assigned to vibrations of the TeO_4 tbps and Te–$_{eq}O_{ax}$–Te linkage. The antisymmetric stretching vibrations of the Te–$_{eq}O_{ax}$–Te linkage have relatively large intensities compared with symmetric stretching vibrations of the same linkage. In pure TeO_2 glass, TeO_4 tbps are formed by most tellurium atom stretching vibrations of Te–$_{eq}O_{ax}$–Te linkages. In pure TeO_2 glass, TeO_4 tbps are formed by most tellurium atoms and connected at vertices by the Te–$_{eq}O_{ax}$–Te linkages. Komatsu et al. (1991) examined the Raman scattering spectra of $(100 - x)$ mol% TeO_2–x mol% $LiNbO_3$ glasses. Their structure was composed of TeO_4 trigonal pyramids (tps), TeO_3 tps, and NbO_6 octahedra. Rong et al. (1992) measured the Raman spectra of binary tellurite glasses containing boron and indium oxides in the form TeO_2–M_2O_3 (M = B and

In) as in Figure 10.20. From the relationship between the M_2O_3 content and the intensity ratios of the deconvoluted Raman peaks, I (720 cm^{-1})/I (665 cm^{-1}) and I (780 cm^{-1})/I (665 cm^{-1}), Rong et al. concluded that In_2O_3 behaves as a network modifier to yield TeO_3 units, and that discrete BO_3 and BO_4 units construct a network of glasses containing boron oxide. Rong et al. (1992) also constructed a structural model for those glasses that involved three-coordination oxygen atoms and TeO_4 units of an intermediate configuration, $O_3Te\delta^+\ldots O\delta^-$. Also in 1992, Sekiya et al. measured the Raman spectra of binary tellurite glasses of the form $TeO_2-MO_{1/2}$ (where M is Li, Na, K, Rb, Cs, and Tl) as shown in the 1st Ed. Depending on their alkali content, these glasses have the following characteristics:

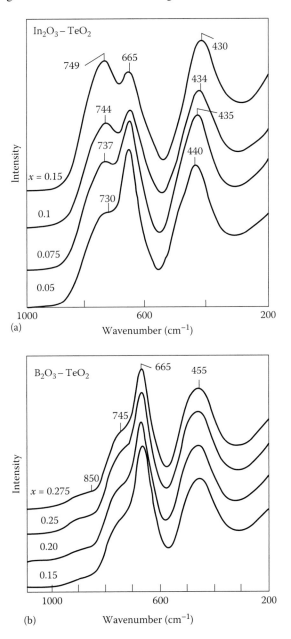

FIGURE 10.20 (a–d) Raman spectra of binary tellurite glasses containing boron and indium oxides in the form $TeO_2-M_2O_3$ (M = B and In). (From Q. Rong A. Osaka, T. Nanba, J. Takada, and Y. Miura, *J. Mater. Sci.*, 27, 3793, 1992. With permission.)

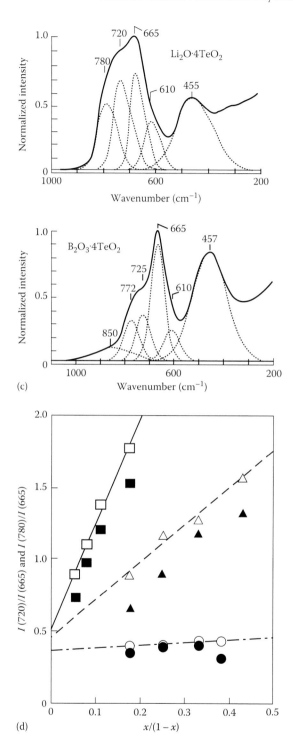

FIGURE 10.20   (Continued)

- Low alkaline content—a continuous network constructed by sharing corners of TeO_4 tbps and TeO_{3+1} polyhedra having one NBO
- 20–30 mol% alkali oxide—TeO_3 tps with NBOs, formed in contiguous network
- >30 mol% alkali oxide—isolated structural units, such as $Te_2O_3^{2-}$ ion, coexist in the continuous network
- >50 mol% alkali oxide—glasses composed of a continuous network comprising TeO_{3+1} polyhedra and TeO_3 tps, and also of isolated structure units, such as Te_2O and $Te_2O_3^{2-}$ ions

The structure of thallium–tellurite glasses having <30 mol% $TlO_{1/2}$ is similar to that of alkali–tellurite glasses containing equal amounts of $MO_{1/2}$. Sekiya et al. (1992) put forward a new hypothesis for the mechanism of the basic structural changes in the tellurite glasses as shown in the 1st Ed. Burger et al. (1992) measured the Raman spectra of the following polycrystalline compounds: α-TeO_2, $Zn_2Te_3O_8$, $ZnTeO_3$, crystallized sample containing 70 mol% TeO_2–30 mol% ZnO, and the binary glasses 80 mol% TeO_2–20 mol% ZnO, 75 mol% TeO_2–25 mol% ZnO, and 70 mol% TeO_2–30 mol% ZnO. The data errors were $\upsilon \pm 0.2$ cm^{-1}; ordinate accuracy and repeatability were better than 0.02%; and the scattering intensity (Figure 10.23, ⇓) indicated the laser plasma line. Tatsumisago et al. (1994) identified the Raman spectra of TeO_2-based glasses and glassy liquids as shown in the 1st Ed. Parameters for initial refinement included Te–O and Te–Te interactions found in experimental D (r) along with other interactions of interest found in 52 mol% Li_2O–48 mol% P_2O_5. The interatomic distance number (n) and temperature factor (b) of each interaction were refined using the method of least squares. Standard deviations for r, b, and n were 0.001 nm, 0.000003 nm^2, and 0.1, respectively. High-temperature Raman spectra revealed that the TeO_4 tbps unite with NBOs as the temperature is increased above the melting temperature, whereas the major structural units in conventional glass-forming systems, such as silicates, are not likely to change with increasing temperature. The structural change with temperature can be described by the structural-change model in which the Te–O (axial) bond cleavage occurs with heating.

Sekiya et al. (1994a) investigated the structure of TeO_2–MO glasses, where M is Mg, Sr, Ba, and Zn, by Raman spectroscopy. The structure of MgO–TeO_2 was different from that of BaO–TeO_2 glasses. The structures of SrO–TeO_2 and ZnO–TeO_2 glasses were the same as those of BaO–TeO_2 and MgO–TeO_2 glasses, respectively. However, Sekiya et al. (1994b) investigated the structure of both $(100 - x)$ mol% TeO_2–x mol% WO_3 glasses (for which x is 0, 5, 10, 15, 20, 25, and 30.1) and tungstate crystals. The agreement between the observed and simulated spectra was quantitatively evaluated by the value of R, which was defined as:

$$R = \sum_i \left| I_i\left(observed\right) - I_i\left(simulated\right) \right| \Big/ \sum_i \left| I_i\left(observed\right) \right| \qquad (10.13)$$

where I_i stands for the intensity at υ_i cm^{-1}. The average and maximal values of R for these binary glasses with the same modifier but of different percentages were 0.02 and 0.03, respectively. Good agreement was obtained between the observed and simulated spectra. The glasses contain clusters composed of corner-shared WO_6 octahedra, and an increase in WO_3 content promotes the growth of the clusters. The coordination state of W^{6+} having a W–O bond was discussed on the basis of crystal chemistry. Tungsten coordination polyhedra occupying the interface of the cluster have a W=O bond.

Dexpert-Ghys et al. (1994) used Raman scattering to investigate tellurite glasses of the form TeO_2–$RMO_{0.5}$, where M is Ag or Tl and $0.22 \leq R \leq 1$, and of the related ionic-conductor tellurite glasses $(TeO_2$–$RMO_{0.5})_{(1-x)}$–AgI_x. The ternary glasses were reported as ionic conductors. The structures of the binary TeO_2–$RMO_{0.5}$ glasses with M as Ag or Tl were investigated by Raman scattering under VV and VH polarizations. The two systems behave in nearly the same way. In the tellurite-rich part, the structure is built up of TeO_4 tps, a distorted form of the paratellurite α-TeO_2. With increasing

R between 0.40 (0.35) and 1.0, an increasing proportion of tricoordinated tellurium (TeO$_3$) is formed in the glass. Also in 1995, Shaltout et al. investigated the same binary tungstate–tellurite glasses of the form (100 – x) mol% TeO$_2$–x mol% WO$_3$, for 5 < x < 50 and discussed the Raman vibrational bands and the different coordination states of the constituent oxides. Shaltout et al. (1995) found other interesting aspects of Raman spectra and differential-scanning calorimetry, which are mentioned in Chapter 5.

Sekiya et al. (1995a) measured the Raman spectra of binary tungstate–tellurite glasses of the form (100 – x) mol% TeO$_2$–x mol% MoO$_3$, for x as 0, 5, 9.4, 15, 20, 25, 30, 35, and 40 mol%. Sekiya et al. (1995a) discussed local structural differences around second-component atoms between TeO$_2$–MoO$_3$ and TeO$_2$–WO$_3$ glasses. Sekiya et al. (1995b) also measured the Raman spectra of binary tellurite glasses containing trivalent or tetravalent cations. The second component oxides were YO$_{3/2}$, InO$_{3/2}$, LaO$_{3/2}$, ZrO$_2$, SnO$_2$, HfO$_2$, and ThO$_2$. Sekiya et al. (1995b) stated that no glasses were obtained in the two systems, TeO$_2$–SnO$_2$ and TeO$_2$–ZrO$_2$, because of the precipitation of ZrTe$_3$O$_8$ and SnTe$_3$O$_8$ crystals. Cuevas et al. (1995) measured the Raman scattering of the tellurite glass system of the form (100 – x) mol% TeO$_2$–5 mol% TiO$_2$–x mol% Li$_2$O, for x = 5, 10, 15, 20 and 25. Tatsumisago et al. (1995) measured the Raman spectra of binary 80 mol% TeO$_2$–20 mol% Li$_2$O and 70 mol% TeO$_2$–30 mol% Li$_2$O at the following stages: rapidly quenched glass samples before heating, crystallized samples during heating, liquid samples after melting, and samples after melting and then cooling.

Komatsu et al. (1996) measured the Raman scattering on local structures of Te^{4+} and Te^{5+} in LiNbO$_3$–TeO$_2$ glasses and basic structure units of TeO$_4$ tbps and TeO$_3$ tps, and Te$_{-eq}$O$_{ax}$–Te bonds in TeO$_2$-based glasses. Muruganandam and Swshasayee (1997) measured the Fourier transform-Raman spectroscopy of 75 mol% LiPO$_3$–25 mol% TeO$_2$, 65 mol% LiPO$_3$–35 mol% TeO$_2$, and 50 mol% LiPO$_3$–50 mol% TeO$_2$ glasses as shown in the 1st Ed. Short Te–Te interaction (compared with α-TeO$_2$) was the major deformation exhibited by TeO$_4$ polyhedra in their chains. Kosuge et al. (1998) proposed from Raman spectral analysis of TeO$_2$–WO$_3$–K$_2$O glasses that TeO$_4$ tbps changed to TeO$_3$ tps with the addition of K$_2$O, and that Te–O–W bonds were formed by the substitution of WO$_3$ for TeO$_2$.

Also in 1998, Ilieva et al. used high-intensity Raman bands of the (100 – x) mol% TeO$_2$–x mol% Ga$_2$O$_3$, for x values of 5–35. They questioned whether the four peaks of glasses containing Ga$_2$O$_3$ arise only from the vibrations of TeO$_n$ polyhedra. In the 600–800 cm^{-1} range, the β-Ga$_2$O$_3$ crystal had intense peaks assigned to vibrations of GaO$_4$ tetrahedra. Ilieva et al. concluded that it is difficult to estimate the gallium contribution because of its low concentration and the band overlap of the modes of GaO$_4$ and TeO$_n$ (n is 3 or 4) polyhedra. Thus, gallium coordination with respect to oxygen remains an open question. Bersani et al. (1998) investigated the structure of x mol% TeO$_2$–(100 – x) mol% V$_2$O$_5$, x mol% ZnO–(100 – x) mol% [2TeO$_2$–V$_2$O$_5$], and x mol% CdO–(100 – x) mol% [2TeO$_2$–V$_2$O$_5$]. In x mol% TeO$_2$–(100 – x) mol% V$_2$O$_5$ glass, the coordination of Te changes from TeO$_4$ to TeO$_3$ with increasing x content. The main effect of the addition of modifier oxides in vanadate and telluro-vanadate glasses is a transformation of the VO$_5$ bipyramids to VO$_4$ tetrahedra. No apparent change in the local structure around the modifier atoms was revealed by x-ray absorption fine structure (XAFS) measurements.

In 1999, Komatsu and Mohri, by measuring the Raman scattering spectra of tellurite glasses and crystalline phases containing PbO and CdO, demonstrated that the number of Te–O–Te linkages in the glasses—and thus the connectivity of this network—decreases with increasing PbO and CdO content. Their glasses were heat treated at 320°C (below the crystallization temperature, 346°C) in nitrogen or 93% Ar–7% H$_2$ atmosphere; they became slightly shiny and black but retained good transparency, and an unknown crystalline phase (i.e., not CdTe or Cd$_2$PbO$_4$) was formed at the surface. Oishi et al. (1999) measured the Raman scattering spectroscopically at room temperature in samples heat-treated for 1 h at temperatures of 390°C, 425°C, 450°C, and 475°C for glass and glass-ceramic forms of 70 mol% TeO$_2$–15 mol% K$_2$O–15 mol% Nb$_2$O$_5$ as shown in Figure 10.21. From Figure 10.21 it can be seen that the peak at >750 cm^{-1} decreases with increasing temperature of heat treatment, implying that the maximum phonon band in the glass-ceramics would be

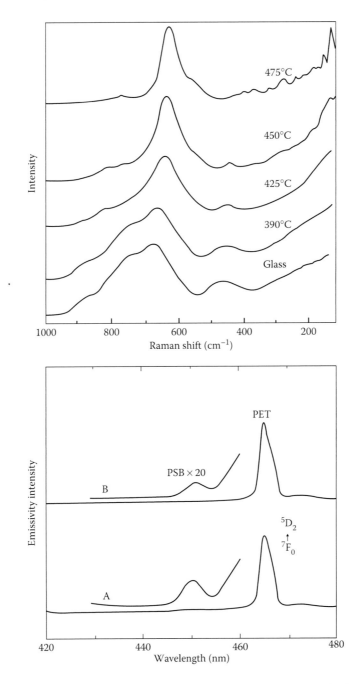

FIGURE 10.21 Raman scattering spectroscopy and PSB at room temperature and for heat-treated samples for 1 h at temperatures of 390°C, 425°C, 450°C, and 475°C for the glass and glass-ceramics of 70 mol% TeO_2–15 mol% K_2O–15 mol% Nb_2O_5. (From H. Oishi, Y. Benino, and T. Komatsu, *Phys. Chem. Glasses*, 40, 212, 1999. With permission.)

smaller than that in the glass. Previously, Sidebottom et al. (1997) measured the phonon sideband spectra (PSB) for zinc oxyhalide–tellurite glasses, as shown in Chapter 8. The PSB of Eu^{3+} and Er^{3+} in tellurite glasses gives a new possibility for optical applications of transparent TeO_2-based glass ceramics with low phonon energies and second harmonic generation. Oishi et al. (1999) suggested that further studies are necessary to clarify the state of the glass-ceramics fabricated in their study. Miyakawa and Dexter (1970) calculated the rate of nonradiative decay caused by multiphonon relaxation with an energy gap to the next lower level (ΔE) of 0 (i.e., $W_p[0]$) at a temperature of 0°C as follows: $W_p(0) = W_o \exp(\alpha \Delta E)$, where p is the number of phonons consumed for relaxation, $\alpha = \hbar \omega^{-1} (\ln p/g - 1)$, $p = \Delta E/\hbar \omega$, and g is the electron phonon coupling strength.

Akagi et al. (1999) studied the structural change of TeO_2–K_2O glasses from room temperature to temperatures higher than the melting point (T_m) by using Raman spectroscopy, as shown in the 1st Ed., XRD, x-ray radial distribution function (RDF), and XAFS. The Raman results indicated that TeO_4 tbp units convert to TeO_3 tp units with increasing temperature and by the addition of more K_2O to the TeO_2–K_2O glasses. The high-temperature XRD and x-ray (RDF) results were in agreement with the results of high-temperature Raman spectroscopy. In 2000, Mirgorodsky et al. studied the structure of TeO_2 glass by using Raman scattering as shown in Figure 10.22. The results were used to interpret the Raman spectra of two new polymorphs (γ and δ) of tellurium dioxide and to clarify their relationships with the spectrum of pure TeO_2 glass.

Data on the Raman spectra of nonoxide–tellurite glasses is a very important area for further research work. As previously described, tellurite glasses are nonlinear solids with high values of the susceptibility term χ^3, which appear by the AC-Kerr effect, optical-Kerr effect, and Raman and Brillouin scattering to be applicable in fast-switching, time-resolved (grating) experiments, and generation of different wavelengths, respectively. Low-frequency Raman spectra of tellurite glasses should be measured at low temperature and compared with the heat capacity data to correlate the Raman intensity with the number of vibrational excitations (density of the vibrational states $g[\upsilon]$, Debye frequency, and Bose factor), as well as to distinguish the scattering caused by acoustic modes from that caused by excess light scattering. Quantitative comparisons between the behavior of Raman scattering and ultrasonic behavior should be determined in the same type of

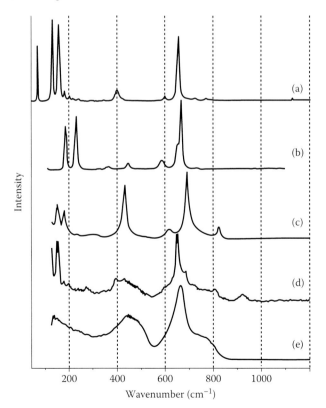

FIGURE 10.22 (a–e) Raman scattering of TeO_2 glass. (From A. Mirgorodsky, T. Merle-Mejean, and B. Frit, *J. Phys. Chem. Solids,* 61, 501, 2000. With permission.)

glasses (i.e., tellurite glasses). It is very important to carry out low-frequency Raman and Brillouin scattering experiments in these strategic solid materials in a wide range of temperatures, from that of liquid helium to the glass transition temperature. Finally, analysis of the Raman data (Stokes and anti-Stokes), together with complementary low-temperature specific heat and ultrasonic velocity, is needed to check whether the light scattering and acoustic relaxation result from different microscopic motions.

10.6 INFRARED AND RAMAN SPECTRA OF NEW TELLURITE GLASSES

As mentioned in Chapter 8, Ovcharenko and Sairnova (2001) prepared transparent TeO_2-based glasses with high refractive indices and were diamagnetic materials with high values of the Verdet constant. Ovcharenko and Sairnova (2001) suggested that the TeO_2-based glasses would become of great interest for use in optical communication systems. Also, the authors prepared tellurite glass of the forms TeO_2–WO_3–Bi_2O_3 and TeO_2–WO_3–PbO. The authors studied both IR reflection spectra in the range 1100–300 cm^{-1} and Raman spectra in the range 1000–200 cm^{-1}. The data proved that the structure of tellurite glasses did not display remarkable changes with variations of the composition. The major structural groups were composed of structural-tolerant bipyramids $[TeO_4]$ and tetrahedra $[WO_4]$ with other participants including $[WO_6]$, $[PbO_4]$ and $[BiO_6]$. In the late 1990s, tellurite glass attracted considerable attention and development effort as a host for broadband erbium-doped fiber amplifiers (EDFA), as mentioned by Oshishi et al. (1998) and Nakai et al. (1998). Shen et al. (2002) studied the Er^{3+}-doped tungsten–tellurite glass as a host for broadband erbium-doped fiber amplifiers (EDFA) in the form: $40WO_3$–$40TeO_2$–$20Li_2O$, $80TeO_2$–$10ZnO$–$10Na_2O$. The authors measured the Raman spectra of tungsten–tellurite and tellurite glasses in the frequency range 300–1000 cm^{-1}. The highest phonon energy peak in tungsten–tellurite glass was at 920 cm^{-1}, which is significantly above that of tellurite at 770 cm^{-1}. Tellurite EDFA cannot be pumped efficiently at 980 nm due to the long lifetime of the $^4I_{11/2}$ level in this glass. Tungsten–tellurite EDFA therefore promised a better performance than tellurite EDFA.

Xu et al. (2006) studied the spectral properties and thermal stability of tungsten–tellurite glass doped with 1.0 wt% Er_2O_3 and 5.0 wt% Yb_2O_3. The authors prepared tellurite glasses and measured the refractive index on a prism by the minimum deviation method, together with the density of the glass by using Archimedes' principle. The tellurite glasses were in the form: $(100 - x)TeO_2$–xWO_3–$10BaO$, with x = 30, 25, and 20 mol%; $(100 - y)TeO_2$–yWO_3–$10La_2O_3$, with y = 25, 20, 15 mol%; $(100 - z)TeO_2$–zWO_3–$10Bi_2O_3$, with z = 30, 25, 20 mol%; and $65TeO_2$–$25WO_3$–$10PbO$, mol%. The absorption spectra of the glasses were recorded with a PERKIN-ELMERLAMBDA 900UV/VIS/NIR spectrophotometer in a spectral range of 400–1700 nm. The fluorescence spectra in the range of 1400–1650 nm were measured with a TRIAX550 spectrofluorimeter upon excitation at 970 nm. The fluorescence at full width and half maximum (FWHM) and the peak emission cross section (σ_{peak}) of the $^4I_{13/2} \rightarrow {}^4I_{15/2}$ transition of Er^{3+} ions in different glass hosts have been compared. The results indicated that the tungsten–tellurite glass is a promising host material for 1.55 μm broadband amplifications. The FWHM and σ peak values are very important for optical amplifiers to realize broadband amplification, and the product FWHM × σ_{peak} is often used to evaluate the gain bandwidth of optical amplifiers (Yang et al. 2003). The σ_{peak} value of tungsten–tellurite glasses was much larger than that of other glass hosts, and the FWHM value is slightly smaller than that of bismuthate glass, but much larger than those of tellurite, silicate, and phosphate glasses, and better than other glasses used as host materials for 1.55 μm broadband amplification.

Moreover, Gao et al. (2009) measured the refractive index of:

1-$70TeO_2$–$20ZnO$–$2.5Na_2O$–$2.5K_2O$–$5La_2O_3$, 2-$70TeO_2$–$10ZnO$–$10ZnF_2$–$2.5Na_2O$–$2.5K_2O$–$(5 - x - y) La_2O_3$–$xHo_2 O_3$–yTm_2O_3, with x = 0, 0.5, 1, 2, 3, 4, and y = 0.5, 1, 1.5, 2, 3, 4 mol%, as explained in Chapter 8. The authors also investigated the mid-IR transmittance properties of glass by using a Fourier transform infrared (FTIR) spectrometer. Demonstration of mid-IR transmittance spectra of glasses with different F^- ions concentration has been achieved. Absorption bands ranging from 2500 to 3700 cm^{-1} were due to stretching vibration of free OH$^-$ groups. It was obvious that the introduction of fluoride compounds is effective to remove OH$^-$ hydroxyl in tellurite glass. The hydroxyl and fluorine ions were isoelectronic and their ionic size is similar; therefore, hydroxyl ions can be easily replaced by fluorine during melting through a reaction OH$^-$ + F$^-$ → HF$^+$ + O^{2-}. Thus, the decrease of OH$^-$ absorption bands in IR transmittance spectra could be attributed to the introduction of fluorine. The results of fluorescence spectra indicated that the maximum emission intensity of Ho^{3+} 2.0 μm emission could be achieved at the concentration of 1.5 mol% Tm$_2$O$_3$ and 1 mol% Ho$_2$O$_3$ in oxyfluoride–tellurite glass. The maximum absorption cross section of Tm^{3+} at 800 nm is 9.8 × 10^{-21} cm^2 at 793 nm, and the absorption and emission cross section of Ho^{3+} near 2.0 μm were 7.0 × 10^{-21} cm^2 at 1950 nm and 8.8 × 10^{-21} cm^2 at 2048 nm, respectively. Tm^{3+}/Ho^{3+} codoped oxyfluoride–tellurite glass was promising material for 2 μm laser due to its wide optical transmission window, high refractive index, good thermal stability, strong 2 μm fluorescence, and long irradiative lifetime (4.4 ms) of the Ho^{3+}; ^{5}I$_7$ level.

As mentioned in Chapter 8, Gebavi et al. (2009) fabricated glass samples based on the following notation: $(100 - x)$ (75 TeO$_2$–20 ZnO–5 Na$_2$O) where $x = 0.36$, 0.72, 1.08, 2.14, 3, 4, 5, 6, 7, and doped with Tm^{3+} up to 10 mol% (0, 8690, 12.993, 25.08, 35.494, 46.920, 58.151, 69.194, 80.052, and 111.564 ppm). The prepared glasses were studied to assess the most suitable rare-earth content for short cavity fiber lasers. Based on the structural investigations (Raman spectra), the addition of Tm^{3+} produced a progressive cleavage of the glass network and the conversion of trigonal bipyramid units (TeO$_4$ polyhedra) into trigonal pyramid (TeO$_3$) polyhedra through the formation of TeO$_{3+1}$ units. Lifetime quenching for the Tm^{3+} concentrations start at higher than 3 mol% with maximum predicted lifetime value of around 3 ms. The best emission intensity ratio for lasing at 1822 nm is at around 5 mol% of Tm^{3+}. Based on optical properties and on thermal analysis, the optimum candidate for a short cavity laser is a sample with 58.151 (ppm) by using the cross-relaxation mechanism higher emission efficiency can be achieved. New barium chloride–oxide–tellurite glasses, BaCl$_2$–BaO–TeO$_2$, were obtained by Sokolov et al. in 2009. The authors studied the structure of these glasses by means of quantum-chemical modeling and Raman spectroscopy, and a model of the glass network was developed. The main results can be stated as follows:

1. Tellurite component of the barium chloride–oxide–tellurite glass network can be formed mainly by structural groups of three types:
   a. TeO$_4$ bipyramid (fourfold coordinated tellurium atom with four Te–O–Te linkages).
   b. O$_2$Te=O trihedral pyramid (threefold coordinated tellurium atom with one nonbridging oxygen atom forming Te=O double bond and two Te–O–Te linkages).
   c. O$_3$Te–O$^-$ tetrahedron (fourfold coordinated tellurium atom with one negatively charged nonbridging oxygen atom, Te–O$^-$).
2. Barium and chlorine atoms do not form any covalent bond in the glass network occurring in interstitial sites such as Ba$_2$ and Cl$^-$ ions.
3. No other structural group, in particular O$_3$Te, is needed to form a continuous network of BaCl$_2$–TeO$_2$ glass.
4. The Raman band near 760 cm^{-1}, which is characteristic of barium chloride–oxide–tellurite glasses is contributed by vibration of three types, Te=O double-bond stretching vibration in nonbridging oxygen atoms in O$_2$Te = O groups and stretching vibrations of both long and short Te–O$^-$ bonds in negatively charged nonbridging oxygen atoms in O$_3$Te–O$^-$ groups.

Ozdanova et al. (2010) prepared the x(PbO)–y(Nb$_2$O$_5$)–$(1 - x - y)$(TeO$_2$) where $x = 0$, 0.1, and 0.2; and $y = 0$, 0.1, and 0.2 mol% glasses from very pure oxides and measured the Raman spectra using an FTIR spectrometer. All Raman spectra were taken at room temperature in the spectral region 50–1000 cm^{-1} using 200 scans of the bulk sample with a flat optical surface. The structure of TeO_2-based glasses, as deduced from Raman spectroscopy, can be assessed using two different approaches:

1. The most typical broad RF feature in the wave number region 600–1000 cm^{-1} was due to an overlap of two Raman frequencies (RF), of which the first was around 650 cm^{-1} (RF650) and is attributed to asymmetric vibrations in symmetric Te–O–Te interchain linkages. Corresponding symmetric vibrations are observed at around 480 cm^{-1}. The second RF was around 750 cm^{-1} (RF 750) and was attributed to asymmetric stretching of intrachain rather than asymmetric Te–O–Te linkage. Relevant symmetric stretching vibrations were located near 430 cm^{-1}.
2. RF in the region 600–1000 cm^{-1} was due to an overlap of four distinct RF, where:
   a. The RF at around 610 cm^{-1} and the RF at around 660 cm^{-1} were attributed to stretching modes of the TeO_4 trigonal bipyramids (tbp) with two nonequivalent oxygen atoms. And
   b. The RF at around 710 cm^{-1} and the RF at around 790 cm^{-1} were attributed to stretching modes of the TeO_3 trigonal units (tp).

Hagar and El-Mallawany (2010) used the melt-quenching technique to prepare the quaternary tellurite glass systems $(70-x)TeO_2$–$20WO_3$–$10Li_2O$–xLn_2O_3, where $x = 0, 1, 3$ and 5 mol%; and Ln are La, Pr, Nd, Sm, Er, and Yb, respectively. The IR absorption spectra of the present glass systems were determined at room temperature over the range of wave numbers 400–1.600 cm^{-1}. Raman spectra of the present glass samples were measured in the range 30–1.030 cm^{-1}. Quantitative interpretations were based on concentration of ions per unit volume of Te, Ln, and O, short distance in nm between ions for (Te–O) of TeO_4, TeO_3 groups, (W–O) of WO_4, WO_6 groups and calculated wave number, v, for TeO_4 and TeO_3, respectively. The average stretching force constant that was present in these quaternary glasses has been calculated to interpret the data obtained. The calculated number of bonds per unit volume of the glasses was increased and the glasses became more rigid. The bond length between the ions of the main structural groups of the glasses was calculated, and the wave number band position of the absorption bands was also calculated. The structure of these glasses is built up from TeO_4 tbp, TeO_3 tp, WO_4 tetrahedral and/or WO_6 octahedral groups. The low frequency Raman absorption band obtained for all the studied samples, which is around 50 cm^{-1}, was identified with the Boson peak, a feature characteristic of the vitreous state. This means that the rare-earth ions were incorporated in the network of the present glasses and act as a network intermediate. Also, the authors suggested that present glasses are good hosts for these rare-earth ions and good for optical properties and optical applications.

Hayakawa et al. (2010) successfully studied the metal oxide doping effects on Raman spectra and third-order nonlinear susceptibilities of thallium–tellurite glasses. The glasses were in the forms: $5TiO_2$–$30TlO_{0.5}$–$65TeO_2$, $10ZnO$–$30TlO_{0.5}$–$60TeO_2$, $10GaO_{1.5}$–$36TlO_{0.5}$–$54TeO_2$, $5PbO$–$28.5TlO_{0.5}$–$66.5TeO_2$, and $5BiO_{1.5}$–$38TlO_{0.5}$–$57TeO_2$. Raman data were collected using a micro-Raman spectrophotometer (JASCO, NRS-2000). An Ar$^+$ laser with a wavelength of 514.5 nm and 20 mW power was employed as an excitation source. The Raman spectra were first calibrated by temperature (Bose–Einstein), refractive index, and instrument corrections. The spectra were then compared with one another by both the frequency position and the relative Raman intensity (RRI) criteria. Also, the authors measured χ^3 using the Z-scan technique; a femtosecond pulse laser from a regenerative Ti: sapphire laser system (Spectra Physics, Harricane) was used at a wavelength of 800 nm with a 1 kHz repetition rate and approximately 170 fs pulse duration. The planar sample with optical flats was moved at 0.5 mm intervals along a 30 mm optical path before and behind the focal position concentrated on by the lens (f = 200 mm). Simultaneously, the optical transmittance was measured with an optical power meter system (Newport, 2930-C/818-SL silicon photodiodes), where the diffraction length $z_R = 11.2$ mm and the beam waist radius $\omega_0 = 53$ μm. The samples were thin enough to satisfy the condition $L \ll n_o z_R$. Close and open Z-scan measurement using the femtosecond laser was applied to estimate the third-order nonlinear optical susceptibilities (the real and imaginary parts of χ^3, respectively) for the glass samples studied. To test the accuracy of our Z-scan measurement, a ZnSe plate was also examined as a reference for both the real and imaginary parts of χ^3. The materials under study were found to consist of TeO_{3+1} units, thus representing intermediate systems between the two types of $TeO_2–TlO0.5–MOY/x$ glasses: those of island-type structures (M = Pb[II], Zn[II], Ga[III], and Bi[III]) and those of framework-type structures (M = Ti[IV]). The real parts of the third-order nonlinear susceptibilities $R(\chi^3)$ of the above glasses were: 5.2 ± 0.7, 2.5 ± 0.3, 2.6 ± 0.4, 4.5 ± 0.6, and 4.4 ± 0.6, respectively, while the Raman band frequencies were 697.5, 730, 732.5, 710, and 722 cm^{-1}, respectively. It was concluded that the incorporation of TiO_2 into $Tl_2O–TeO_2$ glasses is highly favorable for glass polymerization, resulting in an increase in their third-order hypersusceptibilities. Surprisingly, the highest change in refractive index was obtained for M = Ti other than Pb and Bi, which was attributed to the formation of the TeO_2/TiO_2 glass structure without destruction of the initial three-dimensional network, accounting for a strong dielectric response intrinsic to pure TeO_2 glass.

Ardelean et al. (2010) studied both IR and Raman spectroscopies on the $xMnO–(100 – x)[As_2O_3–TeO_2]$ glassy system, with $x = 1, 3, 5, 10, 20, 35$, and 50 mol%. The IR spectra of these glasses suggest that the structure of this system is modified by the addition of MnO content. The spectrum of the glass matrix suggests a structure formed from AsO_3, TeO_4 tbp, and TeO_3 tp units. Then, with the addition of manganese ions, it can be observed that the intensities of the bands present in the spectrum of the glass matrix decrease, thus suggesting that the presence of the manganese leads to the disordering of the structure. The manganese ions enter in the structure by breaking some of the dominant bonds. For 10 mol% MnO, the structure is changing again. A new band specific to Mn–O in MnO_2 units appears, and the intensity of the bands characteristic to AsO_3, TeO_4 tbp, and TeO_3 tp units become more intense. This behavior suggests that the MnO_2 units became distinctive units in a structure formed from AsO_3, TeO_4 tbp, and TeO_3 tp units. The Raman spectrum of the glass matrix suggests a structure formed from TeO_4 tbp and TeO_3 tp units, where the presence of As–O–As bonds cannot be excluded. The addition of manganese ions leads to a disordering of the structure proposed by the spectrum of the glass matrix (behavior that is also suggested by IR measurements). This behavior was suggested by the intensity decrease of the bands. Then, with the increase in manganese content, the intensity of these bands increases, showing that the manganese content favored the formation of TeO_4 tbp and TeO_3 tp units. For the highest MnO content, these units decrease due to the disorder increase and the As–O–As bonds become finally visible.

New tellurite glasses with a large glass-forming domain were elaborated within the $(80 – x)$ $TeO_2–xTl_2O–20ZnO$ (where $x = 10, 20$, and 30 mol%) system have been achieved by Soulis (2010). The impact of the addition of both Tl_2O and ZnO on both the calorimetric and structural characteristics of the glasses was examined. The addition of each oxide modifier clearly provokes a depolymerization of the glass framework, as testified by the Raman spectroscopy data. Optical properties, such as transmission data, were measured as a function of the Tl_2O content, evidencing a red-shift of the absorption/transmission threshold with the increase in the Tl_2O content; that is, a reduction of the optical band-gap. For the tested compositions, second harmonic generation was unambiguously detected after optical poling, with a tendency for the SHG amplitude to be amplified by the Tl_2O addition. Nd^{3+}-doped alkali–niobium zinc–tellurite glasses of molar compositions: $(60 – x)$ $TeO_2–20ZnO–7.5Na_2O–7.5Li_2O–5Nb_2O_5–xNd_2O_3$ ($x = 0.1, 1.0$, and 2.0) were prepared using the conventional melting procedure by Babu et al. in 2010. Thermal properties were measured. The FTIR spectra were used to analyze the functional groups present in the glass. The Judd–Ofelt intensity parameters were derived from the absorption spectrum and used to calculate the radiative lifetime, branching ratio, and stimulated emission cross section of the $^4F_{3/2}–^4I_{9/2, 11/2, 13/2}$, transitions. The alkali–niobium zinc–tellurite glass was found to be thermally stable for rare-earth ion doping, and maximum phonon energy was found to be higher than the zinc–tellurite glasses because of the presence of the niobium oxide. The study of the saturation intensity indicates lower laser threshold power for the studied glass system than for most of the other glasses.

Zhou et al. (2010) studied the effects of TeO_x ($x = 2.6–2.8$) films on temperature coefficients of delay of Love-type wave devices. The TeO_x thin films are deposited onto several kinds of substrates including glass slides, silicon wafers, and $36°$ YX–$LiTaO_3$ crystal substrates by the RF magnetron-sputtering technique at room temperature in different ratios of oxygen and argon gases, in which a 99.99% TeO_2 powder target was used. The deposited films are smooth and transparent, with a light yellow color, and are strongly adherent. The thicknesses of the films are measured using a surface profiler, and the temperature coefficient of delay (TCD) of the Love-type wave devices was investigated. Kaur et al. (2010) prepared tellurite glasses of two systems: $xPbO–(100 – x)TeO_2$ ($x = 13$, 15, 17, 19, and 21 mol%) and $yZnO–(100 – y)TeO_2$, ($y = 18, 20, 22, 25, 30, 33$, and 35 mol%) at two melt-cooling rates and characterized these glasses by measuring their densities by UV-visible and Raman spectroscopy. The Raman measurements were carried out in an unpolarized mode, at room temperature in a backscattering geometry, in the wave number range 50–1500 cm^{-1}, with a spectral resolution of 1 cm^{-1} and the incident laser power of 10 mW. Raman scattering studies found that the addition of glass modifiers (PbO, ZnO) leads to a shift in the relative intensities and frequencies of Raman bands, which clearly indicates the continuous distortion of symmetry of TeO_4 bipyramidal trigonal units, leading to the creation of TeO_{3+1} polyhedron and/or TeO_3 trigonal units. A low-frequency peak at 76 cm^{-1} was observed in the Raman spectra of all lead and zinc–tellurite glasses.

Upender et al. (2010a), measured the Raman spectra of $(100 – x)(TeO_2)–x(WO_3)$ ($x = 10, 20$, 30, and 40 mol%), $60(TeO_2)–(40 – y)(WO_3)–yPbO$ ($y = 0, 10, 20, 30$, and 40 mol%), and $(85 – z)$ $TeO_2–z(WO_3)–15 Ag_2O$ ($z = 0, 10, 20, 30$, and 40 mol%). Raman measurements revealed the dual role of lead atoms in the ternary system $TeO_2–WO_3–PbO$, while the addition of WO_3 was found to promote the formation of TeO_3/TeO_{3+1} units and WO_6 units in all the glasses at the expense of TeO_4 units and WO_4 units. The results showed that in all the systems, the coordination state of the W ion changed from 4 to 6 when WO_3 concentration increases beyond 30 mol%. Also, Upender et al. (2010b) prepared $(90 – x)TeO_2–10GeO_2–xWO_3$ glasses with $7.5 \leq x \geq 30$ mol%. The authors measured the Raman shifts for these glasses, which were: $\delta \pm 1 = 340, 460, 666, 755$, and 930 ($cm^{-1}$) for the glass $82.5TeO_2–10GeO_2–7.5WO_3$, and $\delta \pm 1 = 380, 480, 720$, and 938 ($cm^{-1}$) for the glass $60TeO_2–10GeO_2–30WO_3$. The Raman studies showed that the glass system contains TeO_4, TeO_{3+1}/TeO_3, GeO_6, WO_4, and WO_6 groups as basic structural units. Tara et al. (2010) used the Raman spectroscopy technique to characterize the microstructure and the crystallization properties of the as-cast and heat-treated binary $TeO_2–WO_3$, $TeO_2–CdF_2$, and ternary $TeO_2–CdF_2–WO_3$ glasses and glass-ceramics. The effect of the WO_3 and CdF_2 contents on the TeO_2 glass network and the intensity ratios of the deconvoluted Raman peaks were determined. The Raman results indicated that the glasses were mainly formed by the $[TeO_4]$ and $[TeO_3]$ units. The $[TeO_4]$ units convert to $[TeO_3]$ units with the addition of WO_3 and CdF_2 into tellurite glasses. All the crystalline phases, such as α-TeO_2, δ-TeO_2, and γ-TeO_2 existing in the $TeO_2–WO_3$, $TeO_2–CdF_2$, and TeO_2–$WO_3–CdF_2$ glasses were determined. The transformation of the metastable γ TeO_2 phase into stable α-TeO_2 was observed for the $(1 – x)TeO_2–xWO_3$ (where $x = 0.15, 0.20, 0.25$), $0.90Te_2–0.10CdF_2$, $0.85TeO_2–0.10CdF_2–0.05WO_3$, and $0.80TeO_2–0.10CdF_2–0.10WO_3$ glasses, and the transformation of the metastable δ-TeO_2 phase into the stable α-TeO_2 was also observed for the $TeO_2–CdF_2–WO_3$ glass system. In addition, the authors found an unidentified phase formation, labeled ε.

Ohishi et al. (2010) investigated the effects of Raman spectrum on the relative gain flatness and the effective bandwidth using the $TeO_2–BaO–SrO–Nb_2O_5–WO_3–P_2O_5$ glass with one broad main Raman shift peak. The Brillouin gain coefficients of tellurite fiber were measured. A peak value of Brillouin gain coefficient of 1.70×10^{-10} m/W was obtained for this tellurite fiber, which is about 3.4 times larger than that (5×10^{-11} m/W) of silica fiber. The results have shown that these tellurite fibers can realize a broadband Raman gain spectra covering the S + C + L band. Highly efficient Brillouin slow-light-generation was demonstrated in a tellurite fiber. Tellurite fiber gave the highest value (~3.76 ns/mW) of the time delay per unit power ever reported. The authors have successfully fabricated dispersion-controlled tellurite microstructure fibers with high nonlinearity and obtained broadband SC spectra expanding from 474 to 2400 nm in a tellurite microstructure optical fiber

MOF. The authors have shown that the chromatic dispersion-controlled tellurite microstructure fiber can generate flattened SC spectra expanding from visible to mid-IR; that tellurite fibers have high potential as optical signal processing and coherent light source media; and they expect that tellurite fiber devices will open a new prospect for optical signal processing and coherent light generation.

In 2011, Rada et al. studied the structure of the $Fe(NO_3)_3$–TeO_2 glasses obtained using the sol-gel synthesis. The samples were characterized by x-ray diffraction, FTIR, UV-VIS, and EPR spectroscopy. The results showed that the iron ions located in the glassy matrix are largely in the form of Fe^{3+} ions. The analysis of the IR spectra indicated a gradual transformation of the iron ions from tetrahedral to octahedral sites when the concentration of $Fe(NO_3)_3$ was increased up to 0.8. The increase of $Fe(NO_3)_3$ content in the host matrix induces the growth of the number of effective g values. The authors considered that the orbitals of O^{2-} ions with a large spin-orbit interaction constant will interact with the 3d orbital of Fe^{3+} ion bonded to this O^{2-} ion, thus leading to the appearance of an orbital angular momentum, which contributes to the magnetic moment of Fe^{3+} ion. Wang et al. (2011) prepared a systematic series of $(Ge15Ga10Te75)1 - x(CsI)x$ ($x = 0$, 5,10,15 at%) far IR-transmitting chalcohalide glasses. The most fascinating and original feature of these new glasses was their exceptionally wide transparency range. The optical transmission spectra showed that the transparency of the Ge–Ga–Te–CsI glass system extends from the band gap region at 1.7 mm to the phonon region at about 25 mm, and the maximum average transmission percentage is beyond 70% at a CsI content of 15 at%.

REFERENCES

Abdel-Kader, A., Higazy, A., and Elkholy, M., *J. Mater. Sci.*, 2, 157, 1991a.
Abdel-Kader, A., Higazy, A., and Elkholy, M., *J. Mater. Sci.*, 2, 204, 1991b.
Abdel-Kader, A., El-Mallawany, R., and Elkholy, M., *J. Appl. Phys.*, 73, 71, 1993.
Abdel-Baki, M., Abdel Wahab, F., and El-Diasty, F., *Mater. Chem. Phys.*, 96, 201, 2006.
Abdel-Kader, A., El-Mallawany, R., Elkholy, M., and Farag, H., *Mater. Chem. Phys.*, 36, 365, 1994.
Abdel-Kader, A., Higazy, A., ElKholy, M., and Farag, H., *J. Mater. Sci.*, 28, 5133, 1993.
Adams, R., *Phys. Chem. Glasses*, 2, 101, 1961.
Ahmed, A., Harini, R., and Hogarth, C., *J. Mater. Sci. Lett.*, 3, 1055, 1984.
Ahmed, A., Shamy, T., and Sharaf, N., *J. Non-Cryst. Solids*, 70, 17, 1985.
Ahmed, M., Hogarth, C., and Khan, M., *J. Mater. Sci.*, 19, 4040, 1984.
Akagi, R., Handa, K., Ohtori, N., Hannon, A., Tatsumisago, M., and Umeski, N., *J. Non-Cryst. Solids*, 256, 111, 1999.
Al-Ani, S., and Hogarth, C., *J. Mater. Sci., Lett.*, 6, 519, 1987.
Al-Ani, S., Hogarth, C., and El-Mallawany, R., *J. Mater. Sci.*, 20, 661, 1985.
Amirtharaj, P., *Handbook of Optical Constants of Solids II*, D. Palik (ed.), Academic Press: San Diego, CA, 655, 1991.
Ardelean, I., Lupsor, S., and Rusu, D., *Physica B*, 405, 2259, 2010.
Arnaudov, M., Dimitriev, Y., Dimitrov, V., and Pankov, M., *Phys. Chem. Glasses*, 27, 48, 1986.
Baars, J., and Sorger, F., *Sol. Stat. Commun.*, 10, 875, 1972.
Babu, A., Jamalaiah, B., Kumar, J., Sasikala, T., and Moorthy, L., *J. Alloys Compd.*, 509, 457, 2011.
Babu, S., Rajeswari, R., Jang, K., Jin, C., Jang, K., JinSeo, H., and Jayasankar, C., *J. Lumin.*, 130, 1021, 2010.
Bahgat, A., Shaisha, E., and Sabry, A., *J. Mater. Sci.*, 22, 1323, 1987.
Barady, G., *J. Chem. Phys.*, 27, 300, 1957.
Beall, G., and Duke, D., *J. Mater. Sci.*, 4, 340, 1969.
Bendow, B., In *Experimental Techniques of Glass Science*, C. Simmons and O. El-Bayoumi (eds.), *Am. Ceram. Soc.*, Westerville, Ohio, 33, 1993.
Bersani, D., Antonioli, G., Lottici, P., Dimitriev, Y., Dimitrov, V., and Kobourova, P., *J. Non-Cryst. Solids*, 232, 293, 1998.
Berzelius, J., *Ann. Phys. Chem.*, 32, 577, 1834.
Blair, G., Merry J., and Wylot, J., U.S. Patent No. 4265517, 1981.
Bobovich, Y., and Yakhkind, A., *J. Struct. Chem.*, 4, 851, 1963.

Boiteux, S., Segonds, P., Canioni, L., Sarger, L., Cardinal, T., Duchesne, C., Fragin, E., and Flem, G., *J. Appl. Phys.*, 81, 1481, 1997.

Borrelli, N., Aitken, B., Newhous, M., and Hall, D., *J. Appl. Phys.*, 70, 2774, 1991.

Burger, H., Kenipp, K., Hobert, H., Vogel, W., Kozhukarov, V., and Neov, S., *J. Non-Cryst. Solids*, 151, 134, 1992.

Burger, H., Vogal, W., and Kozhukarov, V., *Infrared Phys.*, 25, 395, 1985.

Chen, Y., Nie, Q., Xu, T., Dai, S., Wang, X., and Shen, X., *J. Non-Cryst. Solids*, 354, 3468, 2008.

Chen, F., Xu, T., Dai, S., Nie, Q., Shen, X., Zhang, J., and Wang, X., *Opt. Mater.*, 32, 868, 2010.

Cheng, J., Chen, W., and Ye, D., *J. Non-Cryst. Solids*, 184, 124, 1995.

Cheremisinov, V., and Zlomanov, V., *Opt. I Spectrsk.*, 12, 110, 1962.

Cherukuri, S., Joseph, I., and Pye, D., In *Experimental Techniques of Glass Science*, C. Simmons and O. El-Bayoumi (eds.), *Am. Ceram. Soc.*, Westerville, Ohio, 185, 1993.

Chiaruttini, I., Fonteneau, G., Zhang, X., and Lucas, J., *J. Non-Cryst. Solids*, 111, 77, 1989.

Chopa, N., Mansingh, A., and Chadha, G., *J. Non-Cryst. Solids*, 126, 194, 1990.

Congshan, Z., Xiaojuan, L., and Zuyi, Z., *J. Non-Cryst. Solids*, 144, 89, 1992.

Cuevas, R., Barbosa, L., Paula, A., Lui, Y., Reynoso, V., Alves, O., Arnha, N., and Cesar, C., *J. Non-Cryst. Solids*, 191, 107, 1995.

Davis, E., and Mott, N., *Phil. Mag.*, 22, 903, 1970.

Denney, R., and Sincleair, R., *Visible and UV Spectra*, Mowthorpe, D. (ed.), John Wiley & Sons: New York, 1987.

Desirena, H., Schulzgen, A., Sabet, S., Ramos-Ortiz, G., E. de la Rosa, E., and Peyghambarian, N., *Opt. Mater.*, 31, 784, 2009.

Dexpert-Ghys, J., Piriou, B., Rossignol, S., Reau, J., Tanguy, B., Videau, J., and Portier, J., *J. Non-Cryst. Solids*, 170, 167, 1994.

Dimitriev, Y., Dimitrov, V., and Arnaudov, M., *J. Mater. Sci.*, 14, 723, 1979.

Dimitriev, Y., Dimitrov, V., Gatev, E., Kashchieva, E., and Petkov, H., *J. Non-Cryst. Solids*, 95/96, 937, 1987.

Dimitriev, Y., *Glass Sci. Technol.*, 67, 488, 1994.

Dimitrov, V. and Komatsu, T., *J. Non-Cryst. Solids*, 249, 160, 1999.

Dimitrov, V., and Sakka, S., *J. Appl. Phys.*, 79, 1736, 1996.

Dimitrov, V., *J. Sol. Stat. Chem.*, 66, 256, 1987.

Dimitrov, V., Komatsu, T., and Sato, R., *J. Ceram. Soc. Japan*, 107, 21, 1999.

Dimitrov, V., Arnaudov, M., and Dimitriev, Y., *Monatshefte Chem.*, 115, 987, 1984.

Dislich, H., Hinz, P., Arfdten, N., and Hussmann, E., *Glastech. Berlin*, 62, 46, 1989.

Donnell, M., Seddon, A., Furniss, D., Tikhomirov, V., Rivero, C., Ramme, M., Stegman, R., Stegman, G., Richardson, K., Stolen, R., Couzi, M., and Cardinal, T., *J. Am.Ceram. Soc.*, 90, 1448, 2007.

Dow, J., and Redfield, D., *Phys. Rev. B*, 5, 597, 1972.

Duclère, J., Lipovskii, A., Mirgorodsky, A., Thomas, P., Tagantsev, D., and Zhurikhina, V., *J. Non-Cryst. Solids*, 355, 2195, 2009.

Duffy, J., *Phys. Chem. Glasses*, 30, 1, 1989.

Efimov, A., *J. Non-Cryst. Solids*, 209, 209, 1997.

Efimov, A., *Optical Constants of Inorganic Glasses*, CRC Press Inc.: Boca Raton, FL, 1995.

Elliott, S., In *Chalcogenide Glasses in Glasses and Amorphous Materials*, J. Zarzycki (ed.), VCH Publishers Inc.: Weinheim, New York, 1991.

El-Mallawany, R., Abdalla, M., and Ahmed, I.A., *Mater. Chem. Phys.*, 109, 291, 2008.

El-Mallawany, R., *Infrared Phys.*, 29, 781, 1989.

El-Mallawany, R., *J. Appl. Phys.*, 73, 4878, 1993.

El-Mallawany, R., *J. Appl. Phys.*, 72, 1774, 1992.

El-Mallawany, R., *Mater. Sci. Forum*, 67/68, 149, 1991.

El-Mallawany, R., *Phys. Non-Cryst. Solids*, L. Pye, W. LaCourse, and H. Stevens (eds.), Taylor & Francis: London, 276, 1992.

El-Samadony, M., Sabry, A., Shaisha, E., and Bahgat, A., *Phys. Chem. Glasses*, 32, 115, 1991.

Eraiahe, B., *Bull. Mater. Sci.*, 29, 375, 2006.

Eyzaguirre, C., Rodriguez, E., Chillcce, E., Osorio, S., Cesar, C., and Barbosa, L., *J. Am. Ceram. Soc.*, 90, 1822, 2007.

Fagen, E., and Fritzsche, H., *J. Non-Cryst. Solids*, 2, 180, 1970.

Fargin, E., Berthereau, A., Cardinal, T., Flem, G., Ducasse, L., Canioni, L., Segonds, P., Sarger, L., and Ducasse, A., *J. Non-Cryst. Solids*, 203, 96, 1996.

Ferreira, B., Fargin, E., Guillaume, B., Le Flem, G., Rodriguez, V., Couzi, M., Buffeteau, T., Canioni, L., Sarger, L., Martinelli, G., Quiquempois, Y., Zeghlache, H., and Carpentier, L., *J. Non-Crys. Solids*, 332, 207, 2003.

Fleming, J., In *Experimental Techniques of Glass Science*, C. Simmons and O. El-Bayoumi (eds.), *Am. Ceram. Soc.*, London, Ohio, 185, 1993.

Gebavi, H., Milanese, D., Liao, G., Chen, Q., Ferraris, M., Ivanda, M., Gamulin, O., and Taccheo, S., *J. Non-Cryst. Solids*, 355, 548, 2009.

George, V. and McIntyre, P., In *Infrared Spectroscopy*, Mowthorpe, D. (ed.), John Wiley & Sons: New York, 1987, p. 68.

Ghosh, G., *J. Am. Ceram. Soc.*, 78, 2828, 1995.

Greenwood, N., and Earnshaw, A., *Chemistry of the Elements*, 2nd Ed., Butterworth-Heinemann: Linacre House, Jordan Hill, Oxford, 1997.

Guojun Gao, G., Wang, G., Yu, C., Zhang, J., Hu, L., *J. Lumin.*, 129, 1042, 2009.

Hager, I., El-Mallawany, R., and Bulou, A., *Physica B: Condensed Matter*, 406, 4, 972, 2011.

Hager, Z., and El-Mallawany, R., *J Mater Sci.*, 45, 897, 2010.

Harani, R., Hogarth, C., Ahmed, M., and Morris, D., *J. Mater. Sci. Lett.*, 3, 843, 1984.

Hassan, M., and Hogarth, C., *J. Mater. Sci.*, 23, 2500, 1988.

Hassan, M., Khleif, W., and Hogarth, C., *J. Mater. Sci.*, 24, 1607, 1989.

Hayakawa, T., Koduka, M., Nogami, M., Duclère, J., Mirgorodsky, A., and Thomas, Ph., *Scripta Materialia*, 62, 806, 2010.

Hayakawa, T., Hayakawa, M., Nogami, M., and Thomas, P., *Opt. Mater.*, 32, 448, 2010.

Heckroodt, R., and Res, M., *Phys. Chem.*, 17, 217, 1976.

Herzfeld, K., *Phys. Rev.*, 29, 701, 1927.

Hilton, A., *J. Non-Cryst. Solids*, 2, 28, 1970.

Hirao, K., Kishimoto, S., Tanaka, K., Tanabe, S., and Soga, N., *J. Non-Cryst. Solids*, 139, 151, 1992.

Hirayama, C., and Lewis, D., *Phys. Chem. Glasses*, 5, 44, 1964.

Hogarth, C., and Kashani, E., *J. Mater. Sci.*, 18, 1255, 1983.

Hong, Ma, Zhang, X., and Lucas, J., *J. Non-Cryst. Solids*, 135, 49, 1991.

Hu, L., and Jiang, Z., *Phys. Chem. Glasses*, 37, 19, 1996.

Hussain, N., Annapurna, K., and Buddhudu, S., *Phys. Chem. Glasses*, 38, 51, 1997.

Ikushima, A., *J. Non-Cryst. Solids*, 178, 1, 1994.

Ilieva, D., Dimitrov, V., Dimitriev, Y., Bogachev, G., and Krastev, V., *Phys. Chem. Glasses*, 39, 241, 1998.

Inoue, S., Nukui, A., Yamamoto, K., Yano, T., Shibata, S. and Yamane, M., *J. Non-Cryst. Solids*, 324, 133, 2003.

Inoue, S., Shimizugawa, Y., Nuki, A., and Maeseto, T., *J. Non-Cryst. Solids*, 189, 36, 1995.

Inoue, S., Shimizugawa, Y., Nukui, A., Maeseto, T., *J. Non-Cryst. Solids*, 189,36, 1995.

Ivanova, I., *J. Mater. Sci.*, 25, 2087, 1990.

Izumitani, T., and Namiki, K., U.S. Patent No. 3,661,600, 1972.

Jeansannetas, B., Blanchandin, S., Thomas, P., Marchet, P., Mesjard, J., Mejean, T., Frit, B., Nazabal, V., Fargin, E., Flem, G., Martin, M., Bousquet, B., Canioni, L., Boiteux, S., Segonds, P., and Sarger, L., *J. Sol. Stat. Chem.*, 146, 329, 1999.

Jose, R., Suzuki, T., and Ohishi, Y., *J. Non-Cryst. Solids*, 352,5564, 2006.

Judd, B., *Phys. Rev.*, 127, 750, 1962.

Kakiuchida, H., Sekiya, E., Shimodaira, N., Saito, K., and Ikushima, A., *J. Non-Cryst. Solids*, 353, 568, 2007.

Kaur, A., Khanna, A., Pesquera, C., Gonzalez, F., and Sathe, V., *J. Non-Cryst. Solids,* 356, 864, 2010.

Kenipp, K., Burger, H., Fassler, D., and Vogel, W., *J. Non-Cryst. Solids*, 65, 223, 1984.

Khan, M., *Phys. Stat. Solids*, 117, 593, 1990.

Kim, H., and Komatsu, T., *J. Mater. Sci. Lett.*, 17, 1149, 1998.

Kim, H., Komatsu, T., Sato, R., and Matusita, K., *J. Ceram. Soc. Japan*, 103, 1073, 1995.

Kim, H., Komatsu, T., Sato, R., and Matusita, K., *J. Mater. Sci.*, 31, 2159, 1996a.

Kim, H., Komatsu, T., Shioya, K., Mazumasa, M., Tanaka, K., and Hirao, K., *J. Non-Cryst. Solids*, 208, 303, 1996b.

Kim, S., and Yoko, T., *J. Am. Ceram. Soc.*, 78, 1061, 1995.

Kim, S., Yoko, T., and Sakka, S., *J. Am. Ceram. Soc.,* 76, 2486, 1993a.

Kim, S., Yoko, T., and Sakka, S., *J. Am. Ceram. Soc.*, 76, 865, 1993b.

Klaska, P., Zhang, X., and Lucas, J., *J. Non-Cryst. Solids*, 161, 297, 1993.

Komatsu, T., Kim, H., and Mohri, H., *J. Mater. Sci. Lett.*, 15, 2026, 1996.

Komatsu T., Onuma J., Kim H., and Kim J., *J. Mater. Sci. Lett.*, 15, 2130, 1996.

Komatsu, T., and Mohira, H., *Phys. Chem. Glasses*, 40, 257, 1999.

Komatsu, T., Kim, H., and Oishi, H., *Inorgan. Mater.*, 33, 1069, 1997b.

Komatsu, T., Shioya, K., and Kim, H., *Phys. Chem. Glasses*, 38, 188, 1997a.

Komatsu, T., Tawarayama, H., and Matusita, K., *J. Ceram. Soc. Japan*, 101, 48, 1993.

Komatsu, T., Tawarayama, H., Mohri, H., and Matusita, K., *J. Non-Cryst. Solids*, 135, 105, 1991.

Kosuge, T., Benino, Y., Dimitrov, V., Sato, R., and Komatsu, T., *J. Non-Cryst. Solids*, 242, 154, 1998.

Kurik, M., *Phys. Stat. Solids* 8, 9, 1971.

Kutub, A., Osman, A., and Hogarth, C., *J. Mater. Sci.*, 21, 3571, 1986.

Lasbrugnas, C., Thomas, Ph. Masson, O., Champarnaud-Mesjard, J., Fargin, E., Rodriguez, V., and Lahaye, M., *Opt. Mater.*, 31, 775, 2009.

Lezal, D., Kasik, I., and Gotz, J., *J. Non-Cryst. Solids*, 90, 557, 1987.

Liao, G., Chen, Q., Xing, J., Gebavi, H., Milanese, D., Fokine, M., and Ferraris, M., *J. Non-Cryst. Solids*, 355, 447, 2009.

Lin, J., Huang, W., Sun, Z., Ray, C., and Day, D., *J. Non-Cryst. Solids*, 336, 189, 2004.

Lines, M., *J. Appl. Phys.*, 69, 6876, 1991.

Lines, M., *Phys. Rev. B*, 41, 3383, 1990.

Lipson, H., Burkmelter, J., and Dugyer, C., *J. Non Cryst. Solids*, 17, 27, 1975.

Liu, Y., Reynoso, V., Barbosa, L., Rojas, R., Frangnito, H., Cesar, C., and Alves, O., *J. Mater. Sci. Lett.*, 14, 635, 1995.

Lucas, J., and Zhang, X., *J. Non-Cryst. Solids*, 125, 1, 1990.

Lucas, J., Chiaruttini, I., Zhang, X., and Fonteneau, G., *Mater. Sci. Forum*, 32/33, 437, 1988.

Maciel, G., Rakov, N., Arajo, C., Lipovskii, A., and Tagantsev, D., *Appl. Phys. Lett.*, 79, 584, 2001.

Malik, M., and Hogarth, C., *J. Mater. Sci. Lett.*, 8, 649, 1989a.

Malik, M., and Hogarth, C., *J. Mater. Sci. Lett.*, 8, 655, 1989b.

Malik, M., and Hogarth, C., *J. Mater. Sci.*, 25, 116, 1990.

Mandouh, Z., *J. Appl. Phys.*, 78, 7158, 1995.

Marchese, D., De Sario, M., Jha, A., Kar, A., and Smith, E., *Conference on Infrared Glass Optical Fibers and Their Applications*, 3416, 177, 1998.

Massera, J., Haldeman, A., Milanese, D., Gebavi, H., Ferraris, M., Foy, P., Hawkins, W. Ballato, J., Stolen, R., Petit, L., and Richardson, K., *Opt. Mater.*, 32, 582, 2010.

McDougall, J., Hollis, D., and Payne, M., *Phys. Chem. Glasses*, 37, 73, 1996.

Mirgorodsky, A., Merle-Mejean, T., and Frit, B., *J. Phys.Chem. Solids*, 61, 501, 2000.

Miyakawa, T., and Dexter, D., *Phys. Rev. B*, 1, 2961, 1970.

Mizuno, Y., He, Z., and Hotate, K., *Opt. Commun.*, 283, 2438, 2010.

Mizuno, Y., Ikeda, M., and Yoshida, A., *J. Mater. Sci. Lett.*, 11, 1653, 1992.

Mochida, N., Sekiya, T., Ohtsuka, A., and Tonokawa, M., *J. Ceram. Soc. Japan*, 96, 973, 1988.

Mochida, N., Takahshi, K., and Nakata, K., *Yogyo-Kyokai-Shi*, 86, 317, 1978.

Montant, S., Freysz, E., and Couzi, M., *Opt. Commun.*, 281, 769, 2008.

Mori, A., Ohishi, Y., and Sudo, S., *Electron. Lett.*, 33, 863, 1997.

Mott N., and Davis, E., *Electronic Process in Non-Crystalline Materials*, 2nd Ed., Oxford, UK, Clarendon Press, 1979.

Mott, N., *J. Non-Cryst. Solids*, 1, 1, 1969.

Murata, T., Takebe, H., and Morinaga, K., *J. Am. Ceram. Soc.*, 81, 1998.

Muruganandam, K., and Swshasayee, M., *J. Non-Cryst. Solids*, 222, 131, 1997.

Nakai, T., Noda Y., Tani, T., Mimura, Y., Sudo, S., and Ohno, S., *OSA TOPS*, 25, 82, 1998.

Narazaki, A., Tanaka, K., Hirao, K., and Soga, N., *J. Appl. Phys.*, 83, 3986, 1998.

Narazaki, A., Tanaka, K., Hirao, K., and Soga, N., *J. Appl. Phys.*, 85, 2046, 1999.

Nayak, R., Gupta, V., Dawar, A., Sreenivas, K., *Thin Solid Films*, 445, 118, 2003.

Neindre, L., Smektala, F., Foulgoc, K., Zhang, X., and Lucas, J., *J. Non-Cryst. Solids*, 242, 99, 1995.

Noguera, O., and Suehara, S., *J. Non-Cryst. Solids*, 354, 188, 2008.

Oermann, M., Ebendorff-Heidepriem, H., Li, Y., Foo, T., and Monro, T., *Opt. Express*, 17, 15578, 2009.

Ofelt, G., *J. Chem. Phys.*, 37, 511, 1962.

Ohishi, Y., Mori, A., Yamada, M., Onon, H., Nishida, Y., and Oikawa K., *Opt. Lett.*, 23, 97, 1998.

Ohishi, Y., Qin, G., Liao, Yan, M., Suzuki, T.; *Optical Fiber Communication (OFC), Collocated National Fiber Optic Engineers Conference, 2010 Conference on (OFC/NFOEC)*, 1–3, 2010.

Oishi, H., Benino, Y., and Komatsu, T., *Phys. Chem. Glasses*, 40, 212, 1999.

Oshishia, Y., Mitachi, S., Kanamori, T., and Manabe, T., *Phys. Chem. Glasses*, 24, 135, 1983.

Ovcharenko, N. and Sairnova, T., *J. Non-Crys. Solids*, 291, 121, 2001.

Ovcharenko, N., and Yakhkind, A., *Optic. Technol.*, 38, 163, 1972.

Ovcharenko, N., and Yakhkind, A., *Sov. J. Opt. Technol.*, 35, 192, 1968.

Ozdanova, J., Helena Ticha, H., and Tichy, L., *Opt. Mater.*, 32, 950, 2010.

Pankova, M., Dimitriev, Y., Arnaudov, M., and Dimitrov, V., *Phys. Chem. Glasses*, 30, 260, 1989.

Prasad,N., Annapurna, K., K., Sooraj Hussain, N., and Buddhudu, S., *Mater. Lett.*, 57, 2071, 2003.

Rada, S., Dehelean, A., Stan, M., Chelcea, R., and Culea, E., *J. Alloys Compd.*, 509, 147, 2011.

Rakhshani, A., *J. Appl. Phys.*, 81, 7988, 1997.

Rasifield, R., *Struct. Bonding*, 13, 53, 1973.

Reddu, C., Ahammed, Y., Reddy, R., and Rao, T., *J. Phys. Chem. Solids*, 59, 337, 1998.

Redman, M., and Chen, J., *J. Am. Ceram. Soc.*, 50, 523, 1967.

Reisfeld, R., Hormadaly, J., and Nuranevich, A., *Chem Phys. Lett.*, 38, 188, 1976.

Reisfeld, R., *J. Phys. Chem. Solids,* 34, 1467, 1973.

Rivera,V., Chillcce, E., Rodriguez, E., Cesar, C., and Barbosa, L., *J. Non-Cryst. Solids*, 352, 363, 2006.

Rockstad, H., *J. Non-Cryst. Solids*, 2, 192, 1970.

Rolli, R., Montagna, M., Chaussedent, S., Monteil, A., Tikhomirov, V., and Ferrari, M., *Opt. Mater.*, 21, 743, 2003.

Romanowski, W., Golab, S., Cichosz, L., and Trazebiatowska, B., *J. Non-Cryst. Solids*, 105, 295, 1988.

Rong, Q., Osaka, A., Nanba, T., Takada, J., Miura, Y., *J. Mater. Sci.*, 27, 3793, 1992.

Rossignol, S., Reau, J., Tanguy, B., Videau, J., Portier, J., Ghys, J., and Piriou, B., *J. Non-Cryst. Solids*, 162, 244, 1993.

Sabadel, J., Armand, P., Herreillat, D., Baldeck, P., Doclot, O., Ibanez, A., and Philippot, E., *J. Sol. Stat. Chem.*, 132, 411, 1997.

Sabry, A., and El-Samadony, M., *J. Mater. Sci.*, 30, 3930, 1995.

Sabry, A., Bahagat, A., and Kottamy, M., *J. Mater. Sci. Lett.*, 7, 1, 1988.

Sahar, M., and Noordin, N., *J. Non-Cryst. Solids*, 184, 137, 1995.

Sakida, S., T. Nanba, T., and Miura, Y., *Mater. Lett.*, 60, 3413, 2006.

Saleem, S., Jamalaiah, B., Jayasimhadri, M., Rao, A., Jang, K., and Moorthy, L., *J. Quant. Spectro. Rad. Trans.*, 112, 78, 2011.

Salem, Sh.M., *J. Alloys Compd.*, 503, 242, 2011.

Sekiya, T., Mochida, N., and Ogawa, S., *J. Non-Cryst. Solids*, 176, 105, 1994b.

Sekiya, T., Mochida, N., and Ogawa, S., *J. Non-Cryst. Solids*, 185, 135, 1995a.

Sekiya, T., Mochida, N., and Ohtsuka, A., *J. Non-Cryst. Solids*, 168, 106, 1994a.

Sekiya, T., Mochida, N., and Soejima, A., *J. Non-Cryst. Solids*, 191, 115, 1995b.

Sekiya, T., Mochida, N., Ohtsuka, A., and Tonokawa, M., *J. Non-Cryst. Solids*, 144, 128, 1992.

Sekiya, T., Mochida, N., Ohtsuka, A., and Tonokawa, M., *Nippon Seramikkusu Kyokai Gakujutsu Rombunshi*, 97, 1435, 1989.

Serényi, M., Lohner, T., Petrik, P., Zolnai, Y., and Khnh, N., *Thin Solid Films*, 516, 8096, 2008.

Shaltout, I., Tang, Y. I., Braunstein, R., and Abu-Elazm, A., *J. Phys. Chem. Solids*, 56, 141, 1995.

Shanker, M., and Varma, K., *J. Non-Cryst. Solids*, 243, 192, 1999.

Sheik-Bahae, M., Said, A., Wei, T., and Vanstryland, E., *I.E.E.J. Quantum Electron.*, QE-26, 760, 1990.

Shen, S., Naftaly, M., and Jha, A., *Opt. Commun.*, 205, 101, 2002.

Shioya, K., Komatsu, T., Kim, H., Sato, R., and Matusita, K., *J. Non-Cryst. Solids*, 189, 16, 1995.

Shivachev, B., Petrov,T., Yoneda, H., Titorenkova, R., and Mihailova, B., *Scripta Materialia*, 61, 493, 2009.

Sidebottom, D., Mruschka, M., Potter, B., and Brow, R., *J. Non-Cryst. Solids*, 222, 282, 1997.

Sigel, G., *Treatise on Materials Science & Technology*, Vol. 12, M. Tomozawa (ed.), Academic Press: London, 5–89, 1977.

Simmons, J., Chen, D., Ochoa, R., and EIbayoumi, O., *The Physics of Non-Crystalline Solids*, L. Pye, W. LaCourse, and H. Stevens (eds.), Taylor & Francis: London, 517, 1992.

Smith, H., and Cohen, A., *Phys. Chem. Glasses*, 4, 173, 1963.

Sokolov, V., Plotnichenko, V., Koltashev, V., *J. Non-Cryst. Solids*, 355, 1574, 2009.

Sooraj, N., Hungerford G., El-Mallawany, R., Gomes, M., Lopes, M., Ali, N., Santos, J., and Buddhudu, S., *J. Nanosci. Nanotechnol.*, 8, 1, 2008.

Soulis, M., Duclère, J., Hayakawa, T., Couderc, V., Colas, M., and Thomas, P., *Mater Res. Bull.*, 45, 557, 2010.

Stanworth, J., *J. Soc. Glass Technol.*, 36, 217, 1952.

Stanworth, J., *J. Soc. Glass Technol.*, 38, 425, 1954.

Strehlow, W., and Cook, E., *J. Phys. Chem. Ref. Data*, 2, 163, 1973.

Sugimoto, N., Kanbara, H., Fujiwara, S., Tanaka, K., Y. Shimizugawa, Y., and Hirao, K., *J. Opt. Soc. Am. B*, 16, 1904, 1999.

Swanepeol, R., *J. Phys. E* 16, 1214, 1983.

Takabe, H., Fujino, S., and Morinaga, K., *J. Am. Ceram. Soc.*, 77, 2455, 1994.

Tanabe, S., and Hanada, T., *J. Appl. Phys.*, 76, 3730, 1994.

Tanaka, K., Kashima, K., Hirao, K., Soga, N., Mito, A., and Nasu, H., *J. Non-Cryst. Solids*, 185, 123, 1995.

Tanaka, K., Kashima, K., Kajihara, K., Hirao, K., Soga, N., Mito, A., and Nasu, H., *SPIE*, 2289, 167, 1994.

Tanaka, K., Narazaki, A., Hirao, K., and Soga, N., *J. Appl. Phys.*, 79, 3798, 1996a.

Tanaka, K., Narazaki, A., Hirao, K., and Soga, N., *J. Non-Cryst. Solids*, 203, 49, 1996b.

Tanaka, K., Narazaki, A., and Hirao, K., *Opt. Lett.*, 25, 251–253, 2000.

Tanaka, K., Yoko, T., Yamada, H., and Kamiya, K., *J. Non-Cryst. Solids*, 103, 250, 1988.

Tantarintsev, B., and Yakhkind, A., *Sov. J. Opt. Technol.*, 39, 654, 1972.

Tantarintsev, B., and Yakhkind, A., *Sov. J. Opt. Technol.*, 42, 158, 1975.

Tatar, D., Ozen, G., Erim, F., and Ovecoglu, M., *J. Raman Spectroscopy*, 41, 797, 2010.

Tatsumisago, M., Kato, S., Minami, T., and Kowada, Y., *J. Non-Cryst. Solids*, 192/193, 478, 1995.

Tatsumisago, M., Lee, S., Minami, T., and Kowada, Y., *J. Non-Cryst. Solids*, 177, 154, 1994.

Ticha, H., and Tichy, L., *J. Optoelectron. Adv. Mater.*, 4, 896, 2002.

Trnovcova, V., Pazurova, T., Sramkova, T., and Lezal, D., *J. Non-Cryst. Solids*, 90, 561, 1987.

Ulrich, D., *J. Am. Ceram. Soc.*, 47, 595, 1964.

Upender, G., Vardhani, C., Suresh, S, Awasthi, A., and Mouli, V., *Mater. Chem. Phys.*, 121, 335, 2010b.

Upender, G., Sathe, V., and Mouli, V, *Physica B.*, 405, 1269, 2010a.

Urbach, F., *Phys. Rev.*, 92, 1324, 1953.

Varshneya, N., *Fundamentals of Inorganic Glasses*, Academic Press: San Diego, 455–505, 1994.

Vithal, M., Nachimuthu, P., Banu, T. and Jagannathan, R., *J. Appl. Phys.*, 81, 7922, 1997.

Vogel, W., Burger, H., Muller, B., Zerge, G., Muller, W., and Forkel, K., *Silikattechnel*, 25, 6, 1974.

Wang, G., Dai, S., Zhang, J., Yang, J., and Jiang, Z., *J. Mater. Sci.*, 42, 747, 2007.

Wang, G., Nie, Q., Barj, M., Wang, X., Dai, S., Shen, X., Xu, T., and Zhang, X., *J. Phys. Chem. Solids*, 72, 5, 2011.

Wang, J., Vogel, E., Snitzer, E., Jackel, J., Silva, V., and Silberg, Y., *J. Non-Cryst. Solids*, 178, 109, 1994.

Wang, Y., Dai, S., Chen, F., Xu, T., and Nie, Q., *Mater. Chem. Phys.*, 113, 407, 2009.

Wang, Z., Tu, C., Li, Y., and Chen, Q., *J. Non-Cryst. Solids*, 191, 132, 1995.

Wang, G., Xu, S., Dai, S., Zhang, J., and Jiang, Z., *J. Alloys Compd.*, 373, 246, 2004.

Weber, M., Myers, J., and Blackburn, D., *J. Appl. Phys.*, 52, 2944, 1981.

Weissenberg, G., Marburg, L., Meinert, N., Munchhausen, M., and Marbug, K., U.S. Patent No. 2,763,559, 1956.

Weng, L., Hodgson, S., and Ma, J., *J. Mater. Sci. Lett.*, 18, 2037, 1999.

Wood, R., *Optical Materials*, Cambridge University Press: Cambridge, UK, 1993.

Xu, J., Yang, R., Chen, Q., Jiang, W., and Ye, H., *J. Non-Cryst. Solids*, 184, 302, 1995.

Xu, T.F., Shen, X., Nie, Q., and Gao, Y., *Opt. Mater.*, 28, 241, 2006.

Yakhkind, A., and Chebotarev, S., *Fiz. I Khim. Stekla*, 6, 485, 1980.

Yakhkind, A., Ovcharenko, N., and Semenov, D., *Opt. Glass*, 35, 317, 1968.

Yakhkind, A., and Loffe, B., *Sov. J. Opt. Technol.*, 33, 1, 1966.

Yakhkind, A., *J. Am. Ceram. Soc.*, 49, 670, 1966.

Yamamoto, H., Nasu, H., Matsuoka, J., and Kamiya, K., *J. Non-Cryst. Solids*, 170, 87, 1994.

Yang, J., Dai, S., Zhou, Y., Wen, L., Hu, L., Jiang, Z., *J. Appl. Phys.*, 93, 977, 2003.

Yasui, I., and Utsuno, F., *Feature*, 72, 65, 1993.

Yu, B., Zhu, C., Gan, F., *J. Appl. Phys.*, 82, 4532,1997.

Zeng, Z., *Kuei Suan Yen Hsueh Pao* (Chinese), 9, 228, 1981.

Zhang, X., Blanchetiere, C., and Lucas, J., *J. Non-Cryst. Solids*, 161, 327, 1993.

Zhang, X., Fonteneau, G., and Lucas, J., *Mat. Res. Bull.*, 23, 59, 1988.

Zhenhua, L., and Frischat, G., *J. Non-Cryst. Solids*, 163, 169, 1993.

Index

A

Absorption, 423
Acoustic relaxation properties of tellurite glasses
    low-coordination-bond networks, 153
    ultrasonic wave velocity, 154
    ultrasound absorption, 154
Acousto-optical properties of tellurite glasses
    acoustic-wave flux intensity, 171–172
    Akhieser acoustic loss, 173–174
    Dixon–Cohen method, 171
    figure of merit, 174–175
    frequency dependence, 172–173
    phonon–phonon interaction, 173–174
    polarization dependence, 172
    sample–referent bonding, 171–172
    specific heat capacity, 173
    temperature dependence, 173
    thermal conductivity, 173
    thermal conductivity and specific heat capacity, 172
    Woodruff–Ehrenreich equation, 174
Additivity, 369
Advanced rheometric expansion system (ARES), 254
Alkali–tungsten–tellurite glass
    Beck's threeparameter hybrid method, 64
    decomposition process, 64
    electronic structure of, 64
Al_2O_3-doped and La_2O_3-doped sodium tellurite glasses, 65
American Society for Testing and Materials, 272
Amorphous chalcogenide films, 451
Anderson-Stuart (A-S) model, 285, 312
ARES. *See* Advanced rheometric expansion system (ARES)
Assyrian hieroglyphic text, 367

B

Barium–tellurite glasses, 37
Base-metal electrode multilayer ceramic capacitors (BME-MLCCs), 354–355
Beam-bending viscometer (BBV), 253
BFS. *See* Brillouin frequency shift (BFS)
Bhatia–Singh's (BS) parameters, 147
Binary and ternary tellurite glasses. *See also* Elasticity and tellurite glasses
    experimental low-temperature properties, 157–158
    gamma-radiation, 120–121
    optical absorption, 440
    oxygen density, 135
    TeO_2–MoO_3, 111–112
    TeO_2–V_2O_5, 115–117
    TeO_2–V_2O_5–Ag_2O, TeO_2–V_2O_5–CeO_2, and TeO_2–V_2O_5–ZnO, 117–120
    TeO_2–WO_3 and TeO_2–$ZnCl_2$, 109–111
    TeO_2–ZnO, 112–115

Binary tellurium oxide glasses, glass-forming ranges
    amorphous thinfilm samples, 8, 10
        distribution of elements, 10–11
        glass produced, 10
        monotectic temperature, 10
        phase diagram and immiscibility, 11–13
        XRD, 9–10
    density and molar volume, 38
    interatomic distances in, 12
        atomic arrangement, 15
        liquification, initial stages, 13
        structure models, 13–17
    ion radius and bond energy, comparison of, 16
    molar and mean atomic volumes, 39–40
    nearest coordination polyhedra, bonding in, 43
    rare-earth metal oxides, 8
    thermal expansion and infrared (IR) spectra, 8
    thin-film synthesis, 8
    TMOs, 8
Born–Lande type of potential, 129–130
Brillouin amplification in tellurite fiber, 70
Brillouin frequency shift (BFS), 417
Brillouin optical correlation-domain reflectometry (BOCDR), 417
Brillouin scattering, 459
Bulk compression model
    Avogadro's number, 89
    bond lengths and angles, 87–90
    bond-stretching force constants, 87–90
    correlation coefficient, 90
    elastic strain, 87–88
    network bonds, 88–89
    three-dimensional network, 88–89

C

Capacitance bridge methods
    capacitive and resistance bridges, 328
    dielectric loss, 328–329
    electrodes, 328
    Hewlett–Packard bridge method, 328–329
    permittivity measurement, 328–329
CBH. *See* Correlated-barrier-hopping (CBH) model
Central force model
    bond energy, 91–92
    deformation potentials, 91–92
    elongation factor, 92
    equilibrium interatomic separation, 91–92
    linear arrangement, 91–92
    longitudinal vibrations, 91–92
    phenomenological theory, 91
    potential energy, 91–92
    transverse vibrations, 91–92
Chalcogenides
    amorphous solids semiconductors, 24

501

classification scheme for, 25
  field of, 25
 mid-IR region, 23
Chalcohalides, 23–24. *See also* Nonoxide-tellurite Glasses
Chen's formula, 256
Clausius–Mossotti equation, 382
Cohen–Fritzsche–Ovshinsky model, 281–282
Compressibility of tellurite glasses, 110
Conduction band minimum (CBM), 454
Copper–constantan (type-T) thermocouple, 272
Copper–tellurite glasses, 158
Correlated-barrier-hopping (CBH) model, 289, 311, 323–324
Couplings
  sound transmission, 93
  test surface and specimen, 93
  types, 93
  ultrasonic energy, 93
Crystallization activation energies of glass, 201
  Johnson–Mehl–Avrami formula, 232–233
  Kissinger formula, 229–230
  Lasocka formula, 229
  Moynihan et al. formula, 230–231
  Ozawa–Chen equation, 234
  thermal stability, 228
Czochrlski method, 97

D

Debye theory, 178, 181
Department of Silicate Technology, 269
Dielectric constant models
  charge density, 331
  conduction losses, 332
  deformation and vibrational losses, 332
  dipole orientations and space-charge contributions, 331–332
  dipole relaxation losses, 332
  electric displacement, 331–332
  electric flux, 331
  electrode materials, 331–332
  energy losses, 332
  polarization and relaxation process
    instantaneous value, 335–336
    transient effects, 335–336
  pressure models, 136–137
  relaxation phenomena
    charge surface density, 333
    Debye equations, 334
    displacement field, 333–335
    electric field strength, 333–335
    electric modulus, 334–335
    loss and power factor, 334
    polarization of, 333
    stretching exponent parameter, 334–335
  temperature and composition
    Clausius–Mossotti equation, 336
    polarization factor, 336
Dielectric properties
  absorption peak, 327–328
  capacitance, 327–328
  constant of, 327–328
  dielectric constant data
    distribution parameter, 354
    equivalent circuit, 339–340

  frequency, temperature, and composition, 337–339
  hydrostatic pressure and different temperatures, 346–348
  loss data in, 350–354
  low-frequency, 340–346
  pressure and temperature effect, 348–350
  electric dipoles, 327
  experimental techniques
    capacitance bridge methods, 328–329
    equivalent circuit, 329–330
    hydrostatic pressure and different temperatures, 330–331
    low-frequency constants, 330
  loss factor and strength, 327–328
  magnetic data
    activation energy, 356–357
    BME-MLCCs, 354–355
    dissipation factor, 356–360
    efficiency of, 358–360
    electrical conductivity, 355–356
    electron spin resonance, 358–359
    electrophoretic deposition (EPD), 360
    LTCCs, 354–355
    luminescent emission spectra, 358–360
    network structure, 355
    optical transmission, 359–360
    reflection resonant-cavity method, 359–360
    refractive index, 359–360
    relative permittivity, 359–360
    structural changes, 358
    tunneling phenomena, 358
  polarization, 327–328
Differential scanning calorimetry (DSC), 25, 204–207, 212, 221–222, 228, 232, 240, 245–248, 259–262
Differential thermal analysis (DTA), 204–208, 215–219, 228, 243, 245–246, 260
Dispersion, 368
Displex Closed-Cycle Refrigeration System CSA-202, 273
DV-Xα cluster method, 304

E

EDFA. *See* Erbium-doped fiber amplifiers (EDFA)
Elasticity and tellurite glasses
  abrasion resistance, 81–82
  BFS, 152
  BOCDR, 152
  bond-bending and bond-stretching forces, 147–148
  bulk compression model, 87–90
  central force model, 91–92
  constants of elasticity, 84–86
  conventional rapid quenching method, 150–151
  couples forces, 82–83
  couplings, 93
  cross-link density, 150
  Debye temperature, 82, 151–152
  elasticity constants, 84
  elastic stiffness constants, 84–85
  far transmission infrared (FTIR), 151–152
  fiber structure, 152
  Gruniesen parameter, 148–149
  hardness measurements, 95–96
  Hooke's law, 81–82, 84

interatomic binding forces, 143, 145–146
K–V relations
  atomic radius, 127
  atomic volume, 127–130
  Born potential, 127–130
  interatomic bonding, 127–130
  network modifiers and formers, 127–130
  network rigidity, 129–130
  thermodynamic property, 127–130
longitudinal and transverse ultrasonic waves, 150
Makishima–Mackenzie model, 86–87
matrix, 151–152
microhardness, 146
modulus data of TeO_2 crystal
  piezoelectric properties, 97
  piezoelectric tensor, 97
  polarized ferroelectric ceramics, 97
  torsional-resonance modes, 97
modulus of elasticity, 81–82
non-Hookean behavior, 81
phase separation, 82
piezoelectric transducers, 93–94
plastic deformation, 81–82
Poisson's ratio, 84–85
pulse-echo technique, 92–93
relative strain, 85
relative volume change, 85
ring deformation model, 90–91
sample holders, 94–95
shearing force, 81
short-range-order structure, 148–149
single-bond strength, 86
SOEC, 81–82
solid deformation, 81–82
stresses, 83–85
stress–strain relationships, 82–85
stress tensor configuration, 83
tensile stress, 82–85
TOEC, 82
transition temperature, 150–151
transverse polarization, 147
ultrasonic wave velocity, 82, 147
uniaxial and hydrostatic pressure measurement, 95
valence force field, 147–148
vibrational spectra, 81–82
wavelength limit, 82
Young's modulus, 83
Electrical conductivity of tellurite glasses
AC data
  frequency-dependent component, 310
  glasses $3TeO_2$–$x$$Li_2O$–$(1-x)$$V_2O_5$, 312, 316
  nonoxide–tellurite glasses, 310
  semiconducting vanadium-tellurite glasses, 310–312
  sodium phosphor–tellurite glasses, 311–312
  sodium–tellurite glasses, 312, 317
  transition metal tellurite glasses, 311, 313–315
in chalcogenide glasses
  amorphous semiconductors properties, 280–281
  Cohen–Fritzsche–Ovshinsky model, 281–282
  8-N bonding rule, 280–281
  small-polaron model, 281–282
current–voltage drop/semiconducting characteristics
  binary TeO_2–WO_3 glasses, 269

chalcogenide glassy semiconductors, 271
crystallization, memory switching by, 270
electrical insulation, 271
impedance spectroscopy, 271
monostable memory effect, 270
phonon-assisted hopping processes, 271
reversible monopolar switching phenomena, 270
semiconductor glasses use, 270
small-polaron theory, 271
switching characteristics and mechanisms, 270
tellurite–vanadate glasses, 269–270
TMI oxide glasses, 270–271
DC conductivity at different temperatures, 291
and activation energy data, 299
binary vanadium–tellurite glasses, 292
conduction process, theoretical values, 292, 297
electron spin resonance analysis, 292, 294
experimental data, 292, 295–296
low temperatures, oxide-tellurite glasses, 298–306
molybdenum–tellurite glasses, 299–300, 301, 304
nonoxide–tellurite glasses, 308–310
oxide–tellurite glasses, 292–294, 296–297
oxide–tellurite glasses containing alkalis, 306–308
rare-earth ions effect, 299, 302
rare-earth tellurite glass system, 303, 305
semiconducting vanadium–tellurite glasses, 299, 303
tellurite glass system, 306
and thermopower, 292–293
in glassy electrolytes
  anionic polarization and covalency, 283
  conducting pathways, creation/loss, 285–286
  electrical resistivity, Rasch-Hinrichsen law, 283
  FICs, 283–284
  ion transport, classical theory, 285
  lattice movement, 282–283
measurement, experimental procedure, 271
  AC and DC conductivity at different temperatures, 272–276
  impedance plane plot, 275
  sample holder, 276–277
  sample preparation, 272
  thermoelectric power, 276–277
new tellurite glasses
  alkali-borate glasses, 324
  alkali–tellurite system, 323
  all-solid-state batteries, 318–319
  CBH model, 322–323
  EPR, 324–325
  experimental data, 319–321
  lithium–vanadium–tellurite glasses, 323
  and mean distance, 322
  NBO, 321
  non-Debye conductivity relaxation, 322
  photonic information technology, 318
  polaronic hopping and other models, 321–322
  silver content increment, 323–324
  superionic glasses, 318–319
  temperature dependence, 320–321
  Triberis–Friedman percolation model, 321
oxide glasses, DC conductivity
  charge-transport process, 277
  low-and high-temperature, 278–280
  room-temperature conductivity, 280

semiconductors, hopping mechanism in, 278
    temperature dependence analysis, 277
    and temperature slope, 279
semiconducting and electrolyte glasses
    Austen–Mott formula, 288
    CBH model, 289
    data presentation methods, 291
    Drude formula, 287
    HOB mechanism, 289
    OLPT model, 291
    polaron distortion clouds overlap, 290
    QMT mechanism, 289
    tellurite glass-ceramics (*see* Tellurite glass-ceramics)
    thermoelectric power
        at high and low temperatures, 287
Electro-deposited films, 453–454
Electron paramagnetic resonance (EPR), 303, 324–325
Equivalent circuit
    admittance and impedance data, 329–330
    oscillation and bias voltages, 329
Erbium-doped fiber amplifiers (EDFA), 413, 489
Erbium-doped tellurite glass
    absorption coefficient, 62
    amplified spontaneous emission (ASE) spectrum, 62
    color of, 60
    density and molar volume, 59–60
    Er^{3+}/Yb^{3+}-co-doped, 60, 62
    Kerr coefficient, 59
    by melt-quenching method, 59
    NBO bonds, 59
    for optoelectronics devices, 60
    photoluminescence study, 59
    Raman spectra, 63
    refractive index, 62
    second-order nonlinearity property, 60
    sol-gel method for magnetic susceptibility, 58
    structural evolutions, 63
    thermal stability and spectroscopic properties, 64
    waveguiding properties, 59
Er^{3+}-doped oxyfluoride–tellurite glasses (EDFA), 248

F

Fast ion conductors (FICs), 283–285
Fiber-to-the-home program, 367
Fluorotellurite, 69
Four-probe method, 273–274
Fragile-glass-forming liquids, 203

G

Gamma-radiation effect
    elastic scattering, 121
    electrons ionization, 121
    interstitial cations, 121
    irradiation, 121
    radiation variation, 121, 122, 123
    γ-rays, 120–121
    source of, 121
Gated-carrier pulse superposition apparatus, 95
Glass-ceramics, transmission spectra
    calculated, 475–476
    lead–barium–tellurite, 467
    measured

        IR absorption spectra, 467–470, 472, 474, 477
        optical absorption, 472, 473
    molybdenum oxide, 467
    tellurite-based materials, 466
Glass, Science and Technology, 178
Glass transformation temperature and structure parameters
    average crosslink density, 234
    average force constant, 235–236
    crystallization activation energies, 235
    quantitative analyses, 237
Glass transition, 201

H

Halide–tellurite glasses, UV properties
    characterization, 451
    chloride ions effect, 450
Hardness measurements
    deformation/fracture pattern, 95–96
    fracture toughness, 96
    indenter constant, 95–96
    transition temperature, 95–96
    Vickers hardness, 95–96
Hashin and Strikman's expression, 180
Heat capacity, 202
High-coordination-bond networks, 153
Higher Institute of Chemical Technology, 269
Hooke's law, 81–82
Hopping mechanism in semiconductors, 278
Hopping-over-barrier (HOB) mechanism, 289, 310
Hydrostatic pressure and different temperatures
    copper–constantan thermocouple, 330–331
    nitrogen storage dewier, 331
    pressure-induced phase transitions, 330–331

I

Impedance spectroscopy, 271
Infrared and Raman spectra of tellurite glasses, 459
    absorption spectra, characteristic, 460
    $BaCl_2$–TeO_2 glass, 490
    barium chloride–oxide–tellurite, 490
    binary rare-earth, 461
    close and open Z-scan, 491–492
    EDFA, 489
    effective bandwidth, 493
    electronic moment, 466
    experimental spectrum, 463–464, 465
    fabricated glass samples, 490
    $Fe(NO_3)_3$–TeO_2 glasses, 494
    flat optical surface, 490
    fluorescence, 489
    Ge–Ga–Te–CsI glass, 494
    glass-ceramics
        agreement, 485
        crystal line, 482
        data, 488, 489
        measured, 482–485, 486–488
        polycrystalline compounds, 485
        spectroscopic studies, 482
        structure of thallium, 485
        studied, 488
        systems, 482

TeO$_2$–MO glasses, 485
TeO$_2$–P$_2$O$_5$, 482
glass-ceramics, transmission spectra (*see* Glass-ceramics, transmission spectra)
large glass-forming domain, 492
major structural groups, 489
melt-quenching technique, 491
metal oxide doping effects, 491
microstructure fibers, 493–494
nonbridging oxygen atoms, 490
normal mode displacement, 465–466
number of vibrational modes, 460–461
optical properties, 463, 492
photon-counting, 460
process, 464–465
reflectivity, 463
refracting and absorbing media, 461
refractive index, 489–490
scattered radiation, 460
spectral data of oxyhalide, 477–480
spectral properties and thermal stability, 489
spectra of chalcogenide glasses
Ge–As–Se, 482
optical materials, 482
spectra of halide, 480–481
optical transmission of Te–Br glass, 480, 481
Te–Se–Br, 480
spectrophotometer for measurements, 460
structure of TeO$_2$-based glasses, 490–491
temperature coefficient of delay, 493
theoretical band position, 462
transmission measurements, 460
transparent TeO$_2$-based, 489
X-ray diffraction, 494

K

Kissinger formula, 229–230
K–Nb–Te glass and glass-ceramics, 141

L

Lasocka formula, 229
Lead–tellurite glasses, 37
Linear and nonlinear optical properties
Abbe numbers, 385
absorption coefficient, 369
acousto-optical materials
broadened fluorescence lines in, 372
configurations of, 371
stationary particles, 372
switching speed of, 371
technology limits, 371
triangular cells with, 371
true two-dimensional deflection, 372
additivity, 369
Assyrian hieroglyphic text, 367
binary and ternary tellurite glasses
relation between density and refractive index for, 387
and chromatic dispersions, variation, 390
coloration of glass, 370
coordination number for some cations and anions
values of, 384

crystalline solids, 370
dense-flint glasses, 368
and density, 391
dielectric materials, 370
Maxwell's equations, 368
dispersion
reciprocal relative, 368
Drude-Voigt relation, 387
electric displacement, 370–371
electric field and polarization, 370
electromagnetic wave oscillation, 371
electron probability density distributions, 369–370
fiber-to-the-home program, 367
fluorescence and thermal luminescence
diode laser use, 375
photoluminescence spectra, 377
spectrophotometer and dye laser, 376
thermal luminescence intensity, 377
fluorides and chalcogenides, investigation, 369
fraction of beam intensity, 369
free-space permittivity, 371
Fresnel's formula, 369
function of frequency of light, 368
glass-ceramic, transparency of, 389
incident radiant flux of photons, 367–368
infrared region, 370
ionic polarizability values of, 384
ionic refraction values of, 384
lightwave, 370
linear refractive index and dispersion of glass
cubic coefficient of expansion, 373
methods, 372–373
quantitative analysis, 377–379, 382
Lorentz–Lorenz plot, 384
of new tellurium glass
Archimedes principle, 413–414
average dispersion and Abbe number of, 415
BOCDR and BFS, 417
Cauchy's equation, 413
core–clad fiber, 418
EDFA, 413
luminescence spectra, 420–421
MMF and SMF, 417
refractive index, 413–414, 419
room temperature values, 418
SHG, 419
spectroscopic and photoluminescence properties, 421
TZNK sample, 416
zinc fluoride content, 416
nonlinear refractive index of bulk glass
Gaussian beam, 374
laser frequency, 374
normalized energy transmittance, 375
photomultiplier (PM), 374
Pockels cell, 377
Q-switched Nd-YAG laser, 373–374
THG method, 373
Z-Scan technique, 374–375
of oxide–tellurite glass and glassceramics
Judd-Ofelt, 404–405
Kerr effect, electrooptic, 406
mechanism, 408–413
multicompnent tellurite glasses, 406

nonradiative multiphonon relaxation, 405
  quantum efficiencies for, 405
  rare earth, optimum concentration, 407
  tellurite glasses in acousto-optic devices and, 406
  thermal luminescence, 407
  up-conversion fluorescence, two-step excitation,
      407
Pauling's electronegativity, 392
photon interactions, types, 367
radiation energy, 370
refraction, 368
refractive index data of tellurite glasses, 391
  and glass-ceramics, 394–395, 397–398–403
  of tellurium nonoxide glasses, 403–404
  of tellurium oxide bulk, 394–395, 397–398–403
  tellurium oxide bulk glasses and glass-ceramics,
      383–390, 392
  thin-film glasses, 394–395, 397–398–403
Sellmeier coefficients and, 389
simple oxides, 392
switches with, 367
tellurium nonoxide bulk and thin-film glasses, 393
  optical-materials investigations, 394
tellurium oxide thin-film glasses, 392
  of single-component oxides and optical basicity, 393
T_eO_2–R_nX_m/R_nO_m, 386
transparency of, 391
  crystalline insulator, 369
  glass, 369
VIS range of the spectrum, 369
Low-temperature co-fired ceramics (LTCCs), 354–355
Low-temperature ultrasonic attenuation and room-
      temperature elastic moduli
  activation energy, 170–171
  cation–anion–cation spacing, 168
  linear regression, 171
  oxygen densities, 169

M

Madelung energy, 85–86
Makishima–Mackenzie model
  bond energy, 86–87
  crystalline oxides, 87
  dissociation energy, 87
  glass structure, 86–87
  hypothetical chain networks, 87, 88
  Madelung energy constant, 86
  Poisson's ratios, 87
  TeO_2, TeO_2–V_2O_5, and TeO_2–MOO_3
    basic unit volume, 130
    bond angles, 132–135
    calculated dissociation energy, 130–134
    cation–cation spacing, 132–135
    crystalline arrangement, 132–134
    longitudinal and transverse double-well potentials,
        132–135
    oxygen density, 134–135
    packing density, 130–132
Melt-quenching technique, 450
Microhardness of tellurite glasses, 140
Microphase separation in tellurite glasses, ultrasonic
      detection
  binary tellurite glasses, 180

  equivalent upper and lower bounds, 180
  Hashin and Strikman's expression, 180
  matrix of volume fractions, 179
  second model, 179
  shear modulus, 179
  variations in moduli, 181
  Young's modulus, 179
Mixed oxyhalide and oxysulfate tellurite glasses
  elastic moduli and compressibility, 36
  glass-forming tendencies, 34–35
  halide–tellurite glass systems, 33–34
  investigated sections in, 36
  optical constants of, 34
  Raman measurements, 37
  study of, 32, 34
  tellurite–sulfate glasses, 35
MMF. *See* Multimode fibers (MMF)
Monatomic (primary) glasses, 4–5
MO–Nb_2O_5–TeO_2 glass, 457
Monostable memory effect, 270
MoO_3 and V_2O_5 glass
  atomic variation, 125–126
  binary transition metal tellurite and phosphate glasses,
      125–126
  crystal structure, 121–130
  nonnetwork bond compression processes, 123–130
  ring diameter with composition, 125–126
  various modifier percentages, 123–125
Mossbauer analysis, 52–53
Moynihan et al. formula, 230–231
Moynihan's formulas., 256
Multicomponent tellurium oxide glasses, glass-forming
      ranges
  conventional methods
    crucibles, 17
    electronegativity of, 17–18
    infrared spectra of, 21
    in quaternary TeO_2–B_2O_3–MnO–Fe_2O_3 system,
        19–20
    six-atom coordination, 17
    of tellurite system, 17
    TeO_2–MoO_3–V_2O_5 system, 18
    in ternary tellurite glasses, 19
    in vitreous ternary TeO_2–MoO_3–CeO_2 system, 19
  derivative of tellurium 2-methyl-2,4-pentanediol,
      hydrolysis, 25
  prepared by sol-gel technique, 22
  thin films, 22–23
  $2TeO_2V_2O_5$–GeO_2 and TeO_2–V_2O5–GeO_5
    phase diagram of, 24
Multimode fibers (MMF), 417

N

Nanoparticles, 72
8-N bonding rule, 280–281
Nonbridging oxygen (NBO), 15, 177, 188, 194, 319, 321,
      323, 325, 455
Non-Hookean behavior, 81
Nonoxide-tellurite glasses
  chalcogenide and halide glasses
    amorphous solids semiconductors, 24–25
    covalent amorphous semiconductors, 25
    DSC, 25

Ge–Se–Te and Ge–Se–Te–Tl systems, 28
  mid-IR region, 23
  optical properties of Ge–Te and Ge–Se–Te, 27
  purification procedure, 28
  qualitative results obtained for, 25–26
  transmission loss characteristics of Ge–Se–Te, 27
composition, 245–246
cycling studies, 246
devitrification behavior, 248
DSC data, 247
fiber preparation
  ampoule for, 29
  antireflection coating, 29
  chalcogenide glass melt, casting, 29
  cladding diameter of, 29–30
  crucible drawing method for, 28–29
  transition loss and mechanical strength, 29
halide-tellurite glasses
  characteristics of, 30
  experimental procedure, 30–31
  for IR transmission, 32
  optical properties of, 31
  structural models, 31
  TeX glasses, 31–32
heat capacity and thermal diffusivity, 246
thermally induced transformations, 246
thermal properties, 247
UV properties of
  amorphous chalcogenide films, 451
  electro-deposited films of CdTe, 453
  film dispersion, 453–454
  heat treatment, 452
  Te-Se alloy film, 452
XRD and DTA data, 246
Nonoxide–tellurite glasses, ultrasonic attenuation properties
  activation energies, 162
  temperature coefficient, 162
  wave frequencies, 162
Normal fluorescence, 375

O

OGS. *See* Oxide glassy semiconductors (OGS)
Optical-absorption-edge data, 451, 452
Optical energy gap, 427
  in amorphous solid material, density of states, 428
  amorphous systems, 429
  chemical compound, 429
  linear relationship, 429
  relation, 429
  transparency of glass, 429–430
Optical limiters (OL), 458
Optical transition, 423–424
Optical transmission, 451
Orthorhombic β-TeO$_2$ (tellurite), 43
  lattice dynamic-model studies, 52
Oxide and nonoxide–tellurite glasses, Debye temperature
  experimental acoustic, 181–185, 187–189
    and calculated optical, 190–193
  radiation effect on, 193–194
Oxide glassy semiconductors (OGS), 277–278
Oxide–tellurite glass ceramics
  absorption bands, 445
  form, 448

heat treatment, 444
optical energy, 444
properties of, 446
Rb-doped crystalline phase, 447–448
TeO$_2$–Nb$_2$O$_5$–K$_2$O, 448
TeO$_2$–PbO–CdO, 449
TeO$_2$–PbO glass, 446
transparency, 444, 447
transparent mixed alkali–tellurite glass, 445
Oxide–tellurite glasses at low temperature, ultrasonic attenuation
  acoustic and optical properties, 161
  activation energy, 155–156
  angular frequency, 155
  Arrhenius-type relaxation, 158–160
  coefficient, 155–158
  deformation potential, 158–161
  logarithm variation, 155–160
  longitudinal absorption, 158–161
  longitudinal waves variation, 155–158
  low-temperature properties, 155–158
  network-forming oxides, 155
  parameters, 161–162
  peaks shape, 158–161
  potential energy and oxygen density, 160–162
  rapid-quenching method, 161
  relaxation strength, 158–161
  transverse waves, 161–162

P

Parallel-plate viscometer (PPV), 253
Phase diagram and immiscibility
  binary tellurium oxide glasses, glass-forming ranges
    borate and silicate systems, 11
    chemical incompatibility between, 11
    curves for RDFs, 11–12
    monotectic temperature, 11
    structural interpretation factors, 12
Piezoelectric transducers, 93
PPV. *See* Parallel-plate viscometer (PPV)
Pulse-echo technique
  oscilloscope screen, 92–93
  polarized shear waves, 92–93
  short sinusoidal electrical wave, 92–93
  sound frequencies, 92–93
  transducer–specimen interface, 92–93
  ultrasonic transducer, 92–93
  ultrasonic vibrations, 93

Q

Q-switched operation, 372
Quantum-mechanical tunneling (QMT) mechanism, 289, 291, 310–311
Quaternary tellurite glass systems, 455
  average stretching force constant, 54
  chemical durability and thermal properties, 57
  compressibility model, 55–56
  cooling rates, 56
  core/clad glass pair, 58
  densities, 55
  fluorescence spectra, 59
  photoluminescence, 59

force constant for, 54
glass composition, 54–55
IR and Raman studies, 55–56
by melt-quenching technique, 53
molar volume, 54–55
Mossbauer effect spectra, 56
nonbridging oxygen (NBO) bonds, 59
volatilization, 58
XRD patterns, 55

R

Radial distribution function (RDF), 3
Radiation effect of tellurite glasses
    Avogadro's number, 164
    double-silver ions, 164–165
    glassy network splitting, 162–163
    ionizing radiation, 164
    photochromic glasses, 162
    two-dimensional representation, 164
Raman scattering, 459
    TeO_2 glass, 488
    theory, 466
    three-step process, 464–465
Rare-earth oxide-tellurite glasses
    Mackishima–Mackenzie model, 144
    parameters adopted from crystal structure of, 141
    quantitative analysis of elasticity moduli, 143
        cation–anion bond, 139
        cross-link density, 139
        dissociation energy, 143
        parameters, 139–143
        structure-sensitive factor, 140
        systematic relationship, 139–140
        ternary and quaternary systems, 139–140
        Young's modulus, 143, 145
    thermal properties of, 215
Rasch-Hinrichsen law of electrical resistivity, 283
Rayleigh scattering, 459
RDF. See Radial distribution function (RDF)
Reflectance, 423
Relaxation phenomena and ultrasonic attenuation
    structural analysis
        interaction with thermal phonons, 166–168
        thermal diffusion, 165
Reversible monopolar switching phenomena, 270
Ring deformation model
    anion vibration, 91
    bond-bending force constant, 90–91
    diameter and perimeter, 90–91
    macroscopic elastic behavior, 90
    nanometers, 90
    three-dimensional structure, 90–91

S

Samarium–phosphate glasses, 137
Second generation glasses, 35–36
Second harmonic generation (SHG), 6–7
Second-order constants of elasticity (SOEC), 81–82
    bond-bending distortions, 108–109
    isotropic compression, 108–109
    isotropic elastic deformation, 108–109
    of pure TeO_2 glass

elastic constants, 97, 98
    longitudinal and shear ultrasonic waves, 97
ring diameter, 108
three-dimensional network, 108
Second order elastic constants of transition metal tellurite
    glasses, 98–99
SHG. See Second harmonic generation (SHG)
Short-range order (SRO), 5
Short wavelength absorption edge (SWAE), 457
Single-mode fibers (SMF), 417
Sinusoidal electromagnetic wave
    intensity of, 368
Small polaron hopping (SPH) model, 323
Small-polaron model, 281–282
SMF. See Single-mode fibers (SMF)
Sodium–zinc–tellurite (TZN), 69
Spectral data of oxyhalide
    absorption spectra for, 478–479, 481
    concentrations of water, 477, 479
    FTIR spectrum of TeO_2–TiO_2, 480
    tellurite–halide systems, 477
    tellurite–phosphate glasses, 479
    transmission, 477
Spectrophotometers. See also UV absorption and
        transmission spectra
    double-beam, 424
    medium, 426
    normal incidence, 426
    percent transmission, 425
    polarizer, 425
    reflective losses, 425–426
    room temperature, 427
    single-beam, 424
    techniques, 425
    transmittance, 426
    two-beam optical bridge, 425
SPH. See Small polaron hopping (SPH) model
SRO. See Short-range order (SRO)
Stabilization, 201
Strong-glass-forming liquids, 203
Structure models
    of binary tellurium oxide glasses, glass-forming ranges
        chains and high deformation, 14–15
        incompatibility model, 15
        microhomogeneous glasses, 13
        nature of atomic arrangement, 15
        NBO ions, 15
        PO–Te bridging oxygen (BO) bonds, 13–14
        second type of phosphorus atoms, 15
        SI model, 15–16
        SRO levels, phases, 15
        structure of atomic arrangement, 14
        TeO_2–GeO_2 binary glass system, 17
SWAE. See Short wavelength absorption edge (SWAE)
Switching effect, 305–306

T

Tail width, 427
    densities, 428
    transparency of glass, 429–430
TBSN glass, 454
TCD. See Temperature coefficient of delay (TCD)
Tellurite glass-ceramics, 239, 241, 243, 244, 245

crystalline phases, 242
crystallization process, annealing effect, 315–316
crystal phases, 243
cubic structural phase, 242
glasses kinds, 315–316
oxyfluoride–tellurite glass-ceramics, 315–316
phase separation and devitrification, 316
second harmonic generation (SHG), 243
Se–Te–Ge, transport properties, 317–318
solid-state synthesis, 240
switching phenomena, 315–316
thermal stability, 240
XRPD, 240
Tellurium dioxide (TeO$_2$) glass
    advantages, 307–308
    applications, 66
        broadband lightwave processing, 70
        fiber cladding, 72
        fiber optic communications, 68–69
        magnetic recording devices, 67
        mid-infrared optical fiber lasers, 71
        optical disk system, 67
        practical functional devices, 70
        slow light generation, 70
        tellurite single-mode fiber, 67–68
    atomic structure, 42
    binding energies, 45
    bonding nature, 42
        chemical shifts of core electron-binding
            energies, 47
        intermediate-range order, 49–53
        oxygen-first (O-first) spectra, 45
        oxygen-first photoelectron spectra, 45–46
        polymorphic forms, 43
        by roller technique, 43
        SRO analysis, 43–44
        valence band spectra, 47–48
    chemical durability, 37
    crystal structure and properties, 1
    definition and research
        benefits, 6
        conventional glasses, characteristics, 7
        disrupted-edge (O) ions, 6
        distribution function, 3
        first reports on, 2–3
        functions and properties, 7
        interatomic distances and areas, 3
        microchemical analysis, 3
        monatomic (primary) glasses, 4–5
        oxygen atom, 5
        RDF in, 3–4
        SHG, 6–7
        SRO structure in, 5–6
        transformation to, 5
        transformation to glass, process, 5–6
    degree of homogeneity, 38
    density and temperature, 38–39
    diffraction pattern of, 3
    distances between components in, 2
    doped with europium ions, 58
    elasticity constants, 146
    elastic properties of, 143, 145
        Brillouin gain, 152
        BS parameters, 147

        bulk and Young's moduli, 149–150
        DeLaunay–Nath–Smith (DNS) equation, 147
        Gruneisen parameter, 149
        longitudinal and shear ultrasonic velocities, 149
        mechanical and optical properties, 151
        Poisson's ratio, 151
        SOEC, 148
        TOEC, 147
        Vickers nanoindentation measurements, 148
    interatomic distances and coordination numbers, 44
    long-wavelength acoustic-mode Gruneisen
            parameters, 239
    Me of, 172
    molar volume, 39
    molecular weight, 39
    praseodymium and neodymium, doped with, 441
    preparation of
        aluminum crucible, 7
        interim stage of cooling, 7–8
        melt, 7
    properties, analyses
        electrical conduction measurements on, 2
    schematic of, 2
    semitransparent with pale lime green color, 38
    spectroscopic analyses, 44–45
    stability of, 1
    thermal analysis, 38
    three-dimensional structure, analyses
        cell dimensions of, 1
    ultrasonic wave propagation, 38
Temperature coefficient of delay (TCD), 493
TeO$_2$–CuO–CuCl$_2$ glass, 450
TeO$_2$–MoO$_3$ glass
    longitudinal and shear ultrasonic wave velocities,
            112, 113
    molar volume, 111–112
    semiconductors, 111–112
    two-dimensional representation, 112
TeO$_2$–Nb$_2$O$_5$–Bi$_2$O$_3$ glass, 456
TeO$_2$–TeCl$_2$, 450, 451
TeO$_2$–V$_2$O$_5$–Ag$_2$O, TeO$_2$–V$_2$O$_5$–CeO$_2$, and TeO$_2$–V$_2$O$_5$–
            ZnO glass
    atomic masses, 117–120
    atomic weights, 118–120
    covalent bonds, 117–120
    cross-link density, 119–120
    elasticity moduli, 117–120
    longitudinal and shear values, 117–120
    NBO, 117–120
    octahedral groups, 118–120
    property data, 119–120
    pulse-echo technique, 117–120
    rigidity, 118–120
    ring diameter, 118–120
    shear ultasonic waves, 117–120
    two-dimensional representation, 118
TeO$_2$–V$_2$O$_2$ glass
    atomic mass and atomic volume, 115–117
    bipolar threshold and memory switching, 115–117
    bond compression model, 116–117
    elasticity moduli, 115–117
    lateral and longitudinal strains, 115
    microhardness value, 116–117
    semiconductors, 115–117

TeO$_2$–WO$_3$ and TeO$_2$–ZnCl$_2$ glass
   adiabatic moduli, 110–111
   compressibility data, 110–111
   dielectric molar volume, 109–111
   elastic stiffness, 109–111
   equation of state, 110–111
   Gruneisen parameter, 110–111
   hydrostatic and uniaxial stress, 109–111
   normal mode frequencies, 111
TeO$_2$–ZnO glass
   average cross-link density, 114
   bond compression model, 114–115
   elasticity moduli, 112–115
   longitudinal modulus, 113–115
   longitudinal sound velocity, 113–114
   multicomponent optical glass synthesis, 112–115
   network rigidity, 113–115
   neutron diffraction patterns, 114–115
   phase separation, 114–115
   structural parameters, 114–115
   three-dimensional vitreous tellurite networks, 114
70T$_e$O$_2$-20ZnO-2.5Na$_2$O-2.5K$_2$O-5La$_2$O$_3$ (TZNK) sample, 416
Te–Se alloy, 452
Tetragonal α-TeO$_2$ (paratellurite), 43
   ionicity of, 48–49
   lattice dynamic-model studies, 52
   structural model, 53
Thallium–tellurite glass, 222–223
Theoretical absorption spectra
   cation and oxide ion, 429–430
   direct transitions, 428
   exponential edges, 428
   glassy materials, 428–429
   linear functions of energy, 428
   Lorentz–Lorentz equation, 429
   transparency of glass, 429–430
   Urbach rule, 427
Theoretical analysis of optical constants
   linear refractive index, quantitative analysis
      area of glass formation, 379, 382
      Clausius–Mossotti equation, 382
      halide-tellurite glass systems and, 380–382
      multicomponent glass, 382
      number of atoms per unit volume, 383
      and polarizability of tellurite glasses, 377–379, 382–383
      tellurite–sulfate glasses, 382
      values of, 378
Thermal analyzer (TMA), 207, 216
Thermal expansion coefficient, 202
   and vibrational properties
      approximation, 238
      cation–anion pair, 238
      isothermal and adiabatic bulk moduli, 238
      Mie–Gruneisen relationship, 239
      modes, 238
Thermal properties of tellurite glasses, 209–211, 249–251
   analysis, scanning calorimetry, 212
   ARES, 254
   BBV, 253
   binary glass system, 260

Brewster angle method, 248
Chen's and Moynihan's formulas, 255–256
cross model accurately, 255
crystallization activation energies (*see* Crystallization activation energies of glass)
DR, 258
EDFA, 248
experimental results, 258
formation and color properties, 261–262
germano-tellurite glasses (GTPC), 256
glass transition, 214, 260
   temperature, 253
measurement, experimental techniques
   Andrade equation, 208
   DSC curve, 205–206
   DTA, 204–205
   Gent's equation, 208
   TMA, 207
melting temperatures, 208–214
Moynihan's formula, 256
oxyfluoride–tellurite glasses, 248
phase transformation, 213
physical properties, 201–202
PPV, 253
quaternary tellurite glass systems, 259–260
RE, 212
sodium zinc–tellurite glass, 253–254
specific heat capacity, 222–225, 227
   anharmonicity of vanadium ions, 226
   atomic vibration, 228
   Gruneisen parameter, 228
stability against crystallization, 214
   devitrification tendency, 219
   DTA curves, 219
   endothermic and exothermic peaks, 216
   glass transformation temperature, 216–219
   Gruneisen parameters, 219–220
   NBO creation, 216
   stretching-force constant, 216
   TMO, 217
   by using dilatometer, 215
stable glass, 255
super-cooled liquids, 202–203, 227
TBSN glass, 261
TBZ system, 253
TDFA, 252
TeO$_2$–Tl$_2$O–ZnO ternary system, 256
TeO$_2$–WO$_3$–PbO system, 257
thermal conductivity, 203, 252
thermal diffusivity, 248
thermal expansion coefficient, 202, 208–214
   and vibrational properties, 238–239
thermal lens spectrometry, 248
thermo-optical coefficient, 252
TOEC, 204
traditional melt-quenching method, 248
transformation temperature and structure parameters, 208–214, 234, 236, 237
   average force constant, 235
viscosity and fragility
   Arrhenius form, 221
   data, 220–221
   glass transition region, 221

mixed alkali–tellurite glasses, 221
  temperature dependence, 222
volume–temperature relationship, 202
YAG laser, 262
Third-order constants of elasticity (TOEC), 82, 204
  experimental, rare-earth tellurite glasses, 138
  of tellurite and other glasses, 100, 102
  and α values of tellurite and other glasses, 103
  and vibrational anharmonicity, 97, 102
    anomalous effects, 107
    bridging oxygen (BO), 105–107
    correlation coefficient, 103
    effective elasticity modulus, 101
    Gruneisen parameters, 101, 103–108
    Hooke's law, 101
    hydrostatic pressure, 101
    kinetic energy, 105
    Lame constant, 103–104
    linear thermal expansion coefficient, 104–106
    physical principles, 101
    third-order constants, 100
    trigonal bipyramids, 105–107
    uniaxial pressure and stress, 100
    vibrational anharmonicity, 105–107
    wave vector, 103–104
TMA. See Thermal analyzer (TMA)
Tm^{3+}-doped fiber amplifiers (TDFA), 252
TMOs. See Transition metal oxides (TMOs)
Transducers
  frequency pulses, 94
  oscilloscope, 94
  piezoelectric, 93
  radiation pattern, 94
  sensitivity and resolution, 94
  ultrasonic flaw detector, 94
  ultrasonic wave velocity, 94
Transition metal ions (TMI), 270–271, 277–279, 287, 294, 311
Transition metal oxides (TMOs), 8, 455
Transmission, 423
Tungsten–tellurite, 69
Two-probe method, 273–274

U

Ultrasonic flaw detector, 94
Ultrasonics applications on tellurite glasses
  analytical method, 177
  Bragg's law, 177
  Brillouin scattering, 177
  creep failure, 179
  Debye temperature of oxide and nonoxide
Ultrasonics applications on tellurite glasses
  Debye temperature of oxide and nonoxide (see
    Oxide-nonoxide, Debye temperature)
  Hashin–Strikman/Kerner elastic modulia curvature, 180, 182
  lattice, 178, 184, 189
  limiting composition, 177
  microphase separation detection, 179–181
  oxide and nonoxide, experimental values, 182, 183–184
  phase separation, 177, 178

static fatigue, 178
temperature of new multicomponent, 185, 186–187
Young's modulus, 177–180, 182, 189
Uniaxial and hydrostatic pressure measurements
  atmospheric pressure, 95
  electrical resistivity, 95
  natural velocity, 95
  ultrasonic longitudinal wave, 95
  ultrasonic wave transit times, 95
UV absorption and transmission spectra
  experimental procedure, 424–427
  measurement of
    absorption bands, 435–437
    binary sodium–borate glasses, absorption spectra, 431
    edge and band gap, 441
    function of wavelength, optical absorbency, 431, 433
    optical transmission spectra, 434
    tellurite glass and binary tungsten, optical absorption, 431, 434–436
    TeO_2–PbO, 442–443, 446
    ternary tellurite glasses, 432–433
    transmission spectra, 442
    UV-VIS cutoff, 441–442
    zinc–tellurite glasses, 435
  optical density, 424
  oxide–tellurite glass ceramics, 444–449
  properties of, 424
    binary lead–tellurite glasses, 444
    B_2O_3, 442–443
    halide–tellurite glasses, 450–451
    new tellurite glasses, 454–458
    nonoxide–tellurite glasses, 451–454
    optical gap, 431
    optical transitions, 438–439
    praseodymium–phosphate glasses, 437–438
    tellurite glasses, data for, 430
    TeO_2–P_2O_5, 439
    transmission spectrum, 444, 447
    Urbach rule, 439
  spectrophotometers (see Spectrophotometers)
UV spectra of ternary tellurite glasses, 432
UV–VIS–NIR absorption, 454

V

Valence band maximum (VBM), 454
Valence force field (VFF) theory, 147
Variable-range hopping (VRH) conduction models, 323
Vickers hardness measurement of glasses, 95, 135
  densities and colors of, 136
  elastic stiffness, 136–138
  microhardness, 138–139, 140
  natural velocity, 137–138
  photochromic properties, 135–136
  second-order constants, 136
Vickers nanoindentation technique, 148–149
Viscosity, 202
Voigt model, 179

W

Wavelength-division-multiplexing (WDM) networks, 58
Weak-electrolyte theory, 285

X

X-cut plate, 93

Y

YAG laser, 262
Y-cut plate, 93

Z

"Z-cut LiNbO$_3$" transducer, 171–172

Printed and bound by CPI Group (UK) Ltd, Croydon, CR0 4YY

18/10/2024

01776271-0015